MODELS OF LUNG DISEASE

LUNG BIOLOGY IN HEALTH AND DISEASE

Executive Editor: **Claude Lenfant**

Director, National Heart, Lung, and Blood Institute
National Institutes of Health
Bethesda, Maryland

MODELS OF LUNG DISEASE

MICROSCOPY AND STRUCTURAL METHODS

Edited by

Joan Gil

Department of Pathology
Mount Sinai Medical Center
New York, New York

CRC Press
Taylor & Francis Group
Boca Raton London New York

CRC Press is an imprint of the
Taylor & Francis Group, an **informa** business

CRC Press
Taylor & Francis Group
6000 Broken Sound Parkway NW, Suite 300
Boca Raton, FL 33487-2742

First issued in paperback 2019

© 1990 by Taylor & Francis Group, LLC
CRC Press is an imprint of Taylor & Francis Group, an Informa business

No claim to original U.S. Government works

ISBN-13: 978-0-8247-8096-8 (hbk)
ISBN-13: 978-0-367-40315-7 (pbk)

Library of Congress Cataloging-in-Publication Data

Models of lung disease : microscopy and structural
methods / edited by Joan Gil.

    p. cm.--(Lung biology in health and disease ; v. 47)
Includes bibliographical references.
Includes index.
ISBN 0-8247-8096-5 (alk. paper)
    1. Lungs--Diseases. 2. Lung--Histopathology. 3. Anatomy,
Pathological. 4. Lungs--Pathophysiology. I. Gil, Joan
II. Series.
    [DNLM: 1. Lung--ultrastructure. 2. Lung Diseases--pathology.
3. Microscopy, Electron--methods. 4. Models, Biological. W1 LU62
v. 47 / WF 600 M689]
RC711.M625 1990
616.2'4075--dc20
DNLM/DLC
for Library of Congress
                                  90-3621
                                      CIP

Visit the Taylor & Francis Web site at
http://www.taylorandfrancis.com

and the CRC Press Web site at
http://www.crcpress.com

To Jerome I. Kleinerman

INTRODUCTION

In many ways, Leonardo Da Vinci may be regarded as a true anatomist. His drawings have given us some of the earliest insights into the structure of the human body. Since then, many investigators from a variety of disciplines have endeavored to uncover the body's detailed construct: they have used their eyes, and they have invented new methodologies to magnify what the eye could not see.

The great pathologist of the nineteenth century, Julius Frederick Cohnheim, wrote: "It is true that if anywhere, then certainly in the domain of respiration, the relations between physiology and pathology are. . .intimate,. . .there is the greatest difficulty in determining the boundary line between them; in fact, it is barely possible to do so." Were Cohnheim alive today, he would recognize that our current knowledge of lung physiology is a mirror image of what we know about the fine structures of the lung's anatomy. Since the end of World War II, we have witnessed a stream of new knowledge about the anatomy and the physiology of the respiratory system. I have no doubt that historians of medicine will attribute these many discoveries to the antecedent or collateral development of new technologies that enabled precise and sophisticated description of the lungs.

Parallel to the increased knowledge in lung physiology has been remarkable progress in lung pathology. These strides would not have been achieved without the fundamental knowledge of anatomy, histology, and subcellular structure that we now have. But this should come as no surprise: without intimate knowledge of an organ, its dysfunction may not be recognized.

Models of Lung Disease: Microscopy and Structural Methods, edited by Joan Gil—a pioneer of modern microscopy, describes very accurately the steps that have served us so well over the years. First, it presents the most modern tools and techniques employed by "anatomists"; then, it shows their applications to the description of the "normal lung"; finally, it illustrates how quantitative pathology is served when these tools and techniques are used. To achieve this tour de force, Dr. Gil has assembled a roster of distinguished contributors from several continents, all well known for their pioneering work and many achievements. This is the first volume of its kind in the series Lung Biology in Health and Disease; undoubtedly it will fill an important gap. I am deeply grateful to Dr. Gil and to his authors for adding such a remarkable contribution to the series.

Claude Lenfant, M.D.
Bethesda, Maryland

PREFACE

This is a book about techniques and morphology which has been rendered necessary because of the decline of anatomic sciences as primary research fields. Although anatomy, histology ultrastructure, and anatomic pathology continue to enjoy a secure existence as necessities in the medical school curriculum and in the hospital practice of pathology, they have ceased to be acceptable principal occupations for innovative investigators, and specialized training in these areas has disappeared or is disappearing for lack of career commitments.

A brief review of the evolution of the position of anatomy as a research discipline in the exciting last 30 years will illustrate the point. In the early 1960s, a relatively new instrument, the electron microscope, gained wide acceptance and popularity as a path-breaking research tool. For one brief moment, an enthralled new generation of scientists perceived electron microscopy more as a discipline than as a plain research tool. The new virgin world of ultrastructure was within reach waiting to be charted. Gone were the days when scientists could conduct research with a scalpel or light microscope. True explanations could come only from the higher resolving power of the new instrument. A major breakthrough was the addition of quantitative methodologies ("morphometry") to histology, an event that can be traced to 1963 with the publication of the momentous book *Morphometry of the Human Lung* by E. R. Weibel, a contributor to this monograph. International Congresses of Electron Microscopy were international events, gatherings of the scientific avant-garde where the leading edge of science was pushed forward. But none of this was to last. By the mid 1970s, cell biology, an ill-defined mixture of ultrastructure and biochemistry had already claimed center stage. Primary anatomists, victims of their own success, heard for the first time the fatal phrase "purely descriptive studies" as a derogatory characterization of their efforts. A long tradition started in Ancient Greece or earlier had come to an end: anatomists were no longer a necessary contingent of scientists. Anatomy Departments either changed their names to better compete in attracting bright graduate students or were dismantled; pathologists, in addition to immunology, always had at their disposal, at least in the United States, the biochemical tools and training of clinical pathology in which to find refuge. Soon, of course, the neverending quest for true explanations steered the new generation (and with it, as always, the vast federal grants, career expectations, spectacular laboratories, and building renovations) away from cell biology and into the fields of molecular biology and genetics: perpetuating a familiar and seemingly unavoidable cycle.

Throughout these painful phases of renewal and progress, anatomy was, is, and will continue to be one of the fundamental disciplines, always needed to show relevance, to put findings to the test or into perspective. The field of experimental pathology is inexorably linked to structural expertise and it is particularly in its connection with experimental procedures that structural methods are invaluable.

This book presents the reader with scholarly presentations of some of the most common techniques of morphological and experimental pathology. It is comprised of three parts: the first deals with general structural procedures that are important to lung workers; the second with methodologies needed in structure-function correlations for the study of the normal and diseased lung; finally, the third section summarizes key experimental procedures.

In a volume of this type and size, truly comprehensive coverage of all related topics would have been impossible. The selection of subjects in Parts I and II reflects the emphasis and trends in the current literature and probably some bias of the editor: items such as lung fixation, morphometry and immunohistochemistry are thoroughly covered. Many authors agreed to emphasize the application of their procedures to studies on human tissues.

The editor is indebted to Jerry Kleinerman (to whom this volume is dedicated) for his decisive help in the early stages of development, particularly in selecting topics and authors for the experimental pathology section (Part III). Here, especially, it was out of the question to cover all experimental methodologies. The editor chose a handful of topics which appeared to be of fundamental interest in the context of this book or that are not widely understood.

This book is addressed to a wide spectrum of lung research workers. It will interest primarily experimentalists, but just about anyone in pulmonary research will sooner or later need some morphologic studies. Hopefully this book will help.

Joan Gil, M.D.

CONTENTS

**Part Two: STRUCTURAL TECHNIQUE IN CORRELATION
BETWEEN NORMAL STRUCTURE AND FUNCTION**

CONTRIBUTORS

Peter S. Amenta, M.D., Ph.D.   Assistant Professor of Pathology and Laboratory Medicine, Department of Pathology, Hahnemann University, Philadelphia, Pennsylvania

H. Bachofen, M.D.   Departments of Anesthesiology and Anatomy, University of Berne, Berne, Switzerland

Marianne Bachofen, M.D.   Departments of Anesthesiology and Anatomy, University of Berne, Berne, Switzerland

Drummond H. Bowden, M.D.   Department of Pathology, University of Manitoba, Faculty of Medicine, Winnipeg, Manitoba, Canada

Peter H. Burri, M.D.   Professor of Anatomy, Department of Anatomy, University of Berne, Berne, Switzerland

Ling-Yi Chang, Ph.D.   Medical Research Assistant Professor, Department of Medicine, Duke University Medical Center, Durham, North Carolina

James D. Crapo, M.D.   Professor, Department of Medicine, Duke University Medical Center, Durham, North Carolina

Paul Davies, Ph.D.   Research Associate Professor, Department of Pharmacology, University of Pittsburgh School of Medicine, Pittsburgh, Pennsylvania

Daphne deMello, M.D.   Department of Pathology, St. Louis University and Cardinal Glennon Children's Hospital, St. Louis, Missouri

Nicolae Ghinea   Institute of Cell Biology and Pathology, Bucharest, Rumania

Joan Gil, M.D.   Professor, Department of Pathology, Mount Sinai Medical Center, New York, New York

Ronald E. Gordon, Ph.D.   Associate Professor and Director of Electron Microscopy, Department of Pathology, Mount Sinai Medical Center, New York, New York

Russell A. Harley, M.D.   Department of Pathology, Division of Histochemical Research, Medical University of South Carolina, Charleston, South Carolina

Keith Horsfield, D.Sc., Ph.D., M.D.   Cardiothoracic Institute, Midhurst, West Sussex, England

Yutaka Kikkawa, M.D.   Professor and Chairman, Department of Pathology, New York Medical College, Valhalla, New York

Charles Kuhn III, M.D.   Professor of Pathology, Department of Pathology, Brown University Program in Medicine, Memorial Hospital of Rhode Island, Pawtucket, Rhode Island

Edward J. Macarak, Ph.D.   Connective Tissue Research Institute, University of Pennsylvania, Philadelphia, Pennsylvania

Antonio Martinez-Hernandez, M.D.   Professor of Pathology and Cell Biology, Department of Pathology, Thomas Jefferson University, Philadelphia, Pennsylvania

Neal Mettler, Ph.D.   Professor and Chairman, Department of Pathology, New York Medical College, Valhalla, New York

Ulrich Mohr, Prof. Dr.   Experimentelle Pathologie, Medizinische Hochschule, Hannover, West Germany

Charles G. Plopper, Ph.D.   Professor, Department of Anatomy, School of Veterinary Medicine, University of California, Davis, California

Lynne M. Reid, M.D.   S. Burt Wolbach Professor of Pathology, Department of Pathology, Harvard Medical School, Pathologist-in-Chief, Department of Pathology, The Children's Hospital, Boston, Massachusetts

Stephen F. Ryan, M.D.   Department of Pathology, St. Luke's-Roosevelt Hospital, New York, New York

Eveline E. Schneeberger, M.D. Associate Professor, Department of Pathology, Massachusetts General Hospital and Harvard Medical School, Boston, Massachusetts

Bradley A. Schulte, Ph.D. Department of Pathology, Division of Histochemical Research, Medical University of South Carolina, Charleston, South Carolina

Sanae Shimura, M.D. First Department of Internal Medicine, Tohoku University School of Medicine, Sendai, Japan

Dennis A. Silage, Ph.D. Department of Electrical Engineering, Temple University, Philadelphia, Pennsylvania

Maya Simionescu, Ph.D. Institute of Cell Biology and Pathology, Bucharest, Romania

Samuel S. Spicer, M.D. Department of Pathology, Division of Histochemical Research, Medical University of South Carolina, Charleston, South Carolina

Shinji Takenaka, M.D.* Experimentelle Pathologie, Medizinische Hochschule, Hannover, West Germany

J. Christopher Wagner, M.D., F.R.C.Path.[†] Pathologist, Medical Research Council, London, England

Margaret M. Wagner, M.D.[†] Medical Research Council, London, England

Nai-San Wang, M.D., Ph.D. Associate Professor, Department of Pathology, McGill University, Montreal, Quebec, Canada

Ewald R. Weibel, M.D. Department of Anatomy, University of Berne, Berne, Switzerland

*Present affiliation: Gesellschaft für Strahlen-und Umweltforschung, Neuherberg, West Germany
[†] Retired.

MODELS OF LUNG DISEASE

Part One

GENERAL TECHNIQUES

1

Controlled and Reproducible Fixation of the Lung for Correlated Studies

JOAN GIL

Mount Sinai Medical Center
New York, New York

I. Introduction

The purpose of fixation is to stabilize and preserve tissue, a goal which is not difficult to achieve unless we are dealing with an organ of continuously changing morphology. When this is the case, the morphologist must attempt to define the prevailing conditions at the instant of fixation and fix in such a way that these conditions are not altered or lost before fixation is complete. Several organs and tissues require this treatment if quantitative correlations between structure and function are to be studied (heart, smooth and striated muscle cells, transitional epithelium), but the lung is particularly vexing. The lung lacks a fixed, or standard volume, and even worse, it collapses following opening of the chest and the morphology of the air spaces changes depending on whether they are reinflated with air or fluid (Gil, 1985; Gil et al., 1979).

Although less spectacular, the situation is equally complex regarding the morphology of the blood vessels: it has been shown that individual capillary segments are open or closed depending on the prevailing zonal pressure conditions; it is in dispute whether increase in flow is linked to vessel recruitment or dilation (Gil, 1980, 1988; Warrell et al., 1972).

The investigator must determine whether the morphology seen under the microscope corresponds to that prevailing in a given point in time when the conditions present in the lung (inflation, volume history, vascular pressures) were known or well defined. If the answer is negative, the investigator will have to make sure that he or she understands how the difference will affect the results of the examination. Another set of difficulties is related to anatomic peculiarities of the lung, such as the unusual physical dimensions of type I cells (large but very thin) which imparts great delicacy to them (Haies et al., 1981); to the abundance of lipids which are easily extracted during chemical fixation and also the ease of development during fixation of fluid accumulations which must be differentiated from edema.

II.   Chemical Fixation

A.   Routes of Administration or the Failed Dream of the Physiological Fixation

It is possible to fix lungs by "in situ" procedures (dripping fluid fixative over the surface, or immersing clamped pulmonary lobes in fluid) (Gil, 1985; Hayat 1981; Kikkawa, 1970). If the specimen is to be studied by electron microscopy, the greatest difficulty is the slow penetration of chemical fixatives across the pleura which allows autolytic damage, but satisfactory results in the preservation of an air-filled lung have been reported. In situ techniques, although not practical for routine use, are useful in providing a comparison to determine the effects of route of fixation on preserved morphology (Kuhn, 1972).

Since the appearance and dimensions of the organ depend on the way it has been fixed, interpretation of lung histology requires an understanding of the physiology of the organ. The lung has no fixed volume and the worker has the option of defining that parameter at will; but even at the same volume, the morphology need not be the same. The relationship between pressure and volume and configuration depends on such elements as the volume history (how many inflation/deflation cycles prior to fixation; whether it was fixed while reducing the pressure down from TLC to the selected range or while increasing it from zero; whether active degasing prior to inflation was or was not carried out). Certain facts are of more obscure significance: in pulmonary mechanics, two different moduli characterize the pulmonary elasticity by relating the pressure to the volume: the "bulk" modulus estimated during the normal expansion of the lung during inflation/deflation and the "shear" modulus which applies to local deformations due, for instance, to localized external compression. These two different moduli should have a different anatomic basis. Nor do the complications end here: the shape of the pressure–volume curve varies drastically depending on whether the lung is being inflated with air, saline, or a detergent (Gil et al., 1979); additionally it has been shown that there are morphological differences between lungs fixed in zones 1, 2, and 3 (i.e., the precise morphology depends on the relations between vascular (arterial and venous) and air pressures, which are adjustable (Gil, 1980, 1988). These problems can be very confusing; therefore it may be helpful to reduce them to different expressions of only two variables: (a) changes in the tissue configuration of the alveolar septal wall at the interface between epithelium and air and (b) changes in the configuration of the alveolar capillaries.

First, the configuration of the *alveolar surface* is variable because of the effect of the surface tension on tissue elements that have only limited connective tissue support. In its semifluid condition, alveolar epithelium evens out whenever exposed to surface tension forces, much in the way the surface of a liquid becomes smooth. This plasticity is limited by the connective tissue scaffolding of the alveolar wall (Sobin et al, 1988) which probably sets limits to the adaptability and imparts to the air spaces a recognizable shape. A remarkable tissue property is the capability of alveolar septa to fold and pleat over itself in the alveolar corners (Gil and Weibel, 1972). The formation of pleats is supposed to be driven by the surface tension which generates a higher recoil force in corners. It is worth mentioning, however, that a surface tension exists only between tissue and air; whenever air is replaced by fluid, as in fixation by filling the airspaces with aldehydes instilled into the trachea, surface tension forces are virtually abolished and the relaxed epithelial surface is free to undulate.

Second, it has been shown that *alveolar capillaries* can be open (recruited) or closed (derecruited). If open, they can be of different diameter and their cross-sectional outlines may be either round, quadrangular, or slitlike (Gil, 1980, 1988; Mazzone et al., 1988; Warrell et al., 1972). In general, it can be said that these changes are related to the intravascular and intra-alveolar pressures.

The worker will have a choice between the two most popular methods of fixing the lung: *instillation*, in which the fluid fixative is poured into the airways following a pneumothorax induced to collapse the lungs and *vascular perfusion*, in which the fixatives are injected into the pulmonary artery, after achieving distension of lung parenchyma by air inflation.

Instillation into the Airways (Fig. 1)

Instillation is the injection of chemical fixatives into the conducting airways (trachea or bronchi) to distend a lung previously collapsed by pneumothorax. It is an easy and widely used procedure that can be started while the heart is still beating and results in full distension of the organ and an excellent preservation at the light and electron microscopic level.

The most important prerequisite is that the instillation pressure with respect to the level of the supine lung should be defined and generally in the order of 20 cm of water. In small rodents, this is best met by vertically attaching a ruler to the table with a 20 cm mark aligned with the level of the fixative contained inside a burette. As the fluid flows into the lung, the pressure can be kept constant either

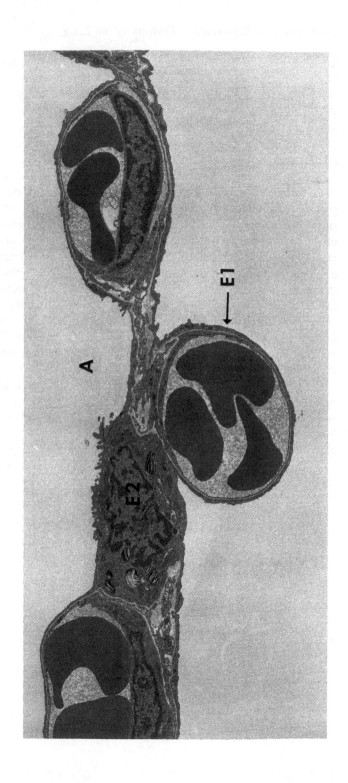

6

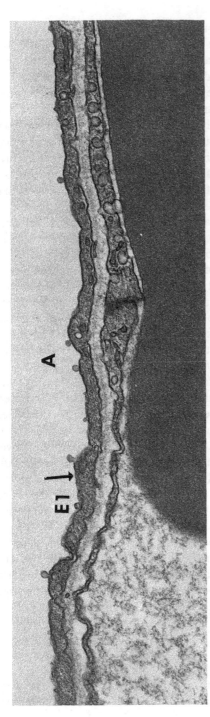

Figure 1 Electron micrographs of rabbit lungs fixed by instillation of glutaraldehyde into the trachea following pneumothorax. (A) An alveolar space, E1 squamous alveolar epithelium, E2 cuboidal alveolar epithelium (5000×). (B) High-power micrographs of rabbit lung fixed by instillation (as above). Note the thinness of the air–blood barrier consisting of epithelium (E1) and endothelium (EN) 30,000×).

by lowering the operating table, if it is mounted on top of a laboratory jack, or by raising the burette so that the 20 cm mark and the fluid level are flush. The fluid is allowed to pour until it spontaneously stops. After several seconds, the airway is clamped to prevent loss of fluid and the lung is dissected from the chest and immersed in a beaker containing the same fixative used to fill the air spaces. Care must be taken to insure that the lungs in the beaker are well covered with fixative: they will sink spontaneously only in the rare cases when no air was trapped inside. The lungs can be turned around inside the fluid with the apex facing the bottom or a lead weight (from a fishing outfit) can be attached to the tracheal ligature to make them sink. The lungs should be allowed to fix for at least 1–2 h; alternatively, they can remain in the fixative without harm over the weekend. To prevent lipid extraction, the beaker should be stored in a refrigerator.

Regarding final volume and external and internal configuration of the lung, one must remember that the organ has no shape of its own. External shape is determined by the chest cage, and depends on whether the chest wall was open and the front plate removed. Whenever part of the ventral thoracic wall is missing prior to the instillation, the tissue will be luxated out of the chest cavity like an overflowing fluid and air spaces outside of the chest in that area will be overdistended. In describing the technique used, it is necessary to mention whether the lung was fixed inside or outside of the chest cavity. As discussed above, additionally the air spaces will acquire the configuration characteristic of the fluid-filled lung with undulating epithelial surfaces and protruding capillaries.

It is commonly believed that a fluid-filled lung inflated at 20 cm H_2O will have a volume equivalent to some two thirds of the total lung capacity. To understand this properly, one must remember the characteristics of the fluid-filled pressure–volume curve where a pressure of 20 cm H_2O on the ascending loop is already located at the plateau of the curve. This is of advantage, since small changes of pressure will not affect the volume. On the other hand, the fixatives impart stiffness to the tissue and aldehydes act rapidly; last but not least, as mentioned above, the final lung shape depends on whether the lung has been fixed inside the closed chest, inside the chest but after removal of the ventral chest plate, or completely outside the chest and isolated.

Regarding fixative, the most commonly used solution is 1% glutaraldehyde in isotonic buffer (phosphate or cacodilate) (Gil, 1977; Hayat, 1981). The lung is then immersed into the same fixative and diced. Fragments of lung must then be washed in isotonic buffer (1 h, or 3 × 10 min) if possible while shaking, and this must be followed by osmification for 1-2 h (1-1.5% osmium tetroxide in isotonic buffer, collidine, or similar).

Vascular Perfusion (Fig. 2)

This technique is appealing because the fixative fills the dense capillary network and allows a very rapid and thorough fixation of the whole organ in record time;

moreover, since the air spaces stay filled with air, it permits a preservation of the airspaces and of alveolar surfactant (Gil, 1985; Gil and Weibel, 1969/70) in their original configuration without abolishing the tissue or fluid–air interfacial tension, which occurs when the alveoli are flooded with fixative. Technically, it is far more demanding than airway instillation because it requires strict control and proper adjustment of both air and perfusate inflow pressures. As a result, the procedure becomes involved and has a substantial rate of failure in unexperienced hands.

The first step after inducing deep anesthesia and immobilizing the animal on its back is to practice a tracheostomy and insert a gauge of appropriate size and length into the trachea. If the cannula fits tightly, it is advisable to sharpen its tip to facilitate insertion. The cannula must be tightened by a ligature externally placed around the whole trachea. Among practical difficulties associated with this first step, the following should be noted: the cannula slips out easily; the ligature collapses the cannula, occluding the trachea; the tip of the cannula perforates the intrathoracic trachea creating a pneumothorax or mediastinal emphysema; and the cannula is too long and becomes enmeshed in one of the stem bronchi resulting in inflation of only one lung.

After the tracheal cannula has been successfully placed, it is necessary to ensure that a valve system appropriate for both artificial ventilation after opening the chest and for holding intrapulmonary air pressure at a constant level during the perfusion is in place. Small animal ventilators cannot normally hold the pressure constant and cannot easily be rebuilt for that purpose. We prefer a constant air pressure source (gas cylinder, institutional air pressure pipe, small motor), but when the pressure source is a mechanical oil pump, air filters are needed. Medicinal gas cylinders are an excellent source.

Fine air pressure regulation requires some ingenuity. We like to immerse a side tube of the air line into an open water container to act as as overflow valve whenever the pressure exceeds the depth by which the tube is immersed. It is best to have at least two fine flow regulators. The final goal is to be able to set up a system that maintains constant air pressure by immediately adding gas whenever volume is lost, but without ever exceeding a preset pressure value. Ventilation can be conveniently accomplished with a three-way stop clock. The user must remember two positions for ventilation: inspiration (open communication between air pressure source and trachea, side opening closed; and expiration (open communication between trachea and side opening to the outside, pressure source closed). Physiological experiments require recording of the air pressure. This is achieved by connecting a side arm of the tracheal cannula to a pressure transducer. Although the specifications of the common transducers require them to be filled with fluid, we have found them to work reasonably well when air filled. Transducers can be connected to any standard amplifying–monitoring system linked to a video display and/or paper recording and equipped with an instantaneous and an averaging channel. This set-up can be used to create a record of the mechanical history and to monitor and document

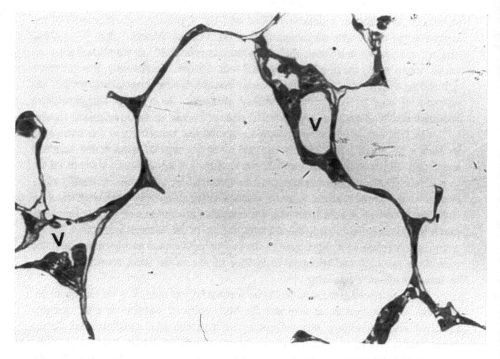

(A)

Figure 2 Electron micrographs of rabbit lungs fixed by instillation of glutaraldehyde into the trachea following pneumothorax. (A) An alveolar space, E1 squamous alveolar epithelium, E2 cuboidal alveolar epithelium cell. Notice bulging, maximally distended capillaries (5000×). (B) High-power micrographs of rabbit lung fixed by instillation (as above). Note the thinness of the air-blood barrier consisting of epithelium (E1) and endothelium (EN) (30,000×).

the stability of the desired pressure during the experiment. This record is particularly important if the adjustment of zonal conditions is contemplated. In general, the simplest way of establishing a volume history is to run a full inflation-deflation cycle and then slowly deflate from full inflation down to the desired range. Since the tissue undergoes some stress relaxation after changing volume, it is often necessary to readjust the pressure after several seconds. By then, the experimentator will have determined whether the lung leaks or not.

Next, vascular surgery can be started. The experimentator performs first a midline skin incision from the neck to the abdomen to dissect free the chest plate. The diaphragm should then be carefully punctured from the abdominal cavity to avoid damage of the lung when cutting the ribs, and the sternum must be removed to expose the heart and greater vessels. The next step is to widely open the

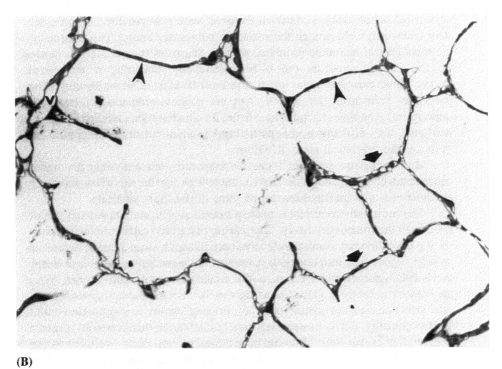

(B)

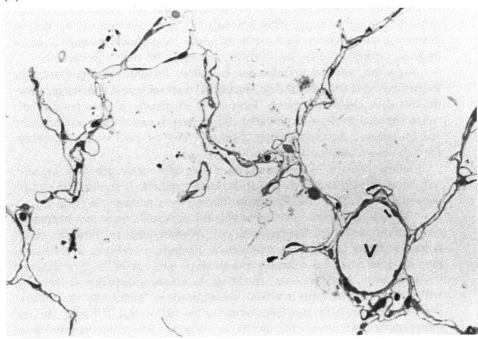

(C)

pericardial sac exposing as much of the great vessels as possible. A strong silk string must be placed around the aorta and pulmonary artery. This is probably the most critical and underestimated step. A sharp instrument is often needed to form a tunnel across the root of both vessels by puncturing the mediastinal fat and fibroconnective tissue; great care must be taken to avoid ligating either pulmonary veins or the left auricle. Any one of these problems will lead to an impairment of outflow, irregularities in the flow distribution, and increased flow resistance. This difficulty occurs particularly in small rodents and represents the most common technical cause of failure.

After the ligature is placed, it must be temporarily left loose while the surgeon concentrates on pulling down the heart to immobilize it in the axis of the pulmonary artery so that a proper incision can be done in the right ventricle.

The cut into the ventricle is made perpendicular to and immediately below the axis of the truncus arteriosus. The plastic (not glass) catheter to be introduced into the pulmonary trunk should have been dilated by heat at the tip to prevent its later slippage. In introducing it, care must be taken not to insert it so deeply that it abuts against the bifurcation, as this would obstruct the outflow path. When the catheter is safely in place, the string can be tied and the left appendage immediately transected to open an outlet for drainage before excess pressure builds up. Occasionally, it may be necessary to adjust left atrial (that is venous) pressure to levels higher than zero. This can be achieved by introducing vertically a cannula into the left appendage, tying a string around it, and devising some simple device to hold it perpendicularly to the operating table. The atrial pressure is adjusted by cutting the height of the cannula to a level equivalent to the desired pressure in a hydrostatic column, beyond which the perfusate coming out of the lungs will overflow. This step is always necessary in zone 3 perfusions.

After this, vascular perfusion can be started. Possible pressure sources for the perfusate could be gravimetry (placing the reservoir at a certain height above the animal) or a peristaltic pump. Flow out of a peristaltic pump is pulsatile and we recommend it without reservation. An air trap is needed between the pump and the catheter. An inverted drop counter of the type used in clinical settings for intravenous infusion is suitable.

Inflow pressure is measured via a side tube connected to a pressure transducer. If adjustment of a particular zonal condition is required, it is again necessary to record all fluid pressures during the experiment (see Fig. 2).

Our experience shows that for reasons difficult to understand, not all pressure combinations are feasible and few work well. When no other specifications are required, we choose the following combination: 10 mmHg air pressure on the descending slope, after two full inflations and an inflow pressure of 12–13 mmHg and 0 atrial pressure. These are Zone 2 conditions, since intracapillary flow (2–3 mmHg) will be the difference between arterial and air pressure. Technically, the effective perfusion pressure in the capillaries should be low, of less than 10 mmHg, but one must consider that pressure may be lost in a very thin intracardiac cannula due to capillarity effects. Therefore, in small animals the effective pressure will be lower than the transducer reading. In small rodents differences due to gravity in the lung can be neglected, but in larger animals they must be considered by measuring the distance between regions of interest in the line between the dorsal and ventral surfaces of the lungs. Zone 1 conditions can be best achieved with high inflation pressures (Gil, 1980, 1988; Rosenzweig et al., 1970; Warrell et al., 1972) and inflow pressures of the order of 10 mmHg. Adjustment of Zone 3 conditions is difficult: in our experience it has required a venous pressure of at least 8 mmHg, an air pressure of 6 mmHg, and a high inflow pressure of over 20 mmHg, (capillary driving pressure is 14, the difference between arterial and venous pressures) and has often resulted in edema formation. Irregular distribution of the fixatives is mostly due to accidental ligation of a pulmonary vein when placing the ligature around the pulmonary artery. The use of aldehydes, in particular glutaraldehyde, which is so common in electron microscopy because of its beneficial effects in the quality of fixation, poses a problem because it induces a vascular contraction which seems to increase the resistance to flow and requires an increase in the inflow pressure to squeeze it through. This makes it difficult to know the effective pressure in the capillaries. Although we prefer primary perfusion of osmium, glutaraldehyde followed by dicing and osmium post fixation has yielded excellent results in other laboratories.

B. Chemical Fixatives

For all practical purposes, our current picture of cells and tissues in the lung and in other organs is that conveyed by chemical fixation with glutaraldehyde and osmium. On the other hand, only osmium is needed when no subcellular studies are planned, because it is in the fixation and visualization of organelles and filaments where the additional use of glutaraldehyde is beneficial.

The mechanism of glutaraldehyde fixation, essentially a protein cross-linkage which probably has far reaching consequences, has been discussed in many technical handbooks of electron microscopy (Hayat, 1981). Osmium tetroxide, on the other hand, is the key fixative that cannot be omitted. By depositing along membranes, it confers electron density (that is, opacity) to them, which yields the outlines of the cell and most of its organelles. The problem with the osmium fixatives is twofold: Its extraordinary toxicity and its cost. For a number of practical reasons, we have repeatedly fixed lungs by vascular perfusion of osmium fixative. First, osmium abolishes membrane semipermeability, therefore reducing osmolarity damage to a minimum; glutaraldehyde, however, does not and the membranes remain semipermeable, which opens the possibility of considerable cellular swelling or shrinkage if the osmolarity of the perfusates is not properly adjusted; second, glutaraldehyde induces muscular contraction, which results in increased resistance to perfusion when introduced into the vasculature; finally, gluraraldehyde seems to induce a far greater level of chemical shrinkage than osmium, and osmium imparts a much higher level of stiffness to the tissue than glutaraldehyde. We have shown that after osmium perfusion of air-inflated lungs there is virtually no pulmonary retraction following removal of the source of air pressure (Gil et al., 1979). This signifies, of course, lack of mechanical recoil due to fixation of all structures, not lack of shrinkage which is a separate issue. If no elastic recoil occurs, one would infer that internal configuration of the tissue-air interface remains unchanged; if recoil, however small, is observed, other microscopic changes cannot be ruled out.

Comparison of Fixatives

When choosing a fixative, it is important to consider whether it is to be used inside the air spaces (instillation) or inside the vasculature (vascular perfusion). In both cases, we use the same concentration (1.5% glutaraldehyde or 1.5% OsO_4 in isotonic 330 mOsm buffer) in one of several buffers: cacodylate, phosphate, or collidin (not for glutaraldehyde). The choice of buffer is of little consequence, except that we showed that an artefact consisting of dark perimembranous granules appears when glutaraldehyde pretreated tissue is fixed with OsO_4 in phosphate buffer; this means that phosphate buffer is to be avoided for the osmification (Gil and Weibel, 1968).

Fluid intended for vascular fixation requires an additional increase of the osmolarity by addition of 1.5% Dextran 40, which empirically has been found to be satisfactory.

If the lung has been perfused with osmium, it is advisable to flush it out of the tissue prior to dicing to reduce the risk of exposure to the experimentor. We attempt this by following the osmium perfusion with a final perfusion with uranyl acetate. There is always a risk of precipitation because uranyl acetate precipitates at the pH of 7.4 used for the osmium. Others have recommended an in situ perfusion of ethanol to reduce elastic recoil (Oldmixon et al., 1985).

III. Physical Fixation

In nonspecialized language the term "to freeze" signifies the immediate arrest of an action or movement. The goal of correlating structure and function would be greatly advanced if it were possible to literally "freeze" the lung in a well-defined condition. In fact, attempts at rapidly freezing the lung were reported many years ago (Staub and Storey, 1962). The other physical fixing agent is heat, which causes protein denaturation and stabilization (as in cooking). A major difficulty with both agents is that they penetrate very slowly, in particular in an organ that contains more air than tissue and is therefore well insulated. The physical techniques of practical use are the rapid freezing followed by freeze substitution and the microwaving of the tissue.

A. Rapid Freezing

Freezing techniques have played a major role in specialized research areas in physiology, ultrastructure, and immunobiology; among them freeze fracturing deserves special mention. We will limit this discussion to methods useful to study whole lungs, ignoring specialized procedures in cell biology. One must distinguish between preservation for light and for electron microscopy, because the degree of technical difficulty is different. In lung research, the goal followed by the choice of rapid freezing is the same as in vascular perfusion: the fixation of alveolar spaces and airways and their contents (surfactant, macrophages, bronchial mucus) in their original condition without filling them with fluid.

Successful studies at the light microscopy level were reported as early as 1962 by Staub and Storey (1962). The main difficulty is ice formation due to the slow reduction of the temperature. If freezing could be achieved fast enough, it would result in vitrification of the whole organ without crystal formation. Unfortunately, the conducting properties of the tissues and the additional insulation provided by the alveolar air, result in a slowly penetrating cold front with a large temperature gradient which causes the formation of large crystals, which tear and disrupt normal structures. In general, there is little hope of achieving vitrification of more than a few microns under the pleural surface and this has discouraged many users. Mazzone et al. (1978, 1979) have recently revived interest in the approach for ultrastructural studies in electron microscopy by adapting and modifying a freeze-substitution procedure originally introduced by Pease 1967 until it yielded satisfactory results. The two-steps procedure involves (1) rapid freezing in situ and (2) freeze substitution.

Rapid Freezing In Situ

Rapid freezing is achieved by pouring a cold fluid over the exposed surface of the lung. Liquid nitrogen is the coldest commonly available fluid, but is not recommended, because in contact with the body, it boils, generating an insulating coat of nitrogen gas which retards the freezing process. Some other fluids, worth mentioning include Freon 22, 70% ethylene glycol in Hank's buffer, isopentane, and propane. Among these, Freon 22 and ethylene glycol are probably the most popular. Freon 22 has a temperature of $-155\,°C$, whereas ethylene glycol can be supercooled to $-80\,°C$ by immersing it in the popular mixture of acetone and dry ice. One would assume that, since the lower the temperature, the better the chances of vitrification, Freon is the better choice, but in biology things do not necessarily work that way. Mazzone et al. (1978) reported that the same results can be achieved with either agent, although the depth of the outer shell of crystal-free glassification is greater with Freon. Even so, there is no expectation of preserving well for electron microscopy a subpleural zone deeper than 100–150 μum. For light microscopy, where the quality expectations are lower, this could be extended to 0.5 mm, although at this depth large ice crystals are already visible.

This matter was studied in detail by Weibel et al. (1982). One could consider the instillation or perfusion of cryoprotective agents (glycerol, DSO), but this would defeat the main purpose of the technique, which is the preservation of the lung in its original condition without chemical treatment. After the temperature has been stabilized, the specimen can be sliced preferably with a precooled sharp knife, for further processing.

Fixation by Freeze-Substitution

Freezing does not constitute fixation; it is still necessary to perform some additional procedure capable of stabilizing the structure during and after thawing. The oldest approach to achieve that goal is *freeze-drying*, where the water is sublimated in vacuum without thawing at −70 °C; this produces acceptable results at the light microscope level, but excessive artefacts in electron microscopy (Chase 1959). For electron microscopy, the recommended procedure is *freeze substitution*, in which the water is replaced slowly by immersing the frozen tissue in a cold solvent such as acetone and a fixative solution (typically osmium in acetone) for a long period (usually several weeks). Mazzone et al. (1978) introduced a useful freeze-substitution method in electron microscopy that yielded acceptable results at least within a narrow subpleural region. The procedure is basically an adaptation of Pease's (1967) approach. After in situ freezing, samples of tissue measuring $1 \times 2 \times 4$ mm are placed in a large volume (50–60 ml) of cold 70% ethylene glycol containing 4% glutaraldehyde and 2% paraformaldehyde. The tissue is kept in a freezer at −50 °C under slow rotation for 24-36 h. The tissue is then rapidly thawed to +4 °C by plunging the blocks into a 60% ethylene glycol solution at that temperature. The ethylene glucol is then removed by successive passses through 50, 30, 20, and 10% ethylene glycol for 10-15 min. The samples are then placed in Hank's balanced salt solution containing again 4% glutaraldehyde and 2% formaldehyde. From here, the standard electron microscopy procedure is followed: washing, post fixation in osmium, block staining, dehydration, and embedding. Weibel and collaborators performed a critical study of that technique and found that it indeed yields well-preserved tissue. There were, however, a number of frequent artefactual problems, such as hemolysis, irregular cell outlines, breaks in cell membranes, and so-called "hemocrystals" (intracellular crystalline hemoglobin precipitates). The overall configuration of septa and preservation of surfactant and other intraalveolar fluids seems acceptable and the gross appearance of tissue and surfactant correlates well with that observed in perfusion-fixed lungs, except that the blood vessels are filled with blood.

Recently Sobin et al. (1988) have reported a method for the processing of the inflated lung intended for studies of the collagen and elastin networks at the light microscopy level. Their method involves double clamping, rapid freezing and freeze substitution with OsO_4 in CCl_4 followed by celloidin embedding.

B. Microwaving

Unquestionably, cooking prevents food decay by stabilizing its molecular structure in a highly complex way and therefore cooking is a form of fixation. However, water boiling is much too slow to be seriously considered. Microwaves, have found some limited applications. In histochemistry, they have been recommended to raise the temperature and rate of penetration of the fixative (Login et al. 1987). The main appeal of all physical procedures is that they have the potential to preserve original structures in their native conditions. To achieve that goal, a very fast fixation is needed and a truly satisfactory microwave source ought to be very powerful, to the point where it probably would create an unacceptable health hazard for the staff. We only know of efforts by Sweeney et al. (1983) in light microscopy who used the technique to study the sites of particle disposition on the airway surface. They placed inflated lungs inside a commercial microwave of the kind intended for domestic use for 20 min and subsequently embedded them externally in a polymerizing foam; the lungs could then be thinly sliced in a commercial meat slicing instrument (as luncheon meats are sliced) and processed for light microscopic evaluation. Although it is probable that the original configuration of the tissue and location of intraluminal particles was well preserved, the general quality of preservation is disappointing.

C. On Artifacts

With the availability of high-quality and high-resolution electron microscopes in the 1960s, the community of the lung scientists witnessed a great surge of interest in pulmonary ultrastructure; indeed for the next decade many important observations were reported, and many structural correlations were established. On certain problems, however, controversial and conflicting interpretations have been reported. Those who are not professionally trained morphologists and do not like a certain line of work will find an easy outlet in uncritical charges of "artifacts" and "technical limitations," which are not always easy to counter. It has always been very difficult to define good fixation, because a good standard for comparison is lacking. In fact, we may well be biased by the effects of osmium fixation and one could sustain the view that everything we see is an artifact.

Weibel et al (1982) defines criteria of goodness based on the compatibility with established findings in other comparable cells of the body and in consistent reproducibility. This writer personally believes in *reproducible beauty*, a poorly definable attribute. Beauty has always been thought to reside in the eye of the beholder. In this case, the qualified beholder is one with reasonable training in biological morphology. But *reproducibility* is of course the leading consideration, because an artifact is either not reproducible at all or its reproducibility is linked to specific circumstances. An example of the latter was an artifact described by Gil and Weibel (1968) linked to the use of phosphate buffer with osmium following glutaraldehyde fixation. This ugly artifact consisted in dark round deposits of electron-dense material located along cell surfaces, in particular the luminal epithelial surface. Its appearance was linked to a reaction between phosphate, osmium, and bivalent cations in the tissue. It can be avoided by using any buffer other than phosphate for the osmium.

Another frequently observed artifact are heavy deposits also along luminal surfaces found when the tissue has been block stained with uranyl acetate. They are due to the fact that uranyl acetate is soluble only at acid pH (around 5.5) and is therefore usually prepared with maleate at that pH. The pH of all other reagents and fixatives is customarily adjusted to 7.4, at which point uranyl acetate, if still present in bulk amounts, will precipitate along surfaces. This can be avoided by vigorously washing the tissue with acid maleate buffer before changing from the uranyl solution to another solution at neutral pH.

Osmolarity is another difficult issue. Grossly hypotonic solutions will cause cellular swelling, cytoplasmic washouts, and even tears of membranes; hypertonic solutions will cause increased cytoplasmic density and loss of membrane contrast. This is particularly true with glutaraldehyde fixation, which (in contrast to osmium, preserves the semipermeable characteristics of the cell membrane). Mathieu et al. (1978) showed the importance of a well-defined osmolarity on the reproducibility of quantitative results.

Conditions for fixation must be clearly defined and applied in a standardized manner. When starting our vascular fixation technique, we noticed unusual osmolarity-related problems. While the time-honored isotonic (330 mOsm) solutions worked very well in lungs fixed by instillation into the airways, they seemed to consistently cause cellular swellings, and intraalveolar edema which was undesired because it causes the surfactant layer to be washed off. As mentioned above, hypertonic solutions (500 mOsm) are not a valid alternative because they lead to loss of contrast and general increase in electron density. Years ago, most people had drawn a practical difference between osmotic pressure due to small solutes and so-called "oncotic" pressure, due to macromolecules. For this reason, we started adding dextran 40 to isotonic 330 mOsm solutions. The amount of dextran was approximately 1.5% to create "oncotic" pressure. This step has repeatedly proven to be useful and necessary for perfusion, although few scientists in our day see substantial differences between the two pressures.

In attempting "physiological" preservation (Gil, 1977, 1985, 1988; Gil and Weibel; 1972; Gil et al., 1979; Mazzone et al., 1978, 1979. Rosenzweig et al., 1970; Warrell et al., 1972), the issue arises frequently as to whether the result is a true depiction of the original prevailing conditions. The word "artifact" is not properly applied to discrepancies in this case. If we were taking photographs of a bird, it would be no artifact to show the bird with its wings open or folded or in between. The issue is whether it shows the wings open when we shoot the picture after making them open and when we believe that they should be open. One could also imagine the difficulty in correlating wind velocity with the bending of the branches of a tree in a photograph. These examples give an idea of the tremendous challenge facing those workers interested in preserving surfactant or specific configurations of the alveolar surface or the capillaries related to respiratory movements or perfusion changes.

From the published literature, it is evident that the degree of filling of the pulmonary capillaries is variable and dependent on the prevailing zonal conditions (Fig.2). But it is, above all, the alveolar surface itself that shows many different configurations. On one hand, we have the glutaraldehyde-filled lung where the septa are completely expanded with capillaries bulging into the fixative-filled air spaces (Fig. 1); on the other hand, the air-filled lung, preserved by either vascular perfusion or rapid freezing, exhibits features such as septal pleats, infoldings, and pools of intraalveolar surfactant (Fig. 2). Regarding the airspace morphology and dimension in our previous comparative study of the inflation patterns of fluid- and air-filled rabbits, we determined that the expansion pattern differs quantitatively depending on whether inflation is with air or water (Gil et al., 1979). Many times this problem can be alleviated by performing morphometry at full inflation, where dimensions are likely to be similar.

IV. Outlook for the Future

Over the years, lung morphology has been plagued by a number of famous problems. Most of them have been solved, a few have not, and still a few remain controversial.

For at least a century, morphologists wondered about the epithelial lining of the alveoli (is there one or not?). What is a "kernlose Platte" (unnucleated plate) (Weibel, 1971)? The electron microscope brought the answers: unnucleated plates are squamous epithelial cells and yes, the epithelial lining is continuous.

Another fundamental question was how to measure items of interest in lung histology, that is, how to obtain from histological sections the same quantitatiave data available from physiological recorders. The now classic book, *Morphometry of the Human Lung* by Weibel explaining how to do this met a need and was eagerly received (Weibel, 1963).

It was precisely the success of morphometry which led to the greatest challenge of our times in morphological research: the so-called "physiological fixation." The goal of the exercise was to adjust well-known physiological conditions and to fix the organ without altering anything. First, emphasis was put on the need for "standard conditions," that is, fix the lung anyway the worker wishes, provided it was always done exactly the same. This was applied consistently to all identifiable single factors: osmolarity, temperature, pH, concentration of fixatives instillation pressure. But soon an additional number of issues arose: Why was surfactant not discernible on routine micrographs? What happened inside the air spaces during inspiration and expiration? What were the anatomic correlates of pulmonary mechanics? What is the blood flow distribution?

These problems prompted the development of alternative techniques such as rapid freezing and perfusion by vascular perfusion. This innovative work proved useful (Gil, 1985; Gil and Weibel, 1969/70). The surfactant turned out to be present mostly in the form of fluid pools in alveolar corners, sometimes lined by an osmophilic line thought to represent a monolayer; the filling degree of capillaries varies depending on circumstances; inflation and deflation appear to take place by pleating and folding, and corrugation and plastic adaptations at many locations of the alveolar surface rather than by any kind of elastic stretching of the walls and the patterns of inflation and deflation can markedly differ. In recent work, we have proposed that functional lung anatomy can be best understood by regarding the organ as an aggregate of tubes (alveolar ducts) which change in length and diameter during respiratory movements (Ciurea and Gil, 1990). This notion deemphasizes somewhat the significance of alveoli, whose lateral walls ("secondary septa") are incompletely developed. Ductal changes in length and width require corrugation and undulation of walls supported by a collagen network (Sobin et al. 1988) with pleating of corners devoid of fibers. Little sketching of the alveolar wall takes place. These exciting findings depart from the usual rigid version of structures as entities of unalterable size and shape (think of the meaning of the common phase "anatomically fixed") and as a whole they are difficult to understand. Additionally, a number of details remained to be filled out. That is where we stand and that is where the outlook for the future begins. May it bring us a full development of the notion of physiologic fixation and a complete quantitative and qualitative description of all changes that take place in the lung during respiratory movements. For instance no true understanding of surfactant function will be possible as long as we do not precisely know what happens in the alveolar surface during respiratory movements.

Acknowledgments

The author is indebted to Dr. D. Ciurea and Mrs. M. Barbee for their skillful help.

References

Chase, W. H. (1959). The surface membrane of pulmonary alveolar walls. *Exp. Cell Res.* **18**:15–28.

Ciurea, D., and Gil J. (1990) Morphometric study of human acini based on viral reactions *J. Appl. Physiol.* (in press).

Gil, J. (1977). Preservation of tissues for electron microscopy under physiological criteria. In *Techiques of Biochemical and Biophysical Morphology* Vol. 3. Edited by D. Glick, and R. M. Rosenbaum. WileyInterscience, New York.

Gil, J. (1980). Organization of microcirculation in the lung. *Ann. Rev. Physiol.* **42**:177–186.

Gil, J. (1985). Histological preservation and ultrastraucture of alveolar surfactant. *Ann. Rev. Physiol.* **47**:753–63.

Gil, J. (1988). The normal lung circulation. State of the Art. *Chest* **93**:80S–82S.

Gil, J., Bachofen, H., Gehr, P., and Weibel, E. R. (1979). The alveolar volume to surface area relationship in air and saline filled lungs fixed by vascular perfusion. *J. Appl. Physiol.* **47**:990–1001.

Gil, J., and Weibel, E. R. (1968). The role of buffers in lung fixation with glutaraldehyde and osmium tetroxide. *J. Ultrastr. Res.* **25**:331–348.

Gil, J. and Weibel, E. R. (1960/70). Improvements in demonstration of lining layer of lung alveoli by electron microscopy. *Resp. Physiol.* **8**:13–36.

Gil, J., and Weibel, E. R. (1972). Morphological study of pressure-volume hysteresis in rat lungs fixed by vascular perfusion. *Respir. Physiol.* **15**:190–213.

Haies, D. M. Gil, J., and Weibel, E. R. (1981). Morphometric study of rat lung cells. I Numerical and dimensional characteristics of parenchymal cell population. *Am. Rev. Resp. Dis.* **123**:533–541

Hayat, M. A. (1981). *Principles and Techniques of Electron Microscopy.* University Park, Baltimore.

Kikkawa, Y. (1970). Morphology of the alveolor lining layer. *Anat. Rec.* **167**:390–400.

Kuhn, C. (1972). A comparison of freeze-substitution with other methods for preservation of the pulmonary alveolar lining layer. *Am. J. Anat.* **133**:495–508.

Login, G. R. Schnitt, S. J., and Dvorak, A. M. (1987). Rapid microwave fixation of human tissues for light microscopic immunoperoxidase identification of diagnostically useful antigens. *Lab. Invest.* **57**:585–591.

Mathieu, O., Claasen, H. and Weibel, E. R. (1978). Differential effect of glutaraldehyde and buffer osmolarity on cell dimensions: a study on lung tissues. *J. Ultrastr. Res.* **63**:20–34.

Mazzone, R. W., Durand, C. M., and West, J. B. (1978). Electron microscopy of lung rapidly frozen under controlled physiological conditions. *J. Appl. Physiol.* **54**:325–333.

Mazzone, R. W., Durand, C. M., and West, J. B. (1979). Electron microscopic appearances of rapidly frozen lung. *J. Microscopy* **117**:269–284.

Oldmixon, E. H., Suzuki, S., Butler, J. P., and Hoppin, F. G., Jr. (1985). Perfusion dehydration fixes elastin and preserves airspace dimensions. *J. Appl. Physiol.* **58**:105–113.

Pease, D. C. (1967). Eutectic ethylene glycol and pure prophylene glycol as substituting media for the hydration of frozen tissues. *J. Ultrastr. Res.* **21**:75–97

Rosenzweig, D. Y., Hughes, J. M. B., and Glazier, J. B. (1970). Effects of transpulmonary and vascular pressures on pulmonary blood volume in isolated lung. *J. Appl. Physiol.* **28**:553–560

Sobin, S.S., Fung Y. C. and Tremer H. M. (1988) Collagen and elastin fibers in human pulmonary alveolar walls. *J. Appl. Physiol* **64**: 1659-1675.

Staub, N. C., and Storey W. F. (1962). Relation between morphological and physiological events in the lung studied by rapid freezing. *J. Appl. Physiol.* **17**:381–390.

Sweeney, T. D., Brain, J. D., Tryka, A. F., and Godleski, J. J. (1983). Retention of inhaled particles in hamsters with pulmonary fibrosis. *Am. Rev. Respir. Dis.* **128**:138–145.

Warrell, D. A., Evans, J. W., Clarke, R. O., Kingaby, G. P., and West, J. B. (1972). Pattern of filling in the pulmonary capillary bed. *J. Appl. Physiol.* **32**:346–356.

Weibel, E. R. (1971). The mystery of "non-nucleated plates" in the alveolar epithelium of the lung, explained. *Acta Anat.* **78**:425–443.

Weibel, E. R. (1963). *Morphometry of the Human Lung*. Academic Press, New York.

Weibel, E. R., Limacher, W., and Bachofen, H. (1982). Electron microscopy of rapidly frozen lungs: evaluation on the basis of standard criteria. *J. Appl. Physiol.* **53**:516–527.

2

Fixation of Human Lungs

MARIANNE BACHOFEN and H. BACHOFEN

University of Berne
Berne, Switzerland

I. Introduction

For investigating the structure of tissue elements and their alterations, routine fixation techniques yield sufficient information for most diagnostic purposes. Either by surgical biopsy or at autopsy lung tissue probes are removed, immersed in a fixative (usually a formalin solution), and then processed by the methods that are most adequate to resolve the diagnostic problem. However, in routine lung preparations the tissue architecture of the expanded lung is not preserved. Neither the determinants of the functional state of the lung, such as the dimensions of peripheral air spaces, the alveolar air liquid interface, the geometric arrangement of fibers, the fluid distribution, the calibers of airways and vessels, nor the extent and distribution of pathologic alterations can be quantitatively evaluated. Furthermore, the preservation of cell structures is not sufficient for electron microscopy.

If morphological data are required that more precisely reflect the state of the living organism, more demanding fixation techniques are required. In the past (Gough and Wentworth, 1949; Blumenthal and Boren, 1959; Tobin, 1952) and more recently (Heard, 1960, 1969; Dailey, 1973; Weibel, 1963, 1970/71;, 1984; Gil, 1977; Gil and Weibel 1969/70; Bachofen et al., 1975) several techniques have been devised to fix human lungs without causing gross distortions of the delicate structural framework. This chapter presents some useful and feasible methods, and discusses their advantages, disadvantages, and limitations.

II. Goals and Limitations

To date, the application of more elaborate fixation methods, which ensure acceptable preservation of tissue architecture and/or tissue elements, has proved to be particularly useful in at least four different research areas.

Gough and Wentworth (1949) fixed whole inflated lungs at autopsy, cut sections from the whole lung, and mounted them on paper. The resulting overviews of the normal and morbid anatomy of lung parenchyma impressively illustrate the extent and distribution pattern of diffuse lung diseases, and may even suggest pathogenetic mechanisms. In addition, whole-lung slices are invaluable tools for teaching. Low-power magnification of thick slices also allowed investigators to visualize the three-dimensional arrangement of peripheral air space structures in healthy and diseased lungs (Heard 1960, 1969).

Whole-lung fixations are required to establish the correlations between radiological and pathological findings (Cureton and Trapnell, 1961; Dailey, 1973; Heitzman, 1973). In combination with contrast injections into vessels and airways, such studies have definitely improved the art of reading chest x-rays (Heitzman, 1973).

Quantitative evaluations of the dimensions of peripheral airways and air spaces yielded important insights into the relations between structural and functional alterations of lungs in patients with obstructive lung disease (Heard, 1960, 1969; Thurlbeck, 1967; Bignon et al., 1969; Depierre et al., 1972; Cosio et al., 1977, Berend et al., 1981). The same methods were applied in epidemiological studies, in particular for estimating the relations between the dose and damaging effect of air pollution and smoking (Sobonya and Kleinerman, 1972; Cosio et al., 1980).

Improvements in techniques for lung tissue preparations (Weibel, 1963, 1970/71; Bachofen et al., 1975; Gehr et al., 1978) and morphometric methods (Weibel 1963, 1979, 1984) allowed for better definitions of the structure–function relations within the framework of physiological studies, and provided more insight into some disease processes and the mechanisms of the corresponding functional impairments of the lung (Weibel, 1963; Thurlbeck, 1966; Butler and Kleinerman, 1970; Kapanci et al., 1972; Bachofen and Weibel, 1977; Gehr et al., 1978).

In parallel with the vast efforts to improve fixation methods and to preserve human lung tissue more usefully for various research purposes, much experimental work has been done on animal lungs, with strong focus on lung fixation (Gil,

1977; Bachofen et al., 1982; Weibel et al., 1982; Weibel, 1984). In particular techniques have been determined to preserve lung tissue under well-controlled physiological conditions in order to establish and test hypotheses on the effect of the structural design on lung mechanics and pulmonary gas exchange. In theory, this knowledge and sophisticated methods can be applied for the fixation of all mammalian lungs, including human lungs. However, the fixation of human lungs raises additional major obstacles owing to the fundamental differences between a dying patient and an experimental animal: fixations of human lungs under optimally controlled conditions with regard to physiology will never be feasible. In addition, different legal barriers are present in different countries and states. In many places autopsies are not allowed immediately postmortem and hence the study of the lungs' ultrastructure may not be possible. Even beyond any legal restrictions, there are ethical issues to be considered carefully. Disregarding the will of the deceased, hurting the feelings of relatives, or neglecting the ideological environment in the hospital not only endangers the research project but may also damage the reputation of clinical research in general.

III. Methods of Fixation

To preserve the architecture of the lung as much as possible, and to allow comparison with other experiments, the lungs have to be fixed in a controlled state of inflation. The choice of the fixation procedure depends on the purpose of the examination and on the particular condition prevailing at the patient's death. Attempts to reinflate edematous lungs or to preserve tissue taken long after death (at autopsy) for ultrastructural studies are wasted efforts.

A. Air Drying

This method is of mere historical interest. Simple air drying does not fix the air-inflated lung, but reversibly stiffens the fibrous skeleton, especially the elastic fibers. Upon contact with water vapor, or with aqueous solutions, a partial elastic collapse will occur. More important, however, is that the cellular elements cannot be fixed by drying, and the extraction of water from lung tissue without replacement by another fluid inevitably results in shrinkage and distortion of the delicate structures. Hence, no currently acceptable level of histologic quality of specimen can be achieved (Tobin, 1952).

B. Formalin Vapor

Lungs excised at autopsy can be fixed in a fairly well controlled state of expansion by formaldehyde vapors. This technique has a real advantage, that the specimen can easily be radiographed for pathological–radiological comparisons, and an apparent advantage: the fixative does not directly alter the alveolar air–tissue interface (Blumenthal and Boren, 1959; Pratt and Klugh, 1961). However, the earlier techniques had several shortcomings. A fixation period of 3–5 days was required, and the resulting specimen was dry, and therefore inadequate for histological analysis. Later, Weibel and Vidone (1961) improved the method by delivering formalin steam to the lung enclosed in a box and inflated by negative pressure. Since artifacts such as shrinkage could readily be accounted for (Weibel, 1968), the procedure produced good results for morphometric evaluation at the light microscopic level. However, the method is not suitable for fixing lungs with edema and tissue consolidations, and even in normal or near-normal lungs the tissue preservation is certainly not sufficient for electron microscopy.

C. Airway Instillation

Instillation of a suitable fixative into the airways of a collapsed lung is the basic procedure for lung fixation. Under normal or near-normal conditions, a fixative solution evenly expands the lung parenchyma because it eliminates the surface forces in the collapsed alveoli and peripheral airways. In most experiments done so far the lungs excised at autopsy were filled with a 10–20% neutral buffered formalin solution. Special care is required to control the inflation of the lung, and to maintain a constant volume during the fixation period (Heard, 1969). Although widely used, formalin is not an ideal fixative. The lungs remain flabby and are difficult to cut, the degree of shrinkage cannot readily be accounted for, and the ultrastructure of lung tissue is not well preserved. In spite of various methodological problems, however, the analysis of lungs fixed by this simple procedure has yielded a wealth of information about the morphology, pathogenesis, and structure–function relations of chronic obstructive lung disease (Thurlbeck, 1967; Bignon et al., 1969; Heard, 1969; Depierre et al., 1972; Cosio et al., 1977, 1980; Berend et al., 1981).

With two improvements, fixation by airway instillation has become a standard method for fixing human and animal lungs (Weibel, 1970/71), 1984; Weibel and Knight, 1964). First, the use of glutaraldehyde instead of formalin is preferable. A solution of 2.5% glutaraldehyde, buffered with 0.03 M potassium phosphate to a pH of 7.4 (total osmolarity about 350 mOsm), hardens the lung to a point where crushing artifacts caused by cutting and handling of blocks are minimal, and ensures excellent preservation of lung tissue (including the capillary contents) for both light and electron microscopy.

Second, it is preferable to leave the lung in the chest to prevent distortions due to the weight of the fluid-filled organ. In the ideal case, within 30 min after confirmed death of the patient both lungs are brought to collapse by a 1 inch incision in the midaxillary lines of the fourth intercostal space. The fixative is then instilled through a cuffed endotracheal tube, using a positive pressure of 20–30 cmH$_2$O above the chest. Great care should be taken that the cuff and the tubings are absolutely tight, to protect the body and in particular the face of the deceased from contamination with the fixative. With the lungs filled completely, the endotracheal tube is occluded. At autopsy (i.e., hours later), the lungs are carefully removed in toto without displacing the endotracheal tube. The main bronchi are ligated and the right and the left lung separated and cleaned of adhering tissues. Their volume can be determined by volumetry (Scherle, 1970; Gehr et al., 1978).

Pressure of about 20–30 cm H$_2$O is necessary to ensure even inflation of all parts of the lung, which is necessary for reliable sampling of lung tissue. It also ensures sufficient unfolding of the alveolar walls in order to estimate reliably the total alveolar surface area. Higher pressures do not improve the result but may crush the capillaries.

Despite the standardized pressure, the degree of lung inflation achieved by this procedure is not quite constant. Usually, the fixation occurs at lung volumes of 50–70% of total lung capacity (Gehr et al., 1978). In most cases the limitation is probably set by the individually varying compliance of the thorax. However, at this point it seems worth mentioning that the glutaraldehyde instillation should be done as swiftly as possible. At least in rat lungs it has been conclusively shown that the degree of lung inflation depends on the rate of fixative infusion (Gholamhossain et al., 1980). With slow instillation the fixation process may decrease the compliance of lung parenchyma enough to limit the degree of the final expansion.

An important prerequisite for the success of this method is that the lungs collapse sufficiently at the thoracic incision, and that a homogeneous reexpansion can be achieved by the subsequent instillation. Therefore, in severe lung diseases such as the adult respiratory distress syndrome the procedure is barely applicable. Also in other conditions with voluminous masses of inflammatory tissue and edema fluid, penetration of the instilled fixative may not be fast enough to prevent autolytic tissue alterations. Finally, the method is not suitable if one intends to quantitate alterations of the airways including the extent of mucous plugging. Niewoehner and Kleinerman (1974) have proposed an alternative: they clamped the airway and inflated the lung by instillation of the fixative via arteries and veins.

D. Transthoracic Injection

If a fixation of the lung by airway instillation is not feasible, and if technical or legal conditions prohibit an autopsy immediately after death, human lung tissue cannot be sufficiently preserved for electron microscopy. This can be overcome by early transthoracic injections of a fixative into the peripheral air spaces of the intact lung without opening the thorax. By adding a dye to the fixative the fixed tissue samples can be retrieved more easily at autopsy (Bachofen et al., 1975).

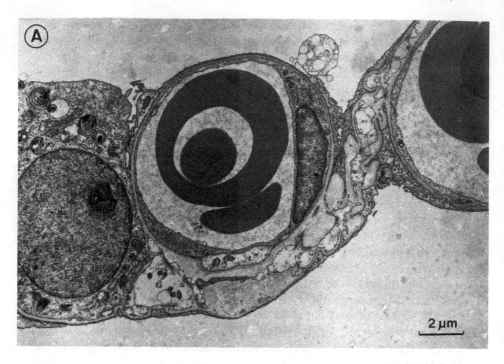

Figure 1   Autolytic lesions dependent on the premortal condition of the lung. A. Healthy lung fixed 8 hr postmortem. Except for minimal autolytic lesions (swelling of pericytes and fibroblasts), the tissue is well preserved. B. *Pneumocystis carinii* pneumonia. Although the lung was fixed 1½ hr postmortem, the autolytic alterations are conspicuous: all cells are swollen and partially destroyed (arrows).

The injection is done within 30 min after confirmed death, using a long thin needle (20 gauge, 10 cm long), which is introduced through an intercostal space. After it passes the rib, which serves as a landmark for the proximity of the pleura, the needle is advanced into the lung to a depth of 2–3 cm. While 5–10 ml of the solution is gently injected, the needle is slowly withdrawn to achieve adequate distribution of the fixative. Experience has shown that the success of the procedure depends partly on the location of the injection. Axillary, lateral, or dorsal injection sites are chosen, the latter not far below the scapular angle. Ventral injection sites pose some pitfalls: the lung is thin toward the anterior margin, and spreading of the fixative may be inadequate because of rapid draining through the bronchial tree. To ensure adequate representation of the tissue samples studied, one should always attempt to fix several probes in different locations; this is particularly important with diseased lungs. In some cases, radiographs of the lung may help in the choice of preferred injection sites.

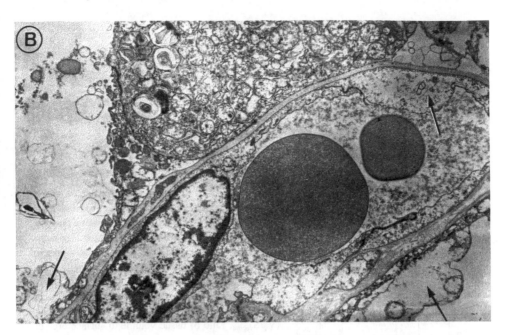

As fixative, we use the buffered 2.5% glutaraldehyde solution recommended for instillation fixation to which we add methylene blue (a small drop of a concentrated solution to 10 ml of fixative). This allows recovery of the fixed samples when the lung is removed from the chest at autopsy. These samples are trimmed from unfixed tissue and stored in the fixative without the dye; they are then processed for light and electron microscopy as for instillation-fixed material.

With this procedure the tissue is fixed in situ and hence the air spaces and capillaries preserve the geometric arrangement of the lung expanded in the intact chest. Gentle injection of the fixative, avoiding the use of excessive force, is important to avoid distortions of the delicate tissue elements. The tissue quality corresponds to that obtained by instillation fixation, but the delay between death and fixative injection is a crucial factor, especially in diseased lungs. Whereas normal lung tissue remains well preserved for hours, early autolytic lesions can be observed in diseased lungs (Bachofen et al., 1975). Probably depending on the type and severity of disease, extensive autolytic alterations may occur as early as 1.5 hr postmortem (Fig. 1).

This method allows only "blind" fixations of small focal samples. In some cases of diffuse lung disease one sample from one fixation site may be representative of the entire lung, both in a qualitative and morphometric sense. In other cases, however, the structural inhomogeneities may be considerable (Fig. 2; Gaensler, 1981). Hence, multiple injections must be performed and samples

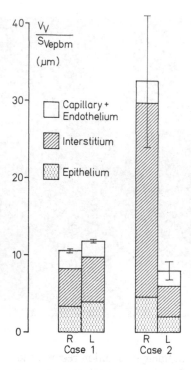

Figure 2 Even in diffuse lung disease, a single tissue probe might not yield results representative for the whole lung, as shown by a comparative analysis of lung samples from the right (R) and left (L), fixed by transthoracic injection. In case 1 with subacute adult respiratory distress syndrom (ARDS), consistent data were obtained. In case 2 with chronic ARDS (right columns), morphometric measurements showed large variations between right and left lung (V_v/S_{vepbm}, ratio of tissue volume to surface area of the epithelial basement membrane; values shown are means $\pm$ SE).

from different locations analyzed. It is also possible, in a second step, to instill the total lung at autopsy with glutaraldehyde or formalin, and to use the secondarily fixed tissue for light microscopic examination in order to account for the regional inhomogeneities.

E. Vascular Perfusion

In order to retain the alveolar surface lining layer, and to preserve the alveoli in their natural air-filled state, lung fixation by vascular perfusion is the method of choice. Special techniques have been developed and tested to fix animal lungs under well-controlled physiological conditions (Gil and Weibel, 1969/70; Gil, 1977; Gil et al., 1979; Bachofen et al., 1982). For obvious reasons, however, these procedures are not readily applicable for the fixation of human lungs, and one must be content with compromises, at best. There are at least three prerequisites for a perfusion fixation to be successful: (1) the lung must be viable (i.e., an autopsy immediately after death must be feasible), in rare cases an intact lung may be obtained from a transplant donor at surgery; (2) the microvasculature has to be intact: pathological increase in capillary permeability results in flooding of the alveoli, and thus annihilates the potential advantages of perfusion fixation;

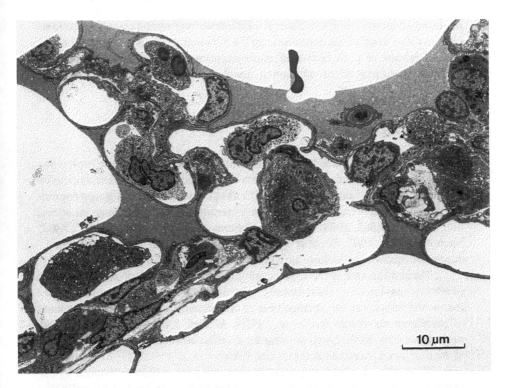

Figure 3 Lung tissue from a patient with neurogenic pulmonary edema, fixed by local perfusion. Note the undisturbed, smooth air–edema fluid interface (the dislocation of the erythrocyte is a processing artifact and serves to demonstrate the high quality of fixation, even of edema fluid and of the air–fluid interface).

(3) the vasculature must be patent: obstruction of vessels due to any cause prevents the fixative from reaching the desired site of action.

All these difficulties and limitations explain the scanty experience with this method. Wierich and Carmanns (1972) succeeded in fixing normal and diseased lungs (respiratory distress syndrome) of newborns. The lungs were excised 30–120 min after death, inflated and ventilated, and then fixed by perfusion with 5% glutaraldehyde. The resulting preparations appeared to be satisfactory for electron microscopic studies.

Information about perfusion fixation of whole adult lungs is not available, whereas attempts have been made to accomplish spatially limited lung fixations by delivering the fixative into a pulmonary artery. After cardiac arrest with a pulmonary catheter in situ, glutaraldehyde with dye was infused under low pressure into the lung periphery via a tightly wedged Swan-Ganz catheter. As with the transthoracic fixation method, the circumscribed pieces of fixed tissue could be retrieved at autopsy. In a few cases good electron micrographs could be produced, showing the septal configuration and the alveolar contents undisturbed by the fixation process (Fig. 3). However, more experiments are needed to evaluate the potential usefulness of the method to resolve some special problems of lung pathological conditions.

IV. Artifacts

No tissue preparation fixed by any of the known techniques reveals a perfect image of the living structure. However, once the artifacts are recognized and defined, the evaluation of the morphological findings can be improved by appropriate corrections. To this end several fixation experiments have been done on animal lungs (Weibel, 1963, 1984; Gil, 1977; Mathieu et al., 1978; Weibel et al., 1982; Bachofen et al., 1982; Brain et al., 1984; Oldmixon et al., 1985), and the accumulated knowledge is fully applicable to the fixation of human lungs.

The introduction of the fixative solution into the peripheral airspaces either by airway instillation or transthoracic injections eliminates the alveolar surface forces and thus alters the arrangement of the alveolar septa and the dimensions of peripheral air spaces (Gil et al., 1979; Weibel, 1984). Furthermore, both methods are unsuitable for preserving the alveolar contents (both fluids and cells) at their original location (Brain et al., 1984).

Much effort has been expended in correcting for shrinkage induced by different fixatives. However, by choosing glutaraldehyde instead of formalin as a fixative, shrinkage can be minimized. Much less attention has been paid to the fact that no fixative is able to fix the elastic fibers irreversible (Oldmixon et al., 1985). Alcohol does stiffen elastin, but not irreversibly. By processing the tissue (i.e., by cutting blocks and sections, by staining with aqueous solutions, or use of other methods), a barely controllable retraction of the tissue samples may occur. This artifact may result in an underestimation of the width of the peripheral air spaces and an overestimation of the alveolar surface area; the error is worst if the lungs were fixed at maximum inflation levels.

If the volume of cells, the interstitial space, and the width of the blood–gas barrier are the focus of interest, particular attention has to be paid to the osmolarity of aldehyde fixatives (Mathieu et al., 1978; Bachofen et al., 1982). In contrast to osmium tetroxide, aldehyde fixatives do not abolish the osmotic properties of cell membranes (Wangensteen et al., 1981), and hence nonisotonic solutions induce changes of cell volumes. Hypotonic solutions may also qualitatively impair the preservation of tissue.

However, even optimally fixed preparations of human lung tissue barely reflect the state of the living organism as well as samples of animal lungs obtained by well-designed experiments. This is true for two reasons. First, human lungs cannot be fixed under physiologically well-controlled conditions, and hence the recognition, correction for, and the prevention of artifacts are much more difficult. Second, one has to make allowance for ill-defined perimortal alterations, such as the relaxation of smooth muscles (Berend et al., 1981), the redistribution of intravascular fluid caused by cardiac arrest, and iatrogenic lesions.

V. Conclusion

Different techniques are available for satisfactorily preserving the architecture and tissue elements of human lungs for qualitative and morphometric studies at the light and electron microscopic level. Obviously, there is no "best recipe"; the specific problem to be resolved determines the method of choice. Due to legal and ethical considerations, which have to be carefully examined, the sophisticated methods developed for the fixation of animal lungs can never be fully exploited, and hence one must accept compromises with regard to the quality of structural preservation and artifacts. However, the knowledge gained from animal experiments may well serve to improve the interpretation of the morphology of less than perfect preparations of human lungs.

Acknowledgments

We are greatly indebted to Professor E. R. Weibel for his expert help in the preparation of the manuscript. This work was supported in part by grants from the Swiss National Science Foundation.

References

Bachofen,H., Amman, A., Wangensteen,D., and Weibel, E. R. (1982). Perfusion fixation of lungs for structure-function analysis: credits and limitations. *J. Appl. Physiol.* **53**: 528–533

Bachofen, M., and Weibel, E. R. (1977). Alterations of the gas exchange apparatus in adult respiratory insufficiency associated with septicemia. *Am. Rev. Respir. Dis.* **116**: 589–615.

Bachofen, M., Weibel, E. R., and Roos, B. (1975). Postmortem fixation of human lungs for electron microscopy. *Am. Rev. Respir. Dis.* **111**: 247–256.

Berend, N., Wright, J. L., Thurlbeck, W. M., Marlin, G. E., and Woolcock, A. (1981). Small airways disease: reproducibility of measurements and correlation with lung function. *Chest* **79**: 263–268.

Bignon, J., Khoury, F., Even, P., André, J., and Brouet, G. (1969). Morphometric study in chronic obstructive bronchopulmonary disease. *Am. Rev. Respir. Dis.* **99**: 669–695.

Blumenthal, B. J., and Boren, H. G. (1959). Lung structure in three dimensions after inflation and fume fixation. *Am Rev. Tuberc.* **79**: 764–772.

Brain, J. D., Gehr, P., and Kavet, R. I. (1984). Airway macrophages. The importance of the fixation method. *Am. Rev. Respir. Dis.* **129**: 823–826.

Butler, C., and Kleinerman, J. (1970). Capillary density: alveolar diameter, a morphometric approach to ventilation and perfusion. *Am Rev. Respir. Dis.* **102**: 886–894.

Cosio, M. G., Ghezzo, H., Hogg, J. C., Corbin, R., Loveland, M., Dosman, J., and Macklem, P. T. (1977). The relations between structural changes in small airways and pulmonary function tests. *N. Engl. J. Med.* **298**: 1277–1281

Cosio, M. G., Hale, K. A., and Niewohner, D. E. (1980). Morphologic and morphometric effects on prolonged cigarette smoking on the small airways. *Am. Rev. Respir. Dis.* **122**: 265–271.

Cureton, R. J. R., and Trapnell, D. H. (1961). Postmortem radiography and gaseous fixation of the lung. *Thorax* **16**: 138–143.

Dailey, E. T. (1973). Preparation of inflated lung specimens. In *The Lung. Radiologic–Pathologic Correlations.* Edited by E. R. Heitzman. St Louis, CV Mosby.

Depierre, A., Bignon, J., Lebeau, A., and Brouet, G. (1972). Quantitative study of parenchyma and small conductive airways in chronic nonspecific lung disease. *Chest* **62**: 699–708.

Gaensler, E. A. (1981). Open and closed lung biopsies. In *Diagnostic Techniques in Pulmonary Disease.* Part II. Edited by M. A. Sackner. New York, Marcel Dekker.

Gehr, P., Bachofen, M., and Weibel, E. R. (1978). The normal human lung: ultrastructure and morphometric estimation of diffusion capacity. *Respir. Physiol.* **32**: 121–140.

Gholamhossain, H., Crapo, J. D., Miller, F. J., and O'Neil, J. J. (1980). Factors determining degree of inflation in intratracheally fixed rat lungs. *J. Appl. Physiol.* **48**: 389–393.

Gil, J. (1977). Preservation of tissues for electron microscopy under physiological criteria. In *Techniques of Biochemical and Biophysical Morphology.* Edited by D. Glick and R. M. Rosenbaum. New York, Wiley-Interscience.

Gil, J., and Weibel, E. R. (1969/70). Improvements in demonstration of lining layer of lung alveoli by electron microscopy. *Respir. Physiol.* **8**: 13–36.

Gil, J., Bachofen, H., Gehr, P., and Weibel, E. R. (1979). Alveolar volume to surface area relationship in air- and saline-filled lungs. *J. Appl. Physiol.* **47**: 990–1001.

Gough, J., and Wentworth, J. E. (1949). The use of thin sections of entire organs in morbid anatomy. *J. R. Microsc. Soc.* **69**: 231–235.

Heard, B. E. (1960). Pathology of pulmonary emphysema. *Am. Rev. Respir. Dis.* **82**: 792–799.

Heard, B. E. (1969). *Pathology of Chronic Bronchitis and Emphysema.* London, Churchill, Livingstone.

Heitzman, R. E. (1973). *The Lung. Radiologic–Pathologic Correlations.* St. Louis, CV Mosby.

Kapanci, Y., Tosco, R., Eggermann, J., and Gould, V. E. (1972). Oxygen pneumonitis in man. Light- and electron-microscopic morphometric studies. *Chest* **62**: 162–169.

Mathieu, O., Claassen, H., and Weibel, E. R. (1978). Differential effect of glutaraldehyde and buffer osmolarity on cell dimensions: a study on lung tissue. *J. Ultrastruct. Res.* **63**: 20–34.

Niewoehner, D. E., and Kleinerman, J. (1974). Morphologic basis of pulmonary resistance in the human lung and effects of aging. *J. Appl. Physiol.* **36**: 412–418.

Oldmixon, E. H., Suzuki, S., Butler, J. P., and Hoppin, F. G. jr. (1985). Perfusion dehydration fixes elastin and preserves lung airspace dimensions. *J. Appl. Physiol.* **58**: 105–113.

Pratt, P. C., and Klugh, G. A. (1961). A technique for the study of ventilatory capacity, compliance, and residual volume of exised lungs and for fixation, drying and serially sectioning in the inflated state. *Am. Rev. Respir. Dis.* **83**: 690–696.

Scherle, W. F. (1970) A simple method for volumetry of organs in quantitative stereology. *Mikroskopie* **26**: 57–60.

Sobonya, R. E., Kleinerman, J. (1972). Morphometric studies of bronchi in young smokers. *Am. Rev. Respir. Dis.* **105**: 768–775.

Thurlbeck, W. M. (1966). The internal surface area of nonemphysematous lungs. *Am. Rev. Respir. Dis.* **95**: 765–773.

Thurlbeck, W. M. (1967). Internal surface area and other measurements in emphysema. *Thorax* **22**: 483–496.

Tobin, C. E. (1952). Methods of preparing human lungs expanded and dried by compressed air. *Anat. Rec.* **114**: 453–466.

Wangensteen, D., Bachofen, H., and Weibel, E. R. (1981). Effects of glutaraldehyde or osmium tetroxide fixation on the osmotic properties of lung cells. *J. Microsc.* **124**: 189–196.

Weibel, E. R. (1963). *Morphometry of the Human Lung.* Berlin–Göttingen–Heidelberg, Springer.

Weibel, E. R. (1968). A note on lung fixation. *Am. Rev. Respir. Dis.* **97**: 463–465.

Weibel, E. R. (1970/71). Morphometric estimation of pulmonary diffusion capacity. I. Model and method. *Respir. Physiol.* **11**: 54–75.

Weibel, E. R. (1979). *Stereological Methods.* Vol. 1: *Practical Methods for Biological Morphometry.* London, Academic Press.

Weibel, E. R. (1984). Morphometric and stereological methods in respiratory physiology including fixation techniques. In *Techniques in the Life Sciences*. Edited by A. B. Otis. Ireland, Elsevier Scientific Publishers.

Weibel, E. R., and Knight, B. W. (1964). A morphometric study on the thickness of the pulmonary air–blood barrier. *J. Cell. Biol.* **21**: 367–384

Weibel, E. R., Limacher, W., and Bachofen, H. (1982). Electron microscopy of rapidly frozen lungs: evaluation on the basis of standard criteria. *J. Appl. Physiol.* **53**: 516–527.

Weibel, E. R., and Vidone, R. A. (1961). Fixation of the lung by formalin steam in a controlled state of air inflation. *Am. Rev. Respir. Dis.* **84**: 856–861.

Wierich, W., and Carmanns, B. (1972). Postmortale Perfusionsfixation der Lungen von Neugeborenen bei Atemnotsyndrom. *Beitr. Pathol.* **146**: 94–98.

3

Electron Microscopy in Lung Research

CHARLES KUHN III

Brown University Program in Medicine
Memorial Hospital of Rhode Island
Pawtucket, Rhode Island

I. Introduction

The limit of resolution of the light microscope is 0.2 μm. Two tenths of a micron is also the thickness of the alveolar–capillary membrane of rodent lungs, and that of the human lung is only slightly greater. No wonder, then, that even such elementary questions of lung structure as the existence of an epithelium covering the alveolar capillaries could not be settled to the satisfaction of all anatomists until the great resolving power of the electron microscope was applied to the lung by Low (1952). Indeed, because of the extreme thinness of the alveolar structures, it is often difficult to distinguish the cells of the walls of airspaces without the use of electron microscopy. In routine paraffin sections stained with hematoxylin and eosin, endothelial cells, pericytes, interstitial macrophages, septal connective tissue cells, and smooth muscle cells are often indistinguishable. Each appears as an elongated nucleus in the interior of an alveolar wall. They are readily identified by electron microscopy. The type II alveolar epithelial cells had been noted by occasional anatomists as long ago as 1871 (Parrot, 1871), but only became generally accepted as a distinct cell when electron microscopy was used to examine the lung (Karrer, 1956; Kisch, 1955; Policard et al., 1957). Electron microscopy is one of several tools that eventually led to the clear differentiation

37

of alveolar macrophages from type II cells. In the conducting airways, the use of electron microscopy (EM) has clarified the interrelationships of secretory epithelial cells and remains the only technique that can be used to identify brush cells, for which we currently know of no unique secretory product or immunological marker (Breeze and Wheldon 1977; McDowell et al., 1978; Reid and Jones, 1979).

The application of a range of powerful histochemical and immunological techniques to the investigation of lung structure supplements but does not replace the use of EM as a tool for the study of the normal and diseased lung, just as EM supplements but does not replace light microscopy. Of course many histochemical and immunohistochemical techniques can also be adapted to the electron microscope. The particular strengths of EM as an approach are, first of all, its resolving power. Within the thickness of an alveolar septum, often no greater than 10 μm, many delicate cell processes and extracellular matrix components are packed, which necessitate excellent spatial resolution for study. Fine intercellular relationships and subcellular organization can only be observed by electron microscopy (EM). The second strength is its broad range of applicability. Unlike histochemical or immunological approaches, it requires no specific reagents and no foreknowledge of the structures that will be fruitful for study. An important limitation is the small volume of tissue that can be studied. Applications of EM to pulmonary research also include studies of cells, organelles, or molecules separated from the lung. Such applications do not differ from similar applications to other tissues and will not be discussed further.

II. Types of Electron Microscope

Two types of electron microscope are widely available: transmission electron microscopes (TEMs) and scanning electron microscopes (SEMs). Each uses a different mechanism of image formation and gives a different kind of information (Fig. 1) (Watt, 1985). In the TEM, a thin specimen, usually less than 0.1 μm thick, is placed in the electron beam. The electrons pass through the specimen and are brought to a focus on a fluorescent screen or photographic film beneath it. Contrast results from scattering of electrons as they pass through the specimen. The unscattered electrons pass through the specimen to interact with the fluor of the screen or photographic emulsion. Scattering of electrons is a function of the atomic number of the atoms making up the specimen. The bulk of biological material is made up of elements of low atomic number (hydrogen, oxygen, carbon, and nitrogen), so most biological samples have little inherent electron contrast. Contrast is usually introduced by the OsO_4 used nearly universally as a fixative, and by the use of heavy metal stains.

In the SEM, the electrons are focused on the surface of the specimen and scanned across it. As they are deflected over the surface of the specimen, atoms

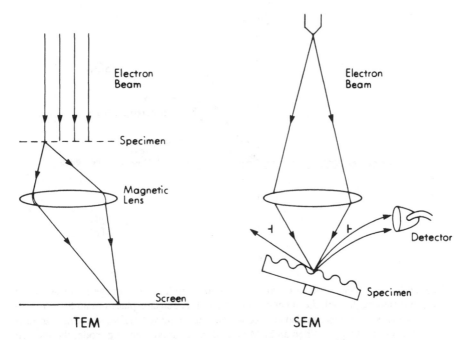

Figure 1 Schematic diagrams contrast the two principal types of electron microscope. In the transmission electron microscope (TEM), the electrons passing through a thin specimen are brought to a focus on the plane of a viewing screen or photographic emulsion. In the scanning electron microscope (SEM), the lens is used to bring the beam to a focus on the surface of a specimen. The focused beam is scanned over the specimen. Emissions from the surface of the specimen are collected by a detector and processed electronically to generate an image.

near the point where the electrons enter the specimen are excited by the absorbed energy. As they decay back to ground state, they give off the absorbed energy by emitting secondary electrons, x-rays, or photons of light (cathodoluminescence) (Fig. 2). These secondary emissions all contain potential information about the specimen. In addition, some of the incident electrons are elastically scattered as a result of interactions with atomic nuclei in the specimen, and these electrons also contain information about the specimen. Ordinarily the secondary electrons are used to generate an image. The other types of emission are used for special applications, particularly analytical microscopy.

Back-scattered electrons are of variable energy, up to the energy of the incident beam as a limit. Secondary electrons are of much lower energy, usually less than 20 ev, and consequently are readily absorbed in the specimen. Thus

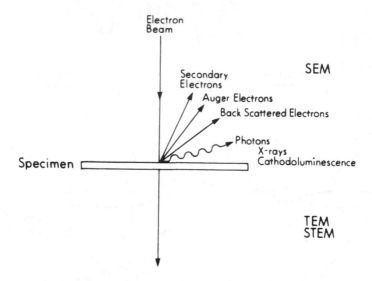

Figure 2 Emissions used in electron microscopy. In TEM the image is formed from electrons passing through the specimen unscattered. A detector placed in an SEM to collect unscattered electrons can also be used to generate an image (STEM, scanning transmission electromicroscopy). When an SEM is used in the usual scanning mode, the focused beam is scanned over the specimen and energy from the incident beam excites atoms in the specimen. As the atoms decay back to ground state, they emit low-energy secondary electrons, x-rays, and photons of light, which are used to create an image of the specimen. High-energy back-scattered electrons can also be used.

only secondary electrons emitted at the surface of the specimen escape. The contour of the surface determines the likelihood of electrons escaping and the image developed displays the surface in relief.

The types of specimen used in the two types of EM are quite different. TEM uses thin slices through tissue embedded in plastic; the images that result are essentially two-dimensional, and spatial resolution is excellent (down to 1 nm in practice with biological material). SEM gives a three-dimensional representation of the surface of the specimen with great depth of field. The thickness of the specimen is irrelevant, because the signal comes from its surface. The resolving power, operating in the secondary electron mode, is typically 10–30 nm, which is better by an order of magnitude than light microscopy but less than that of a TEM.

III. Preparation of Lung for Transmission Electron Microscopy

The basic procedures for TEM include fixation, dehydration, embedding, sectioning, and staining. Fundamental techniques were worked out in the 1950s and 1960s and are now standard. An investigator unfamiliar with this methodology should consult standard texts such as those by Hayat (1986) or Glauert (1972–1980) for an introduction. Nevertheless, the lung presents certain peculiarities compared to most solid tissues. We will point here out some of the choices to be considered when planning a study of lung.

The airspaces of the lung necessarily change in shape and volume during the respiratory cycle and whether control of lung volume is an important concern for the questions to be addressed, must be determined. An additional consideration is that the more the lung is expanded, the less actual volume of solid lung tissue can be studied in a section of fixed area. Therefore, full expansion is not invariably advantageous. The choice of optimal chemical fixatives to be used is influenced by the structures of particular interest. Finally, even the dehydration schema can be varied purposefully.

A. Fixation

Immersion of small tissue blocks in fixative has the advantage of saving fixative and may be the most practical way to handle biopsies or samples from large lungs. Another advantage is that the proportion of tissue (as opposed to empty air space) per unit area is large, since the lung is collapsed. The major drawbacks are that the tissue may be traumatized by mincing while it is unfixed. Penetration of fixative to the cells is slow, and the distortion of airspaces due to collapse and compression during mincing may make interpretation of the histological appearance difficult.

Instillation of fixative through the trachea or bronchi is a widely used procedure with a number of advantages. It is technically simple, rapidly brings fixative in contact with the tissue, expands the alveoli so that the histological appearance is easily interpreted, and is convenient for many types of morphometry. Limitations are that the alveolar lining fluid and airway secretions are dispersed, particles and wandering cells such as macrophages (Brain et al., 1984) can be displaced, and the tissue/air ratio is low. The distending pressure is usually controlled at 20–30 cm H_2O, but Hayatdavoudi et al. (1980) showed that the rate of instillation of fixative is more important than pressure in determining the final volume of the fixed lung. Kuhn and Tavassoli (1976) found no difference in the appearance of airspaces of hamster lung (by SEM) when pressures were increased from 25 to 100 cm of fixative. Hence, for quantitative studies it is important to introduce fixative as rapidly as possible, since the chemical reaction of fixative with tissue reduces the distensibility of the lung, whereas any pressure over 15 cm H_2O suffices to inflate the lung to total lung volume. In

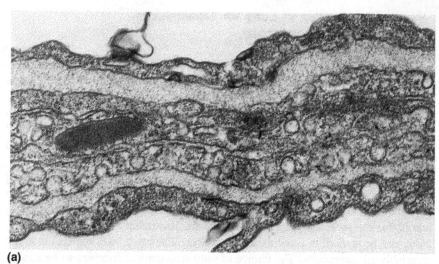

(a)

Figure 3 The appearance of human alveolar basal lamina using different primary fixatives. a. When 3% glutaraldehyde is used, the basal laminae are relatively structureless and the lamina lucida and lamina densa cannot be differentiated. b. When a 1% tannic acid– 2% glutaraldehyde mixture, is used, the basal lamina has a fibrillar texture. A narrow lamina lucida is visible to the left (uranyl acetate–lead citrate stains, x 50,000).

practice most investigators open the chest widely and cannulate the trachea or even excise and degas the lung before cannulation. Under these conditions the lung at 20 cm H_2O or above might expand beyond total lung capacity in vivo since the restraining influence of chest wall compliance is removed. If it is important that lung volume remain physiological, the lung can be collapsed in the animal by making a small puncture in each hemidiaphragm, the chest wall left intact, and the fixative instilled through the trachea until the diaphragm reaches the desired level.

The perfusion of fixative through the right ventricle or pulmonary artery is technically more demanding than airway instillation. It preserves secretions and free alveolar cells in place and, since it leaves the airspaces filled with air, retains the configuration of airspaces as modeled by surface tension. Lung volume can be controlled and fixative can be perfused in sequence to achieve optimal chemical fixation. The method is described in detail by Gil and Weibel (1969) and in Chapter 1.

Choice of fixative formulation is discussed in detail in Chapter 1. Osmolality should approach physiological but details of buffer choice and fixing agent depend on the purpose of the study and, to an extent, personal esthetics. Nevertheless, certain particular problems may influence one's choice of fixative. Retention

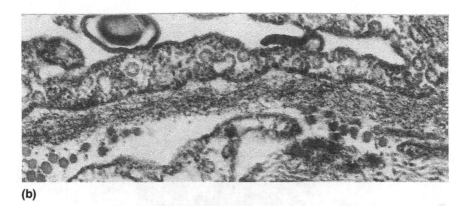

(b)

of the doubly saturated phosphatidyl choline (DSPC) of surfactant has received close attention. Uranyl acetate (Collet, 1979) and tannic acid (Kaline et al., 1977) both aid in the retention of DSPC. Tannic acid (1%) can be added to glutaraldehyde as primary fixative just before use, or it can be used for en bloc treatment after OsO$_4$. Tannic acid also has advantages in visualizing protein components of surfactant (Hassett et al., 1980), elastic fibers (Fukuda et al., 1984; Kajikawa et al., 1975), basal lamina (Vaccaro and Brody, 1981), and aspects of ciliary ultrastructure (Kuhn and Engleman, 1978, Wilsman et al., 1987) (Figs. 3,4). Retention of proteoglycans can be improved by the addition of cetyl pyridinium chloride, alcian blue, or ruthenium red to the primary glutaraldehyde fixative. Inclusion of ruthenium red during the osmium tetroxide postfixative selectively stains proteoglycans as well as preventing their extraction (Luft, 1971a, b; Vaccaro and Brody 1979).

For particular applications, it may be desirable to sacrifice overall fixation to obtain optimal visualization of one particular structure. Excellent visualization of the ultrastructure of ciliary axonemes can be obtained if the tissue sample is treated with a detergent such as triton x-100 or nonidet P-40 to disrupt the plasma membrane prior to fixation with a tannic acid–glutaraldehyde mixture (Fig. 4) (Wilsman et al., 1987). In lung, detergent treatment also is required for visualization of proteoglycans in the extracellular matrix by ruthenium red (Vaccaro and Brody, 1979). In its absence, the ruthenium red only stains the proteoglycans of the epithelial glycocalyx and fails to penetrate to the interstitium.

B. Dehydration

Dehydration is an obligatory step before infiltration with any of the commonly used embedding materials. Because DSPC, the most abundant lipid in surfactant, is not chemically fixed by routine aldehyde–OsO$_4$ fixation, it may be extracted into organic solvents used for dehydration. A variety of approaches have been used to optimize preservation for lamellar bodies, the storage organelles

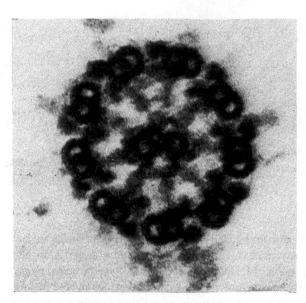

Figure 4 Cross-section of the axoneme of a human nasal cilium. the sample was treated with buffered 0.1% triton X-100 detergent to dissolve the membrane, permitting the soluble proteins to be washed away. Fixation was in tannic acid-glutaraldehyde mixture followed by osmium tetroxide. Radial spokes, dynein arms, and microtubule subunits are easily visible (uranyl acetate–lead citrate stain, x 250,000).

for surfactant. Since the appearance of this organelle in micrographs is highly variable, seemingly dependent on species (Kikkawa and Spitzer, 1969), degree of maturation, and technical factors, it becomes difficult to establish criteria for optimal preservation. Although freeze–fracture has its own artifacts (Schultz et al., 1980), it avoids the problems of fixation and extraction. Therefore, reproducing the appearance of the surface of freeze-cleaved lamellar bodies offers a reasonable "gold standard" (Fig. 5). As mentioned above, the use of tannic acid fixation an en bloc treatment with uranyl acetate improves retention of DSPC. In addition, various dehydration schemes can be used. Surfactant phospholipids have been reported to be precipitated by acetone (Gluck et al., 1966). Askin and Kuhn (1971) and Hallman et al. (1976) found that acetone extracts appreciably less lipid from lamellar bodies than does ethanol. Stratton compared a number of polar dehydration fluids (Stratton, 1976a, b). Epon 812 is more hydrophilic than most epoxy resins and is somewhat water miscible. Stratton combined tannic acid fixation and en bloc uranyl acetate treatment with dehydration with Epon without a transitional organic solvent. He concluded, based on the morphological appearance of the sections, that this method gave excellent lipid retention. Acetone dehydration

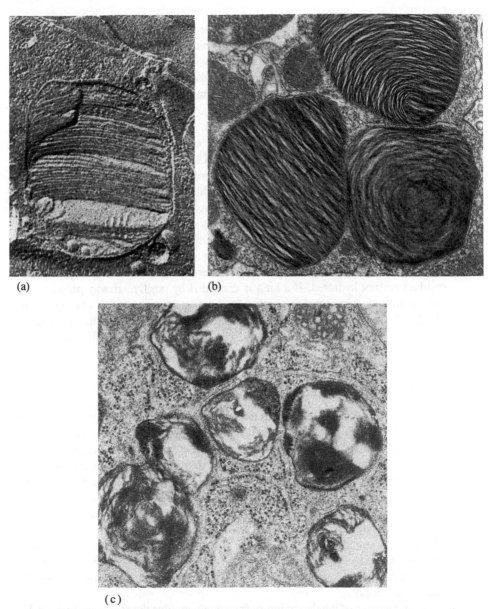

(a)

(b)

(c)

Figure 5 Variation in the appearance of lamellar bodies in rat type II cells. a. Freeze-cleaved lamellar body with regularly spaced, straight lamellae. b. Lamellar body fixed sequentially with glutaraldehyde, osmium tetroxide, and uranyl acetate, and then dehydrated with acetone. Lamellae are well retained and closely spaced. Their wavy appearance may be the result of tissue shrinkage. c. Routine fixation in glutaraldehyde followed by osmium tetroxide and ethanol dehydration. There has been considerable extraction of the contents of the inclusions (a, freeze fracture replica; b and c, uranyl acetate lead citrate; all approximately x 40,000).

is simple and convenient and is probably adequate for most applications. It can be easily incorporated into the routine of any electron microscopy laboratory, since it is as satisfactory as ethanol for routine work. For critical studies, however, Stratton's procedure may be preferred. A biochemically controlled comparison of the two dehydrations has not been reported.

Another situation in which the technique of dehydration needs consideration is when control of airspace size is critical. When the lung is distended, elastic fibers are stretched. Elastin is poorly fixed by chemical fixatives and retains much of its elasticity during fixation (Sobin et al., 1982). When the distending pressure is released, elastic fibers will return to their resting configuration even in a lung that is well fixed by histological criteria. This distorts the shape of alveoli.

The energy for the contraction of elastic fibers depends on the presence of an aqueous environment. Hydrophobic amino acid residues buried in the interior of helical regions of elastin in the resting configuration are exposed to water as the fiber is stretched. When the distending force is released, the molecule will tend to return to the energetically more stable coiled state in which the hydrophobic residues are not hydrated. If a lung is distended by positive airway pressure and then fixed and dehydrated by perfusing the vessels with graded concentrations of ethanol, the elasticity of the elastin disappears and airspace size does not change when airway pressure is subsequently reduced to atmospheric (Oldmixon et al., 1985).

C. Selection of Blocks

Since the volume of tissue that can practically be studied by electron microscopy is a tiny fraction of the whole, selection of tissue is critical and should be guided by knowledge of the relevant histological characteristics at a light microscopic level. Although random sampling is often a goal for morphometric studies when correlations with bulk physiological properties are to be made, it may be inappropriate for study of pathological conditions that involve the tissue sparsely or nonrandomly. When the lesions or structures to be studied are infrequent, it may be desirable to fix the lung by instillation of fixative through the airways and then slice it. The slices can be studied under a dissecting microscopy and small airways, vessels, or lesions identified and dissected out. Perfusion fixation can also be used but differentiation of vessels from airways is more difficult after perfusion because the vessels are empty.

Many types of inhalation injury selectively involve the region of the bronchioloalveolar junction. Stevens and Evans (1973) select this area from rat lung after fixing, dehydrating, and infiltrating thin slices of entire lobes with epoxy. Blocks containing bronchioloalveolar junctions are identified under a dissecting microscope and flat embedded in inverted BEEM capsules. Barry et al. (1985) selectively studied the first alveolar duct bifurcation by punching out blocks containing a complete terminal bronchiole and its alveolar ducts with a trochar and embedding them with the bronchiole in the tip of the capsule. They obtained the bifurcation by trimming through the bronchiole and continuing until the bifurcation was reached. In our laboratory we do histochemical and immunohistochemical tests on pathological material containing small focal lesions. Since the histochemical techniques use frozen sections, it was convenient to adapt methods developed for sampling small nuclei in the CNS. Glass microscopy slides and coverslips are prepared by siliconizing them overnight and drying. The frozen sections, after staining and dehydration, are infiltrated in epoxy resin, transferred to the siliconized slide, covered, and the epoxy is polymerized overnight. The coverglass can be removed with a razor blade, leaving an optically flat surface for microscopy. Lesions of interest can be observed, cut out with a razor, and glued to blank epoxy blocks with cyanoacrylic glue for thin sectioning.

IV. Scanning Electron Microscopy

The first application of SEM to lung was to study particle deposition (Holma, 1969). Because SEM can be used to study large areas of surface at high magnification, and can be equipped with detectors with a capacity for elemental analysis and mapping, it has continued to be a valuable tool both for the study of particle deposition and for the identification and enumeration of particles extracted from lungs. In addition, the morphology of virtually any surface that can be exposed can be studied to advantage: airways, alveoli, pleural mesothelium vessels perfused free of blood, or even connective tissue elements. SEM has proven to be a sensitive way to detect focal abnormalities, selective for one part of the acinus. For example, low–dose ozone injury produces local increases in phagocytic cells in alveoli just distal to the terminal bronchioles. Even when such increases were not detectable as a change in total cells in bronchoalveolar lavage fluid, they could be measured by SEM applied to the bronchiolar–alveolar duct junction (Brummer et al., 1977). Holes in alveolar septa (pores of Kohn and fenestrae) can be seen, measured, and enumerated by SEM (Ranga and Kleinerman, 1980). Epithelial cells can be easily identified and changes in epithelial cell populations due to disease detected. Due to the complex geometry of a three–dimensional image of a tilted specimen, we believe that measurement of airspace size is more conveniently done by simple morphometry of histological sections. For visual impact, however, in illustrating abnormalities in airspace size, SEM has no peer. The new guidelines for recognizing emphysema point out that loss of the orderly pattern of the acinus is an early sign of lung destruction. In the small lung of rodents, SEM is a convenient way to demonstrate this sign (Kuhn and Tavassoli, 1976).

B. Preparing Lungs

Three steps are involved in preparing tissue for SEM: fixation, dehydration, and coating. Fixation must be done by a method that expands the lung to open the airspaces and expose the surfaces. A technique has been described for expanding the airspaces of biopsy specimens (Brody and Craighead, 1975). If possible, however, at least a lobe should be available. The choice is usually between instillation of fixation through the trachea or bronchus and vascular perfusion. With vascular perfusion, the fluids lining air passages are fixed in situ. Perfusion, therefore, is the method of choice for studying particle deposition, because particles and phagocytes are not displaced (Brain et al., 1984). However, many epithelial cells will be covered by the retained secretions (Fig. 6), cell surface specializations may be distorted by surface forces, and pores of Kohn will be filled by alveolar lining material (Kuhn and Finke, 1972; Parra et al., 1978). Therefore, for most applications instillation of fixative via conducting airways at

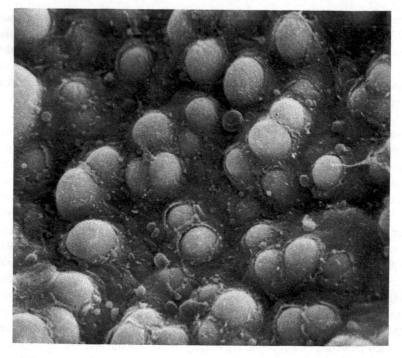

Figure 6   SEM of the lining of rat bronchiole, fixed by perfusion through the vascular system. Secretions line the bronchiole covering the cilia. The apices of Clara cells project through the secretions (x 2,000).

controlled pressure is preferred (Fig. 7). If aldehyde fixatives are used, blocks can be taken from the lung slices for light microscopy and TEM as well as SEM.

Unless the SEM is specially equipped with a cryostage, it will be necessary to remove water before exposing tissue samples to the vacuum of the column. Simply allowing the water to evaporate would allow enormous distortion of the tissue due to the high surface tension of water. Replacing the water with an organic solvent of lower surface tension improves the results, but still distorts delicate structures. To avoid having a surface pass through the tissues as the fluid evaporates, most tissues are prepared either by freeze drying or critical point drying. Since the lung is insulated by entrapped air, lung blocks freeze slowly, allowing ice crystal damage to take place. Critical point drying is consequently the method of choice.

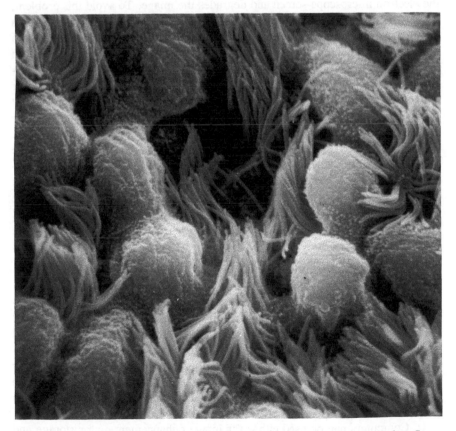

Figure 7 Rat bronchiole fixed by instillation of glutaraldehyde through the trachea. Secretions have been washed away and cilia are visible between the Clara cells (x 4,000).

The critical temperature of a fluid is the temperature above which no amount of pressure will maintain it in the liquid phase (Cohen et al., 1968). By warming a liquid under high pressure through its critical temperature, it is converted to gas. When the phase transition takes place at the critical point in the phase diagram, there is no surface tension between gas and liquid to distort the specimen. Critical point drying is carried out in a pressure chamber and a fluid with critical temperature close to room temperature is used for convenience. Several of the freons and liquid CO_2 give satisfactory results, but owing to environmental concerns about the release of freons into the atmosphere, CO_2 with a critical temperature of 45° is the fluid of choice. Many types of commercial apparatus are available and their manufacturers provide detailed protocols for their use.

Bombarding a tissue sample with an electron beam leads to the build up of electric charge on the sample. The charged area appears as a bright spot on the electron microscope screen and degrades the image. To avoid this problem, tissue samples must be rendered electrically conductive. Two types of approaches can be used: wet chemical reactions applied before drying or evaporation of a thin coating of metal on the sample after it is dried and mounted on a specimen holder (stub). Metal coating, usually gold or gold–palladium alloys, has the advantage of giving a high signal-to-noise ratio. While it is highly satisfactory for cells or simple surfaces, the relief on a lung slice is so deep and complex that one is rarely successful in completely coating the surface even using a "sputter coater" in which the metal alloy is ionized. Wet chemical methods such as the osmiumthiocarbohydrazide (OTC) ligand method (Kelley et al., 1973) give uniform conductivity, but with a lower signal-to-noise ratio, resulting in a slightly softer image than with metal coating. For best results, the two methods can be combined. The OTC procedure is carried out, and then the specimens are dehydrated, critical point dried, mounted on stubs with silver paint, and coated with a thin layer of gold palladium alloy.

V. Future Directions

Although the basic techniques of electron microscopy are now well standardized, the discipline will undoubtedly continue to evolve. Immunohistochemical techniques continue to improve, as discussed by Spicer in Chapter 7, and related specific ligand binding techniques such as lectin histochemistry are already being applied to dissection of the molecular organization of the lung. Other areas in various stages of development include computer technology to improve use of the data obtained by routine techniques, the alternative techniques of tissues preparation, and more advanced instrumentation.

Computers can be used either for image enhancement or for storage and manipulation of large amounts of data. An example of the use of computers for image enhancement is the analysis of the ultrastructure of cilia. Low contrast has inhibited quantitation of many of the components of cilia, such as radial spokes or inner dynein arms. Early studies of normal cilia overcame the difficulty of visualizing these structures by optical enhancement techniques that reinforced periodic structures while blurring out nonperiodic ones (Allen 1968). These approaches cannot be used for quantitative studies of pathological cilia, which require the ability to identify each spoke or arm.

Kennedy et al. (1988) collected electron microscopic images in digitized form and processed them by gray/scale enhancement to accentuate the structures of interest. Although they used scanning transmission (STEM) mode images, current technology also allows the use of conventional TEM to acquire full intensity range video images suitable for computer enhancement.

The large capacity and speed of modern computers permits the storage and manipulation of large numbers of serial sections in digital form. Using noncomputerized techniques, only rare individual lung cells have been serially sectioned and mapped using physical modeling (Adler et al., 1982). Computer reconstruction not only simplifies the reconstruction process but simultaneously provides accurate quantitation of organelles (Young et al., 1985, 1986) Even with computers, serial section reconstruction is technically demanding, and laborious, but it can document the detailed organization of the cell, and intracellular and intercellular relationships that are difficult to show with single random sections. (Parra et al., 1986).

Standard tissue preparation techniques using fixed dehydrated tissues have proven their value. Techniques based on rapid freezing offer some theoretical advantages (Sargent, 1988): chemical fixation is avoided, volume changes are minimized, motile structures are almost instantaneously immobilized, and no tissue components are extracted.

Specimens of rapidly frozen tissue can be directly examined in conventional SEMs equipped with a cryostage (Finch et al., 1987; Sargent, 1988); metal replicas of the natural or fractured surfaces can be prepared and studied by TEM; frozen tissue can be sectioned, stained, and studied by TEM; or the water can be slowly replaced by solvent and then the tissue infiltrated with embedding plastic before thawing (freeze substitution (Feder and Sidman, 1958). To be effective, however, freezing must be extremely rapid to avoid the disruptive effects of ice crystals. Because of its high content of air, lung is a particularly poor conductor of heat, and only a zone a few microns thick at the surface of the specimen vitrifies even with the most efficient means of heat removal (Weibel et al. 1982). Despite this limitation, the technique of freeze fracture has provided information about membranous structures and intercellular junctions that are not obtainable by any other technique (Chapter 4). The removal of water by sublimation from the fractured surface of frozen tissue (etching) before the preparation of replicas exposes nonmembranous tissue components (Heuser, 1981) and has provided powerful high-resolution three-dimensional images of intracellular structures and extracellular matrix components in other tissues. The challenge of adapting this approach to the lung should pay valuable dividends.

The technology of electron imaging continues to evolve. The technique of electron energy loss spectroscopy probably will see its greatest application in analytical microscopy (Chapter 12). The technique of scanning tunneling microscopy (STM), based on quite different principles from standard microscopy, promises to move electron microscopy to the atomic level. The STM probes surface structure by scanning an extremely fine needle over the surface of the sample. Present day technology permits the synthesis of needles with a single atom of metal at the tip. The needle is scanned so close to the surface of the sample that the wave functions of the electrons in the tip overlap the wave functions of the electrons in the sample. If a small voltage is applied between the needle and the surface, electrons tunnel through the gap. The wave functions decrease exponentially with distance, so the tunneling current is extremely sensitive to the width of the gap between the needle tip and surface. Features as small as atoms show up as variations in the tunneling current. The lateral resolution of the STM is 0.1 nm, but Z axis resolution of 0.001 nm is attainable (Quate, 1986). To date, the applications of the STM have been mainly in materials science. Biological samples are poor conductors of electrons and generate little signal. Therefore, new techniques of tissue preparation and improved instrumentation will be required before biological samples can be studied with atomic resolution (Guckenberger et al., 1988). However, the STM has been exploited to study replicas prepared by standard rapid freezing and metal coating methods at vastly superior Z axis resolution than could be obtained with conventional SEM (Zasadzinski et al., 1988). When synthetic phospholipid films have been studied (Zasadzinski et al., 1988) can the study of surfactant be far behind?

References

Adler, K. B., Hardwick, D. H., and Craighead, J. E. (1982). Porcine tracheal goblet cell ultrastructure: A three dimensional reconstruction. *Exp. Lung Res.* **3**: 69–80.

Allen, R. D. (1968). A reinvestigation of cross-sections of cilia. *J. Cell Biol.* **37**: 825–831.

Askin, F. B., and Kuhn, C. (1971). The cellular origin of pulmonary surfactant. *Lab. Invest.* **25**:260–268.

Barry, B. E., Miller, F. J. and Crapo, J. D. (1985). Effects of inhalation of 0.12 and 0.25 ppm ozone on the proximal alveolar region of juvenile and adult rats. *Lab. Invest.* **53**:692–704.

Brain, J. D., Gehr, P., and Kavet, R. I. (1984). Airway macrophages. The importance of the fixation method. *Am. Rev. Respir. Dis.* **129**:823–826.

Breeze, R. G., and Wheeldon, E. B. (1977). The cells of the pulmonary airways. *Am. Rev. Respir. Dis.* **16**:705–777.

Brody, A. R., and Craighead, J. E. (1975). Preparation of human lung biopsy specimens by perfusion-fixation. *Am. Rev. Respir. Dis.* **112**:645.

Brummer, M. E. G., Schwartz, L. M., and McQuillen, N. K. (1977). A quantitative study of lung damage by scanning electron microscopy. Inflammatory response to high-ambient levels of ozone. *Scanning Electron Microsc.* **2**::513–517.

Cohen, A. L., Marlow, D. P., and Garner, G. E. (1968). A rapid critical point method using fluorocarbons (freons) as intermediate and transitional fluids. *J. Microsc.* **7**:331–342.

Collet, A. J. (1979). Preservation of alveolar type II pneumocyte lamellar bodies for electron microscopic studies. *J. Histochem. Cytochem.* **27**:989–996.

Finch, G. L., Bastacky, S. J., Hayes, T. L., and Fisher, A. L. (1987). Scanning electron microscopic studies of frozen hydrated lung exposed to intratracheally instilled particles. *J. Microsc.* **147**:Part 2:193–203.

Feder, N., and Sidman, R. L. (1958). Methods and principles of fixation by freeze-substitution. *J. Biochem. Biophys. Cytol.* **4**:593–602.

Fukuda, Y., Ferrans, V., and Crystal, R. G. (1984). Development of elastic fibers of nuchal ligament aorta and lung of fetal and postnatal sheep: an ultrastructural and electron microscopic immunohistochemical study. *Am. J. Anat.* **170**:597–629.

Gil, J., and Weibel, E. R. (1969). Improvements in demonstration of lining layer of lung alveoli by electron microscopy. *Respir. Physiol.* **8**:13–36.

Glauert, A. M. (1972–1980). *Practical Methods in Electron Microscopy.* Vol. 1–8. New York, Elsevier.

Gluck, L., Kulovich, M. V., and Brody, S. J. (1966). Rapid quantitative measurement of lung tissue phospholipids. *Lipid Res.* **7**:570.

Guckenberger, R., Kosslinger, C., Gatz, R., Breu, H., Levai, N., and Baumeister, W. (1988). A scanning tunneling microscope (STM) for biological applications: Design and performance. *Ultramicroscopy* **25**:111–133.

Hallman, M., Miyai, K., and Wagner, R. M. (1976). Isolated lamellar bodies from rat lung: Correlated ultrastructural and biochemical studies. *Lab Invest.* **35**:79–86.

Hassett, R. S., Engleman, W., and Kuhn, C. (1980). Extramembranous particles in tubular myelin from rat lung. *J. Ultrastruct. Res.* **71**:60–67.

Hayat, M. A. (1986). *Basic Techniques for Transmission Electron Microscopy.* Orlando, Academic Press.

Hayatdavoudi, G., Crapo, J.D., Miller, F. J., and O'Neil, J. J. (1980). Factors determining degree of inflation in intratracheally fixed rat lungs. *J. Appl. Physiol.* **48**:389–93.

Heuser, J. (1981). Preparing biological samples for stereomicroscopy by the quick-freeeze deep-etch, rotary-replication technique. *Methods Cell Biol.* **22**:97-122.

Holma, B. (1969). Scanning electron microscopic observation of particles deposited in the lung. *Arch. Environ. Health* **18**:330-339.

Kajikawa, K., Yamaguchi, T., Katsuda, S., and Miwa, A. (1975). An improved electron stain for elastic fibers using tannic acid. *J. Electron Microsc.* (Tokyo) **24**:287-289.

Kalina, M., and Pease, D. C. (1977). The preservation of ultrastructure in saturated phosphatidyl cholines by tannic acid in model system and type II pneumonocytes. *J. Cell Biol.* **74**:726-741.

Karrer, H. E. (1956). The ultrastructure of mouse lung. General architecture of capillary and alveolar walls. *J. Biophys. Biochem. Cytol.* **2**:241-252.

Kelley, R. O., Dekker, R. A. F., and Bluemink, J. G. (1973). Ligand-mediated osmium binding: Its application in coating biological specimens for scanning electron microscopy. *J. Ultrastruct. Res.* **45**:254-258.

Kennedy, J. R., Dunlap, J. R., Bunn, R. D., and Edwards, D. F. (1988). Utilization of digital image processing to study dynein arms (ATPase) in normal and immotile cilia. *J. Electron Microsc. Tech.* **8**:159-172.

Kikkawa, Y., and Spitzer, R. (1969). Inclusion bodies of type II alveolar cells: Species differences and morphogenesis. *Anat. Rec.* **163**:525-542.

Kisch, B. (1955). Electron microscopic investigation of the lungs: capillaries and specific cells. *Exp.Med. Surg.* **13**:101-117.

Kuhn, C., and Engleman, W. (1978). The structure of the tips of mammalian respiratory cilia. *Cell Tissue Res.* **186**:491-498.

Kuhn, C., and Finke, E. H. (1972). The topography of the pulmonary alveolus: Scanning electronmicroscopy using different fixations. *J. Ultrastruct. Res.* **38**: 161-173.

Kuhn, C., and Tavassoli, F. (1976). The scanning electron microscopy of elastase-induced emphysema: a comparison with emphysema in man. *Lab. Invest.* **34**:2.

Low, F. N. (1953) The pulmonary epithelium of laboratory animals and man, *Anat. Rec.* **17**:241-264.

Luft, J. H. (1971). Ruthenium red and violet. I. Chemistry, purification, methods of use for electron microscopy and mechanism of action. *Anat. Rec.* **171**:347-368.

Luft, J. H. (1971). Ruthenium red and violet. II. Fine structural localization in animal tissues. *Anat. Rec.* **171**:360-416.

McDowell, E. M., Barrett, L. A., Glavin, F., Harris, C. C., and Trump, B. F., (1978). The respiratory epithelium, 1. Human bronchus. *Natl. Cancer Inst.* **61**:539-545.

Oldmixon, E. H., Suzuki, S., Butler, J. P., and Hoppin, F. G., (1985). Perfusion dehydration fixes elastin and preserves airspace dimensions. *J. Appl. Physiol. Respirat. Environ. Exercise Physiol.* **58**:105–113.

Parra, S. C., Burnette, R., Price, H. P., and Takaro, T. (1986). Zonal distribution of alveolar macrophages type II pneumonocytes and alveolar septal connecting tissue gaps in adult human lungs. *Am. Rev. Respir. Dis.* **133**:908–912.

Parra, S. C., Gaddy, L. R., and Takaro, T. (1978). Ultrastructural studies of canine interalveolar pores (of Kohn). *Lab. Invest.* **38**:9–13.

Parrot, M. J., Sur la steatose viscerale que l'on observe a l'etat physiologique chez quelques animaux. *J. Arch. Phys. Norm. Pathol.* **4**:27–47.

Policard, A., Collet, A., Pregermain, S., and Rouet, C. (1957). Proceedings of the Stockholm Conference, Sept. 1956. In *European Regional Conference on Electron Microscopy*. Edited by F. S. Sjostrand and J. Rhoden. New York, Academic Press, 1957.

Quate, C. F. (1986). Vacuum tunneling: a new technique for microscopy. *Physics Today* August: pp. 26–33.

Ranga, V., and Kleinerman, J. (1980). Interalveolar pores in mouse lungs. *Am. Rev. Respir. Dis.* **122**:477–482.

Reid, L., and Jones, R. (1979). Bronchial mucosal cells. *Fed. Proc.* **38**:191–196.

Sargent, J. A. (1988). Low temperature scanning electron microscopy: advantages and applications. *Scanning Microsc.* **2**:835–849.

Schulz, W. M., McAnnaley, W. H., and Reynolds, R. C. (1980). Freeze fracture of pulmonary lamellar body membranes in solid crystal phase. *J. Ultrastruct. Res.* **71**:37–48.

Sobin, S. S., Fung, Y. C., and Tremer H. M. (1982). The effect of incomplete fixation of elastin on the appearance of pulmonary alveoli. *J. Ultrastruct. Res.* **104**:68–71.

Stephens, R. J., and Evans, M. S. (1973). Selection and orientation of lung tissue for scanning and transmission electron microscopy. *Environ. Res.* **6**:52–59.

Stratton, C. J., (1976). The three dimensional aspect of mammalian lung multilamellar bodies. *Tissue Cell* **8**:693–712.

Stratton, C. J. (1976). The high resolution ultrastructure of the periodicity and architecture of lipid-retained and extracted lung multilamellar body laminations. *Tissue Cell* **8**:713–728.

Vaccaro, C. A., and Brody, J. S. (1979). Ultrastructural localization and characterization of proteoglycans in the pulmonary alveolus. *Am. Rev. Respir. Dis.* **120**:901–910.

Vaccaro, C. A., and Brody, J. S. (1981). Structural features of alveolar wall basement membrane in the adult rat lung. *J. Cell Biol.* **91**:427–437.

Watt, I. M. (1985). *The Principles and Practice of Electron Microscopy*. Cambridge, The University Press.

Weibel, E. R., Limacher, W., and Bachofen, H. (1982). Electron microscopy of rapidly frozen lungs: Evaluation on the basis of standard criteria. *J. Appl. Physiol. Respirat. Environ. Exercise Physiol.* **45**:325-333.

Wilsman, N. J., Morrison, W. B., Farnum, C. E., and Fox, L. E. (1987). Microtubular protofilaments and subunits of the outer dynein arm in cilia from dogs with primary ciliary dyskinesia. *Am. Rev. Respir. Dis.* **135**:137-143.

Young, S. L., Fram, E. K., and Craig, B. L. (1985). Three dimensional reconstruction and quantitative analysis of rat lung type II cells: a computer-based study. *Am. J. Anat.* **174**:1-14.

Young, S. L., Fram, E. K., and Randell, S. H. (1986). Quantitative three dimensional reconstruction and carbohydrate cytochemistry of rat non-ciliated bronchiolar (Clara) cells. *Am. Rev. Respir. Dis.* **133**:899-907.

Zasadzinski, J. A. N., Schneir, J., Gurley, J., Elings, V., and Hansma, P. K. (1988). Scanning tunneling microscopy of freeze-fracture replicas of biomembranes. *Science* **259**:1013-1015.

4

Freeze Fracture in Lung Research

EVELINE E. SCHNEEBERGER

Massachusetts General Hospital
and Harvard Medical School
Boston, Massachusetts

I. History

The concept of replicating a frozen surface from which ice had been sublimated was first formulated by Hall (1950). Later Steere (1957) modified the procedure for fracturing by introducing the use of a cold scalpel; this afforded a means of producing surface replicas of viruses, which was the first biological system to be examined. In 1961, Moor et al. incorporated a liquid-nitrogen-cooled ultra-microtome into a vacuum chamber and this prototype has served as a model for all subsequent freeze etch units.

II. Principles and Methods of Freeze Etching

Freeze etching is a specialized ultrastructural technique in which frozen samples are fractured under vacuum with or without subsequent partial sublimation of the surrounding ice (etching). The exposed surface is then replicated with a heavy metal, usually platinum. To strengthen the replica, a layer of carbon is further evaporated onto the surface. The underlying sample is removed by soaking in sodium hypochlorite or a strong acid and the resulting surface replica is mounted on a copper grid and viewed in the electron microscope. The steps involved in freeze etching are briefly outlined below. For greater detail, the reader is referred to the monograph edited by Rash and Hudson (1979).

A. Specimen Preparation

In conventional freeze etching, tissues or cells are fixed in aldehyde fixatives prior to freezing. Since fixation results in the cross-linking of structural proteins, the fixation time should be brief (5–30 min) in order to maximize the number of fracture planes passing through the hydrophobic regions of cell membranes. To freeze biological samples with a minimum of ice crystal formation, two strategies have been devised. In the first, the aldehyde-fixed specimen is soaked in a cryoprotectant such as ethylene glycol, dimethyl sulfoxide, or glycerol. Of these, glycerol at a concentration of 25–30% in buffer, pH 7.3, is most widely used. The cryoprotective action of glycerol is thought to involve its ability to form hydrogen bonds with water, thereby interfering with the tendency of water to crystallize. While these cryoprotectants have been extensively used in conventional freeze fracture, they may be associated; with a variety of artifacts and cannot readily be applied to unfixed samples (Boehler, 1975).

A second strategy to prevent ice crystal formation is the use of rapid freezing, a technique developed by Heuser et al. (1979). This involves the freezing of unfixed samples within milliseconds, it does not require cryoprotectants, and has the added advantage of capturing biological events that are completed within a few milliseconds. This technique will be discussed at greater length at the end of the chapter.

B. Specimen Freezing

After appropriate cryoprotection, the specimen is frozen either in liquid freon 22 cooled to close to its freezing point ($-165\,°C$) with liquid nitrogen or in liquid nitrogen slush produced by putting liquid nitrogen under vacuum, which results in a reduction in its temperature from $-196\,°C$ to $-210\,°C$. The samples are then introduced under vacuum into the freeze etch chamber, onto a precooled specimen stage.

C. Fracturing and/or Etching

The fracturing process may be carried out either with a stainless steel razor blade, in which case a single fractured surface is produced, or a double replica device in which the resulting two-halves of the sample are replicated. The optimal fracturing temperature is between $-110\,°C$ and $-115\,°C$. During freeze fracturing, the fracture plane preferentially passes through the hydrophobic regions of cell membranes, although cross-fractures through the cytoplasm also occur. Additional information can be obtained by allowing some of the ice to sublime away at $-100\,°C$, thereby exposing nonfractured intra- and extracellular surfaces. At $-100\,°C$, the vapor pressure of ice is approximately 1×10^{-5} mmHg, which corresponds to an etching rate of 2.3 nm of depth/sec. Thus nonetchable structures, approximately 100–150 nm above the surrounding ice level, are exposed in 1 min. To reduce contamination during the etching process, it is helpful to have a liquid-nitrogen-cooled surface above the sample to act as a cold trap.

D. Replication

The fractured and/or etched surfaces of the sample are replicated with platinum/carbon through the use of electron beam guns. In conventional freeze fracture, the heavy metal is applied unidirectionally at a 45° angle. However, replication can also be carried out by rotary shadowing at smaller angles, such as 15–25°. After a 2–2.5 nm thick film of platinum is applied, a layer of carbon 20–25 nm thick is applied at 75–90° to the sample.

E. Cleaning the Replica

A variety of cleaning agents have been used to remove the underlying sample from the replica. These include sulfuric acid, dichromate cleaning solution, sodium hydroxide, and sodium hypochlorite (Chlorox). When one is dealing with cell pellets, the replicas will often fragment during the cleaning process. The application of a film of 0.3% parlodion in amyl acetate is helpful to maintain the replica intact. The cleaned replica is then picked up on a Formvar-coated grid and examined in the electron microscope.

F. Interpretation of Freeze Fracture Replicas

When the fracture plane passes through the hydrophobic region of cell membranes, two halves of the membrane result. By convention, the half that remains associated with the cytoplasmic side is known as the protoplasmic or P face, and the external or noncytoplasmic half is designated as the ectoplasmic or E face. When examining electron micrographs of freeze fracture replicas, it is important to note the direction in which the sample was shadowed and to hold the micrograph in such a way that the shadows cast by intramembrane particles point away from the viewer. In this way one will have no difficulties in interpreting what is projecting up from and what is forming a depression in the surface of the replica. When examining tight junctions for possible discontinuities that might account for "leakiness," it is essential to examine *both* fracture faces by the double replica technique, to determine whether the defect on one fracture face is due to the partitioning of tight junction elements to the opposite fracture face (see below).

III. Freeze Fracture in Lung Research

The technique of freeze fracture has afforded a means of examining cellular structures without the need for dehydration and embedding and, with newer rapid freezing techniques, without the need for fixation. Moreover, with freeze fracture it is possible to examine structures within the plane of the membrane that cannot otherwise by visualized. It is, therefore, not surprising that freeze fracture has been used to advantage in a number of important areas of lung cell biology. These include an elucidation of the structure of surfactant within both type II pneumocytes and alveoli (Untersee et al., 1971; Kikkawa and Manabe, 1978; Williams, 1978, 1982), the appearance of specialized arrays of intramembrane particles (Carson et al., 1984; Bartels and Welsch, 1984; Gordon, 1985; Bartels and Miragall, 1986), and the structure of tight junctions and distribution of gap junctions between alveolar and airway epithelial cells (Schneeberger and Karnovsky, 1976; Schneeberger et al., 1978; Schneeberger, 1980; Olver et al., 1981; Schneeberger and McCormack, 1984; Schneeberger and Lynch, 1984) and between endothelial cells (Schneeberger and Karnovsky, 1976; Schneeberger, 1981). A brief review of representative studies in each of these areas is given below.

A. Surfactant

Ultrastructural studies of rat lungs fixed by vascular perfusion indicated an extracellular layer of material lining alveoli at the air-tissue interface (Weibel and Gil, 1968). Because of the solvents used during dehydration and infiltration with plastic, however, the surfactant lining layer was often disrupted. In a subsequent study (Untersee et al., 1971) in which freeze fracture techniques were applied to air-filled lungs that were either unfixed or fixed by vascular perfusion, a continuous smooth layer of surfactant lining alveoli was observed. Evidence was also obtained for the presence of tubular myelin within alveolar spaces. These observations were confirmed by Manabe (1979), who stressed the absence of intramembrane particles in the surface lining layer. Cross fracture images also clearly showed the presence of tubular myelin in the hypophase underneath the smooth surfactant lining layer.

    Freeze fracture studies were extended to examine the structure of lamellar bodies and tubular myelin within type II pneumocytes of both adult (Belton et al., 1971; Roth et al., 1972; Smith et al., 1972, 1973) and fetal animals (Kikkawa and Manabe, 1978; Williams, 1978). These studies clearly showed that while the intracellular lamellar inclusion bodies were enveloped by a unit membrane containing intramembrane particles 15 nm in diameter, the interior, multilayered lamellae were devoid of particles. In a more recent study, in which rapid freezing techniques were used to avoid chemical fixation and cryoprotectants, Williams (1982) provided evidence that both lamellar and tubular myelin were present within multilamellar bodies and that smooth-surfaced lamellae became particulate when they formed tubular myelin.

B. Specialized Membrane Structures

The use of freeze fracture techniques has revealed a variety of specialized structures within the plane of the plasma membrane of a variety of pulmonary epithelial cells. Rod-shaped particles have been described on the P face of the apical membrane and apical vesicles of mitochondria-rich cells in turtle lungs (Bartels and Welsch, 1984). It appeared that each of these particles consisted of two or three globular subunits resulting in particles measuring between 6×12 and 10×12 nm. Such structures were not observed between type I and II pneumocytes. Based on the similarity in appearance of these particles to those observed in some mitochondria-rich cells of other tissues, Bartels and Welsch suggested that they may be involved in the acidification of the surfactant hypophase. They have thus far not been observed in mammalian lungs.

Further specialized membrane structures include orthogonal arrays of particles (OAP) that have been observed in the basal plasma membrane of pneumocytes in turtles and frogs and in human type I pneumocytes (Bartels and Miragall, 1986). Such orthogonal arrays appeared to be absent from pneumocytes in other mammalian species, but were present in the basal plasma membrane of airway epithelial cells in a variety of species (Inoue and Hogg, 1977; Gordon, 1985). The number of particles per OPA ranged from 4 to 40 in lower vertebrates and from 4 to 24 in humans (Bartels and Miragall, 1986). It is interesting that their number appeared to increase substantially following acute or chronic exposure to NO_2 (Gordon, 1985). Although a possible role in the regulation of ion transport has been suggested, their function is at present unclear.

Using rapid freezing and rotary shadowing techniques on hamster airway epithelium, Carson et al. (1984) have confirmed the presence of OAP on the basal plasma membrane of airway epithelial cells. The particles making up these arrays measured 10 nm in diameter and consisted of 10–12 particles per OAP. In addition, a second array of small particles was observed on both the apical and basal surface of these airway epithelial cells. Neither type of particle array was predictably associated with any given cell type nor is their function known. Structure-function correlations as well as isolation and biochemical characterization of these specialized membrane structures will be required to determine their function.

C. Intercellular Junctions

The morphological characterization of intercellular junctions between endothelial and epithelial cells of the lung has relied on the freeze fracture technique. At present, it affords the only means of visualizing these specialized membrane structures en face, in addition to providing useful structural correlates for functional measurements. A variety of intercellular junctions are present between endothelial and epithelial cells of the lung; these include gap junctions (macula communicans), tight junctions (zonulae occludentes), zonula adherans, and desmosomes. Only the first two of these will be discussed.

Gap junctions represent transcellular channels that permit the exchange of small molecules and ions between adjacent cells (Gilula et al., 1972; Potter et al., 1966). While they are ubiquitous, not all epithelial and endothelial cells are linked by them at any given time. By freeze fracture, gap junctions are shown to form patches of closely packed membrane particles (connexons) that are in register with similar particles in the membrane of the adjacent cell (Fig. 1). Each connexon consists of protein subunits, arranged to form a central channel that allows electrical coupling and the exchange of molecules of up to 800–1,200 daltons. Permeability of the gap junction can be modulated by changes in pH, pCo_2, Ca^{2+}, or membrane potential. Studies of their structural organization within the membrane, their regulation, and biochemical composition are underway (for a review see Revel et al., 1985).

Tight or occluding junctions, on the other hand, form a gasketlike seal near the apex of epithelial and endothelial cells, thereby forming a barrier to the passage of molecules along the paracellular pathway (for a review, see Schneeberger and Lynch, 1984). By freeze fracture examination, epithelial tight junctions consist of a continuous network of interconnected strands on the P face and complementary grooves on the E face. Endothelial tight junction particles, by contrast, preferentially partition onto the E face, where they are present in shallow grooves leaving particle-poor complementary ridges on the P face (Simionescu et al., 1975, 1976). There is an approximate correlation between the number of parallel strands in a tight junction and the measured transepithelial electrical resistance (Claude and Goodenough, 1973). Because the isolation of pure preparations of tight junctions has proven to be difficult (Stevenson and Goodenough, 1984), the biochemical composition of the tight junction strands remains to be determined. Although

Figure 1   Endothelial tight junction from an intra-acinar artery in an adult rat lung. Large gap junctions (arrowheads) are intercalated between the rows of particles of the tight junction. P and E faces are indicated (x 72,000).

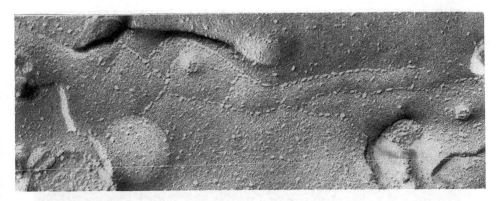

Figure 2 Endothelial tight junction in a pulmonary capillary of adult rat lung. One to two sparsely interconnected rows of particles are present on the E face (95,000).

there is a tacit assumption that they are composed of proteins (Griepp et al., 1983), alternative models, involving inverted lipid micelles, have been suggested (Kachar and Reese, 1982). It is likely that both lipids and proteins are important in the maintenance of normal tight junction structure and function.

Based on early tracer studies (Schneeberger and Karnovsky, 1968, 1971), it appeared that the main permeability barrier to the passage of proteins into the alveolar space resided in the tight junctions of the alveolar epithelium. To evaluate the structural basis for this observation, a freeze fracture study of the mouse air-blood barrier was undertaken (Schneeberger and Karnovsky, 1976). Junctions between pulmonary capillary endothelial cells consisted of one to three interconnected rows of particles present in shallow grooves on the E face and complementary particle-poor ridges on the P face (Fig. 2). A few small gap junctions were associated with the tight junctions in the arteriolar , but not in the venular, segment of the capillary network. By contrast, the tight junctions between type I and II pneumocytes consisted of a continuous band of three to five interconnected strands. These freeze fracture studies supported the observations that epithelial tight junctions constitute the main barrier to the passage of proteins into the alveolar space.

Subsequent studies showed that tight junctions in intra-acinar arteries were more complex and consisted of two to six continuous rows of tight junctions particles that were associated with numerous, large gap junctions intercalated within the tight junction network (Fig. 1). Tight junctions of intra-acinar veins, by contrast, consisted of two to five rows of particles and these were associated with smaller and fewer gap junctions (Fig. 3) (Schneeberger, 1981).

Further freeze fracture studies showed that tight junctions between epithelial cells of airways (Fig. 4) (Schneeberger, 1980) and submucosal gland epithelial cells were heterogeneous (Schneeberger and McCormack, 1984). In addition to

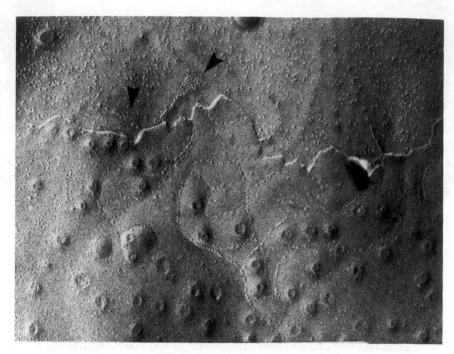

Figure 3 Endothelial tight junction in an intra-acinar pulmonary vein of an adult rat. A few small gap junctions (arrowheads) are present. (x 72,000).

their heterogeneity, the mean number of parallel strands forming the tight junction was significantly higher in extra- than in intrapulmonary airways. A graded increase in the number of tight junction strands was also apparent in submucosal glands. The fewest (3.6±0.4) were observed between serous cells, and their number increased progressively from the mucous to the ductal segments of the gland. These freeze fracture observations correlated well with parallel studies using lanthanum as a tracer. The tight junctions between serous cells were permeable to this tracer, whereas those between mucous and ductal cells were not. Gap junctions were observed between secretory cells of the submucosal glands and extrapulmonary airways but not between Clara cells.

Freeze fracture studies have also been useful in the correlating tight junction structure with functional measurements of epithelial permeability at various stages of fetal development of the lung (Schneeberger et al., 1978; Olver et al, 1981). With the use of double replicas (Fig. 5) and permeability measurements, Schneeberger et al. (1978) showed that the pulmonary epithelium was structurally continuous and functionally "tight" very early in gestation. An interesting

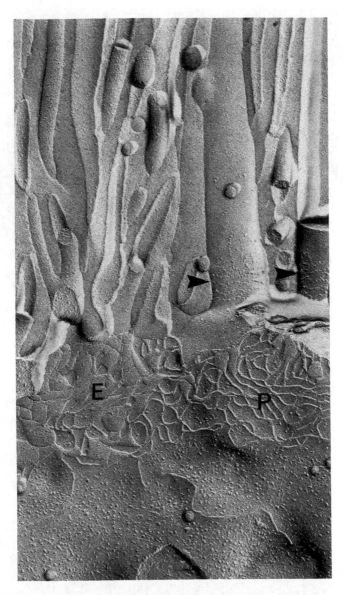

Figure 4   Freeze fracture replica of a ciliated cell from human nasal mucosa. The tight junction forms a continuous, complex, interconnected network of fibrils on the P face and complementary grooves on the E face.   A ciliary necklace of particles is present at the base of the cilia (arrowheads) (x 55,000).

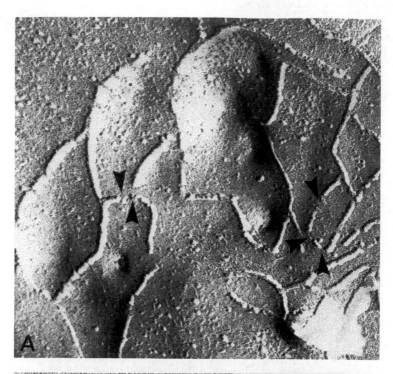

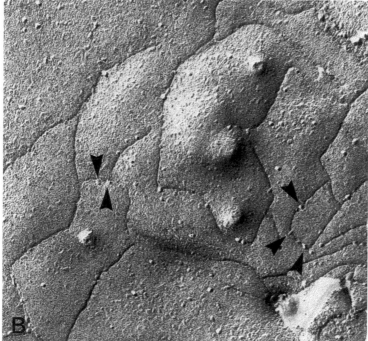

finding was that gap junctions were associated with epithelial tight junctions during the glandular and canalicular stages of development, but they disappeared once alveolar epithelial cells became differentiated.

IV. New Methods

A. Fracture-Labeling

To broaden the usefulness of freeze fracture in cell biology, methods have been devised to combine a variety of labeling techniques with freeze fracture. These have not, thus far, been applied to the study of lung cells. Two of these methods are described below.

Freeze Fracture Autoradiography

A number of attempts have been made to combine autoradiographic labeling techniques with freeze fracture. Most of these methods have been difficult, with loss of morphological detail and a low labeling rate. More recently, however, Carpentier et al. (1985) have described an autoradiographic technique with which radioactively labeled ligands can be detected over large areas of the exoplasmic fracture face. In brief, cells are exposed to the radioactive ligand. They are rinsed, fixed, infiltrated with glycerol, and mounted either as a suspension or on coverslips on replica holders. After fracturing, the replicas are washed several times with distilled water and mounted on Parlodion-coated grids in such a way that the attached cell fragments are sandwiched between the replica and the copper grid. The grids are then coated with radiographic emulsion, developed, and viewed in the electron microscopy. The advantage of the technique is its relative simplicity. Moreover, the use of radioactive ligand interferes less with its handling by the cell than when it is coupled to relatively large, electrondense probes. Its limitations are that the distribution of the ligand can only be examined in the exoplasmic fracture face and the resolution provided by the autoradiographic technique is low.

Figure 5 Double replica of a pulmonary epithelial tight junction from a fetal lamb at 71 days' gestation. Note that wherever portions of the tight junction strand are missing on the P face (arrowhead) an equivalent particle is seen on the complementary E face (arrowhead) (x 149,000).

Colloidal Gold Labeling

Two approaches have been taken with this technique. In the first, labeling of the plasma membrane is carried out in a fashion analogous to that described above for autoradiography, except that cell surfaces are labeled with appropriately coated gold particles before freeze fracturing (Pinta da Silva and Kan, 1984).

In the second, techniques are devised to permit labeling of intracellular structures by colloidal gold particles (Pinto da Silva et al., 1981). In brief, the tissue or cells are fixed, embedded in a gel of 30% cross-linked bovine serum albumin, rapidly frozen in Freon 22 cooled to liquid nitrogen temperature, and fractured in liquid nitrogen with a scalpel. The fractured pieces are thawed, deglycerinated, and labeled with either lectins or antibodies followed by the appropriately coated gold particles. The samples are fixed in 1% osmium tetroxide, dehydrated in ethanol, subjected to critical point drying, and replicated with platinum-carbon. The resulting replicas are cleaned as described for conventional freeze fracture replicas, mounted on Formvar-coated grids, and examined in the electron microscope.

B. Rapid Freezing

The rapid freezing technique (Heuser et al., 1979) was developed to preserve the ultrastructure of biological specimens in the absence of cryoprotectants or fixation. It makes possible the study of biological events that occur in milliseconds and permits the observation of structures previously obscured by nonvolatile cryoprotectants. In essence, the technique is simple and involves bringing a biological specimen, attached to the lower end of a vertical falling rod, rapidly into firm contact with the polished surface of a pure copper block cooled from below to $-260\,^{\circ}C$ with liquid helium. The resulting preparations may either be freeze fractured and replicated by rotary shadowing or prepared for thin sectioning by freeze substitution. For a better perception of the three-dimensional arrangement of cellular structures, the electron micrographs are printed by reverse contrast. This involves photographing the image on a conventional electron micrograph negative with positive film, and then using the positive film to make a print. Except for the studies of Williams (1982) and Carson et al. (1984), rapid freezing techniques have not been applied to the study of the lung.

V. Summary

Freeze fracture, particularly when combined with the newer labeling and rapid freezing techniques, provides a powerful tool with which to examine the structural organization of cells and to correlate these with a number of functional parameters. Although laborious and time-consuming, these techniques provide uni-

References

Bartels, H., and Welsch, U. (1984). Freeze fracture study of the turtle lung. 2. Rod-shaped particles in the plasma membrane of a mitochodria-rich pneumocyte; in *Pseudemys (Chrysemys) scripta. Cell Tissue Res.* **236**:453–457.

Bartels, H., and Miragall, F. (1986). Orthogonal arrays of particles in the plasma membrane of pneumocytes. *J. Submicrosc. Cytol.* **18**:637–646.

Belton, J. C., Branton, D., Thomas, H. V., and Muller, P. K. (1971). Freeze etch observations of rat lung. *Anat. Rec.* **170**:471–486.

Boehler, S. (1975). *Artefacts and Specimen Preparation Faults in Freeze Etch Technology.* Balzers, Liechtenstein, Balzers, A. G.

Carpentier, J. L., Brown, D., Iacopetta, B., and Orci, L. (1985). Detection of surface-bound ligands by freeze-fracture autoradiography. *J. Cell Biol.* **101**:887–890.

Carson, J. L., Collier, A. M., and Smith, C. A. (1984). New observations on the ultrastructure of mammalian conducting airway epithelium: application of liquid propane freezing, deep etching and rotary shadowing techniques to freeze fracture. *J. Ultrastruct. Res.* **89**:23–33.

Claude, P., and Goodenough, D. A. (1973). Fracture faces of zonulae occludentes from "tight" and "leaky" epithelia. *J. Cell Biol.* **58**:390–400.

Gilula, N. B., Reeves, R. O., and Steinbach, A. (1972). Metabolic coupling and cell contacts. *Nature* **235**:262–265.

Gordon, R. E. (1985). Orthogonal arrays in normal and injured respiratory airway epithelium. *J. Cell Biol.* **100**:648–651.

Griepp, E. B., Dolan, W. J. Robbins, E. S., and Sabatini, D. D. (1983). Participation of plasma membrane proteins in the formation of tight junctions by cultured epithelial cells. *J. Cell Biol.* **96**:693–702.

Hall, C. E. (1950). A low temperature replica method for electron microscopy *J. Appl. Physics.* **21**:61–62.

Heuser, J. E., Reese, T. S., and Landis, D. M. D. (1979). Preservation of synaptic structure by freeze fracture. *Cold Spring Harbor Symp. Quant. Biol.* **40**:17–24.

Inoue, S., and Hogg, J. C. (1977). Freeze fracture study of the tracheal epithelium of normal guinea pigs with particular reference to intercellular junctions. *J. Ultrastruct. Res.* **61**:89–99.

Kachar, B., and Reese T.S. (1982). Evidence for the lipid nature of tight junction strands. *Nature* **296**:464–466.

Kikkawa, Y., and Manabe, T. (1978). The freeze-fracture study of alveolar type II cells and alveolar content in fetal rabbit lung. *Anat. Rec.* **190**:627–637

Manabe, T. (1979). Freeze-fracture study of alveolar lining layer in adult rat lungs. *J. Ultrastruct. Res.* **69**:86–97.

Moor, H., Muehlethaler, K., Waldner, H., and Frey-Wyssling, A. (1961). A new freezing ultramicrotome. *J. Biophys. Biochem. Cytol.* **10**:1–13.

Olver, R. E., Schneeberger, E. E., and Walters, D. V. (1981). Epithelial solute permeability, ion transport and tight junction morphology in the developing lung of the fetal lamb. *J. Physiol.* **315**:395–412.

Pinto da Silva, P., Kachar, B., Torrisi, M. R., Brown, C., and Parkison, C. (1981). Freeze-fracture cytochemistry: replicas of critical point-dried cells and tissues after fracture-label. *Science* **213**:230–233.

Pinto de Silva, P., and Kan, F. W. K. (1984). Label-fracture: a method for high resolution labeling of cell surfaces. *J. Cell Biol.* **99**:1156–1161.

Potter, D. D., Furshpan E. J., and Lennox, E. S. (1966). Connections between cells of the developing squid as revealed by electrophysiological methods. *Proc. Natl. Acad. Sci. USA* **55**:325–335.

Rash, J. E., and Hudson, C. S. (1979). *Freeze Fracture: Methods, Artifacts, and Interpretations*. New York, Raven Press.

Revel, J. P., Nicholson, B. J., and Yancey, S. B. (1985). Chemistry of gap junctions. *Annu. Rev. Physiol.* **47**:263–279.

Roth, J. Meyer, H. W. and Winkelmann, H. (1972). Electron microscopic studies in mammalian lungs by freeze etching. II Morphological observations on the metabolism of the surfactant. *Exp. Pathol.* **6**:291–302.

Schneeberger, E. E., and Karnovsky, M. J. (1968). The ultrastructural basis of alveolar-capillary membrane permeability to peroxidase used as a tracer. *J. Cell Biol.* **37**:781–793.

Schneeberger, E. E., and Karnovsky, M. J. (1971). The influence of intravascular fluid volume on the permeability of newborn and adult mouse lungs to ultrastrucutal protein tracers. *J. Cell Biol.* **49**:319–334.

Schneeberger, E. E., and Karnovsky, M.J. (1976). Substructure of intercellular junctions in freeze-fractured alveolar-capillary membranes of mouse lung. *Circ. Res.* **38**:404–411.

Schneeberger, E. E., Walters, D. V., and Olver, R. E. (1978). Development of intercellular junctions in the pulmonary epithelium of the foetal lamb. *J. Cell Sci.* **32**:307–24.

Schneeberger, E. E. (1980). Heterogeneity of tight junction morphology in extrapulmonary and intrapulmonary airways. *Anat. Rec.* **198**:193–208.

Schneeberger, E. E. (1981). Segmental differentiation of endothelial intercellular junctions in intra-acinar arteries and veins of the rat lung. *Circ. Res.* **49**:1102–1111.

Schneeberger, E. E., and McCormack, J. (1984). Intercellular junctions in upper airway submucosal glands of the rat: a tracer and freeze fracture study. *Anat. Rec.* **210**:421–433.

Schneeberger, E. E., and Lynch R.D. (1984). Tight junctions: their structure, composition and function. *Circ. Res.* **55**:723–733.

Simionescu, M., Simionescu, N., and Palade, G. E. (1975). Segmental differentiation of cell junctions in the vascular endothelium. The microvasculature. *J. Cell Biol.* **67**:863–885.

Simionescu, M., Simionescu, N., and Palade, G. E. (1976). Segmental differentiation of cell junctions in the vascular endothelium. Arteries and veins. *J. Cell Biol.* **68**:705–723.

Smith, D. S., Smith, U., and Ryan, J. W. (1972). Freeze-fractured lamellar body membrane of the rat lung great alveolar cell. *Tissue Cell* **4**:457–468.

Smith, U., Smith D. S., and Ryan, J. W. (1973). Tubular myelin assembly in type II alveolar cells: freeze-fracture studies. *Anat. Rec.* **176**:125–128.

Steere, R. L. (1957). Electron microscopy of structural detail in frozen biological specimens. *J. Biophys. Biochem. Cytol.* **3**:45–60.

Stevenson, B. R., and Goodenough, D. A. (1984). Zonulae occludentes in junctional complex enriched fractions from mouse liver. Preliminary morphological and biochemical characterization. *J. Cell Biol.* **98**:1209–1221.

Untersee, P., Gil, J., and Weibel, E. R. (1971). Visualization of extracellular lining layer of lung alveoli by freeze etching. *Respir. Physiol.* **13**:171–185.

Weibel, E. R., and Gil, J. (1968). Electron microscopic demonstration of an extracellular duplex lining layer of alveoli. *Respir. Physiol.* **4**:42–57.

Williams, M. C. (1978). Freeze-fracture studies of tubular myelin and lamellar bodies in fetal and adult rat lungs. *J. Ultrastruct. Res.* **64**: 352–361.

Williams, M. C. (1982). Ultrastructure of tubular myelin and lamellar bodies in fast-frozen adult rat lung. *Exp. Lung Res.* **4**:37–46.

5

Ultrastructural Immunocytochemistry

SAMUEL S. SPICER and BRADLEY A. SCHULTE

Medical University of South Carolina
Charleston, South Carolina

I. Introduction

In view of the vast literature that has resulted from his achievement, a tremendous debt is owed Dr. Albert Coons, who conceived of labeling antibody with a visible marker to localize its antigen microscopically in tissue sections. Although Coon's initial attempt to label antibodies with chromogenic dyes visible with incandescent light failed, the subsequent use of a fluorescent label and viewing in ultraviolet light yielded the requisite sensitivity for detecting antigens present at low concentration in tissues (Coons et al., 1941; Coons and Kaplan, 1950). Intended initially for visualizing specific antigens of invasive microorganisms, the immunostaining method was quickly extended to demonstrating endogenous tissue antigens (Marshall et al., 1959; Emmart et al., 1962, 1963) and offers a potential for revealing the distribution and, to some extent, the amount locally of any antigenic substance in the tissue. The methodology for viewing immunostained sections in visible light was subsequently developed by Nakane and Pierce (1966, 1967), who introduced the immunoenzymatic approach that substituted horseradish peroxidase for the fluorescent marker and achieved sensitivity through catalytic build up of reaction product.

As one of its major goals, the cytochemical approach seeks advantage over conventional biochemical analysis in demonstrating the precise location of a specific tissue component. Histochemical methods suffered in the past in comparison to those of biochemistry in their limited chemical specificity. However, histochemistry's aim of localizing specific entities has been realized with immunocytochemistry. In fact, with monoclonal antibodies immunocytochemistry advances a step farther, by differentiating segments (epitopes) of the peptide backbone or saccharide side chains within a protein or glycoprotein antigen. Immunostaining thus demonstrates in situ differences in expression of the same gene in different species, as evidenced by immunostaining of human but not rodent kidney with monoclonal antibody to human Tamm Horsefall protein (unpublished observation).

Immunostaining offers an advantage over biochemical analysis in allowing the interpretation of the biological significance of a tissue component in terms of the specialized function at the site where it occurs. Light microscopic immunostaining allows one to associate the role of a tissue constituent with the overall specialized activity of its host cells. Extending the method of localization to the electron microscopic level, moreover, permits one to relate the role of the cellular component to the function of a cell organelle at cytological and molecular levels.

A second objective of immunocytochemistry is to extend our understanding of the mechanisms of normal cell function. Immunostaining for glycosyl transferase for example (Roth and Berger, 1982; Roth et al., 1985) has confirmed and extended the results of carbohydrate cytochemistry dependent on staining with lectins and basic cationic reagents (Sato and Spicer, 1982a,b) in showing the Golgi lamellae not to be homogenous but to perform different activities in *cis*, intermediate, and *trans* cisternae. Other immunocytochemical contributions include demonstrating the intracellular distribution of cytoskeletal elements and precursor proteins (Willingham and Pasten, 1985), a glycosidase in granular reticulum (Lucocq et al., 1986), a glycosyl transferase in plasmalemma (Roth et al., 1985), and clathrin in coated pits (Willingham and Pasten, 1985).

In a third area of application, immunocytochemistry aims to gain information about diagnosis, classification, and pathogenesis of disease processes (DeLellis et al., 1979; Spicer, 1987). Cells differ more in chemical composition than in structural features. Chemical analysis can therefore potentially show differences between normal and diseased cells that are not discernible morphologically. Despite its limited quantitative capacity, immunocytochemical examination often surpasses biochemical analysis in the ability to detect chemical characteristics or changes in specific types of diseased cells. By identifying specialized cell constituents, for example, immunoglobulin, light microscopic immunostaining has contributed to the diagnosis and classification of lymphoid neoplasms (Garvin et al., 1974; Lukes et al., 1978; Janossy et al., 1980). Immunostains represent diagnostic probes in diverse areas of pathology including tumor markers (Burns et al., 1978), cutaneous immunmopathology (Harrist and Mihm, 1979) and renal disease (MacIver et al., 1979). Applications of ultrastructural immunostaining to lung are detailed in the final section of this chapter. The prospect for future contributions to pulmonary biology and pathology seems bright, if one extrapolates from past experience.

II. Methods

The wide range of technical variants that have been developed for immunolocalization at the ultrastructural level have been reviewed in a text edited by Polak and Varndell (1984) and a monograph series edited by Bullock and Petruz (1982–1984). The methods are covered more generally here and can be classified into four main approaches as discussed in the following sections.

A. Pre-embedment Immunostaining for Transmission Electron Microscopy

Fixation

Pre-embedment immunostaining begins with fixation, and final results depend critically on this step. Selecting a fixative requires a compromise between retaining the optimal morphology provided by heavy fixation and the optimal chemical reactivity, in this case antigenicity, gained with lighter fixation. The choice depends on the resistance of the antigen to denaturation, and for each antigen a range of fixatives and their concentrations should be tested. For hardy antigens, a near isomolar solution of 0.1–2% glutaraldehyde in 0.1 M phosphate buffer, pH 7.2–7.4, can be used. For less stable antigens, a 4% paraformaldehyde–2% calcium acetate solution is used. Using both glutaraldehyde and formaldehyde sequentially or simultaneously (Karnovsky, 1965) allows a compromise between morphological and chemical restrictions and has the advantage of the better diffusion of formaldehyde into the specimen compared with glutaraldehyde, which acts mainly at the block surface. A solution that was found satisfactory in our experience contains 0.1–0.2% glutaraldehyde and 4% paraformaldehyde in 0.1 M phosphate buffer at pH 7.4.

Fixation times with glutaraldehyde range from 15 min to 2 hr and with formaldehyde from 30 min to 4 hr, although longer intervals can be used for stable antigens. Keeping the fixation as brief as possible is to be recommended for achieving satisfactory sectioning and preserving cell surface constituents.

Preferences for fixing at 4°C and at room temperature have been expressed, and either may be used with compensatory allowance for fixation time. Including detergents such as saponin or Triton X-100 or sucrose in the fixative or rinse before sectioning may enhance the prospect of penetration of fixative or staining reagents.

For some tissues, fixing by immersion is satisfactory. Specimens immersed in fixative should be minced promptly and finely to a range of block sizes down to 1 mm^3 or less. Varying the block sizes makes it easier to obtain areas with different degrees of morphological preservation and chemical denaturation, according to penetration of the fixative from the surface into the block. Areas with the most favorable ratio can be selected in part by examining blocks or sections for the best stained area prior to embedment.

However, many tissues (for example, the central nervous system) require perfusion fixation usually for a brief period followed by more fixation in situ or by immersion. A procedure that can be recommended (Mugnaini, 1985) entails cardiac perfusion at 80 cmH$_2$O pressure with physiological saline containing a vasodilator to clear the blood, approximately 10 min perfusion with a 2% glutaraldehyde–4% paraformaldehyde solution, further perfusion with 4% paraformaldehyde, and retention of the latter fixative in the unopened animal for 1 hr prior to removal of tissue. Exposure to 2% glutaraldehyde even for only 10 min in this procedure may compromise the immunogenicity of the antigen.

For immunocytochemistry of cultured cells, Willingham (1980) developed a method requiring precisely controlled light fixation with glutaraldehyde plus detergents to permeabilize the cell prior to staining and strong fixation after immunostaining. Successful intracellular localization of tubulin, clathrin, sarc protein, and other antigens with this method has been reported (Willingham and Pasten, 1985).

Sectioning

Sections cut in the cryostat from fixed tissue blocks at a thickness of 5–30 μm have been used for pre-embedment ultrastructural immunostaining. However, 10–40 μm sections cut with an oscillating tissue slicer (vibratome) are preferable, since they undergo less tissue disruption from ice crystal formation. Infiltrating the fixed specimen in cryoprotectant solution such as 2.0 M sucrose in 0.1 M phosphate buffer, pH 7.3, prior to freezing lessens the morphological damage by ice crystal formation. Rapid freezing (see below) can protect additionally.

Immunostaining of finely minced, fixed tissue blocks provides an alternative to staining sections. In these miniblocks, only a thin layer at the surface will stain.

Staining Procedures: Linking Marker to Antigen

Immunostaining requires that a labeling moiety be linked to the antigenic site. To avoid the need to conjugate every primary antibody used and to gain sensitivity, immunocytochemical methods generally involve first applying to the section the primary antiserum or antibody against the antigen, engendered most commonly in a rabbit, goat, or mouse. The problem then resolves into one of attaching label to the primary antiserum. This was accomplished initially by chemical conjugation of the label (e.g., an enzyme) to secondary antibody directed against the rabbit, goat, or mouse primary antibody (Nakane and Pierce, 1966, 1967).

The requirement for conjugating antibody to the enzymatic marker was obviated by joining labeling enzyme to antigen through antibody–antigen reactions in the immunoperoxidase bridge method developed in this laboratory (Mason et al., 1969). The bridge procedure, which used rabbit primary antiserum, goat antirabbit secondary antiserum, rabbit antiperoxidase, and then peroxidase, was modified by Sternberger and co-workers (1970). The modification combined the antiperoxidase (AP) antibody and peroxidase (P) of the bridge method into a PAP complex that has become commercially available and is applied in one instead of two steps. The latest modification of the bridge as opposed to the conjugated antibody approach introduced by Hsu and Raine (1981) used the specific high affinity of avidin for biotin and used biotin-derivatized secondary antibody and an avidin–biotin horseradish peroxidase complex (ABC). An available alternative uses labeled protein A, which binds specifically to some types of IgG to reveal sites of binding of the primary antibody in the section.

It is preferable to stain sections free floating in the reagents in each step of the sequential procedure to ensure penetration into the section from all directions. Difficulty can arise in transporting relatively thin frozen sections, in the 5–20 μm range, between steps in the immunostaining sequence. This can be better accomplished by pipetting from one to the next solution rather than by transfer with a fine probe or by sedimentation and resuspension at each step.

In the thinner cryosections and vibratome sections all or most cells will be transected at one aspect and staining reagents need not cross the plasmalemma as is required for miniblocks. Permeation through intracellular membranes remains an obstacle, however, for staining many intracellular components. Use of detergents such as saponin or Triton X-100 on the sections or tissue fragments before or after sectioning (or perhaps preferably in the fixative solution) can be tested for the prospect of enhancing penetration of reagent through plasmalemma and intracellular membranes. Limited experience in our laboratory has indicated that taking the cryostat section through graded steps to absolute ethanol and back to water improves the immunoreactivity. This is presumed to result from removal of membrane lipids.

Diffusion of reagents occurs less readily into tissues fixed with glutaraldehyde at concentrations greater than 0.5% and for times longer than 1 hr. Increased density and impermeability of the cytosol presumably result from cross-linking between amine groups of protein molecules by this bifunctional reagent. Some prospect of enhancing diffusion into the specimen and into cells can be entertained with the use of cryoprotectants such as glycerol or dimethyl sulfoxide in staining solutions at about a 5% concentration.

Staining Procedures: Types of Markers and Intensification

Based on specific binding of a marker to reveal an antigenic site, ultraimmunocytochemistry depends on the availability of electron opaque labeling substances.

Enzymes that generate electron opaque reaction product and intrinsically dense substances have both met this need.

Of the suitable enzymes, horseradish peroxidase (HRP) (Nakane and Pierce, 1966, 1967) has had the widest use because of the electron opacity, insolubility, and minimum diffusability of the reaction product from oxidation of the 3,3′ diaminobenzidine substrate for HRP (Graham and Karnovsky, 1966). After the labeling enzyme has become attached to the antigenic site with one of the above described sequences, its location is revealed by incubation in this substrate medium.

Although immunocytochemistry has used HRP most commonly as a marker enzyme, alkaline phosphatase (Mason and Woolston, 1982) and possible other enzymes can substitute for HRP. In our experience, alkaline phosphatase works less well than HRP in the system for light microscopic localization. However, it could prove satisfactory if not superior for electron microscopy because of the greater density of the lead phosphate reaction product. Alkaline phosphatase and glucose oxidase, the latter of which has proven useful for light microscopic immunostaining, have apparently not been systematically evaluated for applicability at the ultrastructural level.

Antibodies conjugated to ferritin constituted the first approach to ultrastructural immunocytochemistry (Singer, 1959). The 70 A size of ferritin particles restricts use of this label to examination at high magnification. Cell surface antigens or extracellular antigens have been localized by the immunoferritin method, but perhaps because of the high molecular weight of the conjugate the method has proven less amenable to intracellular staining. Staining of intracellular constituents has been accomplished, however, with an immunoglobulin–ferritin bridge method that depends on fixing lightly before and heavily after staining (Willingham et al., 1971). This procedure substitutes ferritin for the usual enzyme marker in the immunobridge approach. The immunoferritin bridge technique applied to lightly fixed, detergent-treated cells in culture works to advantage in its precise location of the antigen compared with the more diffuse distribution of diaminobenzidine reaction product (Willingham and Pasten, 1985).

Dual staining of a single section for two antigens using labeling enzymes with distinguishable reaction product has not been accomplished by electron microscopy as it has by light microscopy (Mason and Woolston, 1982). Achieving such an objective can be envisioned, however, with a sequence of HRP for one and alkaline phosphatase for another antigen.

Intensification of electron opacity at a reactive site could prove advantageous for localizing antigens present in the section at low concentration or low level of immunoreactivity. Postfixation with osmium tetroxide after immunostaining is generally thought to enhance the contrast between positive sites and background. Chelating the HRP produced reaction product of diaminobenzidine with cobalt and other metals changes the color and intensity of the stain at the light microscopic microscopic level and could affect the sensitivity of the electron microscopic stain.

Embedment and Thin Sectioning

Fruitless embedment and further processing of sections lacking positive areas can be avoided by observing sections for their reactivity prior to embedment. Viewing the stained sections or miniblocks with light microscopy for reactive sites permits one to select the most favorable for embedment in epoxy resin. After embedment, thin sections can be taken of known positive areas, possibly without obtaining thick sections prior to electron microscopy.

Flat embedment of cryosections or vibratome sections in epoxy resins becomes highly essential, so that thin sections can subsequently be obtained parallel to the largest area of staining. The importance of flat embedment relates also to the fact that immunoreactivity is often confined to the outer 2 μm or so layer at the surface of the cryosection, vibratome section, or minced fragment. There arises, therefore, a need to conserve and to use the reactive surface layer for electron microscopy. Sections from specimens embedded on edge and thus lying vertical or somewhat tangential to the surface of the tissue will yield a smaller viewing area. They will, however, reveal a gradation of decreasing staining away from the surface, providing an internal control area for comparison with the stained region. Ultrathin sections of immunostained specimens can be viewed with and without uranyl acetate–lead citrate staining, but immunopositivity is detected most sensitively in sections that are not counterstained.

B. Postembedment Immunostaining for Transmission Electron Microscopy

Chemical Fixation and Embedment

Fixation procedures that are recommended for pre-embedment techniques apply in general for postembedment procedures as well.

The physicochemical nature of epoxy resins poses obstacles to immunostaining. First, the epoxy medium resists penetration by most staining reagents, particularly those of high molecular weight, because a medium sufficiently inelastic to yield ultrathin sections lacks the porosity needed to allow diffusion of reagents into it. Hydrophobicity of the medium presumably also plays a part in the impenetrability to reagents in aqueous solution. In addition, chemical reagents in the medium, including the oxidants effecting polymerization, adversely affect the antigenicity of cell constituents. The prolonged high temperature required for polymerization of the resin also contributes to denaturation of the antigen.

However, successful immunolocalizations by staining epoxy thin sections have been achieved (Moriarty, 1976; Baskin et al., 1979; Bauer et al., 1981). These accomplishments are largely restricted to constituents of secretory granules. The explanation for these successes appears to be that antigen sufficiently exposed at the section's surface accounts for staining without penetration of reagents into the section.

Second, antigen molecules are present in these positive sites in such high concentration that a surviving small fraction could account for staining. Third, antigens in the secretory granules resist the denaturing hazards of epoxy embedment through the glycosylation or other stabilizing features characteristic of exocytosed macromolecules. Some improvement in immunostaining of epoxy sections has been reported after the thin section is etched to remove the plastic or increase its porosity. Etching reagents include hydrogen peroxide, xylene, and an appropriately diluted sodium hydroxide–alcohol mixture. Concern persists, however, for the effect these etching reagents may have on the immunogenicity of the tissue component.

Attempts to stain intracellular components in epoxy sections more commonly fail, however. This is exemplified by our experience (unpublished), in which plasma cells in epoxy or methacrylate-embedded mouse spleen failed to stain for IgG, in contrast to the dense reaction product observed in cisternae of granular reticulum in plasma cells of aliquots of the same spleen processed for pre-embedment staining.

A relatively untested variant involves double sectioning and embedding. In this technique, epoxy-embedded tissues are sectioned at 1–2μm and etched with a solution consisting of equal parts ethanol and ethanol saturated with NaOH. The etched sections are then immunostained and re-embedded in Epon for thin sectioning (Mar et al., 1987).

Two recently introduced media, referred to as Lowicryl K4M (Carlemalm et al., 1980, 1982) and LR White (Newman et al., 1983) offer the promise of surmounting the problems encountered with expoxy embedment. Success with ultrastructural immunostaining using these acrylics has been reported (Roth, 1983, 1986; Newman and Jasani, 1984).

Other embedding media that work well for postembedment ultrastructural staining of complex carbohydrates (see Chapter 7) promise to prove useful for immunocytochemistry. These include a mixture of styrene–Vestopal W, styrene–methacrylate or styrene–Spurr's resin (Thomopoulos et al., 1987).

Freeze-Drying or Freeze-Substitution and Embedment

Freeze-drying and embedding potentially eliminates all noxious influences on the antigen except those of the embedding medium (Sjostrand and Baker, 1958) and minimizes displacement of diffusable substances. This therefore appears to constitute a most promising approach (Dudek et al., 1984). However, this procedure is marred by causing structural damage due to the freezing of unfixed tissue. Rapid freezing, however, largely circumvents the latter problem.

A number of fast-freeze techniques have been developed that use the sharp impact of spring-loaded tissue against a polished metal surface chilled with liquid nitrogen (Coulter and Terracio, 1977; Heuser et al., 1979; Van Harreveld et al., 1974; McGuffee et al., 1981; Chiovetti et al., 1985 1987). The "Gentleman Jim" device (Ted Pella Inc., Tustin, CA) described by Phillips and Boyne (1984) has been used with apparent success for fast freezing.

An alternative rapid-freezing technique uses a liquid propane jet at $-188\,°C$ on the exposed organ in the anesthetized animal as well as on excised tissue (Muller et al., 1980). Rapid freezing can also be accomplished by jet propulsion of the specimen into a deep container of propane or an isopentane–propane mixture at -188 to $195\,°C$ (Sitte et al., 1985).

Methods and apparatus for drying frozen specimens have been described by Stirling and Kinter (1967), Stumpf and Roth (1967), and Coulter and Terracio (1977) and modified more recently by McGuffee and co-workers (1981). Frozen dried specimens can then be transferred for embedment in the Lowicryl or LR White acrylic media (McGuffee et al., 1981; Chiovetti et al., 1985, 1987). Immunostaining of thin sections of these blocks has been found superior to staining either of ultrathin cryosections of fixed tissue or of fixed, dehydrated, and embedded specimens (Jorgensen and McGuffee, 1987).

The freeze-substitution procedure merits consideration as a possible means of minimizing fixation-dependent damage to antigenicity while sufficiently preserving morphological structure (Fernandez-Moran, 1960; Harvey, 1982; Sitte et al., 1985). Freeze substitution requires rapid freezing, as with the freeze-drying technique. Frozen tissues are then immersed in absolute acetone at $-188\,°C$ and gradually warmed prior to embedment in Lowicryl K4M at $-70\,°C$ or at $20\,°C$ for other media. Antigens in the thin sections of these blocks suffer only from the exposure to acetone and embedding medium. Their staining potential differs from that of ultrathin cryosections (described later) because of their exposure to acetone and chemicals of the acrylic embedding medium instead of to aldehyde fixatives. Tissue contrast will possibly be better in sections of freeze-substituted or freeze-dried, embedded specimens than in the frozen ultrathin sections.

Staining Plastic Thin Sections

The methods already described for pre-embedment immunostaining, including both systems for linking label to antigen and types of markers, apply also to staining ultrathin sections. In addition, a major advance of importance primarily in staining thin sections resulted from the introduction of colloidal gold as a marker (Geoghegen and Ackerman, 1977; Roth and Wagner, 1977; Roth, 1983).

Taking advantage of the specific affinity of *Staphylococcus* protein A for certain classes of immunoglobulin, Roth (1982) used microspherules of gold coated with protein A to visualize ultrastructurally the sites where a primary antibody applied to a thin section bound to its antigen. Protein-A-coated gold could bind also to endogenous sites of immunoglobulin such as serum and plasma cells, but these can be differentiated by their distribution in control sections and contribute little background interference. This procedure is limited mainly to reacting with primary antibody composed of some members of the IgG class, since protein A does not have affinity for all immunoglobulins of all species.

Colloidal gold-coated antibody to the antibody of the species in which the primary antiserum was generated may provide a more generally applicable method. Silver intensification of immunogold procedures increases sensitivity by several times for light microscopic immunocytochemistry but apparently not for electron microscopy.

A benefit of the colloidal gold staining of thin sections stems from the possibility of making homogenous preparations of spherules of uniform but controllably varying sizes (Roth, 1982; Slot and Geuze, 1984) and using different sized spherules to stain different antigens in the same thin section.

A novel method exists for localizing macromolecules that serve as substrates for hydrolytic enzymes, for example, nucleic acids and specific proteins (Bendayan, 1985). This technique uses colloidal gold coated with the enzyme, for example, ribonuclease or collagenase to bind to and localize its substrate (e.g., ribonucleic acid or collagen) by postembedment staining of thin sections.

C. Immunostaining Ultrathin Cryosections

Staining ultrathin frozen sections now rivals other methods as a means of localizing intracellular antigens. The technique of preparing ultrathin cryosections (Fernandez-Moran, 1950; LeDuc et al., 1967; Tokuyasu, 1973) and immunostaining such sections (Tokuyasu, 1980, 1984) was long considered formidable because of the difficulty of cryoultramicrotomy. This approach has become more feasible through advances in procedure and equipment (Griffiths et al., 1983; Leunissen et al., 1984; Slot and Geuze, 1984; Tokuyasu, 1980).

The method in general uses relatively mild fixation. For example, miniblocks prepared by mincing the specimen in fixative and measuring 1 mm^3 or less are fixed for 30 min–2 hr in chilled buffered 0.2–1% glutaraldehyde or 0.1–0.5% glutaraldehyde plus 2% paraformaldehyde or 2–8% paraformaldehyde; the last is reserved for labile antigens (Slot and Geuze, 1982). Cryoprotection with 0.6–2.3M sucrose precedes ultrathin cryosectioning, which can be accomplished with a diamond or specially prepared glass knife (Griffiths et al., 1983). Satisfactory microtomes for ultracryomicrotomy are now available, for example, the Reichert ultracut E with an FC4D cryochamber. After transfer of sections to gold or nickel grids, immunostaining is performed on the grids with a sequence of primary antibody followed by colloidal gold spherules coated with secondary antibody to the primary antibody or with protein A.

This approach has important advantages. Penetration of staining reagents into a fixed ultrathin cryosection undoubtedly exceeds that into a resin of sufficient hardness for ultramicrotomy. In addition, the frozen sections are spared the adverse action on the antigenicity that results from exposure to catalyzing chemicals and high temperature during polymerization of resin. The cryo method still suffers from the denaturing action of chemical fixatives on the antigen, and only freeze-drying avoids this problem.

A disadvantage of immunostaining ultrathin cryosections compared with sections embedded in the LR White or K4M acrylic resins is the lack of contrast in the frozen section. Even with postfixation in osmium tetroxide and counterstaining in uranyl acetate–lead citrate, cytological detail can be relatively indistinct in ultrathin frozen sections.

D. Immunocytochemistry by Scanning Electron Microscopy

If the antigens are distributed unevenly on the cell surface, the one-dimensional view by transmission electron microscopy allows one to determine the true distribution only with difficulty. Increased information about the distribution of cell surface antigens in two dimensions derives from viewing immunocytochemical results by scanning electron microscopy.

To localize cell surface constituents in the scanning electron microscope, cultured cells, or cells placed in suspension from body fluids or tissues, are fixed usually with glutaraldehyde. Cells can be fixed in suspension, or after sedimentation, or on a coverslip or culture dish. Cryofractured surfaces of tissues may also be suitable for immunocytochemical as they are for morphological examination in the scanning electron microscope. Cells in suspension, pellet form, or on a fixed surface are stained as for pre-embedment immunocytochemistry.

Procedures of immunostaining for scanning electron microscopy have been reviewed by Hodges et al. (1984). Methods of binding marker to antigen for transmission electron microscopy apply here in general. In addition, methods using hybrid antibodies to antigen and marker or antibody to a hapten bound to antigen and to marker (Wofsy, 1979) have been developed for scanning electron microscopy.

A number of markers have been used for immunostaining by scanning electron microscopy (see Hodges et al., 1984). These include the currently preferred colloidal gold particles but also ferritin, viruses, and plastic and other spheres. Enzymatic build up of crystalline reaction product from the oxidation of benzidine by peroxidase also provides a morphologically distinct surface marker (McKeever et al., 1977).

Viewing in the back-scattered electron imaging (BEI) mode affords advantages for immunocytochemical examination with the scanning scope. Because of the different energy of electrons scattered by collisions with different elements, electrons reflected from a given element can be collected and focused selectively to yield an image of the distribution of the one element in the section. Detecting the iron or silver in cytochemical reagents for selective staining of glycoconjugates for example permits the demonstration in the BEI mode of the distribution of the stained glycoconjugates in the section (Bodner et al., 1981). In analogous fashion, gold introduced immunocytochemically at the site of an antigen gives a strong BEI signal and allows the selective demonstration of the distribution of the antigen (de Harven et al., 1984). Electrons scattered by elements in antigens beneath the cell surface can also be imaged in the BEI mode. This penetration of the BEI image into the cell has led to the demonstration of specific antigens in cytoplasmic granules and lysosomes in certain cells, allowing improved differentiation and identification of cell types. Combining immunogold labeling and incubation for intrinsic myeloperoxidase of leukocytes on a single specimen permits one to visualize surface structure, surface antigens, and cytoplasmic granules in a single cell (Soligo and de Harven, 1987).

E. Cytochemical Controls

Immunocytochemistry achieves its objective only when the observed staining demonstrates the location of a specific known constituent of the cell or tissue. False-positive results arise when the antiserum or antibody fraction used contains antibodies against antigens other than that intended.

Monoclonal antibody preparations enjoy freedom from contamination with one or more additional antibodies and, if directed against a known antigen, stain this component specifically. The possible content of a second or more antibodies in addition to that intended in polyclonal antisera requires documentation of the antiserum's specificity.

The specificity of the antiserum depends most on the purity of the isolated antigen. This should be demonstrated by isoelectric focusing and by reaction of antibody with antigen to give a single band in gel immunodiffusion.

False-positive localizations also occur even with well-characterized antibodies because of affinity for immunoglobulin in some sites on a nonimmunospecific basis, for example, in mast cells (Simson et al., 1977) and other histological structures (Spicer et al., 1977). Some of these sites bind antibodies in some species and not others when fixed with one solution and not another and can be differentiated by testing tissues of several species and specimens fixed with different solutions.

A negative result does not conclusively prove absence of an antigen. Evidence against false-negatives attributable to poor antiserum can be obtained by staining sites known biochemically to contain an antigen. For example, immunostaining of macrophages, Paneth cells, and proximal renal tubules with antiserum to lysozyme establishes that the serum gives positive results under the conditions used.

Control procedures serve an important role in ruling out nonspecific staining. These include substituting preimmune serum from the same animal if available or other nonimmune serum for the primary antiserum in the staining procedures. Absorbing the antibody with antigen, using a preparation of antigen not used for generating the antiserum, provides an optimal control.

For ultrastructural immunostaining, it would seem essential that immunostaining be tested with the antiserum first at the light microscopic level to meet these criteria. A problem with the ultrastructural method is background and nonimmunospecific localization. Because such a small cytological area is surveyed ultrastructurally in a section, sufficient tissue should be embedded for wide sampling. It is essential, moreover, to examine sections for randomly distributed label that would indicate nonimmunospecific deposition over the tissue or on medium beyond the specimen's edge. Nonspecific staining can occur in nonrandom distribution, reflecting adsorption of the marker by electrostatic or other forces to selected structural entities. For example, conjugates of ferritin to nonspecific immunoglobulin and even to albumen were found to bind selectively to the A band of skeletal muscle. This artifactual localization discounted the prospect that immunoglobulin from myasthenia gravis patients possessed a specific affinity for the A band in skeletal muscle (Gottlieb et al., 1966). Optimal methods have recently been described for preparing antibody-coated gold spherules and stabilizing the preparation to minimize nonspecific binding of the gold spherules (Birrell et al., 1987). A variety of organs and cell types known to possess or lack the antigen of interest should be assessed for selectivity of localization at the electron microscopic level, and the requisite controls must be included before one interprets a localization as immunospecific.

III. Ultrastructural Immunocytochemical Observations on Lung

The literature on applications of electron microscopic immunocytochemistry to the respiratory tract is small. Lysozyme was localized in type II pneumocytes by light microscopic immunostaining (Fig. 1) and pre-embedment staining at the ultrastructural level (Fig. 2) and showed the enzyme to be in the cytoplasmic lamellar bodies (Spicer et al., 1977). Content of lysozyme and, as previously reported, of acid phosphatase (Goldfischer et al., 1968) likens the lamellar bodies to lysosomes. These bodies, however, function in secretion (Sage et al., 1983), releasing their content by the common exocytic mechanism (Ryan et al., 1975). They appear, therefore, to secrete lysosomelike enzymes. Whether this function relates to resistance to infection, extracellular metabolism, turnover of surface components, or other activity has not been defined.

Secretion of lamellar bodies is said to depend on a mechanism mediated in an unexplained manner by calcium (Mason et al., 1977). Evidence that the calcium action involves calmodulin has been obtained through immunocyto-

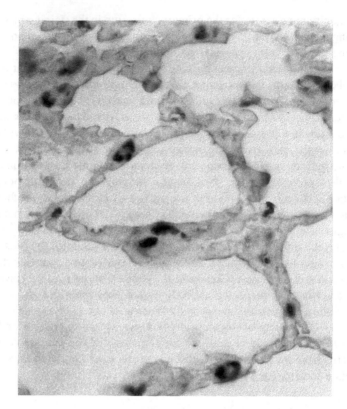

Figure 1   Area from rat lung shows numerous reactive cells with the prevalence and distribution in alveolar septa of type II pneumocytes. Stained with antibody to urinary lysozyme, × 500. (Reproduced with permission from Spicer et al., 1977.)

chemical demonstration of this calcium-binding protein on the surface of lamellar bodies isolated from cultured type II alveolar cells (Fig. 3) (Hill et al., 1984).

    Evidence for a secretory role for Clara cells has been obtained in the rat lung by ultrastructural immunostaining of these cells with monoclonal antibody to an antigen isolated from pulmonary lavage fluid (Bedetti et al., 1987). This antibody stained secretory granules of Clara cells exclusively in rat lung (Fig. 4). Such staining differentiated Clara cells cytochemically from type II pneumocytes, which in some respects, such as immunostaining and lectin staining for blood group antigen (see Chapter 7), share cytochemical properties with Clara cells.

    Ultrastructural immunocytochemistry has been applied mainly to examination of extraepithelial pulmonary structures. Type I collagen has been demonstrated

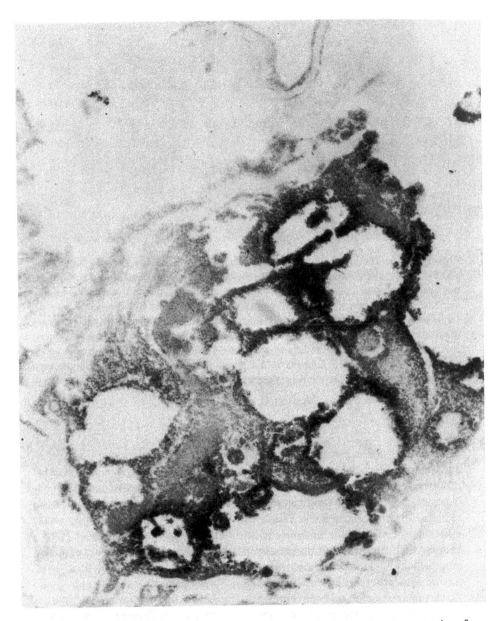

Figure 2 Granular pneumocyte in rat lung shows dense reaction product demonstrative of lysozyme in the rim of lamellar bodies and adjacent cytoplasm. The cytoplasmic reactivity presumably reflects diffusion of enzyme in the lightly fixed cell. More precise localization was not achieved, at least in part, because of loss of immunoreactivity of the antigen after glutaraldehyde fixation. Formalin-fixed cryostat section, immunostained and processed for electron microscopy, × 17,500. (Reproduced with permission from Spicer et al., 1977.)

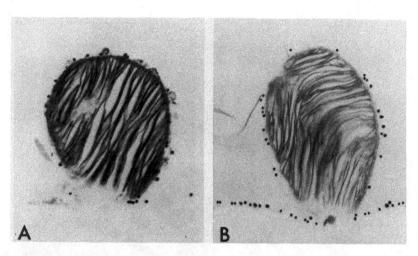

Figure 3 Localization of endogenous calmodulin on lamellar body surfaces. Calmodulin was localized with anticalmodulin IgG and colloidal gold–protein A complexes. A. No treatment prior to calmodulin localization. B. Lamellar bodies treated with 2 mM EGTA prior to calmodulin localization. C. Lamellar bodies treated with colloidal gold–protein A only. D. Lamellar bodies treated with anticalmodulin IgG that had been pretreated with a molar excess of bovine brain calmodulin. Lamellar bodies were subsequently treated with gold-protein A. E. Lamellar bodies treated with 1.5 μM calmodulin in the presence of 50 μM calcium prior to calmodulin localization. F. Lamellar bodies treated with 1.5 μM calmodulin in the presence of 4 mM EGTA prior to calmodulin localization ($\times$ 25,000). (Reproduced by permission from Hill et al., 1984.)

on the surface and in periodic cross-banding of fibers in alveolar septa (Fig. 5) (Gil and Martinez-Hernandez, 1984). This association of type I collagen, considered the more rigid form of collagen, with the mobile and pliable alveolar septum is unexpected. It is not surprising that laminin has been localized exclusively in basal laminae in the lung and found in all basement membranes of bronchiolar and alveolar epithelium, endothelium, and muscle cells (Gil and Martinez-Hernandez, 1984). Fibronectin likewise appears distributed mainly in basement membrane of epithelial endothelial and smooth muscle cells (Torikata et al., 1985) (Fig. 6). It is found also in association with interstitial collagen fibers suggestive of reticulin. Some discrepancies exist in different reports on fibronectin distribution in lung, particularly in alveolar basement membrane. These may be attributed to differences in technique, since it has been shown that fibronectin is associated with glycosaminoglycan (GAG) in some and not other sites. With fixatives that preserve this GAG during tissue processing, the GAG masks fibronectin immunoreactivity in the sites where it occurs (Harrison et al., 1985). The differing results reported could

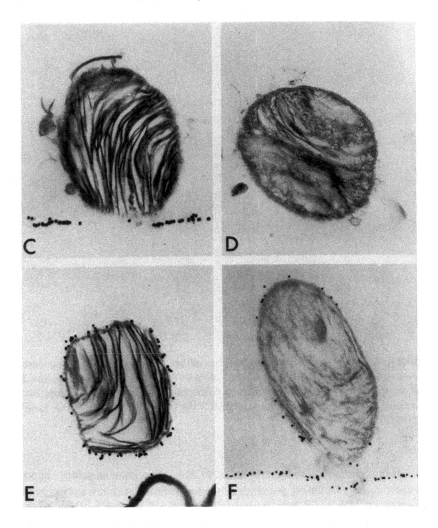

therefore reflect fixation-dependent differences in GAG retention and masking of antigenic sites in the section. This observation substantiates the generalization that a negative immunocytochemical result does not conclusively exclude presence of an antigen at a particular site.

Immunostaining at the electron microscopic level has made visible elastin associated with smooth muscle cells, neighboring endothelium, and epithelium of alveoli and glands (Fukuda et al., 1984). Early in elastogenesis, the microfibrils stained and the amorphous component was immunoreactive throughout. In more mature elastin, only the periphery of the amorphous component was immunopositive; the central

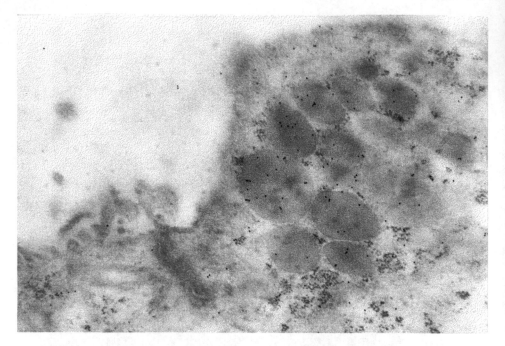

Figure 4   Portions of cytoplasm of rat Clara cell and ciliated cell joined by desmosomes. The secretory granules of the Clara cell are labeled with 10 nm gold particles that show staining with antibody to a protein isolated from alveolar fluid. Immunogold, uranyl acetate–lead citrate stain, original magnification, 40,000. (Reproduced by permission from Bedetti et al., 1987.)

zone of the amorphous component failed to stain. The latter negativity appeared to reflect failure of antibody penetration into the central zone in the pre-embedment immunostain, since the large mature fibers stained uniformly throughout the peripheral to central zones of the amorphous cores with Verhoeff's iron hematoxylin applied to thin sections (Brissie et al., 1975).

Immunocytochemistry has contributed important information concerning diagnosis, classification, and pathogenesis of neoplasms (see for review DeLellis et al., 1979; Spicer, 1987). Although immunocytochemistry has been used for staining bronchopulmonary tumors at the light microscopic level (Blobel et al., 1985) there apparently have been no attempts to extend such examination of pulmonary neoplasms to the ultrastructural level.

Examples of the application of ultrastructural immunocytochemistry to examination of pathological lesions in lung are limited. In one recent report, immunostaining by both light and electron microscopy revealed a dramatic increase in fibronectin associated with pulmonary fibrosis (Torikata et al., 1985) (Figs. 6,7). This

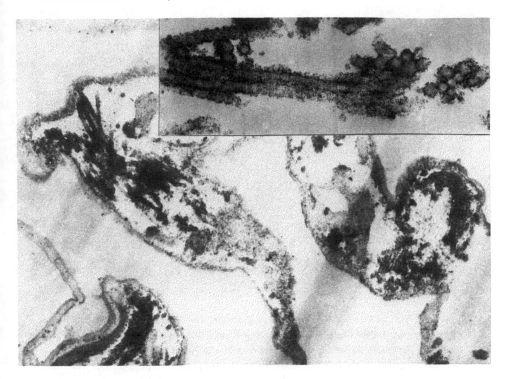

Figure 5   Pulmonary parenchyma. Antitype I collagen antibody. The alveolar epithelium, endothelium, and intervening basement membranes are negative. Type I collagen fibers within the alveolar septum are positive (× 16,400). Inset. Antitype I collagen antibody. Detail of collagen-positive fibers in the alveolar interstitium. The reaction product is largely superficial. However, cross-banding can be seen in the fibers in longitudinal section (× 40,000). (Reproduced by permission from Gil and Martinez-Hernandez, 1984.)

increase occurred in basal lamina of hyperplastic alveolar epithelial cells and of capillaries (Fig. 8). In addition, collagen fibers showed greatly increased immunopositivity for fibronectin on their surface in a 60 nm periodicity. The markedly increased fibronectin of the epithelial basal lamina in fibrotic lung can be interpreted as resulting from greater than normal binding of plasma fibronectin to an often multilayered basal lamina. Since type II pneumocytes synthesize and secrete fibronectin in culture (Sage et al., 1983), the hyperplastic epithelium could, on the other hand, constitute the source of the increased fibronectin in fibrosis. Whether by diffusion of the plasma protein from blood vessels and its binding to basal lamina or by increased production in pneumocytes, the abundant fibronectin in either case could provide a mechanism that favors adherence of epithelial cells and re-epithelialization. Electron microscopic immunocytochemistry offers a promising approach to the future elucidation of this and other problems in pulmonary disease.

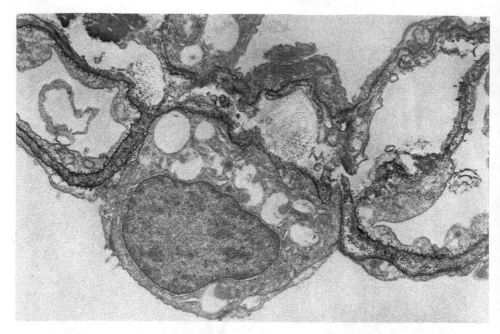

Figure 6   Normal alveolar septa in human lung. Stained with antifibronectin, Fab–peroxidases conjugate. Alveolar and capillary basal lamina and the adjacent basal surface of the epithelial and endothelial cells are stained ($\times$ 10,000). (Reproduced by permission from Torikata et al., 1985.)

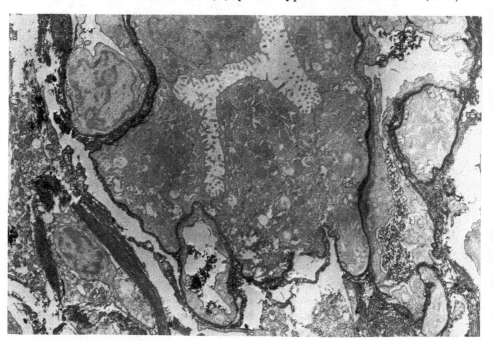

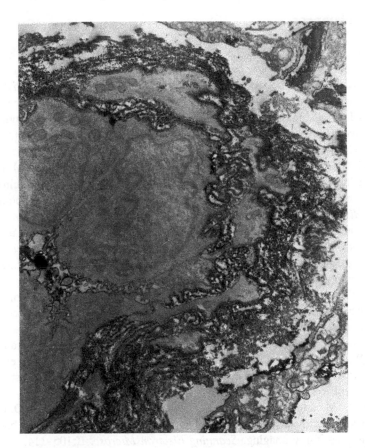

Figure 8 Pulmonary fibrosis. Capillary surrounded by a multilayered basal lamina heavily stained for fibronectin. Perivascular collagen fibers are also stained. At lower right, the process of a fibroblast shows focal dense surface staining similar to that reported for cultured fibroblasts. Antifibronectin Fab–peroxidase conjugate staining, × 10,000. (Reproduced by permission from Torikata et al., 1985.)

Acknowledgments

The authors thank Debra L. Fairfull and Pamela D. Kelley for their editorial assistance.

Figure 7 Pulmonary fibrosis. Stained with antifibronectin, Fab–peroxidase conjugate. Note the heavy staining of the basal laminae of capillaries and beneath the hyperplastic epithelium. Bundles of collagen fibrils are also stained (× 5000). Reproduced by permission from Torikata et al., 1985.)

References

Baskin, D. G., Erlandsen, S. L., and Parsons, J. A. (1979). Influence of hydrogen peroxide or alcoholic sodium hydroxide on the immunocytochemical detection of growth hormone and prolactin after osmium fixation. *J. Histochem. Cytochem.* **27**:1290–1292.

Bauer, T. W., Moriarty, C. M., and Childs, G V. (1981). Studies of immunoreactive gonadotropin releasing hormone (GnRH) in the rat anterior pituitary. *J. Histochem. Cytochem.* **29**:1171–1178.

Bedetti, C. D., Singh, J., Singh, G., Katyal, S. L., and Wong-Chong, M. (1987). Ultrastructural localization of rat Clara cell 10 KD secretory protein by the immunogold technique using polyclonal and monoclonal antibodies. *J. Histochem. Cytochem.* **35**:789–794.

Bendayan, M. (1985). The enzyme-gold technique: a new cytochemical approach for the ultrastructural localization of macromolecules. In *Techniques in Immunocytochemistry*. Volume 3. Edited by G. R. Bullock and P. Petrusz. London, Academic Press, pp. 179–201.

Birrell, G. B., Hedberg, K. K., and Griffith, O. H. (1987). Pitfalls of immunogold labeling: analysis by light microscopy, transmission electron microscopy, and photoelectron microscopy. *J. Histochem. Cytochem.* **35**:843–853.

Blobel, G. A., Gould, V. E., Moll, R., Lee, I., Huszar, M., Geiger, B., and Franke, W. W. (1985). Coexpression of neuroendocrine markers and epithelial cytoskeletal proteins in bronchopulmonary neuroendocrine neoplasms. *Lab. Invest.* **52**:39–51.

Bodner, S. M., Ingram, P., Spock, A., Spicer, S. S., and Shelburne, J. D. (1981). Carbohydrate histaochemistry of surface epithelium of rat trachea; backscatter and X-ray imaging. *Scanning Electron Microsc.* **II**:105–114.

Brissie, R. M., Spicer, S. S., and Thompson, N. T. (1975). The variable fine structure of elastin visualized with Verhoeff's iron hematoxylin. *Anat. Rec.* **181**:83–94.

Bullock, G. R., and Petrusz, P. (1982, 1983, 1984). *Techniques in Immunocytochemistry*. Volumes 1–3. London, Academic Press.

Burns, G. F., Cawley, J. C., Barker, C. R., Worman, C. P., Raper, C. G., and Hayhoe, F. G. (1978). Differing surface marker characteristics in plasma cell dyscrasias with particular reference to IgM myeloma. *Clin. Exp. Immunol.* **31**:414–418.

Carlemalm, E., Garavito, R. M., and Villiger, W. (1980). Proceedings, 7th European Congress, Electron Microscopy, Vol. 2. The Hague, Netherlands, pp. 656–657.

Carlemalm, E., Garavito, R. M., and Villiger, W. (1982). Resin development for electron microscopy and an analysis of embedding at low temperature. *J. Microsc. (London)* **126**:123–143.

Chiovetti, R., Little, S. A., Brass-Dale, J., and McGuffee, L. J. (1985). A new approach for low-temperature embedding: quick freezing, freeze-drying and direct infiltration in Lowicryl K4M. In *The Science of Biological Specimen Preparation for Microscopy and Microanalysis*. Edited by M. Mueller, R.P. Becker, A. Boyde, and J. J. Wolosewick. AMF-O'Hare, Chicago, SEM Inc., p. 155.

Chiovetti, R., McGuffee, L. J., Little, S. A., Wheeler-Clark, E., and Brass-Dale, J. (1987). Combined quick freezing, freeze-drying, and embedding tissue at low temperature and in low viscosity resins. *J. Electron Microsc. Tech.* **5**:1–15.

Coons, A. H., Creech, H. J., and Jones, R. N. (1941). Immunological properties of an antibody containing a fluorescent group. *Proc. Soc. Exp. Biol.* **47**:200–202.

Coons, A. H., and Kaplan, M. H. (1950). Localization of antigens in tissue cells: improvements in a method for the detection of antigen by means of fluorescent antibody. *J. Exp. Med.* **91**:1–13.

Coulter, H. D., and Terracio, L. (1977). Preparation of biological tissues for electron microscopy by freeze-drying. *Anat. Rec.* **187**:477–493.

de Harven, E. P., Leung, R., and Christensen, H. (1984). A novel approach for scanning electron microscopy of colloidal gold labeled cell surfaces. *J. Cell Biol.* **99**:53–57.

DeLellis, R. A., Sternberger, L. A., Mann, R. B., Banks, P. M., and Nakane, P. L. (1979). Immunoperoxidase techniques in diagnostic pathology. *Am. J. Clin. Pathol.* **71**:483–488.

Dudek, R. W., Varndell, I. A., and Polak, J. M. (1984). Combined quick-freeze and freeze-drying techniques for improved electron immunocytochemistry. In *Immunolabelling for Electron Microscopy*. Edited by M. Polak and I. M. Varndell. Amsterdam, Elsevier, pp. 235–248.

Emmart, E. W., Spicer, S. S., Turner, W. A., and Henson, J. G. (1962). The localization of glyceraldehyde-3-phosphate dehydrogenase within the muscle of the roach. *Periplaneta americana*, by means of fluorescent antibody. *Exp. Cell. Res.* **26**:78–97.

Emmart, E. W., Spicer, S. S., and Bates, R. W. (1963). Localization of prolactin within the pituitary by specific fluorescent antiprolactin globulin. *J. Histochem. Cytochem.* **11**:365–373.

Fernandez-Moran, H. (1950). Electron microscopy observations on the structure of the myelimated nerve fiber sheath. *J. Exp. Cell Res.* **1**:143–149.

Fernandez-Moran, H. (1960). Low-temperature preparation for electron microscopy of biological specimens based on rapid freezing with liquid helium II. *Ann. N.Y. Acad. Sci.* **85**:689–713.

Fukuda, Y., Ferrans, V. J., and Crystal, R. G. (1984). Development of elastic fibers of nuchal ligament, aorta, and lung of fetal and postnatal sheep: an ultrastructural and electron microscopic immunohistochemical study. *Am. J. Anat.* **170**:597–629.

Garvin, A. J., Spicer, S. S., Parmley, R. T., and Munster, A. M. (1974). Immunohistochemical demonstration of IgG in Reed-Sternberg and other cells in Hodgkin's disease. *J. Exp. Med.* **139**:1077–1083.

Geoghegan, W. D., and Ackerman, G. A. (1977). Adsorption of horseradish peroxidase, ovomucoid and anti-immunoglobulin to colloidal gold for the indirect detection of concanavalin A, wheat germ agglutinin and goat anti-human immunoglobulin G on cell surfaces at the electron microscopic level: a new method, theory and application. *J. Histochem. Cytochem.* **25**(11):1187–1200.

Gil, J., and Martinez-Hernandez, A. (1984). The connective tissue of the rat lung: electron immunohistochemical studies. *J. Histochem. Cytochem.* **32**:230–238.

Goldfischer, S., Kikkawa, Y., and Hoffman, L. (1968). The demonstration of acid hydrolase activities in the inclusion bodies of type II alveolar cells and other lysosomes in the rabbit lung. *J. Histochem. Cytochem.* **16**:102–109.

Gottlieb, A. J., Douglas, S. D., Strauss, A. J. L., and Spicer, S. S. (1966). Ultrastructural localization of gamma globulin binding to skeletal muscle. *Ann. N.Y. Acad. Sci.* **135**:638–643.

Graham, R. C., Jr., and Karnovsky, M. J. (1966). The early stages of absorption of injected horseradish peroxidase in the proximal tubules of mouse kidney: ultrastructural cytochemistry by a new technique. *J. Histochem. Cytochem.* **14**:291–302.

Griffiths, G., Simons, K., Warren, G., and Tokuyasu, K. T. (1983). Immunoelectron microscopy using thin frozen sections: applications to studies of intracellular transport of semliki forest virus spike gemeoproteins. *Methods Enzymol.* **96**:466–485.

Harrison, F., Van Hoof, J., Vanroelen, C. H., and Foidart, J. M. (1985). Masking of antigenic sites of fibronectin by glycosaminoglycans in ethanol-fixed embryonic tissue. *Histochemistry* **82**:169–174.

Harrist, T. J., and Mihm, M. C. (1979). Cutaneous immunopathology. The diagnostic use of direct and indirect immunoflourescence techniques in dermatologic disease. *Human Pathol.* **10**(6):625–653.

Harvey, D. M. R. (1982). Freeze substitution. *J. Microsc.* **127**:209–221.

Heuser, J. E., Reese, T. S., Dennis, M. J., Jan, Y., Jan, L., and Evans, L. (1979). Synaptic vesicle exocytosis captured by quick freezing and correlated with quantal transmitter release. *J. Cell Biol.* **81**:275–300.

Hill, D. J., Wright, T. C., Jr., Andrews, M. L., and Karnovsky, M.J. (1984). Localization of calmodulin in differentiating pulmonary Type II epithelial cells. *Lab. Invest.* **51**:297–306.

Hodges, G. M., Smolira, M. A., and Livingston, D. C. (1984). Scanning electron microscope immunocytochemistry in practice. In *Immunolabelling for*

Electron Microscopy. Edited by M. Polak and I. M. Varndell. Amsterdam, Elsevier, pp. 189–234.

Hsu, S. M., and Raine, L. (1981). Protein A, avidin, and biotin in immuno-histochemistry. *J. Histochem. Cytochem.* **29**:1349–1353.

Janossy, G., Thomas, J., Bollum, F., Granger, S., Pizzolo, G., Bradstock, K., Wong, L., McMichael, A., Ganeshaguru, K., and Hoffbrand, V. (1980). The human thymic microenvironment: an immunohistologic study. *J. Immunol.* **125**:202–212.

Jorgensen, A. O., and McGuffee, L. J. (1987). Immunoelectron microscopic localization of sarcoplasmic reticulum proteins in cryofixed, freeze-dried, and low temperature-embedded tissue. *J. Histochem. Cytochem.* **35**:723–732.

Karnovsky, M. J. (1965). A formaldehyde-glutaraldehyde fixative of high osmolality for use in electron microscopy. *J. Cell Biol.* **27**:137a–138a.

Leduc, E. H., Bernhard, W., Holt, S. J., and Tranzer, J. P. (1967). Ultrathin frozen sections. II. Demonstration of enzymic activity. *J. Cell Biol.* **34**:773–786.

Leunissen, J. L. M., Elbers, P. F., Leunissen-Bijvelt, J. J. M., and Verkleij, A. J. (1984). An evaluation of the cryosectioning of fixed and cryoprotected rat liver. *Ultramicroscopy* **12**:345–352.

Lucocq, J. M., Brada, D., and Roth, J. (1986). Immunolocalization of the oligosaccharide trimming enzyme glucosidase II. *J. Cell Biol.* **102**:2137–2146.

Lukes, R. J., Taylor, C. R., Chir, B., Phil, D., Parker, J. W., Lincoln, T. L., Pattengale, P. K., and Tindle, B. H. (1978). A morphologic and immunologic surface marker study of 299 cases of non-Hodgkin lymphomas and related leukemias. *Am. J. Pathol.* **90**(2):461–486.

MacIver, A. G., Giddings, J., and Mepham, B. L. (1979). Demonstration of extracellular immunoproteins in formalin-fixed renal biopsy specimens. *Kidney Int.* **16**(5):632–636.

Mar, H., Tsukada, T., Gown, A. M., Wight, T.N., and Baskin, D. G. (1987). Correlative light and electron microscopic immunocytochemistry on the same section with colloidal gold. *J. Histochem. Cytochem.* **35**:419–426.

Marshall, J. M., Jr., Holtzer, H., Finck, H., and Pepe, F. (1959). The distribution of protein antigens in striated myofibrils. *Exp. Cell Res.* **7**:219–233.

Mason, T. E., Phifer, R. F., Spicer, S. S., Swallow, R. A., and Dreskin, R. B. (1969). An immunoglobulin-enzyme bridge method for localizing tissue antigens. *J. Histochem. Cytochem.* **17**:563–569.

Mason, R. J., Williams, M. C., and Dobbs, L. G. (1977). Secretion of disaturated phosphatidyl choline by primary cultures of type II alveolar cells. In *Pulmonary Macrophages and Epithelial Cells,* ERDA Symposium Series, Vol. 43. Springfield, VA, National Technical Information Service, p. 280.

Mason, D. Y., and Woolston, R. E. (1982). Double immunoenzymatic labelling. In *Techniques in Immunocytochemistry* Volume I. Edited by G. R. Bullock and P. Petrusz. London, Academic Press, pp. 135–153.

McGuffee, L.J., Hurwitz, L., Little, S. A., and Skipper, B. E. (1981). A ^{45}Ca autoradiographic and stereological study of freeze-dried smooth muscle of the guinea pig vas deferens. _J. Cell Biol._ **90**:201–210.

McKeever, P. E., Spicer, S. S., Brissie, N. T., and Garvin, A. J. (1977). Immune complex receptors on cell surfaces. III. Topography of macrophage receptors demonstrated by new scanning electron microscopic peroxidase marker. _J. Histochem. Cytochem._ **25**:1063–1068.

Moriarty, G. C. (1976). Immunocytochemistry of the pituitary glycoprotein hormones. _J. Histochem. Cytochem._ **24**:846–863.

Mugnaini, E. (1985). GABA neurons in the superficial layers of the rat dorsal cochlear nucleus: light and electron microscopic immunocytochemistry. _J. Comp. Neurol._ **235**:61–81.

Muller, M., Meister, N., and Moor, H. (1980). Freezing in a propane jet and its application in freeze-fracturing. Mikroskopie (Wien) **36**:129–140.

Nakane, P. K., and Pierce, G. B., Jr. (1966). Enzyme-labeled antibodies: preparation and application for the localization of antigens. _J. Histochem. Cytochem._ **14**:929–931.

Nakane, P. K., and Pierce, G. B., Jr. (1967). Enzyme-labeled antibodies for the light and electron microscopic localization of tissue antigens. _J. Cell Biol._ **33**:307–318.

Newman, G. R., Jasani, B., and Williams, E. D. (1983). A simple post-embedding system for the rapid demonstration of tissue antigens under the electron microscope. _Histochem. J._ **15**:543–555.

Newman, G. R., and Jasani, B. (1984). Post-embedding immunoenzyme techniques. In _Immunolabelling for Electron Microscopy_. Edited by J. M. Polak and I. M. Varndell. Amsterdam, Elsevier, pp. 53–70.

Philipps, T. E., and Boyne, A. F. (1984). Liquid nitrogen-based quick freezing: experiences with bounce-free delivery of cholinergic nerve terminals to a metal surface. _J. Electron Microsc. Tech._ **1**:9–29.

Polak, J. M., and Varndell, I. M. (1984). _Immunolabelling for Electron Microscopy_. Amsterdam, Elsevier.

Roth, J. (1982). The preparation of protein A-gold complexes with 3 nm and 15 nm gold particles and their use in labelling multiple antigens on ultra thin sections. _Histochem. J._ **14**:791–801.

Roth, J. (1983). The colloidal gold marker system for light and electron microscopic cytochemistry. In _Techniques in Immunocytochemistry_, Vol. II. Edited by G. R. Bullock and P. Petrusz. Academic Press, London, pp. 217–284.

Roth, J. (1986). Post-embedding cytochemistry with gold labelled reagents: a review _J. Microscopy_ **143**:125–137.

Roth, J., and Berger, E.G. (1982). Immunocytochemical localization of galactosyl transferase in HeLa cells: codistribution with thiamine pyrophosphatase in trans-Golgi cisternae. _J. Cell Biol._ **93**:223–229.

Roth, J., Lentze, M. J., and Berger, E. G. (1985). Immunocytochemical demonstration of ecto-galactosyltransferase in absorptive intestinal cells. *J. Cell Biol.* **100**:118–125.

Roth, J., and Wagner, M. (1977). Peroxidase and gold complexes of lectins for double labeling of surface-binding sites by electron microscopy. *J. Histochem. Cytochem.* **25**:1181–1184.

Ryan, U. S., Ryan, J. W., and Smith, D. S. (1975). Alveolar type II cells: studies on the mode of release of lamellar bodies. *Tissue Cell* **7**:587–599.

Sage, H., Farin, F. M., Striker, G., and Fisher, A. B. (1983). Granular pneumocytes in primary culture secrete several major components of the extracellular matrix. *Biochemistry* **22**:2148–2155.

Sato, A., and Spicer, S. S. (1982a). Ultrastructural visualization of galactose in the glycoprotein of gastric surface cells with a peanut lectin conjugate. *Histochem. J.* **14**:125–138.

Sato, A., and Spicer, S. S. (1982b). Ultrastructural visualization of galactosyl residues in various alimentary epithelial cells with the peanut lectin-horseradish peroxidase procedure. *Histochemistry* **73**:607–624.

Simson, J. A. V., Hint, D., Munster, A. M., and Spicer, S. S. (1977). Immunocytochemical evidence for antibody binding to mast cell granules. *Exp. Mol. Pathol.* **26**:85–91.

Singer, S. J. (1959). Preparation of an electron dense antibody conjugate. *Nature (London)* **183**:1523–1524.

Sitte, H., Neumann, K., and Edelmann, L. (1985). Cryofixation and cryosubstitution for routine work in transmission electron microscopy. In *The Science of Biological Specimen Preparation for Microscopy and Microanalysis.* Edited by M. Muller, R. P. Becker, A. Boyd, J. J. Wolosewick. Chicago, AMF-O'Hare, SEM Inc., pp. 103–118.

Sjostrand, F. S., and Baker, R. F. (1958). Fixation by freeze-drying for electron microscopy of tissue cells. *J. Ultrastruct. Res.* **1**:239–246.

Slot, J. W., and Geuze, H. J. (1982). Ultracryotomy of polyacrylamide embedded tissue for immunoelectron microscopy. *Biol. Cell* **44**:325–328.

Slot, J. W., and Geuze, H. J. (1984). Gold markers for single and double immunolabelling of ultrathin cryosections. In *Immunolabelling for Electron Microscopy.* Edited by M. Polak and I. M. Varndell. Amsterdam, Elsevier, pp. 129–142.

Soligo, D., and de Harven, E. P. (1987). Simultaneous visualization of myeloperoxidase reactivity and immunogold labeling by backscattered electron imaging with the scanning electron microscope. *J. Histochem. Cytochem.* **35**:267–270.

Spicer, S. S. (Ed.) (1987). *Histochemistry in Pathologic Diagnosis.* New York, Marcel Dekker.

Spicer, S. S., Frayser, R., Virella, G., and Hall, B. J. (1977). Immunocyto-chemical localization of lysozyme in respiratory and other tissues. *Lab. Invest.* **36**:282–295.

Sternberger, L. A., Hardy, P. H., Cuculis, J. J., and Meyer, H. G. (1970). The unlabeled antibody enzyme method of immunohistochemistry: preparation and properties of soluble antigen–antibody complex (horseradish peroxidase–antihorseradish peroxidase) and its use in identification of spirochetes. *J. Histochem. Cytochem.* **18**:315–333.

Stirling, C. E., and Kinter, W. B. (1967). High-resolution radioautography of galactose-^{3}H accumulation in rings of hamster intestine. *J. Cell Biol.* **35**:585–604.

Stumpf, W. E., and Roth, L. J. (1967). Freeze-drying of small tissue samples and thin frozen sections below $-60\,^\circ$C:a simple method of cryosorption pumping. *J. Histochem. Cytochem.* **15**:243–251.

Thomopoulos, G. N., Schulte, B. A., and Spicer, S. S. (1987). Postembedment staining of complex carbohydrates: influence of fixation and embedding procedures. *J. Electron Microsc. Tech.* **5**:17–44.

Tokuyasu, K. T. (1973). Technique for ultracryotomy of cell suspensions and tissues. *J. Cell Biol.* **57**:551–565.

Tokuyasu, K. T. (1980). Immunochemistry on ultrathin frozen sections. *Histochem. J.* **12**:381–403.

Tokuyasu, K. T. (1984). Immuno-cryoultramicrotomy. In *Immunolabelling for Electron Microscopy.* Edited by M. Polak and I. M. Varndell. Amsterdam, Elsevier, pp. 71–82.

Torikata, C., Villiger, B., Khun, C., III, and McDonald, J. A. (1985). Ultrastructural distribution of fibronectin in normal and fibrotic human lung. *Lab. Invest.* **52**:399–403.

Van Harreveld, A., Trubatch, J., and Steiner, J. (1974). Rapid freezing and electron microscopy for the arrest of physiological processes. *J. Microsc.* **100**:189–198.

Willingham, M. C. (1980). Electron microscopic immunocytochemical localization of intracellular antigens in cultured cells: the EGS and ferritin bridge procedures. *Histochem. J.* **12**:419–434.

Willingham, M. C., Spicer, S. S., and Graber, C. D. (1971). Immunocytologic labeling of calf and human lymphocyte surface antigens. *Lab. Invest.* **25**:211–219.

Willingham, M. C., and Pasten, I. (1985). Morphologic methods in the study of endocytosis in cultured cells. In *Endocytosis.* Edited by I. Pasten and M. C. Willingham. New York, Plenum Press, pp. 281–321.

Wofsy, L. (1979). Hapten–antibody conjugates as probes of the lymphocyte surface. *Scanning Electron Microscopy* **III**:565–572.

6

Immunohistochemistry of the Pulmonary Extracellular Matrix

ANTONIO MARTINEZ-HERNANDEZ

Thomas Jefferson University
Philadelphia, Pennsylvania

PETER S. AMENTA

Hahnemann University
Philadelphia, Pennsylvania

I. Introduction

In addition to cells, all organs contain an extracellular matrix. This extracellular matrix (ECM) was initially defined, by morphologists according to its preferential reactions with histological dyes; thus, blue staining matrix was "collagen"; argyrophilic fibers were "reticulin"; those reacting with orcein, elastic fibers; period-acid–Schiff, (PAS) positive material was basement membrane (BM), and the diffuse reactivity with polycations defined the "ground substance." This simple view has been enriched by analytical studies. As of now, 13 distinct collagen types, the 4 major BM components, the protein core of proteoglycans, and the composition of the glycosaminoglycan side chains have been defined (Piez and Reddi, 1984; Cunningham 1987a,b; Philajeniemi et al., 1987; Bentz et al., 1983; Burgeson, 1987; Miller and Gay, 1987). Molecular biology is defining phylogenetically conserved families of ECM components, the multitude of steps regulating ECM synthesis, secretion, assembly, and the basis of some inherited disorders (Piez and Reddi, 1984; Cunningham, 1987a,b; Pihlajaniemi et al., 1987; Cheah, 1985; Pyeritz et al., 1984; Dickson et al., 1984; Jimenez et al., 1986; Nicholls et al., 1984; Pihlajaniemi et al., 1984; Prockop et al., 1985). Knowledge of the molecular charge, conformation, and cross-links provides insights into the functional properties of the molecules; however, to define their specific function in a tissue, their exact distribution and associations in that tissue must be known. The unambiguous identification of the individual components that make up the ECM is beyond the capabilities of classic morphology. At present, only the exquisite specificity of antibodies and immunohistochemical methods can identify these components. The dimensions of most ECM components are beyond the resolving power of the light microscope, a limitation that is particularly obvious in the alveolar ECM, where septal thickness is often less than 2 μm (Gil and Martinez-Hernandez, 1984; Gil, 1978, 1982; Amental et al., 1988). For instance, in the alveolar septum, light microscopy demonstrates fibronectin, but only electron microscopy can resolve the dimensions of individual fibronectin filaments, their associations among themselves, with other ECM components, and with cells (Gil and Martinez-Hernandez, 1984; Amenta et al., 1986).

This chapter describes some immunohistochemical methods and their application to the study of the pulmonary ECM. We will first discuss some theoretical considerations, then specific protocols, and finally review the contributions of immunohistochemistry to our understanding of the pulmonary ECM.

II. Theoretical Considerations

There are many methods and variations available for immunohistochemistry (Martinez-Hernandez, 1987a, b; Amenta and Martinez-Hernandez, 1987; Polak and Varndell, 1984; Bullock and Petrusz, 1982). In any study, choices of fixative sectioning, antibody types, and markers must be based on the possible effects of each treatment on subsequent ones. Therefore, before we present specific protocols, some theoretical considerations are appropriate.

A. Tissue Preparation

The goal of tissue preparation is to stabilize tissue components, which can be achieved by physical or chemical methods. The physical methods include microwave irradiation and quick freezing. Microwave irradiation (Boon and Kok, 1987), although useful for light microscopy, has not been used extensively for electron microscopic studies. Quick freezing can be used as a method to stabilize tissue components, or after fixation, as an alternative to embedding (Martinez-Hernandez, 1987a, b; Amenta and Martinez-Hernandez, 1987). Chemical stabilization is achieved using fixatives (Martinez-Hernandez, 1987a, b; Amenta and Martinez-Hernandez, 1987).

B. Fixation

Optimal fixation is usually achieved by cross-linking tissue components, which may result in loss of antigenicity. Therefore, the immunohistochemist, in choosing a fixative, must accept a compromise between optimal fixation and optimal antigenicity. Given this limitation, only a few fixatives have proven suitable for general use in immunohistochemistry. For example, a commonly used general purpose fixative, glutaraldehyde, has only limited application in immunohistochemical studies. While providing excellent preservation, glutaraldehyde often results in loss of antigenicity, particularly of high-molecular-weight (>20,000 Daltons) antigens (Martinez-Hernandez, 1987b). Given the need to preserve antigenicity, formaldehyde alone, or combined with lysine and periodate, has proven to be dependable for immunohistochemistry (Martinez-Hernandez, 1987b; McLean and Nakane, 1974). This is not to imply that for some antigens other fixatives will not provide adequate antigenicity with superior preservation. Not only the fixative type but also the fixation time is critical for antigenic preservation. For instance, in our experience, the antigenicity of collagen type IV is usually well preserved after fixation in 4% formaldehyde for 2–3 hr. However, fixation for longer than 4 hr often results in complete loss of antigenicity (Amenta et al., 1986)

Whenever feasible, fixation by perfusion is to be preferred over immersion fixation. Vascular perfusion results in rapid delivery of fixative to all areas of the specimen, thus preventing the gradients and autolysis sometimes found with immersion fixation. A further benefit of perfusion is obtained if the fixative is preceded by a short (1 min) perfusion with isotonic buffer, to wash out interstitial plasma (Gil and Martinez-Hernandez, 1984; Martinez-Hernandez et al., 1981b; Courtoy and Boyles, 1983). It is worth emphasizing that due to the need to maintain antigenicity, the structural preservation obtained in immunohistochemistry is often less than optimal.

C. Postfixation Treatments

The goal of postfixation treatments is to quench and remove excess fixative and prepare tissue for the subsequent steps. It is necessary to remove excess aldehyde, which can be accomplished by the addition of sucrose to the buffer to quench the remaining fixative. If the samples are to be frozen, washing with glycerol will minimize ice-crystal formation (Amenta et al., 1983).

D. Embedding

Antibodies have limited tissue penetration (10–12 μm); therefore, only sectioning to less than this thickness will allow complete penetration, thus minimizing false negatives. Lung, being an easily deformable organ, must be embedded prior to sectioning. Multiple embedding media are available for light and electron microscopy (Glauert, 1975). Most embedding procedures involve the replacement of water, either by the embedding media or by an intermediate solvent. In some respects, freezing can be considered as embedding; replacing water by ice. Of the multiple procedures available, freezing, paraffin, and plastics are the most frequently used.

Freezing results in enough tissue hardening to permit cryostat sectioning at 5–10 μm, suitable for light microscopy or, after subsequent embedding in plastic, for electron microscopy (Martinez-Hernandez, 1987a,b; Amenta and Martinez-Hernandez, 1987). With a cryoultramicrotome, thin sections can be obtained directly, obviating the use of organic solvents (Tokuyasu, 1984). Whatever the final use of frozen blocks, it is essential to minimize ice crystal formation. To this end, small blocks, the addition of cryoprotectors, and a fast freezing rate are critical (Martinez-Hernandez, 1987a,b; Amenta and Martinez-Hernandez, 1987). Frozen sections, while not providing the optimal preservation, have some advantages; for instance, the antigenic determinants are directly available. In contrast, with plastic or paraffin, the embedding media remains in the sections (as a barrier to antibody penetration) or must be removed by harsh procedures (which may lessen antigenicity). Vibratome sectioning is an alternative to freezing or embedding and takes advantage of the vibration of an advancing blade to cut tissue blocks immersed in buffer. However, with the vibratome, it is difficult to obtain lung sections thinner than 50–60 μm, which results in incomplete antibody penetration.

E. Prestaining Treatment

Prior to reaction with antibodies, several treatments can be used to facilitate localization of the relevant antigens. These treatments may be used to remove intrinsic enzyme activity, decrease nonspecific antiboy binding, and restore antigenicity.

Intrinsic Peroxidase Activity

Many cells, such as neutrophils and macrophages, contain peroxidatic enzymes and all heme-containing proteins are potential peroxidases. Incubation with diaminobenzidine hydrochloride (DAB) and H_2O_2 in the presence of these peroxidases will yield reaction product unrelated to antibody localization. Oxidation with hydrogen peroxide in methanol or treatment with acids has been used to inactivate the intrinsic peroxidatic activity (Martinez-Hernandez, 1987a,b; Amenta and Martinez-Hernandez, 1987). The lowest effective concentration should be used to minimize any deleterious effects to the tissues (Amenta et al., 1986). If colloidal gold rather than peroxidase is used as the marker, this step is unnecessary (Karkavelas et al., 1988).

Aldehyde Reduction

Free aldehyde radicals will react with ϵ-amino groups; therefore, immunoglobulin molecules will bind to aldehyde fixed tissues by a nonimmune mechanism, resulting in high backgrounds. The ε-amino group in glycine and other amino acids will bind to, and therefore block, free aldehyde radicals, whereas, sodium borohydride reduces the aldehyde groups and also often restores some of the conformation lost in fixation (Martinez-Hernandez et al., 1981b; Boselli et al., Martinez-Hernandez, 1987a,b; Amenta and Martinez-Hernandez, 1987).

Enzyme Digestions

Fixation and embedding often produce some loss of tertiary conformation, thermal denaturation, and aggregation, which may result in loss or masking of antigentic determinants. Partial, controlled enzymatic digestion can often restore some of the lost antigenicity. Glycosidase digestion often decreases the nonspecific background (perhaps by removing lectinlike binding) and sometimes increases the specific staining (perhaps by removing the steric hindrance provided by the charged carbohydrate side-chains) (Amenta et al., 1986; Andrews et al., 1985). Proteolytic digestion often enhances specific staining. This is particularly true in paraffin-embedded (thermally denatured) sections reacted with antibodies to ECM components (Barsky et al., 1984). Pepsin, trypsin, and pronase are the most commonly used enzymes. Enzyme digestion should only be used if there is substantial loss of antigenicity subsequent to tissue processing, and even then, judiciously, to minimize tissue destruction and antigen translocation (Amenta et al., 1986).

Blocking Nonspecific Binding Sites

Immunoglobulins can bind to tissues through several nonimmune mechanisms, such as Fc-binding and nonspecific protein–protein interactions. To minimize this nonimmune binding, treatment of the sections with preimmune serum, from the animal species providing the secondary antibody, is used. Of course, using Fab fragments rather than complete IgG molecules will prevent binding to Fc receptors (Martinez-Hernandez, 1987a,b; Amenta and Martinez-Hernandez, 1987).

F. Antibodies

The choice of antibody to be used in an immunohistochemical procedure is governed by the ultimate objective of the study. Polyclonal antibodies have the advantage of recognizing multiple epitopes in a molecule and having high affinities; however, antibodies directed against contaminants may be present in the same serum. Monoclonal antibodies (MAbs) have the advantage of their unique specificity against a single epitope and the consistent reactivity of different batches. However, they tend to have lower affinities and the likelihood of denaturation of the relevant epitope during processing is greater. If one is using MAbs, those produced in ascites tend to have higher affinity than those raised in tissue culture. Because of their proclivity to aggregate, resulting in precipitation and high backgrounds, IgMs should be avoided (Martinez-Hernandez 1987a,b). Polyclonal antibodies can be used as complete serum, purified IgG, or affinity-purified antibodies. In general, IgGs purified by ammonium sulfate precipitation, followed by chromatography on DEAE, provide satisfactory results (Martinez-Hernandez, 1987a,b). Whole serum can be used, but purification decreases background (hemoglobin has peroxidatic activity) and facilitates storage. Definition of the monospecificity of the antibodies is essential; to this end, radioimmunoassay, ELISA or immunoblotting can be used (Gay and Fine, 1987). If cross-reactivity is found, it can be removed by cross-absorption or affinity chromatography (Gay and Fine, 1987). Antibody penetration in the tissues can be enhanced using Fab fragments, rather than intact antibody molecules (Martinez-Hernandez et al., 1974). As an added benefit, use of Fab fragments eliminates the possibility of interaction with Fc receptors. The working dilution for a particular antibody is determined by "checkerboard" titration on tissue sections.

G. Markers

Numerous markers are available for immunohistochemical procedures. Immunofluorescence was the original method (Coons et al., 1941). The bright green fluorescence, against a black background, results in an esthetically pleasing image. However, several immunohistochemical markers offer advantages over fluorescence, such as greater sensitivity, permanence, and suitability for electron microscopy (Fig. 1).

Particulate Markers

Ferritin and colloidal gold are the most commonly used particulate markers. Ferritin, due to its iron content, can be used as an electron-dense marker (Singer, 1959). Its proclivity to nonspecificx interactions, low sensitivity, high molecular weight (resulting in limited ability to penetrate tissues), and low electron density has limited its use. More recently, colloidal gold has been used as a marker in immunohistochemistry (Faulk and Taylor, 1971). The high atomic weight of gold results in higher electron density than ferritin. The size of the colloidal gold particles can be selected from 3 to 100 nm, permitting dual localizations. Furthermore, the introduction of silver enhancement methods (Geoghegan et al., 1978) has made colloidal gold a practical marker for light microscopy.

Enzyme Markers

With enzyme markers, increased incubation times result in the formation of more reaction product, thus increasing sensitivity. However, prolonged incubation time

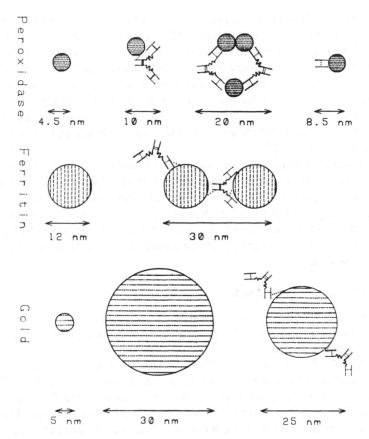

Figure 1 Markers frequently used in electron immunohistochemistry. Horseradish peroxidase has a small radius (4.5 nm) and can be conjugated to IgG molecules by chemical means. The binding of antibodies to horseradish peroxidase with their antigen results in the formation of an immune complex, the peroxidase–antiperoxidase complex (PAP). The penetration problems created by the relative large size of PAP can be minimized using complexes formed by the Fab fragment and horseradish peroxidase.

Ferritin has a large radius (12 nm), low electron density, and a proclivity for nonspecific interactions with other proteins, making it the least desirable of the three markers.

Collodial gold particles of different sizes are available. The commonly used diameters range from 5 to 40 nm. Antibodies labeled with particles of different diameters allow simultaneous localization of more than one antigen. Gold particles have high electron density and are the preferred marker for post-embedding localization. (From Martinez-Hernandez, 1987b.)

may result in diffusion of reaction product, ambiguous localization, as well as the obscuring of underlying structures (Martinez-Hernandez, 1987a; Courtoy et al., 1983). Of the enzyme markers currently available, alkaline phosphatase and horseradish peroxidase are most frequently used (Nakane and Pierce, 1966). They are equally suitable for light and electron microscopy and their relatively small size result in excellent tissue penetration. Compared with gold, peroxidase offers greater sensitivity and penetration, whereas gold provides higher contrast and better resolution (Martinez-Hernandez, 1987a; Faulk and Taylor, 1971; Karkavelas et al., 1988).

Various chromogens are available for peroxidases (Nakane, 1968), offering a choice of colors and permitting simultaneous localization of several antigens. However, some chromogens fade with time. The most frequently used chromogen for peroxidase is DAB (Graham and Karnovsky, 1966). It gives a permanent brown precipitate that is highly osmiophilic, making it a practical marker for light and electron microscopy. In theory, DAB is a potential carcinogen and should be handled with caution.

H. Detection Systems

The different detection systems available offer advantages and disadvantages in terms of reaction time, sensitivity, suitable controls, background staining, and overall complexity (Fig. 2).

The simplest system is the direct method, in which the primary antibody itself is labeled with a suitable marker. In the indirect method, an antibody directed against the primary antibody is labeled with the marker. In the PAP method, the unlabeled secondary antibody is followed by the peroxidase–antiperoxidase complex, in which the antibody is from the same species as the primary. Each subsequent antibody addition increases the method's sensitivity. The direct method requires labeling of each antibody individually, thus the indirect and PAP methods are more desirable, since only the secondary, or the PAP are labeled.

Another alternative is the avidin–biotin system, based on the high-affinity between avidin and biotin (Heitzmann and Richards, 1974). The antibody is labeled with biotin, followed by a avidin enzyme conjugate. Egg-white avidin is highly glycosylated, resulting in relatively high backgrounds (Heitzmann and Richards, 1974; Wooley and Longsworth, 1942). This problem can be overcome using streptavidin, a nonglycosylated protein of bacterial origin (Heitzmann and Richards, 1974; Wooley and Longsworth, 1942). In our hands, the strepavidin–biotin system seems more sensitive than the PAP system.

I. Pre-embedding vs. Postembedding Techniques

Pre-embedding refers to the application of antibodies and markers prior to tissue embedding. In general, postembedding is more desirable because it provides better

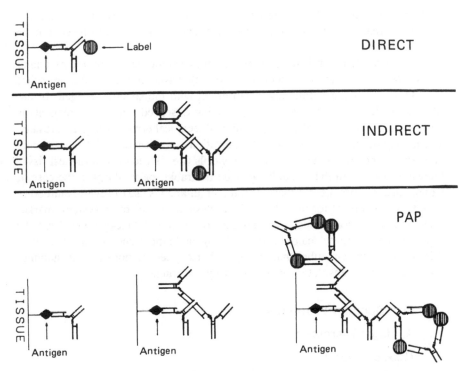

Figure 2 Principles of three commonly used detection methods. The simplest system is the direct method, in which the primary antibody itself is labeled with a suitable marker. In the indirect method, an antibody directed against the primary antibody is labeled with the marker. In the PAP method, the unlabeled secondary antibody is followed by the PAP complex, in which the antibody is from the same species as the primary. Each subsequent antibody addition increases the method's sensitivity. The direct method requires labeling of each antibody individually, thus the indirect and PAP methods are more desirable, since only the secondary, or the PAP are labeled. Multiple variations on these basic principles exist. (From Martinez-Hernandez, 1987b.)

preservation and resolution. However, antigenicity is often decreased and sometimes abolished. Therefore, the choice of either preembedding or postembedding depends essentially on the ability of the relevant antigens to withstand fixation and embedding (Karkavelas et al., 1988).

J. Interpretation of Results

Successful immunohistochemical staining depends on a multitude of factors, such as nonspecific interactions, specificity and integrity of antibodies, preservation of the relevant epitope(s) during processing, and depth of reagent penetration. For these reasons, the need for negative and positive controls cannot be overemphasized.

As negative control, the primary antibody should be substituted by one of the following: preimmune serum (polyclonals), tissue culture media or ascites fluid (MAbs), primary antibody preabsorbed with the specific antigen, or antibodies irrelevant to the tissue. Any staining observed with these reagents is nonspecific. To preclude the interpretation of technical errors as negative results, an antibody of known reactivity and localization in the test cells or tissues should be used as a positive control. Internal controls (i.e., structures in the test tissue known to react with the test antibody) are desirable, particularly in the case of pathological specimens. Furthermore, the simultaneous use of several antibodies makes each a control for the others. Regardless of the detection system, marker used, and precautions taken, some nonspecific staining ("background") is to be expected. Therefore, to determine optimal reagent dilutions and to evaluate results, it is necessary to remember that the signal/noise (specific/non-specific staining) ratio, rather than absolute specific intensity, is crucial.

III. Specific Protocols

A. Light Microscopy

Unfixed Tissues

The procedure used in our laboratory for light microscopy immunohistochemistry on frozen, unfixed lung is as follows:

1. The trachea of the experimental animal is dissected and cannulated following light methoxyfluorane anesthesia. The lungs are ventilated by intermittent positive airway pressure through a three-way stopcock allowing for passive deflation. An 18-gauge catheter is inserted into the pulmonary artery, via the right ventricle, and the left atrium is severed for drainage. A 1 min perfusion with phosphate-buffered saline (PBS) is used to clear the effusate of blood.

2. The lungs are removed and tissue blocks are cut with razor blades. Frozen sections are somewhat difficult to obtain from lung. Injecting cryostat embedding media into bronchi immediately prior to sectioning tissue blocks provides some rigidity to the tissue.

3. Tissue blocks are placed in cryostat embedding media and snap frozen in methylbutane at liquid nitrogen temperature. The freezing device used in our laboratory is a Dewar flask filled with liquid nitrogen and a stainless steel cup containing methylbutane, which is partially immersed in the liquid nitrogen.

4. Cryostat sections, 3–5 µm thick, are collected on slides coated with glutaraldehyde, cross-linked albumin (Amenta and Martinez-Hernandez, 1987).

5. Sections are air dried for 30 min and encircled with a diamond tip pen.

6. Sections are fixed in 100% acetone at 4 °C for 15 min.

7. Sections are hydrated in PBS (Amenta and Martinez-Hernandez, 1987) at 4° for 15 min.

8. To inhibit intrinsic peroxidase activity, the sections are treated with 0.01 M periodic acid in PBS for 15 min at room temperature.

9. The sections are rinsed in PBS for 5 min × 3, at 4°C.

10. The sections are reacted with normal serum of the same species as the secondary antibody (1:10 dilution) for 15 min at room temperature in a moist chamber.

11. The sections are rinsed in PBS for 5 minutes × 3, at 4°C.

12. The sections are reacted either with a suitable dilution of the primary antibody as determined by "checkerboard" titration (experimental) or with the normal serum of the same species (control) at similar dilutions for 1 hr, at 4°C, in a moist chamber. All antibodies and normal sera are applied with enough volume to cover the tissue section, (usually 50–75 µl).

13. The sections are rinsed in PBS for 5 minutes × 3, at 4°C.

14. The sections are reacted with a suitable dilution of the secondary antibody (as determined by "checkerboard" titration) for 1 hr, at 4°C, in a moist chamber.

15. The sections are rinsed in PBS for 5 minutes × 3, at 4°C.

16. The sections are reacted with a suitable dilution of the peroxidase-antiperoxidase (PAP) (as determined by "checkerboard" titration) for 1 hr, at 4°C, in a moist chamber.

17. The sections are rinsed in PBS for 5 minutes × 3, at 4°C.

18. The sections are reacted in DAB 50 mg/150 ml in 0.5 M Tris-buffer, pH 7.2, in the dark for 10 min.

19. Hydrogen peroxide, 20 ml of a 5% solution in water is added to the staining dish, mixed well and the slides incubated for 5–10 m.

20. The sections are rinsed in PBS for 5 minutes × 3, at 4%C.

21. Dehydrate in graded ethanols and mounted with permount.

Fixed Tissues

Frozen Sections

1. Fixation is performed at room temperature (Gil and Martinez-Hernandez, 1984; Amenta et al., 1988). The trachea is dissected and cannulated following light methoxyfluorane anesthesia. The lungs are ventilated by intermittent positive airway pressure through a three-way stopcock allowing for passive deflation. An 18-gauge catheter is inserted into the pulmonary artery through the right ventricle and the left atrium severed. A 1 min perfusion of PBS precedes a 30 min perfusion with 4% formaldehyde in 0.1 M sodium phosphate buffer, at pH 7.4, for 30 min. After 1 min perfusion with fixative, the trachea is injected with fixative.

2. The lungs are removed *in toto* and submerged in a beaker of fixative at 4 °C for 2 hr under gentle agitation with the trachea cannulated and perfused with fixative. Tissue blocks are cut from the lung with new razor blades, placed in scintillation vials, and immersion fixed for an additional 30 min.

3. The fixative is decanted and the tissue blocks washed at 4 °C overnight, under continuous, gentle agitation with multiple changes of PBS–4% sucrose.

4. One hour before being embedded, the tissue blocks are washed in PBS containing 4% sucrose and 7% glycerol. With lung, addition of cryostat embedding media 1:1 to this mixture for the last 10 min of the wash facilitates sectioning.

5. Cryostat sections, 3–5 μm thick, are collected on slides coated with glutaraldehyde, cross-linked, albumin (Gil and Martinez-Hernandez, 1984; Amenta and Martinez-Hernandez, 1987; Amenta et al., 1988).

6. Sections are encircled with a diamond tip pen and air dried for 30 min.

7. Sections are hydrated in PBS at 4 °C for 15 min.

8. To inhibit intrinsic peroxidase activity, the sections are treated with 0.01 M periodic acid in PBS for 30 min at room temperature.

9. Sections are reacted with sodium borohydride (50 mg/ml) in PBS for 45 m at 4 °C for 1 hr to reduce free aldehyde groups.

10. The sections are rinsed in PBS for 5 min × 3, at 4 °C.

11. The sections are reacted with normal serum of the same species as the secondary antibody (1/10 dilution) for 15 min at room temperature in a moist chamber.

12. The sections are rinsed in PBS for 5 min × 3, at 4 °C.

13. The sections are reacted with either a suitable dilution of the primary antibody as determined by "checkerboard" titration (experimental) or with the normal serum of the same species (control) at a similar dilution for 1 hr, at 4 °C, in a moist chamber. All antibodies and normal sera are applied with a micropippetor and enough volume to cover the tissue section (usually 50–75 μl).

14. The sections are rinsed in PBS for 5 min × 3, at 4 °C.

15. The sections are reacted with a suitable dilution of the secondary antibody (as determined by "checkerboard" titration) for 1 hr, at 4 °C, in a moist chamber.

16. The sections are rinsed in PBS for 5 minutes × 3, at 4 °C.

17. The sections are reacted with a suitable dilution of the PAP (as determined by "checkerboard" titration) for 1 hr, at 4 °C, in a moist chamber.

18. The sections are rinsed in PBS for 5 min × 3 at 4 °C.

19. The sections are reacted in DAB 50 mg/150 ml in 0.5 M Tris-buffer pH 7.2 in the dark for 10 min.

20. Hydrogen peroxide, 20 ml of a 5% solution in water, is added to the staining dish, mixed well, and the slides incubated for 5 min.

21. The sections are rinsed in PBS for 5 min ×3, at 4 °C.

22. The sections are dehydrated in graded ethanols and mounted with permount.

Paraffin-Embedded Tissues

Although providing excellent antigenic preservation, freezing introduces artifact, limits the time tissues can be stored, the number of sections that can be handled, and lessens the feasibility of serial sectioning. Methodologies that permit localization of ECM components in paraffin-embedded tissues, resulting in excellent preservation and morphology, are available (Barsky et al., 1984). Optimal results are obtained with tissues, fixed with formaldehyde, embedded in paraffin, deparaffinized, subjected to controlled pepsin digestion, reduced with $NaBH_4$, and reacted with 0.4 M periodic acid. Limited pepsin digestion has been noted to enhance antigenicity of ECM components in paraffin-embedded tissue (Barsky et al., 1984). The protocol is as follows:

1. Fixation is performed as described above.

2. The lungs are removed *in toto* and submerged in a beaker of fixative at 4 °C for 2 hr, under gentle agitation, with the trachea cannulated and perfused with fixative. Tissue blocks are cut with razor blades, placed in scintillation vials, and fixed for an additional 30 min.

3. Tissue blocks are dehydrated in graded ethanols, cleared in xylene, and embedded in paraffin (55–60 °C). Section (3 μm) are placed on coated (Neoprin) slides and kept at 45 °C overnight. The sections will be deparaffinized and rehydrated with xylene, graded ethanols, and PBS (2m changes).

4. The sections are encircled with a diamond-tip pen.

5. Sections are hydrated in PBS at 4 °C for 15 min.

6. Sections are reacted with 0.5M periodic acid in PBS for 20 min to inhibit intrinsic peroxidase activity.

7. The sections are rinsed in PBS for 5 min × 3, at 4 °C.

8. Sections are reduced in $NaBH_4$ (50 mg/ml) in PBS for 45 min at 4 °C for 1 hr to reduce free aldehyde groups.

9. Sections are flooded with 0.25 M pepsin in 0.5 M acetic acid, placed in a moist chamber, and incubated at 37 °C for 30 min.

10. The remainder of the staining procedure is carried out as described above for fixed tissues.

B. Electron Microscopy

Preembedding Methods

For preembedding localization, our preference is to use peroxidase as a marker, with DAB as substrate, followed by osmication (Gil and Martinez-Hernandez, 1984; Amenta and Martinez-Hernandez, 1987; Amenta et al., 1988).

1. Fixation and processing is performed as described above.

2. Frozen sections for electron microscopy (6–8 μm) are cut and collected on coated slides.

3. Sections are air dried for 30 min and encircled with a diamond-tip pen.

4. Sections are hydrated in PBS, at 4 °C, for 15 min.

5. To inhibit intrinsic peroxidase activity, the sections are treated with 0.02 M periodic acid in PBS for 30 min at room temperature.

6. The sections are washed in PBS for 10 min × 3, at 4 °C.

7. To reduce free aldehyde groups, the sections are reacted with 0.05% $NaBH_4$ in PBS at 4 °C for 1 hr.

8. The sections are washed in PBS for 10 min × 3, at 4 °C.

9. The sections are reacted with normal serum of the same species as the secondary antibody (1:10 dilution) for 30 min, at room temperature, in a moist chamber.

10. The sections are washed in PBS for 10 min × 3, at 4 °C.

11. The sections are reacted with a suitable dilution of the primary antibody (as determined by "checkerboard" titration) or with the normal serum of the same species, at a similar dilution, overnight, at 4 °C, in a moist chamber. All antibodies and normal sera are applied with a micropipetor and enough volume to cover the tissue section (usually 50–75 μl).

12. The sections are washed in PBS for 10 min × 3, at 4 °C.

13. The sections are reacted with a suitable dilution of the secondary antibody (as determined by "checkerboard" titration) for 2 hr, at 4 °C, in a moist chamber.

14. The sections are washed in PBS for 10 min × 3, at 4 °C.

15. The sections are reacted with a suitable dilution of PAP (as determined by "checkerboard" titration) for 2 hr, at 4 °C, in a moist chamber.

16. The sections are washed in PBS for 10 min × 3, at 4 °C.

17. The sections are fixed with 2.5% glutaraldehyde in 0.1 M sodium phosphate buffer, pH 7.4, for 30 min, at room temperature.

18. The sections are washed in PBS for 10 min × 3, at 4 °C.

19. The sections are reacted with 0.1 M glycine in PBS for 30 min at room temperature.

20. The sections are washed in PBS for 10 min × 3, at 4 °C.

21. The sections are reacted with DAB 50 mg/150 ml, in 0.5 M Tris-buffer, pH 7.2, in the dark for 10 min.

22. Hydrogen peroxide, 20 ml of a 5% solution in water, is added to the staining dish, mixed well, and the slides incubated for 5–10 min.

23. The sections are washed in PBS for 10 min × 3, at 4 °C.

24. The DAB deposited in the tissues lacks electron density; however, it is highly osmiophilic. Therefore, incubation with OsO_4 results in DAB–OsO_4 complexes suitable for electron microscopy. Following DAB incubation and a PBS wash, the sections are reacted with 1% osmium tetroxide in 0.1 M

sodium phosphate buffer with 0.026 M NaC1, at pH 7.2, for 1 hr at room temperature. Osmium tetroxide vapors are toxic, particularly to corneal tissues: care should be taken to work under a chemical hood. An ampule of osmium tetroxide (1 g/vial) is broken inside a tightly stoppered brown glass bottle and dissolved in 50 ml of distilled water (2%). Complete solubilization may take 24 hr, but it may be accelerated by placing the tightly stoppered bottle in a sonicator bath. This 2% aqueous OsO_4 solution is diluted 1:1 with 0.2 M sodium phosphate buffer containing 0.026 NaC1 (pH 7.2) immediately before use.

25. The sections are washed in PBS for 10 min × 3, at 4°C.

26. The sections are dehydrated in graded ethanols followed by propylene oxide (eight min dehydrations three times).

27. Infiltrate with the following plastic mixture:

Medcast	31,0 ml	28.34%
Araldite	24.0 ml	21.94%
DDSA	50.0 ml	45.70%
DMP-30	4.4 ml	4.02%

28. Invert a capsule containing the plastic mixture on top of the section and allow the mixture to infiltrate overnight at room temperature.

29. The plastic is polymerized at 60°C for 48 hr.

30. The blocks (containing the section) are removed from the slides by being briefly heated over a gas flame and then the block is snapped off quickly.

Postembedding Methods

The ability to localize antigens in tissues with postembedding methods provides some benefits over frozen section methods. As with paraffin-embedded sections, structural preservation is improved and blocks may be preserved indefinitely. The use of immunogold provides for increased resolution, whereas the PAP technique provides greater sensitivity. The combination of the two techniques provides optimal electron microscopic localization, that is, improved preservation, improved resolution (gold), and the sensitivity of the PAP technique (Karkavelas et al., 1988). The procedure we are currently using involves Lowicryl used as embedding media with infiltration and polymerization carried out at −25°C (Carleman et al., 1982).

1. Following perfusion fixation with 4% formaldehyde in 0.2M $NaPO_4$ buffer, the tissues are further fixed by immersion for 1 hr at 4°C with gentle agitation.

2. The tissues are dehydrated in 50% ethanol at 4°C for 15 min, followed by graded ethanols at −20°C.

3. Infiltration is done at −20°C with equal parts of 100% ethanol Lowicryl solution for 2 hr, followed by 1:2 ethanol:Lowicryl solution for 2 hr, and finally in 100% Lowicryl overnight.

4. After infiltration, tissue blocks are placed at the tip of capsules, the capsules filled with Lowicryl, and the resin is cross-linked with ultraviolet (UV) light (370 lambda) at −20°C for 48 hr. The cross-linking is continued at room temperature for an additional 48 hr.

5. Ultrathin sections (60–80 nm) are mounted on uncoated nickel grids (400 mesh).

6. All incubations are carried out in a moist chamber at room temperature. Grids are inverted on a drop of 10% H_2O_2 for 10 min, rinsed ain two changes of PBS, then placed on a drop of PBS for 10 min at room temperature.

7. To reduce free aldehyde groups, grids are placed on a drop of $NaBH_4$ (5 mg/15 ml) PBS for 1 hr, rinsed in PBS, then sequentially exposed to NGS, primary antibody, secondary antibody (goat antirabbit), and PAP for 1 hr, with intervening PBS rinses. Optimal antibody dilutions are determined by serial dilution. The sections are then reacted with DAB–HCl (50 mg/100 ml) in 0.1M Tris-buffer, pH 7.62, with H_2O_2 for 20 min.

7a. For immunogold, the intial steps are as described for the PAP staining. However, instead of the goat antirabbit (GAR), the grids are rinsed with Tris-HCl, pH 8.2, and reacted with GAR–gold; diluted with the Tris-buffer (1:4). The sections are rinsed with Tris-HCl and distilled H_2O, dried, and lightly counterstained with uranyl acetate and lead citrate.

IV. Extracellular Matrix Immunohistochemistry: Normal Lung

The ECM can be defined as the heterogeneous collection of macromolecules surrounding stromal cells and underlying most epithelia (Hay, 1981; Piez and Reddi, 1984; Cunningham, 1987a,b). The ECM consists of three major classes of macromolecules: the collagens, noncollagenous structural glycoproteins, and the proteoglycans (Hay, 1981; Piez and Reddi, 1984; Martinez-Hernandez, 1988; Cunningham, 1987a,b). The ECM has been shown to have a role in modulating cell morphology, growth, differentiation, and migration (Piez and Reddi, 1984; Hay, 1981; Trelstad, 1984; Bernfield and Banerjee, 1982). This modulation is apparently mediated by specific ECM components interacting with or binding to the cell surface (von der Mark and Kuhl, 1985; Ruoslahti et al., 1985; Dedhar et al., 1987; Ruoslahti and Pierschbacher, 1986; Pierschbacher et al., 1983). Many of the ECM proteins contain the cell adhesion tripeptide RGD, first identified on fibronectin (Ruoslahti et al., 1985; Ruoslahti and Pierschbacher, 1986; Pierschbacher et al., 1981, 1983). This tripeptide, and perhaps other sequences, bind to a family of cell surface receptors, the integrins (Ruoslahti et al., 1985; Dedhar et al., 1987; Pierschbacher et al., 1981). Which are heterodimeric proteins consisting of segments spanning the cytoplasmic, plasmalemmal, and extracellular compartments (Urushihara and Yamada, 1986; Ruoslahti and Pierschbacher, 1986; Aumailley and Timpl, 1986). An association of the ECM cell receptors with cytoskeletal filaments has been suggested as a communication route between the matrix and the cellular compartment (Ruoslahti and Pierschbacher, 1986; Urushihara and Yamada, 1986; Hirst et al., 1986).

The form and function of the adult lung are the consequence of the specific integration of a variety of cell types with particular ECM components. That is, since bronchial and alveolar cells represent a particular entodermal differentiation, specific associations and ratios of the ECM components result in an ECM unique to the lung (Rosenkrans et al., 1983a,b; Jaskoll and Slavkin, 1984). The critical role of the ECM in pulmonary physiology is highlighted by the functional derangements that result from pathological matrix alterations (Chen and Little, 1987; Raghu et al., 1985).

The ECM compartment represents approximately 25% of the dry weight of the lung, with collagen making up approximately 40% of that 25% (Crystal et al., 1978; Raghu et al.a, 1985). The pulmonary ECM can be divided into two major compartments: extra-alveolar and alveolar (Amenta et al., 1988). The extra-alveolar ECM envelopes arteries, veins, and conducting airways. The alveolar ECM is confined between the alveolar epithelium and capillary endothelium and is arranged radially around the extraalveolar ECM. Immunohistochemical studies have defined the distribution of various ECM components in the lung, and also provided information on their physiopathology (Adler et al., 1981; Amenta et al., 1988; Barskdyd et al., 1986; Bateman et al., 1981; Madri and Furthmayr, 1980; Raghu et al., 1985; Rosenkrans et al., 1983; Snyder et al., 1987; Torikata et al., 1985; Willems et al., 1986; Woodcock et al., 1984; Konomi et al., 1981).

A. Collagens

This ubiquitous, heterogeneous family (see Fig. 3) of structural proteins is characterized by a repeating amino acid sequence Gly-X-Y (where X and Y are often proline and hydroxyproline), a rodlike shape, and tensile strength (Miller and Gay, 1982, 1987). A collagen molecule consists of three helical polypeptide chains, the α chains. At present, 13 distinct collagen types have been defined with at least 28 genes coding for the distinct α chains (Bentz et al.a, 1983; Miller

TYPE	CHAINS	ASSOCIATION	AGGREGATE	LOCALIZATION
I	$\alpha 1(I)$, $\alpha 2(I)$	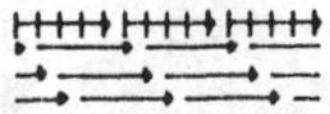		Ubiquitous: Bone Dermis, Tendon, etc
II	$\alpha 1(II)$	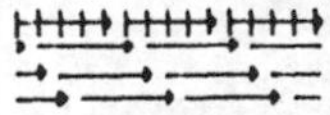		Cartilage Vitreous Humor
III	$\alpha 1(III)$	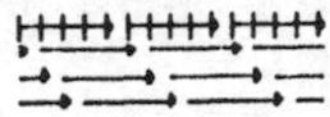		Pliable Organs Vessels, Uterus, etc.
IV	$\alpha 1(IV)$, $\alpha 2(IV)$			Basement Membrane
V	$\alpha 1(V)$, $\alpha 2(V)$, $\alpha 3(V)$	???		Minor Component or Most Tissues
VI	$\alpha 1(VI)$, $\alpha 2(VI)$, $\alpha 3(VI)$			Most Connective Tissues
VII	$\alpha 1(VII)$			Anchoring Fibrils
VIII	$\alpha 1(VIII)$	???		Some Endothelia
IX	$\alpha 1(IX)$, $\alpha 2(IX)$, $\alpha 3(IX)$	???		Minor Component of Cartilage
X	$\alpha 1(X)$	???		Hypertrophic Cartilage

Figure 3 Diagram of 10 collagen types outlines their constituent chains, molecular associations, macromolecular aggregates, and organ distribution. (From Martinez-Hernandez, 1988.)

and Gay, 1987; Myers and Emanuel, 1987; Burgeson, 1987; Pihlajaniemi et al., 1987; Cheah, 1985).

The collagens have recently been categorized into three major groups based on their size and physiochemical properties (Miller and Gay, 1987). Group 1 collagens consist of those collagens with α chains greater than 95 kD, continuous helicity (approximately 300 nm), and the ability to form fibers by lateral aggregation. This group includes collagen types I, II, III, V, and VI and makes up the major structural components of the ECM. Group 2 collagens are similar to group 1, except that they contain intercalated, nonhelical domains that preclude extensive lateral aggregation and therefore large-diameter fiber formation. Group 2 includes collagen types IV, VI, VII, and VIII. Group 3 collagens are characterized by having α chains of less than 95 kD.

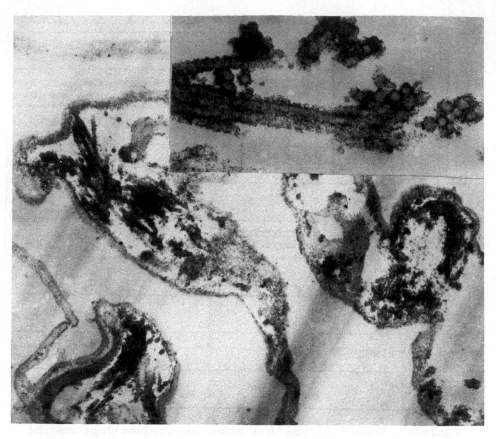

Figure 4 Localization of collagen type I in the alveolar wall. Type I collagen fibers are found in the alveolar septum forming small bundles. Inset: detail of collagen type I fibers in the alveolar interstitium. The reaction product can be seen encircling 30–35nm fibers orthogonally sectioned. In those fibers sectioned longitudinally the characteristic cross-banding of type I fibers is prominent (x 400,00). (From Gil and Martinez-Hernandez, 1984.)

Of the 13 known collagens, types I–IV have been localized in the mammalian lung (Gil and Martinez-Hernandez, 1984; Amenta et al., 1988; Konomi et al., 1981; Madri and Furthmayr, 1980; Bateman et al., 1981). Collagen types I (Fig. 4) and III (Fig. 5) are found in the bronchial and vascular adventitia, lobular septa, and alveolar interstitium (Gil and Martinez-Hernandez, 1984; Amenta et al., 1988; Konomi et al., 1981; Bateman et al., 1981). Type III collagen has a similar distribution to that of type I collagen except that it is more prevalent in

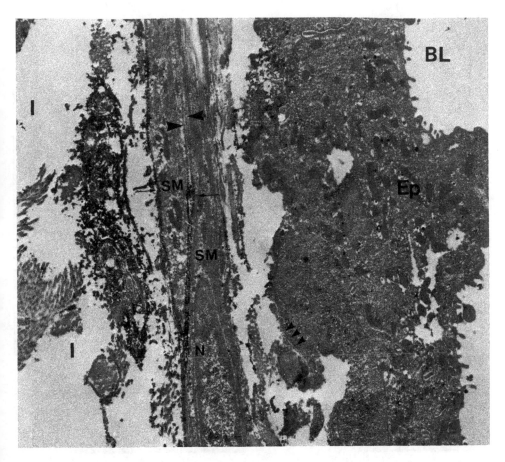

Figure 5 Type III collagen localization. a. Bronchiole. Type III collagen is present in the peribronchiolar interstitium and associated with the adventitial aspect of smooth muscle basement membrane (arrows). The association of collagen type III with the smooth muscle basement membrane ends sharply (arrowheads). The epithelial basement membrane (small arrowheads) and collagen type I fibers are negative **BL**, bronchiolar lumen; **Ep**, epithelial cell; **I**, interstitium; **N**, smooth muscle cell nucleus: **SM**, smooth muscle cell (x 9,630).

the alveolar interstitium (Fig. 5). Electron immunohistochemistry demonstrates type I collagen as cross-banded, 30–35 nm fibers (Fig. 4) present as bundles concentric to the large airways, perivascular adventitia, and lobular septa, or as smaller bundles or even individual fibers in the alveolar instituium (Gil and Martinez-Hernandez, 1984). In contrast, type III collagen forms 10–20 nm, beaded, tortuous

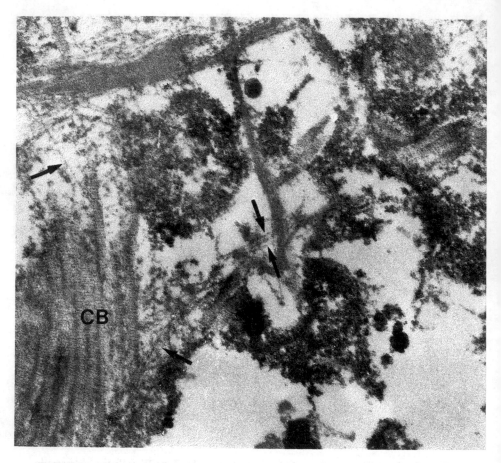

Figure 5 b. Detail of interstitium. Individual, delicate type III fibers (arrows) can be identified emerging from dense bundles where they are obscured by the intense reaction product. Collagen type I fibers (**CB**, collagen bundle) are negative (x 98,069).

fibers (Fig. 5b), often closely associated with type I collagen fibers and bundles (Amenta et al., 1988). Type II collagen is restricted to bronchial cartilage (Konomi et al., 1981). Several studies have localized type V collagen in the lung (Madri and Furthmayr, 1979, 1980; Konomi et al., 1984; Linsenmayer et al., 1983; von Der and Ocalan, 1982). Using immunofluorescence, Madri and Furthmayr found type V collagen in the pulmonary interstitium (Madri and Furthmayr, 1980). Of particular note was their finding of type V collagen in alveolar septa codistributing with collagen type IV. They concluded that basement membranes are composed of more than one collagen type (Madri and Furthmayr, 1979, 1980). However,

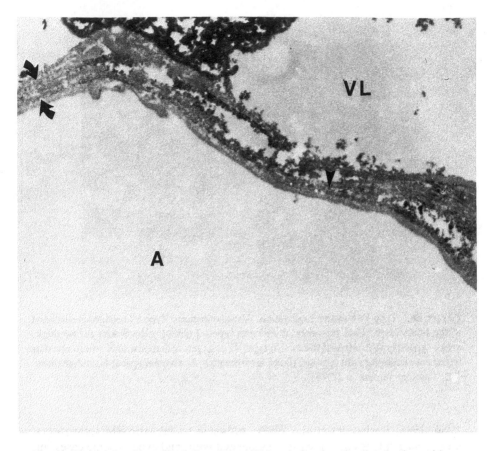

Figure 5 c. Alveolar wall. Collagen type III is restricted to the alveolar interstitium. Thick (30–35 nm) type I fibers are encased by type III fibers (compare wtih Fig. 5b). Some type III fibers can be seen attached to the interstitial aspect of the alveolar basement membranes (arrowheads). The association of collagen type III with the adventitial aspect of basement membranes ends at the fusion point of these basement membranes (curved arrows). A, alveolar space; VL, vascular lumen (x 19,711).

studies by other groups, in other tissues, find different distributions of these two collagen types (Amenta et al., 1986; Martinez-Hernandez et al., 1982; Miller et al., 1982; Linsenmayer et al., 1983; Konomi et al., 1984; von der Mark and Ocalan, 1982). Type VI collagen is ubiquitous, being present in bronchial and vascular adventitia, lobular septa, and alveolar interstitium (Amenta et al., 1988). Electron microscopy demonstrates type VI collagen as 6–10 nm filaments (Fig. 6), often closely associated with both type I and III collagen fibers (Amenta et al.,

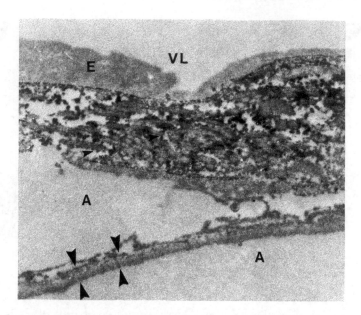

Figure 6a   Type VI collagen localization. Alveolar septum. Type VI collagen is restricted to the interstitium. Thin filaments of collagen type VI (small arrowheads) encase thick, cross-banded type I collagen fibers. No type VI collagen is demonstrable where alveolar basement membranes are opposed (large arrowheads). **A**, alveolar space; **E**, endothelium; **VL**, vascular lumen (x 22,791).

1986, 1988; Karkavelas et al., 1988). Although no intrinsic BM components, these collagen types are all closely associated with, and even in some cases, insert into BM most prominently at the adventitial aspect of bronchial and vascular BM (Fig. 5, 6c) (Amenta et al., 1986, 1988; Karkavelas et al., 1988).

B. Basement Membranes

Basement membranes (BM) are ubiquitous extracellular matrices underlying most epithelia and surrounding mesenchymal cells (Martinez-Hernandez and Amenta, 1983; Kuehn et al., 1982; Timpl and Dziadek, 1986). They represent the heterogeneous assembly of a variety of macromolecules (Fig. 7). At present, there are four well-characterized BM components, two glycoproteins: laminin and entactin; a proteoglycan: heparan sulfate; and a collagen: type IV (Martinez-Hernandez and Amenta, 1983; Timpl and Dziadek, 1986). The functions of the pulmonary BM membranes are not fully defined. However, in other organs, they act as selective filters, serve as points for cell attachment, provide structural support, and modulate cell migration. Given these functions, it is not surprising

Figure 6b Detail of elastic fiber. Type VI collagen filaments are associated with the surface of collagen type I fibers, but not with the elastic fibers (EF) (x 8,230).

that the presence or absence of BM damage is critical in determining the outcome of the pulmonary repair reaction (Martinez-Hernandez, 1988).

Basement membranes are present in the lung, underlying mesothelium, endothelium, airway and alveolar epithelium, and surrounding individual vascular and bronchial smooth muscle cells. The pulmonary distribution of the BM components laminin and type IV collagen (Gil and Martinez-Hernandez, 1984; Amenta et al., 1988; Konomi et al., 1981; Madri and Furthmayr, 1980) has been studied by light and electron microscopy. Both components are restricted to and present throughout the entire thickness of the BM (Fig. 8); however, the preferential localization of laminin in the lamina rara and type IV collagen in the lamina densa is not as obvious in the lung as in other organs. Focally, the alveolar and capillary lamina densa fuse to form a trilaminar BM (Fig. 8b), similar to the glomerular BM. However, unlike the glomerulus, where the epithelial and endothelial BM are fused throughout the entire perimeter of the capillary loop, in the alveolus this fusion is restricted to discrete points (Gil and Martinez-Hernandez, 1984; Amenta et al., 1988).

C. Noncollagenous Structural Glycoproteins

Two of the best-characterized components of this group are elastin and fibronectin. Elastic fibers (Figs. 6b, 8a) are composed of an "amorphous" central core

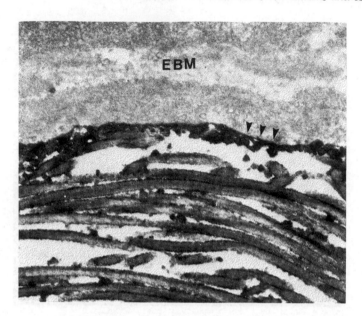

Figure 6c   Detail of a vessel. Type VI Filaments are in close association with thick type I collagen fibers. Occasional type VI filaments can be seen on the interstitial aspect of the vascular smooth muscle basement membrane; however, the endothelial basement membrane (EBM) is negative (x 31,506). (From Amenta 1988.)

surrounded by a system of microfibrils (Cleary and Gibson, 1983; Damiano et al., 1979; Gosline and Rosenbloom, 1984). The central core contains elastin, a highly insoluble protein rich in desmosine and isodesmosine (Gosline and Rosenbloom, 1984); the composition of the peripheral microfibrils is not totally elucidated, but they are known to contain fibrilin (Sakai et al., 1986). The exact function of the peripheral microfibrils is not clear; however, they may serve to integrate elastin with the surrounding ECM (Cleary and Gibson, 1983). Unlike most ECM components, elastic fibers are well defined by conventional microscopy. In the lung, elastic fibers are found in blood vessels, airways, alveolar interstitium, and pleura (Cleary and Gibson, 1983; Damiano et al., 1979; Amenta et al., 1988). Elastin, a true elastomere, provides the recoil properties of elastic fibers (Cleary and Gibson, 1983; Gosline and Rosenbloom, 1984).

Fibronectin (Mr, 440,000), exists as plasma and tissue forms (Hynes et al., 1984; Yamada et al., 1985; Ruoslahti et al., 1981, 1985; Ruoslahti and Pierschbacher, 1986); both of which are derived from a single gene by differential splicing (Oldberg and Ruoslahti, 1986; Kornblihtt et al., 1985; Odermatt et al., 1985; Hynes et al., 1984; Paul and Hynes, 1984; Tamkun et al., 1984; Schwarzbauer et al., 1983). Fibronectin interacts or binds with a variety of macromolecules, bacteria, and cell surfaces via individual globular domains, each having unique binding properties

	CONSTITUENT CHAINS	MOLECULAR COMPOSITION	SUPRAMOLECULAR AGGREGATE	FUNCTION
TYPE IV COLLAGEN	α1(IV), α2(IV)	3 α Chains		Structural
LAMININ	A, B$_1$, B$_2$	1 A and 2 B Chains		Cell Attachment
ENTACTIN	Single Polypeptide Chain	Single Polypeptide Chain		? Unknown
HEPARAN SULFATE PROTEOGLYCAN	Polypeptide Chain Glycosaminoglycans	Polypeptide Chain Glycosaminoglycans	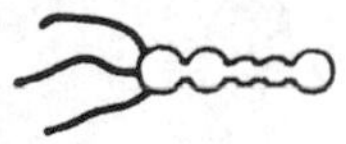	Electrostatic Charge

Figure 7 Diagram of basement membranes four major constituents, highlighting their constituent chains, supramolecular aggregates, and postulates functions. (From Martinez-Hernandez 1988).

(Ruoslahti et al., 1981, 1982, 1985; Yamada et al., 1985). Through these specific interactions, fibronectin exerts its influence on cellular differentiation, development, chemotaxis, opsonization, and hemostasis.

Fibronectin localization in the lung has received considerable attention, since there is evidence suggesting that fibronectin has a central role in the pathogenesis of fibrotic lung disease (Bitterman et al., 1983; Crystal et al., 1976; Rennard and Crystal, 1982; Rennard et al., 1981; Torikata et al., 1985; Rosenkrans et al., 1983a,b; Rosenkrans and Penney, 1986). For example, there are data indicating that patients with pulmonary fibrosis have increased levels of fibronectin in pulmonary lavage fluid (Rennard and Crystal, 1982; Rennard et al., 1981) and increased secretion of fibronectin by alveolar macrophages (Rennard et al., 1981; Laurent, 1985; Hoidal and Niewoehner, 1983; Viljanto et al., 1981; Villiger et al., 1981). This parallels the postulated fibronectin role in other examples of repair, such as dermal wound healing and hepatic cirrhosis, conditions like pulmonary fibrosis, characterized by excessive ECM deposition (Martinez-Hernandez, 1985; Kurkinen et al., 1980; Torikata et al., 1985). In the adult rat lung, fibronectin has been found in the interstitium, particularly in the adventitia of large bronchi, and blood vessels (Gil and Martinez-Hernandez, 1984; Torikata et al., 1985; Rosenkrans et al., 1983a,b). On electron microscopy, fibronectin was present as 6–10 nm filaments coating thick, 30–35 nm type I collagen fibers (Fig. 9) (Gil and Martinez-Hernandez, 1984).

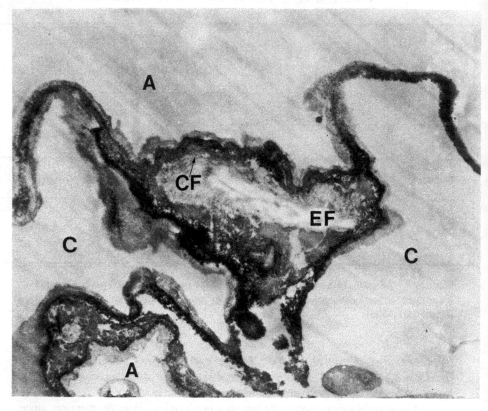

Figure 8a Laminin localization. Epithelial and endothelial basement membranes contain laminin with no perceptible difference. Type I collagen fibers (CF) and elastic fibers (EF) are negative (x 12,300).

The relationship between fibronectin and BM has been the source of considerable debate in a variety of tissues, including the lung. Initial light microscopic immunohistochemical studies with antibodies directed against fibronectin revealed linear staining in the BM zones, and concluded that it was a BM component (Linder et al., 1978). It should be noted that the BM cannot be resolved by light microscopy from the adjacent interstitium (Martinez-Hernandez et al., 1981b; Boselli et al., 1981). Electron immunohistochemical studies in a variety of tissues demonstrate fibronectin in the interstitium immediately adjacent to but not within BM (Martinez-

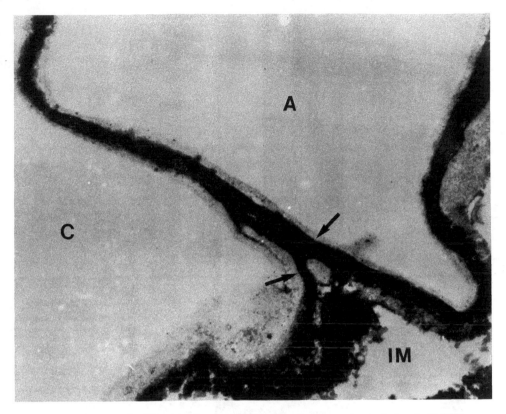

Figure 8b   Detail of the alvolar and endothelial basement membrane fusion (x 10,000).

Hernandez et al., 1981b; Amenta et al., 1986; Courtoy and Boyles, 1983). Similar results are obtained, even after attempts to "unmask" fibronectin antigenic sites with glycosidases, trypsin, pepsin, or acetic acid (Amenta et al., 1986). In those BM with a prominent filtering function, such as the murine parietal yolk sac and the renal glomerulus, fibronectin is occasionally present with an irregular, spotty distribution (Amenta et al., 1983; Martinez-Hernandez et al., 1981a,b; Courtoy et al., 1980). However, brief perfusion with isotonic buffer prior to fixation removes the trace reactivity of these BM with fibronectin antibodies (Martinez-Hernandez et al., 1981b; Courtoy and Boyles, 1983). These findings suggest that the fibronectin occasionally found in some BM is trapped from plasma and should not be considered as "intrinsic" BM component, like laminin or type IV collagen (Amenta et al., 1983, 1986; Martinez-Hernandez et al., 1981b). Electron microscopic localization of fibronectin yields different results, depending on whether or not the lung is perfused with buffer prior to fixation. Thus, in rat

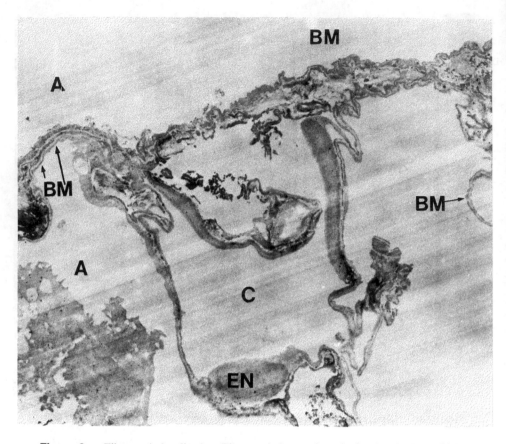

Figure 9a   Fibronectin localization. Fibronectin is prominent in the alveolar interstitium, notably surrounding type I collagen fibers (see Fig. 9b higher magnification inset). Basement membranes (BM) contain only occasional reactivity. A, alveolar space; C, capillary space; EN, endothelial cell (x 4,100).

and hamster, BM of bronchial epithelium, glands, and endothelium of large vessels lack fibronectin; whereas only traces of fibronectin are found in capillary endothelial and alveolar BM (Fig. 9) (Gil and Martinez-Hernandez, 1984). Fibronectin is found consistently at the fusion point of alveolar and capillar BM. In contrast, in human lung fixed by immersion, fibronectin is prevalent not only in the interstitium, associated with collagen fibers but also in alveolar, capillary, and smooth muscle BM (Torikata et al., 1985).

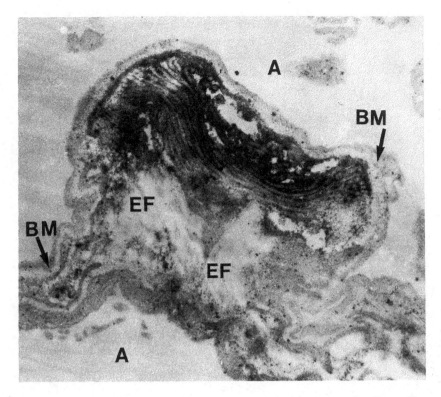

Figure 9b	Detail of alveolar interstitium. Fibronectin encases type I collagen fibers. Basement membranes (BM) and elastic fibers (EF) are devoid of fibronectin (x 19,500). (From Gil and Martinez-Hernandez, 1984.)

Several studies using immersion fixation have defined the distribution of ECM components during pulmonary ontogenesis (Rosenkrans et al., 1983a,b; Snyder et al., 1987). Fibronectin localization has received particular attention. In murine fetuses, on light microscopic examination, fibronectin is found in pulmonary tubular walls, BM zones associated with septal buds, and in the interstitium (Rosenkrans et al., 1983a,b). On ultrastructural examination, fibronectin is identified in the BM along with laminin and heparan sulfate proteoglycan (Jaskoll and Slavkin, 1984).

D. Proteoglycans

The proteoglycans are ubiquitous ECM components consisting of a core protein covalently linked to a variable number (30–100) of glycosaminoglycan (GAGs)

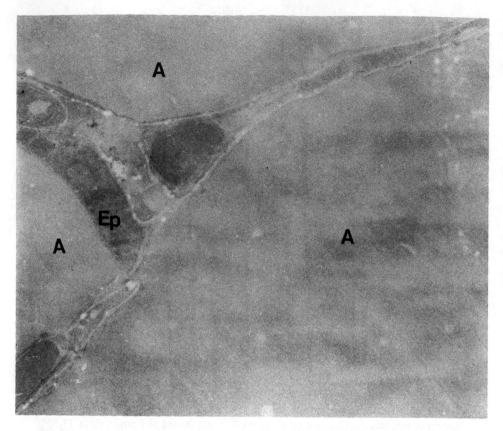

Figure 10   Negative control. Section of lung reacted with normal rabbit serum, goat anti-rabbit IgG antibody, and rabbit PAP. No reaction product is present (x 6,600.)

side chains (formerly referred to as acid mucopolysaccharides) (Iozzo, 1985; Hascall and Hascall, 1981; Brody et al., 1984; Gallagher 1986; Rutten et al., 1988; Takusagawa et al., 1982; Vaccaro and Brody, 1979; van Kuppevelt et al., 1984). There are seven known GAGs: chondroitin-4-sulfate, chondroitin-6-sulfate, dermatan sulfate, heparin, heparan sulfate, hyaluronic acid, and keratan sulfate (Iozzo, 1985; Hascall and Hascall, 1981). The protein cores of different proteoglycans represent distinct gene products, but the attached GAGs are often identical (Iozzo, 1985; Hascall and Hascall, 1981; Heinegard and Paulsson, 1984). In spite of this, and for historical reasons, proteoglycans are named according to the GAG side chains (Iozzo, 1985; Hascall and Hascall, 1981; Heinegard and Paulsson, 1984). Since the same GAGs may be attached to different protein cores, an additional identifier is often used, for example, cell surface heparan sulfate proteoglycan or BM heparan sulfate proteoglycan (Hook et al., 1984).

Heparan sulfate proteoglycan has been localized in the BM zones of fetal lung (Jaskoll and Slavkin, 1984; Vaccaro and Brody, 1979; van Kuppevelt et al., 1984). Most of the available data on pulmonary proteoglycan distribution come from the use of cationic dyes, such as cuprolinic blue, and selective enzymatic degradation (Rutten et al., 1988; Takusagawa et al., 1982; Vaccaro and Brody, 1979; van Kuppevelt et al., 1984). These studies have confirmed the presence of heparan sulfate proteoglycan in human alveolar BM (Takusagawa et al., 1987). Dermatan sulfate proteoglycans were identified in the interstitium associated with banded collagen fibers (Vaccaro and Brody, 1979). Additional information on the localization of proteoglycans in the pulmonary ECM is clearly needed.

The immunohistochemical studies (Amenta et al., 1988; Gil and Martinez-Hernandez, 1984) suggest the following model for the pulmonary ECM (Fig. 11): In the lung, as in other tissues, type I collagen is the apparent major structural buttress, forming a cablelike system surrounding the vascular and bronchial adventitia. Type III collagen fibers, prevalent in pliable tissues (Gay and Miller, 1983), and characterized by their tortuosity may contribute to the pliability of vessels, airways, and alveoli. The other ECM components are closely associated with type I collagen, among themselves, with BM, and perhaps with elastic fibers. These associations suggest that these proteins integrate the pulmonary ECM into a single anatomical and functional unit (Amenta et al., 1986, 1988; Gil and Martinez-Hernandez, 1984).

V. Extracellular Matrix Immunohistochemistry: Pulmonary Diseases

Although a considerable body of information exists on the role of the ECM in pulmonary diseases, few immunohistochemcial studies have been published (Raghu et al., 1985; Bateman et al., 1981; Torikata et al., 1985; Madri and Furthmayr, 1980). Interstitial pulmonary fibrosis (IPF) and emphysema are two pulmonary disorders characterized by prominent changes in the ECM (Bateman et al., 1981; Crystal et al., 1976, 1978; Hance and Crystal, 1975).

IPF is the common endstage of innumerable pulmonary disorders characterized by a disruption of the pulmonary architecture, with initial infiltration of inflammatory cells, followed by proliferation of interstitial cells and excessive, disorderly ECM accumulation (Crystal et al., 1976; Henderson et al., 1983). Several studies have quantitated specific collagen types in IPF (Crystal et al., 1976, 1978; Madri and Furthmayr, 1980; Quinones and Crouch, 1986; Takiya et al., 1983; Last et al., 1983; Reiser and Last 1983; Laurent, 1985). The results indicate that whereas the ratio of collagen types I, III, V, in normal

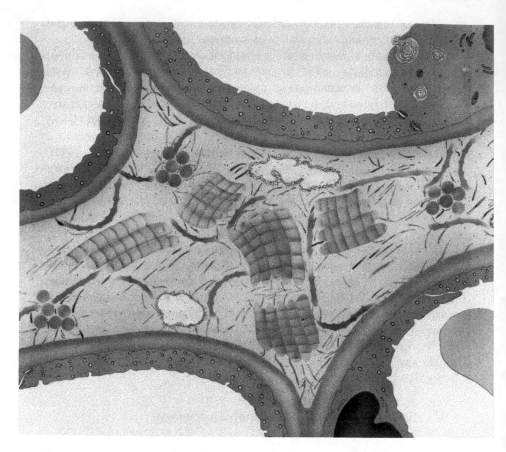

Figure 11   Diagram of the alveolar ECM. Type I collagen fibers are often associated with type III fibers, type V, type VI, and fibronectin filaments. In turn, type III, V, VI, and fibronectin often insert in the alveolar and endothelial basement membranes. Thus the alveolar extracellular matrix is integrated into a functional unit in continuity with that around airways, forming the pulmonary framework. Fibronectin may be attached to the cell membrane of fibroblasts and smooth muscle cells via interactions with specific cell receptors. How elastic fibers are integrated with the rest of the ECM is not defined, although it is likely that the peripheral microfibrils may serve this purpose, interacting with some collagens or fibronectin. Thus the ECM can be viewed as a resilient framework of type I collagen, with type III collagen and elastic fibers providing deformability and recoil properties, whereas the filamentous collagens and fibronectin will act as connecting structures.

lung was 35:58:7, in patients with endstage IPF it was 60:30:10 (Madri and Furthmayr, 1980). It has been suggested that this change in ratio simply represents a shift from the normal predominance of collagen type III to the predominance of collagen type I rather than in increase in total pulmonary collagen (Crystal et al., 1976). However, Starcher et al.(1978) have reported an actual doubling of collagen and elastin content in hamsters with bleomycin-induced IPF. Furthermore, increased prolyl-hydroxylase activity has also been reported in IPF (Chichester et al., 1981; Selman et al., 1986; Gadek et al., 1984). Thus in IPF there seems to be an increase in total collagen content, to which decreased collagen degradation may contribute (Gadek et al., 1984; Laurent and McAnulty,1983; Rich et al., 1983).

Raghu et al. (1985) studied by immunofluorescence the distribution of collagen types I, III, IV, and laminin in human pulmonary disease. Patients with nonfibrotic lung disease had no appreciable change in the pulmonary ECM components. In contrast, in the early stages of IPF,there was an accumulation of collagen types I and III in the interstitium, whereas collagen type I was more prevalent in the latter stages of the disease process. Localization of collagen type IV and laminin distribution showed disruptions of the alveolar basement membranes with subsequent fibrosis. Torikata et al. (1985) studied the ultrastructural distribution of fibronectin in four patients with IPF. All patients' disease was in the early stages of the fibrotic process, with inflammatory infiltrates in the septal walls and had prominent staining of alveolar BM and interstitium with antifibronectin antibodies. These studies suggest a similar sequence of ECM deposition in IPF and other fibrotic conditions, that is, hepatic cirrhosis and dermal wound healing (Martinez-Hernandez, 1985, Kurkinen et al., 1980). In these conditions, collagen type I deposition is preceded temporally and spatially by fibronectin and type III collagen (Martinez-Hernandez, Kurkinen et al., 1980). The role of other ECM components in this apparent common pathway of fibrosis remains to be defined.

In emphysema (Laurent and Tetley, 1984; Hoidal and Niewoehner, 1983; Niewoehner and Hoidal, 1982; Sandberg et al., 1981), the attention has focused on elastin catabolism. Nevertheless, other ECM components may also be involved in the pathogenesis of emphysema. For instance, there are conflicting data on the quantity of collagen in emphysematous lungs (Karlinsky et al., 1983). No immunohistochemical data are available on the localization and distribution of the different collagen types in emphysema.

Our current knowledge of the ECM in pulmonary disease is, at best, rudimentary. For instance, little is known about the deposition sequence of individual ECM components, the cells responsible for their synthesis, or the role of collagen type VI and the more recently defined collagen types in pulmonary diseases. Furthermore, although elastin has received considerable attention, the elastic fiber microfibrils have been ignored. Additional immunohistochemical studies of the pulmonary ECM are needed to define better its role in normal and disease states.

References

Adler, K. B., Craighead, J. E., Vallyathan, N. V., and Evans, J. N. (1981). Actin-containing cells in human pulmonary fibrosis. *Am. J. Pathol.* **102**:427–437.

Amenta, P. S., Clar, C. C., and Martinez-Hernandez,A. (1983). Deposition of fibronectin and laminin in the basement membrane of the rat parietal yolk sac: immunihistochemical and biosynthetic studies. *J. Cell Biol.* **96**:104–111.

Amenta, P. S., Gay, S., Vaheri, A., and Martinez-Hernandez, A. (1986). The extracellular matrix is an integrated unit: ultrastructural localization of collagen types I, III, IV, V, VI, fibronectin, and laminin in human term placenta. *Coll. Relat. Res.* **6**:125–152.

Amenta, P. S., Gil, J., and Martinez-Hernandez, A. (1988). The connective tissue of the rat lung. II: Ultrastructural localization of collagen types III, IVC and VI. *J. Histochem. Cytochem.* (in press).

Amenta, P. S., and Martinez-Hernandez, A. (1987). Specific Methods for Electron Immunohistochemistry. In *Methods in Enzymology. Structural and Contractile Proteins: Extracellular Matrix.* Edited by L. W. Cunningham. New York, Academic Press, pp. 145–173.

Andrews, L. P., Clark, R. K., and Damjanov, I. (1985). Mixed glycosidase pretreatment reduces nonspecific binding on antibodies to frozen tissue section. *J. Histochem. Cytochem.* **33**:695–698.

Aumailley, M. and Timpl, R (1986). Attachment of cells to basement membrane collagen type IV. *J. Cell Biol.* **103**:1569–1575.

Barsky, S. H., Rao, N.C., Restrepo, C., and Liotta, L. A. (1984). Immunocytochemical enhancement of basement membrane antigens by pepsin: applications in diagnostic pathology. *Am. J. Clin. Pathol.* **82**:191–194.

Barsky, S. H., Huang, S. J., and Bhuta, S. (1986). The extracellular matrix of pulmonary scar carcinomas is suggestive of a desmoplastic origin. *Am. J. Pathol.* **124**:412–419.

Bateman, E. D., Turner Warwick, M., and Adelmann Grill, B. C. (1981). Immunohistochemical study of collagen types in human foetal lung and fibrotic lung disease. *Thorax.* **36**:645–653.

Bentz, H., Morris, N. P., Murray, L. W., Sakai, L. Y., Hollister, D. W., and Burgeson, R. E. (1983). Isolation and partial characterization of a new human collagen with an extended triple-helical structural domain. *Proc. Natl. Acad. Sci. USA* **80**:3168–3172.

Bernfield, M., and Banerjee, S. C. (1982). The turnover of basal lamina glycosaminoglycan correlates with epithelial morphogenesis. *Develop. Biol.* **90**:291–305.

Bitterman, P., Rennard, S., Adelberg, S., and Crystal, R. G. (1983). Role of fibronectin in fibrotic lung disease. A growth factor for human lung fibroblasts. *Chest.* **83**:96S–96S.

Boon, M. B., and Kok, L. P. (1987). *Microwave Cookbook of Pathology*. Leyden, Coulomb Press.

Boselli, J. M., Macarak, E. J., Clark, C. C. Brownell, A. G., and Martinez-Hernandez, A. (1981). Fibronectin: its relationship to basement membranes. I. Light microscopic studies. *Coll. Relat. Res.* **1**:391–404.

Brody, J. S., Vaccaro, C. A., Hill, N. S., and Rounds, S. (1984). Binding of charged ferritin to alveolar wall components and charge selectivity of macromolecular transport in permeability pulmonary edema in rats. *Circ. Res.* **55**:155–167.

Bullock, G. R., and Petrusz, P. (1982). *Techniques in Immunocytochemistry*. New York, Academic Press.

Burgeson, R. E. (1987). The collagens of skin. *Curr. Probl. Dermatol.* **17**:61–75.

Carleman, E., Garavito, M. and Villiger, W. (1982). Resin development for electron microscopy and an analysis of embedding at low temperature. *J. Microsc. (London)* **125**:123–143.

Cheah, K. S. (1985). Collagen genes and inherited connective tissue disease. *Biochem. J.* **229**:287–303.

Chen, J. M., and Little, C. D. (1987). Cellular events associated with lung branching morphogenesis including the deposition of collagen type IV. *Dev. Biol.* **120**:311–321.

Chichester, C. O., Palmer, K. C., Hayes, J. A., and Kagan, H. M. (1981). Lung lysyl oxidase and prolyl hydroxylase: increases induced by cadmium chloride inhalation and the effect of beta-aminopropionitrile in rats. *Am. Rev. Respir. Dis.* **124**:709–713.

Cleary, E. G., and Gibson, M. A. (1983). Elastin-associated microfibrils and microfibrillar proteins. *Int. Rev. Connect. Tissues Re.* **10**:97–209.

Coons, A. H., Creech, H. J., and Jones, R. N. (1941). Immunological properties of an antibody containing a fluorescent group. *Proc. Soc. Exp. Biol.* **47**:200–202.

Courtoy, P. J., Kanwar, Y. S., Hynes, R. O., and Farquhar, M. G. (1980). Fibronectin localization in the rat glomerulus. *J. Cell Biol.* **87**:691–696

Courtoy, P. J., and Boyles, J. (1983). Fibronectin in the microvasculature: localization in the pericyte-endothelial interstitium. *J. Ultrastruct. Res.* **3**:258–273.

Courtoy, P. J., Picton, D. H., and Farquhar, M. G. (1983). Resolution and limitations of the immunoperoxidase procedure in the localization of extracullar matrix antigens. *J. Histochem. Cytochem.* **31**:945–951

Crystal, R. G., Fulmer, J. D., Roberts, W. C., Moss, M. L., Line, B. R., and Reynolds, H. Y. (1976). Idiopathic pulmonary fibrosis. Clinical, histologic, radiographic, physiologic, scintigraphic, cytologic, and biochemical aspects. *Ann. Intern. Med.* **85**:769–788.

Crystal, R. G., Fulmer, J. D., Baum, B. J., Bernardo, J., Bradley, K. H., Bruel,

S. D., Elson, N. A., Fells, G. A., Ferrans, V. J., Gadek, J. E., Hunninghake, G. W., Kawanami, O., Kelman, J. A., Line, B. R., McDonald, J. A., McLees, B. D., Roberts, W. C., Rosenberg, D. M., Tolstoshev, P., Von Gal, E. and Weinberger, S.E. (1978). Cells, collagen and idiopathic pulmonary fibrosis. *Lung* **155**:199–224.

Cunningham, L. W. (1987a). Structural and contractile proteins. Part D. Extracellular matrix *Methods Enzymol.* **144**.

Cunningham, L. W. (1987b). Structural and contractile proteins. Part E. Extracellular matrix. *Methods Enzymol.* **145**.

Damiano, V. V., Tsang, A., Christner, P., Rosenbloom, J., and Weinbaum, G. (1979). Immunologic localization of elastin by electron microscopy. *Am. J. Pathol.* **96**:439–455.

Dehar, S., Ruoslahti, E., and Pierschbacher, M. D. (1987). A cell surface receptor for collagen type I recognizes the Arg–Gly–Asp sequence. *J. Cell Biol.* **104**:585–594.

Dickson, L. A., Pihlajaniemi, T., Deak, S., Pope, F. M., Nicholls, A., Prockop, D. J. and Myers, J. C. (1984). Nuclease S1 mapping of a homozygous mutation in the carboxyl-propeptide-coding region of the pro alpha 2(I) collagen gene in a patient with osteogenesis imperfecta. *Proc. Natl. Acad. Sci. USA* **81**:4524–4528.

Faulk, W. P., and Taylor, G. M. (1971). An immunocolloid method for the electron microscope. *Immunohistochemistry* **8**:1081–1083.

Gadek, J. E., Fells, G. A., Zimmerman, R. L., and Crystal, R. G. (1984). Role of connective tissue proteases in the pathogenesis of chronic inflammatory lung disease. *Environ. Health Perspect.* **55**:297–306.

Gallagher, B. C. (1986). Branching morphogenesis in the avian lung: electron microscopic studies using cationic dyes. *J. Embryol. Exp. Morphol.* **94**:189–205.

Gay, S., and Fine, J-D. (1987). Characterization and isolation of poly- and monoclonal antibodies against collagen for use in immunohistochemistry. *Methods Enzymol.* **145**:148–170.

Gay, S. and Miller, E. J. (1983). What is collagen, what is not. *Ultrastruct. Pathol.* **4**:365–377.

Geoghegan, W. D., Scillian, J. J., and Ackerman, G. A. (1978). The detection of man B lympyhocytes by both light and electron microscopy utilizing colloidal gold labelled anti-immunoglobulin. *Immunol. Commun.* **7**:1–12

Gil, J. (1978). Lung interstitium, vascular and alveolar membranes. In *Lung Water and Solute Exchange.* Edited by N. C. Staul. New York, Marcel Dekker, pp. 76–92.

Gil, J. (1982). Alveolar wall relations. *Ann. NY. Acad. Sci.* **384**:31–43.

Gil, J., and Martinez-Hernandez, A. (1984). The connective tissue of the rat lung: electron immunohistochemical studies. *J. Histochem. Cytochem.* **32**:230–238.

Glauert, A. M. (1975). *Fixation, Dehydration and Embedding of Biological Specimens.* Amsterdam, Elsevier North-Holland.

Gosline, M. J., and Rosenbloom, J. (1984). Elastin. In *Extracellular Matrix Biochemistry.* Edited by K. A. Piez and A. H. Reddi. New York, Elsevier, pp. 191–228.

Graham, R. C. and Karnovsky, M. J. (1966). The early stages of absorption of injected horseradish peroxidase in the proximal tubules of mouse kidney: ultrastructural cytochemistry by a new technique. *J. Histochem. Cytochem.* **14**:291–302.

Hance, A. J., and Crystal, R. G. (1975). The connective tissue of lung. *Am. Rev. Respir. Dis.* **112**:657–711.

Hascall, V. C., and Hascall, G. K. (1981). Proteoglycans. In *Cell Biology of Extracellular Matrix.* Edited by E. D. Hay. New York, Plenum Press, pp. 39–64.

Hay, E. D. (1981). In *Cell Biology of Extracellular Matrix*, New York, Plenum Press.

Heinegard, D., and Paulsson, M. (1984). Structure and metabolism of proteoglycans. In *Extracellular Matrix Biochemistry.* Edited by K. A. Piez and A. H. Reddi. New York, Elsevier, pp. 277–322.

Heitzmann, H., and Richards, R. M. (1974). Use of the avidin–biotin complex for specific staining of biological membrances in electron microscopy. *Proc. Natl. Acad. Sci. USA* **73**:3537–3541.

Henderson, A. S., Myers, J.C., and Ramirez, F. (1983). Localization of the human 2(I) collagen gene (COL1AS) to chromosome 7q22. *Cytogenet. Cell Genet.* **36**:586–587.

Hirst, R., Horwitz, A., Buck, C., and Rohrschneider, L. (1986). Interaction of plasma membrane fibronectin receptor with talin—a transmembrane linkage. *Proc. Natl. Acad. Sci. USA* **83**:6470–6474.

Hoidal, J.R., and Niewoehner, D. E. (1983). Pathogenesis of emphysema. *Chest* **83**:679–685.

Hook, M., Kjellen, L., and Johansson, S. (1984). Cell-surface glycosaminoglycans. *Annu. Rev. Biochem.* **53**:847–869.

Hynes, R. O., Schwarzbauer, J. E., and Tamkun, J. W. (1984). Fibronectin: a versatile gene for a versatile protein. *Ciba Found. Symp.* **108**:75–92.

Iozzo, R. V. (1985). Proteoglycans: structure, function, and role in neoplasia. *Lab. Invest.* **53**:373–396.

Jaskoll, T. F., and Slavkin, H. C. (1984). Ultrastructural and immunofluorescence studies of basal-lamina alterations during mouse-lung morphogenesis. *Differentiation* **28**:36–48.

Jimenez, S. A., Williams, C. J., Myers, J.C., and Bashey, R. I. (1986). Increased collagen biosynthesis and increased expression of type I and type III procollagen genes in tight skin (TSK) mouse fibroblasts. *J. Biol. Chem.* **261**:657–662.

Karkavelas, G., Kefalides, N. A., Amenta, P. S., and Martinez-Hernandez, A. (1988). Comparative ultrastructural localization of collagen types III, IV, VI, and laminin in rat uterus and kidney. *J. Ultrastruc. Mol. Struct. Res.* (in press).

Karlinsky, J., Fedette, J., Davidovits, G., Catanese, A., Snider, R., Faris,B., Snider, G. L., and Franzblau, C. (1983). The balance of lung connective tissue elements in elastase-induced emphysema. *J. Lab. Clin. Med.* **102**:151–162.

Konomi, H., Hori, H., Sano, J., Sunada, H., Hata, R., Fujiwara, S., and Nagai, Y. (1981). Immunohistochemical localization of type I, II, III, and IV collagens in the lung. *Acta Pathol. Jpn.* **31**:601–610.

Konomi, H., Hayashi, T., Nakayasu, K., and Arima, M. (1984). Localization of type V collagen and type IV collagen in human cornea, lung, and skin. Immunohistochemical evidence by anti-collagen antibodies characterized by immunoelectroblotting. *Am. J. Pathol.* **116**:417–426.

Kornblihtt, A.R., Vibe, Pedersen, K., and Baralle, F. E. (1984). Human fibronectin: cell specific alternative mRNA splicing generates polypeptide chains differing in the number of internal repeats. *Nucleic Acids Res.* **12**:5853–5868

Kornblihtt, A. R., Umezawa, K., Vibe, Pedersen, K., and Baralle, F. E. (1985). Primary study of human fibronectin: differential splicing may generate at least 10 polypeptides from a single gene. *EMBO J.* **4**:1755–1759.

Kuehn, K., Schoene, H. H., and Timpl, R. (1982). In *New Trends in Basement Membrane Research*, New York, Raven Press.

Kurkinen, M., Vaheri, A., Roberts, P. J., and Stenman, S. (1980). Sequential appearance of fibronectin and collagen in experimental granulation tissue. *Lab. Invest.* **43**:47–51.

Last, J. A., Siefkin, A. D., and Reiser, K. M. (1983). Type I collagen content is increased in lungs of patients with adult respiratory distress syndrome. *Thorax* **38**:364–368.

Laurent, G. J., and McAnulty, R. J. (1983). Protein metabolism during bleomycin-induced pulmonary fibrosis in rabbits. In vivo evidence for collagen accumulation because of increased synthesis and decreased degradation of the newly synthesized collagen. *Am. Rev. Respir. Dis.* **128**:82–88.

Laurent, G. J., and Tetley, T. D. (1984). Pulmonary fibrosis and emphysemsa: connective tissue disorders of the lung. *Eur. J.Clin. Invest.* **14**:411–413.

Laurent, G. J. (1985). Biochemical pathways leading to collagen deposition in pulmonary fibrosis. *Ciba. Found. Symp.* **114**:222–233.

Linder, E., Stenman, S., Lehto, V. P., and Vaheri, A. (1978). Distribution of fibronectin in human tissues and relationship to other connective tissue components. *Ann. NY. Acad. Sci.* **20**:151–159.

Linsenmayer, T. F., Fitsch, J. M., Schmid, T M., Zak, N. B., Gibney, E., Sanderson, R.D., and Mayne, R. (1983). Monoclonal antibodies against

chicken type V collagen: production, specificity, and use for immunocytochemical localization in embryonic cornea and other organs. *J. Cell Biol.*
96:124–132.

Madri, J. A., and Furthmayr, H. (1979). Isolation and tissue localization of type
AB2 collagen from normal lung parenchyma. *Am. J. Pathol.* **94**:323–331.

Madri, J. A., and Furthmayr, Hl. (1980). Collagen polymorphism in the lung.
An immunochemical study of pulmonary fibrosis. *Hum. Pathol.*
11:353–366.

Martinez-Hernandez, A., Nakane, P. K., and Pierce, G. B. (1974). Intracellular
localization of basement membrane antigen in parietal yolk sac cells. *Am.
J. Pathol.* **76**:549–560.

Martinez-Hernandez, A., Marsh, C. A., Horn, J. F., and Munoz, E. (1981a).
Glomerular basement membrane: lamina rara, lamina densa. *Renal Physiol.*
4:137–144.

Martinez-Hernandez, A., Marsh, C. A., Clar, C. C., Macarak, E. J., and
Brownell, A. G. (1981b). Fibronectin: its relationship to basement membranes. II. Ultrastructural studies in rat kidney. *Coll. Relat. Res.* **1**:405–418.

Martinez-Hernandez, A., Gay, S., and Miller, E. J. (1982). Ultrastructural
localization of type V collagen in rat kidney. *J. Cell Biol.* **92**:343–349.

Martinez-Hernandez, A. (1985). The hepatic extracellular matrix. II. Electron
immunohistochemical studies in rats with CC14-induced cirrhosis. *Lab Invest.* **53**:166–186.

Martinez-Hernandez, A., and Amenta, P. S.. (1983). The basement membrane
in pathology. *Lab. Invest.* **48**:656–677.

Martinez-Hernandez, A. (1987a). Methods for electron immunohistochemistry. In
*Methods in Enzymology. Structural and Contractile Proteins: Extracellular
Matrix.* Edited by L. W. Cunningham. New York, Academic Press, p. 145.

Martinez-Hernandez, A. (1987b) Electron immunohistochemistry of the ex-
tracellular matrix. In *Methods in Enzymology. Structural and Contractile
Proteins: Extracellular Matrix.* Edited by L.W. Cunningham. New York,
Academic Press, p. 145.

Martinez-Hernandez, A. (1988). Repair, regeneration, and fibrosis. In *Pathology.*
Edited by E. Rubin and J. L. Farber. Philadelphia, J. B. Lippincott.

McLean, I. W., and Nakane, P. K. (1974). Periodate-lysine-paraformaldehyde
fixative. A new fixation for immunoelectron microscopy. *J. Histochem.
Cytochem.* **22**: 1077–1083.

Miller E. J., and Gay, S. (1982). Collagen: an overview. *Methods Enzymol.*,
82:3–32.

Miller, E. J., Gay, S., Haralson, M. A., Martinez-Hernandez, A., and Rhodes,
R. K. (1982). Chemistry and biology of the type V collagen system. In
New Trends in Basement Membrane Research. Edited by K. Kuehn, H.
H. Schoene, and R. Timpl. New York, Raven Press.

Miller, E. J., and Gay, S. (1987). The collagens: an overview and update. *Methods Enzymol.* **144**:3–41.

Myers, J. C., and Emanuel, B. S. (1987). Chromosomal localization of human collagen genes. *Coll. Relat. Res.* **7**:149–159.

Nakane, P. K. (1968). Simultaneous localization of multiple tissue antigens using the peroxidase-labeled antibody method: a study in pituitary glands of the rat. *Histochem. Cytochem.* **16**:557–560.

Nakane, P. K., and Pierce, G. B. (1966). Enzymes-labeled antibodies: preparation and application for the localization of antigens. *J. Histochem. Cytochem.* **14**:291–302

Nicholls, A. C., Osse, G., Schloon, H.G., Lenard, H. G., Deak, S., Myers, J. C., Prockop, D. J., Weigel, W. R., Fryer, P., and Pope, F. M. (1984). The clinical features of homozygous alpha 2(I) collagen deficient osteogenesis imperfecta. *J. Med. Genet.* **21**:257–262.

Niewoehner, D. E., and Hoidal, J. R. (1982). Lung fibrosis and emphysema: divergent responses to a common injury? *Science* **217**:359–360.

Odermatt, E., Tamkun, J. W., and Hynes, R. O. (1985). Repeating modular structure of the fibronectin gene: relationship to protein structure and subunit variation. *Proc. Natl. Acad. Sci. USA* **82**:6571–6575.

Oldberg, A., and Ruoslahti, E. (1986). Evolution of the fibronectin gene. Exon structure of cell attachment domain. *J. Biol. Chem.* **261**:2113–2116.

Paul, J. I., and Hynes, R. O. (1984). Multiple fibronectin subunits and their post-translational modifications. *J. Biol. Chem.* **259**:13477–13487.

Pierschbacher, M. D., Hayman, E. G., and Ruoslahti, E. (1981). Location of the cell attachment site in fibronectin using monoclonal antibodies and proteolytic fragments of the molecules. *Cell* **26**:259–261.

Pierschbacher, M.D., Hayman, E. G., and Ruoslahti, E. (1983). A synthetic peptide with the cell attachment activity of fibronectin. *Proc. Natl. Acad. Sci. USA* **80**:1224–1227.

Piez, K. A., and Reddi, A. H. (1984) In *Extracellular Matrix Biochemistry*. New York, Elsevier.

Pihlajaniemi, T., Dickson, L. A., Pope, F. M., Korhonen, V. R., Nicholls, Ar., Prockop, D. J., and Myers, J. C. (1984). Osteogenesis imperfecta: cloning of a pro-alpha 2(I) collagen gene with a frameshift mutation. *J. Biol. Chem.* **259**:12941–12944.

Pihlajaniemi, T., Myllyla, R., Seyer, J., Kurkinen, M., and Prockop, D. J. (1987). Partial characterization of a low molecular weight human collagen that undergoes alternative splicing. *Proc. Natl. Acad. Sci. USA* **84**:940–944.

Polak, J. M., and Varndell, I. M. (1984). *Immunolabelling for Electron Microscopy*. Amsterdam, Elsevier.

Prockop, D. J., Chu, M. L., de Wet, W., Myers, J. C., Pihlajaniemi, T., Ramirez, F., and Sippola, M. (1985). Mutations in osteogenesis imperfecta leading to the synthesis of abnormal type I procollagens. *Ann. NY Acad. Sci.* **460**:289–297.

Pyeritz, R. E., Stolle, C. A., Parfrey, N. A., and Myers, J. C. (1984). Ehlers-Danlos syndrome IV due to a novel defect in type III procollagen. *Am. J. Med. Genet.* **19**:607–622

Quinones, F., and Crouch, E. (1986). Biosynthesis of interstitial and basement membrane collagens in pulmonary fibrosis. *Am. Rev. Respir. Dis.* **134**:1163–1171.

Raghu, G., Striker, L. J., Hudson, L. D., and Striker, G. E. (1985). Extracellular matrix in normal and fibrotic human lungs. *Am. Rev. Respir. Dis.* **131**:281–289.

Reiser, K. M., and Last, J. A. (1983). Type V collagen. Quantitation in normal lungs and in lungs of rats with bleomycin-induced pulmonary fibrosis. *J. Biol. Chem.* **258**:269–275.

Rennard, S. I., and Crystal, R. G. (1982). Fibronectin in human bronchopulmonary lavage fluid. Elevation in patients with interstitial lung disease. *J. Clin. Invest.* **69**:113–122.

Rennard, S. I., Hunninghake, G. W., Bitterman, P. B., and Crystal, R. G. (1981). Production of fibronectin by the human alveolar macrophage: mechanisms for the recruitment of fibroblasts to sites of tissue injury in interstitial lung diseases. *Proc. Natl. Acad. Sci. USA* **78**:7147–7151.

Rich, E. A., Seyer, J. M., Kang, A. H., and Mainardi, C. L. (1983). Identification of a type V collagen-degrading enzyme from human sputum. *Am. Rev. Respir. Dis.* **128**:166–169.

Rosenkrans, W. A. Jr., Albright, J. T., Hausman, R. E., and Penney, D. P. (1983a). Ultrastructural immunocytochemical localization of fibronectin in the developing rat lung. *Cell Tissue Res.* **234**:165–177.

Rosenkrans, W. A. Jr., Albright, J. T., Hausman, R. E., and Penney, D. P. (1983b). Light-microscopic immunocytochemical localization of fibronectin in the developing rat lung. *Cell Tissue Res.* **233**:113–123

Rosenkrans, W. A., Jr., and Penney, D. P. (1986). Cell-cell matrix interactions in induced lung injury: III. Long term effects of X-irradiation on basal laminar proteoglycans. *Anat. Rec.* **215**:127–133.

Ruoslahti, E., Engvall, E., and Hayman, E. (1981). Fibronectin: current concepts of its structure and functions. *Coll. Relat. Res.* **1**:85–128.

Ruoslahti, E., Hayman, E. G., Pierschbacher, M. D., and Engvall, E. (1982). Fibronectin: purification, immunochemical properties, and biological activities. *Method Enzymol.* **82**:803–831.

Ruoslahti, E., Hayman, E. G., and Pierschbacher, M. D. (1985). Extracellular matrices and cell adhesion. *Arteriosclerosis* **5**:581–594.

Ruoslahti, E., and Pierschbacher, M. D. (1986). Arg-Gly-Asp: a versatile cell recognition signal. *Cell* **44**:517–518.

Rutten, T. L., van Kuppevelt, T. H., Janssen, H. M., and Kuyper, C. M. (1988). Ultrastructural localization of a chondroitinase-sensitive, cuprolinic blue-positive filament in developing mouse lung. *Eur. J. Cell Biol.* **45**:256–261.

Sakai, L. Y., Keene, D. R., and Engvall, E. (1986). Fibrillin, a new 350-kD glycoprotein, is a component of extracellular microfibrils. *J. Cell Biol.* **103**(1):P 2499–2550.

Sandberg, L. B., Soskel, N. T., and Leslie, J. G. (1981). Elastin structure, biosynthesis, and relation to disease states. *N. Engl. J. Med.* **304**:566–579.

Schwarzbauer, J. E., Tamkun, J. W., Lemischka, I. R., and Hynes, R. O. (1983). Three different fibronectin mRNAs arise by alternative splicing within the coding region. *Cell* **35**:421–431.

Selman, M., Montano, M., Ramos, C., and Chapela, R. (1986). Concentration, biosynthesis and degradation of collagen in idiopathic pulmonary fibrosis. *Thorax* **41**:355–359.

Singer, S. J. (1959). Preparation of an electron dense antibody conjugate. *Nature (London)* **183**:1523–1524.

Snyder, J. M., O'Brien, J. A., and Rogers, H. F. (1987). Localization and accumulation of fibronectin in rabbit fetal lung tissue. *Differentiation* **34**:32–39.

Starcher, B. C., Kuhn, C., and Overton, J. E. (1978). Increased elastin and collagen content in the lungs of hamsters receiving an intratracheal injection of bleomycin. *Am. Rev. Respir. Dis.* **117**:299–305.

Takiya, C., Peyrol, S., Cordier, J. F., and Grimaud, J.A. (1983). Connective matrix organization in human pulmonary fibrosis Collagen polymorphism analysis in fibrotic deposits by immunohistological methods. *Virchows Arch. (B)* **44**:223–240.

Takusagawa, K., Ariji, F., Shida, K., Sato, T., Asoo, N., and Konno, K. (1982). Electron microscopic observations on pulmonary connective tissue stained by Ruthenium red. *Histochem. J.* **14**:257–271.

Takusagawa, K., Asoo, N., Hasuike, M., Sato, T., Nagai, H., Motomiya, M., and Konno, K. (1987). Immunoelectron microscopic observations on proteoheparan sulfate in normal human lung. *Nippon Kyobu. Shikkan Gakkai Zasshi* **25**:738–743.

Tamkun, J. W., Schwarzbauer, J. E., and Hynes, R. O. (1984). A single rat fibronectin gene generates three different mRNAs by alternative splicing of a complex exon. *Proc. Natl. Acad. Sci. USA* **81**:5140–5144.

Timpl, R., and Dziadek, M. (1986). Structure, Development, and molecular pathology of basement membranes. *Int. Rev. Exp. Pathol.* **29**:1–112.

Tokuyasu, T. K. (1984). Immuno-cryoultramicrotomy. In *Immunolabelling for Electron Microscopy.* Edited by J. M. Polak and M. Varndell. Amsterdam, Elsevier, pp. 71–82.

Torikata, C., Villiger, B., Kuhn, C,3rd., and McDonald, J. A. (1985). Ultrastructural distribution of fibronectin in normal and fibrotic human lung. *Lab. Invest.* **52**:399–408.

Trelstad, R. L. (1984). *The Role of Extracellular Matrix in Development.* . New York, Alan R. Liss.

Urushihara, H., and Yamada, K. M. (1986). Evidence for involvement of more than one class of glycoprotein in cell interactions with fibronectin. *J. Cell Physiol.* **126**:323–332.

Vaccaro, C. A., and Brody, J. S. (1979). Ultrastructural localization and characterization of proteoglycans in the pulmonary alveolus. *Am. Rev. Respir. Dis.* **120**:901–910.

van Kuppevelt, T. H., Cremers, F. P., Domen, J. G., and Kuyper, C. M. (1984). Staining in proteoglycans in mouse lung alveoli. II Characterization of the cuprolinic blue-positive, anionic sites. *Histochem. J.* **16**:671–686.

Viljanto, J., Penttinen, R., and Raekallio, J. (1981). Fibronectin in early phases of wound healing in children. *Acta Chir Scand.* **147**:7–13.

Villiger, B., Broekelmann, T., Kelley, D., Heymach, G. J., 3d., and McDonald, J. A. (1981). Bronchoalveolar fibronectin in smokers and nonsmokers. *Am. Rev. Respir. Dis.* **124**:652–654.

von Der, Mark. K., and Ocalan, M. (1982). Immunofluorescent localization of type V collagen in the chick embryo with monoclonal antibodies. *Coll. Relat. Res.* **2**:541–555.

Willems, L. N., Otto Verberne, C. J., Kramps, J.A., ten Have, Opbroek, A. A., and Dijkman, J. H. (1986). Detection of antileukoprotease in connective tissue of the lung. *Histochemistry* **86**:165–168.

Woodcock, M. J., Adler, K. B., and Low, R. B. (1984). Immunohistochemical identification of cell types in normal and in bleomycin-induced fibrotic rat lung. Cellular origins of interstitial cells. *Am. Rev. Respir. Dis.* **130**:910–916.

Wooley, D. W., and Longsworth, L. G. (1942). Isolation of an antibiotin factor from egg white. *J. Biol. Chem.* **142**:285–290.

Yamada, K. M., Akiyama, S. K., Hasegawa, T., Hasegawa, E., Humphries, M. J., Kennedy, D. W., Nagata, K., Urushihara, H., Olden, K., and Chen, W. T. (1985). Recent advances in research on fibronectin and other cell attachment proteins. *J. Cell Biochem* **28**:79–97.

7

Carbohydrate Histochemistry

BRADLEY A. SCHULTE, RUSSELL A. HARLEY, and SAMUEL S. SPICER

Medical University of South Carolina
Charleston, South Carolina

I. Introduction

Macromolecules consisting entirely or partially of carbohydrate are important constituents of all biological systems. In animal tissues, homopolysaccharides made up exclusively of a single monosaccharide such as glycogen serve as primary energy reserves. Heteropolysaccharides consisting of two or more different sugars such as the glycosaminoglycans (GAGs), which are most often found linked to protein to form proteoglycans, are important structural components of connective tissues and of basement membranes. Other glycoconjugates (GCs) composed of saccharide side chains of varying degrees of complexity linked glycosidically to protein (glycoprotein) or lipid (glycolipid) are important elements of cell membranes and macromolecular secretions.

Biochemical analysis has revealed much about the oligosaccharide structure of secretory glycoproteins (mucins) in the respiratory tract (Yeager, 1971; Boat and Cheng, 1980; Sachdev et al., 1980; Woodward et al., 1982; Gallagher and Richardson, 1982; Lamblin et al., 1980, 1984a,b; Chandrasekaran et al., 1984; Rose et al., 1984). Secreted pulmonary GCs are thought to function mainly to prevent desiccation and to lubricate the airways. In contrast, much less is known about the structure and function of GCs, which are integral or peripheral constituents of the plasmalemma of airway epithelial cells.

The epithelial lining of the large and small airways and of the alveolar surface consists of several highly specialized cell types, all of which perform specific functions. Evidence accumulated in recent years suggests that GCs associated with the cell membrane play a key role in the regulation of molecular events occurring at the cell surface. It is probable that many biological processes such as fluid and ion transport, interactions with hormones and other chemical messengers, protection from proteolytic attack, and interactions with potential pathogenic microorganisms are regulated by specific membrane-associated GCs (Spiro, 1969; Nicolson, 1974; Hughes, 1975; Sharon, 1975; Pigman, 1977; Glick and Flowers, 1978; Montreuil, 1980; Lis and Sharon, 1986a,b). The physiological significance of cell and tissue GCs remains unknown in many instances. However, the biological role of GCs of more or less well-known composition conceivably could be better understood through knowledge of their precise location.

The heterogeneous nature of respiratory tract epithelial cells and the difficulties inherent in separating often tightly bound membrane GCs have impeded biochemical investigation into their structure. Histochemical approaches offer advantages that supplement biochemical procedures for elucidating the structure and function of complex carbohydrates. By providing knowledge of the precise location in the tissue or cell without contamination by related substances in tissue extracts, histochemistry yields insights into the biological significance of GCs through relating their chemical structure with possible functional activities in a given site. As two further advantages, these methods often disclose unsuspected heterogeneity of GC structure among morphologically indistinguishable cell types and reveal substances that are unrecognized biochemically.

From the viewpoint of the pathologist, detecting qualitative and/or quantitative changes in specific carbohydrate-containing macromolecules offers a prospect of providing information concerning diagnosis, prognosis, and pathogenesis of disease.

The general methods available for localizing and characterizing carbohydrate-containing components at the light microscopic level (Spicer and Henson, 1967; Spicer et al., 1967, 1983c) and the electron microscopic level (Spicer and Schulte, 1982; Spicer et al., 1983c; Thomopoulos et al., 1987) have been the subject of previous reviews from this laboratory. We describe here the more recently developed lectin methods for characterizing GCs in situ placing emphasis on technical details and application to the study of normal and pathologically altered respiratory tract.

II. Methods

A. Light Microscopy

Fixation

As the initial step in tissue processing, fixation is the single most important element in determining the outcome of histochemical reactions. The fixative is chosen by its ability to immobilize and prevent the extraction of tissue GCs and at the same time to preserve maximum reactivity with the various reagents used. On the other hand, close attention must be paid to the fixative's ability to preserve tissue morphology, since little is gained by optimizing histochemical reactivity at the expense of poor morphological preservation.

Although certain fixative solutions have been deemed advantageous for retaining some complex carbohydrates, solutions of buffered formalin used for 24 hr have long been considered the most satisfactory fixative for the general preservation of basophilia and periodic acid–Schiff (PAS) reactivity (Spicer et al., 1967). However, studies using labeled lectins as histochemical probes for specific sugars or sugar sequences have demonstrated that a number of other fixatives such as Carnoy's fluid or 95–100% ethanol (Stoward et al., 1980; Okusa, 1983; Rittman and Mackenzie, 1983) or in particular, a solution consisting of 6% $HgCl_2$–1% sodium acetate and 0.1% glutaraldehyde (Schulte and Spicer, 1983 a-c; Hennigar et al., 1985) consistently provide better retention of tissue-binding sites for most lectins than do the formaldehyde-based fixatives. This is not to say that formalin does not have its place in lectin histochemical studies, particularly in retrospective studies of routinely fixed pathology specimens. Formalin fixation also may be of use in prospective studies comparing the distribution of specific antigens and GCs in serial sections or on the same section by means of combined immunocytochemical and lectin histochemical methods (Holthofer et al., 1987). It should be emphasized that the interaction between fixative and lectin-binding GCs is variable, with some fixatives affecting the affinity for a given lectin more than others (Allison, 1987). Furthermore, a shorter fixation interval (2–4 hr) using a dilute fixative is generally less destructive to lectin-binding sites, which is also often the case for antigen preservation in immunostaining . This is particularly true in the case of aldehyde-based fixatives. It is recommended that, whenever possible, preliminary studies be performed with a number of different fixation protocols, with a view to optimizing the reactivity of complex carbohydrates and preserving tissue morphology.

It is difficult to obtain good morphological preservation of the lung simply by immersing a piece of lung in fixative. More satisfactory approaches used in this laboratory are vascular perfusion or infusion of fixative directly into the lung via the trachea. The former can be performed using any number of standard protocols usually involving cardiac perfusion. The latter is accomplished in laboratory rodents by placing the animal under deep anesthesia and inserting a small piece of tubing affixed to a syringe through the partially severed trachea. The thorax is opened rapidly, the right atrium is snipped, and the fixative is infused slowly. Excessive force should be avoided to prevent damage to respiratory bronchioles and alveolar ducts. From 20 to 50 ml of fixative solution is infused slowly over 5–10 min, and selected regions of the lung are sectioned with a razor blade and fixed further by immersion with occasional agitation for 3–4 hr at room temperature.

The choice of an appropriate fixation procedure is critical in studies aimed at examining the apical glycocalyx of airway and alveolar epithelial cells, as discussed in the final section of this chapter.

Specimens of resected human lung obtained at surgery can be processed by instilling fixative through major bronchi while immersing the remainder of the lung in fixative. In most instances it is more practical to slice the lung thinly and fix the slices by immersion. Postmortem autolysis can be retarded significantly by injecting an area of lung with fixative containing a visible dye such as eosin or alcian blue immediately after death and long before the autopsy. Well-fixed areas are subsequently identified by the presence of the dye.

Fixation effects on the ability of specific enzymes to digest carbohydrate components for characterizing specific saccharide moieties have not been investigated thoroughly. It will be especially pertinent to use a variety of fixation protocols, possibly including freeze-drying and freeze-substitution in future studies using previously untested but currently available exo- and endoglycosidases for characterizing specific carbohydrate moieties.

Immunohistochemical studies using polyclonal antisera or monoclonal antibodies directed against specific carbohydrate sequences also should be preceded by preliminary investigations using a variety of fixation protocols. This will be especially true for studies that propose the combined use of lectin histochemistry and immunocytochemistry, as mentioned above.

Tissue Processing and Sectioning

Staining with the histochemical procedures outlined in the next section requires stabilization of the tissue to allow generation of sections of different thickness, depending on the objectives of the study. This is accomplished either by embedding the tissue in one of a number of available media or by freezing.

For most investigations, the method of choice is dehydration through graded alcohols and xylene and embedment in paraffin. Paraffin has the advantages of low cost, ease of use, excellent stability over long-term storage, and easy removal prior to staining, thus minimizing interference with staining reagents. Paraffin embedment also allows the rapid generation of large numbers of serial sections over a wide range of section thickness (2–20 μm).

Optimal resolution depends ultimately on section thickness, however, and requires sections thinner than 2 μm. Tissues can be embedded in a variety of plastic resins that are sometimes suitable for both light and electron microscopic histochemical examination. Available embedding media include the more water-soluble acrylic plastics or the more hydrophobic epoxy resins. These media will be discussed in more detail in the section on electron microscopic cytochemistry. Embedment in polymerizable plastics allows sections ranging in thickness from fractions of a micron up to 2 μm to be cut with a glass knife. Although these thinner sections provide excellent morphological detail and increased resolution, they generally must be cut from a relatively small block face, resulting in a greatly reduced tissue sampling area. The use of some plastics also increases specimen preparation time significantly, since many of these monomers require from 24 to 72 hr for complete polymerization. Other potential drawbacks to the use of plastics include, to a variable extent, extraction or denaturation of tissue constituents during prolonged dehydration and infiltration steps, deleterious reactions with components of the embedding medium itself, and perhaps most importantly, severely restricted access of staining reagents to reactive moieties. Heat generated by the exothermic polymerization process, in some cases, may act to denature tissue constituents. Moreover, dissolution of the polymerized resins to improve accessibility to staining reagents, if they can be removed at all, requires extremely harsh chemical treatment (Kaissling, 1973; Baskin et al., 1979), which poses a further serious threat to the preservation of tissue carbohydrates.

In some cases it is advantageous to perform histochemical analysis on thicker sections. This is particularly true for studies of large cells with a complex cytoarchitecture such as neurons. For this purpose, fixed tissue is embedded in a loosely polymerized mixture of albumin and gelatin. Sections ranging from 10 to 100 μm in thickness are generated with an oscillating tissue slicer (Vibratome). These relatively thick sections are normally stained floating free, to promote penetration of reagents from both sides of the section and afford a more intense histochemical reaction. The sections are then affixed to a glass slide, dehydrated, and covered with a cover slip. This procedure is especially adapted to combined light and electron microscopic studies, since it permits staining sections for viewing with the light microscope so that regions of interest can be trimmed, flat embedded in plastic, and finally thin sectioned for electron microscopic examination.

A final method of processing tissues involves rapid freezing of small blocks of fixed or unfixed tissues and sectioning in a cryostat. Cryosections of unfixed tissue are normally fixed briefly prior to histochemical staining. Cryosections ranging in thickness from 5 to 100 μm are affixed to a glass slide for staining or stained free-floating. The use of frozen sections is advantageous when time is crucial, such as in the pathological diagnosis of surgical specimens using labeled lectins. Even in the presence of cryoprotectants such as sucrose or glycerol, the freezing process results in some damage to the cell membrane. This damage, when combined with cutting open the cells and staining free-floating sections, ensures penetration of staining reagents for light and electron microscopic cytochemical studies.

Staining Techniques

The earlier light microscopic techniques of carbohydrate histochemistry were based on the selective reactivity of functional groups found in complex carbohydrates, such as vicinal diols, sulfate esters, and carboxyl groups and on the autoradiographic demonstration of $^{35}SO_4$ and [^3H]sugars selectively incorporated into tissue GCs. These methods have been comprehensively reviewed (Spicer and Henson, 1967; Spicer et al., 1967, 1983c).

Rapid advances in our understanding of the distribution and structure of tissue GCs have depended on the recent availability of a large number of lectins (agglutinins). The lectins are proteins or glycoproteins considered to be of nonimmune origin, which bind to specific sugars or sugar sequences in complex carbohydrates. Lectins with specificities for a wide range of carbohydrate moieties have been isolated and purified from many plant and animal sources and their carbohydrate-binding specificity has been the subject of several reviews (Sharon and Lis, 1972; Goldstein and Hayes, 1978; Gallagher, 1984; Wu and Herp, 1985; Goldstein and Poretz, 1986; Lis and Sharon, 1986b). The use of lectins as probes to study the distribution and structure of complex carbohydrates in situ has expanded rapidly in recent years and much of this work also has been reviewed previously (Spicer et al., 1983c; Alroy et al., 1984; Cooper, 1984; Damjanov, 1987; Thomopoulos et al., 1987).

The methods for lectin histochemistry, with few exceptions, use as a first step a lectin that has undergone chemical modification. The most direct method uses lectins conjugated directly to a fluorescent marker such as fluorescein or rhodamine. Many lectins conjugated to these fluorescent dyes are available commercially. The labeled lectins generally are diluted in a solution of phosphate-buffered saline (PBS), pH 7.0–7.4, and applied directly to a briefly fixed frozen section affixed to a slide. However, this method can also be used to stain deparaffinized sections or free-floating sections cut on a Vibratome or cryostat. The concentration of the lectin conjugate varies, but generally is in the range of 50–100 $\mu g/ml$. Incubation times also vary but usually are less than 1 hr. The slide is then rinsed with PBS and viewed directly under ultraviolet light. The advantages of this technique are its simplicity and rapidity. This method also has the advantage of allowing the simultaneous viewing of binding sites for two lectins of differing specificity on the same section with the use of fluorescent tags that have different emission spectra.

The use of fluorescent labels however, has some disadvantages. First, the fluorescent probes work best on frozen sections and these often result in compromised morphological preservation. Second, the dark background makes difficult the identification of specifically stained cell and tissue constituents. Third, the staining lacks permanence and must be recorded photographically. Finally, the high concentration of labeled lectins used in the majority of studies increases the possibility of introducing staining by contaminating lectins or isolectins of differing specificity. Despite these drawbacks, lectins tagged directly with fluorescent reagents have and will continue to yield valuable histochemical results.

An alternative approach in wide use involves lectins conjugated directly to the enzyme horseradish peroxidase (HRP). This simple and rapid procedure is used most frequently on paraffin sections but also can be applied on plastic sections or free-floating or affixed frozen or Vibratome sections. Lectins conjugated to HRP are available commercially from several sources or are easily prepared in the laboratory. Although many methods of conjugating protein to protein are available, the simplest and most direct procedure is that using the bifunctional reagent glutaraldehyde (Avrameas, 1969). Used extensively in our laboratory, this procedure consists of mixing the lectin and HRP in the presence of glutaraldehyde (Schulte and Spicer, 1983b,c; Hennigar et al., 1985). One mg of lectin and 5 mg of HRP (Type VI, Sigma Chemical Co., St. Louis, MO) are dissolved in 1 ml of 0.1M PB, pH 6.8. A 20 μl aliquot of a 1% solution of glutaraldehyde in distilled water is then added slowly and the solution is stirred for 2 hr at room temperature. The conjugates are used directly or purified further by fractionation through a molecular sieve (Schulte and Spicer, 1983c). The conjugates can be stored at 4°C or aliquoted and frozen for later use.

The staining procedure consists of exposing the sections to labeled lectin followed by an HRP substrate solution. Sections from tissues fixed in buffered $HgCl_2$-glutaraldehyde should be treated with Lugol's solution to remove mercury prior to this step. We have found a 0.1M solution of PBS, pH 7.2–7.4, containing 1.0mM $CaCl_2$, $MnCl_2$, and $MgCl_2$ to be an excellent buffer for lectin histochemistry. The divalent cations are added because many of the lectins require one or more of them for binding (Goldstein and Hayes, 1978). After being equilibrated in buffer, the sections are incubated with a solution of the lectin-HRP conjugate usually ranging in concentration from 5 to 20 μg/ml for anywhere from 30 min to overnight. Incubation for 1 or 2 hr is generally adequate and can be performed at room temperature or 4°C with little difference in results. The sections are rinsed thoroughly in PBS to remove unbound lectin conjugate and exposed to the H_2O_2-diaminobenzidine (DAB) substrate medium of Graham and Karnovsky (1966) or to an alternative substrate for HRP. The duration of this step can be monitored with the light microscope. If the deposited reaction product is insoluble in organic solvents, as with the DAB substrate medium, the sections are dehydrated and a cover slip mounted with Permount. Otherwise cover slips should be mounted with a water-soluble medium such as glycerol.

A variation on this approach that is now used frequently in both immunohistochemical and lectin histochemical studies takes advantage of the very high affinity of biotin for avidin (Hsu and Raine, 1982). The section is first incubated with a biotinylated lectin and then rinsed and exposed in a second step to an avidin–biotin–HRP complex (ABC reagent) for which the biotin on the lectin bound to the section possesses high affinity. Many biotinylated lectins as well as the ABC reagent are available from Vector Labs. (Burlingame, CA), the company that pioneered the use of avidin and biotin in histochemistry. A number of other commercial firms now offer a wide range of biotinylated lectins and avidin labled detection systems. In general, concentrations of reagents, buffer systems, and HRP substrate media for the avidin–biotin system resemble those used for lectin–HRP conjugates. In our experience, these two methods provide identical specificity and similar sensitivity.

A remaining method depends on the capacity of concanavalin A (con A) to bind to the mannose-rich oligosaccharides of HRP, and on the multivalency of con A. In this procedure con A bridges the tissue glycoprotein and HRP. Thus, a two-step procedure that involves sequential exposure to unlabeled con A and then to HRP (Bernhard and Avrameas, 1971; Katsuyama and Spicer, 1978) localizes con A binding sites in tissues. A variation on this method uses first an unlabeled multivalent lectin and then a marker such as colloidal gold that has been combined with a glycoprotein for which the lectin has affinity (Geoghegan and Ackerman, 1977). This indirect technique has been used with con A and HRP-coated colloidal gold spherules as well as with lectins from the slug *Limax flavus* and from wheat germ that bind with high affinity to the glycoproteins fetuin and ovomucoid, respectively (Geoghegan and Ackerman, 1977; Roth, 1983; Roth et al., 1984).

Some lectins can be adsorbed onto the surface of colloidal gold spherules and used for direct staining. The relatively weak color reaction of the colloidal-gold-labeled reagents (rose to light pink) has limited their application at the light microscopic level (Roth et al., 1983; Lucocq and Roth, 1984, 1985), but they have been used widely in electron microscopic cytochemical procedures. The rapidly expanding use of the photochemical silver reaction (Danscher, 1981; Lucocq and Roth, 1985; King et al., 1987; Taatjes et al., 1987), which uses solutions of reduced silver to complex with and intensify gold particles, promises to advance the use of colloidal gold probes in light microscopic studies.

Exposing sections to purified enzymes enhances the specificity of cytochemical procedures by hydrolyzing known sugar linkages and eliminating staining attributable to the sugar moiety removed from the section. Methods for the specific identification of glycogen by obliterating PAS staining with amylase, of sialoglycoconjugates by eliminating characteristic basophilia with neuraminidase (Spicer and Warren, 1960), and of hyaluronic acid and chondroitin 4- or 6-sulfates by loss of basophilia after digestion with testicular hyaluronidase (Leppi and Stoward, 1965) or chondroitinase (Yamada, 1974) have been reviewed previously (Spicer and Henson, 1967). Enzyme cytochemical methods demand optimal temperature and duration of digestion, ionic content of medium, and concentration of enzyme as detailed in the references cited.

A recently introduced method for localizing GC containing terminal N-acetylglucosamine (GlcNAc) depends on loss of affinity for *Griffonia simplicifolia* isolection II (GSA II) resulting from digestion with N-acetylglucosaminidase (Hennigar et al., 1986). Moreover, a loss of staining with GSA II following digestion with amylase improves the sensitivity of glycogen detection obtainable with the amylase–PAS sequence.

Loss of staining after digestion with a specific glycosidase yields information about the location of the sugar hydrolyzed by the enzyme. Moreover, lectin histochemical methods make it possible to determine the location of a given sugar with a specific enzyme digestion by imparting lectin staining for the specific penultimate sugar or sugar sequence exposed after the digestion. Most sialoglyco-conjugates, for example, contain galactose or N-acetylgalactosamine penultimate to the terminal sialic acid and gain reactivity for peanut lectin (Stoward et al., 1980; Schulte and Spicer, 1985) or a lectin from *Dolichos biflorus* (Schulte and Spicer, 1985; Schulte et al., 1985a), respectively, after enzymatic removal of the sialic acid. It is conceivable, that a combination of digestion with one or another specific exoglycosidase to remove the terminal sugars in an oligosaccharide chain at a precise location, followed by a battery of lectin procedures to identify the sugar rendered terminal by digestion, could be performed repeatedly to yield information about the sequence of sugars in the oligosaccharide chains.

B. Pre-embedment Electron Microscopic Cytochemistry

Localizing and characterizing complex carbohydrates by electron microscopy increases the prospect of assigning a biological role to a specific GC, on the assumption that its presence in an organelle or organellar substructure relates to the structure's biological activity. Methods for localizing GCs ultrastructurally are currently being developed and will be reviewed under two headings: staining prior to embedment and staining after embedment.

Fixation

Fixation requirements are more stringent for electron microscopic than light microscopic cytochemistry because of the need to preserve the morphological fine structure of the tissue and ensure the retention and chemical reactivity of the GC. Dilute glutaraldehyde has met the requirement for preserving morphology better than has formaldehyde, but has proven less satisfactory for the retention of staining reactivity. Fixation involves immersing the specimen, perfusing the fixative through the vascular system, installing fixative via the airways, or using a combination of these methods as detailed above.

The extent of tissue fixation influences the rate at which cytochemical reagents diffuse into the interstitial space and cells of the specimen. Diffusion occurs more readily, for example, into formaldehyde- than into glutaraldehyde-fixed tissue. Little effort has been made, however, to compare available fixatives or improve fixation for penetration of the reagents of ultrastructural carbohydrate cytochemistry. Fixation for ultrastructural immunocytochemistry has received considerable attention, however (see Chap. 6). This experience seems applicable to methods for visualizing complex carbohydrates, since diffusion limitations are probably similar for immunoglobulins and reagents for detecting GCs, particularly the lectins.

Postfixing with osmium tetroxide (OsO$_4$) after a primary glutaraldehyde fixation improves fine structural morphology and visualization of cell membranes. However, OsO$_4$ postfixation impairs the reactivity of vicinal glycols, carboxyl groups, and sulfate esters to subsequently applied reagents (Thomopoulos et al., 1983a,b, c) and should only be used after the cytochemical stain as a possible means of improving fixation and accentuating and making permanent the stain product.

Tissue Processing and Sectioning

Penetration of histochemical reagents limits the applicability of the ultrastructural pre-embedment cytochemical methods. The size of the tissue block obviously determines the requisite diffusion distances and, hence, the access of reagent to tissue GCs. Mincing the tissue finely during fixation minimizes diffusion distances.

Staining sections cut at a thickness of 5–20 μm in the cryostat reduces the diffusion problem, and if ice crystal damage is minimized by rapidly freezing tissue soaked with cryoprotectants, the morphology can be sufficiently preserved. Moreover, most cells are cleaved in such cryosections so that staining entities gain direct access to the intracellular space. Sectioning gelatin-embedded tissues on a Vibratome also provides increased access of reagents to interstitial and intracellular components while preserving morphology. Surface epithelial cells in the respiratory tract are favorable sites for staining, as are luminal surfaces of endothelial cells (Skutelsky and Danon, 1976; De Bruyn et al., 1978). These sites are advantageous because the accessibility of apical surfaces of epithelial and endothelial cells obviates the need for penetration into the block or through more than one cell. Cells grown in culture or dispersed into suspension from body fluids or tissues also provide favorable specimens for histochemical staining.

Intracellular staining by pre-embedment methods requires diffusion of staining reagents across the plasmalemma, unless the cell is opened by sectioning before staining. Fixation itself appears to alter plasmalemmal permeability to some but not other reagents, presumably permitting entry of substances of relatively low molecular weight. Freezing and thawing tissue or extracting lipids by treatment with graded alcohols and treatment with detergents such as saponin, Triton X-100, or bile salts have been used to disrupt plasmalemmal continuity and improve the penetration of immunocytochemical staining components (Willingham and Yamada, 1979). Such procedures have found little application as yet to carbohydrate ultracytochemistry, but appear to merit further testing.

Staining Techniques

Exposing reactive sites to cytochemical reagents usually involves immersing tissue blocks or sections in the solution containing the agent. However, in studies of endothelial and epithelial surfaces in the respiratory tract, this has been accomplished by intravascular or intratracheal infusion of the reagent.

A wide range of staining methods exist for detecting with the electron microscope the vicinal glycols, acid groups, and many of the sugar residues characteristic of complex carbohydrates (Spicer and Schulte, 1982; Spicer et al., 1983c; Roth, 1986; Thomopoulos et al., 1987). Procedures for visualizing macromolecular carbohydrates by electron microscopy resemble those used at the light microscopic level, except that the reagent reacting with the carbohydrate obstructs transmission of electrons rather than absorbing visual light. The basis for electron opacity has not been fully explained but, except perhaps for nonmetallic organic reagents such as oxidized diaminobenzidine, depends in part on atomic weight.

Several of the cytochemical methods commonly used for pre-embedment staining use heavy metals in a positively charged state for the selective densification of tissue macromolecular anionic groups. For example, ruthenium red has usually been used by incorporation into the initial fixative fluid. This reagent does not penetrate cells and provides staining confined to extracellular and cell surface components. Although ruthenium red apparently binds to anionic groups of GCs, the specificity of this component has not been established. Since the procedure is conducted at neutral pH, staining of phosphate groups may occur and reaction with phospholipid has been noted.

Colloidal iron suspensions, because of the inherent electron opacity of the iron, serve at the electron microscopic, as well as the light microscopic level, to visualize GCs. As with pH 2.5 alcian blue at the light microscopic level, dialyzed iron has selective affinity for both the carboxyls and the sulfate esters in GCs. From extensive experience, the Abul Haj–Reinhart dialyzed iron preparation diffuses into cells and stains intracellular organelles better than does Muller's colloidal iron prepared by boiling.

Cationized ferritin also binds selectively to acidic mucosubstances, having affinity for both carboxyls and sulfates. This macromolecule of high molecular weight does not penetrate cells and stain intracellular complex carbohydrates. Because of the distinctive structure of cationized ferritin particles, this method has proven valuable for quantitating as well as localizing acidic GCs of cell surfaces.

A recently introduced method (Skutelsky and Roth, 1986) uses gold spherules coated with poly-L-lysine to localize negatively charged components on the surface of erythrocytes. These cationic colloidal gold particles offer a promising alternative to cationic ferritin in quantitative studies of acidic cell surface GCs.

The high-iron diamine solution, which stains sulfated complex carbohydrates selectively for light microscopy, also imparts electron density selectively to sulfated substances. High iron diamine stains sulfomucin in some sites that are negative with dialyzed iron, such as rat tracheal glands (Spicer et al., 1980), and apparently diffuses into cells more readily.

Lectin–HRP conjugates or biotinylated lectins followed by the ABC complex can be used at the electron microscopic level because of the density and electron opacity of the substrate diaminobenzidine after it has been oxidized by HRP. In addition, the diaminobenzidine reaction product becomes more dense after complexing with OsO_4 in the same manner in which it becomes darker brown or black in paraffin sections. Lectins conjugated directly with ferritin and more recently with colloidal gold have also been used extensively in pre-embedment cytochemical studies (Spicer et al., 1983c). The lectin procedures in general fail to stain intracellular components in pre-embedment procedures, except in disrupted cells. However, they provide strong staining of stromal matrix and cell surface GCs.

The advantages and disadvantages of the various stains should be considered in relationship to the specific goal. Two or more similar or complementary stains can be used to gain the advantages of one and overcome the limitations of another procedure.

C. Postembedment Electron Microscopic Cytochemistry

Postembedment techniques for demonstrating complex carbohydrates are evolving rapidly (for review see Roth, 1986; Thomopoulos et al., 1987). This approach offers an important advantage in avoiding the severe limitations imposed on pre-embedment methods from failure of the reagent to penetrate the tissue, the cell, or the organelle. Another advantage of postembedment staining is the potential for comparing reactivity by a variety of cytochemical procedures in a precise cytological location. Thus, reactivity of a single cell organelle, such as a secretory granule or a Golgi cisternae, can be compared in serial sections with periodate methods to demonstrate vicinal diols, cationic reagents to visualize anionic groups, and various lectins to identify different specific sugars or sugar sequences (Thomopoulos et al., 1987)

Problems are encountered, however, in staining the ultrathin sections. Reagents often penetrate the impermeable embedding medium poorly and in some instances deposit background contamination more or less randomly on the thin section. Chemicals in the fixative and embedding medium have a capacity to react with functional groups of tissue carbohydrates, thus altering their chemical structure and cytochemical reactivity. Moreover, solvents to which the tissues are exposed could extract GCs. Comparison of staining with different methods applied to postosmicated and nonosmicated tissues as well as tissues embedded in epoxy and nonepoxy embedding media, has revealed a considerable degree of technique-dependent variability in the intensity and specificity of the staining (Thomopoulos et al., 1987). It is clear that the detection of complex carbohydrates on thin sections is influenced by the mode of fixation and the embedding medium used.

Fixation

As with pre-embedment methods, the fixative must preserve the reactivity of carbohydrate moieties but also must prevent their extraction during the more extensive tissue processing required by embedment prior to staining. The fixative of choice in most postembedment cytochemical studies has been a buffered solution of 1–2% glutaraldehyde, in some cases combined with paraformaldehyde. In our experience, fixation for 1 hr at room temperature in a 1–2% solution of glutaraldehyde in 0.1M PB, pH 7.2, provides good morphology and adequate retention of lectin-binding sites in most tissues. As an alternative, a freshly prepared solution of 4% paraformaldehyde containing no or only a small amount of glutaraldehyde (0.1–0.3%) in the same buffer can be used. Primary or secondary fixation with OsO_4 should be avoided since it is known to destroy the reactivity of carbohydrate functional groups when several histochemical methods are used (Thomopoulos et al., 1983a, b, c). Although glutaraldehyde appears to have little direct effect on carbohydrate functional group reactivity or lectin binding, it does act to block penetration of staining reagents through its capacity to form a tightly cross-linked mesh work of tissue protein and glycoprotein. It is recommended that the concentration of glutaraldehyde and the duration of fixation be held to the minimum necessary to provide adequate morphological preservation. These parameters will of course vary with the tissue under study and the method of fixation (i.e., immersion vs. perfusion) and should, therefore, be determined empirically for each constituent under investigation.

Tissue Processing and Sectioning

The technique of tissue processing and the choice of embedding medium are extremely important factors in retaining the reactivity of carbohydrates in postembedment ultrastructural cytochemistry. The combined efforts of a number of laboratories have demonstrated that light aldehyde fixation and embedment in nonepoxy resins results in the strongest staining of GCs with a wide range of postembedment techniques (Thomopoulos et al., 1987). Moreover, light aldehyde fixation and embedment in beam-stable acrylic resins such as Lowicryl K4M and LR White have provided the best sensitivity with the postembedment lectin histochemical methods. Embedment in these more hydrophillic acrylic plastics avoids the need for prolonged dehydration and infiltration steps, which may minimize extraction of lightly fixed tissue constituents. Furthermore, with Lowicryl the ability to process tissue at progressively lower temperatures and to polymerize the resin with ultraviolet light at low temperatures presumably also helps to prevent heat denaturation or extraction of reactive moieties.

It is clear from the results of immunocytochemical and lectin histochemical studies that the majority of staining, especially with the colloidal gold methods, takes place at the exposed surface of the ultrathin sections. The reason why reactive groups are more accessible to staining reagents at the surface of tissue sections cut from acrylic as opposed to epoxy resins remains unclear. An excellent overview of the use of acrylic resins and the theoretical considerations behind their enhanced efficiency in postembedment cytochemistry can be found in a series of three papers presented at a recent symposium (Kellenberger et al., 1987; Newman and Hobot, 1987; Bendayan et al., 1987)

Staining Techniques

Ultrastructural postembedment methods for demonstrating complex carbohydrates in situ can be grouped into three categories based on the nature of the reactive moiety stained. A first category is composed of techniques aimed at detecting glycosubstances by the reactivity of hydroxyl groups or of vicinal diols present in sugars. All of these methods, references to which can be found in a recent review (Thomopoulos et al., 1987), use as a first step a mineral acid, most commonly and preferentially periodic acid, which acts to oxidize the vicinal diol groups or hydroxyl groups of sugars. The resulting dialdehyde or aldehyde group is then exposed to any one of a number of aldehyde reactive agents, which, if electron opaque, can be visualized directly in the electron microscope or complexed with an electron-opaque heavy metal. The Thiery technique (1967) used extensively in our laboratory involves a periodic acid–thiocarbohydrazide–silver proteinate (PA-TCH-SP) sequence and extends the light microscopic PAS procedure to the electron microscopic level, achieving greatly increased sensitivity. This method deposits electron-opaque silver deposits at the site of periodate-engendered aldehydes. The greatest sensitivity with the PA–TCH–SP method has been observed in tissues fixed with glutaraldehyde and embedded in nonepoxy resins such as a styrene–methacrylate resin mixture (Thomopoulos et al., 1983a) and LR White (unpublished observations).

These methods have proved to be very sensitive and considerable success has been achieved at localizing glycogen, secretory and cell surface glycoproteins, nascent glycoprotein in Golgi cisterna, lysosomal enzyme glycoproteins, and perhaps glycolipids. The presence of glycolipids in thin sections after most standard dehydration and embedding procedures is questionable, however. Unfortunately, these methods provide limited chemical information for differentiating types of GCs, since all sugars contain hydroxyl groups and sugars containing oxidizable vicinal diols are abundant in all classes of complex carbohydrates, with the exception of the GAGs and proteoglycans. Although such reactions have often been claimed as specific for localizing neutral complex carbohydrates, this is a misconception since all sialic acids contain hydroxyl groups and the most frequently occurring sialic acid in mammals, N-acetylneuraminic acid, has an unsubstituted polyhydroxyl side chain containing vicinal diols that are susceptible to even the most mild oxidation procedures. Moreover, glycoproteins containing hexoses with vicinal diols are sometimes acidic by virtue of their sulfate ester content.

A second category of methods are those used for detecting carboxyl groups and sulfate esters of acidic complex carbohydrates such as those described in the section on pre-embedment staining. The application of these methods directly to ultrathin sections has been reviewed recently (Thomopoulos et al., 1987). These techniques rely on complexing electron-opaque cationic reagents with acidic functional groups of glycosubstances including sulfate esters, and carboxyl groups of sialic, glucuronic, and iduronic acid. Early attempts to demonstrate acidic GCs on thin sections were mostly unsuccessful. Progress of later efforts has depended largely on the use of a nonepoxy-embedding medium, such as the methacrylates, which unfortunately provide suboptimal morphological preservation. We have recently achieved some success using dialyzed iron (Thomopoulos et al., 1983b) and high-iron diamine (Thomopoulos et al., 1983c) on thin sections. Successful staining depended on avoiding OsO_4 or using only a very brief (5 min) OsO_4 postfixation step and embedment in nonepoxy resins such as the styrene-methacrylate and styrene-Vestopal W resin mixtures. We have not assessed the reactivity of tissue GCs embedded in Lowicryl or LR White with these methods and are unaware of any other reports describing their use with these embedding media.

The third and currently most fully investigated group of postembedment methods uses lectins for detecting specific sugars or sugar sequences. Although from 50 to 100 lectins with a wide range of sugar-binding specificities are available commercially, a comparatively small number have been applied in postembedment cytochemical studies (Spicer et al., 1983c; Roth, 1986; Thomopoulos et al., 1987).

A major problem encountered in lectin cytochemistry appears to be the limited penetration into the hydrophobic epoxy media of water-soluble labeled lectins used in the direct methods or of the unlabeled lectins and the labeling macromolecules used in the indirect methods. Similar problems are encountered with penetration of immunoglobulins in ultrastructural immunocytochemistry and are probably related to the molecular size of the reagents, including their sphere of hydration.

A number of alternative, less hydrophobic, embedding media have been used in lectin cytochemical studies. To date, the most consistent and satisfactory staining has been obtained by Roth and co-workers (Roth, 1983, 1986; Roth et al., 1983, 1984; Lucooq et al., 1987) using gold particles coated directly with lectin or a two-step procedure using unlabeled lectin followed by gold particles coated with a glycoprotein to which the lectin binds. A comparison of such staining on thin sections of material embedded in Epon or in Lowicryl K4M showed staining intensity to be far superior in the latter resin (Roth, 1983). Excellent results also have been achieved with LR White embedment using lectin–HRP conjugates or biotinylated lectins followed by the ABC complex, as well as lectin-coated colloidal gold particles (Newman and Hobot, 1987). This is in contrast to Lowicryl K4M, in which success has been obtained consistently with the colloidal gold methods but not with lectins labeled with enzymes such as HRP.

III. Interpretation of Results

Support for the valid interpretation of a histochemical result necessarily depends on performing the appropriate method controls. Control procedures for the more generalized light and electron microscopic cytochemical methods have been described elsewhere (Spicer and Henson, 1967; Spicer et al., 1967, 1983c; Spicer and Schulte, 1982; Thomopoulos et al., 1987). Method controls for lectin histochemistry include (1) omission of one or more steps in the staining sequence, (2) elimination of reactivity by periodate oxidation, and (3) competitive inhibition of staining in the presence of the appropriate monosaccharide or more complex sugar sequence. Reactivity in the same histological site with a battery of lectins with similar or identical carbohydrate-binding specificities also serves to increase confidence in the specificity of the staining reaction, as does lack of staining with lectins of different specificity.

Elimination or a pronounced decrease of staining after exposure to a highly specific exo- or endoglycosidase serves in an important way to establish the specificity of the staining method. The validity of the enzyme digestion obviously depends on the purity of the enzyme preparation and its freedom from other enzymatic activities, especially of proteinases and other glycosidases. A control procedure against a false-positive result from digestion requires eliminating simple solubilization and extraction and entails simultaneous exposure to the same solution minus the enzyme or containing heat-inactivated enzyme. A false-negative result on the other hand, should be checked by testing activity of the enzyme preparation on a known positive site.

A positive histochemical result generally reflects true localization if it is not seen in cytochemical controls, whereas lack of reactivity is not necessarily conclusive. For this reason, a known test site containing reactive material should be run in parallel with the tissue of interest as a control against a false negative from a faulty technique. As a tissue control against a false-positive test, an organ that has been shown biochemically to lack the substance of interest should be stained to rule out nonspecific staining or staining artifacts. The control tissues should be collected, fixed, and processed identically to the tissue under study, preferably in the same experimental protocol. For light microscopic studies of rodent tissues, we have used composite paraffin blocks containing a large selection of organs. Such blocks allow the direct comparison of staining in a wide range of both positive and negative histological sites while eliminating technical variability. This approach is obviously not feasible for studies of tissue from larger animals and humans, but it is usually possible to include one or two pieces of control tissue in a composite block along with the tissue of interest. Composite tissue blocks cannot be prepared for electron microscopic studies, but carefully selected control tissues should be included in all ultrastructural cytochemical studies. As an alternative to using a different tissue from the same species, the same tissue from a different species that is known to lack or suspected of lacking the substance of interest can be used as a negative control. As a bonus, this approach has revealed significant information concerning genetic and sex-related differences in the structure of complex carbohydrates (Schulte, 1987; Spicer et al., 1987).

A negative staining result at the ultrastructural level is particularly difficult to interpret. As a rule, preliminary studies at the light microscopic level should precede electron microscopic cytochemistry to establish the presence in the cell of the entity under study. Only the very exceptional cell or tissue constituent cannot be visualized histochemically by careful observation under the light microscope.

Interpretation of the findings rests on the selectivity or specificity of the technique. A technique is said to be selective when it recognizes a range of tissue constituents having in common a particular reactive moiety or functional groups with similar chemical properties. For example, the more general carbohydrate histochemical staining methods, such as the PAS and PA-TCH-SP techniques, are selective for GCs with hexoses containing a vicinal diol function or with sialic acids possessing an unmodified polyhydroxyl side chain. The basic dyes, such as alcian blue, used at the light microscopic level and the electron-opaque cationic reagents, such as dialyzed iron and cationic ferritin, used at the electron microscopic level are selective for GCs with sulfate and carboxyl groups. These reagents are selective for GCs including proteoglycans and glycoproteins because other tissue macromolecules lack residues with vicinal diols or comparable basophilia.

A procedure is considered specific if it can be shown to react with only one functional group in a macromolecule or one species of macromolecule in tissue sections. The high-iron diamine procedure, for example, has been determined to be specific for sulfate esters when used at either the light or electron microscopic levels. Some lectins are highly specific for a given sugar or sugar sequence, whereas other lectins show affinity for more than one terminal sugar or sugar sequence and thus must be considered to be selective.

Used on adjacent sections or in combination on the same section at the light microscopic level, the selective histochemical staining procedures serve to differentiate the various classes of tissue GCs. A positive reaction with the PAS or PA–TCH–SP technique identifies glycogen or glycoproteins since GAGs and proteoglycans fail to react with these methods. Staining in the same sites with alcian blue at pH 1.0 or high iron diamine reveals sulfated glycoprotein. In contrast, sites stained positively with PAS or PA–TCH–SP and with alcian blue at pH 2.5 or dialyzed iron and lacking reactivity with alcian blue at pH 1.0 or high-iron diamine contain nonsulfated sialoglycoprotein. Cell or tissue constituents that fail to stain with the PAS or PA–TCH–SP sequence but that react with basic dyes or electron-opaque cationic reagents are identified as GAGs or proteoglycans. The selectivity or specificity of these staining techniques has been confirmed through application of a wide range of chemical tests discussed in previous reviews (Spicer and Henson, 1967; Spicer et al., 1967, 1983c).

The specificity of the lectin histochemical method is known in some cases and requires further investigation in others. Interpretation of the sugar or sugar sequences to which the lectin is bound is based primarily on biochemical analysis using hemagglutination inhibition, precipitin inhibition, and/or competitive binding assays (Sharon and Lis, 1972; Goldstein and Hayes, 1978; Wu and Herp, 1985; Goldstein and Poretz, 1986; Lis and Sharon, 1986b). The specificity of many of the lectins has been evaluated using highly purified mono-, di-, and trisaccharides and more complex oligosaccharides, glycoproteins, and glycopeptides. Such studies have shown that lectins often have extended binding sites that recognize not only terminal sugars but also penultimate sugars and internal sugar sequences. It is clear that many of the lectins show much greater affinity for their most complementary binding sites in complex oligosaccharides than for the monosaccharides often listed as the sugar with which they bind (Debray et al., 1981). Moreover, some lectins show a higher affinity for glycopeptides and glycoproteins than for isolated oligosaccharides, a difference that is perhaps attributable to the structural rigidity imparted by the glycan–amino acid linkage.

Histochemical studies contribute otherwise unavailable information showing that the interaction between a lectin and its ligand is not simply a function of the presence or absence of a single monosaccharide. Lectins reported to have a similar or identical nominal binding specificity for a particular terminal sugar often show different staining patterns when used on adjacent sections of the same tissue (Hsu and Ree, 1983; Schulte and Spicer, 1983d, 1984; Alroy et al., 1984; Laden et al., 1984; Sato and Muramatsu, 1985; Virtanen et al., 1986). Differences in histochemical reactivity have been shown among groups of lectins represented to be specific for terminal β-D-galactose (Gal), for terminal N-acetyl-D-galactosamine (GalNAc) and for α-L-fucose.

Both biochemical and histochemical studies have thus demonstrated that the lectins are a highly diverse group of macromolecules with different degrees of affinity for individual sugars and their more complex derivatives. In view of the complicated and often incompletely characterized interaction between lectins and GCs, it is recommended that a positive reaction, in the absence of other supporting data, be interpreted as evidence of the presence of the least complex sugar or sugar sequence for which the lectin is known to show affinity. Moreover, lack of staining in a histological site with a single lectin should not be interpreted as showing absence of the sugar with which the lectin binds, but if all lectins available for visualizing the sugar fail to stain, the sugar is probably not present.

IV. Past Contributions to Lung Biology and Future Perspectives

A. Secretory Glycoproteins

Analysis of carbohydrate-containing macromolecules in situ has contributed significantly to our understanding of the distribution and structure of respiratory tract complex carbohydrates. As reviewed previously (Spicer et al., 1983a, b), analysis by histochemistry has been of particular importance in determining the origin of glycoproteins that form the thick coat overlaying the luminal surface of airway epithelial cells. Glycoproteins of this coat have been incompletely defined biochemically in normal pulmonary lavage or abnormal sputum because of limited material for analysis. These diverse glycoprotein secretions derive mainly from serous and mucous glands in the lamina propria and, to a lesser extent, from secretory cells in the surface epithelium of the airway proper.

The secretory GCs are thought to influence the degree of hydration at the cell's luminal surface and in the airspace, countering the desiccation by inhaled air. Moreover, they protect the surface cells from microbial invasion and thermal or chemical injury and may provide lubrication of ciliary motion. In addition, the secretions form a blanket covering the cilia that is swept cephalad by cilia action transporting trapped microorganisms and foreign matter from the lung.

Differences between stored secretions in glandular serous and mucous cells have long been established (Spicer et al., 1983a, b). More recent lectin histochemical studies have revealed differences in secretory glycoproteins among glandular serous cells of mice, rats, and humans (Schulte and Spicer, 1983b, 1985). Serous cells in the mouse produce only neutral glycoproteins lacking basophilia and containing the terminal disaccharides Gal-($\beta1 \rightarrow 3$)-GalNAc and terminal α-GalNAc in all cells and terminal GlcNAc in about 70% of cells. In contrast, serous cells in the rat synthesize acidic glycoprotein with heavily O-acetylated terminal sialic acid and penultimate Gal-($\beta1 \rightarrow 3$)-GalNAc. With use of the lectin methods, human serous tubules and demilunes revealed sialic acid and penultimate Gal-($\beta1 \rightarrow 3$)-GalNAc in all cells.

In the human respiratory tract, serous demilunes and acinar cells vary markedly in the fine structure of their granules (Thaete et al., 1981; unpublished observations). Some cells contain homogeneous granules and others contain granules with a small to large, denser or less dense eccentric core. The granules in one cell differ cytochemically from those in another in their content of high-iron diamine, dialyzed iron, or PA–TCH–SP-reactive GC in the granules' rim or core. The nature of these different carbohydrate-containing substances remains to be investigated.

Lectin histochemistry has also demonstrated differences in mucous cells. These cells were previously thought to form a uniform cell population from their homogeneous PAS staining, but showed variability with the alcian blue–PAS sequence in their content of neutral compared with acidic mucosubstance and, with the high-iron diamine–alcian blue sequence, in their content of sialylated vs. sulfated complex carbohydrate (Chakrin et al., 1972; Spicer et al., 1973). The variability between mucous cells in an animal and mucous tubules of one species compared with another appears much greater, however, with lectin histochemistry. Stored acidic secretions of glandular mucous cells in rodents vary in the chemistry of terminal sialic acids and the penultimate sugar linkage (Schulte and Spicer, 1985). Most mucous cells in the mouse and hamster trachea contain sialic acids with heavily O-acetylated polyhydroxyl side chains linked to penultimate Gal-(β1→3)-GalNAc. Secretions in most mucous cells in the rat trachea contain a similar trisaccharide, but the terminal sialic acid has an additional O-acetyl substituent at carbon 4. In addition, all or most mucous cells in the rat and hamster trachea stain for content of mucins containing sialic acids linked α-2→6 to penultimate GalNAc, whereas this terminal disaccharide is not present in mouse tracheal mucins.

Glycoprotein secretions in mouse mucous cells fail to show any terminal sugar other than sialic acid, except for the presence of fucose in about half the cells. In the rat, some mucous tubule cells contain terminal α-GalNAc, suggesting a similarity in structure to human blood group A antigen or the possible incomplete synthesis of oligosaccharides (Schulte and Spicer, 1983b). Human mucous tubules show differences correlated with the ABO blood group and perhaps with the secretor status of the donors. Thus, blood group A and AB specimens showed staining demonstrative of terminal α-GalNAc, (the terminal sugar in blood group A antigen), in contrast to the nonreactivity of blood group B and O specimens. Blood group B and AB specimens stain for the presence of terminal α-Gal, (the terminal residue in blood group B antigen), contrasting with the nonreactivity of blood group A and O specimens. These differences have since been confirmed using monoclonal antibodies against human ABH blood group antigens (unpublished observation).

In rat and human respiratory tracts, mucous cells of mucous ducts, acini, and surface epithelium differ markedly from serous cells in their fine structure and GC content. The individual mucous cells differ from one another in the fine structure of their granules (Thaete et al., 1981; Mochizuki et al., 1982; Spicer et al., 1983a; unpublished observations). Granules in different cells vary also in the content and distribution of GC, as shown by variable staining of rim and core or even three zones in different granules with the dialyzed iron, high iron diamine, and PA–TCH–SP methods. Nothing is known about the biological significance of such variability.

The observed differences between serous or mucous cells of one compared with another species and between serous and mucous cells in each species presumably relate to specific biological activities in these sites. Having only recently been discovered, the significance of these differences awaits an explanation. However, the relative sparsity of carbohydrate reactivity in serous cell secretions and exceptional abundance of carbohydrate seen cytochemically in mucous cell secretions appear consistent with the view that upstream serous cells produce a nonviscous secretion capable of washing through the duct system the more viscous secretion of the mucous cells downstream. Support for this concept is derived from immunocytochemical demonstration of carbonic anhydrase II in tracheobronchial serous cells (Spicer et al., 1982) since this enzyme functions in fluid and ion transport across epithelial cells and could contribute to production of a watery secretion.

B. Cell Surface Glycoconjugates

Highly specialized cells line the airways and alveoli of the adult respiratory tract. These cells produce diverse GCs that form the glycocalyx at the cell's surface. We have little knowledge about the structure of GCs that form the glycocalyx of respiratory tract epithelial cells, since the glycocalyx of one cell type is not biochemically separable from that of another and glycocalyx GCs cannot be reliably isolated from the more abundant secretory glycoproteins that form the mucous surface coat. Cytochemistry provides advantages over biochemical methods for analyzing cell surface components including (1) revealing differences in the structure of the glycocalyx of histologically different cell types, (2) providing the capacity to show differences in GCs of the apical and the basolateral plasmalemma of a cell, and (3) showing changes in the chemical nature of a given cell's surface during fetal and postnatal development or resulting from pathological change or experimental manipulation.

Recent observations (Schulte and Spicer, 1985, 1986: Schulte et al., 1986) have shown that the surface of the cilia lining the entire tracheobronchial tree in three rodent species and in humans contain GCs with the terminal trisaccharide NeuNAc-($\alpha 2 \rightarrow 3$, 6)-Gal-($\beta 1 \rightarrow 3$)-GalNAc (Figs. 1–3). In addition, some cilia in the proximal trachea in the rat, throughout the trachea to the level of the primary bronchi in the mouse, and throughout the entire tracheobronchial tree to the level of the respiratory bronchioles in humans are stained for the presence of terminal Gal-($\beta 1 \rightarrow 3$)-GalNAc uncapped by sialic acid which suggests incomplete sialylation or the loss of sialic acid after synthesis.

Cilia on the surface of some but not other cells in the C57BL/6J mouse and in Sprague-Dawley rats also express GCs with terminal α-GalNAc residues (Fig. 4). In contrast, a number of other species and strains of mice lacked cilia staining for terminal α-GalNAc, providing evidence that the expression of some cell surface GCs is genetically regulated (Spicer et al., 1987). Similar species differences in the lectin-binding properties of the ciliary glycocalyx have been

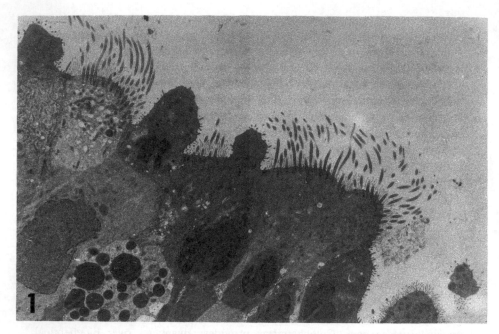

Figure 1 Mouse trachea fixed with 2% glutaraldehyde and incubated with peanut agglutinin (PNA) conjugated to horseradish peroxidase (HRP) prior to embedment. No staining is present on the surface of secretory or ciliated cells (UA-LC counterstain, × 4,000).

reported for the rat, guinea pig, and hamster (Geleff et al., 1986). These findings suggest that the surface glycocalyx of cilia is composed of relatively simple O-glycosidically linked side chains in various stages of completion. The challenge remains of relating the structure of this ciliary GC to the specialized properties required at this surface.

Another problem relates to characterizing the GCs that form the apical glycocalyx of serous and mucous secretory cells lining the luminal surface of the airways. Differences in the staining properties of carbohydrate moieties at the apical surface of ciliated cells and serous secretory cells have been described using more classic histochemical staining reactions and more recently with the lectin methods (Geleff et al., 1986). However, since exocytosed secretions often adhere tightly to the luminal surface of the secretory cells (Fig. 5), it is difficult to ascertain whether staining at their surface is attributable to adherent secretory product or to GCs that are more integrally associated with the apical plasmalemma. Experience with other mucosal surfaces such as that lining the gastrointestinal tract, where the mucous surface coat is more easily removed, suggests that the

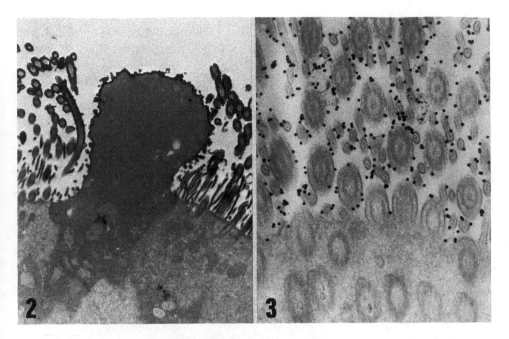

Figure 2 Region similar to that in Figure 1 but exposed to sialidase (neuraminidase) prior to incubation with PNA-HRP conjugate. After cleavage of sialic acid, the newly exposed penultimate β-Gal residues on the apical surface of secretory cells and on cilia are stained intensely (pre-embedment staining, × 10,000).

Figure 3 Surface of cilia and apical microvilli on a ciliated cell bordering the lumen of the mouse trachea are decorated with PNA-coated colloidal gold spherules after digestion with sialidase (pre-embedment staining, × 50,000).

structure of glycoproteins forming the apical cell surface glycocalyx is similar or identical to that of the secretions that fill the cell's secretory granules or vesicles and very likely form the surface coat when exocytosed. This similarity in GC structure is of interest since an important function of cell surface GCs may be the role they play in serving as receptor sites for attachment of potential pathogens. Recent studies, mainly of gastrointestinal tract epithelium, have shown that colonization by bacteria is directly related to their ability to adhere to the glycocalyx of epithelial cells and that this attachment is mediated through lectin-like proteins on the surface of bacteria. That similar mechanisms play a role in colonization and infection of mucosal surfaces of the respiratory tract was demonstrated in a study showing that *S. pneumoniae* interacts with human nasopharyngeal cells by binding to a receptor on the epithelial cell surface containing the terminal disaccharide GlcNAc-(β-1 $\rightarrow$ 3)-Gal (Anderson et al., 1983). Thus, the thick secreted

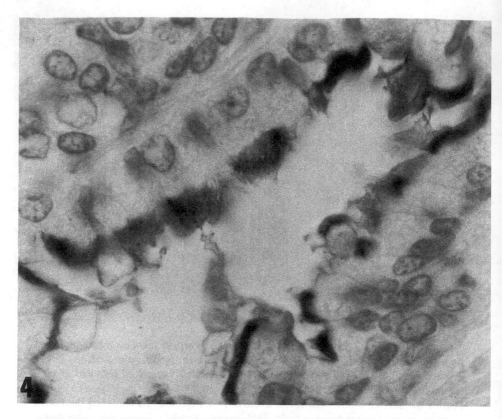

Figure 4   Cilia lining the surface of a bronchiole in a C57BL/6J mouse are stained with *Dolichos biflorus* agglutinin-HRP (DBA-HRP) conjugate (paraffin section fixed with buffered HgCl$_2$–glutaraldehyde, hematoxylin counterstain; × 800).

glycoproteinaceous coat at the surface of the respiratory tract can be envisioned as competing with the apical glycocalyx of epithelial cells for sugar-binding sites on potential pathogens, and trapping them for clearance through ciliary motion. The secretion serves, one could say, to provide receptor-like decoy material to divert pathogens from the cell surface. Differences in the structure of secretory and cell surface GCs among species could relate to known differences in a species' susceptibility or resistance to certain types of bacterial or viral infection.

In contrast to the better characterized GCs in the trachea and upper airways, comparatively little attention has been paid to the structure and function of the glycocalyx and secretory glycoproteins in the distal airways and alveoli.

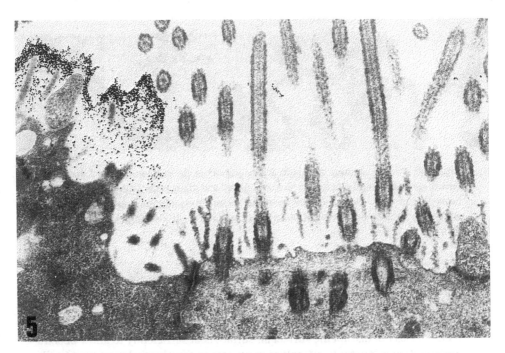

Figure 5 The apical surface coat of the secretory cell at left is labeled heavily with DBA–gold complex, whereas cilia and the apical microvilli of an adjacent ciliated cell are unreactive (pre-embedment staining of rat trachea, × 30,000).

Light and electron microscopic studies have established the presence of a negatively charged surface coat lining the alveolar surface (Luke and Spicer, 1966; Meban, 1972; O'Hare, 1974). Katsuyama and Spicer (1977) showed further that granular pneumocytes (type II cells) in the rat differ from membraneous pneumocytes (type I cells) in containing a layer of sulfated GC overlaying their sialoglycoconjugate-rich surface coat. More recently, a report by Simionescu and Simionescu (1983) has challenged the presumably established presence of anionic sites on the apical surface of type I cells. This discrepancy serves to illustrate the importance of taking into account the fixation and staining methods in the interpretation of results. In all previous studies exposure to cationic reagents was performed after fixation or simultaneously with the fixative in the case of ruthenium red. In contrast, Simionescu and Simionescu (1983) instilled cationic ferritin directly into the trachea after extensive washing with a physiological buffer but prior to fixation. Under these conditions, the cationic ferritin was bound heavily at the apical surface of type II cells but failed to react with type I cells. Together, these results suggest that fixation by immersion, vascular perfusion, or airway infusion,

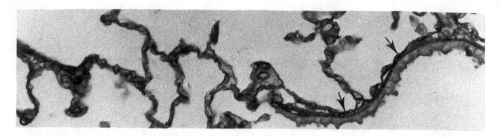

Figure 6 Heavy staining for sialic acid is present at the surface of type I and type II pneumocytes and on the apical surface of ciliated and Clara cells lining a rat respiratory bronchiole at right. Note also staining in the basement membrane region underlying the respiratory bronchiole (arrows) (*Limax flavus* agglutinin–HRP conjugate [LFA-HRP]. × 400).

regardless of the extent to which the procedure preserves the continuous polysaccharide-rich surface coat (Bignon et al., 1976), allows access of staining reagents to the glycocalyx surface coat of all alveolar lining cells. In contrast, it would appear that rinsing with physiological solutions and staining prior to fixation does not disrupt the surfactant surface coat sufficiently to allow penetration of cationic ferritin to sialoglycoconjugates in the glycocalyx of type I cells. The cationic ferritin can, however, react with the sulfated GC or more abundant sialoglycoconjugate at the surface of type II cells, presumably because it extends into and through the epiphase.

Faraggiana et al. (1986) have recently demonstrated the presence of GC with terminal sialic acid and penultimate Gal-($\beta1\rightarrow3$)-GalNAc on the alveolar surface of type I and II cells in the adult human lung. We have confirmed and extended this observation (Schulte and Spicer, 1986; Schulte et al., 1986), showing that in mice, rats, and humans the respiratory surface of Clara cells and type I and II pneumocytes is coated with GCs containing an acidic trisaccharide similar to that found on the surface of cilia (Figs. 6–15). Furthermore, in the rat but

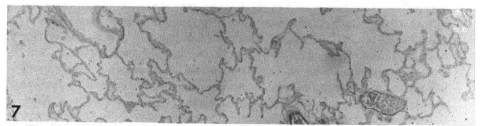

Figure 7 Section similar to that in Figure 6 but incubated with sialidase prior to staining with LFA-HRP. Staining is abolished in all sites (× 400).

Figure 8 An occasional histiocyte in rat lung shows weak reactivity with PNA-HRP (arrows). The respiratory bronchiole at left fails to stain (PNA-HRP, × 400).

not mouse or humans, type II cells differed from type I cells in containing an apically localized GC with heavily O-acetylated sialic acid, thus providing a specific marker for Clara cells and type II pneumocytes in the rat (Fig. 10).

The apical glycocalyx of Clara cells and of type II cells in rats and mice show considerable similarity in their lectin-binding profile (Schulte and Spicer, 1986). The surface of type I cells, however, differs from that of type II cells and Clara cells in reacting strongly with the lectin from soybean (Figs. 16, 17). In contrast, the alveolar surface of type II cells stains much more intensely than that of type I cells with *Pisum sativum* agglutinin (Fig. 18). Brandt (1982) has reported that the lectins from *Maclura pomifera* and *Ricinus communis* bind selectively in vivo to rat type II and type I cells, respectively, and Dobbs et al. (1985) have shown that these lectins are useful for studying the transition of type II cells to those having characteristics more similar to type I cells in culture.

A comparison of lectin-binding sites in human lung is not yet complete because of complexities introduced by the expression of isologous A, B, and H blood group antigens with which many of the lectins bind (Laden et al., 1984; Schulte et al., 1985b). Preliminary studies using lectins specific for the terminal

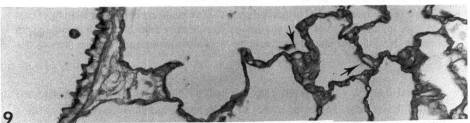

Figure 9 Section adjacent to that in Figure 8, incubated with sialidase prior to staining with PNA-HRP. The alveolar surface of type I and type II pneumocytes (arrows) is stained heavily, as is the surface of ciliated and Clara cells lining the respiratory bronchiole at left. Note also the heavily stained histiocyte adjacent to the respiratory bronchiole (× 400).

Figure 10 Section near to that illustrated in Figure 9 subjected to 1.0 mM periodate oxidation for 15 min prior to incubation with sialidase and staining with PNA-HRP. The alveolar surface of type II pneumocytes and the apical surface of Clara cells in the respiratory bronchiole at left continue to stain selectively because of the presence of periodate-resistant O-acetylated sialic acid (× 400).

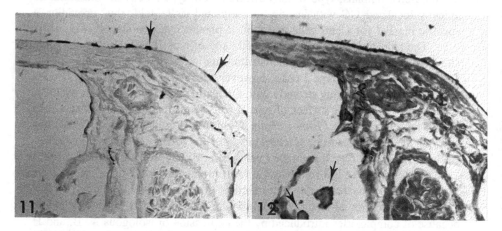

Figure 11 The apical surface of some ciliated cells (arrows) lining a human respiratory bronchiole is stained intensely with PNA-HRP. Intervening Clara cells do not stain (buffered HgCl$_2$-glutaraldehyde, paraffin embedment, × 400).
Figure 12 Section adjacent to that in Figure 11 shows uniform PNA-HRP staining of cilia and at the apical surface of Clara cells lining the respiratory bronchiole after digestion with sialidase. Note also the intense staining at the surface of alveolar macrophages (arrows) (× 400).

sugar residues in A and B antigens, as well as monoclonal antibodies directed against the terminal sugar sequences in A, B, and H antigens, have demonstrated the expression of isologous blood group antigens by Clara and type II cells in the human lung (unpublished results; Figs. 19–25).

    The above observations may prove to be of significance for several reasons. First, they point to the similarity of GC structure on the apical glycocalyx of Clara

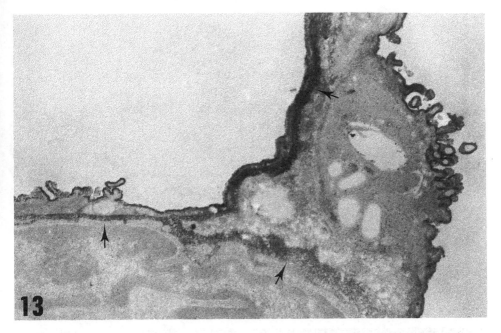

Figure 13 Vibratome section of rat lung stained free-floating with LFA-HRP. The sialoglycoconjugate–rich apical surface of type I and type II pneumocytes is stained intensely. The basement membrane subtending type I and type II pneumocytes (arrows) is also heavily stained ($\times$ 20,000).

and type II cells in a given species, thus providing evidence for an as yet unproven developmental or functional relationship between these two cell types. Second, the lectin cytochemical methods have provided markers selective for type I and type II pneumocytes and Clara cells in rodents and humans. Although the functional significance of structural differences in GCs associated with the alveolar surface of type I and II pneumocytes remains to be determined, selective markers for these cells should aid in investigations of changes associated with differentiation and disease. Third, the finding that vascular endothelial cells in the rodent lung stain specifically with an isolectin from *Griffonia simplicifolia* (GSA I-B$_4$) (Fig. 26) and those in human lung stain specifically with UEA I (Holthofer et al., 1982) has provided a histochemical approach to evaluating angiogenesis during normal lung development and in pathological processes.

Another potentially informative observation is the selective staining of basal cells in rodent and human lung with lectins and in human lung with monoclonal antibodies against isologous A, B, or H antigens. In the mouse and rat trachea and bronchi, the plasmalemma and the Golgi region of basal cells stain intensely

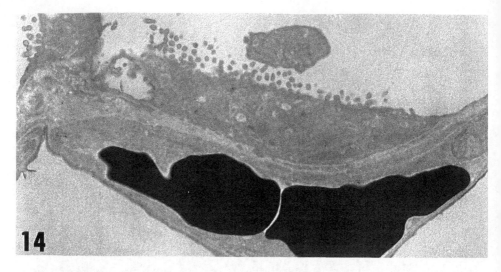

Figure 14   No staining is present in finely minced fragments of rat lung fixed in 2% glutaraldehyde and exposed to PNA-HRP prior to embedment (× 10,000)

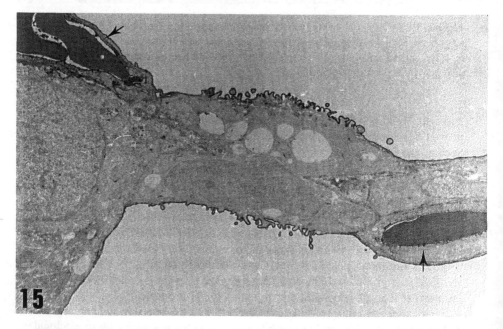

Figure 15   Following incubation with sialidase, the surface of type I and type II cells reacts strongly with PNA-HRP. Sialoglycoconjugates on the surface of endothelial cells (arrows) also react with PNA-HRP after enzymatic cleavage of sialic acid (× 8,000).

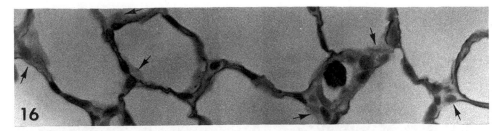

Figure 16 The surface of type I pneumocytes in the mouse lung stains intensely with soybean agglutinin-HRP conjugate (SBA-HRP) whereas type II cells (arrows) fail to stain (× 600).

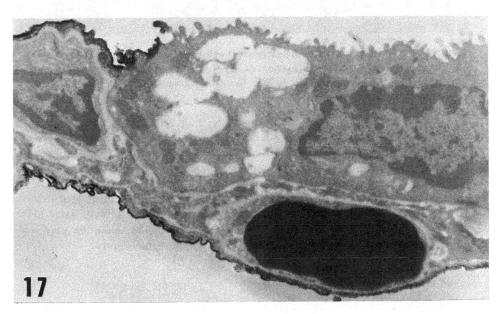

Figure 17 The apical surface of a mouse type II cell is unreactive with SBA-HRP, whereas the surface of adjacent type I cells stains intensely (pre-embedment staining, × 15,000).

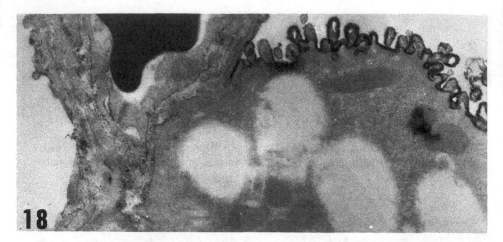

Figure 18 Apical surface of a mouse type II cell is stained intensely with the lectin from *Pisum sativum* in a minced fragment of lung from same mouse as shown in Figure 17. In contrast, staining is much weaker at the surface of adjacent type I pneumocytes (pre-embedment staining, × 25,000).

with GSA I-B$_4$ (Fig. 27) specific for terminal α-Gal. In several human specimens, basal cells in at least some bronchi were stained selectively with antisera against isologous A, B, or H antigens and with blood group specific lectins (Figs. 28–33). Whether this selective staining of basal cells can be related to a developmental or some functional property remains to be determined, but it is clear that the discovery of selective markers for basal cells will be of value in exploring their role in normal lung development and function and in the repair of airway epithelium in response to disease or experimental injury.

C. Extracellular Glycoconjugates

Carbohydrate cytochemistry has provided important information on the presence and distribution of GAGs and glycoproteins in the basement membrane and extracellular connective tissue matrix of the lung. The presence of acidic GCs in the alveolar basement membrane was demonstrated using dialyzed iron and high-iron diamine (Katsuyama and Spicer, 1977). Evidence also has been presented that the alveolar basement membrane underlying type II cells differs from that of type I cells and that the more abundant sulfate esters in the former site may represent a cation-retaining layer (Katsuyama and Spicer, 1977). Staining with ruthenium red and correlative enzyme digestions (Vaccaro and Brody, 1979, 1981) has confirmed and extended these observations, showing distinct differences in the content of heparan sulfate proteoglycan in different laminae of the alveolar basement membrane of the adult rat.

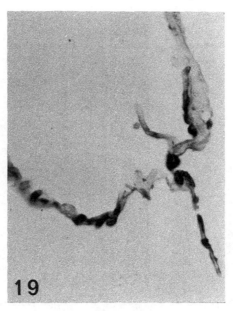

Figure 19 Type II pneumocytes in a blood group A human lung are heavily stained with *Dolichos biflorus* agglutinin-HRP conjugate (DBA-HRP), specific for terminal α-GalNAc. Type I pneumocytes lack reactivity (× 400).

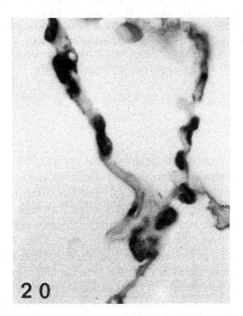

Figure 20 Type II pneumocytes in the same group A specimen as shown in Figure 19 are heavily stained with mouse monoclonal antibody against A antigen (× 400).

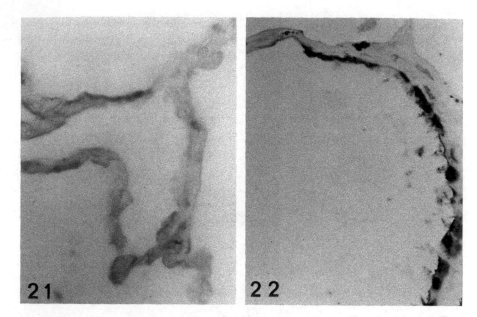

Figure 21 A section from same individual shown in Figures 19 and 20 fails to stain with *Griffonia simplicifolia* isolectin B_4-HRP conjugate (GSA I-B_4-HRP) specific for terminal α-Gal, the terminal sugar in blood group B antigen. This specimen also fails to stain with monoclonal antibody against B antigen ($\times$ 400).

Figure 22 Clara cells lining a respiratory bronchiole in a group A specimen are heavily stained with monoclonal antibody against A antigen. ($\times$ 400).

A study of the mouse lung showed that cuprolinic blue at a critical electrolyte concentration stains filaments in the basement membrane underlying type I cells (Van Kuppevelt et al., 1984 a, b). The basement membrane under type II cells also contained cuprolinic-blue- positive filaments but their length was shorter than that of type I cells. It has also been demonstrated that the different laminae of the basement membrane underlying type I and II cells differ in their content and distribution of high-iron-diamine-reactive sulfate esters (Sannes, 1984). The finding of microdomains in different laminae and regions of the basement membrane underlying different cell types in the adult takes on greater significance from the fact that these microdomains are not present in neonatal rats but develop rapidly after birth (Brody et al., 1982).

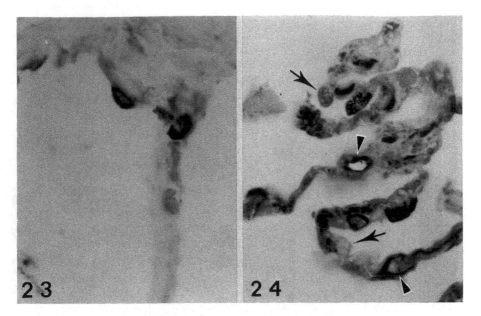

Figure 23 Type II pneumocytes are stained heavily with DBA-HRP in a blood group AB specimen (× 400).

Figure 24 Type II pneumocytes in the same group AB specimen shown in Figure 23 show variable staining with monoclonal antibody against B antigen. Some cells are stained heavily whereas others lack or show only weak reactivity (arrows). Note also heavy staining of vascular endothelium (arrowheads) (× 400).

A significant contribution of the lectin methods may be found in their ability to identify extracellular components of basement membrane and connective tissue. Investigations in this laboratory have shown that a component of the alveolar basement membrane in rodents (Fig. 13) and humans reacts with a sialic-acid-specific lectin from the slug *Limax flavus*. A constituent of the lung connective

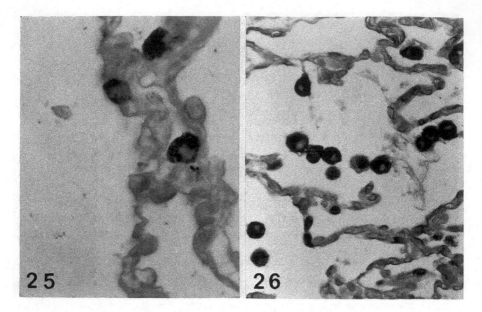

Figure 25   Type II pneumocytes in the same blood group AB specimen illustrated in Figures 23 and 24 show variable reactivity with GSA I-B$_4$-HRP, but vascular endothelium is unreactive ($\times$ 400).

Figure 26   Alveolar macrophages in rat lung react strongly with GSA I-B$_4$-HRP, as do vascular endothelial cells ($\times$ 400).

tissue and/or basement membranes, on the other hand, stains intensely with the lectins from *Pisum sativum* and *Aleuria aurantia* (Fig. 34), indicating the presence of glycoprotein, most probably of the N-glycosidic complex type. Glycoproteins that contain oligosaccharides of the complex N-acetyllactosaminic type and, surprisingly, peptide chains with collagenous and noncollagenous amino acid sequences (Bhattacharyya and Bell, 1984) have been isolated by lung lavage from normal subjects and patients with alveolar proteinosis. It appears that glycoprotein containing or complexed with collagen is substantially altered in alveolar connective tissue disease. Thus, ultrastructural histochemical studies using lectins

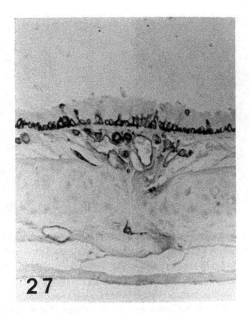

Figure 27 Basal cells lining the mouse trachea are stained heavily with GSA I-B₄-HRP (× 200).

such as *Pisum sativum* specific for the core region of complex type side chains may help in elucidating changes in connective tissue components during pulmonary fibrosis.

D. Pathological Changes and Glycoconjugates

Carbohydrate histochemistry may ultimately be of value in advancing our understanding of molecular events associated with specific disease processes. Little such effort has been made as yet, however, in the lung. In one example, con A and a lectin from *Ricinus communis* have been shown to react more strongly with dysplastic and neoplastic lesions than with normal epithelium or simple squamous metaplasia in both rats and humans (Shiba et al., 1984).

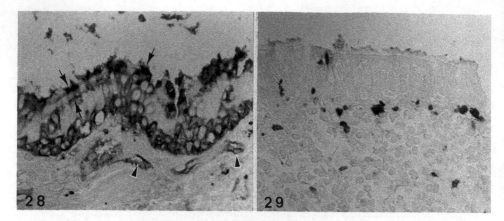

Figure 28   Basal cells undergoing proliferation in a bronchiole of a blood group A human lung are stained intensely with monoclonal antibodies against A antigen. Note also the variable staining in the Golgi region and at the surface of ciliated cells (arrows) and the staining of vascular endothelial cells (arrowheads) (× 400).

Figure 29   Section from a similar region in the same specimen as in Figure 28 shows no reactivity with monoclonal antibody against B antigen. Stromal mast cells are stained intensely because of their affinity for the avidin–biotin–horseradish peroxidase complex (× 400).

Findings in this laboratory have provided evidence for the potential usefulness of the lectin techniques for evaluating histopathological changes in degenerative, inflammatory, and neoplastic disease. Sections of human lung from several patients with centrilobular emphysema showed a generalized loss of sialic acid from the surface of alveolar pneumocytes as evidenced by increased reactivity for PNA. This loss was greater in type I than in type II cells: most type II cells failed to show increased PNA binding (Figs. 35, 36). In adjacent regions of the same specimens undergoing repair, type II cells showed heterogeneity in PNA binding sites, but only a fraction were reactive and the majority of the hypertrophic and hyperplastic type II cells were unstained with PNA in these regions.

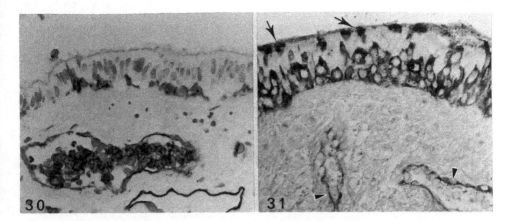

Figure 30 Basal cells in a bronchiole with no obvious signs of hyperplasia from a different group A individual stain heavily with monoclonal antibody against A antigen. Note lack of staining on the cilia surface and intense linear staining of vascular endothelial cells (× 400).

Figure 31 Bronchiole from a blood group O specimen shows intense staining of basal cells in a region of basal cell hyperplasia as well as reactivity in what appears to be stored secretions of surface secretory cells (arrows). Vascular endothelial cells are also stained (arrowheads) (anti-H antigen, × 400).

Changes similar to those observed in emphysema were seen in areas of the lung affected by bronchopneumonia and interstitial pneumonitis (Fig. 37). The selective loss of sialic acid at the surface of type I and not type II cells could be attributed either to a decrease of sialyl transferase activity in the injured type I cells or to cleavage of sialic acid from the surface of only type I cells by bacterial neuraminidase activity. In the latter event, a plausible explanation for the relatively high level of cleavage at the surface of type I but not type II cells may be the protection afforded against bacterial sialidase by a sulfated GC similar to that coating the sialoglycoconjugate-rich glycocalyx of the type II cells in the rat, as discussed earlier.

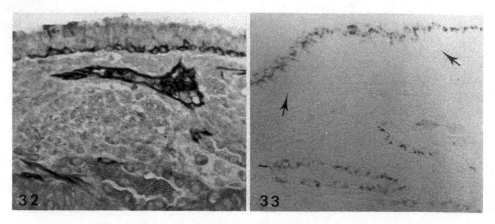

Figure 32   Basal cells in a bronchiole of a type O specimen are stained intensely with *Ulex europeus*-HRP conjugate, as is the surface of the vascular endothelium (× 400).
Figure 33   Section similar to that in Figure 32 taken from same specimen shows punctate staining in the Golgi region of ciliated cells lining a bronchiole whereas basal cells are unreactive (arrows) (DBA-HRP, × 400).

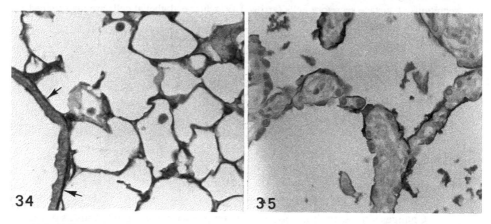

Figure 34   A lectin from *Aleuria aurantia* (orange peel fungus) conjugated to HRP appears to stain connective tissue in the alveolar septa. Note the intense linear staining of the basement membrane region (arrows) subtending a respiratory bronchiole (× 400).
Figure 35   Type II cuboidalization and subpleural scarring in patient with emphysema. Note the intense staining at the surface of type I pneumocytes, whereas adjacent type II cells lack reactivity (PNA-HRP, × 400).

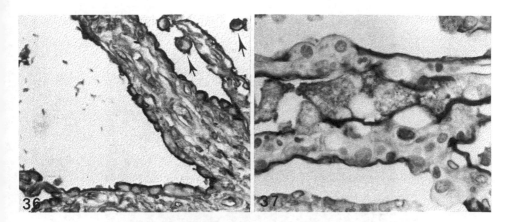

Figure 36 Region similar to that shown in Figure 35 but incubated with sialidase shows intense PNA-HRP staining at the alveolar surface of both type I and type II pneumocytes as well as at the surface of alveolar macrophages (arrows) (× 400).
Figure 37 Type II cells undergoing hypertrophy and hyperplasia in a area of interstitial pneumonitis show little or no reactivity with PNA-HRP. In contrast, type I cells show uniform strong surface staining (× 800).

An apparently comparable observation has established that infection with *P. carinii* in rats is characterized by attachment of this organism to type I alveolar cells and that the attachment is increased in corticosteroid-treated rats. The steroid treatment also caused decreased binding of cationized ferritin and ruthenium red at the surface of type I cells, suggesting that the loss of sialic acids was associated with the increased susceptability to infection (Yoneda and Walzer, 1984). It has also been reported that the surfaces of alveolar cells in premature neonates contain binding sites for PNA (Faraggiana et al., 1986; Meban, 1986), whereas such sites are capped by sialic acid in the normal adult lung. Such findings may be significant to the pathogenesis of infection, since premature infants, as well as adult patients with respiratory distress syndrome, are at much greater risk for infection by *P. aeruginosa, E. coli*, and *Streptococci*. Studies are needed to assess the structure of cell surface GCs that mediate the attachment of these pathogenic microorganisms.

Two cases of squamous cell carcinoma, one of which was relatively undifferentiated and the other highly differentiated, were examined with lectin cytochemistry. All of the tumor cells in both neoplasms expressed sialic acid at their surface, as documented by staining with *Limax flavus* agglutinin and appearance of PNA affinity following digestion with sialidase (Figs. 38, 39). The only PNA binding associated with these carcinomas in undigested sections was

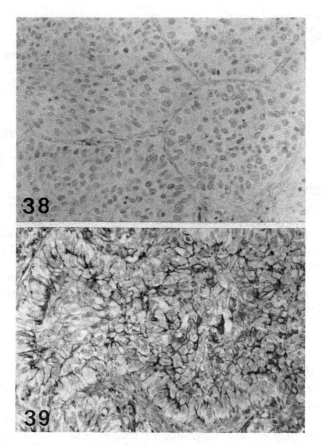

Figure 38 A relatively undifferentiated squamous cell carcinoma lacks reactivity with PNA-HRP (× 160).
Figure 39 Malignant cells in the same tumor as in Figure 38 show reactivity with PNA-HRP following incubation with sialidase (× 160).

variable light to heavy reactivity in regions of epithelial keratinization in the well-differentiated tumor. A striking change occurred in the basal cells lining bronchi adjacent to the tumors. These basal cells differed from normal basal cells in staining intensely with the sialidase–PNA sequence. In addition, areas of squamous metaplasia stained strongly with this procedure.

Lectin cytochemistry showed a distinctive property not otherwise detectable in a mucus-secreting adenocarcinoma. This tumor showed marked differences in the lectin-binding profile between regions of tumor in the solid tumor mass and adjacent papillary projections (Fig. 40). In the papillary area, a generalized

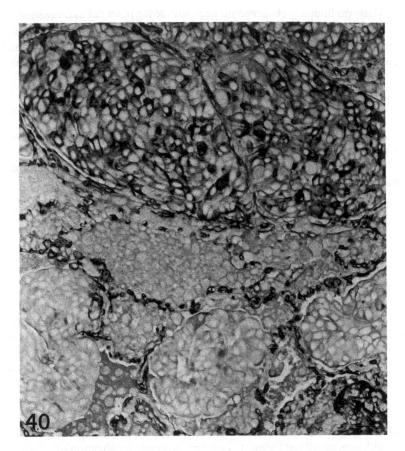

Figure 40 Malignant cells in a mucus-producing adenocarcinoma from a blood group O specimen show striking variability in their staining with UEA I-HRP. Cells in the solid tumor mass at top show variable content of fucosylated GC, whereas those in papillary projections invading alveoli at bottom show an almost uniform lack of UEA-I-reactive secretory or cell surface fucose residues. Capillaries outlining original alveoli are stained intensely (× 160).

loss of affinity for UEA I contrasted with strong staining in the region with a more medullary growth pattern. The loss of binding with UEA I correlated with loss of H substance from papillary but not medullary areas in an adjacent immunostained section from this blood type O specimen. The loss of carbohydrate at the surface of adenocarcinoma cells accords with reports suggesting that loss of isologous A, B, and H blood group antigen is a generalized phenomenon in highly differentiated tumors.

These preliminary observations demonstrate the potential usefulness of lectin histochemistry for evaluating histopathological changes in a wide range of lung diseases. Whether or not the lectin techniques will eventually have diagnostic and/or prognostic applications awaits the outcome of much more comprehensive studies.

Acknowledgments

The authors acknowledge Christina Smith for technical assistance and Debra L. Fairfull and Pamela D. Kelley for editorial assistance.

This work was supported in part by grant HL-29775 from the National Institutes of Health.

References

Allison, R. T. (1987). The effect of various fixatives on subsequent lectin binding to tissue sections. *Histochem. J.* **19**:65–74.

Alroy, J., Ucci, A. A., and Pereira, M. E. A. (1984). Lectins: histochemical probes for specific carbohydrate residues. In *Diagnostic Immunohistochemistry*, Vol 2. Edited by R. A. DeLellis. New York, Masson, p. 67.

Anderson, H., Leffler, H., Magnusson, G., and Svanborg Eden, C. (1983). Molecular mechanisms of adhesions of *Streptococcus pneumoniae* to human oropharyngeal epithelial cells. *Scand. J. Infect. Dis.*, Supp. **39**:45–47.

Avrameas, S. (1969). Coupling of enzymes to proteins with glutaraldehyde. Use of the conjugates for the detection of antigens and antibodies. *Immunocytochemistry* **6**:43–52.

Baskin, D. G., Erlandsen, S. L., and Parsons, J. A. (1979). Immunocytochemistry with osmium-fixed tissue. I. Light microscopic localization of growth hormone and prolactin with the unlabeled antibody method. *J. Histochem. Cytochem.* **27**:867–872.

Bendayan, M., Nanci, A., and Kan, F. W. K. (1987). Effect of tissue processing on colloidal gold cytochemistry. *J. Histochem. Cytochem.* **35**:983–996.

Bernhard, W., and Avrameas, S. (1971). Ultrastructural visualization of cellular carbohydrate components by means of concanavalin A. *Exp. Cell. Res.* **64**:232–236.

Bhattacharyya, S. H., and Bell, D. Y. (1984). Characterization of two glycoproteins isolated from lung lavage fluid of normal humans and from patients with two pulmonary diseases. *Inflammation* **8**:407–415.

Bignon, J., Jaubert, F., and Jaurand, M. C. (1976). Plasma protein immunocytochemistry and polysaccharide cytochemistry at the surface of alveolar and endothelial cells in the rat lung. *J. Histochem. Cytochem.* **24**:1076–1084.

Boat, T. F., and Cheng, P. W. (1980). Biochemistry of airway mucus secretions. *Fed. Proc.* **39**:3067–3074.

Brandt, A. E. (1982). Cell surface saccharides of rat lung alveolar type I and type II cells. *Fed. Proc.* **41**:755 (abs.).

Brody, J. S., Vaccaro, C. A., Gill, P. J., and Silbert, J. E. (1982). Alterations in alveolar basement membrane during postnatal lung growth. *J. Cell Biol.* **95**:394–402.

Chakrin, L. W., Baker, A. P., Spicer, S. S., Wardell, J. R. Jr., DeSanctis, N., and Dries, C. (1972). Synthesis and secretion of macromolecules by canine trachea. *Am. Rev. Respir. Dis.* **105**:368–381.

Chandrasekaran, E. V., Surjit, S. R., Davila, M., and Mendicino, J. (1984). Structures of the oligosaccharide chains in swine trachea mucin glycoproteins. *J. Bio. Chem.* **259**:12908–12914.

Cooper, H. S. (1984). Lectins as probes in histochemistry and immunohistochemistry: the peanut (*Arachis hypogaea*) lectin. *Hum. Pathol.* **15**:904–906.

Damjanov, I. (1987). Biology of Disease: lectin cytochemistry and histochemistry. *Lab. Invest.* **57**:5–20.

Danscher, G. (1981). Localization of gold in biological tissues. A photochemical method for light and electron microscopy. *Histochemistry* **71**:81–88.

Debray, H., Decout, D., Strecker, G., Spik, G., and Montreuil, J. (1981). Specificity of twelve lectins towards oligosaccharides and glycoproteins related to N-glycosyl proteins. *Eur. J. Biochem.* **117**:41–55.

De Bruyn, P. P. H., Michaelson, S., and Becker, R. P. (1978). Nonrandom distribution of sialic acid over the cell surface of bristle-coated endocytic vesicles of the sinusoidal endothelium cells. *J. Cell Biol.* **78**:379–389.

Dobbs, L. G., Williams, M. C., and Brandt, A. E. (1985). Changes in biochemical characteristics and pattern of lectin binding of alveolar type II cells with time in culture. *Biochim. Biophys. Acta* **846**:155–166.

Faraggiana, T., Villari, D., Jagirdar, J., and Patil, J. (1986). Expression of sialic acid on the alveolar surface of adult and fetal human lungs. *J. Histochem. Cytochem.* **34**:811–816.

Gallagher, J. T. (1984). Carbohydrate-binding properties of lectins: a possible approach to lectin nomenclature and classification. *Biosci. Rep.* **4**:621–632.

Gallagher, J. T., and Richardson, P. S. (1982). Respiratory mucus: structure, metabolism and control of secretion. *Adv. Exp. Med. Biol.* **144**:335–364.

Geleff, S., Brock, P., and Stockinger, L. (1986). Lectin binding affinities of the epithelium in the respiratory tract. A light microscopical study of ciliated epithelium in rat, guinea pig, and hamster. *Acta Histochem.* **78**:83–95.

Geoghegan, W. D., and Ackerman, G. A. (1977). Absorption of horseradish peroxidase, ovomucoid and anti-immunoglobulin to colloidal gold for the indirect detection of concanavalin A, wheat germ agglutinin and goat anti-human immunoglobin G on cell surfaces at the electron microscopic level:

a new method, theory and application. *J. Histochem. Cytochem.* **25**(11):1187–1200.

Glick, M. C., and Flowers, H. (1978). Surface membranes. In *The Glycoconjugates*, Vol. 2. Edited by M. I. Horowitz. New York, Academic Press, pp. 337–384.

Goldstein, I. J., and Hayes, C. E. (1978). The lectins: carbohydrate binding proteins of plants and animals. *Adv. Carbohydrate Chem. Biochem.* **35**:127–340.

Goldstein, I. J., and Portez, R. D. (1986). Isolation, physiochemical characterization, and carbohydrate-binding specificity of lectins. In *The Lectins: Properties, Functions and Applications in Biology and Medicine*. Edited by E. I. Liener, N. Sharon, and I. J. Goldstein. Orlando, FL, Academic Press, p. 35–247.

Graham, R. C., Jr., and Karnovsky, M. J. (1966). The early stages of absorption of injected horseradish peroxidase in the proximal tubules of mouse kidney: ultrastructural cytochemistry by a new technique. *J. Histochem. Cytochem.* **14**:291–302.

Hennigar, R. A., Schulte, B. A., and Spicer, S. S. (1985). Heterogeneous distribution of glycoconjugates in human kidney tubules. *Anat. Rec.* **211**:376–390.

Hennigar, R. A., Schulte, B. A., and Spicer, S. S. (1986). Histochemical detection of glycogen using *Griffonia simplicifolia* agglutinin II. *Histochem. J.* **18**:589–596.

Holthofer, H., Virtanen, I., Kariniemi, A.-L., Hormia, M., Linder, E., and Miettinen, A. (1982). *Ulex europaeus* I lectin as a marker for vascular endothelium in human tissues. *Lab. Invest.* **47**:60–66.

Holthofer, H., Schulte, B. A., and Spicer, S. S. (1987). Expression of binding sites for *Dolichos biflorus* agglutinin at the apical aspect of collecting duct cells in rat kidney. *Cell Tissue Res.* **249**:481–485.

Hsu, S.-M., and Raine, L. (1982). Versatility of biotin-labeled lectins and avidin-biotin-peroxidase complex for localization of carbohydrate in tissue sections. *J. Histochem. Cytochem.* **30**:157–161.

Hsu, S.-M., and Ree, H. J. (1983). Histochemical studies on lectin binding in reactive lymphoid tissues. *J. Histochem. Cytochem.* **31**:538–546.

Hughes, R. C. (1975). The complex carbohydrates of mammalian cell surfaces and their biological roles. In *Essays in Biochemistry* Vol 2. Edited by P. N. Campbell and W. N. Aldridge. New York, Academic Press, pp. 1–36.

Kaissling, B. (1973). Histologische und histochemische untersuchungen un semidunned Schnitten. *Gegenbaurs. Morphol. Jahrb.* **119**:1–13.

Katsuyama, T., and Spicer, S. S. (1977). A cation-retaining layer in the alveolar-capillary membrane. *Lab. Invest.* **36**:428–435.

Katsuyama, T., and Spicer, S. S. (1978). Histochemical differentiation of complex carbohydrates with variants of the Concanavalin A–horseradish peroxidase method. *J. Histochem. Cytochem.* **26**:233–250.

Kellenberger, E., Durrenberger, M., Villiger, W., Carlemalm, E., and Wurtz, M. (1987). The efficiency of immunolabel on lowicryl sections compared to theoretical predictions. *J. Histochem. Cytochem.* **35**:959–969.

King, T. P., Brydon, L., Gooday, G. W., and Chappell, L. H. (1987). Silver enhancement of lectin–gold and enzyme–gold cytochemical labelling of eggs of the neomatode *Onchocerca gibsoni*. *Histochem. J.* **19**:281–287.

Laden, S. A., Schulte, B. A., and Spicer, S. S. (1984). Histochemical evaluation of secretory glycoproteins in human salivary glands with lectin–horseradish peroxidase conjugates. *J. Histochem. Cytochem.* **32**:965–972.

Lamblin, G., Lhermitte, M., Boersma, A., and Roussel, P. (1980). Oligosaccharides of human bronchial glycoproteins. *J. Biol. Chem.* **255**:4595–4598.

Lamblin, G., Lhermitte, M., Klein, A., Roussel, P., Van Halbeek, H., and Vliegenthart, J. F. G. (1984a). Carbohydrate chains from human bronchial mucus glycoproteins: a wide spectrum of oligosaccharide structures. *Biochem. Soc. Trans.* **12**:599–600.

Lamblin, G., Boersma, A., Lhermitte, M., Roussel, P., Mutsaers, J. H. G. M., Van Halbeek, H., and Vliegenthart, J. F. G. (1984b). Further characterization by a combined high-performance liquid chromatography/^{1}H-NMR approach, of the heterogeneity displayed by neutral carbohydrate chains of human bronchial mucins. *Eur. J. Biochem.* **143**:227–236.

Leppi, T. J., and Stoward, P. J. (1965). On the use of testicular hyaluronidase for identifying acid mucins in tissue sections. *J. Histochem. Cytochem.* **13**:406–407.

Lis, H., and Sharon, N. (1986a). Lectins as molecules and as tools. *Annu. Rev. Biochem.* **55**:35–67.

Lis, H., and Sharon, N. (1986b). Biological properties of lectins. In *The Lectins: Properties, Functions and Applications in Biology and Medicine*. Edited by E. I. Liener, N. Sharon, and I. J. Goldstein. Orlando, FL, Academic Press, pp. 266–291.

Lucocq, J. M., Berger, E. G., and Roth, J. (1987). Detection of terminal N-linked N-acetylglucosamine residues in the Golgi apparatus using galactosyltransferase and endoglucosaminidase F/peptide N-glycosidase F: adaption of a biochemical approach to electron microscopy. *J. Histochem. Cytochem.* **35**:67–74.

Lucocq, J. M., and Roth, J. (1984). Applications of immunocolloids in light microscopy. III. Demonstration of antigenic and lectin-binding sites in semithin sections. *J. Histochem. Cytochem.* **32**:1075–1083.

Lucocq, J. M., and Roth, J. (1985). Colloidal gold and colloidal silver-metallic markers for light microscopic histochemistry. In *Techniques in Immunocytochemistry*, Vol. 3. Edited by G. R. Bullock and P. Petrusz. London, Academic Press, pp. 203–236.

Luke, J. L., and Spicer, S. S. (1966). Histochemistry of surface epithelial and pleural mucins in mammalian lung. The demonstration of sialomucin in alveolar cuboidal epithelium. *Lab. Invest.* **14**:2101–2109.

Meban, C. (1972). An electron microscopic study of acid mucosubstances lining the alveoli of hamster lung. *Histochem. J.* **4**:1–8.

Meban, C. (1986). Lectin binding sites on the surfaces of the pneumocytes in human neonatal lung. *Histochem. J.* **18**:196–202.

Mochizuki, I., Setser, M. E., Martinez, J. R., and Spicer, S. S. (1982). Carbohydrate histochemistry of rat respiratory glands. *Anat. Rec.* **202**:45–59.

Montreuil, J. (1980). Primary structure of glycoprotein glycans: basis for the molecular biology of glycoproteins. *Adv. Carbohydrate Chem. Biochem.* **37**:157–223.

Nicholson, G. L. (1974). The interaction of lectins with animal cell surfaces. *Int. Rev. Cytol.* **39**:90–190.

Newman, G. R., and Hobot, J. A. (1987). Modern acrylics for post-embedding immunostaining techniques. *J. Histochem. Cytochem.* **35**:971–981.

O'Hare, K. H. (1974). Fine structural observations of ruthenium red binding in developing adult rat lung. *Anat. Rec.* **178**:267–288.

Ookusa, Y., Takata, K., Nagashima, M., and Hirano, H. (1983). Distribution of glycoproteins in normal human skin using biotinyl lectins and avidin-horseradish peroxidase. *Histochemistry* **79**:1–7.

Pigman, W. (1977). General aspects. In *The Glycoconjugates Vol 1, Mammalian Glycoproteins and Glycolipids*. Edited by M. I. Horowitz and W. Pigman. New York, Academic Press, pp. 1–14.

Rittman, B. R., and Mackenzie, I. C. (1983). Effects of histological processing on lectin binding patterns in normal mucosa and skin. *Histochem. J.* **15**:467–474.

Rose, M. C., Voter, W. A., Brown, C. F., and Kaufman, B. (1984). Structural features of human tracheobronchial mucus glycoprotein. *Biochem. J.* **222**:371–377.

Roth, J. (1983). Application of lectin–gold complexes for electron microscopic localization of glycoconjugates on thin sections. *J. Histochem. Cytochem.* **31**:987–999.

Roth, J. (1986). Post-embedding cytochemistry with gold labelled reagents: a review. *J. Microsc.* **143**:125–137.

Roth, J., Brown, D., and Orci, L. (1983). Regional distribution of N-acetyl-D-galactosamine residues in the glycocalyx of glomerular podocytes. *J. Cell. Biol.* **96**:1189–1196.

Roth, J., Lucocq, J. M., and Charest, P. M. (1984). Light and electron microscopic demonstration of sialic acid residues with the lectin from *Limax flavus*: a cytochemical affinity technique with the use of fetuin-gold complexes. *J. Histochem. Cytochem.* **32**: 1167–1176.

Sachdev, G. P., Myers, F. J., Horton, F. O., Fox, O. F., Wen, G., Rogers, R. M., and Carubelli, R. (1980). Isolation, chemical composition and properties of the major mucin component of normal human tracheobronchial secretion. *Biochem. Med.* **24**:82–94.

Sannes, P. N. (1984). Differences in basement membrane-associated microdomains of type I and type II pneumocytes in the rat and rabbit lung. *J. Histochem. Cytochem.* **32**:827–833.

Sato, M., and Muramatsu, T. (1985). Reactivity of five N-acetyl-galactosamine recognizing lectins with preimplantation embryos, early postimplantation embryos and teratocarcinoma cells of the mouse. *Differentiation* **29**:29–38.

Schulte, B. A. (1987). Genetic and sex-related differences in the structure of submandibular glycoconjugates. *J. Dent. Res.* **66**:442–450.

Schulte, B. A., Harley, R. A., and Sens, M. A. (1986). Histochemical characterization of glycoconjugates in normal and diseased human lung. *Proc. R. Microsc. Soc.* **21**:577 (abs).

Schulte, B. A., Miller, R. L., and Spicer, S. S. (1985a). Lectin histochemistry of secretory and cell surface glycoconjugates in the ovine submandibular gland. *Cell. Tissue Res.* **240**:57–66.

Schulte, B. A., Rao, K. P. P., and Spicer, S. S. (1985b). Histochemical examination of glycoconjugates of epithelial cells in the human fallopian tube. *Lab. Invest.* **52**:207–219.

Schulte, B. A., and Spicer, S. S. (1983a). Light microscopic histochemical detection of terminal galactose and N-acetylgalactosamine residues in rodent complex carbohydrates using a galactose oxidase–Schiff sequence and peanut lectin–horseradish peroxidase conjugate. *J. Histochem. Cytochem.* **31**:19–24.

Schulte, B. A., and Spicer, S. S. (1983b). Light microscopic histochemical detection of sugar residues in secretory glycoproteins of rodent and human tracheal glands with lectin–horseradish peroxidase conjugates and the galactose oxidase–Schiff sequence. *J. Histochem. Cytochem.* **31**:391–403.

Schulte, B. A., and Spicer, S. S. (1983c). Light microscopic detection of sugar residues in glycoconjugates of salivary glands and the pancreas with lectin-horseradish peroxidase conjugates. I. Mouse. *Histochem. J.* **15**:1217–1238.

Schulte, B. A., and Spicer, S. S. (1983d). Histochemical evaluation of mouse and rat kidneys with lectin–horseradish peroxidase conjugates. *Am. J. Anat.* **168**:345–362.

Schulte, B. A., and Spicer, S. S. (1984). Light microscopic detection of sugar residues in glycoconjugates of salivary glands and the pancreas with lectin-horseradish peroxidase conjugates. II. rat. *Histochem. J.* **16**:3–20.

Schulte, B. A., and Spicer, S. S. (1985). Histochemical methods for characterizing secretory and cell surface sialoglycoconjugates. *J. Histochem. Cytochem.* **33**:427–438.

Schulte, B. A., and Spicer, S. S. (1986). Differences in the structure of glycoconjugates on the surface of various cell types in the mouse and rat lung. *J. Histochem. Cytochem.* **34**:1363 (abs.).

Sharon, N. (1975). *Complex Carbohydrates: Their Chemistry, Biosynthesis, and Function*. Massachusetts, Addison-Wesley.

Sharon, N., and Lis, H. (1972). Lectins: cell-agglutinating and sugar specific proteins. *Science* **177**:949–959.

Shiba, M., Ohiwa, T., and Klein-Szanto, A. J. P. (1984). Lectin-binding sites in preneoplastic and neoplastic lesions of human and rodent respiratory tracts. *J. Natl. Cancer Inst.* **72**:43–51.

Simionescu, D., and Simionescu, M. (1983). Differentiated distribution of the cell surface charge on the alveolar–capillary unit. *Microvasc. Res.* **25**:85–100.

Skutelsky, E., and Danon, D. (1976). Redistribution of surface anionic sites on the luminal front of blood vessel endothelium after interaction with polycationic ligand. *J. Cell Biol.* **71**:232–241.

Skutelsky, E., and Roth, J. (1986). Cationic colloidal gold—a new probe for the detection of anionic cell surface sites by electron microscopy. *J. Histochem. Cytochem.* **34**:693–696.

Spicer, S. S., and Henson, J. G. (1967). Methods for localizing mucosubstances in epithelial and connective tissues. In *Series on Methods and Achievements in Experimental Pathology*, Vol. 2. Edited by E. Bajusz and G. Jamin. Basel, S. Karger, pp. 78–112.

Spicer, S. S., and Schulte, B. A. (1982). Ultrastructural methods for the localization of complex carbohydrates. *Hum. Pathol.* **13**:343–354.

Spicer, S. S., and Warren, L. (1960). The histochemistry of sialic acid containing mucoproteins. *J. Histochem. Cytochem.* **8**:135–137.

Spicer, S. S., Horn, R. G., and Leppi, T. J. (1967). Histochemistry of connective tissue mucopolysaccharides. In *The Connective Tissue*. Edited by Bernard M. Wagner. Baltimore, Williams & Wilkins, pp. 251–303.

Spicer, S. S., Chakrin, L. W., and Wardell, J. R., Jr. (1973). Respiratory mucous secretion. In *Sputum-Fundamentals and Clinical Pathology*. Edited by M. J. Dulfano. Springfield, IL, Charles C. Thomas, pp. 22–68.

Spicer, S. S., Mochizuki, I., Setser, M. E., and Martinez, J. R. (1980). Complex carbohydrates of rat tracheobronchial surface epithelium visualized ultrastructurally. *Am. J. Anat.* **158**:93–109.

Spicer, S. S., Sens, M. A., and Tashian, R. E. (1982). Immunocytochemical demonstration of carbonic anhydrase in human epithelial cells. *J. Histochem. Cytochem.* **30**:864–873.

Spicer, S. S., Chakrin, L. W., and Schulte, B. A. (1983a). Ultrastructural and cytochemical observations of respiratory epithelium and glands. *Exp. Lung Res.* **4**:137–156.

Spicer, S. S., Schulte, B. A., and Thomopoulos, G. N. (1983b). Histochemical properties of the respiratory tract epithelium in different species. *Am. Rev. Respir. Dis.* **128**:S20–S26.

Spicer, S. S., Schulte, B. A., Thomopoulos, G. N., Parmley, R. T., and Takagi, M. (1983c). Cytochemistry of complex carbohydrates by light and electron microscopy: Available methods and their application. In *Connective Tissue Diseases. International Academy of Pathology Monograph.* Edited by B. M. Wagner, P. Fleischmajer, and N. Kaufmen. Baltimore, Williams & Wilkins, pp. 163–211.

Spicer, S. S., Erlandsen, S. L., Wilson, A. C., Hammer, M., and Schulte, B. A. (1987). Genetic differences in the histochemically defined structure of oligosaccharides in mice. *J. Histochem. Cytochem.* **35**:1231–1244.

Spiro, R. G. (1969). Glycoproteins: their biochemistry, biology and role in human disease. *N. Engl. J. Med.* **281**:991–1001.

Stoward, P. J., Spicer, S. S., and Miller, R. L. (1980). Histochemical reactivity of peanut lectin-horseradish peroxidase conjugate. *J. Histochem. Cytochem.* **28**:979–990.

Taatjes, D. J., Schaub, U., and Roth, J. (1987). Light microscopical detection of antigens and lectin binding sites with gold-labelled reagents on semi-thin Lowikryl K4M sections: usefulness of the photochemical silver reaction for signal amplification. *Histochem. J.* **19**:235–245.

Thaete, L. G., Spicer, S. S., and Spock, A. (1981). Histology, ultrastructure and carbohydrate cytochemistry of surface and glandular epithelium of human nasal mucosa. *Am. J. Anat.* **162**:243–263.

Thiery, J. P. (1967). Mise en évidence des polysaccharides sur coupes fines en microscopie électronique. *J. Microsc.* **6**:987–1018.

Thomopoulos, G. N., Schulte, B. A., and Spicer, S. S. (1983a). The influence of embedding media and fixation of the post-embedment ultrastructural demonstration of complex carbohydrates. I. Morphology and periodic acid–thiocarbohydrazide–silver proteinate staining of *vicinal* diols. *Histochem. J.* **15**:763–783.

Thomopoulos, G. N., Schulte, B. A., and Spicer, S. S. (1983b). The influence of embedding media and fixation of the post-embedment ultrastructural demonstration of complex carbohydrates. II. Dialyzed iron staining. *Histochemistry* **79**:417–431.

Thomopoulos, G. N., Schulte, B. A., and Spicer, S. S. (1983c). The influence of embedding media and fixation of the post-embedment ultrastructural demonstration of complex carbohydrates. III. High iron diamine staining for sulfated glycoconjugates. *J. Histochem. Cytochem.* **31**:871–878.

Thomopoulos, G. N., Schulte, B. A., and Spicer, S. S. (1987). Postembedment staining of complex carbohydrates: influence of fixation and embedding procedures. *J. Electron Microsc. Tech.* **5**:17–44.

Vaccaro, C. A., and Brody, J. S. N. (1979). Ultrastructural localization and characterization of proteoglycans in the pulmonary alveolus. *Am. Rev. Respir. Dis.* **120**:901–910.

Vaccaro, C. A., and Brody, J. S. (1981). Structural features of the alveolar basement membrane in the adult rat lung. *J. Cell Biol.* **91**:427–437.

Virtanen, I., Kariniemi, A.-L., Holthofer, H., and Lehto, V.-P. (1986). Fluorochrome-coupled lectins reveal distinct cellular domains in human epidermis. *J. Histochem. Cytochem.* **34**:307–315.

Van Kuppevelt, T. H. M. S. M., Domen, J. G. W., Cremers, F. P. M., and Kuyper, C. M. A. (1984a). Staining of proteoglycans in mouse lung alveoli. I. Ultrastructural localization of anionic sites. *Histochem. J.* **16**:657–669.

Van Kuppevelt, T. H. M. S. M., Domen, J. G. W., Cremers, F. P. M., and Kuyper, C. M. A. (1984b). Staining of proteoglycans in mouse lung alveoli. II. Characterization of the cuprolinic blue-positive, anionic sites. *Histochem. J.* **16**:671–686.

Willingham, M. C., and Yamada, S. S. (1979). Development of a new primary fixative for electron microscopic immunocytochemical localization of intracellular antigens in cultured cells. *J. Histochem. Cytochem.* **27**:947–960.

Woodward, H., Horsey, B., Bhavanandan, V. P., and Davidson, E. A. (1982). Isolation, purification, and properties of respiratory mucus glycoproteins. *Biochemistry* **21**:694–701.

Wu, A. M., and Herp, A. (1985). A table of lectin carbohydrate specificities. In *Lectins*, Vol. 4. Edited by T. C. Bog-Hansen and G. A. Spengler. Berlin, Walter de Gruyter, pp. 629–636.

Yamada, K. (1974). The effect of digestion with chondroitinases upon certain histochemical reactions of mucosaccharide-containing tissues. *J. Histochem. Cytochem.* **22**:266–275.

Yeager, H. (1971). Tracheobronchial secretions. *Am. J. Med.* **50**:493–509.

Yoneda, K., and Walzer, P. D. (1984). The effect of corticosteroid treatment on the cell surface glycocalyx of the rat pulmonary alveolus: relevance to the host–parasite relationship in *Pneumocystis carinii* infection. *Br. J. Exp. Pathol.* **65**:347–354.

8

Morphometry: Stereological Theory and Practical Methods

EWALD R. WEIBEL

University of Berne
Berne, Switzerland

I. Introduction

Pulmonary physiologists have often felt the need to interpret their data in terms of structural models of some sort, more often than not based on crude visions of the morphological design of the lung, its airways, blood vessels, and tissue elements. Morphologists have, on the other hand, used their microscopes to unravel the lung's complicated tissue design, describing some 24 cell types in great detail, for example.

The virtue of a morphometric approach to the study of the lung is that design and function can be brought into a common perspective. Any attempt to analyze the interdependence of structure and function requires a set of physiological and morphological data that are quantitative and sufficiently reduced to essential and compatible parameters, in order to permit statistical correlations and model calculations. In general, physiological data are of this nature; taking a morphometric approach allows us to generate such data also for the structural characteristics of the lung.

The basis for a sound morphometric approach is threefold. First, a very thorough knowledge of the fine structure of the organ of interest, the lung, is essential (Weibel, 1984a, 1985, 1986; Weibel and Taylor, 1987; Burri, 1986).

This means that any morphometric study must be preceded by an in-depth qualitative study of the structure; this is particularly important when approaching alterations due to disease or experimental manipulation (Bachofen and Weibel, 1987). Second, the reduced data set that one wants to obtain by morphometry must be defined on the basis of models that describe the structure–function relationship, lest the effort be without real meaning and value. Third, methods must be available that produce morphometric data that describe the true structure: it does not suffice to take some measurements on histological sections, even if it is done by computers; such measurements must be transformed to be meaningful for the spatial structure. This chapter presents this approach, particularly to demonstrate and discuss some useful models and the stereological methods needed to estimate the structural parameters of these models.

II. Morphology and Models: Defining Functionally Relevant Morphometric Parameters

The first morphometric question that is commonly asked is about the number of alveoli making up the lung. Not only is this, seemingly basic, figure not easy to obtain, it is also not very useful in attempts to find a morphometric basis for lung function. To find the morphometric variables of physiological interest one must consider the functional situation with which the study is concerned. This is best done by attempting to draw up a model that establishes the relations between physiological and design characteristics.

A. Model for Pulmonary Gas Exchange: Diffusing Capacity

In respiration physiology, the capacity of the lung for O_2 transfer from air to blood is measured by the pulmonary diffusing capacity for O_2, D_{LO_2}, which has been defined by Bohr (1909) as

$$D_{LO_2} = \dot{V}_{O_2}/(PA_{O_2} - \overline{Pc_{O_2}}) \tag{1}$$

where $\dot{V}_{O_2}$ is O_2 consumption, and PA_{O_2} and Pc_{O_2} are O_2 partial pressures in alveolar air and capillary blood, respectively. Since D_{LO_2} is the conductance of the lung for O_2 diffusion it is often thought to be directly proportional to alveolar surface area, S_A, and inversely proportional to barrier thickness, τ, according to Fick's law:

$$D_{LO_2} = K_{O_2} \cdot S_A/\tau \tag{2}$$

where K_{O_2} is the permeation coefficient for O_2 in the barrier tissue. However, this model is too primitive. Oxygen uptake by diffusion across the barrier will evidently also be influenced by the presence and arrangement of blood in the

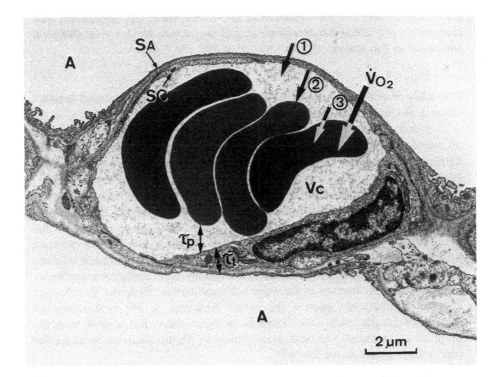

Figure 1 Electron micrograph of human lung capillary explains morphometric model for calculating diffusing capacity (D_{LO_2}). Flow of O_2 ($\dot{V}_{O_2}$) has to traverse in sequence three resistances: (1) tissue barrier, (2) plasma barrier, and (3) erythrocyte interior. Scale marker = 2 μm.

alveolar capillaries: the capillary surface area should be matched as closely as possible to the alveolar surface area for this large surface to be truly usable as a gas-exchanging surface. Furthermore the quantity of blood present on the unit area of alveolar surface will determine the amount of O_2 that can be bound per unit time (Roughton and Forster, 1957). To account for this we have chosen to subdivide the path of an O_2 molecule from alveolar air to its hemoglobin-binding site into three parts (see Fig. 1):

1. The air–blood barrier made of tissue
2. The plasma layer separating the erythrocyte from the tissue
3. The path within the erythrocyte

These three paths offer three distinct resistances to O_2 diffusion, and since they are in series the total resistance is their sum. For convenience we shall write below expressions for the conductance (D) in the tissue (t), plasma (p), and erythrocyte (e); since the resistance is 1/D we obtain for the total diffusion resistance in this system

$$D_L^{-1} = D_t^{-1} + D_p^{-1} + D_e^{-1} \tag{3}$$

It is easily seen that the reciprocal of this sum, namely D_L, is the total conductance of the lung or its "diffusing capacity" (Weibel, 1970/71).

It now remains to find expressions for the three partial conductances based on morphometric information obtainable on histological specimens. The tissue barrier and the plasma layer are sheets of a certain thickness τ separating two adjoining spaces over an area S; D_t and D_p can hence be found by Fick's law, that is, by expressions similar to Eq. (2). To account for the possibility that the alveolar and capillary surface areas, S_A and S_c, which bound the tissue barrier, do not match exactly we use their average, $(S_A + S_c)/2$, as an estimate of the tissue barrier surface. The barrier thickness is variable; since at each point of the barrier the flow of O_2 is inversely proportional to the local barrier thickness, the overall conductance is proportional to the average local conductance and hence proportional to the average of the reciprocal local barrier thickness; this average is called the harmonic mean of the barrier thickness, τ_h. We further need an estimate of the permeation coefficient K for O_2 in tissue; this is taken from the literature and assumed to be a constant (Table 1). On this basis we write for the conductance of the tissue barrier:

$$D_t = K_t(O_2) \cdot (S_A + S_c)/2\tau_{ht} \tag{4}$$

We similarly derive the conductance of the plasma layer as

$$D_p = K_p(O_2) \cdot S_c/\tau_{hp}. \tag{5}$$

Note that we take the capillary surface area, S_c, as an estimator of plasma barrier surface and do not consider the erythrocyte surface as a counterpart; the reason is that part of the erythrocyte surface is "hidden" from the capillary surface by neighboring erythrocytes so that the "available" erythrocyte surface is found to be similar to S_c.

Because of the unknown complexity of the diffusion conditions within the erythrocyte, we use only one morphometric parameter to find an expression for D_e, namely, the capillary volume, V_c; the complex physical events within the red cell are considered in a physical coefficient, Θ_{O_2}, the rate of O_2 binding by whole blood:

$$D_e = \Theta_{O_2} \cdot V_c \tag{6}$$

This approach has been introduced by Roughton and Forster (1957), who also proposed that Θ_{O_2} can be measured in vitro in stopped-flow rapid reaction devices: a suspension of desaturated erythrocytes is rapidly mixed with an oxy-

Table 1 Coefficients for Calculating Pulmonary Diffusing Capacity for O_2 from Morphometric Data

---

Conventional values

Diffusion coefficients $K_t = K_p$ $3.3 \cdot 10^{-8}$ cm^2/min/mmHg

O_2 binding rate θ_{O_2} 1.5 ml O_2/ml/min/mmHg

New values for θ_{O_2}

Eq. (7) with $k'c = 440$ mM/sec

Species	Hemoglobin concentration (g/100 ml)	Body temperature (exercise, °C)	θ_{O_2} (ml O_2/ml/min/mmHg)
Man	15	38	2.99
Dog[a]	18.8	40	3.85
Goat[a]	10.7	40	2.19
Pony (horse)[a]	17.0	38	3.26
Calf[a]	10.3	40	2.10

[a]From Weibel et al. (1987).

genated solution; the initial O_2-hemoglobin reaction rate, $k'c$, is measured spectrophotometrically and Θ_{O_2} is calculated from this rate. With this type of approach, Staub and Storey (1962) found that $\Theta_{O_2} = 2.7$ ml O_2/ml/min/mmHg, whereas Mochizuki (1966) obtained a value of 0.9. Using an improved method, Holland et al. (1977) determined Θ_{O_2} for human blood to be 1.5 ml O_2/ml/min/mmHg, a value we have used until recently (Weibel, 1984a) and which is given in Table 1.

In recent years these estimates of Θ_{O_2} have been questioned. It is now generally believed that the measurements of the initial reaction rate $k'c$, as obtained in the rapid reaction devices, reflects not only the chemical reaction between O_2 and hemoglobin but also the effect of an unstirred diffusion barrier of plasma surrounding the red cells (Huxley and Kutchai, 1983; Vandegriff and Olsen, 1984). Such a diffusion barrier may be larger in the in vitro situation of the rapid reaction device than in the pulmonary capillary. On the basis of current evidence it appears that the estimates of $k'c$ are too slow by a factor of about 2 due to such diffusion barriers (Holland et al., 1985; Yamaguchi et al., 1985). We have recently discussed the consequences of this effect on the calculation of "valid" values of Θ_{O_2} (Weibel et al., 1987), and have concluded that the "best educated guess'' is to accept a constant value of $k'c = 220$ mM/sec, the value measured by Holland et al. (1977) for human blood at 37 °C and 60% saturation, and to double it to 440 to account for the effect of the unstirred boundary layer. This value appears to apply, in first approximation, also to all mammals. Following Holland et al. (1977), we then find Θ_{O_2} by

$$\Theta_{O_2} = k'c \cdot f(T) \cdot (0.0587 \cdot \alpha_{O_2}) \cdot (1 - S_{O_2}) \cdot 0.01333 \cdot [Hb] \quad (7)$$

where $f(T)$ is a body temperature correction factor (which is 1.0 at 37 °C and 1.13 at 40 °C), α_{O_2} is O_2 solubility (0.0227 at 37 °C and 0.0219 at 40 °C), S_{O_2} is initial O_2 saturation of blood (fractional), and [Hb] is hemoglobin concentration in g/100 ml of blood (about 15 g/100 ml in man). Values of Θ_{O_2} calculated by this procedure are given in Table 1; for human blood Θ_{O_2} is 3 ml/ml/min/mmHg.

The estimation of D_{LO_2} by this approach requires estimates of total alveolar and capillary surface areas, of total capillary volume, and of the harmonic mean thickness of tissue and plasma barrier. These estimates are obtained from measurements performed on electron micrographs of lung tissue; the preparation method most commonly used for that purpose is instillation fixation, in which the blood is retained in the capillaries (Fig. 2a). To preserve the alveolar surface in the air-inflated conditions, one may also fix by vascular perfusion (Fig. 2b), but this eliminates the blood from the capillaries (Bachofen et al., 1982). The measurements are done on small tissue samples using stereological methods, as described below; the data thus obtained are surface or volume densities. To calculate the total surface or volume, the densities must be multiplied by the total lung volume, which must be measured independently before sampling. Details of this approach are given below.

B. Lung Mechanics: Surface Forces and Fiber Tension

The main mechanical problem to be solved in lung design is how to keep the alveolar walls with the capillaries well expanded in the air-filled lung, since this is one of the main determinants of gas exchange. Expansion of the alveolar septa depends essentially on the arrangement of the two main force-bearing elements of lung structure with respect to the acinus, the basic structural unit of lung parenchyma; the acinar extensions of the axial and peripheral systems of connective tissue fibers on which the septal fibers are anchored and on which the other forces, particularly the surface force, also act (Weibel and Gil, 1977).

An idealized and greatly reduced structural model of this interrelation of fiber system and surface forces in the acinus is shown in Figure 3. The separation of axial and peripheral fiber systems depends on the level of lung inflation because the pleural pull is transmitted to the acinar periphery through the hierar-

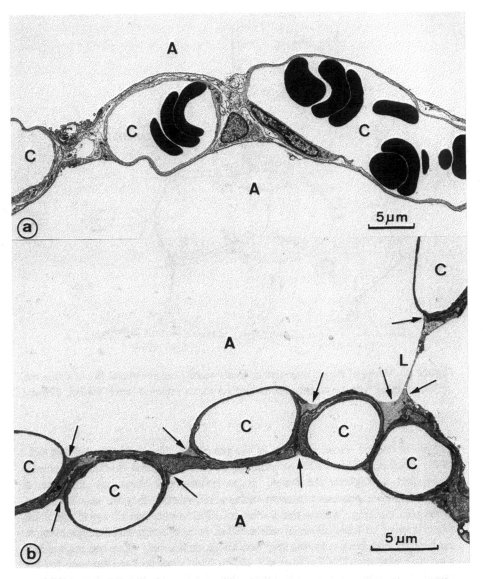

Figure 2 Transmission electron micrographs of alveolar septa. a. Fixation by instillation retains blood in capillaries (C). b. Fixation by vascular perfusion removes blood but retains surface lining layer, which forms pools in depressions of the alveolar surface (arrows); note lamella of lining layer (L) that spans pore of Kohn between alveoli (A) (from Gil et al., 1979).

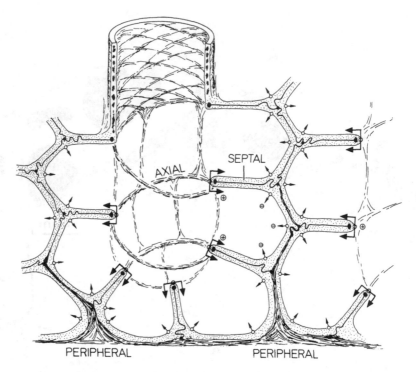

Figure 3 Model of the arrangement of axial, septal, and peripheral fibers within the acinus, together with the distribution of surface forces (arrows) (from Weibel, 1984a).

chy of connective tissue elements of the peripheral fiber hull (Weibel and Gil, 1977; Weibel, 1986). This volume-dependent peripheral traction is evenly distributed throughout the lungs; in perfusion-fixed lungs one can note a remarkably homogeneous expansion of the acini, whether they are near the pleura or deep in the lung. Wilson and Bachofen (1982) noted that all septal and axial fibers appear to hang loose in saline-filled or instillation-fixed lungs even at moderately high lung volumes; they become stretched only when one approaches total lung capacity (TLC) (Gil et al., 1979). This is particularly the case for the septal fibers, which appear to be longer than the distance beween their supports. This changes when surface tension is added in the air-filled lung: the acinar extensions of the axial fiber system now form smooth (helical) loops around the alveolar duct, and the more or less smooth alveolar walls are formed about a flat midplane. There is one additional remarkable feature, namely that alveolar ducts are wider in air-filled than in the saline-filled lungs expanded to the same volume and extremely dilated if surface tension is abnormally high (Bachofen et al., 1979).

This change in acinar structure is essentially the result of surface tension balanced by counteracting fiber tensions. As shown in the idealized model (Fig. 3), the acinar surface forms alternating positive and negative curvatures. The positive curvatures are over the axial fibers and have a small radius of curvature; the negative curvatures are in the alveoli. Surface forces tend to minimize free surface energy by reducing alveolar surface and evening out the curvatures. As a result, the free alveolar surface of air-filled lungs is smaller than that measured in fully unfolded fluid-filled lungs (Bachofen et al., 1979, 1987).

The mechanical properties of lung parenchyma are related to some measurable morphometric quantities characterizing its structure. The local surface force is related to the surface tension coefficient γ (dynes/cm) of the lining layer and the local mean curvature K. By the formula of Gibbs, the overall pressure differential due to surface forces, P_γ, is related to the average mean curvature, $\bar{K}$, by

$$P_\gamma = \gamma \cdot \bar{K} \tag{8}$$

if one assumes that γ is homogeneous throughout the lung. $\bar{K}$(units/cm) can be estimated on sections by obtaining an intersection and a tangent count in relation to a stereological test system as described below. The average mean air space curvature has been shown to decrease with increasing lung inflation in air-filled rat (Gil and Weibel, 1972) and rabbit lungs (Bachofen et al., 1979, 1987). In detergent-rinsed rabbit lungs, $\bar{K}$ measured at TLC was similar to that in normal air-filled lungs, but it decreased as inflation was reduced, in keeping with the formation of larger air spaces (Bachofen et al., 1979). It is, however, difficult to estimate the prevailing surface force from such measurements because the contribution of tissue tension to total recoil pressure is not easy to appreciate (Bachofen et al., 1987).

It has also been shown (Hoppin and Hildebrandt, 1977; Wilson, 1981) that P_γ can be expressed as a function of the surface-to-volume ratio of the air spaces, $(S/V)_A$, by

$$P_\gamma = 2/3 \cdot \gamma \cdot (S/V)_A \tag{9}$$

The surface-to-volume ratio can be measured on sections by stereological methods, as described below. Such measurements have been obtained on rat and rabbit lungs fixed by vascular perfusion (Bachofen et al., 1979; Gil et al., 1979; Gil and Weibel, 1972). Wilson and Bachofen (1982) have recently exploited these data in an attempt to establish a relation between structure and mechanics of lung parenchyma on the basis of a new model that accounts for the surface forces as well as for the contribution of the alveolar duct fibers to total lung recoil. They estimate that the contribution of "tissue line elements" (a physical equivalent to some unit fiber) must be proportional to the length density J_{Vf} of such unit fibers in the unit volume of parenchyma. With T as the tension of the line element, the contribution of this fiber system to recoil pressure is

$$P_T \cdot (1/3)T \cdot J_{Vf} \tag{10}$$

J_V could again be estimated on sections as described below. No such measurements have yet been performed. Although the stereological method as such is very simple, the complexity of the integral fiber strands, made of three different fiber types of varying thickness, does not permit a simple measurement before the theoretical term "line element" is translated into histological terms (Weibel, 1986).

Despite this lack of information, Wilson and Bachofen (1982) were able to draw some conclusions about the interaction of forces and structure in a stable lung by using the physiological and morphometric data set obtained on perfusion-fixed rabbit lungs. They were showed that their model, based on the combination of Eqs. (9) and (10), was able to describe the relation between measured alveolar surface area and the recoil pressure difference between air-filled and saline-filled lungs inflated at 40–80% TLC. They could also estimate the surface tension γ prevailing in the air-filled lungs, obtaining values in very close agreement with those measured directly in rat lungs by Schürch et al. (1978). Combining their approaches, Bachofen and Schürch could recently improve on the interpretation of these findings and calculate local surface tension as well as the surface forces prevailing at different levels of inflation (Bachofen et al., 1987). The morphologic approach from which such data were derived clearly has some limitations, mainly because the specimen reflects a more or less static condition of the lung.

C. Ventilation and Perfusion: Morphometric Model of Branched Tubes

The structures determining air and blood distribution to the gas exchange units—represented approximately by alveoli—are designed as branched hierarchic trees in which the dimensions of the branches decrease gradually towards the periphery (Weibel, 1963, 1984a). The models describing such trees must first account for the branching pattern, which is commonly found to be dichotomous with irregularities in the dimensions of the branches (Fig. 4). Deviations from dichotomy appear to occur mainly when small branches originate from a larger stem, a situation one calls "monopodial branching." Although some parts of larger bronchi appear to follow monopodial branching, this is not a typical condition for airways that, on the whole, quite clearly branch by dichotomy, from the trachea to the acinar airways (Weibel, 1987; Haefeli-Bleuer and Weibel, 1988). The basis for this lies in the branching pattern of airway tubes in the fetal lung (Burri, 1986), which is definitely dichotomous. Irregularities in the dimensions are the secondary results of an irregular distribution of peripheral lung volume that develops into the available pleural space.

The advantage of dichotomy as a branching pattern is that it allows generations of branching to be defined as the basic reference for ordering morphometric

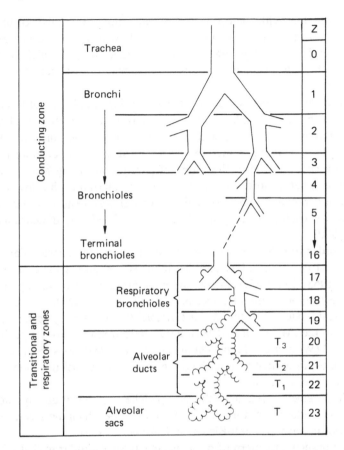

Figure 4 Airway branching in human lung by regularized dichotomy from trachea (generation z = 0) to alveolar ducts and sacs (generations 20–23). The first 16 generations are purely conducting; transitional airways lead into the respiratory zone made of alveoli (after Weibel, 1963).

data with respect to the hierarchy of the tree (Weibel, 1963, 1984a, 1987). The alternative would be to order the data according to "orders up" from the most peripheral airway (Horsfield, 1986); this is more difficult to treat also in physical models, so I shall limit this discussion to dichotomous models with "generations down," starting with the trachea as generation 0. The number of (parallel) branches in generation z is

$$N(z) \ . \ 2^z \tag{11}$$

which means that with each progressive generation the number of branches is doubled, until the final generation is reached, which, for the airways of the human lung, is the 23rd generation, on the average. Irregular dichotomy causes the terminations to occur in different generations: from about 18 to 30 in the human lung.

Many descriptors of the airway tree dimensions can be useful when describing the mechanics of ventilation, but they must always be structured according to the hierarchy of branching, and also with respect to the functional properties of airways: conducting down to the terminal bronchiole, transitional and respiratory beyond. The total volume of purely conducting airways relates to dead space, whereas total airway cross-section determines the flow velocity of air in each generation upon ventilation. Total airway resistance must again be estimated for each generation and is a function of airway dimensions (Pedley et al., 1970).

In order to calculate such overall parameters, the airway diameters and lengths must be estimated as size distributions and means for each generation (Weibel, 1963, 1989). Generalization on the progression of airway dimensions with increasing generation number is evidently very useful for the construction of analytical models. It is found, for example, that the airways can be described as a fractal tree (Mandelbrot, 1983), a generalization that has not yet been really exploited (West et al., 1986). On the other hand, deviations from such general laws are most important to describe design effects on the distribution of ventilation to peripheral units.

The parameters describing the most peripheral or acinar airways are similar, but one must add descriptors of the extent of gas-exchanging tissue associated with these airways (Haefeli-Bleuer and Weibel, 1988).

The branching pattern of blood vessels is very similar to that of airways. Both pulmonary arteries and veins basically adhere to dichotomous branching, but there are additional "supernumerary" branches that emanate from larger vessels in the sense of apparent monopody (Fig. 5). Accordingly, one finds that the terminal vascular branches that lead into the capillary network are reached after about 28 generations in humans (Weibel, 1963, 1984a).

The basic information that allows such hierarchic branched trees to be characterized in terms of physical or analytical models thus comprises an analysis of the branching pattern, the allocation of critical structures (such as terminal or transitional bronchioles leading into the gas exchange units, or alveolar sacs as terminations of the airways) to specific generations, and estimates of the diameter and length distribution of airway (or vessel) segments in each generation. Averages are not adequate; this type of modeling requires a carefully structured data set.

Figure 5 Resin cast of bronchial tree (B) and pulmonary arteries (A) of human lung. The arrowheads point out the sequence of branchings of small pulmonary arteries, which are more frequent than those of the associated bronchioles.

D. Lung Cell Biology: Morphometric Characteristics of Cell Population

The morphometric characterization of the lung cell population is of great importance, particularly when one is considering metabolic properties of the lung, as well as pathological changes (Weibel, 1984a, 1986; Crapo et al., 1982; Evans et al., 1975; Kapanci et al., 1969; Bachofen and Weibel, 1977; Montaner et al., 1986). Its complexity does not permit a detailed discussion of all the relevant morphometric characteristics; these will greatly vary with the goals of the studies.

For a general morphometric description of the lung cell population one will, first of all, have to consider the location of the cell types, that is, their distribution to different levels in the hierarchy of airways and blood vessels with which they are associated. Some basic parameters would include the distribution of the number and volume of the various cell types, their relative number and volume, and the fraction of the surface (of airways) lined by each cell type. One may also want to estimate the total cell mass, for each cell type or for each tissue compartment, epithelial vs. interstitial or endothelial cells, for example.

Then it may be useful to estimate certain cell type characteristics, such as the average cell volume, average cell membrane area, or their composition in terms of cytoplasmic organelles. For further details, the reader is referred to the literature on morphometric cell biology (Weibel, 1986; Weibel et al., 1969; Crapo et al., 1982).

III.   Stereological Methods: Elementary Background

The structural properties of the lung required to define the models discussed in the preceding section are all spatial parameters; volumes, surfaces, lengths, curvatures, and numbers of internal components contained within the space of the lung. The study of these, mostly microscopic, structures can only be done on flat sections through lung tissue; accordingly the "picture" presented to the investigator is two-dimensional: volumes appear as areas, surfaces as linear contours, and so on. Measurements obtained on such sections need to be transformed into data pertaining to the original three-dimensional structure.

This can be achieved through stereological methods. These are based on considerations of the (geometric) probability of obtaining a certain "profile" of a structural component when the structure is cut by a random section. The theory behind these methods has been worked out in great depth over the past few decades. The theoretical foundations of these methods and the practical procedures for their application have been extensively reviewed elsewhere (Weibel, 1979, 1980), so that this discussion is limited to a brief outline of their particular use in lung research.

Figure 6 shows the system of basic stereological formulae by which we can estimate, in particular, volumes, surfaces, lengths, curvature, and number of the components in three-dimensional structures from measurements done on sections. The nature of these stereological methods is that the measurements are obtained relative to a reference system, which is defined with respect to the volume (V) of the containing space. For this purpose, the structure is cut, and a test grid is placed randomly over the cut surface, or on a micrograph. The test grid defines the reference system, which is the area of the microscope field (A), the length of a set of test lines (L), or the number of test points (P), applied to a section of the structure.

Various methods are available by which one can now estimate the structure parameters with respect to these reference systems, but there is, for each parameter, one that is particularly simple and efficient because the "measure-

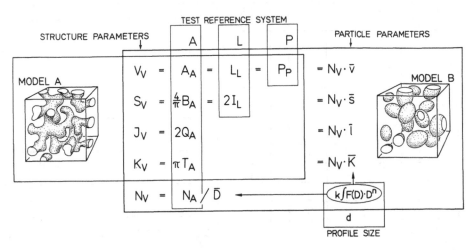

Figure 6 Synopsis of basic stereological formulae (from Weibel, 1979).

ment'' is reduced to a simple count. These so-called point-counting methods turn out to be the most efficient stereological methods and should therefore be given preference over the other methods (Mathieu et al., 1981). The approach to point counting is to overlay the section or micrograph with a suitable *coherent test system* made of test points P_T and test lines L_T enclosed in a test area A_T marked by a frame; three variants of such test systems are shown in Figure 7. The test systems are called coherent because points, lines, and area are in a precisely defined relation to each other (Weibel, 1979) such that

$$A_T = P_T \cdot k_2 \cdot d^2 \tag{12}$$

$$L_T = P_T \cdot k_1 \cdot d \tag{13}$$

where d is the length of the unit test line or the distance between test points, and k_1 and k_2 are constants. In a simple square grid $k_2 = 1$ and $k_1 = 2$; this is realized in Figure 7a for the thick lines with the test points marked by a dot: there are 16 such test points so that the test area is 16 d^2 and the length of all the thick lines is 32 d. The test system of Figure 7a is an example of a double lattice test system in which a second set of points and lines is superimposed on the primary set, again in a coherent fashion: for each primary (thick) line there are two fine lines, and thus the *total* set of test lines (thick and fine) is three times as long. If one takes the intersection of test lines as test points it is easily seen that there are nine total test points for each primary test point. Writing P_T' and L_T' for the total number of test points and the total length of test line in the test system of Figure 7a, we can write

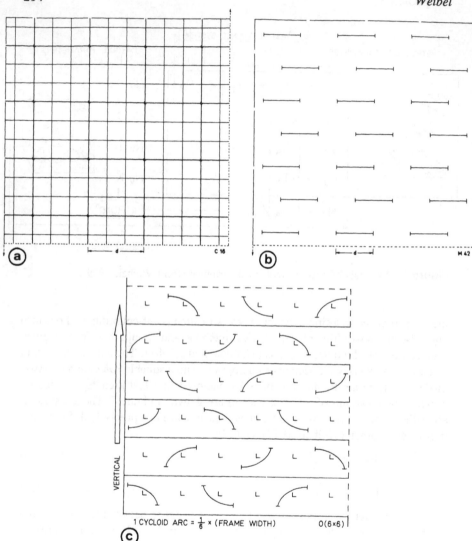

Figure 7 Sample test systems for point-counting stereology. (a, b, from Weibel, 1979; c, according to Baddeley et al., 1986).

$$P_T' = 9 \cdot P_T \tag{14}$$

$$L_T' = 3 \cdot L_T = 3 \cdot 2d \cdot P_T \tag{15}$$

The test system shown in Figure 7b is made of 21 disconnected line segments; taking their end points as test points we have $P_T = 42$, $k_1 = 1/2$, and $k_2 = \sqrt{3}/2$.

The number of different coherent test systems that can be described is almost unlimited (see Weibel, 1979). As an example, Figure 7c presents a test system with the test lines in form of cycloids; such test systems are useful when the section though the material is a "vertical section," as it is obtained when one cuts the bronchial epithelium perpendicular to the surface, for example (Baddeley et al., 1986).

The counting frame, which defines the area A_T of the test system, is bounded by two types of lines that determine the counting rules (Gundersen, 1977; Weibel, 1979): any profile wholly within the frame or intersecting the boundary marked by the solid line is counted, whereas any profile intersecting the dashed line, called the "forbidden line," is rejected. Note that the "forbidden line" extends beyond the counting frame in the direction of the short arrows.

The test systems must be calibrated whenever the test line unit d enters a calculation, as is the case in estimating surface densities. The length of the test line unit d must be expressed in terms of *real* dimensions, that is, its apparent length on the test grid must be reduced by the magnification of the micrograph on which it is applied. For that purpose, the real magnification must be estimated by means of a suitable test graticule (Weibel, 1979).

A. Estimating Volume Density by Point Counting

If a test system is applied to a micrograph (or a section) at random, the probability that a test point hits the profile of a given component is proportional to the volume of that component. The volume density of component [x] in the containing space [y] is therefore directly given by the ratio of the number of test points hitting [x] to the number of test points hitting [y]:

$$V_V(x,y) = P(x) / P(y) = P_p(x,y) \tag{16}$$

To demonstrate the procedure, a variant of the test system of Figure 7b has been superimposed on an electron micrograph of an alveolar septum in Figure 8. The volume density of capillaries (c) in the septum (s) is estimated by counting the points falling on the capillaries, P(c), and those falling on all septal structures, P(s), so that

$$V_V(c,s) = P(c) / P(s) \tag{17}$$

In Figure 8 we find 47 points on capillaries and 36 points on tissue, giving 83 points on the septum, so that $V_V(c,s) = 0.57$.

Double lattice test systems are convenient for volume density estimations if the volume of the component of interest is relatively small. This is, for example, the case at level III where we need to estimate the volume density of alveolar septa (s) in lung parenchyma (p_2). As seen in Figure 9 the alveolar septa occupy only a small fraction of the section area; the chance that 1 of the 16 primary test

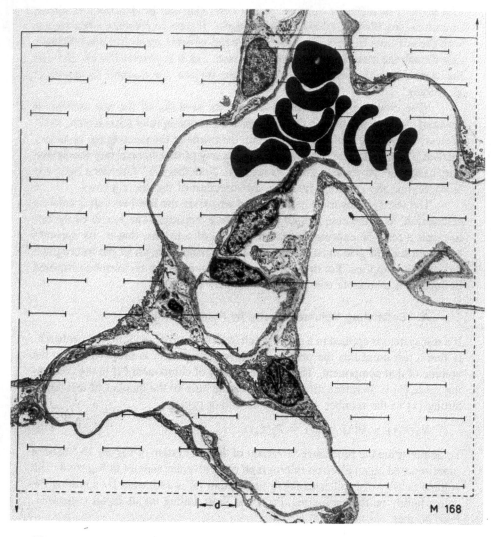

Figure 8 Electron micrograph of alveolar septum with test system that allows estimation of alveolar and capillary surfaces as well as capillary and tissue volumes (from Weibel, 1979).

points of grid C16 (Figure 7a) hits a septum is small, whereas several points of the total grid hit the septum. On the other hand, most of the points hit the parenchyma, with the exception of those that fall on nonparenchymatous components (the small vessel encircled by a thick line in Fig. 9). The particular features of

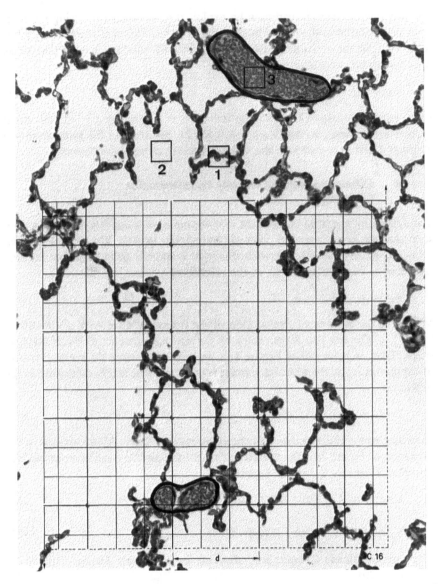

Figure 9 Medium-power light micrograph of lung as used for level III. Elements of fine nonparenchyma (small vessels) are surrounded by black line. Squares indicate size of quadrats (micrographs recorded by electron microscopy) for level IV; note that some fields do not contain alveolar septa. In the lower part a double-lattice test system is superimposed, with which the septal volume density is estimated (level III) (200 ×).

this coherent test system can now be exploited to advantage by using the *total* point grid to estimate the point hits on alveolar septa, $P'(s)$, and the primary point set to estimate the containing space, $P(p_2)$; since the total point set is nine times the primary set we obtain

$$V_V(s, p_2) = P'(s) / 9 \cdot P(p_2) \tag{18}$$

In the particular case of Figure 9 we find $P'(s) = 17$ and $P(p_2) = 15$ (one point on nonparenchyma), so that $V_V(s, p_2) = 0.126$. We can use the same approach at levels I and II to estimate the volume density of nonparenchyma.

B. Estimating Surface Density by Intersection Counting

The surface of a spatial component (for example, of a capillary) appears on a section as the linear boundary trace of the capillary profile. The probability that a test line of given length intersects this trace is directly proportional to the surface density of the component in the containing volume, so that

$$S_V = 2 \cdot I/L_T = 2 \cdot I_L \tag{19}$$

where I is the number of intersection of the boundary trace with a test line of length L_T. The test line length must be defined with respect to the containing space. Thus, in reference to Figure 8, if we wish to estimate the surface density of capillaries (c) in the alveolar septum (s), the test line length contained in the septum is

$$L_T(s) = P(s) \cdot (1/2)d \tag{20}$$

according to the characteristics of the coherent test system M168 applied to the micrograph (Figure 7b). We therefore find for the capillary and alveolar surface densities, respectively, in the septum:

$$S_V(c,s) = 4 \cdot I(c) / [P(s) \cdot d] \tag{21}$$

$$S_V(A,s) = 4 \cdot I(A) / [P(s) \cdot d] \tag{22}$$

where the coefficient 4 results from dividing the coefficient 2 in Eq. (19) by 1/2 from Eq. (20).

Note that we can now calculate the alveolar (or capillary) surface density in lung parenchyma by multiplying Eq. (18) with Eq. (22):

$$S_V(A,p_2) = V_V(s, p_2) \cdot S_V(A,s) \tag{23}$$

One could, however, also estimate $S_V(A,p_2)$ directly on specimens as shown in Figure 9 by counting $I(A)$ on the test grid C16, using, for example, only the thick lines of the primary grid. Noting that, with this test system $L_T \cdot P_T \cdot 2d$, we would have

$$S_V(A,p_2) = I(A) / [P(p_2) \cdot d] \tag{24}$$

The result will, however, not be the same because estimates of surface density depend on the resolution of the micrograph (Paumgartner et al., 1981). Which approach is appropriate depends on the model in which the data are to be used. In the model for gas exchange, the surface is defined with respect to the tissue barrier and we therefore need the resolution offered by electron microscopy.

In contrast, the alveolar surface estimate required in characterizing the lung's mechanical properties is defined with respect to the alveolus; in this case an estimate obtained by light microscopy may be acceptable. Note that the surface-to-volume ratio that appears in Eq. (9) is the ratio of alveolar surface to air space volume; this can be obtained as the ratio of alveolar surface density to volume density in parenchyma:

$$(S/V)_A = S_V(A,p_2) / V_V(A,p_2) \tag{25}$$

Using a coherent test system, we have

$$S_V(A,p_2) = 2\, I(A) / P(p_2) \cdot k_1 \cdot d \tag{26}$$

$$V_V(A,p_2) = P(A) / P(p_2) \tag{27}$$

so that

$$(S/V)_A = [2/(k_1 \cdot d)] \cdot I(A) / P(A) \tag{28}$$

where $P(A)$ means test points that fall on all acinar air spaces, alveoli, and alveolar ducts taken together.

Some morphometric studies on the lung still use the "mean linear intercept" (Lm) as a characteristic of acinar air spaces, following Weibel (1963), Thurlbeck (1967), and others. This parameter is, in some ways, obsolete and should only be used if this is really what one wants to know. In fact, Lm is nothing else than the reciprocal of $(S/V)_A$, but the latter is a much more useful parameter.

C. Estimating Length Density by Transection Counting

The length density of linear features in space, for example of fibers, is reflected on a random section by the density of their transections on the unit section area. Denoting the number of transections by Q we can estimate the length density of fibers (f) in lung parenchyma (p_2) by

$$J_V(f,p_2) = 2 \cdot Q(f) / A_T(p_2) = 2 \cdot Q_A(f,p_2) \tag{29}$$

or, using a coherent test system,

$$J_V(f,p_2) \ . \ 2 \cdot Q(f) \ / \ [P(p_2) \cdot k_2 \cdot d^2] \tag{30}$$

In counting transections we must note that $Q(f)$ is all transections that are totally within the frame of the test grid plus those that lie on two boundary lines of the frame shown in long lines (Figure 7), whereas those that intersect the frame boundary marked by short line segments are excluded, as described above.

D. Estimating Mean Curvature by Tangent Counts

The density of integral mean curvature of a spatial surface, such as the alveolar surface, is reflected on a random section by the density of curvature in the unit area of the section. The simplest method for estimating mean curvature density is to sweep a line across the section and to count the number of tangents that this line forms with the boundary trace of the alveolar surface (De Hoff, 1968; Gil and Weibel, 1972; Weibel, 1979). Figure 10 shows that one must differentiate between tangents with positive (T_+) and negative (T_-) curvatures, that is, with convexities and concavities of the surface. The mean curvature density K_V is directly proportional to the net number of tangents per unit area:

$$K_V = \pi \cdot T_{net} \ / \ A_T = \pi \cdot (T_+ - T_-) \ /A_T = \pi \ T_{Anet} \tag{31}$$

In characterizing the lung's mechanical properties, we found a need to estimate the *mean* mean curvature density, $\bar{K}$, which is the integral mean curvature $\bar{K}$ averaged over the alveolar surface. It is hence plausible the $\bar{K}$ must be obtained by a combination of tangent count (to estimate k_V) and of intersection counts (to estimate S_V), so that

$$\bar{K} = (\pi \ / \ 2) \cdot T_{Anet} \ / \ I_L \tag{32}$$

or, in using a coherent test system,

$$\bar{K} = (\pi \ / \ 2) \cdot [k_1 \ / \ (k_2 \cdot d)] \cdot T_{net} \ / \ I \tag{33}$$

The practical application of this method is shown in Figure 11. A test system M168 made of short line segments is used to count the intersections with the alveolar surface; the sweeping line is represented by a graphic triangle that is swept over the section from top to bottom of the test frame.

E. Estimating the Number of Elements

The numerical density of elements (cells, alveoli) is their number contained in the unit volume of tissue, N_V. The number of profiles seen on the section area, N_A, is evidently proportional to N_V, but not directly because larger elements have a higher probability to be hit by the section. This is intuitively plausible. The basic formula is

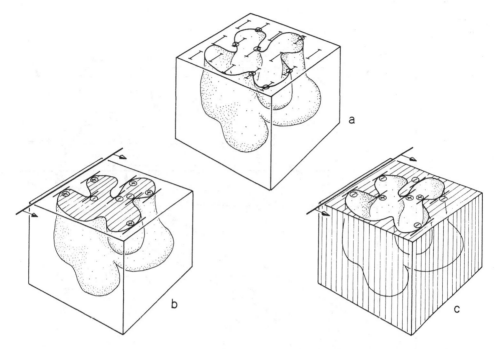

Figure 10 Estimating mean curvature of alveolar surface from (a) counting intersections with a test grid and (b, c) tangents with a sweeping test line (from Weibel, 1979).

$$N_V = N_A/\overline{D} \tag{34}$$

where $\overline{D}$ is the mean tangent diameter of the elements. To calculate N_V, one must therefore first count the number of profiles within the area of the test grid by the counting rule described above. It is then also necessary to obtain an independent estimate of $\overline{D}$, not a trivial task, particularly if the elements are not simple spheres of nearly constant diameter (see Weibel, 1979).

We have therefore proposed an alternative method (Weibel and Gomez, 1962), by which all measurements are obtained on the same section by stereological methods:

$$N_V = (1/\beta) \cdot N_A^{3/2}/V_V^{1/2} \tag{35}$$

The coefficient β is a dimensionless geometric factor that is 1.38 for spheres and increases to about 1.5 for ellipsoids (see Weibel, 1979). As well as a count of N_A, as described above, we need a measurement of V_V by point counting. The method has been used widely thanks to its easy applicability. However, its use is limited to cases in which the elements are convex bodies of well-defined shape and narrow size distribution.

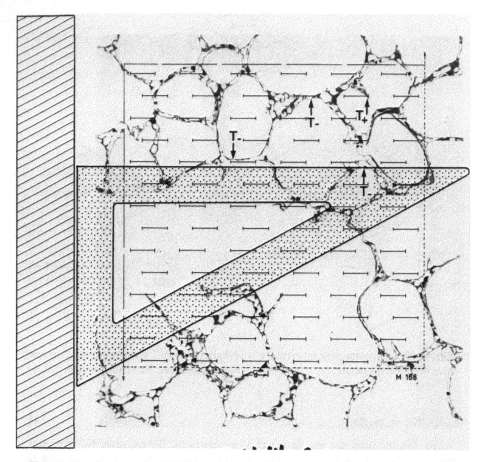

Figure 11   Practical method for estimating surface curvature in lung fixed by vascular perfusion. Sliding ruler is used for tangent count (from Weibel, 1979).

Recent years have brought a new approach that is applicable to particles of arbitrary, even complex shape: the dissector or selector methods (Sterio, 1984; Gundersen, 1986; Cruz-Orive, 1987). With these methods, one uses two parallel sections: on one section a counting frame is applied, the other is a "look-up" section. One counts all profiles that are contained in the counting frame but do not appear on the "look-up" section. This means that the particle "ends" in the space between the sections.

The dissector method is illustrated in Figure 12. The two sections are separated by a known distance, h, which should not be too large, optimally about ⅓ or ¼ of the mean particle height, but certainly not larger than the minimum

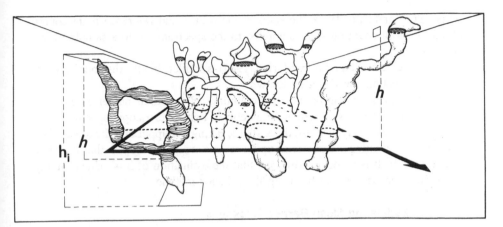

Figure 12 The disector (a two-dimensional sampling frame and a parallel section plane at a distance h) generating sets of planar profiles by cutting through particles. An unbiased counting frame of area A_T is positioned on the lower section. One counts all particles that are not intersected by the upper (infinite) plane and that have profile sets completely or partially inside the frame, and that are not intersected in any way by any fully drawn edge or (infinite) extension line. Two particles are counted in the example. The total height, h, of a particle in the front of the figure is indicated; that particle is not counted because its set of planar profiles is intersected by a fully drawn edge of the frame (from Gundersen, 1986).

height of a particle. On the "counting section" a test grid is applied according to uniform sampling, in practice by systematic random sampling. The test grid has a counting frame of area A_T that contains a set of P_T test points (Fig. 12). The volume of the dissector is

$$v(\text{dis}) = A_T \cdot h \qquad (36)$$

The test points are used to estimate the fraction of the reference space of the specimens that is contained in the disector by counting the points P(ref) that fall on reference space.

By comparing the "counting" and the "look-up" sections, one now counts those particle profiles Q^- contained in the counting frame and *not* found in the look-up section. If h is about ⅓ of the mean particle height, one will count about every third profile. Note that the counting rule is the same as above: all profiles that intersect the "forbidden line" are not counted. The numerical density of particles in the reference space is now obtained by:

$$N_V(\text{particles, ref}) = [\Sigma Q^- / \Sigma(P(\text{ref})] \cdot [P_T / v(\text{dis})] \qquad (37)$$

where ΣQ^- and $\Sigma P(ref)$ are the sums of the profile and reference point counts obtained on all counting frames applied to the spectrum, or to a sample of the specimen.

The selector method (Cruz-Orive, 1987) is similar, but the distance between the paired sections need not be known.

These methods can be used for counting cells or their nuclei on electron micrographs as well as for counting alveoli in light microscopic specimens. In the latter case one may also use thick slices, taking the upper and the lower slice surface as "counting" and "look-up" sections, respectively, so that the slice thickness is h. It is evident, however, that the estimation of h is critical in any case. This can be avoided by using the selector method.

F. Estimating Mean Barrier Thickness

The arithmetic mean thickness, $\overline{\tau}$, of the tissue barrier between air and blood can also be estimated by point counting, since it can be defined as the ratio of tissue volume per barrier surface. Counting points hitting the tissue barrier (t) and intersections with the two bounding surfaces (A and c) on a specimen such as that shown in Figure 8 and using a coherent test system, we derive

$$\overline{\tau} = 2 \cdot V_V(t,s) / \{S_V(A,s) + S_V(c,s)\}$$

$$= 2 \cdot k_1 d \cdot P(t) / \{I(A) + I(c)\} \tag{38}$$

This parameter is useful in characterizing changes in tissue mass, as they occur in edema, for example. It is sometimes useful to use the epithelial basement membrane as reference surface, particularly in cases of tissue alterations (Bachofen and Weibel, 1977).

In attempting to characterize the lung's gas exchange properties we need to estimate the harmonic mean barrier thickness, and this cannot be obtained by point counting. The approach is to superimpose at random a set of long straight lines onto a micrograph (Fig. 13) and to measure the length of the intercepts, l, of the test lines with the barrier, whereby l is the distance between the intersections of the test line with the alveolar and the capillary surface, respectively. Note that intercepts that do not extend from alveolar to capillary surface do not count. This measurement is best performed using a ruler with a logarithmic scale (Weibel, 1979, p. 209). To calculate the harmonic mean barrier thickness, τ_h, we need to calculate the harmonic mean of the intercept lengths, l_h, as the mean of the reciprocal lengths:

$$l_h = \{\Sigma n_i\} / \{\Sigma n_i \cdot l_i^{-1}\} \tag{39}$$

where n_i is the number of intercepts of length l_i. We then obtain (Weibel and Knight, 1964; Gundersen et al., 1978)

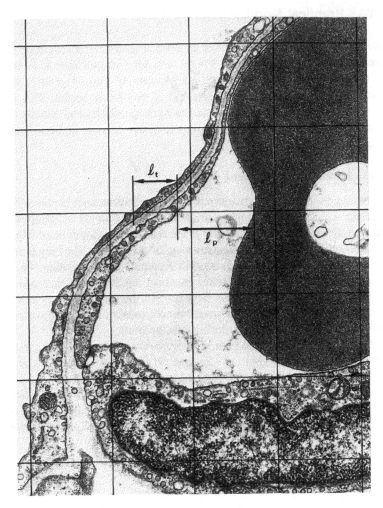

Figure 13 Square lattice of test lines used to measure intercept lengths in tissue (l_t) and plasma (l_p). Note erythrocyte (EC) in capillary (C), and that barrier separating blood from alveolar air (A) is made of the three layers epithelium (EP), interstitium (IN), and endothelium (EN) (from Weibel, 1970/71).

$$\tau_h = (\tfrac{2}{3})l_h \qquad\qquad (40)$$

Note that the model for calculating pulmonary diffusing capacity requires τ_h to be estimated for both the tissue barrier and the plasma layer between the capillary surface and the red cell membranes. This must be done on sufficiently magnified electron micrographs.

IV. Practical Morphometric Methods

To calculate morphometric data such as the alveolar surface or the capillary volume
we need two pieces of information: (1) the total lung volume and (2) an estimate
of the density of the parameter in the unit volume of lung. The latter must be
obtained on a sample that is representative of the entire organ. This chapter
therefore reviews first the method for measuring lung volume, and then the detailed
sampling procedures used in our laboratory to obtain reliable, unbiased estimates
of lung morphometry.

A. Measuring Total Lung Volume

The total lung volume is best measured by a water displacement method (Scherle,
1970). To avoid distortion, it is important that tissue fixation be complete; with
glutaraldehyde instillation fixation this is the case after a minimum of 24 hr. One
must also carefully avoid mechanical crushing or distortions. The lung is carefully
dissected free of the other chest organs. With the lung floating on saline, the trachea
is opened to equilibrate transpulmonary pressure; the trachea is then tied as near
to the bifurcation as possible. With large lungs it may be necessary to separate
the various lobes; in this case, the main bronchi must be tightly ligated to avoid
leakage of intrapulmonary fluid during manipulation.

A sufficiently large container with isotonic saline or buffer is placed on
a balance and the initial weight, W_O, is read off. The lung (or lobe) is now completely submerged by lowering it below the surface either by means of a soft wire
loop or by a clamp grasping the hilum, both fastened to a laboratory stand; this
compensates for buoyancy. The lung may not touch the container wall. The final
weight, Wf, is read off and the weight of the fluid volume displaced by the lung,
W_L, is

$$W_L = Wf - W_O \tag{41}$$

If the specific gravity of the fluid is g, the lung is

$$V_L = W_L / g \tag{42}$$

Since the specific gravity of isotonic saline is 1.0048 we can, without undue error, disregard g and set $V_L \approx W_L$.

It is advisable to measure lung volume shortly before sampling; this ensures that V_L reflects the same degree of lung expansion as the samples processed
for microscopic measurement. It is found that larger lungs may lose some volume
($\sim 5\%$) during the first hours after instillation, probably due to some residual
elastic recoil or because of some trapped air. After that, lung volume is found
to remain constant for weeks if the lung is stored in ample fluid. Glutaraldehyde-
fixed tissue is also very resistant to distortions: if it is crushed (for example

during the sampling procedure) it will restore to its original dimensions when submerged in fluid.

An alternative method is to use the "Cavalieri method," which may be suitable mainly for small lungs. With this method the lung is cut into slices of equal thickness, t, using a random start position for the first section. The slices are then laid out flat on a tray and the area of all the cut surfaces is estimated by point counting on a square lattice test system with period d, which covers all slices. The volume of the lung is then found by

$$V_L = t \cdot d^2 \cdot \Sigma P(l) \tag{43}$$

where $P(l)$ is the number of test points that fall on sectioned lung tissue. These points are summed over all slices.

B. Sampling for Morphometry

General Considerations

To design an appropriate sampling scheme, we must first understand the nature of the microscopic measurements obtained by the stereological methods described above (Weibel, 1979). The principle of *measuring with respect to a reference space* is the basis of all stereological methods: each parameter is obtained as the *ratio of two measurements*, one estimating the size of the objects under investigation, the other the size of the space in which they are contained, which we may call the reference space. To give an example, we cannot measure the capillary surface, S(c), directly, but we can estimate the capillary surface contained in the unit volume of lung parenchyma, V(p), which we call the *surface density of capillaries in lung parenchyma*:

$$S_V(c,p) = S(c) / V(p) \tag{44}$$

Likewise, we can estimate by stereological methods the *volume density of lung parenchyma in the entire lung*,

$$V_V(p,L) = V(p) / V(L) \tag{45}$$

so that we can calculate the absolute capillary surface as the product of Eqs. (44) and (45) with the lung volume, as can easily be seen:

$$S(c) = V(L) \cdot V_V(p,L) \cdot S_V(c,p) \tag{46}$$

The sampling problem we are faced with is that the structures of interest with respect to gas exchange are concentrated in the alveolar walls whose dimensions are such as to require fairly high magnifications ($\sim 10,000 \times$) for adequate resolution (Fig. 8). This therefore limits the sample available for microscopic analysis to a very small size (Fig. 9): since the spacing of the alveolar septa is larger than the size of the microscope field, the chance that a *random* field would fall into an empty space (or into a larger vessel or airway) is rather high. Furthermore, the composition of the lung is rather inhomogeneous if we consider that about 10–20% of the lung space is made up of larger structures, such as conducting airways or vessels (Fig. 14). A rational sampling design must therefore account for the hierarchical design of the lung, as it is shown schematically in Figure 15. In this scheme, the gas exchange structures are concentrated in the *alveolar septum*, which is decomposed into capillaries and tissue barrier. Septa, air in alveoli, and alveolar ducts taken together are defined as *lung parenchyma*. We lump into the term *nonparenchyma* all other components, such as major airways (from large bronchi to bronchioles), blood vessels larger than capillaries, thicker connective tissue septa, and the pleura.

For the measurements to be representative of the lung, the sample on which they are performed must be unbiased. One approach to achieving this is to obtain a simple random sample by cutting the lung into a large number of small tissue blocks and picking out a few of them by a suitable randomization procedure. The problem with this approach is that it is liable to what has been called "laboratory bias," because tissue blocks that do not contain alveolar septa are simply rejected. To avoid this, and also to simplify the task we propose to proceed through a multiple-stage sampling procedure (Cruz-Orive and Weibel, 1981; Müller et al., 1981; Weibel et al., 1981).

Cascade Sampling

The logic of this sampling design is to estimate the parameters of the gas exchange apparatus with respect to the tissue space in which they are confined, that is, to the volume of alveolar septa, and then to obtain the overall parameter estimates by measuring the volume density of septa in the lung volume. This again requires restriction of the sampling space to parenchyma, excluding nonparenchymatous structures that are inhomogeneously dispersed in the lung and may assume a size and spacing larger than the microscope fields required for estimating septal volume density (Fig. 14)

We therefore adopted a sampling design in a cascade of four levels whose theoretical basis has been worked out by Cruz-Orive and Weibel (1981). As shown in Figure 15, the four levels of restricted sampling spaces are defined as follows with the specific notations used in parentheses:

Ω_4 : phase of interest: volume and surface of capillary (c), tissue barrier (t), alveolar (external) surface of barrier (A)

Ω_3 : alveolar septum (s) = capillary + tissue

Ω_2 : "fine" or true parenchyma (p_2) = septa + alveolar air + duct air

Ω_1 : "coarse" parenchyma (p_1) = fine parenchyma + intermediate nonparenchymatous structures (< 1 mm)

Ω_0 : whole lung (L)

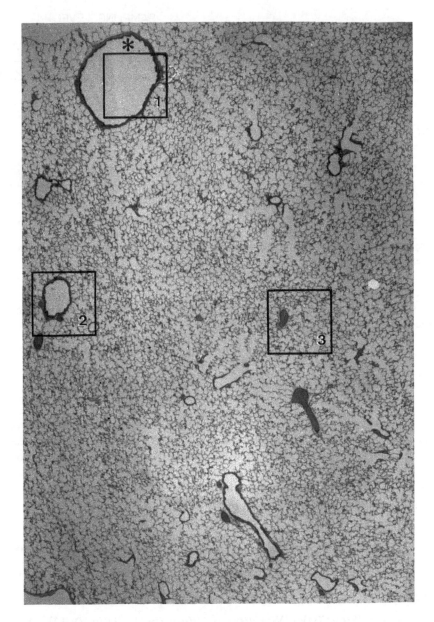

Figure 14 Low-power light micrograph of lung as used for level II showing distribution of nonparenchymatous structures. Bronchiole marked by asterisk belongs to coarse nonparenchyma. Squares indicate size of tissue blocks prepared for sampling levels III and IV (16 ×; from Weibel et al., 1981).

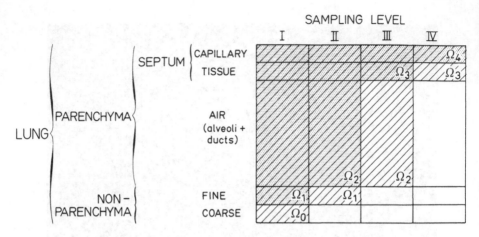

Figure 15 Subdivision of lung into sequential compartments and sampling scheme (from Weibel et al., 1981).

It is also important to note the meaning of the difference sets of some of these spaces:

$\Omega_0 - \Omega_1$: coarse nonparenchyma (np_1) = structures with diameter d > 1 mm

$\Omega_1 - \Omega_2$: fine nonparenchyma (np_2) = structures of size [20 μm < d < 1 mm]

The characteristic of this approach is that the reference space of one level becomes the space of interest (object space) at the next lower level. The parameter of interest $Y(\Omega_4)$, where Y can stand for surface, S, or volume, V, and so on, can be estimated from its density in the third reference space, $Y_V(\Omega_4,\Omega_3) = Y(\Omega_4)/V(\Omega_3)$, by considering the volume densities of the reference spaces in the cascade, and the lung volume:

$$Y(\Omega_4) =$$

$$\{Y(\Omega_4)/V(\Omega_3)\} \cdot \{V\{\Omega_3)/V(\Omega_2)\} \cdot \{V(\Omega_2)/V(\Omega_1)\} \cdot \{V(\Omega_1)/V(\Omega_0)\} \cdot V(\Omega_0)$$

$$(47)$$

or specified for capillary surface, for example,

$$S(c) = \{S(c)/V(s)\} \cdot \{V(s)/V(p_2)\} \cdot \{V(p_2)/V(p_1)\} \cdot \{V(p_1)/V(L)\} \cdot V(L)$$

$$= S_V(c,s) \cdot V_V(s,p2) \cdot V_V(p2,p1) \cdot V_V(p1,L) \cdot V(L) \qquad (48)$$

$$\text{IV} \qquad \text{III} \qquad \text{II} \qquad \text{I}$$

where the roman numerals indicate the level at which the stereological estimate is obtained.

This sampling protocol should ensure that, at each level, the variance of the parameter estimates is as low as possible, because their distribution within the restricted reference space is as homogeneous as possible.

Stratified Sampling

Because of the possibility that regional differences might occur in the size of alveoli and capillaries, particularly in larger lungs due to gravity (Glazier et al., 1967; Gehr and Weibel, 1974), it may be advisable to apply a stratified sampling strategy in addition to the multilevel sampling, by dividing the lung into several segments from cranial to caudal. For these dissections, the lungs are kept submerged in the fixation solution in order to lose as little fluid as possible in the course of cutting. Each stratum should then be weighed while floating in the fixative to estimate the relative stratum volume.

The way the strata are set up depends on the size of the lung, and on the specific questions asked. In our work on medium-sized to large lungs (Weibel et al., 1987), we have divided the lung into three strata using the natural lobar subdivisions as guides. If one attempts to make all strata of about the same size, namely roughly a third of lung volume, and if one observes that regional differences are expected to occur between apical and caudal parts of the lung, the three strata are composed as follows:

UL : right and left upper lobes and right middle lobe

LL : left lower lobe

RL : right lower (and cardiac) lobes

In autopsied patients or in experiments where the animal was kept in the supine position, one may want to separate the dorsal (dependent) and ventral parts of the lung, perhaps in addition to lobar stratification.

Practical Sampling Procedures

This sampling scheme may appear complicated, but it is quite simple to perform in practice, as described in detail by Müller et al. (1981) and Weibel et al. (1981). Above all, adhering to strict sampling rules simplifies the work and ensures confidence in the data obtained by measurements on very small samples.

The procedure of four-level cascade sampling we use is as follows. The lobes of each of the strata are cut into about 10 slices of equal thickness (for larger lungs, 10–20 mm, depending in lung size) using very sharp blades. The slices are assembled on a rectangular tray and photographed for subsequent estimation of coarse nonparenchyma (level I). By means of a pair of crossed rulers and a

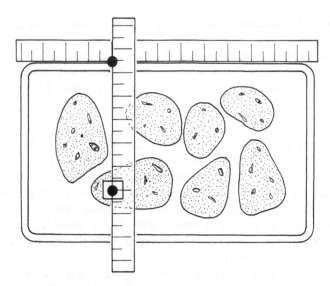

Figure 16 Diagram of tray with lung slices. Random sample location (square) is found by means of two random numbers (dots) using two crossed rulers (from Weibel, 1984b).

random number table, a sample of three tissue blocks 1–3 cm^3 is picked from these slices (Fig. 16). From each block a slice is cut into small dices about 2 mm in diameter, of which six dices are picked and processed for embedding in Epon for high-power light microscopy (level III) and electron microscopy (level IV). Additional blocks are obtained by the same procedure for low-power microscopy (level II) and embedded in either paraffin or methacrylate.

Levels I and II: Estimation of Volume Fraction of Parenchyma in the Whole Lung, $V_V(p_2, L)$

The volume fraction of parenchyma in the lung is determined indirectly by estimating the volume occupied by components of nonparenchyma that occur in a large spectrum of sizes (from several millimeters for the major airways and blood vessels down to 20–40 μm for the smaller conducting blood vessels) and are inhomogeneously distributed. We estimate their volume density at two levels: (I) macroscopically down to structures about 1 mm in size, and (II) by light microscopy for smaller structures. With small lungs (rat and smaller animals), level I cannot be done because most nonparenchyma structures are smaller than 1 mm. In this case, the volume density of nonparenchyma must be estimated by light microscopy (level II) using step sections.

   Level I. This is the volume density of nonparenchymatous structures larger than 1 mm (np_1) estimated macroscopically on the photographic records of the slice surfaces. By means of a test grid, the fraction of points falling on nonparenchymatous components larger than 1 mm is estimated. This step leads to estimation of the volume density of "coarse parenchyma," that is, parenchyma proper plus the finer nonparenchymatous components:

$$V_V(p_1,L) = 1 - V_V(np_1,L) \tag{49}$$

Level II. The finer components of nonparenchyma (smaller than 1 mm) are estimated by point counting at low magnification in the light microscope on sections of paraffin- or methacrylate-embedded blocks, offering a surface of about 2 cm². We use an automatic sampling stage microscope (Weibel, 1970) to generate a square point lattice of approximately 0.5 mm spacing. Part of such a section is shown in Figure 14. Coarse nonparenchyma components ($\geq$ 1 mm), such as the bronchiole marked by an asterisk in Figure 14, are excluded from the reference space because they were already accounted for at level I. By counting the number of points falling on nonparenchymatous structures [20 μm < d < 1mm] = (np_2), we estimate a volume fraction of "true parenchyma" in "coarse parenchyma": $V_V(p_2,p_1) = 1 - V_V(np_2,p_1)$.

Level III: Estimation of the Volume Density of Alveolar Septa in Parenchyma, $V_V(s,p_2)$

The third sampling level is performed by light microscopy at a magnification of about 250. From each of two or three small tissue dices per stratum, a section 1 μm thick is cut. Figure 9 shows such a section on which we estimate the volume density of septa in parenchyma, $V_V(s,p_2)$, by differential point count. Such sections contain some elements of nonparenchyma, such as the vessels encircled by a line in Figure 9; since these were already accounted for in level II, points falling onto such structures are excluded from the reference point set. At this level one can also obtain some additional parameters such as the septal boundary length density, the mean septal width, and the length distribution of linear intercepts of the air spaces.

Level IV: Estimation of Internal Septal Structure

At this level, performed by transmission electron microscopy at about 10,000 magnification, we estimate the surface densities of alveolar and capillary surface within the septum, $S_V(A,s)$ and $S_V(c,s)$, respectively, the capillary volume density, $V_V(c,s)$, and the harmonic mean thickness of tissue, $\tau_h(t)$, and plasma barriers, $\tau_h(p)$.

Since the reference space is the alveolar septum, the sampling strategy at this level must provide a set of electron micrographs that contain as much of the alveolar septa as possible while still being unbiased. To achieve this a two-stage procedure is required because, at the required magnification, only a small fraction of the section is observed. The first consists of the selection of four to six

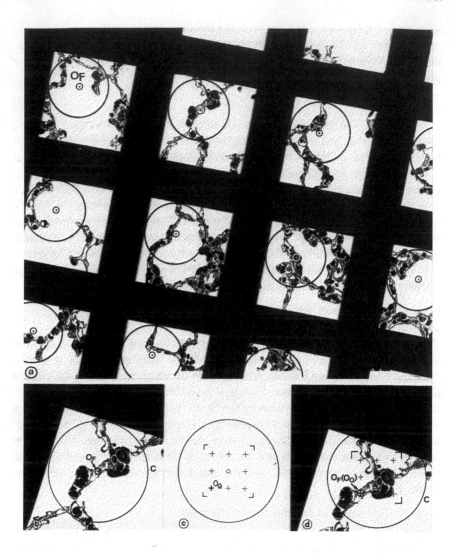

Figure 17   Procedure for systematic subsampling of thin sections to obtain random electron micrographs of septa. a. Low-power view of lung section on·copper grid. The EM screen (large circle) is placed in one corner of the grid squares. With systematic quadrat (SQ) sampling, every field that contains septum is photographed; empty fields are recorded. b. With systematic area-weighted quadrat (SAWQ) sampling, a micrograph is recorded if the center point of the screen (O_F) hits a septum (b). c. By random numbers, a point (O_Q) on the screen is identified (c). d. The specimen is shifted until O_Q overlies the point O_F on the septum (from Müller et al., 1981).

tissue samples from the set of random samples obtained by the procedure of stratified random sampling described above; one section is cut from each sample. In the second stage each section is subsampled to obtain eight micrographs (quadrats). There are two possible procedures (Müller et al., 1981). One can obtain a systematic quadrat (SQ) subsample by placing the screen of the electron microscope into the corners of the supporting copper grid (Fig. 17a), and record a micrograph whenever the quadrat contains septum; the amount of septum contained in the quadrat will greatly vary. Alternatively, one can use a *systematic area weighted quadrats* (SAWQ) subsampling as described by Müller et al. (1981), which yields more septum per micrograph and is still unbiased. In this procedure a micrograph is only recorded if the center point of the screen hits a septum (Fig. 17b), but the recording is preceded by a random shift of the section, as shown in Figure 17c and d.

The SAWQ procedure is superior to SQ subsampling in theory, but in practice the difference is small (Müller et al., 1981). With SQ subsampling the amount of septum contained in each micrograph varies more than with SAWQ. The resulting greater "noise" in the primary data collected (point hits and intersection counts) is, however, attenuated in the subsequent calculation procedures. With SAWQ the ratios of points on structure to points on the septum must be calculated for each quadrat and then averaged over the section (this is imposed by the sampling procedure); with SQ, however, the primary counts of points on structure and on septum are summed and the ratios are obtained as ratio of the sums. For example, the volume density of capillary in septum, $V_V(c,s)$, is obtained from the points on capillary, $P(c)$, and on septum, $P(s)$, counted on the quadrats, j, as follows:

$$SAWQ : V_V(c,s) = (1/n) \cdot \sum_{j=1}^{n} [P(c) / P(s)]j \tag{50}$$

$$SQ : V_V(c,s) = \left[\sum_j P(c)j \right] / \left[\sum_j P(s)j \right] \tag{51}$$

Whereas the SAWQ method has "esthetic" appeal and is perhaps more rigorous sampling procedure, it is also more laborious and may therefore be less efficient. The choice between the two is largely a matter of taste; we tend to prefer the SQ method at present for its greater overall efficiency.

V. A Model Case of Lung Morphometry: Estimating Pulmonary Diffusing Capacity

We have shown here that the pulmonary diffusing capacity for oxygen, D_{LO_2}, can be estimated from morphometric information. The model assumes that O_2 uptake occurs in two steps: O_2 diffusion from alveolar air into blood, followed by the O_2

binding reaction of hemoglobin in the erythrocytes. The model requires the morphometric estimation of the gas exchange surfaces, the capillary blood volume, and the harmonic mean thickness of the tissue and plasma barrier (Eqs. 3–6). Since D_{LO_2} is a major parameter of respiratory physiology, I would now like to illustrate the use that can be made of morphometric data in interpreting lung function by elaborating on the study of D_{LO_2}. I shall base this first on the estimate of D_{LO_2} in the normal human lung (Gehr et al., 1978), which is a study of particular interest because it relates to humans, but poses a number of methodological problems, partly due to the fact that it is difficult to obtain and properly prepare "normal" human lungs under well-controlled conditions. Also some of these specimens were collected and studied over 10 years ago; in the mean time some aspects of the methods have been refined, but not to the point of invalidating the old data. I would then like to illustrate the "state of the art" by discussing a recent study of pairs of mammals with different levels of O_2 needs (dogs and goats, ponies and calves) in which the morphometric analysis was combined with a thorough investigation of a variety of physiological variables (Taylor et al., 1987).

A. The Study of the Human Lung

In the early 1970s we collected, over a period of a few years, seven intact human lungs that could be termed "normal" (Gehr et al., 1978). They were from patients who had died of severe cerebral injuries, except for one who died of a sudden cardiac arrest in connection with his congenital heart disease. Their age ranged from 19 to 40 years. The lungs were fixed by instillation of a glutaraldehyde solution through an endotracheal tube into the collapsed lung in situ, as quickly after confirmed death as possible. Early fixation is important; in a previous study we had found that postmortem alterations occur in lung tissue as early as 1 hr after death, if some disease or inflammatory response was present in the lung (Bachofen et al., 1975).

Samples from all lungs were examined by light and electron microscopy to check for such changes; only lungs free of damage were retained for the morphometric study. Figure 18 documents, on electron micrographs taken from one of these lungs, the good quality of tissue and cell preservation that could be achieved. It must be stressed that this "quality control" is necessary, but that it must be done *before* the morphometric study is undertaken. Items found unsuitable for the study, for whatever reason, must be excluded a priori on the basis of independent criteria (such as presence or absence of inflammation); an a posteriori exclusion on the basis of "unsatisfactory" data is not permissible, or must be thoroughly justified.

The lungs were sampled according to a scheme of stratified random sampling, with samples of dorsal and ventral regions taken from three horizontal slices cut at three levels (apical, medial, basal) in the right and left lung each. The total sample thus consisted of 12 stratified subsamples; from these, small tissue blocks

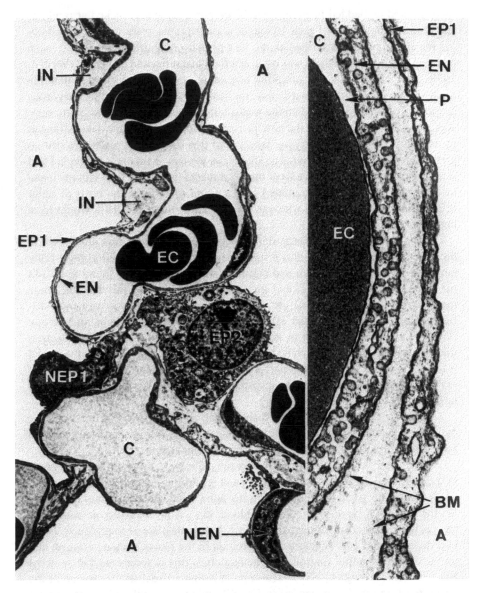

Figure 18 Electron micrographs of thin sections of human lung fixed by instillation of glutaraldehyde show good preservation of capillaries (C) with erythrocytes (EC) and plasma (P), endothelial (EN) and epithelial cells of type 1 (EP1) and type 2 (EP2). The interstitial space (IN) is bounded by basement membranes (BM) (figure on left is 3500×; figure on right is 45,000×; from Gehr et al., 1978).

were cut and embedded in Epon. One block per sample was picked at random, and on one thin section of each 16 electron micrographs were recorded, following the SQ procedure described above, at a primary magnification of 1120, such that the stereological analysis was done at a final magnification of 11,200 (Weibel, 1979). It is evident that this sampling procedure was less sophisticated than the method described above. In particular, the samples were not picked by a rigorous randomization procedure and all the measurements were done in one single step, corresponding to level IV of the new procedure, with the primary data estimated as densities per lung parenchyma. Because of this simpler approach, we had no good estimate of nonparenchyma available, as we now estimate it in levels I and II. It is therefore likely that we have underestimated the part of lung volume taken up by nonparenchyma: we assumed it to amount to 10% on the basis of older information (Weibel, 1963), whereas we now know that about 18% would have been a better guess.

The original morphometric data obtained on this population are shown in the first column of Table 2. For a mean body mass of 74 kg and a mean body height of 177 cm, the alveolar and capillary surface areas were found to be 143 m^2 and 126 m^2, respectively, and the capillary volume 213 ml. The harmonic mean barrier thickness is 0.62 μm. It must be noted that the gas exchange surface areas are larger than the estimates obtained earlier by light microscopy (Weibel, 1963). This is due to the higher resolution afforded by the electron microscope. In deciding which is the relevant value, we must note that the model for D_{LO_2} is based on the tissue barrier as the basic element (Fig. 1) so that the resolving power of the electron microscope is required. We can thus accept the newer data.

The second column of Table 2 shows corrected estimates of these data by assuming lung parenchyma to be 82% of lung volume; this changed only the surface and volume estimates but not the barrier thicknesses.

The physical coefficients needed to calculate D_{LO_2} by Eqs. (3–6) are given in Table 1. The diffusion coefficients K_t and K_p pose no major problem; we can accept the value given in Table 1a. On the other hand, the coefficient determining the rate of O_2 binding by capillary blood, Θ_{O_2}, is problematic; no consensus has yet been reached about its real meaning. Two points are of importance. First, Θ_{O_2} depends on the hemoglobin concentration in the blood, on temperature, and on the degree of initial saturation of hemoglobin; this is accounted for by using Eq. (7) in calculating Θ_{O_2}. The second point is that a "realistic" estimate of the initial reaction rate $k'c$ is needed in Eq. (7) and this is a difficult measurement to make, as discussed above. Table 1a reports the value of Θ_{O_2} as estimated by Holland et al. (1977), whereas Table 1b lists newer values that account for hemoglobin concentration differences in several species and for the possible effect of an unstirred boundary layer on the estimate of $k'c$. These newer values are generally higher than older estimates.

Table 2 Morphometric Data on Human Lung Allowing an Estimate of Pulmonary Diffusing Capacity.

Variable	Original data (1978)	Corrected data (1987)
Body mass (kg)	74 ± 4	74
Body height (cm)	177 ± 3	177
Lung volume (ml)	$4{,}341 \pm 285$ ml	4,341
V_V parenchyma (%)	90	82
Alveolar surface area (m^2)	143 ± 12	130
Capillary surface area (m^2)	126 ± 12	115
Capillary volume (ml)	213 ± 31	194
Barrier thickness (harmonic mean)		
Tissue (μm)	0.62 ± 0.04	0.62
Plasma (μm)	0.15 ± 0.01	0.15
Diffusing capacities: ml O$_2$/min/mmHg		
D_tO_2	716	652
D_pO_2	2,772	2,530
D_eO_2	320[a]	582[b]
D_{MO2}	569	518
D_{LO2}	205[a]	274[b]

Original data according to Gehr et al. (1978), corrected to account for 18% nonparenchyma.
[a]With $\theta_{O_2} = 1.5$ ml/ml/min/mmHg (Holland et al., 1977).
[b]With θ_{O_2}(CST) $= 3$ ml/ml/min/mmHg (Eq. 7).

    The diffusing capacities calculated for the human lung with these coefficients are shown at the bottom of Table 2. Using the estimate of Θ_{O_2} given by Holland et al. (1977) with the original data set (first column), D_{LO_2} is found to be 205 ml O^2/mm/mmHg. Using the most recent estimates of Θ_{O_2} with the corrected morphometric data (second column), we find D_{LO_2} to be 274 ml O^2/min/mmHg.

    It is evident that these morphometric estimates of D_{LO_2} are much larger than those obtained by physiological methods, which range from 20–30 at rest to about 100 in heavy exercise (Hammond and Hempleman, 1987). If we assume that the diffusing *capacity* is best measured physiologically when the lung is fully exploited in heavy exercise, we still find the morphometric estimate to be "too large" by at least a factor of 2. We can propose several explanations on the basis of available evidence. First, the gas exchange surface measured is probably not totally available in vivo: in the air-filled lung the molding effect of surface forces causes part of the barrier to fold up, so that only about 75% of the total alveolar surface is available (Gil et al., 1979). Second, the value of Θ_{O_2} used in these calculations refers to desaturated blood; we shall see below that Θ_{O_2} falls as O_2 is loaded onto the blood, so that the average Θ_{O_2} along the capillary path is about half the desaturated value. If these two effects are considered, the value of D_{LO_2} obtained by morphometry becomes lower, but still remains larger than the largest physiological estimates. But then some "reserve D_{LO_2}" is certainly desirable, because it sets the upper limit of O_2 uptake and some safety margin is important and because ventilation and perfusion of the gas exchange units are not completely homogeneous, which reduces the efficiency of the gas exchanger.

We conclude that the morphometric estimate of pulmonary diffusing capacity is larger than what is found by physiological methods; but this approach studies "ideal conditions" which are never, or at best rarely, fulfilled in life.

B. State of the Art: Improved Lung Morphometry and Respiratory Physiology Combined

In recent years we have extended the scope of our scientific inquiry to encompass structure–function correlation in the entire respiratory system, from the lung through circulation to the mitochondria in muscle cells (Weibel and Taylor, 1981; Taylor et al., 1987). We have furthermore studied the physiology of O_2 consumption and transfer under limiting conditions (i.e., at $\dot{V}_{O_2}$max, the capacity to perform endurance exercise fueled by oxidative metabolism) in order to assess the possible role of structural design in limiting O_2 consumption, and we have looked at different variations of O_2 demand. To summarize the main result of these studies: $\dot{V}_{O_2}$max is, in all instances, proportional to the total mitochondrial volume of locomotor muscles (Hoppeler et al., 1987), whereas at all other levels of the respiratory system adaptation to different levels of O_2 demand involved a combination of structural and functional changes (Taylor et al., 1987, 1988).

These studies became more demanding, also with respect to lung morphometry, where an estimate of D_{LO_2} was most important. The problem was also that we wanted to compare lungs of widely varying size: from 0.1 ml in the Etruscan shrew to 120 ml in the rabbit and 40 L in the horse (Gehr et al., 1981). This made it necessary to develop a new sampling scheme, because the final amount of lung tissue subjected to actual stereological analysis was quite similar in all cases, irrespective of lung volume, namely vanishingly small. The reliability of the morphometric estimate therefore depends essentially on a good, unbiased sampling procedure. The sampling method developed (Cruz-Orive and Weibel, 1981; Müller et al., 1981; Weibel et al., 1981) is the one presented above.

In a recent series of studies on adaptive variation in the respiratory system, we combined this improved sampling method with a thorough physiological investigation (Taylor et al., 1987). The rationale of the study was as follows. Healthy mammals, including humans, can increase their O_2 consumption by about 10-fold above rest before reaching their maximal rates. However, highly trained athletes (e.g., endurance runners) and some mammalian species are able to increase their rates of O_2 consumption by 20–30 times. These species can be thought of as nature's "elite athletes," and they include horses and dogs. We therefore exploited this fact by comparing "athletic" dogs (28 kg body mass) and ponies (175 kg) to "normal" animals of the same size class, goats and calves, and found that $\dot{V}_{O_2}$max of the "athletes" was about 2.5 times that of the two other species, as shown in Table 3 (Taylor et al., 1987). With respect to the lung, we asked to what extent D_{LO_2} was increased in the athletic species, and what other factors may contribute to the 2.5 times higher O_2 uptake rate in the lung (Weibel et al., 1987).

After the physiological study was completed, the animals were sacrificed. In deep anesthesia the lung was fixed by instillation of glutaraldehyde solution (2.5% phosphate buffered to pH 7.4, total osmolarity 350 mOsm) through a tracheal cannula, at a constant head pressure of 25 cm above the highest point of the sternum in the supine animal. Before instillation, a snare was placed around the base of the heart to stop circulation by closing the snare prior to instillation; this avoids the hemoconcentration in the pulmonary capillaries when the heart continues to pump blood into the lung during instillation fixation (Bur et al., 1985). During lung fixation, samples were taken from the locomotor muscles, heart, and diaphragm (Hoppeler et al., 1987). Then the lung was removed from the chest together with heart and esophagus, and kept immersed in a large volume of the fixative for at least 24 hr.

The lung volume was measured by the water displacement method described above. With large lungs, one may want to measure the right and left lungs separately, keeping the main stem bronchi clamped or tied. The lungs were then divided into three strata of about equal size: stratum 1 consisted of both upper lobes together with the right middle and the cardiac lobe. Strata 2 and 3 consisted of the right and left lower lobe, respectively. Note that for ponies this subdivision had to be approximated because horse lungs have no interlobar fissures. The relative stratum weights were estimated by weighing the lobes. On each stratum we then performed the four-level sampling cascade described above. The stereological densities thus obtained for each stratum were combined by multiplying with the stratum volume (lung volume times relative stratum weight) and then added to yield the overall morphometric estimates.

The most important morphometric data are shown in Table 3; they are expressed as mass-specific values per kg body mass to facilitate interspecies comparison. In absolute terms, the dog's alveolar surface area amounts to 78 m^2, that of the pony to 325 m^2. The capillary volume in the dog and goat lungs is about 120 ml, but it is 500 and 380 ml in pony and calf, respectively, if one assigns

Table 3 Morphometric Estimation of Mass-Specific D_{LO_2} in Dogs, Goats, Ponies, and Calves (from Weibel et al., 1987), compared to $\dot{V}_{O_2}$max (Taylor et al., 1987).

Species	Number	Body mass (M_b, kg)	Lung volume (ml)	$S(A)/M_b$ (m²/kg)	$S(c)M_b$ (m²/kg)	$V(c)/M_b$ (ml/kg)	τ_{ht} (μm)	Capillary hematocrit (%)	D_{MO_2}/M_b ml/min/mmHg/kg	D_{LO_2}/M_b ml/min/mmHg/kg	$\dot{V}_{O_2}$max/M_b (ml/min/kg)
Dog	3	28.2	1587	2.75	2.17	4.18	0.51	55	12.60	7.07	136.2
Goat	5	27.7	1742	2.43	2.13	4.38	0.52	25	9.78	4.78	54.0
Ratio dog/goat				1.13	1.02	0.95	0.98	2.2*	1.29*	1.48*	2.52*
Pony	3	175.5	9777	1.85	1.76	3.12	0.51	36	8.76	4.71	88.8
Calf	3	149.0	5397	1.43	1.38	2.39	0.42	27	7.50	2.99	36.4
Ratio pony/calf				1.29*	1.28*	1.31*˙	1.21	1.3	1.17*	1.57*	2.44*

Means only are given here for clarity.

*Ratio significant at p < 0.05.

an average body mass of 160 kg to each. If one considers that the hematocrit is greatly different between the species, one can calculate the total mass of red cells in these lung capillaries. One finds the dog and goat lungs to contain about 65 and 30 ml of erythrocytes, the pony and calf 180 and 103 ml, respectively. The athletic species thus have twice as much red cells present in their lungs for O_2 uptake, but much of this difference is due to their blood having a higher erythrocyte concentration (see also Table 1 for hemoglobin concentration).

When one is calculating the diffusing capacities, these blood differences come into play in the calculation of Θ_{O_2} by Eq. (7), as shown in Table 1. We then find D_{LO_2} to be about 1.5 times greater in the athletic species. Since $\dot{V}_{O_2}$max was 2.5 times larger, it is evident that the pulmonary diffusing capacity is only partly adapted to the higher O_2 uptake rate in the athletic species. Since the O_2 flow rate is determined by the diffusion conductance, D_{LO_2}, and the difference between alveolar and capillary P_{O_2} as driving force (Eq. 1) we must conclude that this driving force is larger in the athletes: about 19 mmHg in dog and pony as compared to 12 mmHg in the goat and calf (Weibel et al., 1987).

To interpret this finding that the driving force (a physiological parameter) is partly responsible for the differences in O_2 uptake rate, we have to call on the physiological data collected in this study (Taylor et al., 1987). In particular, we would like to estimate the actual values of alveolar and mean capillary P_{O_2} (Karas et al., 1987b). The value of PA_{O_2} can be estimated from the blood gas data collected, and we conclude that it is *lower* in athletic (about 100 mmHg) than in the other species (about 115 mmHg). Thus, their higher driving force is not due to a higher pressure head; we must rather suspect that an even larger difference must exist in the mean capillary P_{O_2} to cause the driving force to be larger in the athletic species.

The mean capillary P_{O_2} can be calculated by the so-called Bohr integration (Bohr, 1909; Hill et al., 1973). The procedure assumes a (linear) capillary path from the arteriole to the venule (Fig. 19). The blood enters the capillary at the arteriolar end with a P_{O_2} corresponding to mixed venous blood, $P\bar{v}_{O_2}$; when it leaves the capillary the P_{O_2} of arterial blood, Pa_{O_2}, is reached. Note that if the blood equilibrates with alveolar air, we have $PA_{O_2} = Pa_{O_2}$. In order to calculate $\overline{Pb}_{O_2}$ we must calculate the stepwise increase in capillary P_{O_2} as the blood moves through the lung; this depends on the capillary transit time and on the (local) diffusing capacity, which determines the transfer rate.

The mean capillary transit time, t_c, is obtained from the ratio of capillary volume, measured morphometrically (Table 3), to cardiac output, measured physiologically (Karas et al., 1987a); at $\dot{V}_{O_2}$max we find t_c to be on the order of 0.3 sec with some interspecies variation (Fig. 19). The diffusing capacity is also taken from the morphometric data (Table 3); however, it cannot be assumed to be constant. From Eq. (7) we see that Θ_{O_2} depends on the degree of O_2 saturation of the blood; in particular, it would tend to fall rapidly as O_2 becomes loaded

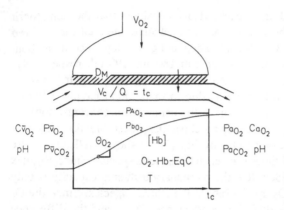

Figure 19 Model to calculate P_{O_2} profile in capillary by Bohr integration from morphometric and physiological information.

into hemoglobin and saturation exceeds 75%, as is the case for most of capillary transit time. Figure 20 shows how Θ_{O_2} decreases in the four species considered. Performing Bohr integration, we found that $\overline{Pb_{O_2}}$ is about 80 mmHg in the athletic and about 105 mmHg in the other species, so that we confirm the P_{O_2} differences deduced above.

One of the important observations now is that the blood becomes equilibrated with alveolar air while it still flows in the capillary (Fig. 20). Indeed, it appears that the last 25% of the capillary path in the athletic species does not contribute to O_2 uptake and is thus redundant; in the goat and calf the redundant part is about 60% of the path or of transit time (Karas et al., 1987b). This would seem to explain the observation made above with respect to the human lung that about half its morphometric $D_{L_{O_2}}$ is redundant: equilibration of capillary blood and alveolar air is completed before the blood reaches the end of the capillary path.

One question we had to face was whether the calculations performed were a valid description of what happened in the lung. In order to test this we performed an additional experiment by varying inspired and, by this, alveolar P_{O_2} (Karas et al., 1987b). The reasoning was as follows. If the goat lung allows O_2 uptake to be completed after 40% of the transit time at $\dot{V}_{O_2}$max when inspired P_{O_2} corresponds to the O_2 concentration, FI_{O_2}, of 21% in ambient air, the lung is not limiting O_2 upake, and the goats should still be able to reach $\dot{V}_{O_2}$max when FI_{O_2} is lowered. We ran some goats on the treadmill at the same speed at which they had reached $\dot{V}_{O_2}$max in ambient air, and let them breath a gas mixture with lowered FI_{O_2}. As shown in Figure 21, the PA_{O_2} became lowered accordingly, but the time needed to equilibrate the capillary blood increased until it reached nearly 100% of transit time at an FI_{O_2} of about 15%. We concluded that in hypoxia the goat can utilize its reserve diffusing capacity to maintain $\dot{V}_{O_2}$max despite of the reduced pressure head (Karas et al., 1987b).

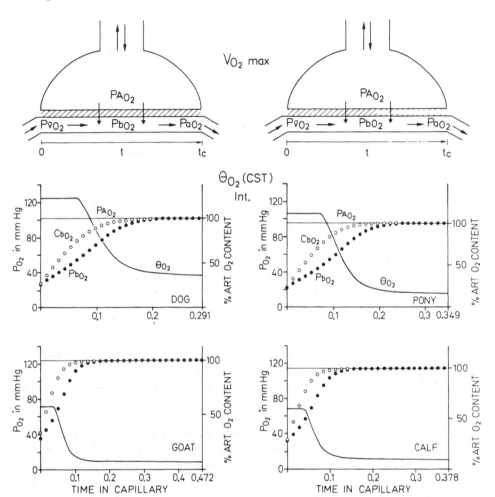

Figure 20 The increase in O_2 concentration (C_bO_2) and P_{O_2} of the blood during the time it transits the pulmonary capillary bed (t_c) when animals exercise at $\dot{V}_{O_2}$max (from Karas et al., 1987b).

The further conclusion we can draw is that the morphometric model for D_{LO_2} presented previously yields a realistic estimate of the "true" pulmonary diffusing capacity. In contrast, the physiological measurement of D_{LO_2} yields an estimate of that part of D_{LO_2} that is actually utilized, excluding the redundant part that the physiological methods do not "see." At rest, less than 10% of D_{LO_2} is probably enough, but in maximal aerobic exercise conditions, up to or more

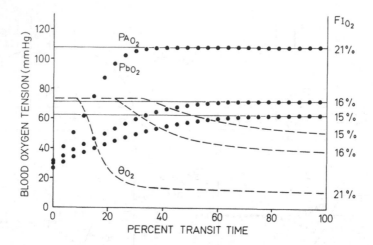

Figure 21 Hypoxia test of model for calculating O_2 transfer in the lung of the goat at $\dot{V}_{O_2}$max. The increase in P_{O_2} and decrease of Θ_{O_2} of the blood in the pulmonary capillaries is plotted as a function of transit time in the lung for inspired O_2 of 21%, 16%, and 15% (from Karas et al., 1987b).

than half of D_{LO_2} is needed. It is also possible that high-performance athletes, whose $\dot{V}_{O_2}$max is over twice that of normal humans, may have to utilize most of the D_{LO_2} available in their lungs, since it is not established that the lung can increase its diffusing capacity with intense exercise training.

We must, however, realize that factors that tend to reduce the efficiency of pulmonary gas exchange, such as inhomogeneities in the ventilation–perfusion ratio (Torre-Bueno et al., 1985), may also make it necessary for the lung to maintain some excess diffusing capacity. In that respect it would be important to study in greater detail the design features of airways and blood vessels as they may affect the distribution of ventilation and perfusion to the gas exchange units in the alveolar walls (Weibel, 1984a). Some progress has recently been made (Rodriguez et al., 1987; Haefeli-Bleuer and Weibel, 1988), but the state of the art is still at a relatively low level, both in terms of the models and the methods for obtaining reliable morphometric information.

Acknowledgments

I gratefully acknowledge the devoted assistance of Ms. Gertrud Reber, Mr. Karl Babl, and Ms. Barbara Krieger in preparing this article. The work at the basis of this review has been supported by grants from the Swiss National Science Foundation (3.036.84).

References

Bachofen, M., and Weibel, E. R. (1977). Alterations of the gas exchange apparatus in adult respiratory insufficiency associated with septicemia. *Am. Rev. Respir. Dis.* **116**:589–615.

Bachofen, M., and Weibel, E. R. (1987). Sequential changes in morphology of the adult respiratory distress syndrome. In *Pulmonary Diseases and Disorders*. Edited by A. P. Fishman. New York, McGraw-Hill.

Bachofen, M., Weibel, E. R., and Roos, B. (1975). Postmortem fixation of human lungs for electron microscopy. *Am. Rev. Respir. Dis.* **111**:247–256.

Bachofen, H., Gehr, P., and Weibel, E. R. (1979). Alterations of mechanical properties and morphology in excised rabbit lungs rinsed with a detergent. *J. Appl. Physiol.* **47**:1002–1010

Bachofen, H., Ammann, A., Wangensteen, D., and Weibel, E. R. (1982). Perfusion fixation of lungs for structure–function analysis: credits and limitations. *J. Appl. Physiol.* **53**:528–533.

Bachofen, H., Schürch, S., Urbinelli, M., and Weibel, E. R. (1987). Relations among alveolar surface tension, surface area, volume, and recoil pressure. *J. Appl. Physiol.* **62**:1878–1887.

Baddeley, A. J., Gundersen, H. J. G., and Cruz-Orive, L. M. (1986). Estimation of surface area from vertical sections. *J. Microscopy* **142**:259–276.

Bohr, C. (1909). Ueber die spezifische Tätigkeit der Lungen bei der respiratorischen Gausaufnahme. *Scand. Arch. Physiol.* **22**:221–280.

Bur, S., Bachofen, H., Gehr, P., and Weibel, E. R. (1985). Lung fixation by airway instillation: effects on capillary hematocrit. *Exp. Lung Res.* **9**:56–66.

Burri, P. H. (1986). Development and growth of the human lung. In *Handbook of Physiology*. Section 3: The Respiratory System. Chapter 1. Edited by A. P. Fishman and A. B. Fisher. Bethesda, MD, American Physiological Society, pp. 1–46.

Cruz-Orive, L. M. (1987). Arbitrary particles can be counted using a disector of unknown thickness: the selector. *Acta Stereol.* **6**:129–135.

Cruz-Orive, L. M., and Weibel, E. R. (1981). Sampling designs for stereology. *J. Microscopy* **122**:235–257.

Crapo, J. D., Barry, B. E., Gehr, P., Bachofen, M., and Weibel, E. R. (1982). Cell numbers and cell characteristics of the normal human lung. *Am. Rev. Respir. Dis.* **125**:332–337.

De Hoff, R. T. (1968). Curvature and topological properties of interconnected phases. In *Quantitative Microscopy*. Edited by R. T. De Hoff and F. N. Rhines. New York, McGraw-Hill, pp. 291–325.

Evans, M. J., Cabral-Anderson, L. J., and Freeman, G. (1975). Transformation of alveolar type 2 cells to type 1 cells following exposure to NO_2. *Exp. Mol. Pathol.* **22**:142–150.

Gehr, P., and Weibel, E. R. (1974). Morphometric estimation of regional difference in the dog lung. *J. Appl. Physiol.* **37**:648–653.

Gehr, P., Bachofen, M., and Weibel, E. R. (1978). The normal human lung: ultrastructure and morphometric estimation of diffusion capacity. *Respir. Physiol.* **32**:121–140.

Gehr, P., Mwangi, D. K., Ammann, A., Maloiy, G. M. O., Taylor, C. R., and Weibel, E. R. (1981). Design of the mammalian respiratory system. V. Scaling morphometric pulmonary diffusing capacity to body mass: wild and domestic mammals. *Respir. Physiol.* **44**:61–86.

Gil, J., and Weibel, E. R. (1972). Morphological study of pressure-volume hysteresis in rat lungs fixed by vascular perfusion. *Respir. Physiol.* **15**:190–213.

Gil, J., Bachofen, H., Gehr, P., and Weibel, E. R. (1979). Alveolar volume-surface area relation in air- and saline-filled lungs fixed by vascular perfusion. *J. Appl. Physiol.* **47**:990–1001.

Glazier, J. B., Hughes, J. M. B., Maloney, J. E., and West, J. B. (1967). Vertical gradient of alveolar size in lungs of dogs frozen intact. *J. Appl. Physiol.* **23**:694–705.

Gundersen, H. J. G. (1977). Notes on the estimation of the numerical density of arbitrary profiles: the edge effect. *J. Microscopy* **111**:219–223.

Gundersen, H. J. G. (1986). Stereology of arbitrary particles. *J. Microscopy* **143**:3–45.

Gundersen, H. J. G., Jensen, T. B., and Osterby, R. (1978). Distribution of membrane thickness determined by lineal analysis. *J. Microscopy* **113**:27–43.

Haefeli-Bleuer, B., and Weibel, E. R. (1988). Morphometry of the human pulmonary acinus. *Anat. Rec.* **220**:401–414.

Hammond, M. D., and Hempleman, S. C. (1987). Oxygen diffusing capacity estimates derived from measured $\dot{V}_A/\dot{Q}$ distributions in man. *Respir. Physiol.* **69**:129–147.

Hill, E. P., Power, G. G., and Longo, L. D. (1973). Mathematical simulation of pulmonary O_2 and CO_2 exchange. *Am. J. Physiol.* **224**:904–917.

Holland, R. A. B., Van Hezewijk, W., and Zubzanda, J. (1977). Velocity of oxygen uptake by partly saturated adult and fetal human red cells. *Respir. Physiol.* **29**:303–314.

Holland, R. A. B., Shibata, H., Scheid, P., and Piiper, J. (1985). Kinetics of O_2 uptake and release by red cells in stopped-flow apparatus: effects of unstirred layer. *Respir. Physiol.* **59**:71–91.

Hoppeler, H., Kayar, S. R., Claassen, H., Uhlmann, E., and Karas, R. H. (1987). Adaptive variation in the mammalian respiratory system in relation to energetic demand. III. Skeletal muscles: setting the demand for oxygen. *Respir. Physiol.* **69**:27–46.

Hoppin, F. G., Jr., and Hildebrandt, J. (1977). Mechanical properties of the lung.

In *Bioengineering Aspects of the Lung*. Edited by J. B. West. New York, Marcel Dekker, pp. 150–153.

Horsfield, K. (1986). Morphometry of airways. In *Handbook of Physiology*. Section 3. Vol. III, Part 1. Edited by P. T. Macklem and J. Mead. Bethesda, MD, American Physiological Society, pp. 75–88.

Huxley, V. H., and Kutchai, H. (1983). Effect of diffusion boundary layers on the initial uptake of O_2 by red cells. Theory versus experiment. *Microvasc. Res.* **26**:89–107.

Kapanci, Y., Weibel, E. R., Kaplan, H. P., and Robinson, F. R. (1969). Pathogenesis and reversibility of the pulmonary lesions of oxygen toxicity in monkeys. II. Utrastructural and morphometric studies. *Lab. Invest.* **20**:101–118.

Karas, R. H., Taylor, C. R., Rösler, K., and Hoppeler, H. (1987a). Adaptive variation in the mammalian respiratory system in relation to energetic demand. V. Limits to oxygen transport by the circulation. *Respir. Physiol.* **69**:65–79.

Karas, R. H., Taylor, C. R., Jones, J. H., Lindstedt, S. L., Reeves, R. B., and Weibel, E. R. (1987b). Adaptive variation in the mammalian respiratory system in relation to energetic demand. VII. Flow of oxygen across the pulmonary gas exchanger. *Respir. Physiol.* **69**:101–115.

Mandelbrot, B. B. (1983). *The Fractal Geometry of Nature*. New York, W. H. Freeman.

Mathieu, O., Cruz-Orive, L. M., Hoppeler, H., and Weibel, E. R. (1981). Measuring error and sampling variation in stereology: comparison of the efficiency of various methods for planar image analysis. *J. Microscopy* **121**:75–88.

Mochizuki, M. (1966). Study on the oxygenation velocity of the human red cell. *Jpn. J. Physiol.* **16**:635–648.

Montaner, J. S. G., Tsang, J., Evans, K. G., Mullen, J. B. M., Burns, A. R., Walker, D. C., Wiggs, B., and Hogg, J. C. (1986). Alveolar epithelial damage. *J. Clin. Invest.* **77**:1786–1796.

Müller, A. E., Cruz-Orive, L. M., Gehr, P., and Weibel, E. R. (1981). Comparison of two subsampling methods for electron microscopic morphometry. *J. Microscopy* **123**:35–49.

Paumgartner, D., Losa, G., and Weibel, E. R. (1981). Resolution effect on the stereological estimation of surface and volume and its interpretation in terms of fractal dimensions. *J. Microscopy* **121**:51–63.

Pedley, T. J., Schroter, R. C., and Sudlow, M. F. (1970). The prediction of pressure drop and variation of resistance within the human bronchial airways. *Respir. Physiol.* **9**:387–405.

Rodriguez, M., Bur, S., Favre, A., and Weibel, E. R. (1987). The pulmonary acinus: geometry and morphometry of the peripheral airway system in rat and rabbit. *Am. j. Anat.* **180**:143–155.

Roughton, F. J. W., and Forster, R. E. (1957). Relative importance of diffusion and chemical reaction rates in determining rate of exchange of gases in the human lung, with special reference to true diffusing capacity of pulmonary membrane and volume of blood in the lung capillaries. *J. Appl. Physiol.* **11**:290–302.

Scherle, W. (1970). A simple method for volumetry of organs in quantitative stereology. *Mikroskopie* **26**:57–60.

Schürch, S., Goerke, J., and Clements, J. A. (1978). Direct determination of volume- and time-dependence of alveolar surface tension in excised lungs. *Proc. Natl. Acad. Sci. USA* **75**:3417–3421.

Staub, N. C., and Storey, W. F. (1962). Relation between morphological and physiological events in lung studied by rapid freezing. *J. Appl. Physiol.* **17**:381–390.

Sterio, D. C. (1984). The unbiased estimation of number and sizes of arbitrary particles using the disector. *J. Microscopy* **134**:127–136.

Taylor, C. R., Karas, R. H., Weibel, E. R., and Hoppeler, H. (1987). Adaptive variation in the mammalian respiratory system in relation to energetic demand. *Respir. Physiol.* **69**:1–127.

Taylor, C. R., Weibel, E. R., Hoppeler, H., and Karas, R. H. (1989). Matching structures and functions in the respiratory system: allometric and adaptive variations in energy demand. In *Comparative Pulmonary Physiology: Current Topics*. Edited by S. C. Wood. New York, Marcel Dekker pp. 27–65.

Thurlbeck, W. M. (1967). The internal surface area of non-emphysematous lungs. *Am. Rev. Respir. Dis.* **95**:765.

Torre-Bueno, J. R., Wagner, P. D., Saltzman, H. A., Gale, G. E., and Moon, R. E. (1985). Diffusion limitation in normal humans during exercise at sea level and simulated altitude. *J. Appl. Physiol.* **58**:989–995.

Vandegriff, K. D., and Olsen, J. S. (1984). Morphological and physiological factors affecting oxygen uptake and release by red blood cells. *J. Biol. Chem.* **259**:12619–12627.

Weibel, E. R. (1963). *Morphometry of the Human Lung*. Berlin, Springer.

Weibel, E. R. (1970). An automatic sampling state microscope for stereology. *J. Microscopy* **91**:1–18.

Weibel, E. R. (1970/71). Morphometric estimation of pulmonary diffusion capacity. I. Model and method. *Respir. Physiol.* **11**:54–75.

Weibel, E. R. (1979). *Stereological Methods*. Vol. 1: Practical Methods for Biological Morphometry. London, Academic Press.

Weibel, E. R. (1980). *Stereological Methods*. Vol. 2: Theoretical Foundations. London, Academic Press.

Weibel, E. R. (1984a). *The Pathway for Oxygen*. Cambridge, MA, Harvard University Press.

Weibel, E. R. (1984b). Morphometric and stereological methods in respiratory physiology including fixation techniques. In *Techniques in the Life Sciences*. Respiratory Physiology. County Clare, Elsevier Scientific Publishers Ireland Ltd., pp. 1–35.

Weibel, E. R. (1985). Lung Cell Biology. In *Handbook of Physiology*. Section 3: The Respiratory System. Vol. I. Edited by A. P. Fishman. Bethesda, MD, American Physiological Society, pp. 47–91.

Weibel, E. R. (1986). Functional morphology of lung parenchyma. In *Handbook of Physiology*. Section 3: The Respiratory System. Vol. III, Part 1. Edited by P. T. Macklem and J. Mead. Bethesda, MD, American Physiological Society, pp. 89–111.

Weibel, E. R. (1989). Lung morphometry and models in respiratory physiology. In *A Quantitative Approach to Respiratory Physiology*. Edited by H. K. Chang and M. Paiva, Marcel Dekker, New York pp. 1–56.

Weibel, E. R., and Gil, J. (1977). Structure–function relationships at the alveolar level. In *Bioengineering Aspects of the Lung*. Edited by J. B. West. New York, Marcel Dekker, pp. 1–81.

Weibel, E. R., and Gomez, D. M. (1962). A principle for counting tissue structures on random sections. *J. Appl. Physiol.* **17**:343–348.

Weibel, E. R., and Knight, B. W. (1964). A morphometric study on the thickness of the pulmonary air-blood barrier. *J. Cell Biol.* **21**:367–384.

Weibel, E. R., and Taylor, C.R. (1981). Design of the mammalian respiratory system. *Respir. Physiol.* **44**:1–164.

Weibel, E. R., and Taylor, C. R. (1987). Design and structure of the human lung. In *Pulmonary Diseases and Disorders*. Edited by A. P. Fishman, New York, McGraw-Hill.

Weibel, E. R., Stäubli, W., Gnägi, H. R., and Hess, F. A. (1969). Correlated morphometric and biochemical studies on the liver cell. I. Morphometric model, stereologic methods and normal morphometric data for rat liver. *J. Cell Biol.* **42**:68–91.

Weibel, E. R., Gehr, P., Cruz-Orive, L. M., Müller, A. E., Mwangi, D. K., and Haussener, V. (1981). Design of the mammalian respiratory system. IV. Morphometric estimation of pulmonary diffusing capacity: critical evalutation of a new sampling method. *Respir. Physiol.* **44**:39–59.

Weibel, E. R., Marques, L. B., Constantinopol, M., Doffey, F., Gehr, P., and Taylor, C. R. (1987). Adaptive variation in the mammalian respiratory system in relation to energetic demand. VI. The pulmonary gas exchanger. *Respir. Physiol.* **69**:81–100.

West, B. J., Bhargava, V., and Goldberger, A. L. (1986). Beyond the principle of similitude: renormalization in the bronchial tree. *J. Appl. Physiol.* **60**:1098–1097.

Wilson, T. A. (1981). The relations among recoil pressure, surface area and surface tension in the lung. *J. Appl. Physiol.* **50**:921–926.

Wilson, T. A., and Bachofen, H. (1982). A model for mechanical structure of the alveolar duct. *J. Appl. Physiol.* **52**:1064–1070.

Yamaguchi, K., Nguyen-Phu, D., Scheid, P., and Piiper, J. (1985). Kinetics of O_2 uptake and release by human erythrocytes studied by a stopped-flow technique. *J. Appl. Physiol.* **58**:1215–1225.

9

Computerized Interactive Morphometry

DENNIS A. SILAGE

Temple University
Philadelphia, Pennsylvania

I. Introduction

With the availability of the first microcomputer-compatible video signal digitizer in 1984, an image analysis laboratory could be established without the expense and difficulty encountered in earlier minicomputer-based systems (Silage and Gil, 1982). Although some of the conceptual problems of digital image processing are not yet resolved, planimetric, densitometric, and morphometric measurements now can be performed efficiently and routinely. However, to understand the enormous capabilities but practical limitations of such digital image processing systems requires some discussion of the design of the specialized hardware, the methodology of the processing software, and the factors that determine an efficient human interface.

The specialized hardware consists of computer interface called a video frame memory (VFM) that is peripheral to a standard microcomputer system, which was originally introduced by IBM in 1981 and now called simply the personal computer (PC). In 1984 and again in 1987 IBM produced improved microcomputer systems, the PC/AT and the personal system/2 (PS/2). Although the PC/AT is somewhat compatible with the PC, the PS/2 is architecturally different from both the PC and the PC/AT. Peripherals such as a VFM designed for a PC or a PC/AT will not operate with the advanced models of the PS/2.

Digital image processing software requires routines that are extremely rapid in execution because of the large amount of information in a typical scene. One standard for a digitized image requires over 250,000 storage locations (bytes). An interesting observation is that if a typical word consists of 12 characters, even 1000 words would only need 12,000 bytes for its storage. This large amount of image information implies that most of the software should be written in the native or machine language of the microprocessor, and not in a conversational language such as BASIC. However, it is difficult to develop a complete application in such a machine or low-level language. Therefore, a compromise must be reached between the execution speed of the low-level language and the ease of programming in a conversational or high-level language.

Finally, an efficient interface between the user and the digital image processing computer system requires that the interactive software be skillfully crafted and that the required pointing and inputting device be a natural extension of some human trait. A high-resolution touch screen periperal (Silage and Gil, 1984a) seems to provide this bridge in many instances and is a solution to the tedium of the task of acquiring morphometric data from tens or hundreds of scenes.

Each of these concepts will be discussed and the contemporary design of a digital image processing system will emerge from that experience. However, an assessment of the state of progress in image processing perhaps can only be tempered by a comparison to those advances made for signals such as the voice.

With an understanding of the biological vocal tract, computer systems can now be configured not only to acquire and measure the characteristics of the voice, but also to recognize words. Although image understanding and recognition remain at issue, in a similar way computational techniques based on neural networks are being applied to recognize objects in a scene. Ultimately such a capability should have more impact, just as voice recognition has more utility than its mere signal measurement.

II. Digital Image Processing Hardware

A. Video Frame Memory

The single computer peripheral that defines digital image processing is the video frame memory (VFM). The VFM is an assembly of electrical subsystems that have been used extensively in the computer processing of analog signals. Electrical analog signals are rendered into a series of numbers that can be read and recorded by the computer by a device called the analog-to-digital converter (ADC). Conversely, a series of numbers can be transformed to an electrical analog signal by the digital-to-analog converter (DAC). The storage of these numbers is usually in the random access memory (RAM) of the computer host, unless the conversions to capture the signal accurately must occur so rapidly that the numbers cannot be transferred to the host in the time available. In this case, RAM is provided on the conversion peripheral for the local storage of the values. These conversions occur equally spaced in time, at a rate called the sampling rate, which is measured in hertz.

A high-resolution video image that conforms to an accepted protocol (the RS-170 standard) consists of 480 horizontal scan lines. Each line is sampled at a rate that results in 512 numbers in each scan line. The ADC subsystem of the VFM produces numbers that range from 0 to 255, which can be stored in a single location (a byte). Since the analog video image is sent 30 times per second, the 480 × 512 or 245,760 picture element (pixel) values cannot be transferred to the host computer in the time available. RAM is provided on the VFM to store the pixels for this reason and because the early host microcomputers (the PC) usually did not have the requisite amounts of RAM. To view the current pixel contents of the VFM, a DAC transforms the values to an analog video image for display on a conventional video monitor.

In this configuration, the VRM behaves as though the analog video input signal is connected directly to the monitor. However, the VFM has a control register (CR) that can be read from and written to by the host computer. The contents of the CR determine, among other functions, whether the image is continuously acquired by the VFM (the function is called GRAB) or if only one image is acquired, stored in RAM but continuously displayed (the function is SNAP).

The analog video monochrome input image is converted to pixels that range in value from 0, corresponding to absolute black, to 255, for intense white. Each of these pixels is stored in a single location or byte. A byte is an ensemble of eight sublocations (a bit) that can only assume the value 0 or 1. The relative position within the byte of those bits that contain a 1 add a weight to the number stored in the byte. The weights, from left to right, are 128, 64, 32, 16, 8, 4, 2, 1, or descending powers of the base 2. Thus a pixel that in bit form is 01010011 is 64 + 16 + 2 + 1, or 83.

Although 256 discrete pixel values are available, only 128 or fewer levels are generally observed in an arbitrary monochrome image. These 128 valves require only 7 of the 8 bits in a byte and the 1 remaining bit is used to provide a graphic overlay for cursors and alphanumerical text by the use of the input and output look-up table (LUT). The LUT is a construct of an additional 256 locations in the memory of the VFM that indexes one pixel value to another pixel value. A LUT can be read from or written to by the host computer system and a VFM can have several sets of input and output LUTs that can be selected from the CR.

When a graphic overlay capability is not required, both of the LUTs are linearized, that is a value of 0 is in location 0, 1 in 1, 2 in 2, and similarly in each location until a value of 255 is in location 255. If a LUT is linear, essentially no translation of a pixel value occurs.

The input LUT translates the pixel value before storage in the RAM of the VFM. The output LUT translates the stored pixel to an output display value. The input and output LUTs can provide an overlay graphics capability. All the odd locations of the input LUT are set to zero and all the even locations have linear values, that is a value of 0 is in location 0, 0 in 1, 2 in 2, 0 in 3, until a value of 254 is in location 254 and 0 in 255. All the odd locations of the output LUT contain a value of 255 and all the even locations likewise have linear value. To evoke a white dot for the graphic overlay only the least significant bit, which has a weight of 1 and determines if a pixel value is odd or even, need be set. The input LUT configuration ensures that no stored pixel value can be an odd number.

B. Video Camera and Monitor

The analog video camera acquires the electrical image signal that is digitized by the VFM. The camera must be of the high-resolution type capable of producing 480 vertical scan lines per image in 2 interlaced fields in which first all the odd lines are sent in 1/60 of a second, then the even lines in the next 1/60 of a second. If the camera is monochrome, the electrical signal is a composite of the image information and the horizontal and vertical sychronization (sync) signals. This composite signal has been standardized and called the RS-170 video signal.

A monochrome video monitor in the digital image processing system displays the analog output of the VFM. The monitor must also be capable of displaying a high-resolution, composite video signal. The performance of a monitor is usually described in terms of its electrical frequency bandwidth in hertz. A bandwidth of at least 20 MHz is recommended for the monitor.

A red–green–blue (RGB) color digital image processing system requires a high-resolution color video camera, three VFMs to acquire the complete image, and an RGB color monitor. Both the color video camera and monitor are expensive and with three VFMs the color system is at least three times as expensive as a monochrome system. Apart from that, however, there is also three times the video information stored in the VFM that must be analyzed by the same host computer system, which can significantly affect the execution speed of the analysis program.

C. Host Computer

The host computer system interacts with the VFM in analyzing the digitized video information. The VFM usually does not have a microprocessor, but because of its elaborate digital hardware, such as the CR and LUT, the VFM does reduce the computation burden on the host computer. For example, the LUT operation could be performed by the host on each pixel in each image by first reading a pixel, transforming it, and then writing it back to the VFM. In a complete digital image this would require over 250,000 storage location transfers. However, the LUT operation is more efficient, requiring only 256 such transfers. In addition, if the LUT operation is fixed, it would only be done once for any image. If the host were to perform the same operation on the pixel data, it would have to be done on each image.

The host computer system is required to have a standard peripheral structure known as either the PC or PC/AT bus. A VFM designed for the PC bus will function on the PC/AT bus, but the reverse is not true.

The RAM memory of the host computer and all other peripherals compete for the available address space of the microprocessor. In the PC and many of the versions of the PC/AT the available space is defined as 1024K storage locations, or 1024×1024 bytes, where the system K represents a multiplication of the index by 1024. The digital image data in the VFM is 256K bytes, which is 25% of the available address space of the host computer system. To be efficient, the image data of the VFM is mapped as four 64K byte contiguous segments and each segment is addressed by first setting digital registers in the CR of the VFM.

While a nonremovable magnetic disk storage device is an option in the host computer system, a numerical data processor (NDP) is a necessity for efficient planimetric and densitometric measurements. However, the mere presence of the NDP does not ensure that it will be used in the analysis of image data. The computer language in which the application software was written must specifically include routines that utilize the NDP.

D. Interactive Peripheral

A high-resolution interactive cursor peripheral is desirable as part of the digital image processing system so that the investigator can mark structures, regions of interest, and select menu items for controlling the executing of the program. A mouse has been used for this application, but a touch screen peripheral mounted directly on the video monitor has a distinct advantage. This mounting eliminates the so-called correspondence image that occurs when the input device is manipulated on a surface other than that where the cursor appears.

Touch-sensitive screens are available that operate by acoustic, infared, capacitive, or resistive sensors. However, of these only the resistive membrane touch screen seems to provide a stable calibration and high resolution. These interactive peripherals are available from several manufacturers and have an integral controller that provides a convenient data output. In the most common case, a serial data port is provided and a stream of characters is sent to the host computer system to indicate a current cursor position.

The assembly language routine CALL TSS decodes the character string to a set of numbers that can be manipulated by the applications software written in BASIC. CALL TSS also provides a repeatable cursor location by averaging several current readings and eliminates the occasional false indication of position inherent in this device.

A touch-sensitive screen is also a random input device that, unlike a mouse peripheral, does not have to be "dragged" to the region of interest. With it even the four corners of the screen can be designated in rapid succession. This facility is especially useful in point and intersect counting morphometry, where the same structures can now be identified at their random locations without the tedium of the traditional random structure identification at sequential locations (Silage and Gil, 1984a).

III. Digital Image Processing Software

The morphometric and planimetric analysis software described here has been under development for several years with the close cooperation and assistance of several research laboratories.

A. Assembly Language Subroutines

A standard set of assembly or machine language subroutines for efficient digital image processing has been developed and linked to the conversational computer language BASIC. The digital image analysis software can be distributed to laboratory systems that utilize VFMs from different manufacturers by providing only the assembly language module for that specific hardware. The BASIC language analysis software is identical and essentially performs the same on each system. The subroutines are described by their CALLed name and the integer BASIC variables that are used as the arguments of the function. In the following description of the functions, the variables I, J, K, JX, KY, and IV are BASIC integer variables and if no arguments are listed, none are required.

CALL VINIT

The VFM is INITialized and two sets of LUTs are defined. The LUT 1 set is a linear function that does not translate either an input or output pixel value. The LUT 2 set provides the graphics overlay capability. The other LUTs of the VFM are not defined and the LUT 1 set is selected (see VLLUT).

CALL SLUT(K)

The single input or output LUT K is selected for all further readings (see RLUT), or writings (see WLUT), or VFM image acquisition functions (see SNAP and GRAB). An input or output LUT remains selected until another LUT of the same type is specified.

CALL VLLUT

The input and output linear LUT set 1 is selected and is useful in returning to this pixel transform without executing several CALL SLUT operations.

CALL VGLUT

The input and output graphical overlay LUT set 2 is selected.

CALL SNAP

The current video image is SNAPed into the VFM using the currently selected input LUT to translate the pixel values before storage. Each execution of CALL SNAP acquires one image.

CALL GRAB

The current video image is continuously GRABbed into the VFM, again using the currently selected input LUT to translate the pixel value. CALL GRAB is essentially a repetitive CALL SNAP operation.

CALL WLUT(I,J)

The previously selected LUT is written with the data J, in the range of 0–255, at the location I, also in the range of 0–255. To write input LUT 3 as an inverse pixel manipulation (negative to positive transform), execute the following BASIC sequence:

```
100 K = 3:CALL SLUT(K)
110 For I = 0 TO 255
120 J = 255–I:CALL WLUT (I,J)
130 NEXT I
```

CALL RLUT(I,J)

The data at location I, again in the range of 0–255, of the previously selected LUT is returned as the BASIC integer variable J. CALL RLUT is provided to verify the values currently stored in the LUTs, but it is not necessary to read the LUT.

CALL WPIX (JX, KY, IV)

The pixel value IV, in the range of 0–255, is stored in the VFM at horizontal (x) location JX and vertical (y) location KY. The range of JX is 0 (left) to 511 (right). By a convention of digital image processing, the range of KY is 0 (note that this is the top) to 479 (bottom). The following BASIC sequence produces a horizontal black line at vertical location 50, which is near the top of the video display:

```
100 KY = 50: IV = 0
120 FOR JX = 0 TO 511
130 CALL WPIX (JX, KY, IV)
140 NEXT I
```

CALL RPIX (JX, KY, IV)

The pixel value stored in the VFM at the horizontal location JX and vertical location KY is returned as a BASIC integer variable IV, again in the range 0–255. The following BASIC sequence arbitrarily adds 50 units to all the pixels in a portion of the digital video scene bounded by the vertical lines at x equals 50 and 100:

```
100 FOR JX = 50 TO 100
110 FOR KY = 0 TO 479
120 CALL RPIX (JX, KY, IV)
130 IV = IV + 50
140 IF IV < = 255 THEN 160
150 IV = 255
160 NEXT KY
170 NEXT JX
```

CALL VMOVE (JX, KY)

The graphics cursor (GCUR) is positioned at the JX, KY as described for the RPIX and WPIX functions. CALL VMOVE is used for overlay graphics to move the GCUR without drawing (see VGLUT, VDRAW, and VERASE).

CALL VDRAW (JX, KY)

An overlay graphics line is drawn by setting bit 0 to 1 (see VGLUT) for every pixel from the last position of GCUR to the new position JX, KY as first described for the RPIX and WPIX functions. The last position of GCUR is determined by the execution of the VMOVE, VDRAW, VPOINT, VERASE, or VTEXT function. After CALL VDRAW is completed, GCUR is at the location JX, KY. As an example of the interaction between the assembly language subroutines and the BASIC language program, the following sequence initializes the VFM, acquires a single digital video image using the graphical overlay LUT set, and draws a rectangular box near the center of the video display:

100 CALL VINIT: CALL VGLUT: CALL SNAP

110 JX = 100: KY = 100: CALL VMOVE (JX, KY)

120 JX = 400: CALL VDRAW (JX, KY)

130 KY = 400: CALL VDRAW (JX, KY)

140 JX = 100: CALL VDRAW (JX, KY)

150 KY = 100: CALL VDRAW (JX, KY)

CALL VERASE (JX, KY)

The overlay graphics line is erased (if one was drawn) from the last position of GCUR to the new position JX, KY and after completion GCUR is then at JX, KY.

CALL VTEXT (T$)

The BASIC character string T$ is formed as a graphic overlay character string, each character of size 8 horizontal and 14 vertical pixels. The upper left corner of the first character is the last position of GCUR. GCUR can be positioned by the CALL VMOVE function.

CALL VCLEAR

The entire overlay graphics bit 0 plane of the VFM is cleared, or set to 0, which removes all the lines and text seen on the video diaplay.

CALL CURS (JX, KY, IV)

A cross-hair cursor of size 9 horizontal by 11 vertical pixels with a pixel value IV is positioned at location JX, KY. This is not an overlay graphics function because the function does not use the bit 0 plane of the VFM. The 19 affected original digital image pixels are first read from the VFM and stored in the RAM of the host computer (see ECURS), then 19 pixels of value IV are stored in the VFM to display the cross-hair cursor.

CALL ECURS

The last cross-hair cursor generated by the function CALL CURS is effectively removed by replacing it with the original 19 pixels of the digital image that were stored in the RAM of the host computer. The following BASIC sequence alternates a cross-hair cursor of pixel value 200 at two locations on the video display:

100 JXA = 100: KYA = 100: JXB = 200: KYB = 200: IV = 200

110 CALL CURS (JXA, KYA, IV)

120 CALL ECURS

130 CALL CURS (JXB, KYB, IV)

140 CALL ECURS: GOTO 110

CALL TSS (JX, KY)

The horizontal and vertical position of a small, passive stylus on the interactive touch-sensitive screen peripheral is returned as the BASIC integer variables JX and KY. If the touch-sensitive is not depressed, a value of 0 is returned for both JX and KY. If JX equal -1, the keyboard of the host computer has been struck and KY contains the key code.

CALL MPIC

The 480 vertical by 512 horizontal pixels stored in the VFM are moved to the RAM of the host computer system. Only one such digital image can be stored in the RAM.

CALL RPIC

The digital image stored in the RAM of the host computer is returned to the VFM, effectively replacing the current scene in the VFM.

CALL ZOOM (JX, KY, IV)

The digital image stored in the RAM of the host computer is zoomed by the factor IV at the original location JX, KY to the center of the video display, or location 256,240. Before this function is called, the original digital image is moved once to RAM by using CALL MPIC.

B. Conversational Language Application Software

The assembly language routines described in the preceding section can be evoked by any conversational or high-level computer language by passing the appropriate parameters. Most research laboratories favor BASIC and have adopted the advanced versions of this language that have recently been developed. BASIC now is not merely an interpreted language that operates slowly but has been markedly improved in both syntax and execution.

An applications program is then designed to manipulate the digital image, perform certain calculations, and store results. Programs have been developed for morphometric and planimetric measurement with features to reduce significantly the tedium such as the digital image reversal of a negative with enhancement (Silage and Gil, 1984b) and the zooming of a scene to facilitate the identification of structures (Silage and Gil, 1985).

The modularity of this approach, which is independent of the specific digital image processing hardware in the laboratory, provides the capability to develop new application programs rapidly and to exchange software with other laboratories. There are many morphometric analyses and current research continues to develop new approaches. (Gil et al., 1986). Only this concept in digital image processing hardware and software would seem to continue to support these advances.

References

Gil, J., Marchevsky, A., and Silage, D. A. (1986). Applications of computerized interactive morphometry in pathology. *Lab. Invest.* **54**:222–227.

Silage, D. A. and Gil, J. (1982). Digital image analyzer for the morphometric reconstruction of biological tissue. *Med. Comput. Sci.* (IEEE/CS) 1:456–459.

Silage, D. A., and Gil, J. (1984a). The use of a touch sensitive screen in interactive morphometry. *J. Microsc.* **134**:315–321.

Silage, D. A., and Gil, J. (1984b). Digital reversal and enhancement of negatives in video based interactive morphometry. *Lab. Invest.* **51**:112–116.

Silage, D. A., and Gil, J. (1985). Digital image tiles: a method for the processing of large sections. *J. Microsc.* **138**:221–227.

10

Methods for Casting Airways

KEITH HORSFIELD

Cardiothoracic Institute
Midhurst, West Sussex
England

I. Introduction

A. Development of Airway Casting

Although the ancients knew that the lung contains a system of branching airways, it was not until late last century that a detailed study of the bronchial tree was made by Aeby (1880). He used metal casts to demonstrate the pattern of branching in the lungs of various animals. The next advance in the technique of casting was made by Schummer (1935), who introduced methyl methacrylate. This was followed the next year with the use of vinyl compounds by Narat et al. (1936). Latex has been used extensively for vascular injection studies, notably by Lieb (1940), because of its property of flexibility. Pump (1964) adapted this technique for airway casting, using moulage, another latex. The use of polyester resins for making lung casts was reported by Scales (1950), and the technique was perfected by Tompsett (1970). These resins make excellent rigid casts. Frank and Yoder (1966) introduced silicone rubber as a casting medium; its use was developed by Phalen et al. (1973) for making airway casts in situ. It is a very satisfactory material for making flexible casts that hold their shape better than latex.

B. Purpose of Airway Casting

Airway casts may be classified into two main groups: negative casts, which are solid and represent the lumen of the tube; and positive casts, which are hollow and represent the walls of the tube.

Negative casts are used for the following purposes: (1) to obtain information on the general arrangement and types of airway to be found in the tree; (2) to make measurements of the dimensions of the airways and their branching angles; (3) to count their numbers; (4) to study the pattern of branching; (5) to demonstrate the normal anatomy of the bronchial tree; (6) to demonstrate pathological changes. A selection of studies on these topics are described in Chapter 11. Methods for studying different parts of the normal airways are described below.

Positive casts are used for experimental purposes involving fluid flow, including the study of flow regimes (West and Hugh-Jones, 1959; Dekker, 1961; Olsen et al., 1973; Snyder and Jaeger, 1983), gas mixing (Horsfield et al, 1977, 1980), and particle deposition (Schlesinger et al., 1977).

II. Methods of Casting

A. Materials for Negative Casts

A variety of materials have been used for making casts; only the more important will be described here. They have a wide range of physical properties, their value depending very much on the requirements of the technique being used. A general need is to remove the tissues once the cast has been made, and this is commonly done with corrosive chemicals. It follows that the material from which the cast is made must be able to withstand attack from such chemicals, whether acid or alkali. For negative casts, the material must be durable and robust enough to withstand handling, pruning, and measuring, and should therefore either be strong and rigid, or flexible. Other desirable properties of an ideal injection medium include the possibility of adjusting the viscosity, an adequate working time, a constant viscosity during the practical working time, and no change in volume on setting. It should not be miscible with water nor permeate the tissues. For positive casts it is necessary to make a negative first, and then use this as a mold for the positive. The negative then has to be removed, which is most commonly achieved by melting it, and so a material with a relatively low melting point is required.

Metals

Metallic alloys can be made with remarkably low melting points, but probably the most often used is Woods metal, which has a melting point of 70 °C. It has to be poured into a dry and fixed lung to avoid deformation by its weight and the problems that might result from a water/molten metal interface. It can pass all the way down to alveoli, where a heavy cast that is extremely difficult to prune can result. To prevent this, careful control of temperatures is required. In the past it has been used to make final negative casts, but the more modern materials are better suited to this purpose. It is still used to make the negative cast mold for positive casts, since it is strong and rigid but can be melted out of the positive.

Wax

A wide variety of waxes are available, with a range of melting points. Some workers have used dental waxes, but paraffin wax with a melting point of 60° is satisfactory. Negative wax casts are made in the same way as metal casts, and are similarly used to make positive casts. They are much lighter and much more fragile than metal casts, and need to be handled with great care, but can give excellent results.

Methyl Methacrylate

Schummer (1935) introduced this material for casting purposes. Its use was investigated in some detail by Bugge (1963), who described how to adjust its viscosity and working life as required for different uses. It can cast structures as fine as capillaries, but it has been superseded by the polyester resins for bronchial tree casting.

Vinylite

This was introduced by Narat et al. (1936) as a material for making casts and was used by Liebow et al. (1947) to demonstrate pulmonary pathology. The technique was improved by Tucker and Krementz (1957a, b) for making vascular and bronchial casts. Vinylite is dissolved in acetone, and suffers considerable shrinkage as the solvent evaporates. In addition, it is rather brittle and is easily broken. It has been superseded by the more modern polyester resins.

Polyester Resins

The potential for using these compounds to make casts was recognized by Scales (1950), but the detailed methods were developed by Tompsett (1970). Casts made from polyester resins are strong and stable, but rather brittle where thin branches have been cast. This actually makes pruning easier, as long as it is done with care. A quick twist with a narrow-ended forceps is sufficient to sever a branch. The resins can be self-colored, for example, when making combined vascular and airway casts, or can be painted, for example, to identify orders. A wide range of resins are available, with many different physical properties. It may be necessary to experiment to find one with the right properties for any given application.

Latex

Pump (1964) used moulage for airway casting, especially for demonstrating the intra-acinar branches. The material can be self-colored, and may also be used for vascular injections. It has the useful property of being instantly solidified by strong acid, so an injected specimen may be immersed for maceration in hydrochloric acid immediately.

Silicone Rubber

Many different kinds of silicone rubber are available and, as with the polyester resins, it may be necessary to experiment to find one to suit a particular application. Their use for airway casting was introduced by Frank and Yoder (1966). They give very accurate reproductions, are strong, stretchable, flexible, and return to their original shape. They undergo very little volume change on setting, may be pruned by cutting, and are immiscible with water. They have the disadvantage of being rather viscous, and require either a high pressure or, preferably, a long time for injection to be completed.

B. Materials for Hollow Casts

These need to be stable and strong enough to retain their shape, and be able to withstand the trauma that might result from the removal of the negative cast. In some cases, especially where flow regimes are being studied, transparency might also be important.

Electroplated Metal

Woods metal negative casts were electroplated directly with copper by Pedersen et al. (1983). The Woods metal was then removed by boiling water, steam, and a propane torch. The casts were made for the purpose of studying fluid flow mechanics. Timbrell et al. (1970) used a wax negative cast coated with colloidal silver to make it electroconductive. They used silver plating, which is quicker than copper and more evenly deposited. Wax was removed by combined heating and blowing. The cast was used for studies of gas mixing in the airways.

Latex

A metal alloy negative cast was coated with two layers of latex by Eisman (1970). The metal was removed by melting, squeezing, using fine forceps, making small incisions that could be easily repaired, and finally by immersion in hydrochloric acid. The cast was used for airflow studies.

Silicone Rubber

Schlesinger et al. (1977) used silicone rubber to coat a wax negative cast by repeatedly painting it on and curing each coat. The wax was removed by heating, squeezing, and blowing. The cast was used for studying particle deposition.

Plexiglas

A metal alloy negative cast was coated with 30 layers of Plexiglas dissolved in chloroform by Snyder and Jaeger (1983). The metal was removed by heating and by amalgamating the residue with mercury. This technique produced a clear cast, which was then used to study the lobar distribution of flow in the central airways.

Polyester Resin Block

West and Hugh-Jones (1959) used a complicated technique to obtain a positive cast of the upper airways. They obtained a negative polyester resin cast using Tompsett's method (see later) and from this made a positive mold. A wax negative was then made, and this was embedded in a solid block of resin. The wax was melted out, the ends of the "airways" accessed by drilling holes in the block, and the block finally faced and polished to give a transparent finish. The cast was used to study patterns of gas flow.

A cast of the trachea was made by Dekker (1961). He started with a wax negative and embedded this in polyester resin. The wax was removed by heating, leaving a transparent block that was used to study flow regimes.

C. Preparation of the Lungs for Casting

In general, the lungs should be obtained as fresh as possible. While there is merit in proceeding with the casting while they are still fresh, consideration should be given to the fact that smooth muscle maintains the ability to contact for some hours after the death of the whole animal. Injection media may stimulate contraction of this muscle and thereby affect the dimensions of the cast. This is probably more important for arteries than for airways. When using excised lungs, they can be sealed in a plastic bag and kept at 4 °C overnight, or stored for longer periods before use at −20 °C in a freezer.

Air-Filled Lungs

With some casting materials, such as wax and Woods metal, it is necessary to use dry, air-filled lungs. These are prepared by blowing air down the trachea from a steady supply, gently at first to inflate the lungs, and then at a constant pressure, usually 30 cm H_2O. The time taken to dry and fix the tissues varies. Periods of 3–7 days are used by different authors, though Timbrell et al. (1970) used warm air for only 3 hr. Human lungs take longer than those from small animals. When thoroughly dry, they are ready for use.

Water-Filled Lungs

In Situ in the Thorax

The technique is described by Phalen et al. (1973, 1978). Animals should be anesthetized, the trachea intubated, and cardiac arrest obtained with intravenous sodium pentobarbitone or potassium chloride. With human cadavers the procedure should be carried out as soon as possible postmortem, after having first dealt with any legal and ethical questions. After intubation, the lungs are ventilated with carbon dioxide. Three or more ventrolateral incisions are carefully made through the chest wall on each side, without damaging the lungs, at the level of the apex, near the diaphragm, and midway between them. This causes a pneumothorax and the lungs collapse. The pneumothorax is then filled at a pressure of 5–10 cm H_2O with normal saline that has been degassed by exposure to a partial vacuum. The expanding lungs displace the saline from the pleural cavity. Several lung volumes of saline are instilled over a period of 15–30 min. The saline passes through the pleura and takes with it the remaining carbon dioxide in solution. In this way all the gas is removed from the lungs, which are expanded to fill the thoracic cavity. They are now ready for injection of the casting medium.

Excised Lungs

Great care must be taken not to damage the lungs during excision. After a pneumothorax is induced, the thorax should be opened widely and the entire mediastinal contents removed, including a length of trachea. If a pulmonary vascular cast is also required the heart should remain in situ (see Tompsett, 1970) but if not it may be removed along with the esophagus and aorta. The trachea is intubated and the tube tied firmly in place. The lungs may now be ventilated with carbon dioxide, and then inflated with degassed normal saline, as decribed for lungs in situ. They can be supported in a water bath while this is being carried out. Animal lungs collapse on removal from the thorax and therefore contain very little air. Because of this it is possible to obtain good casts without washing out first with carbon dioxide. When making resin casts it may help to prevent excessive peripheral filling of the cast if the lungs are next filled with a solution of gelatin. The gelatin also helps to maintain their shape while the resin is setting. A solution of 12 g/L at 40 °C is satisfactory (Tompsett, 1970), and it may help to warm the saline in the supporting bath. The casting medium may now be injected into the saline- or resin-filled lung.

Tompsett (1970) recommends fixing the lungs prior to injecting the casting medium. This can be done in a tank of spirit or 4% formaldehyde solution into which the lungs are placed. The fixative is injected into the tracheal cannula at a constant pressure from a reservoir with an overflow at a fixed height, the fixative being recirculated by a suitable pump. Fixed lungs are difficult to macerate, requiring hydrochloric acid, and good casts can be obtained without fixing thetissues first. For these reasons I prefer not to fix lungs prior to casting.

D. Injecting the Casting Medium

Only two media will be described here in detail; references to papers describing the use of other materials have been given above.

Polyester Resin

Anyone wishing to make airway casts using polyester resins would be well advised to refer to Tompsett (1970), but the main points of the technique will be given here. The original resin recommended by Tompsett, Crystic resin 700, is no longer available, and I have been unable to find one as good for all purposes. However, by experimenting with different products, it is usually possible to find one that will suffice. A gelling time of 10–30 min after adding the catalyst is desirable, to allow sufficient working time without allowing the resin too much opportunity to diffuse into the tissues. It is essential to know how the chosen resin behaves, and to this end some preliminary experiments are necessary. An appropriate quantity of catalyst, according to the instructions, is added to a quantity of activated resin at room temperature (20 °C) and the time is noted. The temperature of the mixture is recorded at regular intervals and a temperature-time graph plotted. In addition, the time and temperature at which gelling occurs are also noted. The reaction is exothermic, and the temperature rises increasingly rapidly. If one needs to reduce the viscosity of the mixture, this can be done by adding monomer before the catalyst, in which case the experiment must be repeated with the diluted mixture. Since about 4 min is an adequate working time, the temperature at which to start injecting the resin is read off the graph 4 min before gelling time. The reaction time is affected by the proportions of the constituents, especially of the catalyst, the presence of dyes for giving color to the casts, and the temperature. It is therefore good practice always to work at the same temperature, and since the materials should be stored in a refrigerator, they must be allowed to warm up before being used.

The quantity of resin to use is determined by a combination of knowledge, experience, and judgement. As a guide, a volume equal to 5% of the expanded lung volume should be injected, and half as much again should be made up to allow for the tubing and other factors. When using colored resin, the material should be just visible under the pleura after injection.

The resin may be injected using a large syringe, or more easily run in by gravity from a funnel held at a predetermined height above the water level in the bath. It may be necessary to control the quantity entering by clamping the tube. With either method, all the tubing should have any air bubbles meticulously excluded first. Resin can be poured straight onto the saline or gelatine, which it easily displaces. When using gelatin it is best to allow some to run out of the tracheal tube first, before injecting the resin, to allow for expansion of the lung as the resin enters. With saline this is not usually necessary, because water passes through the pleura fairly easily. When the resin injection is finished, the gelatin should be solidified by cooling rapidly with the addition of cold water or ice to the bath. Because the reaction is exothermic, any resin remaining in a syringe, funnel, or beaker after the injection can become hot enough to melt a plastic container. For this reason it is advisable to immerse equipment containing resin in cold water as soon as the injection is finished. Thermometers and any other equipment with resin on it should be wiped with acetone as soon as possible, to remove any resin before it sets hard. If one wishes to reuse the tracheal cannula, this can be disconnected from the tubing after a few minutes and much of it cleared of resin while it is still a gel. The preparation should then be left for 1 or more days for the resin to harden, after which the tissues can be macerated.

Silicone Rubber

A suitable material is Silastic E made by Dow Corning (Phalen et al., 1973, 1978). The quantity required is determined in a similar manner as for resin casts. It is thoroughly mixed with the correct quantity of catalyst using an electric stirrer, and then degassed under a partial vacuum to remove any air bubbles. The mixture is then injected slowly via the tracheal tube using a syringe pump, which takes about an hour. Self-curing takes about 12 hr, so the preparation can conveniently be left until the following day. If the cast has been made in situ in the thorax, the lungs may then be carefully removed and are ready for maceration.

E. Macerating the Tissues

Several methods have been used for removing the tissues. These include digestion with pancreatic enzymes, leaving the tissues exposed outside to decompose, and immersion in acidic or alkaline solutions. The first two are slow and smelly, and are not to be recommended. Tompsett (1970) used hydrochloric acid, and there is no doubt that this is effective. It does, however, have several disadvantages. The acid is potentially dangerous, and gives off corrosive fumes that attack every piece of metal in the laboratory. To counteract this it is necessary to work with ammonia fumes at the same time, in order to neutralize the acid fumes. Because of these complications, I do not recommend the use of hydrochloric acid.

Sodium hypochlorite solution (bleach) is effective for macerating small lungs, and is not as troublesome to use as hydrochloric acid. With large lungs, it works rather slowly.

The best general-purpose corrosive is a solution of sodium or potassium hydroxide. The sodium salt may form hard soaps with fatty tissue, so if this is a problem the potassium salt can be used. A 3M solution (120g/L of sodium hydroxide or 168 g/L of potassium hydroxide) is usually adequate, though weaker (1 M) and stronger (6 M) solutions can be used. Small lungs are macerated in a day or 2 while larger lungs may take a week or 2. It may help to remove the lungs after a day or 2 and wash them gently with a jet of water to remove partially digested tissues, and then replace them in the macerating solution for a further period. Small lungs can be handled fairly easily, but larger ones, such as human lungs, may distort on lifting with consequent breaking of the cast. Tompsett (1970) recommends the use of a Plexiglas tray in the shape of the posterior thoracic wall to support them during lifting, moving, or washing. When maceration is complete, the cast should be thoroughly washed in water, and may be rinsed in a dilute acid to remove any traces of alkali. Great care should be taken when using any of these corrosive agents. Avoid splashing, and wear goggles, gloves, and aprons. Any material that accidently gets on to the skin or clothing, or in the eyes, must be rinsed off immediately with copious cold water.

F. Region of Airways Demonstrated by Casting

Negative casts have been used to demonstrate the whole range of hollow airway structures; for example, the nasal passages, mouth, pharynx, and larynx (Olson et al., 1970; Schreider and Raabe, 1980, 1981b), trachea (Dekker, 1961), lobes and segments (Rahn and Ros, 1957; Tucker and Krementz, 1957b), bronchial tree (Ross, 1957; Weibel, 1963; Horsfield and Cumming, 1968, 1976; Tompsett, 1970; Phalen et al., 1973; Raabe et al., 1976), the acinus (Willson, 1922; Pump, 1964, 1969; Horsfield and Cumming, 1968; Schreider and Raabe, 1981a), and channels of collateral ventilation (Raskin and Herman, 1975; Henderson et al., 1968/1969; Andersen and Jespersen, 1980). This is not a complete list but it is sufficient to provide a guide to anyone wishing to make casts of these regions of the airways.

III. Discussion

A. Positive Casts

It is possible to obtain faithful reproductions of the airway lumen with hollow casts, from the nasal passages down to airways 1.0 mm in diameter. Transparent casts have been used successfully to study flow regimes using dyes in liquid, or smoke in air, and pressures and flows have been measured in transparent and opaque casts. Gas mixing in airway casts has also been studied, as has particle deposition. These results have provided a good deal of useful information that would have been difficult to obtain in real lungs because of the problems of access. In life, the airways from the larynx downwards are capable of considerable changes in dimension, varying normally with the phase of respiration. In the larynx, the vocal cords move together and apart with respiration, and have a considerable effect on the flow regime. Studies of flow based on casts from the trachea downwards miss this effect, and may be open to some difficulty of interpretation. Others have included the larynx modeled with a variable-sized opening of the cords to allow for this. Flow in the airways is reciprocating and it is known that the flow regimes differ in the two directions. Thus modeling constant flow in one direction will only demonstrate part of what is happening during respiration. If airflow is induced in a bronchial tree cast by applying a pressure gradient between the trachea and the open ends of the cast, the distribution of flow is determined by the resistance of each pathway. This can be overcome by attaching a resistance on each open ending, although this can be a big task if there are many peripheral airways. Thus, while casts have been helpful in developing our understanding of airway flow, the technique is inevitably limited in its ability to represent events occurring in the airways during life.

B. Negative Casts

For obtaining information on numbers, dimensions, branching angles, and branching patterns, airway casts have proved to be especially valuable. The method is subject to the same criticism as positive casts, namely that living airways vary in dimensions and branching angles during respiration. In general, airway casts have given information on lungs at 3/4 to full inflation, and a scaling down of the dimensions is required to model the situation at functional residual capacity. For comparisons between individuals or species, this matters less, because the lungs need to be inflated to the same degree in each case. The morphometric data that casting have made available to us would have been difficult to obtain by any other means, and for this purpose the method must be considered to have been extremely successful.

Silicone rubber casts have the advantage of being unbreakable; anyone who has dropped a precious resin cast on the floor will understand the enormous value of this. They are not as easy to prune because each branch must be cut, whereas small branches of resin casts are brittle enough to snap off easily. Resin casts are easier to paint, for recording or demonstrating orders. When measuring rubber casts, branches can be bent out of the way to gain access to the interior of the cast, whereas with resin casts it may be necessary to cut a branch off to achieve this. Of course, the branch can be glued back on after the measurements have been made.

IV. Future Developments

A better material for casting the negative mold, prior to making a positive cast, is required. Wax is very fragile and Woods metal is very heavy. Both can set rapidly after pouring, and getting the material to run into all the branches is difficult. What is required is a material with the strength, rigidity, lightness, and easy working properties of resin, but that can be melted out of the positive as wax can.

Making negative casts at lower lung volumes, with the volume controlled and known, is surprisingly difficult. A method needs to be developed to achieve this, so that the changes in airway dimensions with lung volume can be studied in detail.

References

Aeby, C. (1880). *Der Bronchialbaum der Saugethiere und des Menschen. Nebst Bemerkungen uber den Bronchialbaum der Vogel und Reptilien.* Leipzig, Engelman.

Andersen, J. B., and Jespersen, W. (1980). Demonstration of intersegmental respiratory bronchioles in normal human lungs. *Eur. J. Respir. Dis.* **61**:337–341.

Bugge, J. (1963). A standardized plastic injection technique for anatomical purposes. *Acta Anat.* **54**:177–192.

Dekker, E. (1961). Transition between laminar and turbulent flow in human trachea. *J. Appl. Physiol.* **16**:1060–1064.

Eisman, M. M. (1970). Lung models: hollow, flexible reproductions. *J. Appl. Physiol.* **29**:531–533.

Frank, N. R., and Yoder, R. E. (1966). A method of making a flexible cast of the lung. *J. Appl. Physiol.* **21**:1925–1926.

Henderson, R., Horsfield, K., and Cumming, G. (1968/1969). Intersegmental collateral ventilation in the human lung. *Respir. Physiol.* **6**:128–134.

Horsfield, K., and Cumming, G. (1968). Morphology of the bronchial tree in man. *Respir. Physiol.* **24**:373–383.

Horsfield, K., and Cumming, G. (1976). Morphology of the bronchial tree in the dog. *Respir. Physiol.* **26**:173–182.

Horsfield, K., Davies, A., and Cumming, G. (1977). Role of conducting airways in partial separation of inhaled gas mixtures. *J. Appl. Physiol.* **43**:391–396.

Horsfield, K., Davies, A., Mills, C., and Cumming, G. (1980). Effect of flow oscillations on the stationary concentration front in a hollow cast of the airways. *Lung* **157**:103–111.

Lieb, E. (1940). Demonstration of vascular tree with neoprene. *Bull. Int. Assoc. Med. Mus.* **20**:48–56.

Liebow, A. A., Hales, M. R., Lindskog, G. E., and Bloomer, W. E. (1947). Plastic demonstrations of pulmonary pathology. *Bull. Int. Assoc. Med. Mus.* **27**:116–129.

Narat, J. K., Loef, J. A., and Narat, M. (1936). On the preparation of multicolored corrosion specimens. *Anat. Rec.* **64**:155–160.

Olsen, D. E., Sudlow, M. F., Horsfield, K., and Filley, G. F. (1973). Convective patterns of flow during inspiration. *Arch. Intern. Med.* **131**:51–57.

Pedersen, C. J., Watson, J. W., and Jackson, A. C. (1983). Technique for making hollow central airway casts. *J. Appl. Physiol.* **55**:254–257.

Phalen, R. F., Yeh, H. C., Raabe, O. G., and Velasquez, D. J. (1973). Casting the lungs *in situ. Anat. Rec.* **177**:255–263.

Phalen, R. F., Yeh, H. C., Schum, G. M., and Raabe, O. G. (1978). Application of an idealized model to morphometry of the mammalian tracheobronchial tree. *Am. Rev. Respir. Dis.* **190**:167–176.

Pump, K. K. (1964). The morphology of the finer branches of the bronchial tree of the human lung. *Dis. Chest* **46**:379–398.

Pump, K. K. (1969). Morphology of the acinus of the human lung. *Dis. Chest* **56**:126–134.

Raabe, O. G., Yeh, H. C., Schum, G. M., and Phalen, R. F. (1976). *Tracheobronchial Geometry: Human, Dog, Rat, Hamster. LF-53.* Albuquerque, NM, Lovelace Foundation for Medical Education and Research.

Rahn, H., and Ross, B. B. (1957). Bronchial tree casts, lobe weights and anatomical dead space measurements in the dog's lung. *J. Appl. Physiol.* **10**:154–157.

Raskin, S. P., and Herman, P. G. (1975). Interacinar pathways in the human lung. *Am. Rev. Respir. Dis.* **111**:489–495.

Ross, B. B. (1957). Influence of bronchial tree structure on ventilation in the dog's lung as inferred from measurements of a plastic cast. *J. Appl. Physiol.* **10**:1–14.

Scales, J. T. (1950). Use of polyester resins in medicine. *Lancet* **1**:796–799.

Schlesinger, R. B., Bohning, D. E., Chan, T. L., and Lippmann, M. (1977). Particle deposition in a hollow cast of the human tracheobronchial tree. *J. Aerosol Sci.* **8**:429–445.

Schreider, J. P., and Raabe, O. G. (1980). Replica casts of the entire respiratory airways of experimental animals. *J. Environ. Pathol. Toxicol.* **3**:427–435.

Schreider, J. P., and Raabe, O. G. (1981a). Structure of the human respiratory acinus. *Am. J. Anat.* **162**:221–232.

Schreider, J. P., and Raabe, O. G. (1981b). Anatomy of the nasopharyngeal airway of experimental animals. *Anat. Rec.* **200**:195–205.

Schummer, A. (1935). Ein neues Mittel (Plastoid) und Verfahren zur Herstellung korrosionsanatomischer Praparate. *Anat. Anz.* **81**:177–224.

Snyder, B., and Jaeger, M. J. (1983). Lobar flow patterns in a hollow cast of canine central airways. *J. Appl. Physiol.* **54**:749–756.

Timbrell, V., Bevan, N. E., Davies, A. S., and Munday, D. E. (1970). Hollow casts of lungs for experimental purposes. *Nature* **225**:97–98.

Tompsett, D. H., (1970). *Anatomical Techniques,* 2nd edition. London, Churchill Livingstone.

Tucker, J. L., and Krementz, E. T. (1957a). Anatomical corrosion specimens. I. Heart-lung models prepared from dogs. *Anat. Rec.* **127**:655–665.

Tucker, J. L., and Krementz, E. T. (1957b). Anatomical corrosion specimens. II. Bronchopulmonary anatomy in the dog. *Anat. Rec.* **127**:667–676.

Weibel, E. R. (1963). *Morphometry of the Human Lung.* Berlin, Springer-Verlag.

West, J. B., and Hugh-Jones, P. (1959). Patterns of gas flow in the upper bronchial tree. *J. Appl. Physiol.* **14**:753–759.

Willson, H. G. (1922). The terminals of the human bronchiole. *Am. J. Anat.* **30**:267–295.

Part Two

STRUCTURAL TECHNIQUE IN CORRELATION BETWEEN NORMAL STRUCTURE AND FUNCTION

11

Morphometric Analysis of Casts of Branching Structures

KEITH HORSFIELD

Cardiothoracic Institute
Midhurst, West Sussex
England

I. Introduction

Casts of the pulmonary airways were originally used to obtain qualitative data on the organization of lung structures and their dimensions (Aeby, 1880; Ogawa, 1920; Willson, 1922). With the advent of chest surgery, detailed knowledge of the bronchopulmonary segments of the lung became necessary. The study of lung casts made an important contribution in this field, especially from the work of Brock (1946) and Boyden (1955). Ross (1957) was the first to make an analysis of the bronchial tree as a whole, basing his approach on path lengths as measured on a cast, and relating both structural and functional variables to the position of branches in the tree in terms of distance from the carina. This study emphasized the asymmetry of the bronchial tree and attempted to assess the effects of asymmetry on lung function. Casts have also been used to study the smaller branches within an acinus, that is, distal to the terminal bronchiole (Pump, 1964, 1969), while McLaughlin et al. (1961) used them to study the interrelation between airway and vascular trees in the lung. Weibel (1963), in his study of the human bronchial tree, estimated the number of alveolar ducts from samples of sections, and calculated the number of generations down to the ducts. He expressed the results in terms of a symmetrical model, thereby excluding asymmetry. The problem of how to express asymmetry in terms of the position of branches in the hierarchy was broached by Horsfield and Cumming (1968). They counted orders up from the smaller branches to the trachea and expressed asymmetry in terms of both path length and orders. Subsequently (Horsfield et al., 1971) they developed a method of expressing asymmetry at a bifurcation in terms of orders.

For many years geomorphologists have been studying the branching patterns of rivers (Hagett and Chorley, 1969), which bear considerable similarity to the branching of the airways. Research in the two fields was brought together by Woldenberg (Woldenberg et al., 1970) who introduced the method of ordering of Strahler (1952, 1957) to lung biology. Much of the subsequent work using this methodology (e.g., Singhal et al., 1973; Horsfield and Cumming, 1976: Horsfield, 1978; Horsfield and Thurlbeck, 1981a,b) owes a debt to Woldenberg's foresight and effort, which have gone largely unacknowledged.

II. Objectives of the Method

A complete cast of the conducting airways of the human lung consists of about 50,000 branches. If identification number, order, diameter, length, branching angles, connectivity of daughter and parent branches, and number of terminal bronchioles supplied by each are recorded, there may be half a million units of information. The handling, understanding, and use of such a data set requires that it be simplified to a form that brings out the underlying trends and relationships between its parts. This facilitates the calculation of physiological variables, comparison between trees of the same and different species, and the formulation of models, which are in themselves summary statements of simplification processes. This approach not only contributes to our knowledge of normal structure and function of the lung but also provides a baseline from which changes occurring in diseased lungs may be better understood.

III. Methods of Analyzing Branching Structures

A. Definitions of the Parts of a Tree

This account is concerned primarily with binary trees, that is, trees in which branching is by division of a parent branch into two daughter branches (dichotomy) and in which rejoining of branches to form loops or networks does not occur. Occasionally, division into three daughter branches (trichotomy) may be observed in the airways, and in the arterial tree division into four or more daughter (polychotomy) may also occur. The simplest such tree, shown in Figure 1A, consists of a stem that divides to give rise to two daughter branches, termed *links*. A node or vertex is situated at each end of a link, an interior vertex where three

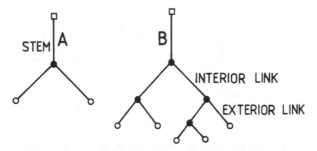

Figure 1 A. The smallest binary tree, consisting of three links meeting at an interior vertex or node. Each link has an exterior vertex at its free end, the stem link a root vertex, and each of its two daughter links a pendant vertex. B. A more complex tree, with interior links between interior vertices, and exterior links terminating in exterior vertices (see Fig. 2).

links meet, and an exterior vertex on the free end. That on the stem is the "root" vertex (although the figure is the wrong way up for this) and those on the free ends of the other links are pendent vertices. Figure 1B shows a more complex tree with examples of the various features, including interior links between interior vertices, and exterior links terminating with pendent vertices. Figure 2 shows the different types of vertex. Primary vertices give rise to two exterior links distally (i.e., away from the root), secondary vertices give rise to one interior and one exterior link, while tertiary vertices give rise to two interior links.

B. Grouping or Classification of Tree Structures

The first step in reducing the data set to manageable proportions is to group elements of the tree by some form of classification. There are two broad ways of doing this, one based on classifying branches and the other on classifying vertices. Many different ways of classifying branches of rivers have been described (Haggett and Chorley, 1969), but only three of these are in general use for classifying lung structures.

Counting Branches Centrifugally from the Stem

In this method, the stem is defined as generation 0, its two daughter links constitute generation 1, their daughters generation 2, and so on out to the exterior links. In symmetrical trees (Fig.3A), in which the numbers of links along every pathway are equal, the number of branches in successive generations doubles up to the last generation. In asymmetrical trees (Fig. 3B), the number of branches doubles until the exterior link on the shortest pathway has been reached, but thereafter the numbers decrease with increasing generation.

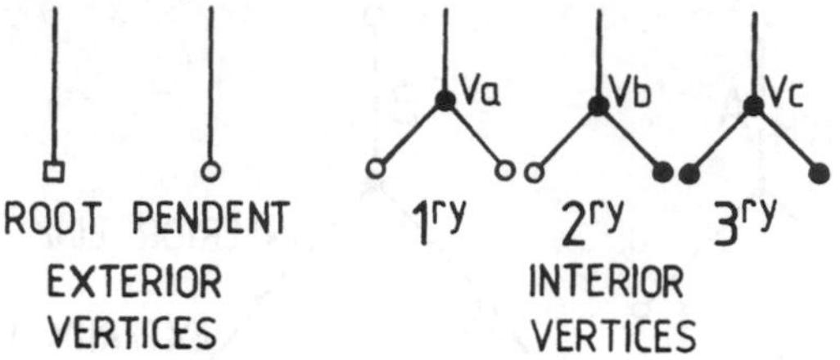

Figure 2 Types of vertex. Interior vertices are of three kinds: primary vertices, Va, which give rise to two exterior links distally; secondary vertices, Vb, which give rise to one interior and one exterior link; and tertiary vertices, Vc, which give rise to two interior vertices.

Counting Branches Centripetally Toward the Stem

Horsfield's Method

With this method (Horsfield and Cumming, 1968), exterior links are counted as order 1. Two of these meet to form an order 2 link, and at each successive node the order of the parent link is one more than the higher ordered of the two daughter links. In symmetrical trees (Fig. 4A), the result is similar to that obtained for generations, except that the numbering is inverted. In asymmetrical trees (Fig. 4B) the result is quite different from generations, except that the number of generations and orders along the longest pathway are equal.

Inverse Orders

It is sometimes mathematically more convenient for the numerical value of orders to increase down the tree, as it does in the case of generations. This can be achieved by the use of inverse orders, which simply correspond to Horsfield orders with the numbering inverted. Thus the tree must be ordered first, as in Figure 4B, and then the numbers changed as indicated

| Horsfield orders | 1 2 3 4 5 |
| Inverse orders | 5 4 3 2 1 |

to give Figure 4C.

Strahler's Method

This method (Strahler, 1952, 1957), which was developed for the study of rivers, is carried out in two stages. In the first stage, exterior links are numbered order

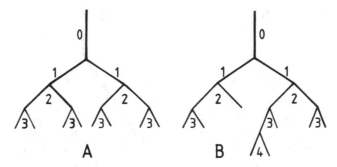

Figure 3 Classifying links by generation. A. A symmetrical tree, in which all the pathways have the same number of generations. B. An asymmetrical tree, in which the number of generations down the different pathways varies.

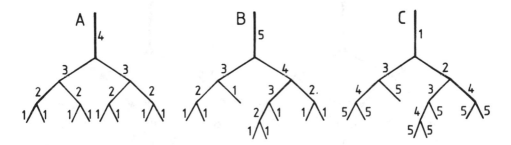

Figure 4 Classifying links by Horsfield order. A. A symmetrical tree. B. An asymmetrical tree. C. Inverse orders.

1, but subsequently order number only increases when two similarly ordered links meet (Fig. 5A). When dissimilarly numbered links meet, the order of the parent link continues unchanged from the higher ordered daughter. In the second stage, contiguous links of the same order are considered to constitute just one branch (Fig. 5B), so that Strahler-ordered branches may consist of one or more contiguous links. With symmetrical trees the result is identical to that obtained by Horsfield ordering (Fig. 4A), but with asymmetrical trees the results are different. This is because the formation of branches from contiguous links results in fewer branches and fewer orders (compare Figure 4B with Fig. 5B). Strahler orders may be inverted if required.

Counting Vertices

Figure 6A shows a symmetrical tree with the types of vertex labeled. Although there are four primary (Va) vertices, there are no secondary (Vb) vertices. Sadler

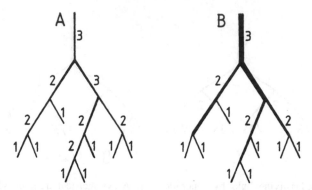

Figure 5 Classifying links by Strahler orders. A. First stage B. Second stage.

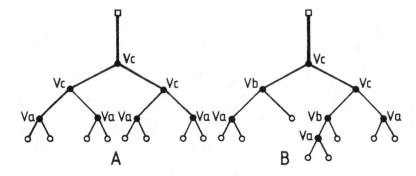

Figure 6 Types of vertix in A. a symmetrical tree, and B. an asymmetrical tree (see text and Fig. 2).

and Berry (1983) use the ratio of primary to secondary vertices, Va/Vb, as an index of symmetry, and in Figure 6A this ratio is infinite. In Figure 6B the Va/Vb ratio is 3/2, that is, the tree is less symmetrical. Horsfield and Woldenberg (1986) suggested that the ratio would be better expressed in its inverse form, Vb/Va, as an index of asymmetry. Thus in Figure 6A the ratio is zero, corresponding to no asymmetry, while in Figure 6B the ratio is 2/3, which is more asymmetrical.

C. Grouping Data by Structural Hierarchy

Generations

Weibel (1963) obtained dimensions of branches of the human bronchial tree, partly from measurement of a cast, partly from measurement of microscopic sections, and partly by interpolation between these two data sets. A dichotomously branching

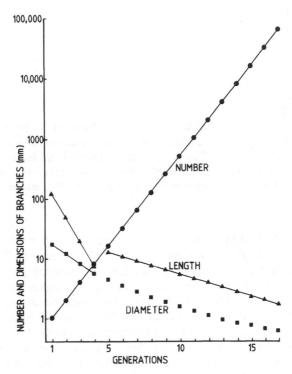

Figure 7 Numbers and dimensions of branches in a model of the human pulmonary airways, by Weibel (1963). Abscissa, generations, ordinate, number and mean dimensions on a logarithmic scale. The points for the mean diameters of generation 5 downwards do not lie on a straight line.

tree was assumed, which if symmetrical would need to have 23 generations to account for the observed number of alveolar ducts. Mean values of diameter and length were then ascribed to each generation. Figure 7 shows a plot of Weibel's model, in which the number of branches doubles in successive generations. Path length is identical for all the pathways, as is the number of terminal units supplied by each branch within any given generation. If we assume that flow is proportional to the number of terminal units supplied, flow is identical in all the branches of any given generation. Furthermore, the connectivity of every branch is defined, such that a branch of generation w arises from a parent of generation $w-1$, and gives rise to two daughter branches of generation $w+1$. Thus the path taken by a molecule of gas flowing through the tree can be determined. This symmetry greatly simplifies the calculation of physiological variables and as a result the model has been used extensively.

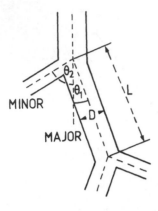

Figure 8 The model used by Raabe et al. (1976) for recording morphometric data from an asymmetrical tree. The major daughter branch is the larger, and branches at the smaller angle θ_1, while the minor daughter branches at the larger angel θ_2. D, diameter; L, length.

Raabe et al. (1976) produced a massive data set based on the measurement of airway casts from four species: humans, dog, rat, and hamster. They organized their data in terms of generations, using a binary numbering system to identify each branch. Just as Weibel based his data organization on a model concept, so did Raabe et al. Figure 8 shows the model they used, based on asymmetrical dichotomy with a major and a minor daughter branch at each bifurcation, defined in terms of their diameters. When, as sometimes happens, the two daughters are equal in diameter, the minor branch has the greater branching angle. If the two angles are also equal, or they have not been measured, the minor branch has the smaller length. Using these criteria, it is usually possible to make an unambiguous attribution of major and minor between the two daughters. In the binary numbering system the digit 1 was used for the major daughter branch and the digit 2 for the minor daughter branch. The trachea is 1, the right and left main bronchi 11 and 12, respectively, and their daughters 111, 112, 121, and 122. This system can be continued indefinitely, and can be adapted for handling on a computer. The identification of the parent or the daughters of any branch is simple; for example, 1121 arises from 112 and gives rise to 11211 and 11212. Thus the connectivity is known, and every pathway can be defined. Given this, and that the data set is complete, the length of every pathway can be calculated and the number of terminal units supplied by each branch determined.

Diameter, and branching angle were measured and recorded for each branch, and in addition, in of most the casts, the angle that the branch would make to the direction of gravity in the living animal. It is to the credit of these authors that they made their data set available to other workers, both in the form of a printed book and on magnetic tape. The main purpose of their work was to produce a model that would permit calculation of how inhaled particles deposit in the airways. Their model was developed in the following way. If we consider Figure 3A, it is obvious that the number of branches doubles in successive generations, and that the number Nn in generation n is given by

$$Nn = 2^n \tag{1}$$

Some authors prefer to start counting at the stem with generation 1, and to allow for this Eq. (1) can be expressed as

$$Nn = 2^{n-w} \tag{2}$$

where w is the generation of the stem. It is possible to have a model based on this equation in which the number of branches increases by a (constant) noninteger number in successive generations. The branching pattern of such a model could not be drawn on a piece of paper, but mathematically it presents no problems. Thus Eq. (2) becomes

$$Nn = Rb^{n-w} \tag{3}$$

where Rb is the ratio between the number of branches in successive generations. If we take Nn as the number of respiratory bronchioles, and n as the median number of generations down to respiratory bronchioles in the data, a value for Rb can be calculated from Eq. (3). This value then determines the number of generations in the model. Finally, average values for diameter, length, and branching angles, determined from pathways within generations in the data, were used to give a "typical path lung model," the numbers in each generation being calculated from Eq. (3). The method can also be applied to individual lobes.

The original account of this method by Yeh (1979) is not very clear because he seems to apply Eq. (3) to lung data, whereas in fact it is only applicable to symmetrical trees in which all the pathways have equal numbers of generations. With respect to pulmonary airways, this means in practice that it can only be applied to models. What Yeh and co-workers (Yeh, 1979; Yeh et al., 1979; Yeh and Schum, 1980) did was to produce a model in which the number of generations down to terminal bronchioles is equal to the median number in the data, and to calculate a value for the branching ratio which, in a symmetrical tree, would give the same number of terminal bronchioles as in the data. The result for the human lung (Yeh and Schum, 1980) was very similar to Weibel's model, with a branching ratio of 2.0, but with some difference in dimensions. In the rat, the branching ratio was 1.68 (Yeh et al., 1979); the model is summarized in Figure 9. When the branching ratio is noninteger, as in this case, the model has a kind of connectivity; for example, 38 branches of generation 8 join on to 23 branches of generation 7. Although it is impossible to draw the branching pattern, nevertheless the flow can be equally distributed between the branches of each generation in turn and calculations made of physiological variables.

Orders

The use of orders for classifying branches of the bronchial tree was reported by Horsfield and Cumming (1968). Their data were about 90% complete down to

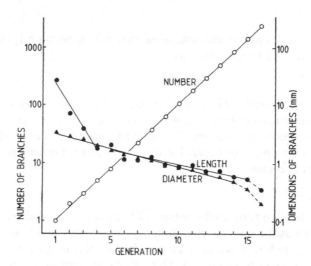

Figure 9 Numbers and dimension of branches in a model of the pulmonary airways of the rat, by Yeh et al. (1979). Axes as in Figure 7.

0.7mm diameter, with the addition of data from samples of the more peripheral parts of the cast. In the basic data the connectivity is preserved, but in the model, shown in Figure 10, the connectivity is lost. This is because only half of the branches of generation n are required by definition to join on to parents of generation n-1 (see Fig. 4B); the other half may or may not do so. Thus, inspection of Figure 4B will show that not all of the flow must traverse a branch in each successive generation: for example, that coming to or from the shortset pathway will miss out generations 2 and 4. Consequently, calculations of physiological variables are more difficult to make than when a symmetrical model is used.

   With Strahler orders (Strahler, 1952, 1957), some individual links and their connectivity are lost. As a result there are fewer orders and fewer branches than when the same tree is ordered by Horsfield's method. Thus physiological calculations are even more difficult when using Strahler orders. Figure 11 shows a Strahler-ordered model derived from the same tree as the model shown in Figure 10. The reduced numbers of orders and branches are obvious, as is the resulting increased simplicity. It will also be noticed that the number, diameter, and length plots give relatively straight lines with the semilogarithmic scale. The antilogs of the absolute values of the slopes of these lines, determined from regression equations by least squares best fit, give three useful parameters: the branching ratio, Rb, the factor by which the number of branches increases in successive orders down the tree; and the diameter ratio, Rd, and the length ratio, Rl, the

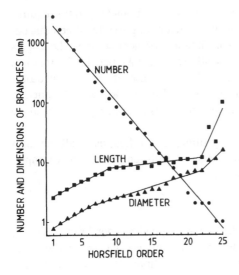

Figure 10 Numbers and dimensons of branches in a model of the pulmonary airways of humans, based on Horsfield orders, by Horsfield and Cumming (1968). Abscissa, orders, ordinate as in Figure 7 (from Horsfield, 1986, by permission).

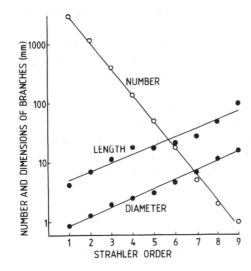

Figure 11 Numbers and dimensons of branches in a model of the pulmonary airways of humans, based on Strahler orders, by Horsfield (1981). Axes as in Figure 10 (from Horsfield, 1986, by permission).

factors by which the mean values of these dimensions increase in successive orders up the tree. These parameters can also be calculated for Horsfield orders, but the lines are often less straight (Fig. 10) and to that extent less satisfactory. Sometimes a line can be divided into relatively straight sections, each with its own ratios (Fig. 10). It is quite often found that the point for the number of branches in the highest order (normally 1) is noticeably above or below the line through the other values (see Fig. 13). This arises because the branching ratio requires a noninteger number, and this is obviously physically impossible. When this occurs, the point should not be used in the calculation of branching ratio. The lengths of the upper three or so orders often lie on a different line (Weibel, 1963; Horsfield and Cumming, 1976; Horsfield, 1981), perhaps because of spatial constraints around the hilum of the lung. The concept of Rb, Rd, and Rl can be applied to generations in symmetrical models, provided the plots are linear, and indeed we have already used Rb in this context.

Path length is determined simply by summing the lengths of all the links along the pathway; the frequency distribution of path lengths to, for example, terminal bronchioles can then be plotted.

Terminal Branches

Calculation of the number of terminal branches supplied by each branch is easy, and can be done by hand or by computer for any complete data set with defined connectivity, whether classified by orders or generations. It is obtained simply by summing the numbers for each of the two daughter branches, starting at the terminals and working up the tree. If the number supplied by a branch is t, and the total (i.e., the number for the stem) is T, t/T gives a first-order estimate of the fraction of the total flow carried by the given branch. This assumes that all the terminal units are of the same size and empty synchronously at the same rate. It is obvious that all the branches of any given generation in a symmetrical tree will supply the same number of terminals.

McBride and Chuang (1985) found a large variation in airway dimensions within a given order in the ferret's lung, which made the method unsuitable for the precise definition of the hierarchical position of an airway. Instead they demonstrated a linear log–log (i.e., power) relation between the cross-sectional area of a branch and the number of terminal bronchioles supplied by it. They also showed a similar relation between the cross-sectional area of a branch and the summed area of the terminal bronchioles supplied by it, with much less variation between branches than when orders were used. They could then use both of these relations to make a better estimate of expected diameter from the number and summed cross-sectional area of the terminal bronchioles supplied by it.

D. Sampling Techniques

Because these casts may be comprised of so many branches, measurement of every branch may be impracticable. The human lung cast that Horsfield measured (Horsfield and Cumming, 1968) took 14 months, while the rat lung cast in the data of Raabe et al. (1976) took 1 year (Yeh et al., 1979). Much time can be saved by the use of sampling methods, and these are also useful when the cast is incomplete peripherally.

Peripheral Segment Sampling

When starting out with a cast, the first decision should be how far down to make complete measurements. This decision can be based on inspection and is easier if one has some experience. In general, the cut-off point, defined as a diameter, will be a compromise between the desire for complete data and the resources available for its collection. Because of the asymmetry of the cast, the definition of order 1 branches has to be carefully worded. For example, an order 1 branch could be defined as the first branch encountered, on tracing up any pathway, equal to or greater in diameter than the cut-off diameter. A selection of such branches, along with the segments of lung subtended by them, are broken off and studied separately. Since the branches in the samples are likely to be small, this may involve working with a dissecting microscope. Ideally, the sample for study should be randomly selected, but in practice this is remarkably difficult to achieve. The main reasons for this are the large number of potential samples and the inaccessibility of many of them on the intact cast. In most studies, either a lobe or a segment has been used as a sample (Hogg et al., 1970; Hislop et al., 1972; Plopper et al., 1983), or individual broken off portions have been chosen from different parts of the lung (Horsfield, 1978). In either case, the most important criterion for selection is the completeness of the cast in the selected segment.

On studying the samples it is again necessary to define an order 1 branch. Where a structural definition, such as a terminal bronchiole, can be used, this is the most satisfactory. But often it may be necessary to define a cut-off diameter, which is done in just the same way as for the main part of the cast. When using generations, the sample data are simply a continuation of the main cast data and are dealt with in a similar manner. When using orders, the data obtained from the different samples may be pooled, and joined onto the main cast data.

This is usually done in the following manner. The data for the main cast are ordered, and the mean diameter of the branches in each order determined. Log mean diameter is then plotted against order, which with Strahler orders invariably gives a straight line. Each tree in the sample data is similarly ordered, the data are pooled, the mean diameter of branches in each order determined, and log mean diameter plotted against order. The two plots now have to be lined up, as shown in Figure 12, to determine how the sample orders relate to those in the main cast. Sometimes there will be an overlap of one order, or the largest diameter in the sample data may be one order lower than order 1 on the main cast. If the samples have been carefully taken from order 1 branches of the main cast, the two plots will line up without difficulty, because of the very constant linear relation of log mean diameter to Strahler order. Once diameter plots have been lined up, the length plots can be lined up on orders corresponding to those in the diameter plots (Horsfield et al., 1987).

This sampling method gives good results for mean dimensions, but the number of branches in the samples is too low. An estimate of the true numbers

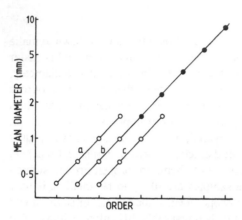

Figure 12 Lining up plots of diameter data from an airway cast. ●, main cast data; ○, peripheral sample data. Three possible positions (a), (b), and (c), for the peripheral data plot are shown. That at (b) lines up with the main cast data.

in the periphery of the tree can be obtained by plotting log number of branches against order for the main cast and for the sample data, as shown in Figure 13. Ideally, the value of Rb determined from the main cast is the same as determined from the sample data, in which case the plot for the main cast can simply be extended for the appropriate number of orders. In practice the two values are usually slightly different, in which case the data for the samples are multiplied by a constant factor to bring them into line with the main cast data. When there is an overlap of one order, this factor is Nm_1/Np_w, where Nm_1 is the number of branches in order 1 of the main cast calculated from the regression equation, and Np_w is the number of branches in the highest order of the peripheral data, calculated from the regression equation. When there is no overlap, the number plot for the main cast data should be extended by one order, to give an extrapolated value of Nm_0 for the number of branches in the next lower order. The factor is now given by Nm_0/Np_w. Sometimes the value for Np_w is obviously too low, clearly lying below the line for the other data points; when this occurs, the point is best ignored and the calculations carried out using the data for the next order, Np_{w-1}, instead.

Pathway Sampling

An entirely different method of reducing the labor involved in measuring a cast is to sample pathways (Horsfield, 1984), which is best done using the axial paths in lobes or segments. These can be used to compare a named lobe or segment in different individuals (Hogg et al., 1970; Hislop et al., 1972; Plopper et al., 1983) or to make an estimate of the total data using pooled data from several lobes or segments (Horsfield, 1984). The axial pathway is defined by starting at the lobar or segmental branch, defining its major daughter, and repeating the process down the tree, always choosing the major branch (see above for definition of the major branch.) This method naturally follows generations down the

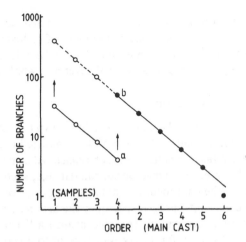

Figure 13 Numbers of branches from a cast from peripheral samples. Abscissa, Strahler orders, ordinate, number of branches in each order on a logarithmic scale; • , main cast data; ○ , peripheral sample data. Continuous lines are the regression lines, that for the main cast data having been calculated excluding the point for the highest order. The dashed line has been drawn parallel to the line for the peripheral sample data, starting from the point where the regression line for the main cast data cuts the vertical line for order 1 on the main cast. thus point a is moved up to point b, and other peripheral data points are moved up by an equal distance. This is equivalent to multiplying the peripheral sample data by a constant factor.

tree, and these can be conveniently counted at the same time. It is of course necessary to define unambiguously the last branch on the pathway. For airways this might be the terminal bronchiole, or the first branch less than 0.7mm diameter, for example. Usually, but not invariably, the axial pathway defined in this way is the longest in terms of the number of generations. As required, the diameter and length of each branch along the path can be measured, and comparison made between individual lobes or segments.

Where one needs to make an estimate of the total data from the sample, several axial pathways should be measured and mean values for diameters and lengths determined. To do this it is essential to work with Horsfield orders rather than generations, unless by chance all the sampled pathways happen to have the same number of generations, in which case both methods give the same result. This is because generations do not classify like branches together (see below). Comparison of Figures 3B and 4B will show that the longest pathway in terms of generations is also the longest in terms of orders, and that the longest path consists of one branch of each order. Therefore, having defined the axial path by generations, it is reasonable to start at the distal end and start counting orders. When this has been done for the required number of pathways, mean values of diameter and length for branches in each order can be calculated. This gives a reasonable estimate for the mean values in the tree as a whole (Horsfield, 1984).

E. Methods of Measuring

If it has been decided to record data on every link down to some stated diameter, it may be worthwhile to make a drawing of the complete branching pattern, annotated with the identification number of each branch. At some later date, perhaps when the original cast has sustained accidental damage, such records can prove invaluable. They also help morphometrists to find their way on a cast and to know how far measurements have progressed, since the task may take many months. As data are acquired, they can be recorded against a list of the identification numbers. It is only too easy for an occasional error to creep in, and the ability to check a measurement with certainty is most reassuring. Also helpful, especially when Strahler ordering, is to paint the branches of the cast according to a color code. In order to do this, the order 1 links must first be identified. This may entail measuring diameters in order to define the first link up every pathway equal to or greater than the defined diameter. Ordering can then proceed as far as the first broken branches (if any), which must then be dealt with as described below. This process is continued until the stem is reached. Once the branches have been ordered and painted, counting and measuring by order is easy.

The simplest way to measure diameter and length is with a pair of sharp pointed dividers placed across the part to be measured and then read against a suitable scale, which may be a ruler or an eyepiece grid in a microscope. For diameter, the middle of the length of the branch is usually a suitable site to measure. When the cross-section is oval rather than round, two measurements at right-angles may be made and the mean value used. Length can be measured between the junction points situated at each end of the branch, as illustrated on the model shown in Figure 8. In practice, these points are difficult to determine accurately. An easier but theoretically less accurate method is to put the points of the dividers right into the apex of the branching angles. A better instrument for the job is a caliper for measuring external diameter, with a rotary scale that can be read direct to, for example, 0.05 mm, and estimated to 0.01 mm. This should have sharp points attached, like those on a pair of dividers (Fig. 14).

For microscopic work, as is needed for small lungs or for peripheral samples, a calibrated scale in the eyepiece of a dissecting microscope is satisfactory. This can be read direct to 0.1 mm or less, depending on the magnification. With this technique it is important to hold the cast so that the branch being measured is at right angles to the line of sight, particularly for measurement of length.

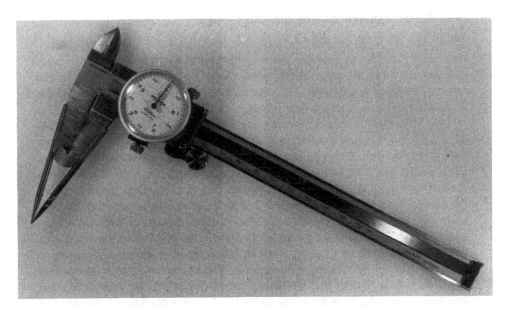

Figure 14 External diameter calipers with a circular scale and two sharp points attached to the jaws, used for measuring diameter and length of branches.

Branching angles are particularly difficult to measure accurately. One reason for this is that a branch may curve away from the line of the parent, making its new direction difficult to define. Even when the two branches concerned are straight, their center lines and the point where they meet (Fig. 8) are not easy to visualize. A simple and practical method is to construct a series of templates, shaped as sectors of a circle, at 5 degree intervals, and fix these to a simple handle. The sectors can then be placed on the branches at their junction and the best fit judged by eye. There is no point in using smaller angle intervals on the sectors because the method is not precise enough to yield meaningful answers. Phalen et al. (1978) do not give a clear account of how they measured angles, but do mention that they used sector templates. The between-observer differences for their angle measurements had a standard deviation of 10 degrees.

Another method that is easy in principle, but difficult in practice, is to mark a point A on the parent, a point B on the major daughter branch, a point C on the minor daughter branch, and a point D at the meeting of the center lines of the three. The lengths AB, AC, BC, AD, BD, and CD are measured, and the angles are then calculated by trigonometry (Woldenberg and Horsfield, 1983).

F. Broken and Missing Branches

The techniques so far described for analyzing casts have been based on the assumption that the cast, or at least that part of it being measured, is complete. In practice this often not the case, so that some means of correcting the data for the missing branches is required. The most important need is to estimate the order of a broken branch, which would normally be determined by the pattern of the missing branches. It should be noted that when generations are used this need does not arise, because generation is determined by the remaining intact branches up to the stem. A simple way is to order the cast up to branches of diameter equal to that of the broken branch and determine the mean order of those branches. The nearest integer value to the mean can then be taken as the order of the broken branch. A more complex, but perhaps mathematically more acceptable way, is to plot the frequency distributions of diameters of branches in each order (Fig. 15) and to draw a vertical line where the distributions overlap such that the area belonging to order w on its left is equal to the area belonging to order w − 1 on its right. Order is then attributed according to whether the diameter lies to the left or right of the line. This procedure equalizes the number of times that branch of order w is attributed order w − 1, and a branch of order w − 1 is attributed order w. A practical difficulty arises from the fact that diameters are usually measured at discrete intervals (e.g., 0.1 mm), so that the distributions are better represented by histograms. The dividing line may then fall through a discrete diameter value, so that one must decide how to deal with broken branches of that diameter. They can either be randomly distributed in proportion to their frequency in the data, or the dividing line moved to include all broken branches of that diameter in one or other order according to which is the most frequently occurring. In carrying out this method, which is best done on a computer, broken branches are provisionally allocated order 1, the tree is ordered, and the above procedure completed. This gives a new provisional order for the broken branch, which is used to reorder the tree, and the whole procedure is repeated until a stable value for the order is obtained. This generally requires between one and four iterations. The method was developed by Alison Thurlbeck (Thurlbeck and Horsfield, 1980)

The method of compensating for the number of missing branches arising from a broken branch (Horsfield et al., 1982) depends on the assumption that the corrected plot of log number of branches against order will be linear. As an example, it will be assumed that there are three broken order 3 branches and five broken order 2 branches. First, Rb is determined provisionally from the available data by calculating the least-squares regression of log number on order; let us assume that a value of 2.5 is obtained. Each broken branch of order 3 is attributed 2.5 order 2 branches and this would give 7.5 missing order 2 branches, which number is rounded to eight, the nearest integer. Adding this to the

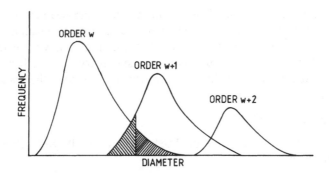

Figure 15 Frequency distribution of diameters of branches in three succesive orders. Vertical line drawn so that the area to the left of the line under the curve for order $w+1$ is equal to the area to the right of the line under the curve for order w.

five broken order 2 branches gives an estimate of 13 missing order 2 branches, which in turn would give rise to $2.5 \times 13 = 32.5$ order 1 branches, rounded to 33. Now eight order three branches and 33 order 2 branches are added to the data, branching ratio is recalculated using these revised values, and the whole procedure repeated using the new value of Rb thus obtained. This procedure is repeated until a stable value for Rb is obtained, which normally takes from one to six iterations. When this point has been reached, the value of Rb and the estimated numbers of branches are taken to be the corrected values.

G. Expression of Asymmetry

Perhaps the most straightforward way of expressing the asymmetry of a tree is to plot the distribution of generations down from the stem to the terminal branches, as was done for the human bronchial tree by Horsfield and Cumming (1968) and for the rat by Yeh et al. (1979). Similar plots can be made for the distribution of path lengths (Ross, 1957; Horsfield and Cumming, 1968). Branching ratio Rb is directly related to asymmetry. In a symmetrical tree, in which the number of generations is the same down every pathway, $Rb = 2.0$, and this is true of both Horsfield and Strahler orders. In an asymmetrical tree, $Rb < 2.0$ for Horsfield orders, and in the limit $\rightarrow 1.0$. With Strahler orders, $Rb > 2.0$, and in the limit $\rightarrow \infty$. Thus in the human airways, Rb lies between 2.5 and 2.8, while in the more asymmetrical airways of the dog, Rb lies between about 3.0 and 3.6.

The vertex ratio Vb/Va is a useful indicator of asymmetry, even though it can give a value of zero in some artificially generated asymmetrical trees. In practice it is closely, but not absolutely, related to branching ratio (Horsfield and Woldenberg, 1986). It has been related to the pattern of growth, especially by Berry and co-workers (Berry et al., 1975; Berry and Bradley, 1976).

Asymmetry at an individual junction can be expressed in terms of the difference, Δ, between the orders or inverse orders, of the two daughter branches (Horsfield et al., 1971). At a symmetrical bifurcation $\Delta = 0$, with the least degree of asymmetry $\Delta = 1$, and with increasing asymmetry $\Delta > 1$. If a model tree is constructed in which the value of Δ is constant throughout, the branching ratio Rb(H) for Horsfield orders is given by

$$1/rb(H)^x + 1/Rb(H) = 1 \tag{4}$$

where $x = \Delta + 1$ (Horsfield, 1976). The branching ratio Rb(S) for Strahler orders is then given by

$$Rb(S) = Rb(H)^x \tag{5}$$

The value of Δ, and hence the value of x, is an integer, but there is no mathematical reason why x in Eq. (4) should not have a noninteger value, just as Rb in Eq. (3), which would then imply a noninteger value for Δ. This cannot exist at a single bifurcation, but may exist as an average value for the whole tree. Thus, once one has measured Rb(H) for a tree, the asymmetry can be expressed as an average value of x, and hence Δ, which satisfies Eq. (4). The derivation of Eq. (4) is given by Horsfield (1976). This approach has been used by Horsfield et al. (1971). Fredberg and Hoenig (1978), Sidell and Fredberg (1978), and Yeates and Aspin (1978).

IV. Discussion

A. Generation Versus Orders

The stem branch (trachea in the airways) is usually the most accessible, and this is especially true in the clinical situation. It is therefore natural to count down from the trachea to locate a branch in terms of generations, rather than count upwards in orders, which would be nearly impossible to do. Also, as already described, the distribution of generations down to terminal branches is a good way to define asymmetry. Thus, generations are particularly useful for describing the location of parts of a tree in relation to the stem.

To use orders it is necessary to locate the terminal branches, and in practical terms this can only be done on a preparation such as a cast. Orders are best used for classifying and grouping branches, and then relating their classification to other morphometric variables such as diameter, or comparing groups between trees.

When dealing with a symmetrical tree, there is no important difference between using generations and orders for grouping the branches, the one being simply the inverse of the other. But with asymmetrical trees important differences emerge. Inspection of Figure 3B shows that branches in generation 2, for example, include three functionally different branches. One supplies one terminal, two supply two terminals each, and one supplies three terminals. In the airways of the lung these would carry different flows and have different diameters. Thus generations do not classify like with like. In contrast, inspection of Figure 4B shows that the branches of order 2 each supply two terminals, while the branches of order 3 each supply three terminals. Thus, this method does classify like with like. Of course, in a real tree, the branches of any one order would not be identical, but the grouping is much better than it would be using generations. Therefore, if the purpose of the study is to make comparison between different trees, it is better to use orders since these give better average data for each level in the tree. The use of generations for obtaining average dimensions tends to suppress information related to asymmetry, whereas with orders this information is retained.

B. Horsfield Orders vs. Strahler Orders

The choice between these two methods of ordering depends on the objectives of the study. Perhaps the most important point is that with Horsfield orders the connectivity of the tree can be retained, whereas with Strahler orders it is lost. Mean values can be obtained for the pooled data from every link in a Horsfield-ordered tree, but with Strahler orders a considerable proportion of the data is excluded. Strahler's method gives fewer branches and fewer orders than Horsfield's, and of particular value is the fact that Strahler orders give linear plots of log number and log mean diameter against order. This greatly facilitates comparison between trees, for example, when studying growth of the airways (Horsfield, 1977; Horsfield et al., 1987), and simplifies the expression of the parameters of the tree, Rb, Rd, and Rl.

C. Inverse Orders

These are used purely for mathematical convenience. With the usual ordering methods the larger number of branches are associated with the lower order values, whereas with inverted orders smaller numbers are associated with lower order values, and this often simplifies the use of mathematical expressions (Horsfield, 1976; Horsfield and Woldenberg, 1986). It should be stressed that inverse orders can only be determined after a tree has been ordered in the usual way, starting with the terminal branches.

D. Morphometric Parameters Rb, Rd, and Rl

These represent perhaps the greatest degree of simplification of a data set for a tree, since these three numbers summarize the information obtained from many thousands of measurements. They are useful for comparing trees from different individuals, different species, and even totally different origins (Woldenberg et al., 1970). They contain information on asymmetry, can be used to relate structure to function in a general way (Horsfield, 1980; Horsfield and Thurlbeck, 1981a,b), and may change their values during growth (Horsfield et al., 1987)

E. Vertices

Since every vertex except the root vertex is associated with a parent branch, it is not surprising that vertex analysis has close affinities with ordering (Horsfield and Woldenberg, 1986). Vertex analysis has been used particularly in the study of growth, certain modes of growth giving rise to known probable values of Va/Vb (Berry et al., 1975; Berry and Bradley, 1976).

F. Sampling Methods

Measurement of peripheral samples can save a great deal of time and effort with most casts, but the difficulty is in being sure that the samples are representative of the whole tree. I know of no study that has formally demonstrated this to be so, nor any method that can be used in practical circumstances to ensure that the samples are truly randomly chosen. The best indication that it gives reliable results is the fact that the diameter plot for the sample nearly always is a straight continuation of the diameter plot for the main cast data.

Pathway sampling methods can also save a great deal of time, and these have been demonstrated to give results representative of the tree as a whole. In human lungs it is best to sample at least one pathway from each bronchopulmonary segment (about 20 pathways) to obtain a reliable result.

G. Correction for Broken Branches

Estimating the order of a broken branch can never be totally reliable; a degree of uncertainty must always remain. In the great majority of cases an error of 1 in the estimation of the order of a broken branch (it would be unlikely to be greater than this using Strahler orders) makes no practical difference to the results for the whole cast. Just occasionally, however, the broken branch is situated in a critical position such that altering its order affects the orders of all the branches up the pathway way to the stem branch. This, in turn, changes the number of orders in the tree and can thus alter the values of Rb, Rd, and Rl. With practice, it is remarkably easy to estimate the Strahler order of the stem branch of a cast by observation and counting, and hence to determine whether the presence of a broken branch might alter the order so obtained. In such a case either the cast can be discarded or great care taken with the correction. Very often the approximate value of Rb to be obtained from the cast is known, perhaps from measurement of similar casts made previously. In this case an error in the number of orders is glaringly obvious because of the large effect it has on the value of Rb.

For example, in a tree with 2000 terminal branches and 6 orders, Rb $= 2000^{1/5}$ $= 4.57$, but in a similar tree with 7 orders, Rb $= 2000^{1/6} = 3.55$. Thus, by one means or another, this potentially serious problem can be contained.

Estimating the number of branches missing distal to a broken branch is likewise always subject to uncertainty, but in this case the effects of any error are minimal. Given the underlying assumption for the iterative method described, namely that the plot of log numbers against orders is linear, the error must be small. This is because the method lines up the corrected data points with the rest of the data in a straight line.

H. Airway Dimensions on the Cast and During Life

The modern materials used for making casts, such as polyester resins and silicone rubber, are capable of producing very accurate replicas of hollow structures. However, the casts are prepared postmortem, and as such may not represent the structures as they were during life. Furthermore, most biological tubular structures can change diameter to a marked degree, so that one measurement can only represent the living structure at one point in a spectrum of possible dimensions.

Casts may be made from excised lungs, or from lungs in situ in the thorax. In either case, it is the responsibility of the author to state clearly the conditions under which the cast was made, and the responsibility of the user of the data so obtained to take account of these conditions. In general, airways change dimensions approximately as cube lung root of volume (Hughes et al., 1972), so an appropriate correction to the dimensions obtained from measuring a cast can be made as required for lung volume. This, or course, does not take into account bronchoconstriction.

I. Models of the Airways

In addition to their use for describing trees and making comparisons between them, these methods are frequently used to obtained a data base from which to develop a model. Such models are usually conceptual rather than physical, although physical models based on data have occasionally been used to study flow, for example by Pedley et al. (1970). In developing them, the first requirement is to formulate how the branches will be arranged in series, for example, by generation. Then the mean values for dimensions are calculated for each group of branches, and possibly other data such as branching angle. Estimates of flow are usually based on the number of terminal units supplied by each branch, and this follows from the pattern of branching.

Models developed in this way can be classified into one of four groups: symmetrical with and without connectivity, and asymmetrical with and without connectivity. Weibel's model A (Weibel, 1963) is the best known symmetrical model with defined connectivity, and it, or variants of it, have been the most frequently used for the calculation of physiological variables. Its great merit is its ease of use, and its disadvantage is that it excludes asymmetry. In addition, the number of terminal bronchioles in the model is probably too high by a factor of about 2.5 (Horsfield, 1981) and the number of alveolar ducts too low (Hansen and Ampaya, 1974).

The model of the rat lung by Yeh et al. (1979) is a good example of a symmetrical model without connectivity, in the sense of being able to reconstruct the pattern of branching. However, since all of the branches in generation w connect with all of the branches in generations w − 1 and w + 1, the flow can be divided equally between the branches of any given generation. Since all pathways are identical, calculations can be made for one pathway, as in the case of Weibel's model, but asymmetry is similarly excluded. The model is the only one to include branching angles and gravity angles, which may be useful for calculating particle deposition, as intended by the authors.

The first asymmetrical model with connectivity was that of Horsfield et al. (1971) for the human bronchial tree. It was based on the average value of Δ at each order. Connectivity was completely defined, with an average degree of asymmetry. The principle was used by Fredberg and Hoening (1978) in their model of the human bronchial tree, and by Horsfield et al. (1982) for a model of the dog's airways. Because the pattern of branching arising from every branch of a given order is identical, the number of terminal branches supplied can easily be calculated. More complex calculations using the model are best handled with a computer.

Ross (1957) was the first to attempt to describe the asymmetry of the airways as a whole, using a cast of a dog lung. Horsfield and Cumming (1968) produced an asymmetrical model of the human bronchial tree without connectivity, using Horsfield orders, and a similar one using Strahler orders (Horsfield, 1981). These models are good for purposes of description, but difficult to use for calculation because of asymmetry combined with the near absence of connectivity. However, the number of terminal branches supplied by each branch is defined, as is the connectivity for the longest pathway, which by definition includes one branch of each order. Calculations can therefore be made for this one pathway, which represents the extreme of the range of path lengths. This can sometimes be useful to give one limit to the range of possible results.

V. Prospects for Future Developments

The main problem with the study of lung casts is the time and effort required to measure them. If it were possible to speed up this aspect, more could be achieved. If the trees with which we are concerned were two-dimensional, they could be scanned and the data fed directly into a computer, which could then calculate the dimensions and even order the tree. However, with these three-dimensional structures the problem is much more difficult. Perhaps computerized axial tomography could be applied to the problem, although the small size of the branches would present difficulties in definition and resolution. A useful step in this direction has been made by Block et al. (1984). They used computed tomography to study the three-dimensional anatomy of pulmonary vessels in vivo, and tested their system by scanning a glass model of part of the arterial tree. They obtained dimensions and branching angles of branches down to 2 mm diameter, and were able to produce simulated three-dimensional images.

Sampling peripheral segments of the cast greatly reduces the work of measuring, but so far there is no easy method of randomly choosing the samples. The development of a randomized technique would improve the scientific veracity of this approach, but the difficulties are considerable. It is probable that the peripheral lung units are not randomly distributed in three-dimensional space, they are packed such that the inner ones are hidden from external view, and the sampling procedure would have to be nondestructive.

Models are simplifications of the data. They have been used to ease the burden of calculation where large amounts of data are available, or to substitute for data where there are none. Now that large data sets including connectivity (even complete sets in some cases) of mammalian conducting airways are available, and with the advent of modern computers, perhaps the anatomical data themselves could be used in calculations. We could then dispense with the use of dimensional models for this purpose.

Although some qualitative information is available from casts of diseased airways (Leibow et al., 1947; Wyatt et al., 1961; Horsfield et al., 1966; Pump, 1973), there are no published quantitative data of which I am aware. Since one of the most important motives for studying lung anatomy is to gain insight into the normal functioning of the lung, in order to understand better what goes wrong in disease, it would seem that the time is right for the application of these casting techniques to the quantitative study of diseased lungs. The difficulties are considerable, the most frequently encountered being the negative artifact. This bedevils attempts at quantification, so that some way of overcoming the problem would greatly enhance the value of the technique in pathological studies.

Casting of anatomical tree-like structures, especially in the lung, has added enormously to our knowledge of anatomy, physiology, and surgical technique. I would like to see its further development applied to the enhancement of our knowledge of pulmonary pathology and internal medicine.

References

Aeby, C. (1880). *Der Bronchialbaum der Säugethiere und des Menschen. Nebst Bermerkungen über den Bronchialbaum der Vogel und Reptilien.* Leipzig, Engelman.

Berry, M. T., and Bradley, P. M. (1976). The application of network analysis to the study of branching patterns of large dendritic fields. *Brain Res.* **109**:111–132.

Berry, M., Hollingworth, T., Anderson, E. M., and Flinn, R. M. (1975). Application of network analysis to the study of branching patterns of dendritic fields. In *Advances in Neurology*, Vol.12. Edited by G. W Kreutzberg. New York, Raven Press, pp. 217–245.

Block, M., Liu, Y., Harris, L. D., Robb, R. A., and Ritman, E. L. (1984). Quantitative analysis of a vascular tree model with the dynamic spatial reconstructor. *J. Comput. Assist. Tomogr.* **8**:390–400.

Boyden, E. A. (1955). *Segmental Anatomy of the Lungs*. New York, McGraw-Hill.

Brock, R. C. (1946). *The Anatomy of the Bronchial Tree with Special Reference to Lung Abscess*. Oxford, University Press.

Fredberg, J. J., and Hoenig, A. (1978). Mechanical response of the lungs at high frequencies. *J. Biomech. Eng.* **100**:57–66.

Haggett, P., and Chorley, R. J. (1969). *Network Analysis in Geography*. London, Arnold.

Hansen, J. E., and Ampaya, E. P. (1974). Lung morphometry: a fallacy in the use of the counting principle. *J. Appl. Physiol.* **37**:951–954.

Hislop, A., Muir, D. C. F., Jacobsen, M., Simon, G., and Reid, L. (1972). Postnatal growth and function of the pre-acinar airways. *Thorax* **27**:265–274.

Hogg, J. C., Williams, J., Richardson, J. B., Macklem, P. T., and Thurlbeck, W. M. (1970). Age as a factor in the distribution of lower-airway conductance and in the pathologic anatomy of obstructive lung disease. *N. Engl. J. Med.* **282**:1283–1287.

Horsfield, K. (1976). Some mathematical properties of branching trees with application to the respiratory system. *Bull. Math. Biol.* **38**:305–315.

Horsfield, K. (1977). Postnatal growth of the dog's bronchial tree. *Respir. Physiol.* **29**:185–191.

Horsfield, K. (1978). Morphometry of the small pulmonary arteries in man. *Circ. Res.* **42**:593–597.

Horsfield, K. (1980). Are diameter, length and branching ratios meaningful in the lung? *J. Theor. Biol.* **87**:773–784.

Horsfield, K. (1981). The structure of the tracheobronchial tree. In *Scientific Foundations of Respiratory Medicine*. Edited by J. G. Scadding and G. Cumming. London, William Heinemann.

Horsfield, K. (1984). Axial pathways compared with complete data in morphologaical studies of the lung. *Respir. Physiol.* **55**:317–324.

Horsfield, K. (1986). Morphometry of the airways. In *Handbook of Physiology, Respiration*, Sect. III, vol. III. Edited by P. Maklem and J. Mead. Washington, D.C., American Physiological Society, p. 7.

Horsfield, K.,and Cumming. G. (1968). Morphology of the bronchial tree in man. *J. Appl. Physiol.* **24**:373-383.

Horsfield K., and Cumming,G. (1976). Morphology of the bronchial tree in the dog. *Respir. Physiol.***26**:173-182.

Horsfield, K., and Thurlbeck, A. (1981a). Volume of the conducting airways calculated from morphometric parameters. *Bull. Math. Biol.* **43**:101-109.

Horsfield, K., and Thurlbeck, A. (1981b). Relation between diameter and flow in branches of the bronchial tree. *Bull. Math. Biol.* **43**:681-691.

Horsfield, K., and Woldenberg, M.J. (1986). Comparison of vertex analysis and branching ratio in the study of trees. *Respir. Physiol.* **65**:245-256.

Horsfield, K., Cuming, G., and Hicken, P. (1966). A morphologic study of airway disease using bronchial casts. *Am. Rev. Respir. Dis.* **93**:900-906.

Horsfeld, K., Dart, G., Olson, D. E. Filley, G. F., and Cumming, G. (1971). Models of the human bronchial tree. *J. Appl. Physiol.* **31**:207-217.

Horsfield, K., Kemp., W., and Phillips, S. (1982). An asymmetrical model of the airways of the dog lung. *J. Appl. Physiol.***52**:21-26.

Horsfield, K., Gordon, W. I., Kemp, W., and Phillips, S. (1987). Growth of the bronchial tree in man. *Thorax*, **42**:383-388.

Hughes, J. M. B., Hoppin, F. G., and Mead, J. (1972). Effect of lung inflation on bronchial length and diameter in excised lungs. *J. Appl. Physiol.* **32**:25-35.

Liebow, A. L., Hales, M. R., Lindskog, G. E., and Bloomer, W. E. (1947). Plastic demonstrations of pulmonary pathology. *Bull. Int. Assoc. Med. Mus.* **27**:116-129.

McBride, J. T., and Chuang, C. (1985). A technique for quantitating airway size from bronchial casts. *J. Appl. Physiol.* **58**:1015-1022.

McLaughlin, R. F., Tyler, W. S., and Canada, R. O. (1961). Subgross pulmonary anatomy in various mammals and man. *J.A.M.A.* **175**:148-151.

Ogawa, C. (1920). The finer ramifications of the human lung. *Am. J. Anat.* **27**:315-332.

Pedley, T. J., Schroter, R. C., and Sudlow, M. F. (1970). Energy losses and pressure drop in models of human airways. *Respir. Physiol.* **9**:371-386.

Phalen, R. F., Yeh, H. C., Schum, G. M., and Raabe, O. G. (1978). Application of an idealized model to morphometry of the mammalian tracheobronchial tree. *Anat. Rec.* **190**:167-176.

Plopper, C. G., Mariassy, A. T., and Lollini, L. O. (1983). Structure as revealed by airway dissection. *Am. Rev. Respir. Dis.***128**:S4-S7.

Pump, K. K. (1964). The morphology of the finer branches of the bronchial tree of the human lung. *Dis. Chest* **46**:379-398.

Pump, K. K. (1969). Morphology of the acinus of the human lung. *Dis. Chest* **56**:126-134.

Pump, K. K. (1973). The pattern of development of emphysema in the human lung. *Am. Rev. Respir. Dis.* **108**:610–620.

Raabe, O. G., Yeh, H. C., Schum, G. M., and Phalen, R. F. (1976). *Tracheobronchial Geomtry: Human, Dog, Rat, Hamster, LF–53.* Albuquerque, NM, Lovelace Foundation for Medical Education and Research.

Ross, B. B. (1957). Influence of bronchial tree structure on ventilation in the dog's lung as inferred from measurements of a plastic cast. *J. Appl. Physiol.* **10**:1–14.

Sadler, M., and Berry, M. (1983). Morphometric study of the development of Purkinje cell dendritic trees in the mouse using vertex analysis. *J. Microsc.* **131**:341–354.

Sidell, R. S., and Fredberg, J.J. (1978). Noninvasive inference of airway network geometry from broadband lung reflection data. *J. Biomech. Eng.* **100**:131–138.

Singhal, S., Henderson, R., Horsfield, K., Harding, K., and Cumming, G. (1973). Morphometry of the human pulmonary arterial tree. *Circ. Res.* **33**:190–197.

Strahler, A. N. (1952). Hypsometric (area-altitude) analysis of erosional topography. *Bull. Geol. Soc. Am.* **63**:1117–1142.

Strahler, A. N. (1957). Quantitative analysis of watershed geomorphology. *Trans. Am. Geophys. Union* **38**:913–920.

Thurlbeck, A., and Horsfield, K. (1980). Branching angles in the bronchial tree related to order of branching. *Respir. Physiol.* **41**:173–181.

Weibel, E. R. (1963). *Morphometry of the Human Lung* Berlin, Springer-Verlag.

Willson, H. G. (1922). The terminals of the human bronchiole. *Am. J. Anat.* **30**:267–295.

Woldenberg, M.J., Cumming, G., Harding, K., Horsfield, K., Prowse, K., and Singhal, S. (1970). *Law and Ordere in the Human Lung. Geography and Properties of Surfaces.* Selries No. 41, Cambridge, MA, Harvard University Press.

Woldenberg, M. J., and Horsfield, K. (1983). Finding the optimal lengths for three branches at a junction. *J. Theor. Biol.* **104**:301–318.

Wyatt, J. P., Fischer, V. W., and Sweet, H. (1961). Centrilobular emphysema. *Lab. Invest.* **10**:159–177.

Yeates, D. B., and Aspin, N. (1978). A mathematical description of the airways of the human lungs. *Respir. Physiol.* **32**:91–104

Yeh, H. C. (1979). Modeling of biological tree structures. *Bull. Math. Biol.* **41**:893–898.

Yeh, H. C., and Schum, G. M. (1980). Models of human lung airways and their application to inhaled particle deposition. *Bull. Math. Biol.* **42**:461–480.

Yeh, H. C., Schum, G. M., and Duggan, M. T. (1979). Anatomic models of the tracheobronchial and pulmonary regions of the rat. *Anat. Rec.* **195**:483–492.

12

Methods for the Morphological Study of Tracheal and Bronchial Glands

SANAE SHIMURA

Tohoku University School of Medicine
Sendai, Japan

I. Introduction

By the coordination of ciliary movement, airway mucus secretion plays a primary role in mucociliary clearance and thus in the defense mechanisms against inhaled particles and pathogens as well as reflexes (bronchoconstriction and cough) and local immune responses. Secretory cells exist at the airways both in the surface epithelium that contains mucuous cells (goblet cells), serous cells, and Clara cells, and in the submucosal glands, which consist of serous and mucous cells. The epithelial serous cell has only recently been found in human and rat, and in human it is frequently found during the fetal stage and little is known of its fate postnatally. It resembles the serous cell of the human and rat submucosal gland, and, compared to the surface mucous cell (goblet cell), it is very rare. Clara cells are found only in the terminal bronchiolus of many species, including humans, and the nature of their secretion is still obscure. In healthy adult humans the volume of the submucosal glands down to the fifth generation in bronchi was calculated to be as much as 4 ml and submucosal glad cells, in volume, exceed the surface mucous cells (goblet cells) by a ratio of about 40:1 (Reid, 1960). Furthermore, autonomic nerves do not regulate secretion from surface goblet cells; the submucosal glands probably make the greater contribution to the production of respiratory tract mucus in human lungs.

In the human respiratory tract the submucosa consists of glands and cartilage, the two occurring together, with the gland mainly internal to the cartilage lying between it and the epithelium but also between the plates of cartilage. Glands are most numerous in the trachea, progressively decreasing distally, along with the cartilage. They are absent from airways smaller than 1 mm in diameter (i.e., bronchiolus).

The total volume and airway distribution of submucosal glands is species related (Table 1). In some species, such as the goose, rat, and rabbit, submucosal glands are scarce or absent, therefore goblet cells are presumably the primary source of mucus. In the rat, cartilage is absent from any intrapulmonary airway and glands are concentrated centrally in the upper trachea. In the rabbit, glands are absent while cartilage extends into the small intrapulmonary bronchi. In cat, ferret, dog, and human trachea, submucosal glands are abundant and tracheas from these species are usually used for the experimental study of submucosal gland secretion. To date, site-related physiological differences (tracheal vs. bronchial or central vs. peripheral) in submucosal gland secretion have not been found despite the differences in the amount of mucus secreted and in glandular tissues examined.

Almost all the tools and methods available for morphological experimental research in other tissues have also been applied to tracheobronchial gland studies. These include light microscopy, electron microscopy (transmission and scanning), histochemistry, immunocytochemistry, morphometry, fluorescence, and freeze-fracture. Since each has its own limitations as well as advantages in morphological examinations, it is important to know which method or tool is most suitable for gaining an understanding of the mechanism operating in the secretory response of the airway submucosal gland.

II. Tridimensional Morphology of the Tracheobronchial Gland

Little knowledge of the tridimensional structure of tracheobronchial glands has been gained because they are embedded in the submucosal gland layer. Meyrick et al. (1969) obtained a graphic reconstruction of a submucosal gland from a normal human main bronchus using serial paraffin sections, and established that the submucosal gland consists of four distinct regions: the ciliated duct, collecting duct, mucous tubules, and serous tubules. They have determined the tridimensional relations of these regions to each other.

Applying the separation technique originally developed for isolating kidney tubules (Burg et al., 1966), Shimura et al. (1986) have been successful in isolating single submucosal glands from trachea. The method for isolating single submucosal glands from trachea will be outlined here (Fig. 1). Submucosal glands are isolated mechanically from the membranous portion under a stereoscopic microscope. The

Table 1 Species-Related Quantity and Distribution of Tracheobronchial Glands

Species	Trachea		Bronchus
	Proximal	Mid and Distal	
Human	+ +	+ +	+ +
Monkey	+	+	+
Dog	+ +	+ +	+
Pig	+ +	+ +	+
Cow	+ +	+ +	+
Rat	+	−	−
Mouse	+	−	−
Hamster	−	−	−
Guinea pig	−	−	−
Ferret	+	+	+
Horse	+	+	−
Oppossum	+ +	+ +	+ +

+ +, Abundant; +, present; −, absent or scarce.
Source: Jeffery, 1983.

tracheas are kept cold (4 °C) in a 5% CO_2–95% O_2 saturated modified Krebs–Ringer bicarbonate (KRB) solution until use. The external surface of the trachea is cleaned of fat and connective tissues, cut into rings 3–4cm long, and fixed by pins in KRB at 4 °C with the posterior (membranous) wall side up. Light through a flexible fiber bronchoscope placed inside the tracheal ring is used to transilluminate the membranous portion of the trachea. The outermost layer and thick smooth muscle layer are carefully removed. The glands can then be easily differentiated from the surrounding connective tissues under a stereoscopic microscopy × 60–80). Fresh, unstained submucosal gland is isolated using two pairs of tweezers and microscissors. To avoid tissue damage during isolation, care is taken to isolate the gland by picking up some of the connective tissues surrounding the gland. The entire procedure is performed in a KRB solution bubbled with 5% CO_2–85% O_2 gas at pH 7.4 and at 4 °C. For the observation of morphology alone, a slight stain with 1% indocian green or 1% neutral red is recommended, by which the submucosal gland in situ can be observed more easily and clearly. Isolated glands from canine trachea have the appearance of a flattened and elongated bunch of graphs (Fig. 2), similar to that obtained from histological reconstruction by serial section of human bronchial wall (Meyrick et al., 1969). The size of the gland from canine trachea ranged from 300 μm to 1.5 mm in maximum length with a collecting duct 50–70 μm in external diameter in the fresh

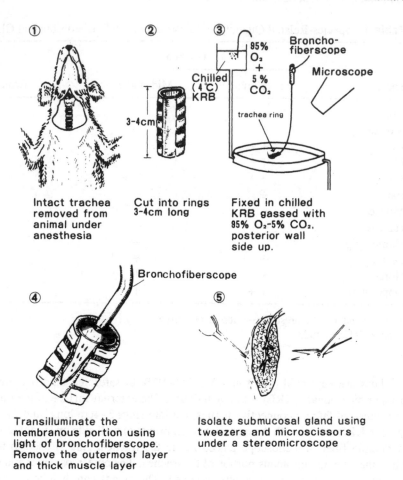

① Intact trachea removed from animal under anesthesia

② Cut into rings 3-4cm long

3-4cm

③ Fixed in chilled KRB gassed with 85% O_2-5% CO_2, posterior wall side up.

85% O_2 + 5% CO_2

Chilled (4°C) KRB

Broncho-fiberscope

Microscope

trachea ring

Bronchofiberscope

④ Transilluminate the membranous portion using light of bronchofiberscope. Remove the outermost layer and thick muscle layer

⑤ Isolate submucosal gland using tweezers and microscissors under a stereomicroscope

Figure 1 Diagrammatic illustration of method used in this laboratory for isolating submucosal glands from animal trachea. See text for details.

and unstained state. The isolated gland preparation enables us to observe directly the glandular contraction in response to cholinergic agents through the activity of myoepithelial cells (Fig. 3) (see also section VIII) (Shimura et al., 1986; 1987a,b). In addition, using this preparation we can examine secretory responses of airway submucosal glands in a well-defined condition, excluding possible effects of surrounding tissues, that is, epithelial inhibitory action (Sasaki et al., 1986, 1987; Shimura et al., 1987b; 1988). For morphological study, isolated glands are stained with Mercurochrome for 10 sec to increase visibility and, after the fixation with glutaraldehyde, they are embedded in molten 3% Bacto-agar to prevent loss

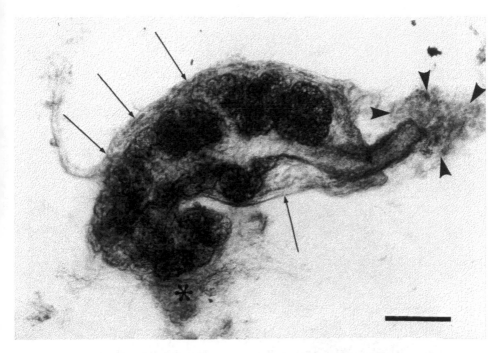

Figure 2 Single submucosal gland in the fresh and unstained state, isolated from canine trachea that is secreting mucus from the tip of a duct (arrowheads). The gland has the appearance of a "bunch of grapes." Arrows indicate small vessels attached to gland, which contain some blood cells (bar, 100 μm).

during postfixation and dehydration. Staining with mercurochrome and embedding in Bacto-agar are techniques that have been used for handling such very small samples as single isolated alveolar walls and are known to have no effects on the ultrastructure (Shimura et al., 1985a). For light microscopic studies, specimens are embedded in resin (Epon 812), mounted on aluminum stubs. A longitudinal section (1 μm thick) can then be cut with a glass knife on an ultramicrotome. This has revealed a tubuloacinar pattern in submucosal gland from feline and canine trachea. After 1 μm sections are obtained, ultrathin sections are cut for electron microscopic examination. It is difficult to obtain such a longitudinal profile of the submucosal gland by conventional histological methods using airway walls. An example from canine trachea is shown in Figure. 4.

Knowledge of tridimensional structure contributes to an understanding of the developmental patterns, hyperplasia, and morphometrical evaluation of submucosal glands.

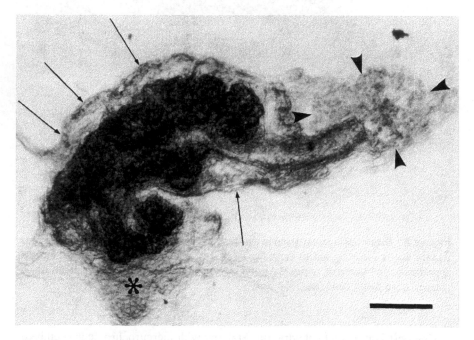

Figure 3   Submucosal gland shown in Figure 2 contracted 16% of its length in response to 10^{-4} M methacholine, simultaneously squeezing mucus into medium (arrowheads). The collecting duct outside the glandular tissue shows no change in diameter. Small vessels (arrows) and connective tissues (asterisk) attached to the gland seem neither to contract nor to play a role in contraction of the gland (bar, 100 μm).

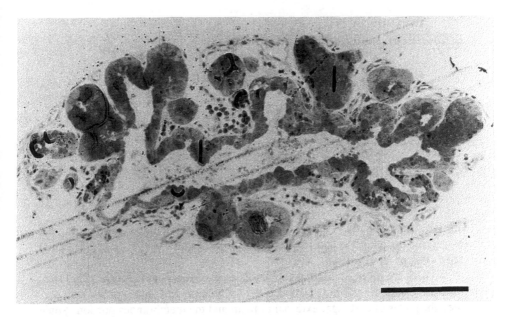

Figure 4   A longitudinal 1 μm section of isolated submucosal gland from canine trachea with toluidine blue stain. Visible are a tubuloacinar pattern containing a small amount of connective tissue, a few small vessels, migratory cells, but no smooth muscle (bar, 100 μm).

III.   Measurement of Tracheobronchial Gland Size

The fundamental morphological change in chronic bronchitis is enlargement of the tracheobronchial glands brought about by hyperplasia of the acinar secretory cells. Methods of morphologically estimating or quantifying tracheobronchial gland size or volume have been developed to investigate chronic obstructive pulmonary disease, particularly chronic bronchitis.

Reid (1960) described measurement the gland to wall ratio, which is generally referred to as the Reid index, and its increase in chronic bronchitis has been confirmed by most observers. When A represents the internal wall thickness (the distance between the basement and the inner aspect of cartilage), and B, the thickness of the gland at the same point, the Reid index is expressed as ratio of B to A as shown in Figure 5 (Reid, 1960). Reid reported that the ratio ranges from 0.14 to 0.36 (mean 0.26) normally, 0.41 to 0.799 (mean 0.59) in bronchitis, and 0.28 to 0.43 (mean 0.33) in emphysema. In her series, there was no overlap in the Reid index values of patients with and those without bronchitis.

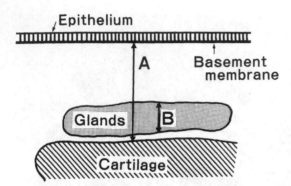

Figure 5 Illustration of the method used for measuring the wall to gland ratio (Reid index). See text for detail (from Reid, 1960).

The Reid index retains a useful place both as a quick method of measurement for routine hospital use and in epidemiological (Scott, 1973), or experimental study (Phipps et al., 1986).

There are inherent problems with this assessment (Reid index) of gland size, since the glands lie not only between the epithelium and cartilage but also between the plates of cartilage, external to them and the membranous portion. Some investigators have cited this fact as the reason that there is some overlap between normal and bronchitic groups and that the Reid index is not significantly related to sputum volume (Thurlbeck and Angus, 1964; Mckenzie et al., 1969; Jamal et al., 1984). Some attempts have been made by several investigators to overcome this criticism.

Alli (1975) has described a radial intercepts method for quantitating bronchial gland size, in which one takes measurements along radii at sectoral intervals of bronchial transection including all portions of the wall. The sum of all the measurements of gland thickness is taken together with the sum of all the radial measurements of wall thickness and expressed as a ratio thus: sum of gland thickness ÷ sum of wall thickness. They reported the high correlation of this ratio with planimetric and point-count techniques for quantifying bronchial wall components. However, the Reid index and radical intercepts method are essentially one-dimensional approaches.

The point-count technique (Hale et al., 1968; Dunhill et al., 1969; Macleod and Heard 1969; Wheeldon and Pirie, 1974; Lennox, 1975; Oberholzer et al., 1978; Berend et al., 1979; Hayashi et al., 1979) and weighing of paper cut-outs of drawing (Restrepo and Heard, 1963a; 1963b; 1964), both reflecting two-dimensional attempts at greater precision, have been applied to glandular and other bronchial constituents. To determine the relative accuracy and efficiency of these methods, Bedrossian et al. (1971) compared results obtained by Reid index, the point-count methods, and the weighing of paper cut-outs with "true" value derived by planimetry from the bronchi of 30 subjects and showed that results obtained with Reid indices correlated poorly with those of planimetry. They concluded that weighing paper cut-outs seemed to be the most economical from the standpoint of time saved. This, however, seems to be a crude method of estimation because weights of paper vary largely, depending on the humidity and nonuniformity of thickness. On the other hand, Takizawa and Thurlbeck (1971a,b) have reported a close correlation between Reid index and point-counting method. For these morphometries, tracheobronchial images are usually projected on to a sheet of white drawing paper fixed to a vertical wall or to a point-count grid, since it is more expensive when gland fields are photographed and printed at a fixed magnification. Stained sections are also examined at a moderate magnification with a Wild sampling microscope (Model M510, Heerbrugg, Switzerland) incorporating a projection head containing a Weibel multipurpose coherent lattice with 42 points used for areal estimation of tracheobronchial glad (Hayashi et al., 1979). Point counting may be inaccurate when the component under investigation is only a small percentage of the whole, such as glandular tissues in the airway wall (Lennox, 1975). Edwards and Carlile (1982) measured the cross-sectional area of the bronchial wall and the area occupied by gland, muscle, and cartilage in human lung by means of a projecting microscope and a Reichert-Jung electronic planimeter. Gland areas expressed as a percentage of the total cross-sectional area were 6.74, 8.1–8.4, and 6.1–6.2 in main, lobar, and segmental bronchi, respectively. Recently, absolute glandular area and volume proportion measurements have been made by projecting images onto a computer-assisted graphics tablet (Jamal et al., 1984; Nagai et al., 1985). In addition to the measurements of whole gland size or volume, gland acinar counts per light microscopic field (Reid, 1960) and the ratio of mucous to serous glad acini (Glynn and Michaels, 1960) have also been suggested as indices of bronchial glandular hypertrophy.

In practice, there are some difficulties with the measurements of tracheobronchial gland size or volume. Although almost all investigators have used paraffin sections for gland morphometry, some have embedded samples in a resin (for example, glycol methacrylate mixture), and sectioned with a glass knife on an ultramicrotome (Hayashi et al., 1979). Resin embedding has the advantage that artifacts such as deformities and clefts due to cartilage in the tracheobronchial wall are avoided during sectioning, as well as the accuracy of thickness, compared with paraffin embedding. However, resin embedding has the disadvantage of being more expensive and painstaking than paraffin embedding. Furthermore, when the area or volume proportion or glands to bronchial wall is obtained there often remains some doubt regarding the outer boundary of the wall during measurement. Specifically, the imperceptible blending of the peribronchial connective tissue with the outer perichondrium makes it difficult to decide precisely the line of cleavage between the outer perichondrium and the areola connective tissue. The artifacts are minimized by separating the lamina propria from the perichondrium (Jamal et al., 1984). Also, Whimster et al. (1984), to overcome this problem,

Table 2 Methods for Measurement of Tracheobronchial Gland Size or Volume

Method	Advantages	Disadvantages	References
1 Gland- to wall ratio (Reid index)	Simple and inexpensive	Does not contain glands between cartilage or membranous portions	Reid (1960; Hayes (1969); Burton and Dixon (1969); Mckenzie et al. (1969); Douglas (1980)
2 Radial intercepts method	Inexpensive	One dimensional; tool is complicated	Hale et al. (1968; Dunhill et al. (1969); Macleod and Heard (1969); Wheeldon and Pirie; (1974); Lennox (1975); Oberholzer et al. (1978); Berend et al. (1979); Hayashi et al. (1979)
4 Weighting of paper cut-outs of drawing	Simple and inexpensive	Crude estimation	Restrepo and Heard (1963a, b, 1964)
5 Planimetry	Same as point counting	Complicated manipulation or expensive	Dunhill et al. (1969; Bedrossian et al. (1971); Edwards and Carlile (1982)
6 Absolute area or volume estimation by a computer-associated graphic tablet	Provides accurate data on area or volume proportion	Expensive; complicated manipulation	Jamal et al. (1984); Nagai et al. (1985); Aikawa et al. (1987)

expressed human tracheobronchial gland amounts volumetrically as the gland volume per unit luminal surface area of the airways. Since the bronchial glands produce secretions that are required for the luminal surface of the airway, it seems appropriate to express gland volume in terms of the luminal area, as these authors asserted in their report. Finally, it should be noted that estimation of volume proportions requires that the structures to be measured are randomly dispersed throughout the organ. This is clearly not the case with bronchial wall components, and thus the theoretical base for the volume proportion of mucus gland is not sound. Despite this, it works empirically and this is the important criterion.

Thurlbeck and Angus (1964) have shown that there is an overlap in the measurements of the Reid index obtained in bronchitic and in nonbronchitic subjects. Furthermore, some investigators have reported that the gland/wall ratio (Reid index) is less applicable to clinicopathological correlations (chronic bronchitis or sputum volume) than the absolute area or the volume density (or proportion) of glands established by planimetry, point-counting, or computer-assisted digitization. The Reid index has been recommended by a number of investigators because it requires less skill and offers a significant saving in time and cost. Now the area or volume proportion of glands can be easily and quickly measured using a computer-assisted digitizer. However, it is not clear that area or volume assessment of glands requires and expensive machine such as a computer. Since more is being learned about the tridimensional structure and the distribution of tracheobronchial submucosal gland, it seems reasonable to attempt to devise, in place of the Reid index, a new yardstick or index that is simple and reflects true gland volume.

Methods for the measurement of tracheobronchial gland size are summarized in Table 2, including their advantages and limitations.

Tracheobronchial gland volume has been known to show developmental (Leigh et al., 1986a), regional (trachea, main bronchi, or segmental bronchi) (Niewoeher et al., 1972), sexual (Hayashi et al., 1979) or species-related (Jeffery, 1983) differences that have been confirmed by morphometric methods. Furthermore, tracheobronchial gland hyperplasia is associated with some conditions or diseases of the lung in addition to chronic bronchitis. These include mainly tobacco smoke (Ryder et al., 1971), air or environmental pollution (Hayes, 1969; Scott, 1973), pneumoconiosis (Mackenzie, 1973; Douglas, 1980; Douglas et al., 1982), bronchial asthma (Dunhill et al., 1969; Takizawa and Thurlbeck, 1971b), and cryptogenic fibrosing alveolitis (Edwards and Carlile, 1982).

IV. Secretory Products

Secretory cells in tracheobronchial submucosal glands consist of mucous and serous cells that contain secretory granules. Much attention has been focused on these secretory granules, which enclose the intracellular product prior to discharge into the duct and airway lumen. Various morphological methods have been applied to these granules in order to allow us to understand the nature of the secretory product in response to various stimuli.

A. Histochemical Methods

A variety of cytochemical procedures are used for localizing and characterizing carbohydrate-rich components of secretory granules by light microscopy.

With the use of hematoxylin and eosin (HE) stains, two secretory cell types, mucous and serous, have been identified in submucosal glands of the respiratory tract. Mucous cells contain many pale droplets on HE stainings and the nucleus is sometimes invisible. Serous cells contain a multitude of small, highly refractile secretory granules in the cytoplasm. Mucous cells are supposed to elaborate a viscid secretion that consists almost exclusively of mucin, while serous cells secrete a watery liquid that lacks mucin but contains electrolytes and proteins.

The presence of both neutral and acid glycoproteins has been demonstrated in the secretory cells of the submucosal glands. Of the various staining techniques available, the periodic acid-Schiff (PAS) and alcian blue (AB) techniques are much used; PAS is used to demonstrate the presence of neutral glycoprotein and AB for acid glycoprotein. Alcian blue, used at pH 2.6, stains both sialic acid and sulfate radicals bright blue. The control of AB staining, by varying the pH between 0.5 and 2.6, has been found by Jones and Reid (1973a, 1973b) to be satisfactorily selective, differentiating between sialo- and sulfomucin. With a combination of AB (pH 2.6) and PAS stains, mucous cells stain blue for acid glycoprotein. The mucous cells of the human tracheobronchial glands have been shown by Lamb and Reid (1972) to produce four groups of acid glycoprotein: sialomucin susceptible to sialidase; sialomucin resistent to sialidase; a sulfate identified by AB staining after acid hydrolysis (which removes all the sialomucin) but that is not stained by stains specific for sulfate; and a sulfate that stains after hydrolysis and stains with the specific stains for it. Ferret tracheal glands develop from intraepithelial cellular aggregates devoid of secretory granules at birth into complex, submucosal tubuloacinar structures composed predominantly of cells containing nonacidic (staining with PAS but not AB) secretory granules at 28 days (Leigh et al., 1986b), and thereafter acidic histochemical staining properties increase in secretory cells (Leigh et al., 1986a).

The high-iron diamine (HID) and low-iron diamine (LID) methods have been used to demonstrate the presence of acid glycoconjugates at the light microscopic level. The HID method appears to stain sulfate groups specifically, as evidenced by its staining of sites incorporating radiosulfate and its failure to stain alcian blue positive carboxyl groups (Spicer, 1965; Sorvari, 1972). On the other hand, LID appears to stain both caroxyl and sulfate groups that demonstrate

Table 3 Major Histochemical Methods for Staining Mucus Glycoproteins at the Light Microscopic Level

		Acid	
	Neutral	Sialylated	Sulfated
Periodic acid -Schiff (PAS)	+	+	+
Alcian blue (AB)	−	+	+
High-irion diamine (HID)	−	−	+
Cell type	Serous	Mucous	

alcian blue reactivity (Spicer, 1965). The combined HID-AB procedure stains some mucous cells black, demonstrative of sulfomucin; stains approximately an equal number blue, demonstrative of sialomucin; and imparts blue and black coloration to a portion of the secretory cells (Spicer et al., 1983).

Since PAS, AB, and HID are most popular among the stains available, interpretation of these techniques is summarized in Table 3.

Recently, lectins conjugated to fluorescence or horseradish peroxidase have been used to localize specific terminal or internal sugars of tissue complex carbohydrates. The specificity of the lectin binding for a single terminal sugar and specific susceptibility of a given linkage to lysis by an exoglycosidase are also utilized for cytochemical analysis of submucosal gland cells. Mucous tubule cells uniformly bind the peanut lectin–horseradish peroxidase (PL-HRP) conjugate after, but not before, sialidase digestion. They, therefore, uniformly contain penultimate alactose and terminal sialic acid (Spicer et al., 1983). As noted above, the mucous cells invariably stain thoughout for sulfomucin or sialomicin. Furthermore, Itoh et al. (1981) have demonstrated positive staining for fucose mainly in mucous cells of human bronchial glands using a peroxidase-labeled lectin from *Ulex europaeous* (UEA-I), which is thought to be specific for fucose (Fig. 6). Terminal sugars, identified by lectin–peroxidase, are known to vary between species and in human with ABO type (see also section on Immunohystological Methods, below).

Serous tubules in the human respiratory tract stain moderately for complex carbohydrate at the light microscopic level. With the AB-PAS method, some serous cells in human submucosal gland evidence red staining indicative of neutral mucosubstance and others blue-purple coloration demonstrative of acidic glycoconjugate. The acidic groups are identifiable as sulfate esters from their gray to black staining with the HID–AB sequence. The latter method rarely reveals blue staining demonstrative of neuraminic acid (sialic acid) in serous cells, in contrast with

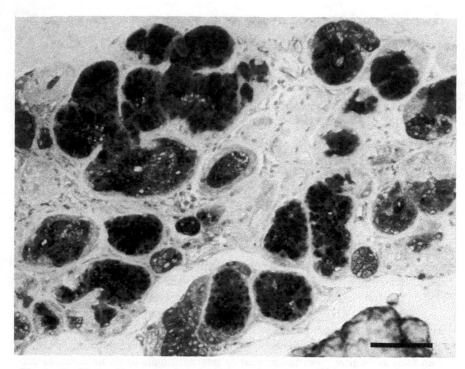

Figure 6 Light micrograph of human bronchial submucosal glands stained with the peroxidase–antiperoxidase (PAP) method using a peroxidase-labeled lectin from *Ulex europaeus* (UEA-I) that is thought to be specific to fucose. Positive staining is seen mainly in mucous cells (bar, 25 μm; photo courtesy of Dr. M. Itoh).

the abundant sialic acid demonstrable histochemically in mucous cells. Meanwhile, serous tubules in mouse trachea have been found to lack affinity for cationic reagents, including AB and iron diamine, and to stain red with the AB-PAS method; they are, therefore, uniformly judged to produce a neutral complex carbohydrate. The peanut lectin–horseradish peroxidase (Stoward et al., 1980) and galactose oxidase–Schiff methods concur in demonstrating terminal galactose in glycoprotein devoid of neuraninic acid (Spicer et al., 1983). Beyond an inconstant degree of affinity for HID, demonstrative of a variable content of sulfate esters in secretory mucosubstance, serous cells differ little from one another at the light microscopic level.

B. Ultrastructural Methods

At the electron microscopic (transmission) level, mucous cells have flat, basal nuclei and lucent granules (300–1800 nm in diameter) and the limiting membranes of many of these granules are fused in human lung. In contrast, serous cells have round, centrally located nuclei and granules (300–1000 nm in diameter) with dense contents, and the membranes of most serous cells are not fused (Meyrick and Reid, 1970) (Figure 7). Although transmission electron microscopy has been used for ultrastructural examination of secretory granules by almost all investigators, Kawada et al. (1981) used scanning electron microscopy (SEM) to observe secretory granules in human bronchial gland using a freeze-fracture method that provides tridimensional images of secretory granules. However, the term *electron microscopy* used in this chapter refers to transmission electron microscopy by ultrathin sectioning except where stated otherwise.

Mucous tubule cells vary in fine structure and the ultrastructual cytochemical properties of their secretory granules. Mucous cells, however, differ from one another less than do serous cells. HID staining at the electron microscopic levels shows variability between the granules within a given cell. HID staining with the ultrastructural level has also demonstrated three different zones with granule profiles in some mucous cells in rat trachea (Bensch et al., 1965). The zones can be differentiated on the basis of an absence of staining and light and heavy staining with both the HID method and the dialyzed iron (DI) procedure. The DI stain reacts with sialated and sulfated glycoprotein (Wetzel et al., 1966). The ultrastructural cytochemical methods, therefore, confirm results obtained by light microscopic cytochemistry and extend these results to show differences between granules within a cell or between zones within a granule profile.

LID method has also been used by Takagi et al. (1982) in ultrastructural studies of secretory granules in mucous tubules of rat tracheal glands, which are known to contain sulfated glycoprotein. Thiocarbohydrazide silver proteinate (TSH-SP) staining of thin sections variable enhanced LID reactive sites. They then suggested that the LID and LID-TCH-SP methods would be useful for ultrastructural localization of carboxylated and sulfated glycoconjugates (Takagi et al., 1982).

Although serous cells differ little from one another at the light microscopic level, when examined ultrastructurally the several cells in a profile of serous tuble differ moderately in the morphology and cytochemistry of their granules. Cytoplasmic granules in some but not other serous cells generally contain a variable-sized, often eccentric core and variably thick rim. The cortex reacts with the periodic acid–thiocarbohydrazide (PA–T–SP) method and, to a varying degree, evidences HID affinity, whereas the core is consistently devoid of complex carbohydrate (Spicer et al., 1978; 1983). Little information is available on the composition of the cores. Since serous cells contain lysozyme, their cores could consist in part of this enzyme.

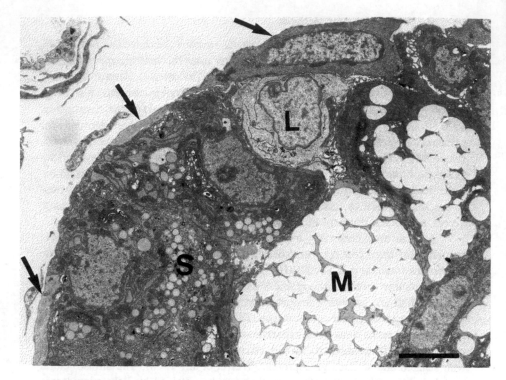

Figure 7 Electron micrograph of acinus in isolated gland from canine trachea. Myoepithelial cells (thick arrows) surround acinus. S, serous cells; M, mucous cells; L, lymphocyte. Thin arrow indicates gland lumen (bar, 10 μm).

The substructure of serous cells from porcine tracheal submucosal glands was studied by Turek et al. (1982). In tissue fixed and processed for TEM, and stained with uranyl acetate and lead citrate, the condensing granules of serous cells occasionally processed hexagonal and sometimes a lamellar substructure. Tissue fixed in paraformaldehyde–glutaraldehyde and stained with periodic acid–thiocarbohydrazide–silver proteinate (PTS) or with phosphotungstic acid (PTA) showed secretory granules stained for complex carbohydrates, revealing a substructure similar to that noted in the condensing granules. The PTS sequence stains for neutral glycoproteins since the serous cell granules do not react with DI stain for acidic glycoproteins (Spicer et al., 1980) and PTA at acidic pH is thought to stain highly polymerized carbohydrates (Hayat, 1975). Treatment of periodic acid-oxidized thin sections with pronase or pepsin prior to thiocarbohydrazide and silver proteinate treatment decreased the intensity of the PTS staining, but did not digest away any components of the granules. The substructure revealed by the carbohydrate stains may be a reflection of the mechanism of packaging or the macromolecular structure of the glycoproteins in the serous cell granules (Turek et al., 1982

It is usually assumed that membrane-bound glycolipids are poorly preserved with commonly used fixatives and, thus, are solubilized during subsequent tissue processing. Although highly probable, this assumption has not been proved and this should be borne in mind, especially when one attempts to interpret histochemical reactivity of the cell surface and intracellular membranes, since staining in these sites could result from the presence of residual glycolipid (Spicer et al., 1983).

C. Immunohistological Methods

Among a population with a given blood group type, their specific blood group substance should be contained in the respiratory tract secretory cells of those that are secretors. Residues other than galactose are known biochemically to terminate glycoprotein side chains, since type A blood group substances contain terminal N–acetylgalactosamine and type O substance contains terminal fucose. These blood-group-reactive substances occur in respiratory secretions and have been localized in mucous cells using immunocytochemical techniques (Glynn and Holborow, 1959). The lectins with specific binding affinity for terminal α-N-acetyl-D-galactosamine (type A) or α-galactose (type B) have been used to characterize the cell population from which these blood-group-reactive substances are derived (Schulte and Spicer, 1982; Spicer et al., 1983).

Serous cells of the human respiratory tract evidence lysozyme and lactoferrin content at light microscopic level by immunocytochemical methods using rabbit antihuman lysozyme or lactoferrin (Mason and Taylor, 1975; Spicer et al., 1977; Bowes and Corrin, 1977; Itoh et al., 1981) (Fig. 8).

Ultrastructural immunostaining localizes lysozyme in the granule rim in some bronchial glands, indicating that the rim consists in part of components other than glycoprotein (Mason and Taylor, 1975). That ultrastructural HID staining localizes sulfated complex carbohydrate in the rim of many bizonal serous cell granules suggests the possibility of complex formation beteween the strongly cationic lysozyme and a sulfated glycoconjugate in the granule rim. Tom-Moy et al. (1983) reported that electron microscopic and immunocytochemical analysis of incubated ferret tracheal segments revealed a loss of serous granules and lysozyme immunoreactivity in response to cholinergic, α- and β-adrenergic drugs. Measurement of lysozyme assayed from the incubating medium indicated that these drugs stimulate lysozyme release. The biological significance of this glycosidase, lysozyme, which is widely encountered in epithelial cells, has not been precisely defined but its presence is possibly related to an antimicrobial function.

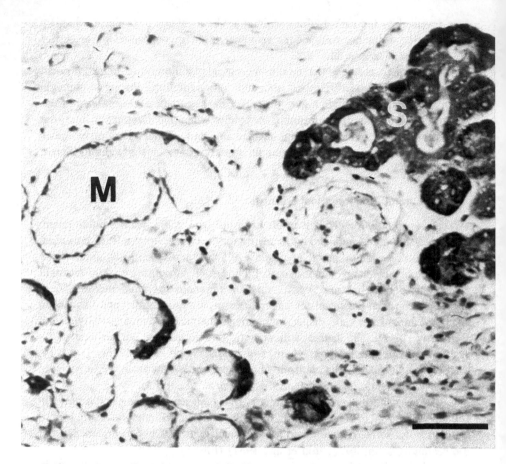

Figure 8 Light micrograph of human bronchial submucosal glands stained immunohistochemically using horseradish-labeled antilactoferrin. Positive staining is seen in serous cells. S, serous cells; M, mucous cells (bar, 25 μm; photo courtesy of Dr. M. Itoh).

The ultrastructural distribution of lactoferrin in human bronchial glands has been studied by an immunoperoxidase method using rabbit antihuman lactoferrin and soluble rabbit antihorseradish peroxidase/horseradish peroxidase complex (Bowes et al., 1981). The distribution of lactoferrin coresponds to that of lysozyme (Bowes and Corrin, 1977). Lactoferrin and lysozyme have both been confirmed in serous acini where the granules show a variable pattern of staining. Some serous granules are filled uniformly with lactoferrin, some lack lactoferrin in a small central core or a thin peripheral rim, and some are competely devoid of lactoferrin (Spicer et al., 1983). Another conspicuous component demonstrated immunocytochemically in human bronchial serous cell is carbonic anhydrase (Spicer et al., 1982). The presence of lactoferrin and carbonic anhydrase is presumably related to an antimicrobial function in airway lumen as with lysozyme.

The distribution of immunoglobulins, IgA, IgG, and IgM in developing human lungs, ranging in age from 12 gestational weeks to 8 years, was studied using the indirect peroxidase-labeled antibody method of Takemura and Eishi (1985). The IgA- and IgM-containing cells were present around the bronchial glands in normal lungs, whereas IgG-containing cells were not associated with the glands. In normal lungs, Ig-A containing cells were most prominent. The apical portions of serous epithelial cells of bronchi and bronchial glands were positive for IgA and IgM. In bacterial and viral pulmonary infections, Ig-A containing cells increased in number in the bronchial glands.

Secretory IgA is the major immunoglobin in bronchial secretions and other external body fluids in humans. IgA and the secretory component in human bronchial mucosa have been localized by immunoelectron microscopy (Goodman et al., 1981). Antibodies to human IgA and secretory component were prepared by hyperimmunization of rabbits with purified myeloma Iga and free secretory component derived from colostrum. IgA was identified on plasma cells located near glandular epithelial cells and on the basolateral plasmalemma and endocytic vesicles of epithelial cells, especially of mucous cells. The secretory component was localized to the perinuclear spaces, endoplasmic reticulum, Golgi complexes, basolateral plasmalemma, and endocytic vesicles of glandular epithelial cells, as well as ciliated epithelial cells (Goodman et al., 1981). These findings suggest that IgA dimers, synthesized in plasma cells, are complexed with the secretory component on the basolateral plasmalemma of epithelial cells in bronchial glands and transported across the cells in endocytic vesicles to the gland lumen.

Using antibodies to basic and acidic proline-rich protein (PRP) of salivary origin, Warner and Azen (1984) have detected PRP immunoreactivity in serous cells of human tracheobronchial glands by an immunoperoxidase technique. Immunoreactive PRP, detected by immunoblotting from sodium dodecyl sulfate (SDS) gels, was also found in culture media from tracheal explants. They hypothesize that PRP, by interacting with glycoproteins of mucous, as other proteins do, may be necessary for maintaining the appropriate viscoelastic properties of respiratory secretions.

Recently, monoclonal antibodies have been used for immunohistological characterization of secretory products from submucosal glands (St. George et al., 1984, 1985; Basbaum et al., 1984b; de Water et al., 1986). Respiratory secretion, which is a complex mixture of large glycoproteins and other materials, has been difficult to isolate and analyze because of its biochemical heterogeneity. Therefore, a large panel of monoclonal antibodies to respiratory tract mucin is feasible. Using a series of 12 monoclonal antibodies made against mucuous and serous components of monkey trachea, St. George et al. (1985) have reported that the intracellular mucous products of tracheal secretory cells exhibit greater heterogeneity than is detectable by conventional histochemistry.

Bronchial secretion contains two proteinase inhibitors: the plasma-derived alpha 1-proteinase inhibitor and antileukoprotease (ALP) (a bronchial mucous proteinase inhibitor), which is not present in the bloodstream. By immunoelectron microscopy, ALP was demonstrated in secretory granules of serous cells inhuman bronchial glands (Mooren et al., 1982). Further, de Water et al. (1986) examined the ultrastructural localization of ALP in human airways by using monoclonal ALP-specific antibodies in a two-step gold-labeling procedure. The monoclonal antibody was produced using spleen cells from purified human ALP-immunized Balb/c mouse and cell fusion procedures. For immunoelectron microscopic studies, ultrathin sections were cut from blocks polymerized at −40°C by the method of Roth et al. (1981). In the serous cells of bronchial glands, ALP could be demonstrated in secretory granules among which four phenotypes could be differentiated morphologically in ultrathin sections. Sometimes gold particles could be observed in the rough endoplasmic reticulum and nuclear envelope, which suggest that ALP is also present in these cell organelles.

D. Morphometric Analysis

Morphometric analyses of secretory cell granules in submucosal glands have been performed at both light and electron microscopic levels. By use of morphometric techniques, we can observe which cell types in glandular tissues respond to stimulation.

Kollerstrom et al. (1977) obtained the percentage area of the bronchial glands taking up AB (pH 2.6) stain for acid mucin by light microscopy using paraffin sections from the trachea and from each generation down the inferior lingular axial bronchial pathway in five nonsmokers and five cigarette smokers. Areas of the submucosal mucous glands what were AB-negative (g) and areas that were AB-positive (m) were estimated separately by counting the number of points superimposed by the intersection of an eyepiece graticule with a 0.5 mm^2 lattice. The percentage of acid mucus in the total gland was estimated by the ratio = m/(m+g) × 100. The mean percentage value for the smokers (59.5% was significantly less than for the nonsmokers (74.4%). The percentage increased between successive generations from the trachea by an average of 6.5. This gradient did not differ significantly between the smoking and nonsmoking groups. Furthermore (Hayashi et al. (1979) reported a sexual difference in the composition of rat tracheal gland secretory granules using quantitative morphologic analyses by light microscopy. Although the relative proportion of gland cells and lumina were equivalent between the sexes, male rats had a larger quantity of mucin. The volume proportion in the glands of AB-positive mucin was 80–84% greater in males and PAS-positive mucin was 18–19% greater in males than in females, both in control animals and those exposed to tobacco smoke. This study demonstrated that sexual differences were present in the mucin composition of the tracheal glands of normal male and female rats, that the female glycoprotein varied during the estrous cycle, and that these differences were retained after exposure to tobacco smoke.

In interpreting such morphometric data related to secretory granules and products of submucosal gland, it should be noted that the following possibilities may account for the differences. First, there is more connective tissue or inflammatory or other nonsecretory cells in the more proximal human glands or in the female rat glands, so the proximal or female secretory cells as a whole must have contained relatively less acid mucin and relatively more cytoplasm. Second, the individual secretory granules in the mucous cell will vary in their rates of production and excretion and their initial volumes. For example, Phipps et al. (1986) have studied effects of O_3 exposure on sheep tracheal mucus secretion by both measuring radiolabeled glycoprotein release and morphometry of submucosal glands. Although chronic exposure produced glandular hypertrophy, it also induced a decrease in glycoprotein release and the gland cells were devoid of secretory material (granules). With recovery, basal glycoprotein secretion was greatly increased above control and gland cells were full of secretory material.

Electron microscopic morphometric analysis has been made of the alteration in secretory cell structure induced by adrenergic and cholinergic agonists and neuropeptides. Basbaum et al. (1981) examined degranulation serous cells induced by adrenergic and cholinergic agonists by electron microscopic observation and stereological analysis of incubated ferret tracheal segments. The jugular vein was cannulated and served as a route for supplemental anesthetic over a 3 hr period in which the animals remained supine prior to removal of the trachea. This step established a consistent resting state in which serous cells were filled with secretory granules. This also allowed the animals to recover the stress of handling and thereby reduce the levels of circulating catecholamines that could provoke secretion (Basbaum et al., 1981; Gashi et al., 1986). The investigators reported that the volume density of serous cell granules was measured using a coherent multipurpose test system (Weibel, 1973) laid over electron micrographs and was significantly reduced by α-adrenergic and cholinergic agonist. This suggests that serous cell granules are discharge by both α-adrenergic and cholinergic, but not by β-adrenergic, stimulation.

Using a similar morphometric analysis, Gashi et al. (1986) examined the effect of neuropeptides (substance P and vasoactive intestinal peptide) on the gland cell morphology of ferret trachea. Serous cells of tracheal segments treated with neuropeptides were markedly degranulated and also these two neuropeptides increased significantly the release of $^{35}SO_4$-labeled macromolecules from tracheal segments. They have not obtained such evidence for mucous cells. However, the presence of cytoplasmic tags and cytoplasmic swelling suggested that mucous cells may be stimulated by the peptides in a manner not made clear by use of these morphometric techniques.

E. Autoradiographic Methods

To investigate glycoprotein (or macromolecule) secretion from airway submucosal glands, radiolabeled glycoconjugates released into the medium have been measured using [^{3}H] (or ^{14}C]) glucosamine or sodium sulfate [^{35}S as precursors. In vitro autographic studies in canine trachea and human bronchus has revealed a time-dependent incorporation of either [^{3}H]glucosamine or [^{35}S]sulfate into mucus-producing cells. Distribution of radiolabeled glycoconjugates has been visualized autoradiographically. Autoradiographs of submucosal glands show silver grains demonstrative of mucosubstance containing [^{35}S]sulfate over the cytoplasm mainly of the mucous cells, whereas those after incubation with [^{3}H]glucosamine were found overlying mucous and serous cells as well as in secretions in the glandular lumen of human and canine tracheobronchial glands (Lamb and Reid, 1970; Chakrin et al., 1972). Three different mucin sugars (N-acetylglucosamine, N-acetylgalactosamine, and sialic acid) may be labeled with tritium after incubation with [^{3}H]glucosamine, whereas [^{35}S]sulfate is incorporated into sulfomucin (acid glycoprotein). Developmental changes in glycoconjugate secretion by ferret trachea have been examined by Leigh et al. (1986b). Incorporation of [^{35}S]sulfate into mucins relative to that for [^{3}H]glucosamine increased with age, consistent with the increasingly acidic histochemical staining properties of secretory cells (Leigh et al., 1986a).

V. Intracellular Secretory Process

A. Cytoskeletal System

From data obtained with various secretory tissues, microfilaments and microtubules have been shown to play an obligatory role in the intracellular process of protein and/or granule secretion of many exocrine glands. Microtubules are 24 nm diameter structural elements and microfilaments are 4–8 nm structures found as parallel arrays or latticelike networks forming the cell web adjacent to the plasma membrane (Figs. 9, 10). Coles and Reid (1981) examined the effect of colchicine (an antimicrotubule agent) and cytochalasin B (an antimicrofilament agent) on glycoconjugate secretion from the submucosal gland of human airways. They suggested that microtubules and microfilaments may be important in secretagogue-induced but not in baseline cellular glycoconjugate discharge, which implies that the mechanics of the two process differ significantly. To our knowledge, however, no attempts have been made to examine the microtubule and microfilament system in airway submucosal gland cells from a morphological

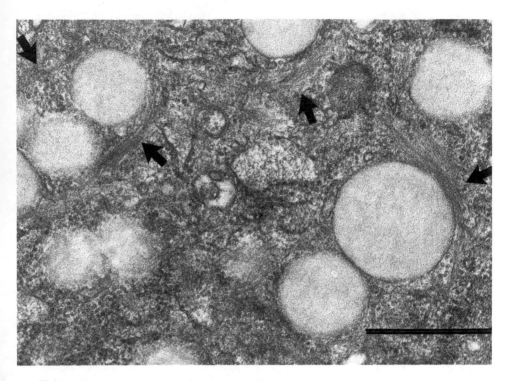

Figure 9 Electron microscopy of serous cell from an isolated gland of canine trachea. Microfilaments (arrows) are seen close to the granules in the cytoplasm (bar, 0.5 μm).

approach. Brown et al. (1985) examined the effects of colchicine on microtubules of cultured type II cells by indirect immunofluorescence using the method of Lazarides and Weber (1974). Attached cells were incubated with colchicine and then processed with the primary antibody of sheep antitubulin and the secondary antibody of fluorescein-conjugated rabbit antisheep immunoglobin G. Coupled with results obtained from incorporation of [H³]glycerol into disaturated phosphatidylcholine, they reported that an intact microtubule-microfilament system is obligatory for enhanced surfactant secretion. Culp et al. (1983) have developed a procedure for isolating submucosal gland cells from cat trachea. Using these isolated cells, it seems possible to examine morphologically the role of microfilaments or microtubules in the secretory responses of submucosal glands. Alder et al. (1982) have shown a specific anatomical relationships between microtubules and microfilaments and granules of porcine tracheal goblet cells by means of transmission electron microscopy of serially prepared ultrathin sections. Busson-Mabillot et al. (1982) examined the pathway and kinetics of secretory protein transport in rat lacrimal glands in an in vitro time-course radioautographic study

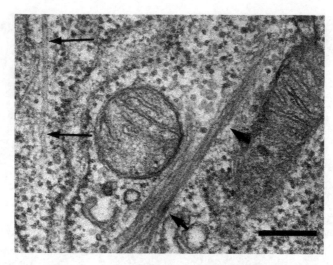

Figure 10 Microtubules (arrows) in the cytoplasm of serous cell from isolated gland of canine trachea (bar, 0.3 μm).

of pulse-labeled protein secretion, coupled with ultrastructural observations of intracellular microfilaments and microtubules. Such morphological investigations of the cytoskeletal system, as in alveolar type II cells, airway goblet cells, and lacrimal glands, to clarify their roles in the synthesis, transport, and release of mucus and/or fluid in acinar cells of tracheobronchial submucosal glands, remain to be performed.

B. Cyclic AMP and Ca Ions in Secretory Cells

Stimulus-secretion coupling'' proposed in other exocrine glands is known to involve mainly these two cellular mechanisms: cyclic AMP and/or calcium ions. Many secretory tissues require cyclic AMP and/or calcium ions in the intra- and extracellular fluid for the secretion of ions (or water) and macromolecules, and the effects of secretagogues are mediated by the intracellular adenylate cyclase–

cyclic AMP system and/or by intracellular calcium ion concentration. In airway submucosal glands, few morphological investigations supporting this hypothesis have been performed despite the following physiological studies. Exogenous dibutyryl cyclic AMP is known to produce macromolecule secretion in airway submucosal glands (Sasaki et al., 1986). The uptake of precursors into mucus glycoprotein decreases in calcium-deficient solution, whereas the release of sulfate incorporated in mucus glycoprotein into the acinar lumen has been shown to be independent of calcium in the external bathing media (Marin et al., 1982). Furthermore, Coles et al. (1984a) have reported that extracellular calcium depletion induces the release of the glycoprotein from canine tracheal glands by increasing the rate of flow of mucus in the duct lumens.

Lazarus et al. (1984, 1986) have used an immunocytochemical probe for the intracellular localization of cyclic AMP to identify the specific cell types that respond to prostaglandins and vasoactive intestinal peptide (VIP) in dog, ferret, and cat trachea. Cryostat sections were stained for cyclic AMP using the indirect immunofluorescence and unlabeled antibody enzyme (PAP) methods (Spruill and Skiner 1979). Anti-cyclic AMP antiserum was prepared by immunizing rabbits with 2'-0-monosuccinyl-cyclic AMP conjugated to human serum albumin. They found that both prostaglandin E_1 and VIP increase stains for immunoreactive cyclic AMP in tracheal submucosal gland cells. However, morphological and physiological estimates of intracellular cyclic AMP in airway submucosal gland in response to other stimuli have not yet been performed, although we know that second messengers in secretory cells are important for understanding the intracellular mechanisms governing submucosal gland secretion.

C. Intracellular Membrane System in Secretory Cells

The intracellular membrane system (membrane apparatus, smooth and rough endoplasmic reticulum, and mitochondria) is known to have the functions of synthesis and transport of secretory protein and phospholipids in other exocrine gland (Shimura et al., 1983, 1985b). In airway submucosal glands, however, little attention has been focused on this intracellular membrane system. The freeze-fracture replication method would be more suitable because it provides a means of obtaining the tridimensional images. Furthermore, autoradiography would be suitable for the understanding of intracellular process of synthesis of secretory materials.

VI. Morphological Approaches to the Innervation of Tracheobronchial Glands

Autonomic nerves regulate secretion from airway submucosal glands but not from surface goblet cells. Thus the glands are under parasympathetic, sympathetic, and

Table 4 Methods for Visualizing Innervations of Tracheobronchial Glands

Parasympathetic nerves		
Acetylcholinesterase technique	Frozen sectioning; light microscopy	El-Bermani et al. (1970; 1975); Murlas et al. (1980); Partanen et al. (1982)
Sympathetic nerves		
Sucrose-potassium phosphate-glyoxlic acid (SPG) technique	Frozen sectioning; fluorescence; light microscopy	de la Torre (1980)
Formaldehyde-induced fluorescence	Same as above	Falck et al. (1962); Pack and Richardson (1984)
6-Hydroxydopamine method	Selective degeneration of adrenergic nerves; light and electron microscopy	Silva and Ross (1974)
5-Hydroxydopamine method	Adrenergic vesicles with intensely osmiophilic material on electron microscopy	Murlas et al. (1980)
Neuropeptide-containing nerves		
Substance P and VIP	Immunocytochemistry; indirect immunofluorescence; peroxidase–antiperoxidase method; avidin–biotin–peroxidase complex method; light and electron microscopy	Uddman et al. (1978); Lundberg et al. (1979); Dey et al. (1981); Shimura et al. (1987b)

noncholinergic, nonadrenergic control. Morphological examination has confirmed autonomic innervation to tracheobronchial glands including neuropeptide-containing nerves. The methods used for investigating this area are summarized in Table 4.

A. Parasympathetic Nerves

The parasympathetic nerve supply is via the vagus nerve, which is observed by electron microscopy to end at varicosities containing small agranular vesicles close to acinar cells. The acetylcholine released from these nerves acts through muscarinic receptors to stimulate the glands. No ganglia are observed to be present in glandular tissues (Shimura et al., 1986) and they are found to be separate from submucosal glands (Knight, 1980; Suzuki 1984; Baluk et al., 1985). Like other exocrine glands, the tracheobronchial glands are innervated with autonomic postganglionic axons (Fig. 11).

Histological staining with supravital methylene blue (El-Bermani et al., 1970) and methods for silver impregnation (Fitzgerald, 1964; Namba et al., 1967) have been used to visualize the general distribution of nerves in monkey bronchial glands (El-Bermani and Grant, 1975). To demonstrate the presence of acetylcholinesterase (AChE) positive nerve fibers, a modification (El-Badawi and Schenk, 1967) of the thiocholine technique of Karnovsky and Roots (1964) has been used with submucosal glands from monkey and human airways (El-Bermani and Grant, 1975; Partanen et al., 1982; Suzuki et al., 1984) (Fig. 12). Frozen sections are dried in a cool air stream, fixed in ice-cold neutral formalin, and incubated with a medium containing acetylthiocholine iodide. The unspecific cholinesterase staining is suppresssed by tetra isopropyl phosphoramide (iso OM-PA) (Partanen et al., 1982). A combination of methylene blue and silver staining with the histochemical method for specific AChE provides a good picture of the general pattern of innervation (El-Bermani et al., 1975).

B. Sympathetic Nerves

The catecholamine-containing nerve fibers in submucosal glands of mammalian airways have been demonstrated by both a modification of the sucrose–potassium phosphate–glyoxylic acid (SPG) technique (de la Torre, 1980) and the formaldehyde-induced fluorescence technique of Falck-Hillarp (1962) (Pack and Richardson, 1984). Doidge and Satchell (1982) have visualized adrenergic fibers in preparations of guinea pig, rat, and rabbit by the combination of these two techniques. The tissues were cut and placed in 2% glyoxylic acid for 30 min (Björklund et al., (1972). The tissues were then frozen in liquid propane cooled in liquid nitrogen and freeze, dried for 48 hr at $-40\,°C$ and 10^{-3} mmHg. Tissues were then brought to room temperature and transferred to a dessicator containing paraformaldehyde at a relative humidity of 70%. The tissue was incubated at

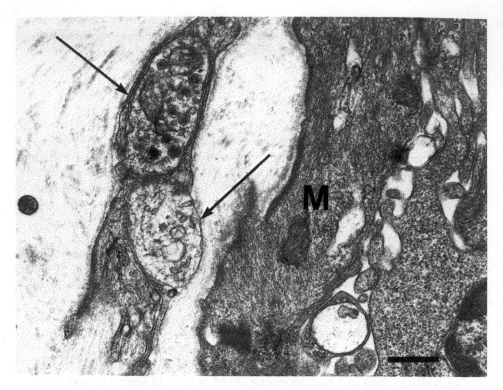

Figure 11 Electron micrograph of nerve endings (arrows) in canine tracheal gland. M, myoepithelial cell (bar, 0.5 μm).

80° for 1 hr and the pieces were then vacuum embedded with paraffin wax for 25–30 min at 62 °C. Sections were made at 10 μm. In monkey and human tissues glyoxylic acid caused a high background fluorescence; therefore, the formaldehyde-induced fluorescence technique alone was used.

Silva and Ross (1974) used ultrastructural and fluorescence histochemical methods to examine the innervation of the tracheobronchial glands of normal cats and cats treated with 6-hydroxydopamine 7 days or more before sacrifice. Administration of 6-hydroxydopamine produces selective degeneration of adrenergic nerve fibers with no specific fluorescence indicative of catecholamines. They found that in the animals treated with 6-hydroxydopamine, the surviving nerves, presumed to be cholinergic, had the same distribution as the adrenergic nerves.

More recently, labeling of nerve endings with false precursors of neurotransmitters (e.g., 5-hydroxydopamine) has been used to demonstrate adrenergic nerves on electron microscopic examination (Murlas et al., 1980). 5-Hydroxy-

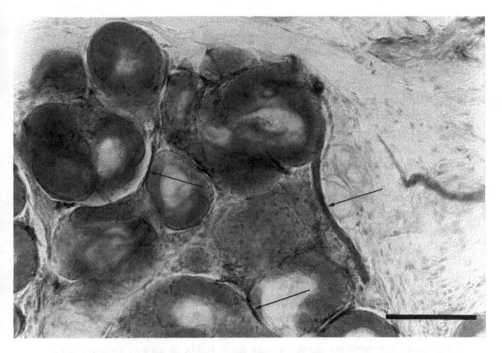

Figure 12 Light micrograph of human bronchial wall stained by the Karnovsky-Roots method (1964) to demonstrate acetylcholinesterase- (AChE) containing nerves. AChE-positive postganglionic nerves (arrows) are seen close to and in acini of submucosal glands (bar, 25 μm; photo courtesy of Dr. T. Suzuki).

dopamine (100 mg/kg in 0.9 NaCl with 1.0% ascorbic acid was injected intravenously 10 min before the removal of trachea or bronchi from cat. The 5-hydroxydopamine injection method, which intensifies the cored vesicles, is highly specific for monoamine-containing vesicles. Cholinergic or peptide-containing vesicles do not become more electron-dense with 5-hydroxydopamine. As described in other systems, varicosities of postganglionic neurons in airway submucosal glands contains small (40–60 nm) and large (80–120 nm) synaptic vesicles, mitochondria, and smooth endoplasmic reticula (Fig. 11). Adrenergic varicosities are identified by the presence of small dense-cored vesicles, a feature that is particularly prominent because of the 5-hydroxydopamine injection. Varicosities containing small agranular vesicles and large dense-cored vesicles, but no small dense-cored vesicles, are presumed to be cholinergic. Thus, Murlas et al. (1980) examined by an electron microscopic morphometric method the distribution of adrenergic and cholinergic axon varicosities to serous and mucous cells of cat tracheal glands after treatment with 5-hydroxydopamine. Of all varicosities identified within 10 μm of the glands, 90% were cholinergic and 10% were adrenergic. No differential innervation of serous and mucous cells was observed. Because sympathetic nervous stimulation apparently produces as much secretion as vagal stimulation (Borson et al., 1980), it seems that the potency of neural control over effector cells depends mainly on the density of receptors in cells rather than on the numbers of nerve endings. (Nadel et al., 1985).

The supply of adrenergic fibers to human submucosal glands is mainly from the plexus around bronchial arteries, whereas AChE-containing fibers are mainly from extrachondrial and/or subchondrial microganglia and from the corresponding nerve plexuses, and represent cholinergic innervation of the gland (Partanen et al., 1982).

The innervation of tracheobronchial glands differs between mammals. In sheep and humans, the glands recieve both AChE- and catecholamine-containing nerve fibers; in goat only AChE-positive fibers; and in calf, pig, and rabbit no AChE-positive or catecholamine-containing nerves exist (Mann, 1971).

C. Neuropeptide-Containing Nerves

It has been speculated that vasoactive intestinal peptide (VIP) is a neurotransmitter in noncholinergic, nonadrenergic nerves. Substance P (SP) is present in vagus nerves as well as in a network of fine nerve fibers supplying the airways and is released from sensory nerves via an axon reflex in airway inflamation.

Immunocytochemical analyses have revealed that VIP- and SP-containing nerves are localized not only in smooth muscle but also in submucosal glands of large airways in dog, cat, ferret, and human tissues (Dey et al., 1981; Uddman et al., 1978; Lazarus et al., 1986; Shimura et al., 1987b). Both indirect immunofluorescence and peroxidase-antiperoxidase PAP methods have usually been used. Furthermore, Shimura et al. (1987b) have reported the use of isolated submucosal glands from feline trachea fixed with 10% phosphate-buffered formaldehyde (pH 7.4), embedded in paraffin, and cut into 5 μm thick sections for SP immunostaining, according to the avidin–biotin–peroxidase complex (ABC) method described by Hsu et al. (1981). Antiserum against SP conjugated to bovine serum albumin was produced in rabbits, and the specificity of the SP antiserum was confirmed by radioimmunoassay and absorption tests. Dot- and fiberlike SP-immunoreactive structures were observed throughout the isolated glands. The positive dots or fibers were seen close to the acini and secretory tubules, and between them. The diameter of the dot- or fiberlike reaction products in the micrographs varied between 2 and 6 μm, as seen in Figure 13. SP is reported by Coles et al. (1984b) to increase tracheal submucosal gland secretion with an initial short time response. To elucidate how SP produces submucosal gland secretion, Shimura et al. (1987b) examined the effects of SP on the glandular contractile response and ^{3}H-labeled glycoprotein release using isolated gland preparation

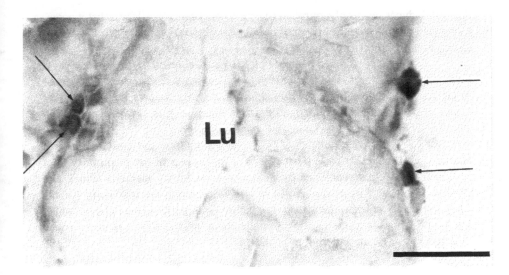

Figure 13 Light micrograph of paraffin section of isolated submucosal gland from feline trachea after incubation with antiserum to substance P (SP). Many SP-immunoreactive dot- and fiberlike structures (arrows) are seen close to secretory tubules. Lu, lumen of secretory tubule (bar, 25 μm).

from feline trachea. The findings obtained indicate that (1) SP induces glandular contraction, which is related to the squeezing of mucus in the ducts and secretory tubules; (2) SP stimulates glycoprotein release in isolated gland, involving mucus synthesis and/or cellular secretion; and (3) these two actions are mediated by a peripheral cholinergic action.

Gashi et al. (1986) have reported that VIP stimulates acinar cell degranulation of ferret tracheal submucosal glands using electron microscopic morphometry, in findings comparable with those of Peatfield et al. (1983), who examined [$^{35}SO_4$]-mucin release from ferret trachea in vitro. However, these findings are are at variance with those of Coles et al. (1981), who used [^{14}C]glucosamine as the macromolecule precursor and reported that VIP caused inhibition of glycoconjugate and lysosome secretion by human bronchi in vitro. By a combination of the indirect immunofluorescence technique using antiserum to VIP with AChE staining, it has been shown that VIP is present in cholinergic neurons in the exocrine glands of cat, such as the salivary gland, nasal mucosa, tongue, and bronchi (Lundberg et al., 1979). At the electron microscopic level, VIP-like immunoreactivity was also observed in postganglionic cholinergic neurons in exocrine glands of cat by PAP methods (Johansson and Lundberg, 1981). Using isolated gland preparations from feline trachea, Shimura et al. (1988) have shown that VIP induces mucus glycoprotein release from secretory cells and that it potentiates the secretion induced by cholinergic stimulation, probably through an interaction with muscarinic receptors in gland secretory cells.

Calcitonin gene-related peptide (CGRP) immunoreactivity was investigated in rat lung by Lauweryns and Ranst (1987). Immunoreactive material was observed in nerves surrounding bronchi, bronchioli, and blood vessels, suggesting that CGRP plays a part in the regulation of airway function. It is, however, unknown whether CGRP regulates submucosal gland secretion.

Other regulatory peptides; bombesin, cholecytokinin, and somatostatin; and neuron-specific enolase have been shown to be present in the respiratory tract of humans and other mammals by immunocytochemistry methods (Polak and Bloom, 1982). However, their roles in airway secretion are unknown. Details concerning antibodies to the airway regulatory peptides described above, which are used for immunocytochemistry to understand localizations, are summarized in a review by Polak and Bloom (1982).

VII. Receptors of Secretory Cells in Tracheobronchial Glands

Recently, morphological methods to demonstrate muscarinic, adrenergic receptors, and receptors for neuropeptides (SP and VIP) have been applied to secretory cells in airway submucosal glands. Localization of these receptors over airway submucosal glands has been examined by autoradiographic method mainly at the light microscopic level using various ligands with binding assay. These are summarized in Table 5.

Studies of the regulation of airway secretion indicate that submucosal glands are under muscarinic cholinergic control. Recently, in a study using autoradiographic techniques, muscarinic receptors were localized to both serous and mucous cells of ferret tracheal gland (Basbaum et al., 1984). The muscarinic receptor distribution in ferret tracheal glands was determined using [^{3}H]propylbenzylcholine mustard ([^{3}H]PrBMC) binding and autoradiography. Specific, atropine-sensitive [^{3}H]PrBMC binding was quantified autoradiographically in submucosal glands. Serous and mucous cells in the gland did not differ in the receptor density. Binding sites on gland cells were associated with basolateral membranes. Functionally, Culp and Marin (1986) utilized the muscarinic antagonist [1-^{3}H]quinuclidinyl benzilate ([^{3}H] QHB) to characterize glandular receptors. They found that muscarinic receptors on tracheal gland cells are of high affinity and density. The finding of muscarinic receptors on both serous and mucous cells confirms the capacity of both cell types to contribute to the secretory response to vagal or cholinomimetic stimulation.

Barnes et al. (1982, 1983a,b) investigated the distribution of adrenergic receptors in ferret trachea using autoradiography. [^{3}H]Dihydroalprenolol, used

Table 5 Receptors Visualized Autoradiographically in Tracheobronchial Glands

Receptors	Ligands	Localization		References
		Mucous cell	Serous cell	
Muscarinic	[³H]Propylbenzyl-choline-mustard	+ +	+ +	Basbaum et al. (1984a)
Adrenergic				
α_1 receptor	[³H]Prazosin	−/+	+ +	Barnes et al. (1983; Barnes and Basbaum (1983)
β-receptor	[³H]Dihydroalprenolol [³H]Iodocyanopindol	+ +	−/+	Barnes et al. (1982); Barnes and Basbaum (1983); Carstairs et al. (1985)
Neuropeptides				
Substance P	¹²⁵I-Bolton-Hunter-labeled SP	+	?	Carstairs and Barnes (1986)
VIP	[¹²⁵]VIP	+ +	?	Leys et al. (1986)

to identify β-receptors, revealed a high density of specific binding sites over surface epithelium and submucosal glands. Specific binding of [^{3}H]prazosin to lung section was used for autoradiography of the localization of $\alpha_1$receptors present in airway submucosal glands and epithelium. Comparison of adrenergic receptor densities in tracheal sections from the same animals showed a rank order for submucosal glands of α_1 greater than β. Within submucosal glands, α_1 and β-adrenergic receptors were differentially distributed, with α_1-receptors being significantly more numerous over serous than mucous cells and β-receptors being significantly more numerous over mucous than serous cells. These findings are consistent with the watery secretion measured by micropipettes in response to α-adrenergic stimulation and with higher protein and sulfur concentrations in secretion induced by β-adrenergic stimulation (Leikauf et al., 1984). Carstairs et al. (1985) examined the localization of β-adrenoreceptors in human lungs by autoradiographic methods, using [^{125}I]iodocyanopindolol (ICYO) to label β-receptors in tissue sections in competition with selective β-receptor antagonists ICI 118, 551 (β_2-selective) and betaxolol (β_1-selective) to obtain the ratio of β_2 to β_1-receptors in the sections. In bronchial submucosal gland, the ratio of β_2 to β_1-receptor was 9:1, although the significance of the distribution of β-receptor subtypes remains to be determined.

Shimura et al. (1987b,c) have shown that both exogenous SP and VIP evoke ^{3}H-labeled mucus glycoprotein release from isolated submucosal gland and thus gland cells are suggested to have their own receptors for SP or VIP. To date, few morphological findings regarding the neuropeptide receptors in airway submucosal glands have been obtained. Recently, localization of specific receptors for SP in the submucosal glands of human airways has been demonstrated by autoradiography using ^{12}I-Bolton-Hunter-labeled SP [^{125}I]BH-SP) (Carstairs and Barnes, 1986). Leys et al. (1986) have examined autoradiographic localization of VIP receptors in human lung using [^{125}I]VIP and reported a lower density of VIP binding sites over bronchial epithelium and submucosal gland than in pulmonary artery smooth muscle and alveolar walls.

It has been suggested that secretory cells in submucosal glands of canine and feline trachea have no receptors for various chemical mediators (histamine, leucotrienes, serotonin, etc.), because chemical mediators failed to produce significant increases in radiolabeled glycoprotein release from isolated submucosal glands, or glandular contraction, which means "mucus squeezing," whereas they augment the secretory responses to neural stimulation (Shimura et al., 1986; Sasaki et al., 1986).

VIII. The Contractile System in Submucosal Gland

Although myoepithelial cells are known to exist in submucosal glands of airways, their precise morphology and function in secretion are still debated. Myoepithelial cells in some exocrine glands are known to be contractile. However, the possibility is often overlooked or ignored in studies of the secretion of airway submucosal glands despite the fact that the presence of the contraction may have a profound effect on the secretory response. Contractions in submucosal glands attributed to the activity of myoepithelial cells have been observed directly using isolated submucosal gland preparation (Shimura et al., 1986). The glandular contraction results in the squeezing of mucus that is secreted from secretory cells into secretory tubules and ducts, and exhibits an initial short (a few minutes) response time in submucosal gland secretion. The contraction is mediated mainly by cholinergic nerves via muscarinic receptors and in small part by adrenergic nerves via α-receptors (Shimura et al., 1987a,b). We have thus postulated that the secretory response of submucosal glands consists of two actions: (1) discharge of mucus from secretory cells into tubules and ducts and (2) squeezing (expulsion) of mucus into airway lumen through glandular contraction. Acinar cell secretion that exhibits a long-term response (up to a few hours) may be mediated by cholinergic, adrenergic, and noncholinergic, nonadrenergic nerves (Shimura et al., 1987a,b). These two actions in submucosal gland secretion are under separate nervous control and not all neural stimuli may be involved in both actions. In fact, reflexly increased secretion in canine and feline tracheal submucosal gland in vivo has been shown to involve only a cholinergic efferent pathway. Reflexly increased secretion has been proven to involve mainly the mucus-squeezing action through myoepithelial cell contraction, representing the initial short-term response with lack of a β-adrenergic contribution (Shimura et al., 1986, 1987a,b).

Myoepithelial cells surround acini and secretory tubules in airway submucosal glands (Figs. 7, 14). The myoepithelial cell is spindle-shaped, sometimes divided at the tail portion. It lies on the basement membrance and is filled with masses of myofilaments (Figs. 7, 14, 15). The cells show a separation of their cytoplasm into two main compartments; one is nonfilamentous and the other contained filaments. The latter forms up to 90% of the cytoplasmic area in the myoepithelial cell. The nonfilamentous part contains the nucleus and a few of the cytoplasmic organelles. The nucleus tends to be flattened in the direction of the stromal basal lamina and organelles are usually situated close to the nucleus. A few rough endoplasmic reticuli, unattached ribosomes, and lysosome-like bodies are present (Fig. 15). Dense bodies are seen in the filamentous part and small vesicles (caveolae) under the cell membrane (Shimura et al., 1986). These ultrastructural findings are similar to those from smooth muscle cells, suggesting that myoepithelial cells is contractile. Intracellular calcium ions stored in such organelles as well as extracellularly play the major role in the contraction of myoepithelial cells (Shimura et al., 1986). Using potassium pyroantimonate, the intracellular localization of calcium ion has been demonstrated in the contracted and resting states of muscular cells (Legato and Langer, 1969). However, such an observation has not been made in myoepithelial cells. Myoepithelial cells have

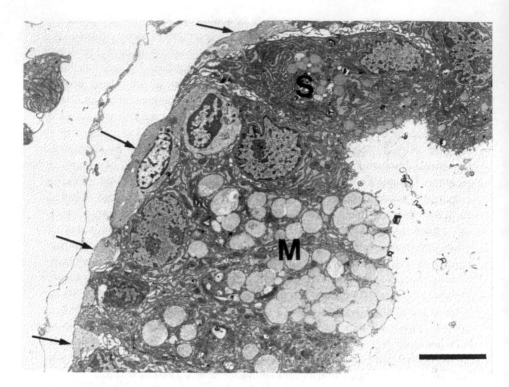

Figure 14 Electron micrograph of secretory tubule in an isolated gland from canine trachea. Myoepithelial cells (arrows) exist between basement membrane and secretory cells. S, serous cells; M, mucous cells (bar, 5 μm).

a desmosomal attachment to myoepithelial and/or secretory cells (Fig. 15), thus differing from smooth muscle. Although they are not frequent, elongated mitochondria as seen in the smooth muscle cell are observed between myofilaments in the filamentous part. Bundles of unmyelinated axons are found between and in acini. Terminal axons (nerve endings) have been seen in the acinar portion, between secretory cells, between myoepithelial cells, and close to myoepithelial cells (Fig. 11).

Although contractile proteins (actin and myosin) in salivary myoepithelial cells have been demonstrated by the use of fluorescence immunohistochemistry (Archer et al., 1971; Puchtler et al., 1974; Drenckhahn et al., 1977), we have no direct evidence that they are also present in the myoepithelial cells of airway submucosal glands.

From examination of radiolabeled glycoprotein release in isolated glands from canine and feline trachea (Shimura et al., 1986, 1987a,b), myoepithelial

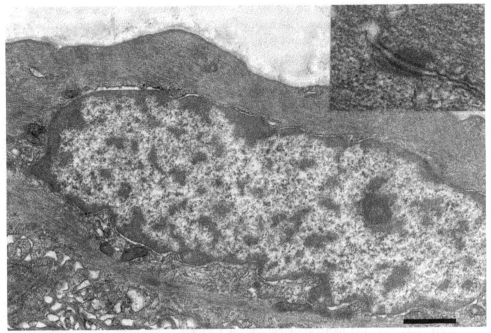

Figure 15 Electron micrograph of myoepithelial cell from canine tracheal gland. A few organelles are seen around nucleus (bar, 1 μm). Inset, desmosome between myoepithelial cells.

cells in submucosal glands have been suggested to possess receptors for muscarinic cholinergic agonist, α-adrenergic agonist, substance P, and VIP. At present, however, no morphological investigations have not demonstrated the presence of these receptors in myoepithelial cells of tracheobronchial glands.

IX. Electrolyte (Fluid) Secretion and the Duct System

The exocrine glands generally consist of two functional units: acinar cells that synthesize, store, and secrete granules; and centroacinar and terminal duct cells that are responsible for electrolyte secretion and reabsorption (i.e., the transport of ions accompanied by water) (Case, 1973). The former have been extensively investigated in airway submucosal glands both by measurement of radiolabeled mucin release and by morphological methods (histochemistry, immunohistology, morphometry, etc.). In contrast, there has been little investigation of electrolyte secretion from submucosal glands because of the lack of a suitable experimental tool, although ion transport across the airway mucosa, which contains submucosal glands as well as epithelium, has been studied using Ussing's chamber. The role of submucosal glands in ion and fluid secretion across the bronchial wall is not clearly understood. Phipps et al. (1986) have examined the effects of ozone exposure on ion and fluid transport across sheep bronchial wall. The concomitant changes observed in net ion and water movement across bronchial wall and in submucosal gland size estimated by Reid index suggest that submucosal glands may contribute to the increased ion fluxes and water secretion with chronic O_3 exposure. More directly, Culp et al. (1987) studied the electrical properties of cat submucosal gland cells in primary culture and suggested a marked potential for regulated active ion transport by airway submucosal glands. Using isolated gland preparation, Sasaki et al. (1987) have found ouabain-sensitive Na^+-K^+ATPase activity and K^+-channels in the acinar cells of feline tracheal submucosal glands, which are known to play major roles in electrolyte secretion in other exocrine glands. By electron microscopy, Mills and Quinton (1981) and Basbaum et al. (1981) have observed that cholinergic stimulation evokes the formation of intracellular vacuoles in serous cells of cat and ferret tracheal submucosal glands, which was inhibited by various factors known to reduce fluid transport, such as ouabain. Thus, it has been suggested that vacuolation is a possible indication of active ion and water transport. Further investigation of this area by both functional and morphological means remains to be done.

Ciliated respiratory epithelium dips into the gland opening and leads to the first part of the duct, the ciliated duct, and then gives way in the collecting duct to an epithelium composed of tall, columnar, eosinophilic cells containing numerous large mitochondria. This cell structure suggests that the collecting duct controls ionic and water concentration (Meyrick et al., 1969). In fact, an abundant supply of blood capillaries in the region of the collecting duct is observed. However, no direct functional or morphological evidence supporting this speculation has been obtained. From the collecting duct arise secretory tubules (mucous and serous). Recently, Shimura et al. have been successful in isolating collecting duct from canine trachea in the same manner in which submucosal glands are isolated (unpublished data). This preparation will be useful for the functional and morphological investigation of ion and water control in the collecting duct of airway submucosal glands.

X. Migratory Cells

Lymphocytes, mast cells, and globule leukocytes (immature mast cells?) have been observed infrequently in the lining of bronchial gland acini (Fig. 7). Although the function of these migratory cells is speculated to be immunoresponsive, their significance in submucosal gland secretion is still unknown. Furthermore, their distribution and subpopulations (T lymphocyte or B lymphocyte, mucosal or connective tissue mast cell) have not yet been examined. Immunohistological or ultrastructural studies in this area remain to be performed.

XI. Future Direction

In the future, an understanding of morphology at the gene and molecular levels will become necessary to allow us to understand airway submucosal secretion, and will require new methods and approaches to the study of airway submucosal glands. For the present, we have selected the following 10 morphological areas in tracheobronchial gland research that require further study and that may be approached by the routine methods described above.

1. The relationship of submucosal gland secretion to airway luminal mucus layer or the amount of mucus. Aikawa et al. (1987) have evaluated the amount of mucus in the airways of autopsied lung using a computer-assisted morphometric method. However, we have as yet no direct information concerning how submucosal gland secretion is related to the amount of mucus in the airway lumen or airway obstruction in the diseased lung.

2. Morphometric analysis of secretory granules in the mucous cells of glands. Changes in serous cell granules in response to secretagogues have been examined by ultrastructural morphometry, whereas those in mucous cells have yet been evaluated by morphometric methods.

3. Microtubule and microfilament system in secretory cells. Their localization and significance in the secretory process of airway submucosal glands are still unknown. Deep freeze–etching techniques with scanning electron microscopy that have been applied to morphological visualization of microtubules and microfilaments in other tissues (Hirokawa and Heuser, 1981) will be among the methods useful for investigating this area.

4. Localization and changes in intracellular second messengers. Morphological studies of the behavior of intracellular Ca ions and cyclic AMP in response to secretagogues remains to be performed.

5. Morphological evidence for ion (or fluid) transport across the submocosal glands. Localization of Na^+-K^+ATPase, intracellular Na^+, K^+, Cl^-, and their changes in the secretory process remain to be examined morphologically.

6. Morphological approaches to examination of the duct system as well as related physiological studies. Isolated collecting duct from canine trachea (unpublished data) may be a useful experimental tool in this area.

7. Innervation and receptors in myoepithelial cells. Shimura et al. (1986, 1987a,b) have demonstrated autonomic innervation and suggested various receptors in submucosal myoepithelial cells using isolated gland preparation. However, no morphological methods have been applied to the investigation of myoepithelial cells in airway submucosal glands to confirm their results.

8. Vessel supply to submucosal glands. The casting method with scanning electron microscopy may be a useful approach.

9. Distribution and subpopulation of migratory cells (lymphocyte or mast cell) in glandular tissues and morphological approaches to the study of "local immune responses" in submucosal gland secretion, that is, changes in the secretory component, IgA, and IgM in response to given stimuli. Such knowledge is important for understanding mucus hypersecretion in bronchial asthma.

10. The morphology of phospholipid production or surfactant in airway submucosal glands. Pulmonary surfactant is known to be present in the airways as well as alveolar regions. The origin, however, is not clear and whether the submucosal glands secrete airway surfactant remains unknown.

Acknowledgment

I gratefully acknowledge Professor Tamotsu Takishima for his helpful advice and discussions, Dr. Ronald Scott for reading, and Ms. Naomi Suzuki for typing the manuscript. This study was in part supported by Scientific Grants from the Ministry of Education, Science and Culture of Japan (No. 59570321, 60570342, 62570340, and 01570422).

References

Aikawa, T., Shimura, S., Sasaki, H., and Takishima, T. (1987). Mucus occupying ratio in small airway in chronic obstructive pulmonary disease. *Am. Rev. Respir. Dis.* **135**:A 146.

Alder, K. B., Hardwick, D.H., and Craighead, J. E. (1982). Porcine tracheal goblet cell ultrastructure: a three-dimensional reconstruction. *Exp. Lung Res.* **3**:69–80

Alli, A. F. (1975). The radial intercepts method for measuring bronchial mucous gland volume. *Thorax* **30**:687–692.

Archer, F. L., Beck, J. S., and Melvin, J. M. D. (1971). Localization of smooth muscle protein in myoepithelium by immunofluorescence. *Am. J. Pathol.* **63**:109–118.

Baluk, P., Fujiwara, T., and Matsuda, S. (1985). The fine structure of the ganglia of the guinea-pig trachea. *Cell Tissue Res.* **239**: 51–60.

Barnes, P. J., and Basbaum C. B. (1983) Mapping of adreneregic receptors in the trachea by autoradiography. *Exp. Lung Res.* **5**: 183–192.

Barnes, P. T., Basbaum, C. B., Nadel, J. A., and Roberts, J. M. (1982) Localization of β-adrenoreceptors in mammalian lung by light microscopic autoradiography. *Nature* **299**:444–447.

Barnes, P. J., Basbaum, C. B., Nadel, J. A., and Roberts, J. M. (1983). Pulmonary α-adrenoceptors: autoradiographic localization using [³H]-prazosin. *Eur. J. Pharmacol.* **88**: 57–62.

Basbaum, C. B., Ueki, I., Brezina, L., and Nadel, J. A. (1981). Tracheal submocosal gland serous cells stimulated in vitro with adrenergic and cholinergic agonists. A morphometric study. *Cell Tissue Res.* **220**:481–489.

Basbaum, C. B., Grillo, M. A. and Widdicombe, J. H. (1984a). Muscarinic receptors: evidence for a nonuniform distribution in tracheal smooth muscle and exocrine glands. *J. Neurosci.* **4**:508–520.

Basbaum, C. B., Mann, J. K., Chow, A. W., and Finkbeiner, W. E. (1984b). Monoclonal antibodies as probes for unique antigens in secretory cells of mixed exocrine organs. *Proc. Natl. Acad. Sci. USA* **81**:4419–4423.

Bedrossian, C. W.M., Anderson, A. E., Jr., and Foraker, A. G., (1971). Comparison of methods for quantitating bronchial morphology. *Thorax,* **26**:406–408.

Bensch, K. G., Gordon, G. B., and Miller, L. R. (1965). Studies on the bronchial counterpart of the Kultschitzky (argentaffin) cell and innervation of bronchial gland. *J. Ultrastruct. Res.* **12**:558–686.

Berend, N., Woolcock, A. J., and Marlin, G. E. (1979). Correlation between the function and structure of the lung in smokers. *Am. Rev. Respir Dis.* **119**:695–705.

Björklund, A., Lindvall, O., and Svensson, L. A. (1972). Mechanisms of fluorophore formation in the histochemical glyoxylic acid (GA) method for monoamines. *Histochemie* **32**:113–131.

Borson, D. B., Chinn, R. A., Davis, B, and Nadel, J. A. (1980). Adrenergic and cholinergic nerves mediate fluid secretion from tracheal glands of ferrets. *J. Appl. Physiol.* **49**:1027–1031

Bowes, D., and Corrin, B. (1977). Ultrastructural immunocytochemical localization of lysozyme in human bronchial glands. *Thorax* **32**:163–170

Bowes, D., Clark, A. E., and Corrin, B. (1981). Ultrastructural localization of lactoferring and glycoprotein in human bronchial gland. *Thorax* **36**:108–115.

Brown, L. A. S., Pasquale, S. M., and Longmore, W. J. (1985). Role of microtubules in surfactant secretion. *J. Appl. Physiol.* **58**:1899–1873

Burg, M., Grantham, M., Abramow, J., and Orloff, J. (1966). Preparation and study of fragments of single rabbit nephrons. *Am. J. Physiol.* **210**:1293–1298.

Burton, P. A., and Dixon, M. F. (1969). A comparison of changes in the mucous glands and goblet cells of nasal, sinus, and bronchial mucosa. *Thorax* **24**:180–185.

Busson-Mabillot, S., Chambaut-Guerin, A.-M., Ovtracht, L., Muller, P., and Rossignol, B. (1982). Microtubules an portion secretion in rat lacrimal glands: localization of short-term effects of colchicine on the secretory process. *J. Cell Biol.* **95**:105–117.

Carstairs, J. R., Nimmo, A. J., and Barnes, P. J. (1985). Autoradiographic visualization of beta-adrenoceptor subtypes in human lung. *Am. Rev. Respir. Dis.* **132**:541–547.

Carstairs, J.R., and Barnes, P. J. (1986). Autoradiographic mapping of substance P receptors in lung. *Eur. J. Pharmacol.* **127**:295–296.

Case, R. M. (1973). Review. Cellular mechanisms controlling pancreatic exocrine secretion. *Acta Hepato-Gastroenterol.* **20**:435–444.

Chakrin, L. W., Baker, A. P., Spicer, S. S., Wardell, J. R., DeSanctis, N., Jr., and Dries, C. (1972). Synthesis and secretion of macromolecules by canine trachea. *Am. Rev. Respir. Dis.* **105**:368–381.

Coles, S. J., and Reid, L. (1981). Inhibition of glycoconjugate secretion by colchicine and cytochalasin B. An in vitro study of human airway. *Cell Tissue Res.* **214**:107–118.

Coles, S. J., Said, S. I., and Reid, L. M. (1981). Inhibition by vasoactive intestinal peptide of glycoconjugate and lysozyme secretion by human airways in vitro. *Am. Rev. Respir. Dis.* **124**:531–537.

Coles, S. J., Bhaskar, K. R., O'Sullivan, D. D., and Reid, L. M. (1984a). Extracellular calcium ion depletion induces release of glycoproteins by canine trachea. *Am. J. Physiol.* **246**:C494–C501.

Coles, S. J., Neil, K. H., and Reid, L. M. (1984b). Potent stimulation of glycoprotein secretion in canine trachea by substance P. *J. Appl. Physiol.* **7**:1323–1327.

Culp, D. J., and Martin, M. G. (1986). Characterization of muscarinic receptors in cat tracheal gland cells. *J. Appl. Physiol.* **61**:1375–1382.

Culp, D. J., Penny, D. P., and Martin, M. G. (1983). A technique for the isolation of submucosal gland cells from at trachea. *J. Appl. Physiol.* **55**:1035–1041, 1983.

Culp, D. J., Kershl, W. C., McBride, P. K., and Martin, M. G. (1987) Electrical properties of cat submucosal gland cells in primary culture. *Fed. Proc.* **46 (4)**:1237.

de la Torre, J. C. (1980). An improved approach to histofluorescence using the SPG method for tissue monoamines. *J. Neurosci. Methods* **3**:1–5

de Water, R., Willens, L. N. A., Van Muijen, G. N. P., Franken, C., Fransen, J. A. M., Dijkman, J. H., and Kramps, J. A. (1986). Ultrastructural localization of bronchial antileukoprotease in central and peripheral human airways by a gold-labeling technique using monoclonal antibodies. *Am. Rev. Respir. Dis.* **133**:882–890.

Dey, R. D., Shannon, Jr. W. A., and Said, S. I. (1981). Localization of VIP-immunoreactive nerves in airways and pulmonary vessels of dogs, cats, and human subjects. *Cell Tissue Res.* **220**:213–238.

Doidge, J. M., and Satchell, D. G. (1982). Adrenergic and nonadrenergic inhibitory nerves in mammalian airways. *J. Auton. Nerv. Syst.* **5**:83–99.

Douglas, A. N. (1980). Quantitative study of bronchial mucous gland enlargement. *Thorax* **35**:198–201.

Douglas, A. N., Lamb, D., and Ruckley, V. A. (1982). Bronchial gland dimensions in coalminers: influence of smoking and dust exposure. *Thorax* **37**:760–764.

Dreckhahn, D., Gröschel-Stewart, U., and Unsicker, K. (1977). Immunofluorescence microscopic demonstration of myosin and action in salivary glands and exocrine pancreas of the rat. *Cell Tissue Res.* **183**:273–279.

Dunhill, M. S., Massarella, G. R., and Anderson, J. A. (1969). A comparison of the quantitative anatomy of the bronchi in normal subjects, in status asthmatics, in chronic bronchitis, and in emphysema. *Thorax* **24**:176–179.

Edwards, C. W., and Carlile, A. (1982). The larger bronchi in crypatogenic fibrosing alveolitis: a morphometric study. *Thorax* **37**:828–833.

El-Badawi, A., and Schenk, E. A. (1967). Histochemical methods for separate, consecutive and stimultaneous demonstration of acetylcholinesterase and norepinephrine in cryostat sections. *J. Histochem. Cytochem.* **15**:580–588.

El-Bermani, A.-W. I., and Grant, M. (1975). Acetylcholinesterase-positive nerves of the rhesus monkey bronchial tree. *Thorax* **30**:162–170

El-Bermani, Al.-W., McNary, W. F., and Bradley, D. E. (1970). The distribution of acetylesterase and catecholamine containing nerves in the rat lung. *Anat. Rec.* **167**:205–212.

Falck, B., Hillarp, N.-A., Thieme, G., and Torp, A. (1962). Fluorescence of catecholamines and related compounds condensed with formaldehyde. *J. Histochem. Cytochem.* **10**:348–354.

FitzGerald, M. J. T. (1964). The double-impregnation silver technique for nerve fibers in paraffin sections. *Q. J. Microsc. Sci.* **105**:354–361.

Gashi, A. A., Borson, D. B., Finkbeiner, W. E., Nadel, J. A., and Basbaum, C. B. (1986). Neuropeptides degranulate serous cells of ferret tracheal glands. *Am. J. Physiol.* **251**:C223–C229.

Glynn, L. E., and Holborow, E. J. (1959). Distribution of blood-group substances in human tissues. *Br. Med. Bull.* **15**:150–153.

Glynn, L. E., and Michaels, L. (1960). Bronchial biopsy in chronic bronchitis and asthma. *Thorax* **15**:142–153.

Goodman, M. R., Link, D. W., Brown, W. R., and Nakane, P. K. (1981). Ultrastructural evidence of transport of secretory IgA across bronchial epithelium. *Am. Rev. Respir. Dis.* **123**:115–119.

Hale, F. C., Olsen, C. R., and Mickey, M. R. Jr. (1968). The measurement of bronchial wall components. *Am. Rev. Respir. Dis.* **98**:978–987.

Hayashi, M., Sornberger, G. C., and Huber, G. L. (1979). Morphometric analyses of tracheal gland secretion and hypertrophy in male and female rats after experimental exposure to tobacco smoke. *Am. Rev. Respir. Dis.* **119**:67–73.

Hayat, M. A. (1975). *Positive Staining for Electron Microscopy.* New York, Von Nostrand-Reinhold.

Hayes, J. A. (1969). Distribution of bronchial gland measurements in a Jamaican population. *Thorax* **24**:619–622.

Hirokawa, N., and Heuser, J. E. (1981). Quick-freeze, depth visualization of the cytoskeleton beneath surface differentiations of intestinal epithelial cells. *J. Cell Biol.* **91**:399–409.

Hsu, S-M., Raine, L., and Fanger, H. (1981). Use of Avidin–Biotin–Peroxidase Complex (ABC) in immunoperoxidase techniques: a comparison between ABC and unlabeled antibody (PAP) procedures. *Histochem. Cytochem.* **29**:577–580.

Itoh, M., Tamada, Z., and Aoki, M. (1981). Studies on the structure and function of the bronchial gland. Significance of the bronchial gland as a bronchial defense mechanism. *Jpn. J. Thorac. Dis.* **19**:926–931.

Jamal, K., Cooney, T. P., Fleethan, J. A., and Thurlbeck, W. M. (1984). Chronic bronchitis. Correlation of morphologic findings to sputum production and flow rates. *Am. Rev. Respir. Dis.* **129**:719–722.

Jeffery, P. K. (1983). Morphologic features of airway surface epithelial cells and glands. *Am. Rev. Respir. Dis.* **128**:S14–S20.

Johansson, O., and Lundberg, L. M. (1981). Ultrastructural localization of VIP-like immunoreactivity in large dense-core vesicles of cholinergic type nerve terminals in cat exocrine glands. *Neuroscience* **6**:847–862.

Jones, R., and Reid, L. (1973a). The effect of pH on Alcian blue staininbg of epithelial glycoproteins. I. Sialomucins and sulphomucins (singly or in simple combinations). *Histochem. J.* **5**:9–8.

Jones, R., and Reid, L. (1973b). The effect of pH on Alcian blue of epithelial glycoproteins. II. Human bronchial submucosal gland. *Histochem. J.* **5**:19–27.

Karnovsky, M. J. and Roots, L. (1964). A direct-coloring thiocholine technique for cholinesterases. *J. Histochem. Cytochem.* **12**:219–221.

Kawada, H., Kawakami, M., Nagai, A., and Takizawa, T. (1981) Scanning electron microscopic study of bronchial glands in cases without bronchial disease and those with bronchiectasis. *Jpn. J. Thorac. Dis.* **19**:873–880.

Knight, D. S. (1980). A light and electron microscopic study of feline intrapulmonary ganglia. *J. Anat.* **131**:413–428.

Kollerstron, N., Lord, P. W., and Whimster, W. F. (1977). Distribution of acid mucus in the bronchial mucous glands. *Thorax* **32**:160–162.

Lamb, D., and Reid, L. (1970). Histochemical and autoradiographic investigation of the serous cells of the human bronchial glands. *J. Pathol.* **100**:127–138.

Lamb, D., and Reid, L. (1972). Quantitative distribution of various types of acid glycoprotein in mucous cells of human bronchi. *Histochem. J.* **4**:91–102.

Lauweryns, J. M., and Ranst, L. V. (1987). Calcitonin gene-related peptide immunoreactivity in rat lung: light and electron microscopy. *Thorax* **42**:183–189.

Lazarides, E., and Weber, K. (1974). Actin-antibody: the specific visualization of actin filaments in non-muscle cells. *Proc. Natl. Acad. Sci. USA* **71**:2268–2272.

Lazarus, S. C., Basbaum, C. B., and Gold, W. M. (1984). Prostaglandins and intracellular cyclic AMP in respiratory secretory cells. *Am. Rev. Respir. Dis.* **130**:262–266.

Lazarus, S. C., Basbaum, C. B., Barnes, P. J., and Gold, W. M. (1986). cAMP immunocytochemistory provides evidence for functional VIP receptors in trachea. *Am. J. Physiol.* **251**:C115–C119.

Legato, M. J., and Langer, G. A. (1969). The subcellular localization of calcium ion in mammalian myocardium. *J. Cell Biol.* **41**:401–423.

Leigh, M. W., Gambling, T. M., Carson, J. L., Collier, A. M., Wood, R. E., and Boat, F. (1986a). Postnatal development of tracheal surface epithelium and submucosal glands in the ferret. *Exp. Lung Res.* **10**:153–169.

Leigh, M. W., Cheng, P-W., Carson, J. L., and Boat, T. F. (1986b). Developmental changes in glycoconjugate secretion by ferret trachea. *Am. Rev. Respir. Dis.* **134**:784–790.

Leikauf, G. D., Ueki, I. F., and Nadel, J. A. (1984). Autonomic regulation of viscoelasticity of cat tracheal gland secretions. *J. Appl. Physiol.* **56**:426–430.

Lennox, B. (1975). Observations on the accuracy of point counting including a description of a new graticule. *J. Clin. Pathol.* **28**:99–103

Leys, K., Morice, A. H., Madonna, O., and Sever, P. S. (1986). Autoradiographic localization of VIP receptors in human lung. *FEBS Letts.* **199**:198–202.

Lundberg, J. M., Hökfelt, T., Schultzgerg, M., Uvnäs-Wallenstein, K., Köhler, C., and Said, S. I. (1979). Occurrence of vasoactive intestinal peptide (VIP)-like immunoreactivity in certain cholinergic neurons of the cat: evidence from combined immunohistochemistry and acetylcholinesterase staining. *Neuroscience* **4**:1539–1559.

Macleod, L. J., and Heard, B. E. (1969). Area of muscle in tracheal section in bronchitis, measured by point-counting. *J. Pathol.* **97**:157–161.

Mann, S. P. (1971), The innervation of mammalian bronchial smooth muscle: the localization of catecholamines and cholinesterases. *Histochem. J.* **3**:319–331.

Marin, M. G., Estep, J. A., and Zorn, J. P. (1982). Effect of calcium on sulfated mucous glycoprotein secretion in dog trachea. *J. Appl. Physiol.* **52**:198–205.

Mason, D. Y., and Taylor, C. R. (1975). The distribution of muramidase (lysozyme) in human tissue. *J. Clin. Pathol.* **28**:124–132.

McKenzie, H. I., Glick, M., and Outhred, K. G. (1969). Chronic bronchitis in coal miners: ante-mortem/post-mortem comparisons. *Thorax* **24**:527–535.

Meyrick, B., and Reid, L. (1970). Ultrastructure of cells in the human bronchial submucosal glands. *J. Anat.* **107**:281–299.

Meyrick, B., Sturges, J. M., and Reid, L. (1969). A reconstruction of the duct system and seretory tubules of the human bronchial submucosal gland. *Thorax* **24**:729–736.

Mills, J. W., and Quinton, P. M. (1981). Formation of stimulus-induced vacuoles in serous cells of tracheal submucosal glands. *Am.J. Physiol.* **241**:C18–C24.

Mooren, H. W. D., Meyer, C. J. L. M., Kramps, J. A., Franken, C., and Dijkman, J. H. (1982). Ultrastructural localization of the low molecular weight protease inhibitor in human bronchial glands. *J. Histochem. Cytochem.* **30**:1130–1134.

Murlas, C., Nadel, J. A., and Basbaum, C. B. (1980). A morphometric analysis of the autonomic innervation of cat tracheal glands. *J. Auton. Nerv. Syst.* **2**:23–37.

Nadel, J. A., Widdicombe, J. H., and Peatfield, A. C. (1985). Regulation of airway secretions, ion transport, and water movement. In *Handbook of Physiology*. Section 3. The respiratory system. Volume I. Edited by A. P. Fishman and A. B. Fisher. Washington, D.C., American Physiological Society, pp. 416–445.

Nagai, A., West, W. W., Paul, J. L., and Thurlbeck, W. M. (1985). The National Institutes of Health intermittent positive pressure breathing trial: Pathology studies. I. Interrelationship between morphologic lesions. *Am. Rev. Respir. Dis.* **132**:937–945.

Namba, T., Nakamura, T., and Grob, G. (1967). Staining for nerve fiber and cholinesterase activity in fresh frozen sections. *Am. J. Clin. Pathol.* **47**:74–77.

Niewoehner, D. E., Kleinerman, J., and Knoke, J. D. (1972). Regional chronic bronchitis. *Am. Rev. Respir. Dis.* **105**:586–593.

Oberholzer, M., Dalquen, P., Wyss, M., and Rohr, H. P. (1978). The applicability of the gland/wall ratio (Reid Index) to clinicopathological correction studies. *Thorax* **33**:779–784.

Pack, R. J., and Richardson, P. S. (1984). The aminergic innervation of the human bronchus: a light and electron microscopic study. *J. Anat.* **138**:493–502.

Partanen, M., Laitinen, A., Hervonen, A., Toivanen, M., and Laitinen, L. A. (1982). Catecholamine- and acetylcholinesterase-containing nerves in human lower respiratory tract. *Histochemistry* **76**:175–188.

Peatfield, A. C., Barnes, P. J., Bratcher, C., Nadel, J. A., and Davis, B. (1983). Vasoactive intestinal peptide stimulates tracheal submucosal gland secretion in ferret. *Am. Rev. Respir. Dis.* **128**:89–93.

Phipps, R. J., Denas, S. M., Sielczak, M. W., and Wanner, A. (1986). Effect of 0.5 ppm ozone on glycoprotein secretion, ion and water fluxes in sheep trachea. *J. Appl. Physiol.* **60**:918–927

Polak, J. M., and Bloom, S. M. (1982). Regulatory peptides and neuron-specific enolase in the respiratory tract of man and other mammals. *Exp. Lung. Res.* **3**:313–328.

Puchtler, H., Waldrop, F. S., Carter, M. G., and Valentine, L. S. (1974). Investigation of staining, polarization and fluorescence microscopic properties of myoepithelial cells. *Histochemistry* **40**:281–289.

Reid, L. (1960). Measurement of the bronchial mucous gland layer: a diagnostic yardstick in chronic bronchitis. *Thorax* **15**:132–141.

Restrepo, G. L., and Heard, B. E. (1963a). The size of bronchial glands in bronchitis. *J. Pathol. Bacteriol.* **85**:305–310.

Restrepo, G. L., and Heard, B. E. (1963b). Mucous gland enlargement in chronic bronchitis: extent of enlargement in the tracheo-bronchial tree. *Thorax* **18**:334,–339.

Restrepo, G. L., and Heard, B. E. (1964). Air trapping in chronic bronchitis and emphysema. *Am. Rev. Respir. Dis.* **90**:395–400.

Roth, J., Bendayan, M., Carlemalm, E., Villiger, W., and Garavito, M. (1981). Enhancement of structural preservation and immunocytochemical staining in low temperature embedded pancreatic tissue. *J. Histochem. Cytochem.* **29**:663–671.

Ryder, R. C., Dunnill, M. S., and Anderson, J. A. (1971). A quantitative study of bronchial mucous gland volume, emphysema and smoking in a necropsy population. *J. Pathol.* **104**:59–71.

Sasaki, T., Shimura, S., Okayama, H., Sasaki, H., and Takishima, T. (1986). Regulatory effect of airway epithelium on secretion of isolated submucosal gland from feline trachea. *Am. Rev. Respir. Dis.* **133**:A295.

Sasaki, T., Okayama, H., Shimura, S., Yashima, K., Sasaki, H. and Takishima, T. (1987). Na^+-K^+ATPase activity of isolated submucosal gland from feline trachea. *Am. Rev. Respir. Dis.* **135**:A365.

Schulte, B. A., and Spicer, S. S. (1982). Light microscopic histochemical detection of terminal galactose and N-acetylgalactosamine residues in rodent complex carbohydrate using a galactose oxidase-Schiff sequence and peanut lectin-horseradish peroxidase conjugate. *J. Histochem. Cytochem.* **31**:19–24.

Scott, K. W. M. (1973). An autopsy study of bronchial mucous gland hypertrophy in Glasgow. *Am. Rev. Respir. Dis.* **107**:239–245.

Shimura, S., Aoki, T. and Takishima, T. (1983). Connection of lamellar body to interacellular membrane system in type II alveolar cells of human lungs. *Am. Rev. Respir. Dis.* **127**:498–499.

Shimura, S., Martin, C. J., Boatman, E. S., and Dhand, R. (1985a). A role for interstitial matrix in tissue tension of alveolar wall. *Respir. Physiol.* **62**:293–303.

Shimura, S., Aoki, T., Tomioka, M., Shindoh, Y., and Takishima, T. (1985b). Concentrically arranged endoplasmic reticulum containing some lamellae (bar-like structure) in alveolar type II cells of rat lung. *J. Ultrastruct. Res.* **93**:116–128.

Shimura, S., Sasaki, T., Sasaki, H., and Takishima, T. (1986). Contractility of isolated single submucosal gland from trachea. *J. Appl. Physio.* **60**:1237–1247.

Shimura, S., Sasaki, T., Okayama, H., Sasaki, T., and Takishima, T. (1987a). Neural control of contraction in isolated submucosal gland from feline trachea. *J. Appl. Physiol.* **62**:2404–2409.

Shimura, S., Sasaki, T., Okayama, H., Sasaki, H., and Takishima, T. (1987b) Effect of substance P on the mucus secretion of isolated submucosal gland from feline trachea. *J. Appl. Physiol.* **63**:646–653.

Shimura, S., Sasaki, T., Ikeda, K., Sasaki, H., and Takishima, T. (1987c). Vasoactive intestinal peptide augments glycoconjugate secretion induced by cholinergic in tracheal submucosal glands. *J. Appl. Physiol.* **65**:2537–2544.

Silva, D. G., and Ross, G. (1974). Ultrastructural and fluorescence histochemical studies on the innervation of the tracheo-bronchial muscle of normal cats and cats treated with 6-hydroxydopamine. *J. Ultrastruct. Res.* **47**:310–328.

Sorvari, T. E., (1972). Histochemical observations on the role of ferric chloride in the high-iron diamine technique for localizing sulfated mucosubstances. *Histochem. J.* **4**:193–204.

Spicer, S. S. (1965).Diamine methods for differentiating mucosubstances histochemically. *J. Histochem. Cytochem.* **13**:211–234.

Spicer, S. S., Frayser, R., Virella, G., and Hall, B. J. (1977). Immunocytochemical localization of lysozymes in respiratory and other tissues. *Lab. Invest.* **36**:282–295.

Spicer, S. S., Hardin, J. H., and Setser, M. E. (1978). Ultrastructural visualization of sulphated complex carbohydrates in blood and epithelial cells with the high iron diamine procedure. *Histochem. J.* **10**:435–452.

Spicer, S. S., Mochizuki, I., Setser, M. E., and Martinez, J. R. (1980). Complex carbohydrates of rat tracheobronchial surface epithelium visualized ultrastructurally. *Am. J. Anat.* **158**:93–109.

Spicer, S. S., Sens, M. A., and Tashian, R. E., (1982). Immunocytochemical demonstration of carbonic anhydrase in human epithelial cells. *J. Histochem. Cytochem.* **30**:864–873.

Spicer, S. S., Schulte, B. A., and Charkin, L. W. (1983). Ultrastructural and histochemical observations of respiratory epithelium and gland. *Exp. Lung Res.* **4**:137–156.

Spruill, W. A., and Steiner, A. L. (1979). Cyclic nucleotide and protein kinase immunocytochemistry. In *Advances in Cyclic Nucleotide Research,* Vol. 10. Edited by G. Brooker and P. Greengard. New York, Raven Press, pp. 169–186.

St. George, J. A., Plopper, C. G., Etchison, J. R., and Dungworth, D. L. (1984). An immunocytochemical/histochemical approach to tracheobronchial mucus characterization in the rabbit. *Am. Rev. Respir. Dis.* **130**:124–127.

St. George, J. A., Cranz, D. L., Zicker, S. C., Etchison, J. R., Dungworth, D. L., and Plopper, C. G. (1985). An immunohistochemical characterization of rhesus monkey respiratory secretions using monoclonal antibodies. *Am. Respir. Dis.* **132**:556–563.

Stoward, P. J., Spicer, S. S., and Miller, R. L. (1980). Histochemical reactivity of peanut lectin-horseradish peroxidase conjugate. *J. Histochem. Cytochem.* **28**:979–990.

Suzuki, T. (1984). The innervation to lung. *Respir. Res. (Kokyu)* **3**:1309–1315.

Takagi, M, Parmley, R. T., Spicer, S. S., Denys, F. R., and Setser, M. E. (1982). Ultrastructural localization of acidic glycoconjugates with the low iron diamine method. *J. Histochem. Cytochem.* **30**:471–476.

Takemura, T., and Eishi, Y. (1985). Distribution of secretory component and immunoglobulins in the developing lung. *Am. Rev. Respir. Dis.* **131**:125–130.

Takizawa, T., and Thurlbeck, W. M. (1971a). A comparative study of four methods of assessing the morphologic changes in chronic bronchitis. *Am. Rev. Respir. Dis.* **103**:773–783.

Takizawa, T., and Thurlbeck, W. M. (1971b). Muscle and mucous gland size in the major bronchi of patients with chronic bronchitis, asthma and asthmatic bronchitis. *Am. Rev. Respir. Dis.* **104**:331–336.

Thurlbeck, W. M., and Angus, G. E. (1964). A distribution curve chronic bronchitis. *Thorax* **19**:436–442.

Tom-Moy M., Basbaum, C. B., and Nadel, J. A. (1983). Localization and release of lysozyme from ferret trachea: effects of adrenergic and cholinergic drugs. *Cell Tissue Res.* **228**:549–562.

Turek, J. J., Sheares, B. T., and Carlson, D. M. (1982). Substructure of glandules from serous cells of porcine tracheal submucosal gland (1982). *Anat. Rec.* **203**:329–336.

Uddman, R., Alumets, J., Densert, O., Hakonson, R., and Sundler, F. (1978). Occurrence and distribution of VIP nerves in the nasal mucosa and tracheabronchial wall. *Acta Otolaryngol.* **86**:443–448.

Warner, E. R. (1973). Stereological techniques for electron microscopic morphometry. In *Principles and Techniques of Electron Microscopy* Edited by M. A. Hayat. New York, Van Nostrand Rheinhold Co.

Weibel, E. R. (1973). Sterological techniques for electron microscopic morphometry. In *Principles and Techniques of Electron Microscopy*. Edited by M. A. Hayat. New York, Van Nostrand Rheinhold Co.

Wetzel, M. G., Wetzel, B. K., and Spicer, S. S. (1966). Ultrastructural localization of acid mucosubstances in the mouse colon with iron-containing stains. *J. Cell Biol.* **30**:299–315.

Wheeldon, E. B., and Pirie, H. M. (1974). Measurement of bronchial wall components in young dogs, adult normal dogs, and adult dogs with chronic bronchitis. *Am. Rev. Respir. Dis.* **110**:609–615.

Whimster, W. F., Lord, P., and Biles, B. (1984). Tracheobronchial gland profiles in four segmental airways. *Am. Rev. Respir. Dis.* **129**:985–988.

13

The Use of Tracers in Transport Studies

MAYA SIMIONESCU and NICOLAE GHINEA

Institute of Cell Biology and Pathology
Bucharest, Romania

I. Introduction

To uncover intricate cellular processes requires that one enter the cell in various ways without perturbing its normal functions. One of the least traumatic ways to penetrate the intimacy of a cell is to use probes (or tracers) that mimic real-life molecules, that enter to stay and be used by the cell, or are ferried across the cell to the neighboring compartments. To follow its pathways, the tracer must give a signal of its location. For localization at the electron microscopic level, the probes should be electrondense either by their own chemistry (ferritin) or be rendered electron opaque by special staining (dextrans), or via a cytochemical reaction by virtue of their peroxidatic activity (hemoproteins). Another possibility is to radiolabel the tracers and detect them indirectly by autoradiography. In addition, tracers can be tagged to specific molecules of biological interest (lectins, plasma proteins and lipoproteins, hormones, etc.) and used to follow the fate of the latter within a cell. The main advantage of using probe molecules is that they can be applied to living cells either in vivo, in situ, or in culture, reasonably close to the physiological conditions. Thus, the most valuable tracers would be physiological molecules, or at least inert probes that do not interfere with cell activity.

The ability to uncover a cellular mechanism in this way largely depends on the use of the right tracer. That implies an appropriate knowledge of the physicochemical characteristics of the probe to be used, its interaction(s) with cells, possible side effects, and great caution in the choice of the general experimental conditions. Tracers may be taken up by the cells in fluid phase, adsorbed nonspecifically to surface components, or they can bind specifically to membrane-binding sites or receptors. Upon binding, the probe is generally internalized in a closed compartment, usually a vesicle (of clathrin-coated or uncoated variety), which carries the ligand that presumably does not interfere in the vesicle's normal traffic. Whether, this is always true must be checked for each tracer using alternative methods. Once internalized, the probe-carrying vesicle is usually either routed to the compartments involved in endocytosis or may cross the cell, transporting and discharging the ligand extracellularly by transcellular transport or transcytosis (Simionescu, 1979a). Both processes, endocytosis and transcytosis, begin by the same basic mechanisms: fluid phase ingestion or nonspecific or specific (receptor-mediated) uptake (Simionescu et al., 1987). In various cells, the existence and fractional contribution of each of these two processes may be evaluated to some degree by the appropriate selection of the tracer molecule. In addition, the probes that bind to the cell surface can be used for studying the retrieval and intracellular traffic of various membrane domains.

II. Tracers as Molecules

A proper tracer should be (a) nontoxic and physiologically inert; (b) composed of uniform particles of known size (molecular weight and effective molecular diameter) and molecular charge; (c) detectable in low concentration; (d) immobilized in its in vivo position by fixation; and (e) capable of being accurately localized when viewed with the electron microscope.

A. Physicochemical Characteristics

The use of the ultrastructural tracers in permeability studies assumes a thorough characterization of their molecular weight (M_r), size, shape, net electrical charge, and chemistry.

Molecular Weight and Effective Molecular Diameter

As shown in Tables 1–4, the tracers available for transport studies cover a large range of molecular weights and dimensions. In general, the choice of a tracer by size should consider the compatibility between its effective molecular diameter and the presumed dimensions and geometry of the pathways to be explored. Molecular weight and size of a probe molecule can be determined by gel filtration (on Sephadex, Sephacryl, or Sepharose) under appropriate conditions of

Table 1 Particulate Tracers

Tracer	Molecular weight (kD)	Molecular diameter (nm)	Molecular charge (pI)
Dextrans	20	5.5–6.5	Neutral
	40	8–9	Neutral
	75	11.6	Neutral
	250	~22.5	Neutral
	500	—	Neutral
	2,000	—	Neutral
Dextran sulfate	500	—	Negative
DEAE–dextran	500	—	Positive
Iron–dextrans			
Imferon	73		Neutral
Imposil	200		
Gleptaferon	2,000		
Ferritins			
Native	~960	~13	4.5
Anionized	~960	~13	3.7–4.1
Cationized	~960	~13	8.4[a]
Gold–protein conjugates	Varies[b]	Varies[b]	Varies[c]
Shellfish glycogen		20	Neutral
Rabbit liver glycogen		30	Neutral
Colloidal carbon		250 × 300	Inert

[a]Ferritin cationized to various pIs (>4.5) can be obtained (see text).
[b]According to the size of gold particles and the size of the proteins adsorbed.
[c]According to the charge of the protein adsorbed.

temperature, pH, and ionic strength against standard proteins. The distribution coefficients (K_{av}) for the probe and standards are calculated as follows: $K_{av} = V_e - V_o/V_t$, where V_e = elution volume for each molecule, V_o = void volume, and V_t = total bed volume. The distribution coefficients of the control probes are plotted against the known molecular radius on semilogarithmic graph paper, and the effective molecular diameter of the probe is calculated from the curve (Rennke and Venkatachalam, 1979).

Table 2   Anionic Peroxidatic Tracers

Tracer	Molecular weight (kD)	Molecular diameter (nm)	Molecular charge (pI)	Optimal pH[a]
Microperoxidases				
Hemeoctapeptide	1.55	1.6	5.4	12
Hemenonapeptide	1.63	1.6	4.95	12.5
Hemeundecapeptide (H11P)	1.88	1.66	4.85	9
Anionized H11P	1.91	1.7	3.5	9.5
Horseradish peroxidases (HRP)				
Acidic isoenzymes	44	6.0	4.3–5.8	4.5
Succinilated HRP	44	6.0	4	4.3
Catalase	240	10.4	5.7	10.5

[a]For visualization by cytochemical reaction using 3,3′-diaminobenzidine as hydrogen donor.

Molecular Shape

The molecule's shape and flexibility or rigidity are also important parameters that determine the permeability behavior of the tracers. For example, high molecular weight dextran unravels while passing through the fenestral diaphragm in visceral endothelium; conversely myoglobin is a particularly rigid molecule (Simionescu et al., 1972, 1973).

Molecular Charge

The net electric charge of a probe molecule at physiological pH affects its accessibility and interactions with the cell surface and thus its permeation. Among the available tracers some are anionic and others neutral or cationic (Tables 1–4). The isoelectric point (pI) of a tracer can be determined either by isoelectric precipitation or by isoelectric focusing on polyacrylamide or agarose gels using ampholines as generators of pH gradient. The latter method is superior by having a higher resolution capability.

To study the effect of charge on uptake and transport of macromolecules with similar molecular weight, the pI of proteins can be chemically modified so as to obtain cationic and anionic derivatives.

Cationization of proteins is based on the covalent blocking of the carboxyl groups by small neutral or cationic molecules. Acid-catalyzed esterification and coupling with nucleophils mediated by a water-soluble carbodiimide are the reactions most commonly used. The method involves the activation of the carboxyl

Table 3 Neutral Peroxidatic Tracers

Tracer	Molecular weight (kD)	Molecular diameter (nm)	Molecular charge (pI)	Optimal pH[a]
Myoglobin	16.9	3.6	7.0	5
Horseradish peroxidase	44	6.0	7.0	3.5
Hemoglobin	65	6.8	6.8	7

[a]For visualization by cytochemical reaction using 3,3′-diaminobenzidine as hydrogen donor.

groups by carbodiimide and the subsequent reaction of the activated carboxyl groups with a nucleophil (Hoare and Koshland, 1967). By the reaction of carboxyl groups with carbodiimide an *O*-acylisourea derivative is formed:

$$\text{Protein—(COOH)}_m + n \underset{\underset{N-R_2}{\parallel}}{\overset{\overset{N-R_1}{\parallel}}{C}} \xrightarrow{+nH^+} \text{Protein} \begin{matrix} /(COOH)_{m-n} \\ +NH-R_1 \\ \backslash (COO-\underset{HN-R_2}{\overset{\parallel}{C}})_n \end{matrix}$$

tracer protein   carbodiimide   *O*-acylisourea derivative

In the second step, *O*-acylisourea derivative reacts with the nucleophil to obtain the cationic derivative of the protein:

$$\text{Protein} \begin{matrix} /(COOH)_{m-n} \\ +NH-R_1 \\ \backslash (COO-\underset{HN-R_2}{\overset{\parallel}{C}})_n \end{matrix} \xrightarrow[-nH^+]{+nHX} \text{Protein} \begin{matrix} /(COOH)_{m-n} \\ \\ \backslash (CO-X)_n \end{matrix} + n \ HO-\underset{\underset{HN-R_2}{\mid}}{\overset{\overset{N-R_1}{\parallel}}{C}}$$

O-acylisourea   cationized tracer
derivative   derivative

To avoid protein polymerization, the nucleophil is introduced in high excess. The degree of modification of the carboxyl groups can be controlled by varying the pH of the coupling reaction or by changing the concentration of carbodiimide reagent. When the nucleophil is a small molecule containing a positive

Table 4 Cationic Peroxidatic Tracers

Tracer	Molecular weight (kD)	Molecular diameter (nm)	Molecular charge (pI)	Optimal pH[a]
Hemeundecapeptide (H11P)				
Cationized H11P$_t$[b]	2.1	1.7	9	8
Cationized H11P$_p$[c]	2.1	1.7	10.6	8
Cytochrome c	12.4	3 × 3.4 × 3.4	10.6	2.5
Horseradish peroxidase (HRP)				
Basic isoenzymes	44	6	8.7–9.2	3.5
Cationized HRP	44	6	8.4–9.2	4.3
Lactoperoxidase	82	7.2	8	7
Myeloperoxidase	160	8.8	10	7.5

[a]Using 3,3'-diaminobenzidine as hydrogen donor in cytochemical reaction.
[b]Contains predominantly tertiary amino groups.
[c]Contains primary amino groups only.

charge at the unattached end (i.e., 1,6-diaminohexane, N,N-dimethyltrimethyl-enediamine), the effect is a change of two unit charges per modified carboxyl group (Danon et al., 1972).

Anionization consists in the covalent blocking of the primary amino groups of the protein by small molecules (electrophilic substitution). The most frequently used electrophilic agents are acidic anhydrides (i.e., acetic anhydride, succinic anhydride) and α-dicarbonyl compounds (i.e., 1,2-cyclohexanedione) (Fraenkel-Conrat, 1957):

$$\text{Protein}-(\text{NH}_2)_p + v\ O\begin{array}{c} \overset{O}{\underset{\parallel}{\ }} \\ C-R_1 \\ \\ C-R_2 \\ \underset{\parallel}{\overset{\ }{O}} \end{array} \xrightarrow[-v\ R_1\ \text{COOH}]{\text{pH 8–9}} \text{Protein}\begin{array}{c} (\text{NH}_2)_{p-v} \\ \\ (\text{NHCOR}_2)_v \end{array}$$

tracer	acidic	anionized tracer
protein	anhydride	protein

When succinic anhydride is used, the modification of protein primary amino groups effects a change of two unit charges per modified amino group:

$$\text{Protein}-(NH_2)_p + v \quad O \begin{array}{c} \overset{\displaystyle O}{\underset{\displaystyle \parallel}{C}}-CH_2 \\ | \\ \underset{\displaystyle \overset{\parallel}{O}}{C}-CH_2 \end{array} \quad \xrightarrow{\text{pH 8-9}} \quad \text{Protein} \begin{array}{c} (NH_2)_{p-v} \\ \diagdown \\ (NHCOCH_2-CH_2COOH)_v \end{array}$$

| tracer | succinic | anionized tracer |
| protein | anhydride | protein |

In general, the procedures for protein cationization and anionization do not significantly change the molecular weight; however, some conformational modifications may occur. In addition, one should keep in mind that derivatives of proteins cationized to an isoelectric point below 7.0 used at physiological pH still behave as anionic probes.

B. Classes of Tracers

Based on the method of detection, the tracers used for macromolecular transport fall into four major classes: particulate tracers, peroxidatic tracers, radiolabeled tracers, and tracers tagged to biological molecules.

Particulate Tracers

To this category belong probes that appear as discrete particles of different sizes and pIs (Table 1), which can be identified either by their electron density (i.e., ferritin, colloidal gold, colloidal iron, carbon) or by special staining procedures (e.g., dextrans) (Simionescu et al., 1972). The degree of tracer dispersion in solution can be assessed by negative staining (Fig. 1).

Ferritin

An anionic protein in its native form, horse spleen ferritin consists of a spherical apoferritin shell ($\sim$ 3 nm thick) with a diameter of $\sim$ 13 nm (Fischbach et al., 1969) and a core of ferric hydroxyphosphate micelles that constitute up to 40% of its dry weight (Crichton, 1971). The prevalence of acidic residues impart to ferritin its pI of 4.5. By chemical modifications of protein functional groups (i.e., amino and carboxyl groups), anionic and cationic ferritin derivatives have been obtained (Danon et al., 1972; Rennke et al., 1975; Kanwar and Farquhar, 1979; Ghinea and Hasu, 1986). Ferritin (and its derivatives) has an M_r of $\sim$ 960 kD,

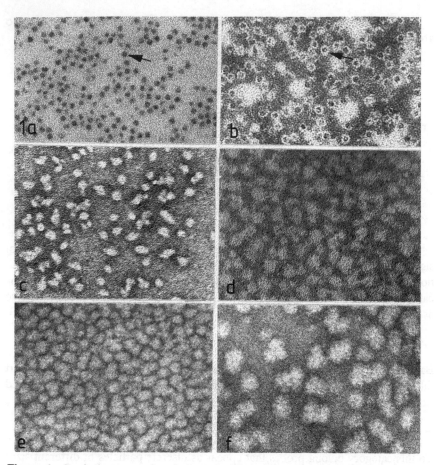

Figure 1   Particulate tracers in solution to be injected: negative staining with 2.5% uranyl acetate, pH 4.2 (except a). a. Electron-dense core of unstained ferritin particles. Negative staining of ferritin (b); dextran, M_r 40,000 (c); dextran, M_r 75,000 (d); shellfish glycogen (e); and rabbit liver glycogen (f). Note the size difference between the ferritin core (a) and the whole molecule (b) (arrows) (a, b, × 228,000; c–e, × 160,000).

to which the apoferritin shell (composed of 24 subunits) contributes an M_r of ~ 445 kD (Panitz and Ghiglia, 1982). Ferritin is electron dense by virtue of its 5.5 nm iron core (Fig. 1a). The whole molecule (the iron core and the apoferritin shell) can become visible upon staining of sections with metallic bismuth, which binds to and stains the apoferritin shell, thus exposing the effective diameter of the molecule (Ainsworth and Karnovsky 1972). In addition to its value for its intrinsic electron opaqueness, ferritin was used successfully as tracer because of its relatively homogeneous size. When cadmium-free it is well tolerated by animals and may be coupled to active molecules to give conjugates that maintain appreciable biological activity (i.e., low-density lipoprotein–ferritin [Anderson et al., 1976], histamine–ferritin [Heltianu et al., 1983]).

Dextrans and Glycogens

Dextrans are long polysaccharide chains of α-D-glucopyranose units linked by $\alpha(1 \to 6)$, $\alpha(1 \to 4)$, and $\alpha(1 \to 3)$ glucosidic bonds, the majority of them being $\alpha(1 \to 6)$.

Glycogens are branched polysaccharides made up of chains of α-D-glucopyranose units linked by $\alpha(1 \to 4)$ glucosidic bonds with branching points arising from additional $\alpha(1 \to 6)$ linkages. Due to the variable degree of polycondensation, dextrans and glycogens have different molecular weights and sizes (Table 1). Dextran and glycogen powders are soluble in 0.15 M NaCl and in various buffer solutions; the dissolution can be accelerated by sonication.

Being uncharged, stable, and well tolerated when intravenously injected into experimental animals, these polysaccharides have been used extensively as probes in physiological experiments and in studies on capillary permeability. The two-pore system concept was derived primarily from experiments performed with graded dextrans (Grotte, 1956; Mayerson et al., 1960). Using the staining procedure of Simionescu et al. (1972), the earlier observations of dextran transport were extended to electron microscopic study of the structural equivalent of the pore system. By this technique, using a one-step fixation procedure (see Addendum) a sharp and high contrast of the tracer particles can be achieved. Iron-dextrans are electron opaque tracers that, due to their iron part, do not require special staining enhancement.

Charged Dextrans

Charged derivatives are polyanionic dextran-sulfate and the polycationic diethylaminoethyl–dextran (DEAE–dextran). The former is usually used as fluid-phase tracer for capillary permeability and the latter as a nonspecific adsorptive membrane marker (the cationic probe binds strongly to cell membranes). Like their neutral forms, the charged dextran derivatives become visible on electron microscopic examination upon fixation by the one-step technique mentioned above.

Peroxidatic Tracers

These probes (Tables 2–4) are nonparticulate molecules that become visible on electron microscopic examination after a cytochemical reaction based on their peroxidatic activity. Sensitivity is increased because of the enzymatic amplifying effect. The peroxidatic reaction is performed by allowing a hemoprotein to act on hydrogen peroxide (H_2O_2), which is coordinated at Fe^{3+} of the heme; subsequently, the complex (tracer–H_2O_2), in the presence of a hydrogen donor (AH_2), yields an insoluble oxidation product (A):

$$
\text{AH}_2 \quad + \ \text{H}_2\text{O}_2 \xrightarrow{\begin{array}{c}\text{peroxidatic}\\\text{tracer}\end{array}} \text{A} \quad + \quad 2\text{H}_2\text{O}
$$

hydrogen   substrate   oxidation
  donor   product

The reagent 3,3-diaminobenzidine (DAB) introduced by Graham and Karnovsky (1966a) has been widely used as a hydrogen donor. In the presence of the peroxidatic tracers, this aromatic compound forms a polymer that, upon treatment with osmium tetroxide gives an electron-opaque complex product known as osmium black. According to the accessibility of the H_2O_2 to the Fe^{3+} of the heme, the peroxidatic activity may vary, being the highest in horseradish peroxidase (HRP) and the lowest in cytochrome c. The detection of tracers with low peroxidatic activity is possible when relatively large amounts of the probe remain in the tissue after fixation and tissue processing for electron microscopy. However, if only small amounts are retained, the sensitivity of the peroxidatic reaction becomes critical. The sensitivity may be increased by varying the concentrations of DAB and H_2O_2, the temperature, and pH or adding nitrogenous compounds to the incubation medium. Among the latter, imidazole increases the peroxidatic activity for HRP from 478 to 713 units, for hemeundecapeptide from 1.9 to 3.4 units and for hemeoctapeptide from 2.9 to 5.5 units when 0-dianisidine is used as hydrogen donor (Tu et al., 1968). Each peroxidatic tracer has an optimum pH for visualization by cytochemical reaction (Tables 2–4); for HRP, staining at an acid pH is more intense than staining at a neutral pH (Weir et al., 1974).

A wide range of tracer sizes is desirable for obtaining structure–function correlations of macromolecular transport into and across cells. The hemoproteins used to study endothelial permeability cover a broad spectrum of sizes and pIs (Tables 2–4). These include hemepeptides (Feder, 1971; Kraehenbuhl et al., 1974; Simionescu et al., 1975; Plattner et al., 1977; Wissig and Williams, 1978), cytochrome c (Karnovsky and Rice, 1969), myoglobin (Anderson, 1972; Simionescu et al., 1973), hemoglobin (Pietra et al., 1969), lactoperoxidase (Graham and Kellermeyer, 1968), horseradish peroxidase and myeloperoxidase (Graham and Karnovsky, 1966b), and catalase (Venkatachalam and Fahimi, 1969).

Horseradish Peroxidase

This is a polymannose glycoprotein that has been widely used as a tracer because of its high peroxidatic activity. It contains at least seven major isoenzymes, all having ferric protoporphyrin IX as the prosthetic group with the nitrogen atom of a histidine residue as the proximal ligand. These various isoenzymes have different pIs (Tables 2–4) due to differences in the number of arginine residues; the acidic pI is characterisitic for the isoenzymes with lowest content of arginine

(Shannon et al., 1966). The main isoenzyme, horseradish peroxidase-C (HRP-C), has a molecular weight of 44 kD and consists of 308 amino acid residues and 8 neutral carbohydrate side chains (Welinder, 1979) containing mannose, mannosamine, galactosamine, fucose, and xylose (Shannon et al., 1966). By chemical modifications, anionic and cationic derivatives of HRP have been produced (Rennke and Venkatachalam, 1979).

Cytochrome c

Cytochrome c (Cyt c) is a cationic globular protein (molecular size $3 \times 3.4 \times 3.4$ nm) containing 104 amino acid residues. The heme moiety is covalently attached to the polypeptide chain by two thioether bridges between the vinyl side chains of the heme and cysteinyl residues 14 and 17. Due to the embedded position of its heme group, Cyt c has a low peroxidatic activity.

Hemepeptides

By enzymatic digestion of Cyt c, a number of low molecular weight hemepeptides (microperoxidases) have been obtained: hemeundecapeptide (H11P), hemenonapeptide (H9P), and hemeoctapeptide (H8P) (Table 2). These hemepeptides have a peroxidatic activity higher than Cyt c due to the significant reduction of the polypeptide chain (from 104 amino acids in Cyt to 11, 9, and 8 in H11P, H9P, and H8P, respectively) and consequently better exposure of heme.

From native H11P, which contains two primary amino groups and four carboxyl groups, three derivatives have been prepared: anionized H11P, pI 3.5, and two cationized H11P with pIs of 9.0 and 10.6. The anionized H11P (aH11p) was prepared by covalent blocking of the primary amino groups of H11P with acetic anhydride. The synthesis of cationic H11P derivatives was based on the activation of carboxyl groups of H11P with carbodiimide. Two nucleophils were used: N,N-dimethyltrimethylenediamine to obtain a cationic derivative that contains predominantly tertiary amino groups (cH11P$_t$, pI 9) and 1,6-hexanediamine to produce a cationic derivative with primary amino groups only (cH11P$_p$, pI 10.6) (Ghinea and Simionescu, 1985). No significant changes in molecular weight, molecular diameter (Tables 2, 4), or adsorption spectrum were noticed, which make these derivatives useful as probes with similar M_r but different pIs.

Radiolabeled Tracers

Virtually all biological molecules can be radiolabeled by an appropriate method and their presence within tissues detected by autoradiography. Most methods for radiolabeling proteins use either iodination or reductive methylation.

Iodination

This can be performed by using oxidants in aqueous solutions (e.g., chloramine T, iodine monochloride, lactoperoxidase) or in solid phase (e.g., iodogen, iodobeads, enzymobeads). The redox reactions produce from sodium iodide an electrophilic iodine species [$^{125}I^+$] that reacts with any aromatic moiety of the protein to be labeled (tyrosine being the major site of insertion). Solid-phase iodination has a number of advantages over aqueous or enzymatic methods, among which are that reducing agents to stop the reaction are not necessary and the time for removal of the free iodine (desalting) is shorter (David, 1972; Markwell, 1983).

The labeling methods using oxidative technique are not suitable for proteins that lack tyrosine residues. In addition, the oxidation of thiol groups, methionine, tryptophan, and histidine, may damage some proteins, which consequently may lose their biological activity; in such cases, reductive methylation provides a useful alternative method.

Reductive Methylation

Reductive methylation of amino groups can be achieved by exposing proteins in alkaline solution to low concentrations of [^{14}C]formaldehyde and sodium borohydride. Both α- and ϵ-amino groups are modified:

$$\text{Protein}-NH_2 + [^{14}C]H_2O \xrightarrow{\quad NaBH_4 \quad} \text{Protein}-NH-[^{14}C]H_3$$

radiolabeled protein
tracer

The physical properties of alkylated protein derivatives are very similar to those of the native protein.

Tracers Tagged to Biological Molecules

Proteins Covalently Bound to Tracers

In addition to their use as general tracers, particulate and peroxidatic probes may be covalently bound to various biological molecules (ligands) and used to follow the fate of these specific molecules. Usually the chemical groups available to couple ligands to tracers are the primary amino groups (terminal α-amino and ϵ-amino groups of lysine) and, to a lesser extent, carboxyl and thiol groups. Homo- and heterobifunctional reagents are most commonly used for coupling. The pH of the reaction plays a major role in successful cross-linking: the correct pH will serve to minimize the potential hydrolysis of the cross-linker and to create the proper environment for the reaction to proceed smoothly. Buffers containing primary amino groups (e.g., glycine or Tris) should be avoided since the cross-linkers may react preferentially with them.

Peroxidatic and particulate tracers have been linked to various ligands using as reagents either p,p'-difluoro-m,m'-dinitrophenyl sulfone (Nakane and Pierce, 1967), water-soluble carbodiimide, or glutaraldehyde (Kishida et al.,

1975). The reactions are usually carried out in a mixture of ligands and tracers so that in addition to the desired tracer–ligand conjugates, other heterogeneous products are produced. These include tracer–tracer aggregates, ligand–ligand conjugates, and high polymers of tracer–ligand conjugates. To avoid the formation of undesired products, a two-step procedure for linking tracers to ligands was proposed (Kishida et al., 1975). The method consists of tracer activation by glutaraldehyde or carbodiimide followed by ligand conjugation to the activated tracer. The degree of conjugation depends on activation and coupling reactions; for maximum efficiency it is important to adjust the quantity of activation agent (first step) and the concentration of ligand (second step) as well as the time, pH of the buffer, and the temperature at which reactions are performed. The residual activated groups must be quenched with appropriate reagent (i.e., aldehyde groups are inactivated with $NaBH_4$). Finally, tracer-ligand conjugates are separated from nonconjugated reagents by gel filtration. Whether the conjugation of a biological molecule to a tracer alters its biological activity and behavior must be routinely checked. In addition, the size of the whole complex must be estimated by gel chromatography or negative staining.

Proteins Adsorbed to Gold Sol

Colloidal gold is an electron-dense, nondegradable, negatively charged hydrophobic sol whose stability is maintained by electrostatic repulsion. There are various methods of producing gold sols of different sizes from 5 to 100 nm. All methods are based on the controlled reduction of an aqueous solution of $HAuCl_4$ using various reducing agents (i.e., white phosphorus, trisodium citrate, citrate–tannic acid) (Slot and Geuze, 1981). Addition of electrolytes to the gold sol results in the compression of the ionic double layers and the reduction of the radius of repulsive forces' action; as a result, gold sols flocculate. Flocculation of gold particles may be prevented by adding a protective coat of macromolecules (e.g., proteins, glycoproteins, polysaccharides). These macromolecules immobilized on gold granules preserve most of their specific properties. The optimal amount of macromolecules necessary to stabilize the gold sol is determined by exposing constant amounts of gold to increasing concentrations of macromolecules (Horisberger and Rosset, 1977). After 5 min, 10% NaC1 is added (to a final concentration of 1% NaC1) to flocculate the unstabilized gold particles. The lowest concentration of macromolecules that prevents the shift in color from red to blue is taken as the saturation point. To ensure that all gold particles are stabilized, 0.006 mg/ml of polyethylene-glycol is added (Horisberger, 1979).

Colloidal gold was first proposed as electron opaque tracer for electron microscopy by Faulk and Taylor in 1971. The spectacular growth in interest in this particulate tracer is due to the following advantages: the preparation method of gold sol is simple, rapid, and efficient; its coupling to proteins and polysaccharides implies adsorption and not covalent binding; the method is suitable for double labeling experiments using gold particles of different sizes; and the contrast of gold particles is higher than that of other particulate tracers such as ferritin. Low-density lipoprotein (LDL) can be conjugated to gold sol (Au) to give an LDL–Au complex. When coupled to 5 nm gold, usually one or two gold particles label one LDL (Fig. 2a). The complex does not change significantly the size of native LDL particle and is more electron dense than LDL–ferritin conjugates (Fig. 2b). When conjugated to 20 nm gold sol, most complexes are made of ~ 8 LDL particles per one Au particle (Handley et al., 1981); in this case the size of the complex is greatly increased. Various ratios of LDL/Au particles may also be obtained (Fig. 2c–i).

III. Tracers as Tools: Choosing A Tracer

According to the purpose of the inquiry, for each particular experimental system one should select from the available tracers those that fit best and produce the least side effects.

A. Fluid-Phase Probes

The tracers used to study the fluid-phase uptake must be uncharged or neutral molecules, without affinity for surface components, and sufficiently dense to scatter electrons or be rendered (indirectly) electron dense for ultrastructural visualization. The most frequently used fluid-phase markers are native ferritin, dextrans of various molecular weights, glycogens, and some hemoproteins.

Ferritin

Ferritin was used successfully as a fluid-phase marker in studies of transendothelial transport (Farquhar and Palade, 1961; Bruns and Palade, 1968). A relatively high concentration (1 mg/ml) is usually required. It can be injected in vivo, providing that the preparation is devoid of impurities, of which trace amounts of cadmium are the most common and toxic. To prevent this, ferritin should be dialysed extensively (48 hr) before use.

Dextrans

Neutral dextrans of various molecular weights and sizes were used as fluid-phase markers in studies of transendothelial transport in fenestrated (Simionescu et al., 1972) and glomerular capillaries (Caulfield and Farquhar, 1974). Dextrans remain as individual particles at high concentrations (10% w/v) in physiological solutions (such as 0.9% NaCl); dissolution of the tracer is accelerated by 50 min sonication at 20 °C (Simionescu and Palade, 1971). To increase the contrast and definition of dextran and glycogen particles, a one-step fixation of the tissue using a mixture containing glutaraldehyde, formaldehyde, OsO_4, and lead citrate in arsenate or phosphate buffer was devised (Simionescu at al., 1972). This mixture can be used in various buffers and gives good results as a general fixative for electron microscopy for tissues, cell suspensions, or isolated membranes or organelles (for details of the method, see Addendum). Iron-dextrans have been used as fluid phase markers by Herzog and Miller (1981). The advantage of using dextran for permeability studies stems from its biological nature, availability in a wide spectrum of sizes, and direct visualization. Of equal importance is that dextrans are well tolerated upon intravenous injection in vivo and do not induce vascular leakage (of histamine type) in mice and rats.

Glycogens

Rabbit liver glycogen and shellfish glycogen are large molecules that can be used when a wide spectrum of neutral molecules of various sizes are needed (Table 1). Both are polysaccharide particles that cannot be "fixed" by the usual reagents for electron microscopic studies; however, they can be well immobilized by fixation of the proteins in their surroundings (plasma or interstitial fluid proteins).

Glycogens were used for permeability studies of intestinal (fenestrated) capillaries as 15% solution (w/v) in 0.9% NaCl, sonicated for 30 min at 25 °C. The state of dispersion of particles and the size distribution was assessed by negative staining (using 2% potassium phosphotungstate, pH 7.4, or 2.5% uranyl acetate, pH 4.2) of a drop of tracer solution placed on carbon-coated Formvar grids (Fig. 1e, f) (Simionescu et al., 1972). The tracers remained mostly as dispersed particles in the plasma.

Hemoproteins

This group of proteins, which are visualized via a cytochemical reaction, range in size from 1.6 nm (H8P) to 10.4nm (catalase) effective molecular diameter (Tables 2, 4). The peroxidatic tracers can be used at low concentrations due to the amplifying effect of the reaction product: few tracer molecules generate a strong signal. The latter characteristic is also a limitation, since little quantitative information can be obtained.

Horseradish peroxidase has been used predominantly in permeability studies and for quantification of fluid-phase pinocytosis (Steinman and Cohn, 1972). However, in some cells, like rat alveolar macrophages, HRP (a polymannose glycoprotein) is adsorbed to the plasma membrane, presumably to a mannose receptor; its binding was specifically inhibited by $10^{-3}M$ mannose or $10^{-3}M$ mannan (Sly and Stahl, (1978). Binding to the plasma membrane was followed by adsorptive endocytosis. In addition, HRP binds strongly to the cell surface of several cultured mammalian cells. Evidence was obtained that HRP may be bound to a glycosyltransferase at the cell surface since its presence was suppressed by 10 mM chitotriose and 10 mM UDP-galactose, an acceptor and donor, respectively, of galactose for the reactions catalyzed by galactosyltransferase (Straus and Keller, 1987). Thus, in some systems HRP is an adsorptive marker rather than a fluid-phase probe; the generality of these results must be tested for each cell under study.

Hemepeptides are the smallest peroxidatic probes used in studies of capillary permeability (Table 2). Since they contain charged residues (i.e., H11P contains four carboxyl groups and two amino groups), they can theoretically bind to both anionic and cationic sites on the endothelial cell surface. Yet in experiments performed in situ, H11P behaves as an anionic molecule and binds to the endothelial surface in a fashion comparable to the anionized H11P (pI 3.5). When used as a fluid-phase marker, especially in studies of endothelial permeability, its ingestion by plasmalemmal vesicles should be considered, to a degree, a combined process of adsorptive (electrostatic) and fluid-phase uptake.

Pitfalls of Fluid-Phase Markers

In general, the probes used are heterologous substances which may induce side effects. This possibility should be checked upon tracer administration to animals. It has been reported that HRP produces hypotension in rats (Deimann et al., 1976) and opening of junctions between endothelial cells in postcapillary venules (Cotran and Karnovsky, 1967; Vegge and Haye, 1977). Like HRP, myoglobin can also induce opening of junctions in postcapillary (pericytic) venules, probably via a histamine-mediated effect (Simionescu, 1979b). For fluid-phase uptake experiments, the probes have to be applied at relatively high concentrations, a condition particularly detrimental for studies on endothelial permeability since modification in blood volume and hydrostatic or colloid osmotic pressure may significantly alter the results. In addition, at high tracer concentration, the amount of impurities and contaminants can reach levels that may be toxic to the cells (i.e., cadmium in ferritin solutions). Dextran endocytosed by the exocrine glandular cells can produce swelling of the Golgi cisternae, possibly due to an osmotic side effect (Herzog and Farquhar, 1977; Herzog and Reggio, 1980). Due to the enzymatic amplification a major limitation of the peroxidatic tracers is the diffusion of the reaction product frequently far beyond the actual location of the tracer molecule per se.

B. Adsorptive Nonspecific Probes

To this category belong mostly probes that interact electrostatically with plasma membrane or its specialized domains, that is, cationic and anionic tracers that adsorb to the opposite charged groups of the plasmalemma. Since such interactions are usually strong, these probes can be used to advantage for mapping the distribution of charge on cell surfaces and to follow the traffic of membrane domains to various intracellular compartments. The most useful tracers are those that can be used at physiological pH and ionic strength and do not require prefixation.

Cationized Ferritin

Cationized ferritin was introduced by Danon et al. (1982) to study the distribution of acidic sites on the surface of endothelial cells. Because of its high isoelectric point (pI 8.4), CF interacts electrostatically with anionic residues of the cell surface components. The strong ionic interaction allows the use of CF at low concentration and for short incubation time. Under these conditions, microdomains of different charge or charge density have been detected on the luminal surface of various capillary endothelia (Simionescu et al., 1981a,b, 1985; Ghitescu and Fixman, 1984) including lung (Pietra et al., 1983; Simionescu and Simionescu, 1983). Information on the molecules that generate these microdomains can be obtained indirectly, by exposing the cell surface to specific enzymes (i.e., neuraminidase, heparinase, etc.) prior to CF. This "in situ" biochemistry allows the identification and localization of various acidic components of the cell membrane.

The strong CF binding to the plasmalemma was used to identify the retrieval of the apical cell membrane in secretory cells of the thyroid follicle (Herzog and Miller, 1979), anterior pituitary (Herzog and Farquhar, 1983), and type II epithelial cells of the lung (Simionescu, 1985a). With the use of ferritin cationized to various isoelectric points (pIs 5–9) more subtle information can be obtained on the role of charge of the molecules in transport.

Cationized and Anionized Hemeundecapeptides

These markers were introduced as probes for cell surface charge and permeability studies by Ghinea and Simionescu (1985). Their access to the surface components is facilitated by the tracer small size (~ 1 nm molecular radius). Other cationic tracers such as DEAE–dextran (Herzog and Miller, 1979) and cationized HRP (Davies et al., 1981) have been used as membrane markers.

Pitfalls of Adsorptive Nonspecific Markers

The major limitation of these probes is the possibility that upon binding to surface molecules conformational changes and, consequently, functional perturbations may occur. Such changes (not always present) have to be checked for each experimental model, but this is not easy. Cationic ferritin was reported to produce redistribution and lateral diffusion of anionic sites on the surface of endothelial cells in large vessels (Skutelsky and Danon, 1976), whereas no such phenomenon was observed in capillary endothelium (Simionescu et al., 1981b). The difference can be explained in part by the fact that the large vessels were used in vitro (cut out ringlike segments of aorta and vena cava) whereas the observations on capillary endothelium were obtained in vivo or in situ. Cationic tracers injected in vivo or added to the medium of cultured cells may form aggregates with anionic serum proteins, in which case changes in size and charge of the probe take place. Care should be exercised in the adequate choice of experimental conditions when charged probes are used. Cationic tracers that initially bind to the cell membrane and are internalized sometimes detach from the container membrane when reaching a certain intracellular compartment. In the lung, CF binds to the plasmalemma of type II epithelial cells and is internalized in vesicles. When reaching the multivesicular bodies or multilamellar bodies, CF detaches from the membrane, possible due to an acidic pH at this level or the digestion of the apoprotein shell of ferritin by the lysosomal enzymes. The same behavior was previously observed in thyroid follicle cells (Herzog and Miller, 1979) and secretory granules of the anterior pituitary (Farquhar, 1978).

C. Adsorptive Specific Probes

These are biological molecules that interact selectively with specific constituents of the cell surface. These strong, chemically specific interactions are particularly useful for studying major physiological cell processes such as receptor-mediated endocytosis, receptor-mediated transcytosis, and the traffic and retrieval of membrane domains involved in these events. When using such ligands, the first requirement is to establish by multiple tests the specificity of the binding, such as, selective competition, saturability. Among the plasma proteins for which specific binding sites have been detected in various endothelial cells are lectins, peptide hormones, low-density lipoproteins (LDL), albumin, and transferrin. These biological molecules are not electron dense by themselves and to be visualized they must be conjugated or adsorbed to an electron dense probe (for standard electron microscopy) or radiolabeled (for autoradiography). The most frequently used tags are ferritin and HRP for covalent conjugation, and gold particles of different dimensions for noncovalent adsorption (Horisberger and Rosset, 1977). Radiolabeling of proteins (considered the mildest labeling procedure) followed by autoradiography of tissues exposed to the ligand allows the detection of the routes and kinetics of native proteins.

Lectins

These are proteins or glycoproteins of plant or animal origin that bind restrictively and specifically to one or, rarely, two sugar residues of oligosaccharides and oligopeptides. They have been routinely defined by their ability to agglutinate erythrocytes. Lectin specificity was established according to the monosaccharide that is the most effective inhibitor of hemagglutination or by precipitation of carbohydrate-containing polymers (for review, see Lis and Sharon, 1986). Conjugated to ferritin, HRP, or gold, lectins have been used extensively for the localization of various sugars on cell surfaces and can be useful for tracking the internalization of lectin-binding surface domains.

Low-Density Lipoproteins

Low-density lipoproteins conjugated to electron-dense particles (ferritin or gold) (Fig. 2) have been used to identify the structural features involved in receptor-mediated uptake of LDL (Anderson et al., 1976; Handley et al., 1981; Vasile et al., 1983; Nistor and Simionescu, 1986).

Albumin

Albumin adsorbed to gold particles has allowed the identification of a novel mechanism for the transport of albumin through capillary endothelium, namely receptor-mediated transcytosis (Simionescu and Simionescu, 1984; Ghitescu et al., 1986a) (see pg. 395).

Pitfalls in the Use of Adsorptive Specific Markers

Since these biological molecules are visualized indirectly by the marker to which they are conjugated, one must measure the overall size of the complex, to check for possible false or diverted routing of the whole complex and to test the nonspecific adsorption of the electron-dense tracer used. When using conjugates, in general, one should be aware that the normal traffic of a protein may be altered to a certain degree when it is carrying or being carried by another molecule. In addition, the conjugation procedure may diminish the biological activity of the protein of interest, in which case changes in binding specificity may occur.

The major limitation in autoradiography studies is the low resolution that can be achieved (90–150 nm) in electron microscopy.

D. General Methodological Requirements

To obtain relevant information and to be able to compare the results obtained in various experimental conditions, a thorough characterization of each tracer used is an absolute requirement. Numerous contradictory results could be avoided if the probe, its interaction with a specific cell, and the conditions used were carefully tested for each experimental model.

Physicochemical characterization of the tracers has to be done both for the commercially available markers and for those prepared in the laboratory. In particular for the fluid phase probes (applied to permeability studies), the molecular weight, size, and charge have to be checked before the experiment.

The pI of the tracers and their derivatives must be precisely determined by isoelectric focusing (Fig. 3) since minor variations may give different results.

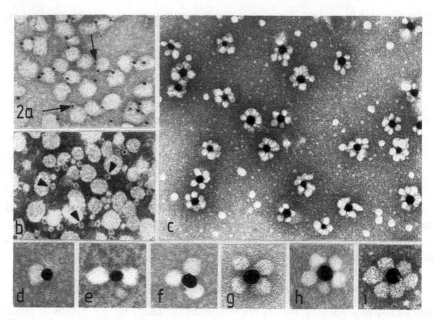

Figure 2 Low-density lipoproteins (LDL) coupled to gold (Au) and ferritin: negative staining with 1% phosphotungstic acid. When conjugated to (a) 5 nm Au (arrows) or (b) ferritin (arrowheads), one or two particles bind to each LDL particle: (c) if adsorbed on 20 nm Au, usually eight LDL particles attach to each Au particle, but other Au/LDL ratios (d–i) may also be obtained (a, b, × 310,000; c, × 95,000; d–i, × 230,000; c–i provided by Dr. R. Mora).

This becomes evident when using the same probe (i.e., ferritin or H11P) prepared at different pIs. For example, CF pI 8.4 labels a small fraction (~10%) of endothelial plasmalemmal vesicles in muscle capillaries, but this proportion increases (to ~25%) as the pI of the tracer decreases to 6.8 (Simionescu et al., 1985). These results indicate the preferential uptake of anionic proteins by the plasmalemmal vesicles. Cationized H11P$_t$ pI 9.0 does not label plasmalemmal vesicles and transendothelial channels, whereas the same structures are labeled by cationized H11P$_p$ of pI 10.6 (Ghinea and Simionescu, 1985). The data suggested that in the endothelia examined, the membrane of plasmalemmal vesicles and channels contains weak acids of high pK$_a$ values, and that in these structures electrostatic restriction operates up to pI 9.0. In addition, the observations convey the idea that at least in some capillary endothelia there is a charge limit in the electrostatic restriction that can be determined only if the precise pI of a tracer is known.

The purity and the state of dispersion of the probe molecules in the tracer solution, in the blood plasma (for in vivo studies), or in the culture medium (for

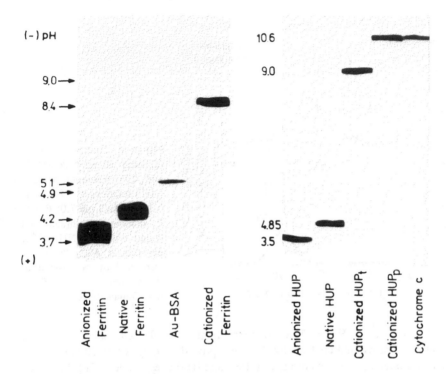

Figure 3 Isoelectric focusing of various tracers of different pIs. Gel conditions: 1% agarose IEF, 1 mm thick gel, Pharmalyte pH 3–10, 1:16 dilution. Focusing conditions: 1500 V, 10 mA, 10°C. Interelectrode distance: 10 cm.

in vitro studies) must be evaluated if the results are to be expressed in terms of effective molecular dimensions as is the case for capillary permeability. For example, gel filtration on Biogel P6 followed by spectrophotometry of the eluate was used to determine the purity, size, and dispersion of H11P and H8P in (1) the tracer solution (0.5% wt/vol in 0.154 M NaC1, pH 7.0), (2) the tracer solution treated with aldehydes for 2 hr at 24°C, (3) the plasma or serum collected 10 min after tracer injection, and (4) plasma samples (as in item 3) treated for 2 hr at 24°C with the aldehyde mixture (Simionescu et al. 1975). The results of these experiments showed that in the tracer solution H8P elutes as a single symmetrical peak at the expected position of the monomer whereas H11P elutes as a slightly asymmetrical peak. This indicated that H11P (obtained in the laboratory) contained ~5% aggregates (Fig. 4). In the plasma (Fig. 5), both hemepeptides were found to occur essentially (H8P) or mostly (H11P) in monomeric form and to bind to a small extent to the plasma proteins: 12% for

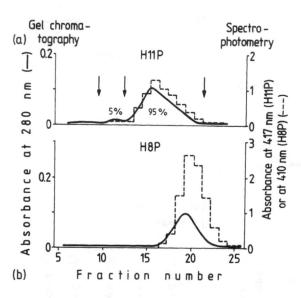

Figure 4 Elution profiles of (a) hemeundecapeptide (H11P) and (b) hemeoctapeptide (H8P) solutions (0.5% w/v) in 0.15 M NaCl pH 7.0; Biogel P_6 column equilibrated and eluted with 0.1 NH_4HCO_3, pH 8.0 at 4°C, 20 ml/hr (reprinted by permission of Rockefeller University Press from Simionescu et al., 1975).

H8P and 16% for H11P. Two hour treatment of similar plasma samples with aldehyde fixative and analysis by the same procedure (Fig. 6) results in ~47% and 94% binding to plasma proteins of H8P and H11P, respectively. The difference between the two hemepeptides can be explained by the existence of an additional $-NH_2$ group (lysine) in H11P. Taken together the findings indicate that both hemepeptides remain predominantly in monomeric form in tracer solution and circulating plasma, H8P being more uniformly dispersed and H11P more efficiently retained by fixation. The general state of dispersion of the molecules in the tracer solution and in plasma samples can also be assessed by negative staining.

Particularly for permeability studies performed in vivo, it is of crucial importance to choose the right animal species and strains that are known to be genetically resistant to histamine-releasing agents, such as Wistar-Furth rats (Cotran et al., 1968) or C3HeB mice. In many other species HRP and myoglobin induce junctional openings at the level of the endothelium of postcapillary venules. Under appropriate control, other laboratory animals can be found suitable for a given experiment.

To check for possible tracer-induced leakage at the level of endothelial junctions, "the cremaster test" can be applied (Majno and Palade, 1961). The test consists of the injection of tracer (negative control), or histamine, in isotonic saline solutions (positive control) subcutaneously on the anterior face of the scrotum after small amounts of colloidal carbon (0.1 ml/100 g body weight) have been injected in the circulation. After 1 hr, the cremaster is excised, stretch-fixed in

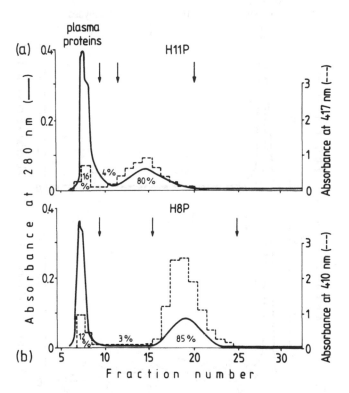

Figure 5 Gel chromatography (Biogel P_6) of 0.25 ml plasma collected 10 min after intravenous injection of 1 ml/100 g body weight of 0.5% hemeundecapeptide (H11P) or hemeoctapeptide (H8P). a. the protein absorbing at 417 nm is monomeric H11P in fractions 9–11; fractions 6–8 contain some H11P and possibly hemoglobin bound to plasma proteins. b. Absorbance at 410 nm is due to monomeric H8P (fractions 15–25) and hemoglobin in fractions 6–9 (reprinted by permission of Rockefeller University Press from Simionescu et al., 1975).

10% formaldehyde, and examined by transillumination under a dissecting microscope. It is scored for the presence or absence of vascular tattoo (carbon leakage) at the site of injection (Simionescu and Palade, 1971).

The concentration and the volume of the injected tracer should be controlled to produce minimal perturbations of the blood volume, hydrostatic and colloid–osmotic pressures. In the mouse lung, HRP injected intravascularly in vivo does not permeate capillary endothelial junctions, except when large volumes are used (Schneeberger and Karnovsky, 1971). Such change in hydrostatic pressure

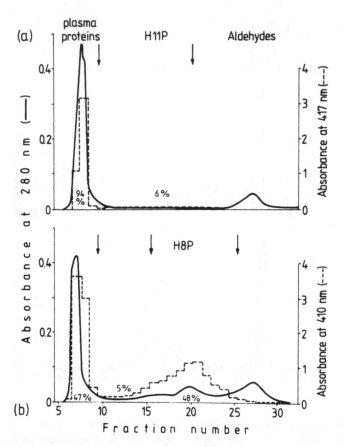

Figure 6 Gel chromatography of 0.25 ml plasma (obtained as described in Fig. 5) treated with aldehydes 2 hr at 24 °C. Figures give percentages of absorbance at 417 nm for H11P and 410 nm for H8P for the fractions indicated by vertical arrows. After fixation 6% H11P does not bind to plasma proteins (a) whereas 53% of H8P remains unbound in the same conditions (b) (reprinted by permission of Rockefeller University Press from Simionescu et al., 1975).

alters the delicate balance of the physical forces at the level of microvasculature, one result of which could be a nonphysiological opening of endothelial junctions. Checking the osmolality of the plasma before and after tracer injection may give an indication whether changes have occurred.

Control of the effect of the fixation procedure used was exemplified for hemepeptides. Leakage of hemoglobin from circulating red blood cells during fixation with aldehydes was reported (Thiessen et al., 1970). This effect is usually limited to the immediate vicinity of red blood cells, and the associated reaction product is generally weak. Yet in the case of peroxidatic tracers, the suitable capillary profiles to be examined are those devoid of circulating cells or in which plasma was well preserved in the lumen.

The artefactual diffusion of the peroxidatic tracers or their reaction product into the tissue blocks (Novikoff et al., 1970; Fahimi, 1973) can be checked by controls in which the enzyme is added to the fixative or to the incubation medium. The time for best fixation should be adjusted according to the probe, the experimental model (shorter time for cultured cells, longer for fixation by perfusion, and even longer when the fixative is topically applied on tissues), and the concentration of aldehydes used. The optimal time and temperature of fixation should be evaluated in preliminary experiments.

Conditions for visualization of hemoproteins via peroxidatic reaction should be adjusted so that for each tracer optimal pH (Tables 2–4), temperature, and time of incubation should be compatible with good preservation of tissue ultrastructure. To increase the peroxidatic activity of some hemoproteins, addition of 0.1 M imidazole to the incubation medium was found useful (Harbury and Loach, 1960). However, this effect must be checked for each tracer since it has been reported that, in vitro, addition of imidazole has a great effect on the enzymatic activity of HRP, H8P, and H11P (~80% increase), whereas no increase was reported for cytochrome c (Tu et al., 1968).

Low molecular weight galloylglucoses (tannic acid) (Simionescu and Simionescu, 1976) used in DAB-cytochemistry has several advantages, such as a better preservation of the general ultrastructure, increased contrast of all membranes (plasma membrane, vesicles, etc.), and an enhancement of the reaction product without a concomitant increase in the background. Stepping up the contrast tallows the use of tissue fragments without treatment with heavy metals and of sections that are not stained prior to their examination with the electron microscope. Tannic acid (used after osmication of tissues) acts as a mordant, preventing to a certain degree the extraction of the osmium black during dehydration.

IV. Experimental Procedures for Studies on Lung Cells

Transport studies in lung, can be performed in vivo, in situ, or on cells in culture. Each experimental model has advantages and limitations, and whenever possible a combination of these three approaches may substantially validate the results.

A. In Vivo

The major advantage of these studies is that the probe is injected into the living animal, in conditions close to normal. Fixation of tissues is supposed "to freeze" the tracer in its location at the moment when the fixative was applied. Yet the time required for complete fixation and the possible artefacts induced by the chemical fixation of tissues must be carefully considered.

To study the permeability characteristics of the lung capillary endothelium, a fluid-phase marker under appropriate conditions of concentration, temperature, pH, volume, and other factors, is injected into a directly accessible vein (e.g., the saphenous vein), allowed to circulate, and at different time intervals the tissue is fixed in situ, then collected and processed for electron microscopy. The in vivo intravascular route allows neither the use of tracers that interact with circulating plasma proteins (e.g., cationic tracers which may bind to anionic plasma proteins) nor the study of binding of plasma proteins to specific sites (or receptors) present on the endothelial luminal surface because these may be occupied by the same endogenous plasma molecules.

To study the transport of molecules from the alveolar space to the interstitia, the tracer can be administered via trachea into the air space, without or with gentle lavage of the surfactant and the alveolar macrophages.

B. In Situ

This is the experimental model of choice when charged tracers or other nonspecific or specific adsorptive membrane markers are to be used. In the lung, these probe molecules can be injected intravascularly or intratracheally after extensive washing out of the blood or the surfactant, respectively. For vascular perfusion, a gel electrophoretic analysis of the last eluate is a desirable control to evaluate the completeness of vessel clearance of plasma proteins. When accurately performed in situ, such procedures expose the cell surfaces, offering unmediated access of the tracers to plasma membrane components and its specific features (vesicles open to the lumen, coated pits, uncoated pits).

Intravascular Administration of Tracers

After general anesthesia and laparotomy, a catheter (or fine needle) connected to polyethylene tubing is introduced into the abdominal aorta to perfuse a physiological solution such as oxygenated phosphate-buffered saline (PBS) supplemented with 1 mM $CaCl_2$ and 14 mM glucose, at 37 °C at a flow rate of ~3 ml/min. For adequate pressure, it is preferable to monitor the flow rate with a manometer. The vena cava is punctured to be used as outlet for the perfusate. At the end of this general perfusion (~20 ml in mice), while the heart is still beating, thoracotomy and exposure of the heart and lungs allow the introduction into the pulmonary artery of a second catheter that is fixed in place with a ligature. The left atrium is punctured and serves as outlet for perfusion. In this system, the probe can be perfused or recirculated through the lung. At different time intervals the unbound tracer is removed by perfusing PBS (~5 ml for mouse lung) followed by the fixative for electron microscopy, which is left in situ for a minimum of 10 min (the time depends largely on the strength of the fixative used). In our experience, excellent results can be obtained with the one-step fixation procedure (see Addendum). At the completion of in situ fixation, the tissue is cut into small ($\sim$ 1mm^3) fragments and the fixation continued in vials for 60–90 min (according to the fixative used) followed by the standard procedure for electron microscopy.

Intratracheal Administration of Tracers

After anesthesia is induced, a polyethylene tubing connected to a syringe is introduced through a small incision placed between two tracheal cartilage rings. If required by the experiment, the lung airways can be gently washed four to six times with $\sim$0.5 ml oxygenated PBS to remove the surfactant and free cells present in the air spaces. The tracer of interest introduced into the alveolar space is allowed to interact with the cells, after which the excess is removed by several gentle washings with PBS. Fixation of the lungs is performed by instillation into the airways of a small amount of fixative ($\sim$1 ml for mouse lung), which can also be applied topically. When using the one-step mixture-fixative, within 10 min the well-fixed areas are completely black (due to the presence of OsO$_4$), which allows the collection of the best fixed lung fragments for further processing.

C. In Vitro

When available, cell culture (pure population) may represent a good alternative model for collecting information on the behavior of a given cell type. For transport studies, such a model is especially useful to investigate the endocytic pathway; however, the results should be thoroughly corroborated with in vivo or in situ studies, since culturing of cells may affect their functions.

Cultured endothelial cells from lung large vessels have been obtained primarily from the pulmonary artery (Ryan et al., 1976). Fragments of bovine, pig, and rabbit pulmonary artery were rinsed in sterile PBS pH 7.4 supplemented with calcium and magnesium, penicillin, streptomycin, and fungizone. The vessel was cannulated and perfused with PBS containing 0.2% collagenase, clamped shut at both ends, and maintained at 37 °C for 15 min in a water bath. The collected effluent was centrifuged at 2500 rpm at 4 °C for 10 min, the pellet washed with medium 199 supplemented with 20% fetal calf serum and plated in plastic tissue culture flasks in the same medium to which 2 mM L-glutamine was added. Endothelial cells from the same source have been obtained by perfusing collagenase directly through the pulmonary circulation after washing out the blood in a heart–lung preparation. The cells collected are primarily of large vessel (arterial and veins) origin. Cultured endothelial cells have been used successfully for localization of various enzymes associated with the endothelial cell surface (Ryan et al., 1976).

Type I epithelial cells have recently been isolated from rat lung using an enzymatic digestion with 0.2% collagenase, 0.05% typsin, 0.008% elastase, and 0.005% DNAse followed by density gradient centrifugation (Weller and Karnovsky, 1986). Reportedly, the isolated cells retained their in vivo morphologic characteristics.

Type II epithelial cells have been isolated using trypsin (Kikkawa and Yoneda, 1974) or elastase (Dobbs et al. 1980) as dissociating agents. Isolated cells were used especially to study their synthetic properties and metabolic activities.

Primary cultures and subsequent passages should be thoroughly characterized for the presence of specific functions as known from in situ or in vivo studies. Thus, for endothelial cells in culture, the presence of von Willebrand factor, angiotensin-converting enzyme activity, and synthesis of prostacyclin, are some of the specific characteristics. In contrast to previous belief, endothelial cells isolated from various species and different anatomical sites exhibit different morphology in culture. The expression of differential phenotype in culture may be due to the influence of the matrix on which the cells are plated or to their site of origin (Williams, 1987).

In the experiments performed in culture, the general procedure involves the removal of culture medium, gentle but thorough washing of the cells with a physiological solution to eliminate any adsorbed proteins with which the probe may nonspecifically interact, followed by exposure of the cells to the tracer of interest. At the end of the incubation period, the excess tracer is discarded, the cells are washed with a physiological solution, and fixed in the culture dish. These experiments can be performed at 37 °C, when the requirements are for physiological conditions, or can be carried out at 4 °C, when surface phenomena (without internalization) are to be observed.

V. Facts and Findings Uncovered Using Tracers in Lung Research

A. Fluid Phase Markers Revealed Pathways for Transport of Macromolecules Across Capillary Endothelium

The use of enzymatic tracers of different molecular weight and size in the study of endothelial permeability has brought interesting information on the routes and structures involved in the transport of macromolecules in lung capillaries. It was reported that 1 min after intravascular administration, at normal hydrostatic pressure, small molecules such as cytochrome c (~ 3 nm diameter) appear in the capillary lumen, plasmalemmal vesicles, and some (but not all) endothelial junctions (Fishman and Pietra, 1976). A similar distribution has been found for HRP (~ 6 nm diameter) except that it was absent at the level of interendothelial junctions. However, injection of HRP in large volumes of saline solution opens the junctions, possibly due to an increase in the intravascular volume and a certain lability of the intercellular junctions (Schneeberger and Karnovsky, 1968, 1971). To some extent, this may also represent side effect (i.e., increased vascular permeability) induced by HRP in certain species (Karnovsky and Cotran, 1966). Tracers bigger than HRP do not permeate endothelial junctions, even when injected in large volumes. The exact location of the labile endothelial junctions in lung microvasculature has yet to be determined, since precise localization, especially of small postcapillary venules, is difficult.

These observations suggest that, at normal intravascular pressure, plasmalemmal vesicles are effective in transendothelial transport of water-soluble macromolecules across the endothelium (Schneeberger, 1978). Under normal conditions, interendothelial junctions do not represent a major pathway for molecules of the size so far tested. However, at increased pressure some junctions may open at least up to the dimension of HRP molecule. Endothelial junctions seem to be more labile than the corresponding junctions between the type I epithelial cells (Schneeberger, 1976). Increased permeability of alveolar epithelial cells to cytochrome c was noticed after rabbits were exposed to 100% oxygen; such phenomena were interpreted as a manifestation preceding pulmonary edema (Nickerson et al., 1981).

B. Radiolabeled 5-Hydroxytryptamine is Removed from Circulation by Pulmonary Endothelial Cells

Removal of the biogenic amine 5-hydroxytryptamine (5-HT) by the pulmonary circulation is extensive ($\sim 95\%$). Radioautographic experiments performed on isolated rat lung perfused with [^{3}H]5-HT in the presence and absence of monoamine oxidase inhibitor have shown convincingly that 90% of silver grains were present over capillary endothelium (Strum and Junod, 1972). The transport of 5-HT by endothelial cells is Na dependent and carrier mediated.

C. Charged Probes Identified the Existence of Microdomains of Different Charge on the Cell Surface of the Alveolar-Capillary Unit

It is generally accepted that all cells expose on their surface anionic and cationic residues; however, overall they generate a net negative charge. The use of charged probes, (i.e., cationic ferritin, cationized and anionized hemeundecapeptides) at low concentration revealed that there are domains of different charge or charge densities on the surface of endothelial cells, generated by the preferential distribution of some glycoconjugates (Simionescu et al., 1981 a,b; Ghinea and Simionescu, 1985). These studies have been extended to the cells that constitute the air–blood barrier (Pietra et al., 1983; Simionescu and Simionescu, 1983, 1987).

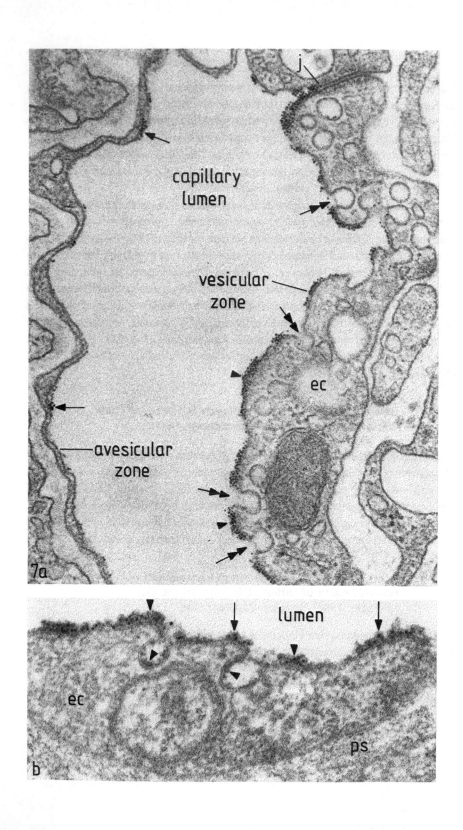

Alveolar Capillary Endothelium

After blood vessels were cleared of blood, CF was perfused in situ into the pulmonary artery, and at various time intervals lung fragments were collected and processed for electron microscopy (as described above for in situ experiments). The inhomogeneous distribution of the probe over the endothelial surface defined macro- and microdomains of different charge and charge density. In the vesicular zone of the endothelium CF marked almost continuously the plasma membrane. whereas the membrane of most plasmalemmal vesicles, channels (when present), and their diaphragms were devoid of CF-binding sites. The latter were present at low density exposing large ligand-free areas on the attenuated avesicular zone of the endothelium (Fig. 7a). In double-labeling experiments in which CF pI 8.4 and HUP pI 4.85 were successively perfused through the lung vasculature, the cationic marker maintains its characteristic distribution, whereas the anionic probe labeled the plasmalemma as well as the vesicle membrane (Fig. 7b). The existence along the endothelial luminal surface of macrodomains (the avesicular zone) and microdomains (the vesicles) of different charge or charge density may express a functional differentiation of the two zones. It is possible, yet not proved, that at the level of the avesicular zone (poor in anionic sites) the circulating red blood cells (negatively charged) face a diminished repellence by the capillary endothelium, which may facilitate the gas exchange. The plasmalemmal vesicles devoid of acidic sites may represent a preferential pathway for the uptake and transport of anionic molecules that is the case with most plasma proteins. It appears that vesicles are able to discriminate the molecules not only by size but also by their charge. In comparable experiments, Pietra et al. (1983) found similar results, except that a larger number of vesicles and their diaphragms were labeled

---

Figure 7 Visualization of anionic and cationic sites on the surface of alveolar capillary endothelium. a. Cationized ferritin (CF) pI 8.4 binds almost continuously to the plasmalemma of the vesicular zone (arrowheads) but fails to label plasmalemmal vesicles and their diaphragms (double arrows). On the avesicular zone of the endothelial cell (ec), CF binding is scarce (arrows). CF binds to the infundibula leading to the junction (j) but does not go beyond it (reprinted by permission of Academic Press from Simionescu and Simionescu, 1983). b. Double labeling with CF and anionic hemeundecapeptide (H11P) pI 4.85 perfused successively through the vasculature. CF has the same labeling pattern as in a (arrows), whereas H11P decorates homogeneously the plasma membrane and the vesicle membrane (arrowheads). ps, Pericapillary space (a, × 61,300; b, × 114,000).

by CF. The difference may be due to the use of CF with a pI between 8 and 10 as opposed to CF of pI 8.4 used in our experiments (Simionescu and Simionescu, 1983; Simionescu, 1985b). These data also stress the importance of the pIs of the molecules used.

Alveolar Epithelial Lining

Analysis of the charge distribution over the alveolar epithelium was carried out according to the experimental procedure described above for in situ experiments, in which CF was injected intratracheally. Marked differences were observed in the CF decoration of the epithelium: while type I epithelial cells were very little labeled, the neighboring type II cells were heavily marked by two or three tightly packed rows of CF attached to the luminal contour up to the level of the junctions with type I cells (Fig. 8). At 3 min, CF labeled the introits and membrane of the multilamellar bodies open to the air space as well as the lamellae exocytosed in the air space. This pattern may be due to the polycationic nature of the probe, which may form bridges between adjacent anionic surfaces.

D. Adsorptive Nonspecific Markers Indicated the Retrieval of Apical Plasma Membrane in Type II Epithelial Cells

Type II epithelial cells are bona fide secretory cells that produce and actively secrete surfactant components by exocytosis into the air space. Lamellar body membrane fuses with the apical cell plasmalemma and upon fission the lamellae are extruded extracellularly. To maintain the balance in the distribution of membranes among the the participants in the exocytotic process, an almost comparable amount of membrane must be internalized from the cell surface.

Cationized ferritin, which binds heavily to the surface of type II cells, was used as probe to follow the route and fate of various membrane domains. After intratracheal injection (as described above) CF decorates the plasma membrane of type II cells, including the microvilli and numerous open vesicles or coated pits at their base. At 20 min, the probe is found in internalized coated and uncoated vesicles (Fig. 9a). After a longer exposure (30–60 min) the multivesicular bodies and forming lamellar bodies are filled with CF (Fig. 9b). Occasionally, the probe is found in some of the Golgi cisternae (Fig. 9c). Tracking the CF within various compartments of the cell suggests that recycling of luminal membrane domains may follow two alternative routes: (1) from the luminal plasma membrane to multilamellar bodies via vesicles and multivesicular bodies; and (2) from the luminal plasma membrane to Golgi complex, probably via another fraction of vesicles. The second route is rarely encountered in in situ experiments; the fractional proportion of the two pathways followed in vivo is yet to be determined. Retrieval of luminal surface domains via vesicles may represent a mechanism

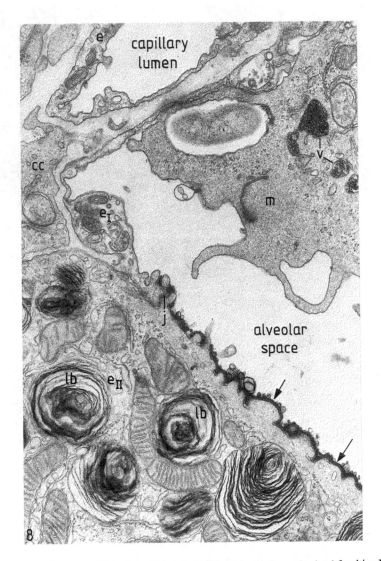

Figure 8 Differentiated labeling of alveolar epithelial cells by cationized ferritin pI 8.4, 3 min after intratracheal administration. Note the extensive labeling (arrows) of type II epithelial cells (e$_{II}$), up to the level of the junction (j) with type I cells (e$_I$). The latter is completely devoid of CF-binding sites. An alveolar macrophage (m) displays CF-loaded vacuoles (v). e, Endothelium; cc, contractile cell; lb, lamellar body ($\times$ 29,000; reprinted by permission of Academic Press from Simionescu and Simionescu, 1983).

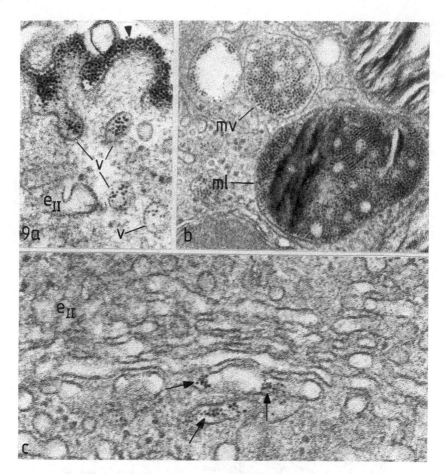

Figure 9 Internalization of cationized ferritin (CF) pI 8.4 from the alveolar space into the type II epithelial cell (e_{II}). a. At 20 min after intratracheal administration, CF labels the apical plasmalemma (arrowhead) and vesicles (v) in various stages of internalization. b. At 60 min, multivesicular bodies (mv) and forming multilamellar bodies (ml) appear studded with CF. c. At 80 min, CF appears occasionally in Golgi cisternae (arrows) (a, c, × 83,000; b, × 55,000).

by which the cells economize on a continuous production of membrane components. The same route may be used for the recovery of surfactant components from the alveolar space to be reutilized by the cell. This was also suggested by experiments using ^{32}P-labeled surfactant (Hallman et al., 1981), CF or lectins (Williams, 1984a,b). Accumulated data based on experiments with various tracers indicate that numerous secretory cells reutilize domains of their apicalplasmalemma (Farquhar, 1983).

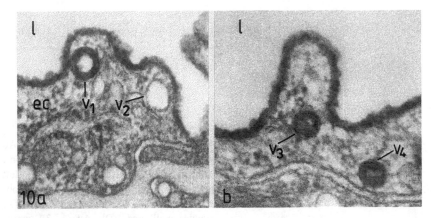

Figure 10 Binding of peroxidase-conjugated wheat germ agglutinin to endothelial cell (ec) surface after intravenous perfusion. a. The labeled vesicle (v_1) is probably in continuity with the lumen (1). The unlabeled vesicle (v_2) was presumably internalized at the time of perfusion. b. Labeled vesicles (v_3 and v_4) appear to be in various stages of internalization, one of them (v_4) reaching the abluminal membrane ($\times$ 90,000).

E. Lectins Detected the Oligosaccharides Exposed on the Cell Surface and Associated Vesicles

Lectins of well-defined specificity were used to detect glycoconjugates with various terminal monosaccharides residues present on endothelial and epithelial cells. The lectins conjugated with HRP were perfused intravascularly or intratracheally, and their presence was visualized indirectly via a peroxidatic reaction. Moieties such as N-acetylglucosaminyl and sialyl residues (detected with wheat germ agglutin), α-L-fucosyl (identified with lotus tetragonolobus lectin), and β-D-galactosyl (revealed by peanut agglutinin) were found on both endothelial and epithelial surfaces as well as in open vesicles (Simionescu, 1985a). Concanavalin A does not bind to alveolar epithelial cells (Dixon and Jersild, 1982). Since lectins are specific markers for membrane components, they can also be used in time-course experiments, to follow the traffic of membrane domains or its special features such as endothelial plasmalemmal vesicles. Vesicles in continuity with the lumina of the vessel are marked by peroxidase-conjugated lectins whereas the cytoplasmic vesicles are not labeled (Fig. 10a). In Figure 10b, a labeled vesicle can be observed close to the abluminal front, which suggests that after the lectin labeled the vesicle and prior to its fixation, that particular vesicle moved toward the abluminal front of the cell. Type II cells plasma membrane binds specifically ferritin-labeled lectin from *Maclura pomifera* (an α-galactose-binding molecule). Upon binding, the lectin-labeled vesicles are internalized and, in time, the ligand decorates multivesicular bodies

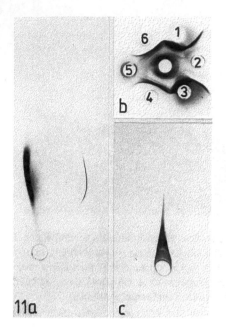

Figure 11 Characterization of LDL–Au conjugate. a. Immunoelectrophoresis against antihuman LDL of LDL–Au conjugate (left well) and native LDL (right well). b. Double immunodiffusion with antihuman LDL (center well) and LDL–Au (1, 3), native LDL (4, 6), and antihuman LDL-IgG (2, 5). c. Laurell immunoelectrophoresis using antihuman LDL in the agarose bed. All tests indicate immunological similarity between LDL-Au and native LDL (reprinted by permission from Nistor and Simionescu, 1986).

and multilamellar bodies, which suggests that domains of luminal plasmalemma are routed intracellularly (Williams, 1984b).

F. Low-Density Lipoprotein–Gold Conjugate Revealed Pathways of Lipoprotein Uptake in Capillary Endothelium

The main sources of cholesterol for surfactant biosynthesis are the plasma lipoproteins (Hass and Longmore, 1980), such as low-density lipoproteins (LDL). To find out how circulating LDL is transported through endothelium, the lipoprotein was adsorbed to 5 nm colloidal gold (LDL–Au). The conjugate retains the same immunoreactivity as native LDL as shown by immunoelectrophoresis, Ouchterlony double immunodiffusion, and Laurrell immunoelectrophoresis against antihuman LDL (Fig. 11). By negative staining, one or two colloidal gold particles appeared attached to an LDL particle (Fig. 2a). The LDL–Au conjugate was recirculated in situ through the pulmonary vasculature, and its presence on and within the endothelium was recorded at 7, 30 and 60 min. (Nistor and Simionescu, 1986). At 7 min, LDL–Au labeled a large number of coated pits and plasmalemmal vesicles open to the luminal front or apparently internalized (Fig. 12a,b). After 30 and 60 min of recirculation, in addition to the features mentioned, the conjugate appeared associated with coated vesicles, plasmalemmal vesicles opened to the abluminal front, and in structures resembling endosomes and lysosomes (Fig. 12c,d). LDL–Au could occasionally be found in the pericapillary space. Similar results were obtained in experiments in which native untagged LDL was perfused and visualized by tissue treatment with 1% tannic acid previous to osmication (Nistor and Simionescu, 1986). The results showed that the features involved in uptake and transport of LDL in alveolar endothelium are those known to be effective in receptor-mediated endocytosis (i.e., coated pits and vesicles, endosomes, and lysosomes) and fluid phase transcytosis (i.e., plasmalemmal vesicles). These data, together with those from experiments in which radiolabeled LDL was used, showed that LDL is partially taken up from the circulation by a receptor-mediated process that is saturable and time and concentration dependent. When heparin was added to the washing buffer to remove receptor-bound LDL (Goldstein et al., 1976), it was found that 50% of LDL uptake is receptor independent. The fractional contribution of each of these processes for uptake and transport of LDL by the lung in vivo remains to be established.

G. Albumin–Gold Complex Binds to Specific Binding Sites on Endothelial Vesicle Membrane and is Transported by Receptor-Mediated Transcytosis

Albumin is the main contributor to the maintenance of the colloid–osmotic pressure of plasma and interstitial fluid and a carrier for fatty acids, hormones, some drugs, etc. within the circulation. The amount of albumin passing from the plasma to the interstitial fluid varies in different organs, and no correlation could be established between the rate of albumin transport and the type of endothelium present. Thus, other functional characteristics of the endothelial cells may account for the differentiated transport of this protein in various locations. Using an albumin–gold conjugate (Alb–Au) perfused in situ in the lung vasculature, intense binding of the complex was found restricted to the membrane of plasmalemmal vesicles and some uncoated pits (Simionescu, 1985a) (Fig. 13a). Morphometric analysis of Alb–Au binding at different tracer concentrations revealed the low affinity of plasma membrane for the conjugate (approximately six times lower than plasmalemmal vesicle membrane). The degree of vesicle labeling increased with ligand concentration and quickly attained saturability (Ghitescu et al., 1986). Starting at 3 min, some of the ligand-labeled vesicles were internalized and with increasing time vesicles were detected discharging their content on the abluminal front (Fig. 13b,c). The Alb–Au labeling pattern was abolished when native albumin was perfused prior to conjugate, or when Alb–Au was injected in vivo, in the presence of endogenous plasma albumin. The results indicate that endogenous and untagged albumin compete with similar affinity for the limited number of binding sites available on the surface of endothelial cells. Heparin or high ionic strength buffers failed to prevent adsorption of Alb–Au

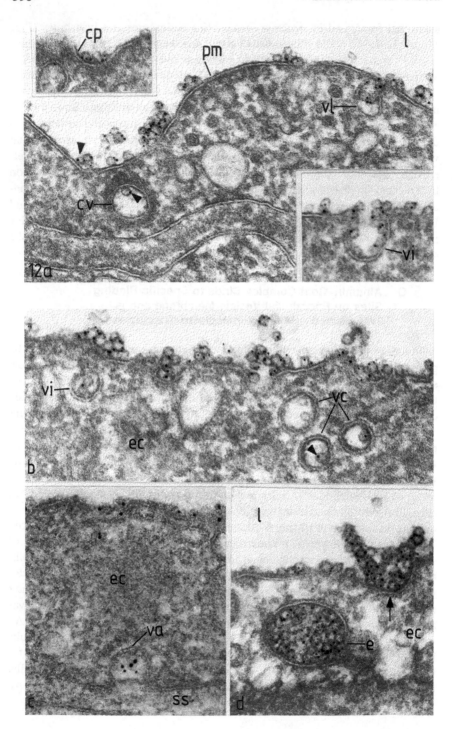

to vesicle membrane, indicating that binding is not electrostatic in nature. Colloidal gold stabilized with other plasma proteins (fibrinogen, fibronectin), with glucose oxidase, or polyethylene glycol did not yield a decoration comparable to that of Alb–Au (Fig. 13d). The labeling pattern of Alb–Au to endothelial surface and the control experiments indicate that albumin has specific binding sites or receptors restricted to vesicle membrane. Upon binding to its receptor, transport of albumin across lung capillaries takes place via endothelial vesicles by receptor-mediated transcytosis. Modulation in the density and affinity of endothelial albumin binding sites may account, at least in part, for the differences in the site-specific transport of albumin. In the lung, the chemically specific uptake of albumin may represent a mechanism for transport of albumin per se, as well as for the transport of fatty acids that are essential for the lung metabolic activities. Experiments using oleic acid–albumin complex showed that endothelial receptor for albumin has a higher affinity for albumin carrying fatty acids, which is transcytosed by the same receptor-mediated mechanism but at a higher rate than defatted albumin (Galis et al., 1987). Albumin-specific binding sites have been found in the capillary endothelium of heart, lung, skeletal muscle, and adipose tissue.

Recently, membrane-associated albumin binding proteins represented by two pairs of polypeptides of M_r 31 and 18 kD were identified in endothelial cells (Ghinea et al., 1988, 1989).

The use of actual plasma proteins as probes in studies of endothelial permeability has led to a new concept of receptor-mediated transcytosis (Simionescu and Simionescu, 1984) and a better understanding of the diversity and magnitude of the active role of the endothelial cells in the transport across the vessel walls.

Figure 12 Binding and transport of LDL–Au conjugate by the alveolar capillary endothelium. a, b. After 7 min of recirculation, LDL–Au (arrowheads) binds to plasma membrane (pm) and labels vesicles open to the lumen (vl), apparently located in the cytoplasm (vc), coated pits (cp, upper inset), and coated vesicles (cv). c. After 30 min of recirculation, LDl–Au appears in vesicles open to the abluminal front (va). d. At 60 min of recirculation, LDL–Au still heavily labels some pits of the plasmalemma (arrow) and endosome-like structures (e) of endothelial cell (ec). ss, Subendothelial space (a, b, × 105,000; c, × 124,000; d, × 131,000).

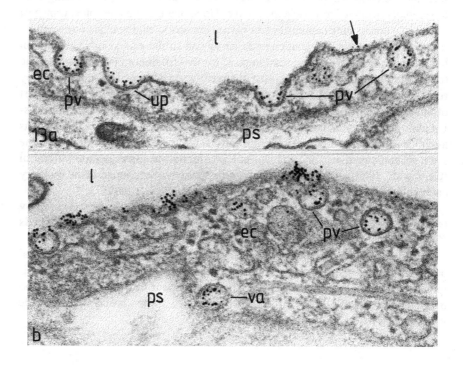

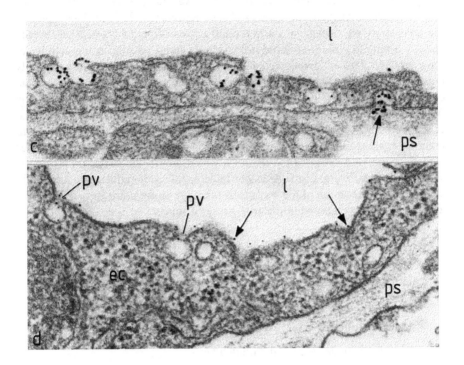

VI. Concluding Remarks

It is an understatement to say that the use of tracers has helped our understanding of cell functions. With the use of tracers we were able to enter the most intimate compartments of a cell and to uncover (with various degree of certainty) the route(s) available for transport either within or across cells. The data have revealed the complexity of mechanisms and the intricacies of pathways present in each cell, as a function of its own physiological state and its role in the activity of the surrounding tissue. Provided that the limitations in using tracers are carefully considered, the experiments performed as close as possible to the physiological conditions, and controlled by alternative procedure (especially biochemical and functional tests), tracers can document and uncover profound cellular processes. If appropriate probes are used in gentle experiments, the accumulated data will continue to show the skill of cells in functioning properly and making themselves useful for the well-being of the tissue to which they belong and, on a broader scale, to the whole organism. A deeper understanding of cellular processes will ease our comprehension of the ways cells function; as Brancusi said, "we arrive at simplicity as we approach the true sense of things."

Addendum: One-Step Fixation for Electron Microscopy

Stock Solutions

A : 3% formaldehyde + 5% glutaraldehyde
B : 2% or 3% OsO_4
C : saturated solution of lead citrate filtered through 0.45 μm Millipore filters when prepared.

The solutions are prepared freshly in different buffers (i..e., 0.1 M arsenate or 0.1 M phosphate or 0.1 M HC1-Na-cacodylate buffers.

The fixative mixture is prepared just before use by mixing in a cold graduated tube: 3 volumes of solution A, 2 volumes of solution B, and 1 volume of solution C.

Throughout the procedure the vials, stock solutions, and mixture are kept in an ice bath.

The mixture should stay clear but may turn slightly brown after 1 hr.

The method provides good preservation and adequate contrast of all cell structures and ensures intensive staining of polysaccharides; when necessary, the staining in block can be avoided.

Acknowledgments

We are grateful for the excellent assistance of G. V. Ionescu and E. Stefan (photographs), C. Neacsu (graphics), and D. Neacsu and L. Barbulescu (word processing and editing).

This chapter is based on work supported by the Ministry of Education, Romania and by the National Institutes of Health (USA) Grant HL-26343.

References

Ainsworth, S. K., and Karnovsky, M. J. (1972). An ultrastructural staining method for enhancing the size and electron opacity of ferritin in thin sections. *J. Histochem. Cytochem.* **20**:225–229.

Anderson, W. A. (1972). The use of exogenous myoglobin as an ultrastructural tracer. Reabsorption and translocation of protein by the renal tubule. *J. Histochem. Cytochem.* **20**:672–684.

Anderson, R. G. W., Goldstein, J. L., and Brown, M. S. (1976). Localization of low density lipoprotein receptors on plasma membrane of normal human fibroblasts and their absence in cells from a familiar hypercholesterolemia homozygote. *Proc. Natl. Acad. Sci. U.S.A.* **73**:2434–2438.

Bruns, R. R., and Palade, G. E. (1968). Studies on blood capillaries. II. Transport of ferritin molecules across the wall of muscle capillaries. *J. Cell Biol.* **37**:277–299.

Caulfield, J. P., and Farquhar, M. G. (1974). The permeability of glomerular capillaries to graded dextrans. *J. Cell Biol.* **63**:883–903.

Cotran, R. S., and Karnovsky, M. J. (1967). Vascular leakage induced by horseradish peroxidase in the rat. *Proc. Soc. Exp.Biol. Med.* **126**:557–561.

Cotran, R. S., Karnovsky, M. J., and Goth, A. (1968). Resistance of Wistar-Furth rats to the mast cell-damaging effect of horseradish peroxidase. *J. Histochem. Cytochem.* **16**:382–383.

Crichton, R. R. (1971). Ferritin: structure, synthesis and function. *N. Engl. J. Med.* **284**:125–131.

Danon, D., Goldstein, L., Markovsky, I., and Skutelsky, E. (1972). Use of cationized ferritin as a label of negative charges on cell surfaces. *J. Ultrastr. Res.* **38**:500–510.

David, G. S. (1972). Solid state lactoperoxidase: a highly stable enzyme for simple, gentle iodination of proteins. *Biochem. Biophys. Res. Commun.* **48**:464–471.

Davies, P. F., Rennke, H. G., and Cotran, R. S. (1981). Influence of molecular charge upon the endocytosis and intracellular fate of peroxidase activity in cultured arterial endothelium. *J. Cell Sci.* **49**:69–86.

Deiman, W., Taugner, R., and Fahimi, H. D. (1987). Arterial hypotension induced by horseradish peroxidase in various rat strain. *J. Histochem. Cytochem.* **24**:1213–1217.

Dixon, M. T., and Jersild, R. A. Jr. (1982). Influence of maternal diabetes on lectin binding to the surface of alveolar epithelial cells. 22nd Ann. Meeting Amer. Soc. Cell Biol. 110a.

Dobbs, L. G., Geppert, E. F., Williams, M. C., Greenleaf, R. D., and Mason, R. J. (1980). Metabolic properties and ultrastructure of alveolar type II cells isolated with elastase. *Biochim. Biophys. Acta* **618**:510–523.

Fahimi, H. D. (1973). Diffusion artefacts in cytochemistry of catalase. *J. Histochem. Cytochem.* **21**:999–1009.

Farquhar, M. G. (1978). Recovery of surface membrane in anterior pituitary cells. Variations in traffic detected with anionic and cationic ferritin. *J. Cell.Biol.* **78**:R35–R42.

Farquhar, M. G. (1983. Multiple pathways of exocytosis, endocytosis and membrane recycling. Validation of a Golgi route. *Fed. Proc.* **42**:2407–2413.

Farquhar, M. G. and Palade, G. E. (1961). Glomerular permeability. I. Ferritin transfer across the glomerular capillary wall in nephrotic rats. *J. Exp. Med.* **114**:699–716

Faulk, W., and Taylor, G. (1971). An immuncolloid method for the electron microscope. *Immunochemistry.* **8**:1081–1083.

Feder, N. (1971). Microperoxidase: an ultrastructural tracer of low molecular weight. *J. Cell Biol.* **51**:339–343.

Fischbach, F. A., Harrison, P. M., and Hay, T. G. (1969). The structural relationships between protein ferritin and its mineral core. *J. Mol. Biol.* **39**:235–238.

Fishman, A. P., and Pietra, G. G. (1976). Permeability of pulmonary vascular endothelium. *Ciba Found. Symp.* **38**:29–36.

Fraenkel-Conrat, H. (1957). Methods of investigating the essential groups for enzyme activity. *Methods Enzymol.* **4**:247–269.

Galis, Z., Ghitescu, L., and Simionescu, M. (1988). Fatty acids binding to albumin increases its uptake and transcytosis by the lung capillary endothelium. *Eur. J. Cell Biol.* **47**:358–365

Ghinea, N., Eskenasy, M., Simionescu, M., Simionescu, N. (1989). Endothelial albumin binding proteins are membrane associated components exposed on the cell surface. *J. Biol. Chem.* **264**:4755–4758.

Ghinea,, N., Fixman, A., Alexandru, D., Popov, D., Hasu, M., Ghitescu, L., Eskenasy, M., Simionescu, J., Simionescu, N. (1988). Identification of albumin-binding proteins in capillary endothelial cells. *J. Cell Biol.*, **107**:231–239.

Ghinea, N., and Hasu, M. (1986). Charge effect on binding, uptake and transport of ferritin through fenestrated endothelium. *J. Submicrosc. Cytol.* **18**:647–659.

Ghinea, N., and Simionescu, N. (1985). Anionized and cationized hemeundecapeptides as probes for cell surface charge and permeability studies. differentiated labeling of endothelial plasmalemmal vesicles. *J. Cell Biol.* **100**:606–612.

Ghitescu, L., and Fixman, A. (1984). Surface charge distribution on the endothelial cell of liver sinusoids. *J.Cell Biol.* **99**:639–647.

Ghitescu, L., Fixman, A., Simionescu, M., and Simionescu, N. (1986). Specific binding sites for albumin restricted to plasmalemmal vesicles of continuous capillary endothelium: receptor-mediated transcytosis. *J. Cell Biol.* **102**:1304–1311.

Goldstein, J. L., Basu, S. K., Brunshede, G. Y., and Brown, M. S. (1976). Release of low density lipoprotein from its cell surface receptor by sulfated glucosaminoglycans. *Cell* **7**:85–95.

Graham, R. C., and Karnovsky, M. J. (1966a). The early stages of adsorption of injected horseradish peroxidase in the proximal tubules of mouse kidney: ultrastructural cytochemistry by a new technique. *J. Histochem. Cytochem.* **14**:291–302.

Graham, R. C., and Karnovsky, M. J. (1966b). Glomerular permeability. Ultrastructural cytochemical studies using peroxidases as protein tracers. *J. Exp. Med.* **124**:1123–1134.

Graham, R. C., and Kellermeyer, R. W. (1968). Bovine lactoperoxidase as a cytochemical protein tracer for electron microscopy. *J. Histochem. Cytochem.* **16**:275–278.

Grotte, G. (1956). Passage of dextran molecules across the blood–lymph barrier. *Acta Chir. Scand.* **211** (Suppl.):1–84.

Hallman, M., Epstein, B. L., and Gluck, L. (1981). Analysis of labeling and clearance of lung surfactant phospholipids in rabbit. *J. Clin. Invest.* **68**:742–751.

Handley, D. A., Arbeeny, C. M., White, L. D., and Chien, S. (1981). Colloidal gold-low density lipoprotein conjugates as membrane receptor probes. *Proc. Natl. Acad. Sci. USA* **78**:368–371.

Harbury, H. A., and Loach, P. A. (1960). Interaction of nitrogenous ligands with heme-peptides from mammalian cytochrome c. *J. Biol. Chem.* **235**:3646–3652.

Hass,M. A., and Longmore, W. J. (1980). Regulation of lung surfactant cholesterol metabolism by serum lipoproteins. *Lipids* **15**:401–406.

Heltianu, C., Simionescu, M., and Simionescu, N. (1983). Histamine receptors of microvascular endothelium revealed in situ with a histamine-ferritin conjugate: characteristic high-affinity binding sites in venules. *J. Cell Biol.* **93**:357–364.

Herzog, V., and Farquhar, M. G. (1977). Luminal membrane retrieved after exocytosis reaches most Golgi cisternae in secretory cells. *Proc. Natl. Acad. Sci. U.S.A.* **74**:5073–5077.

Herzog, V., and Farquhar, M. G. (1983). Use of electron-opaque tracers for studies on endocytosis and membrane recycling. *Methods Enzymol.* **98**:203–225.

Herzog, V., and Miller, F. (1979). Membrane retrieval in epithelial cells of isolated thyroid follicles. *Eur. J. Cell Biol.* **19**:203–215.

Herzog, V., and Miller, F. (1981). Structural and functional polarity of inside-out follicles prepared from pig thyroid gland. *Eur. J. Cell Biol.* **24**:74–84.

Herzog, V., and Reggio, H. (1980). Pathways of endocytosis from luminal plasma membrane in rat exocrine pancreas. *Eur. J. Cell Biol.* **21**:141–150.

Hoare, D. G., and Koshland, D. E. Jr. (1967). A method for the quantitative modification and estimation of carboxylic acid groups in proteins. *J. Biol. Chem.* **242**:2447–2453.

Horisberger, M. (1979). Evaluation of colloidal gold as a cytochemical marker for transmission and scanning electron microscopy. *Biol. Cell* **36**:253–258.

Horisberger, M., and Rosset, J. (1977). Colloidal gold, a useful marker for transmission and scanning electron microscopy. *J. Histochem. Cytochem.* **25**:295–305.

Kanwar, Y. S., and Farquhar, M. G. (1979). Anionic sites in the glomerular basement membrane. *J. Cell Biol.* **81**137–153.

Karnovsky, M. J., and Cotran, R. S. (1966). The intercellular passage of exogenous peroxidase across endothelium and mesothelium. *Anat. Rec.* **154**:365a.

Karnovsky, M. J., and Rice, D. F. (1969). Exogenous cytochrome c as an ultrastructural tracer. *J. Histochem. Cytochem.* **17**:751–753.

Kikkawa, Y., and Yoneda, K. (1974). The type II epithelial cell of the lung. I. Method of isolation. *Lab.Invest.* **30**:76–84.

Kishida, Y., Olsen, B. R., Berg, R. A., and Prockop, D. J. (1975). Two improved methods for preparing ferritin–protein conjugates for electron microscopy. *J. Cell Biol.* **64**:331–339.

Kraehenbuhl, J. P., Galardy, R. E., and Jamieson, J. D. (1974). Preparation and characterization of an immunoelectron microscope tracer consisting of a heme-octapeptide coupled to Fab. *J. Exp. Med.* **139**:208–223.

Lis, H., and Sharon, N. (1986). Lectins as molecules and as tools. *Annu. Rev. Biochem.* **55**:35–67.

Majno, G., and Palade, G. E. (1961). Studies on inflammation. I. The effects of histamine and serotonin on vascular permeability: an electron microscopic study. *J. Biophys. Biochem. Cytol.* **11**:571–606.

Markwell, M. A. K. (1983). A new solid state reagent to iodinate proteins. I. Conditions for the efficient labeling of antiserum. *Anal. Biochem.* **125**:427–432.

Mayerson, H. S., Wolfram, C. G., Shirley, H. H. Jr., and Wasserman, K. (1960). Regional differences in capillary permeability. *Am. J. Physiol.* **198**:155–160.

Nakane, P. K., and Pierce, G. P. Jr. (1967). Enzyme-labeled antibodies for the light and electron microscopic localization of tissue antigens. *J. Cell Biol.* **33**:307–318.

Nickerson, P. A., Matalon, S., and Farhi, L. E. (1981). An ultrastructural study of alveolar permeability to cytochrome c in rabbit lung. *Am. J. Pathol.* **102**:1–9.

Nistor, A., and Simionescu, M. (1986). Uptake of low density lipoproteins by the hamster lung. Interaction with capillary endothelium. *Am. Rev. Respir. Dis.* **134**:1266–1272.

Novikoff, A. B., Novikoff, P. M., Quintana, N., and Davis, C. (1970). Diffusion artefacts in 3,3′-diaminobenzidine cytochemistry. *J. Histochem. Cytochem.* **20**:745–749.

Panitz, J. A., and Ghiglia, D. C. (1982). Point projection imaging of unstained ferritin molecules on tungsten. *J. Microsc.* **127**:259–264.

Pietra, G. G., Szidon, J. P., Leventhal, M. M., and Fishman, A. P. (1969). Hemoglobin as a tracer in hemodynamic pulmonary edema. *Science* **166**:1643–1645.

Pietra, G. G., Samson, P., Lanken, P. N., Hansen-Flaschen, J., and Fishman, A. P. (1983). Transcapillary movement of cationized ferritin in the isolated perfused rat lung. *Lab. Invest.* **49**:54–61.

Plattner, H., Wachter, E., and Grobner, P. (1977). A heme-nonapeptide tracer for electron microscopy. Separation, characterization and comparison with other heme-tracers. *Histochemistry* **53**:223–242.

Rennke, H. G., Cotran, R. S., and Venkatachalam, M. A. (1975). Role of molecular charge in glomerular permeability: tracer studies with cationized ferritins. *J. Cell Biol.* **67**:638–646.

Rennke, H. G., and Venkatachalam, M. A. (1979). Chemical modifications of horseradish peroxidase preparation and characterization of tracer enzymes with different isoelectric points. *J. Histochem. Cytochem.* **10**:1352–1353.

Ryan, U. S., Ryan, J. W., Whitaker, C., and Chiu, A. (1976). Localization of angiotensin converting enzyme (kininase II). II. Immunocytochemistry and immunofluorescence. *Tissue Cell* **8**:125–145.

Schneeberger, E. E. (1976). Ultrastructural basis for alveolar-capillary permeability to protein. *Ciba Found. Symp.* **38**:3–28.

Schneeberger, E. E. (1978). Structural basis for some permeability properties of the air-blood barrier. *Fed. Proc.* **37**:2471–2478.

Schneeberger, E. E., and Karnovsky, M. J. (1968). The ultrastructural basis of alveolar-capillary membrane permeability to peroxidase used as a tracer. *J. Cell Biol.* **37**:781–793.

Schneeberger, E. E., and Karnovsky, M. J. (1971). The influence of intravascular fluid volume on the permeability of newborn and adult mouse lung to ultrastructural protein tracers. *J. Cell Biol.* **49**:319–334.

Shannon, L., Kay, E., and Lew, J. Y. (1966). Peroxidase izoenzymes from horseradish roots. I. Isolation and physical properties. *J. Biol. Chem.* **241**:2166–2172.

Simionescu, D., and Simionescu, M. (1983). Differentiated distribution of the cell surface charge on the alveolar-capillary unit. Characteristic paucity of anionic sites on the air-blood barrier. *Microvasc. Res.* **25**:85–100.

Simionescu, M. (1985a). Cellular organization of the alveolar-capillary unit: structural-functional correlations. In *The Pulmonary Circulation and Acute Lung Injury.* Edited by S. I. Said. Mount Kisco, NY, Futura Publishing, pp. 13–36.

Simionescu, M. (1985b) Regional differentiation of the surface charge distribution in the continuous endothelium of the microvasculature. In *Glomerular Dysfunction and Biopathology of the Vascular Wall.* Edited by A. L. Copley, Y. Hamashima, S. Seno, and M. A. Venkatachalam. Tokyo, Academic Press, Japan, pp. 3–11.

Simionescu, M., and Simionescu, N. (1987). Endothelial surface domains in pulmonary alveolar capillaries. In *The Pulmonary Endothelium in Health and Disease.* Edited by U. S. Ryan. New York, Marcel Dekker, pp. 35–62.

Simionescu, M., Simionescu, N., Silbert, J. E., and Palade, G. E. (1981a). Differentiated microdomains of the luminal surface of the capillary endothelium. II. Partial characterization of their anionic sites. *J. Cell Biol.* **190**:614–621.

Simionescu, M., Simionescu, N., Santoro, F., and Palade, G. E. (1985). Differentiated microdomains of the luminal plasmalemma of murine muscle capillaries: segmental variations in young and old animals. *J. Cell Biol.* **100**:1396–1407.

Simionescu, M., Ghitescu, L., Fixman, A., and Simionescu, N. (1987). How plasma macromolecules cross the endothelium. *News Physiol. Sci.* **2**:97–100.

Simionescu, N. (1979a) The microvascular endothelium: segmental differentiations, transcytosis, selective distribution of anionic sites. In *Advances in*

Inflammation Research, vol.1. Edited by G. Weissmann, B. Samuelson, and R. Paoletti. New York, Raven Press, pp. 61–70.

Simionescu, N. (1979b) Enzymatic tracers in the study of vascular permeability. *J. Histochem. Cytochem.* **27**:1120–1130.

Simionescu, N., and Palade, G. E. (1971). Dextrans and glycogens as particulate tracers for studying capillary permeability. *J. Cell Biol.* **50**:616–624.

Simionescu, N., and Simionescu, M. (1976). Galloylglucoses of low molecular weight as mordant in electron microscopy. I. Procedure and evidence for mordanting effect. *J. Cell Biol.* **70**:608–621.

Simionescu,, N. and Simionescu, M. (1984). Fluid phase and adsorptive transcytosis in the endothelial cell. In International Symposium on Membrane Biogenesis and Recycling, Kannami. Abstract volume VI-1.

Simionescu, N., Simionescu, M., and Palade, G. E. (1972). Permeability of intestinal capillaries. Pathway followed by dextrans and glycogens. *J. Cell Biol.* **53**:365–392.

Simionescu, N., Simionescu, M., and Palade, G. E. (1973). Permeability of muscle capillaries to exogenous myoglobin. *J. Cell Biol.* **57**:424–452.

Simionescu, N., Simionescu, M., and Palade, G. E. (1975). Permeability of muscle capillaries to small heme-peptides. Evidence for the existence of patent transendothelial channels. *J. Cell Biol.* **64**:586–607.

Simionescu, N., Simionescu, M., and Palade, G. E. (1981b). Differentiated microdomains on the luminal surface of capillary endothelium. I. Preferential distribution of anionic sites. *J. Cell Biol.* **100**:1396–1407.

Skutelsky, E., and Danon, D. (1976). Redistribution of surface anionic sites on the luminal front of blood vessel endothelium after interaction with polycationic ligand. *J. Cell Biol.* **71**:232–241.

Slot, J. W., and Geuze, H. J. (1981). Sizing of protein A–colloidal gold probes for immunoelectron microscopy. *J. Cell Biol.* :533–536.

Sly, W. S., and Stahl, P. (1978). Receptor mediated uptake of lysosomal enzymes. In *Transport of Macromolecules in Cellular Systems*. Edited by S. C. Silverstein. Berlin, Dahlem Konferenzen, pp. 229–244.

Steinman, R. M., and Cohn, Z. A. (1972). The interaction of soluble horseradish peroxidase with mouse peritoneal macrophages in vitro. *J. Cell Biol.* **55**:186–204.

Straus, W., and Keller, J. M. (1987). Binding sites for horseradish peroxidase on the cell surface. Suppression of binding by gangliosides and effects of some bivalent cations. *Histochemistry* **86**:453–457.

Strum, J. M., and Junod, A. F.)1972). Radioautographic demonstration of 5-hydroxytryptamine-[^{3}H] uptake by pulmonary endothelial cells. *J. Cell Biol.* **54**:436–437.

Thiessen, G., Thiessen, H., Dowidat, J. H., Lucian, L., and Reale, E. (1970). Die Diffusion des Fe markierten Hamoglobins, ein Artefakt der glutaraldehyde-fixierung. *Histochemie* **23**:1–12

Tu, A. T., Reinosa, J. A., and Hsiao, Y. Y. (1968). Peroxidative activity of hemepeptides from horse heart cytochrome c. *Experientia* **24**:219–221.

Vasile, E., Simionescu, M., and Simionescu, N. (1983). Visualization of the binding, endocytosis and transcytosis of low density lipoproteins in the arterial endothelium in situ. *J. Cell Biol.* **96**:1677–1689.

Vegge, T., and Haye, R. (1977). Vascular reactions to horseradish peroxidase in the guinea pig. *Histochemistry* **53**:217–222.

Venkatachalam, M. A., and Fahimi, D. H. (1969). The use of beef liver catalase as a protein tracer for electron microscopy. *J. Cell Biol.* **42**.:480–489.

Weir, E. E., Pretlow, T. G., Pitts, A., and Williams, E. E. (1974). A more sensitive and specific histochemical peroxidase stain for the localization of cellular antigen by the enzyme–antibody conjugate method. *J. Histochem. Cytochem.* **22**:1135–1140.

Welinder, K. G. (1979). Amino acid sequence studies of horseradish peroxidase. Amino and carboxyl termini, cyanogen bromide and tryptic fragments, the complete sequence and some structural characteristics of horseradish peroxidase C. *Eur. J. Biochem.* **96**:483–502.

Weller, N. K., and Karnovsky, M. (1986). Isolation of pulmonary alveolar type I cells from adult rats. *Am. J. Pathol.* **124**:448–456.

Williams, M. C. (1984a). Endocytosis in alveolar type II cells: effect of charge and size of tracers. *Proc. Natl. Acad. Sci. U.S.A.* **81**:6054–6058.

Williams, M. C. (1984b) Uptake of lectins by pulmonary alveolar type II cells. subsequent deposition into lamellar bodies. *Proc. Natl. Acad. Sci. U.S.A.* **81**:6383–6387.

Williams, S. K. (1987). Isolation and culture of microvessel and large-vessel endothelial cells: their use in transport and clinical studies. In *Microvascular Perfusion and Transport in Health and Disease*. Edited by J. McDonagh. Basel, Karger, pp. 204–245.

Wissig, S. L., and Williams, M. C. (1978). Permeability of muscle capillaries to microperoxidase. *J. Cell Biol.* **76**:341–359.

14

Structural Methods in the Study of Development of the Lung

PAUL DAVIES

University of Pittsburgh
School of Medicine
Pittsburgh, Pennsylvania

LYNNE M. REID

Harvard Medical School and
The Children's Hospital
Boston, Massachusetts

DAPHNE deMELLO

St. Louis University and
Cardinal Glennon Children's Hospital
St. Louis, Missouri

I. Historical Background

It is strange to realize that it is not for much more than 100 years that dissection combined with the microscope has been used to analyze cells and tissues in an organ. With this combination of techniques, the development of organs in the human embryo and fetus was established, and comparative embryology demonstrated what we now know; that ontogeny recapitulates phylogeny. At about the turn of the century, with the addition of experimental studies in animals, much else was discovered; how these changes came about was then explored. Chemical "organizer" substances were recognized, as well as special "zones" in which they operate. A signal must be delivered, but the cell or tissue must be prepared to receive it. This double requirement gives us the window to the perspectives we will mention at the end of this chapter.

Normal embryology also offers a way to interpret and then study disease. Because congenital abnormalities are the most common causes of death in the newborn, their diagnosis and understanding represent not only major scientific challenges, but also ones that must be met if the clinical burden is to be eased.

Growth represents increase in size and weight as well as differentiation. Although function matures with age, our simple concept of growth is of a slide projector: the objects get steadily bigger. In lung this is certainly true, but analysis of the growth pattern, both before and *after* birth, shows that growth is not more of the same. The newborn lung is not the adult in miniature. The template to which growth conforms changes dramatically. Airways, alveoli, and blood vessels each dance to a different tune. There is harmony, but also counterpoint.

The broad framework for lung growth and development is summarized in the following three laws. It is more usual for the physicist to make laws than for the biologist, but the laws have similar use. They summarize facts and provide the basis for interpretation and manipulation of events. These laws serve well in analyzing lung disease and offer a critical term of reference even for experimental techniques (see Perspectives).

Law I. The bronchial tree is developed by the 16th week of intrauterine life.

Law II. Alveoli develop after birth, increase in number until the age of 8 years and in size until growth of the chest wall finishes in adulthood.

At birth the future alveoli are represented by primitive saccules, about 20 million of them. Typical alveoli appear soon after birth and by 8 years about 300 million are present (within the adult range).

Law III. The preacinar vessels (arteries and veins) follow the development of the airways, the intraacinar vessels follow that of the alveoli. Muscularization of the intraacinar arteries does not keep pace with the appearance of new arteries.

II. Methods

A. General Methods, Including Lung Preparation

The methods appropriate to structural analysis of lung tissue depend on a number of factors, but perhaps the most critical, for several reasons, is the species to be studied. Species' size determines the feasibility of certain procedures. The small size of rodent lungs, particularly during fetal development, precludes perfusion fixation or other techniques involving manual manipulation. Second, although the sequence of four stages of lung development is characteristic of all

Table 1 Stages of Lung Development

	Human (weeks)	Sheep (days)
Embryonic	0–5	0–60
Pseudoglandular	5–16	60–80
Canalicular	16–28	80–120
Saccular/alveolar	28–term	120–150 (term)

mammalian species (Table 1), the relative duration of each stage varies among species. This is particularly important during late gestation and early postnatal periods, because it affects the degree of maturation reached at term. For example, in the sheep the alveolar stage begins in utero, whereas in the rat it does not occur until some days after birth. Studying the initial stages of alveolarization in the sheep involves delivering the fetus and using methods appropriate to a fetal lung.

In the list of techniques that follows, some are mandatory for a proper appreciation of lung growth; others are optional, and related to special questions.

Age

The age of the fetus or infant is an essential reference point. For studies on human fetal development, determining gestational age can be difficult. Other indices that normally correlate with age include body length (Usher et al., 1966), crown–rump length, and bone age (Langston et al., 1984). In humans determining postnatal age is straightforward. It can be a problem in experimental animals such as rodents and lagomorphs that often give birth at night or in the early hours of the morning. In these animals, to study the perinatal period if may be necessary to induce parturition.

Gender

It is increasingly clear that gender influences the rate of lung development (Nielsen and Torday, 1987), and so wherever possible, it should be determined. In humans, this is feasible throughout development. In fetuses of small mammals, careful dissection or histological sectioning may be necessary.

Body Weight

Since the degree of lung growth can only be judged by reference to somatic growth, this is a fundamental measurement in all studies.

Lung Weight

To determine their weight, the lungs must be excised from the chest. If this is done *before* fixation, as is often the case for autopsies, the lungs are weighed in the collapsed state after the heart is removed. This is not the method of choice if the lungs are to be fixed under conditions typical of normal ventilation. If the air-inflated lung is fixed by vascular perfusion (see below), its weight in air indicates its tissue mass (plus the weight of perfusate). Whether the lungs are collapsed or inflated, weight is increased by factors not necessarily related to growth such as edema, hemorrhage, and fibrosis.

Lung Volume and Fixation

An estimate of lung volume is basic to all studies on lung development and demands a consistent technique. It is determined by the water displacement method (Scherle, 1970). If the lungs are already excised, a prefixation lung volume can be determined and later compared with the postfixation volume to assess collapse and shrinkage, but inflation conditions must be consistent. These are determined by the nature of the study and the method of fixation selected: satisfactory fixation can be achieved by use of one of the two following methods.

Airway Instillation

Airway instillation of fixative at a head or pressure sufficient to inflate the lung is a simple way to achieve good preservation. A disadvantage is that it removes much of the surface lining material and destroys the surface tension of postnatal air-filled lungs so that the microscopic architecture is altered (Gil et al., 1979). Fetal lungs, being already filled with fluid, may be less affected. Choice of instillation pressure can be difficult, especially for those species about which little is known of in utero lung mechanics. A pressure sufficient to inflate the lung fully is generally chosen, and is often similar to that used for adult lungs, but could be excessive for young lungs. When instilled with formalin at 25 cmH$_2$0, lungs of children achieved higher volumes relative to physiological total lung capacity (TLC) than lungs from adults (Berend et al., 1980). The same was found in a comparative study of adult mammals: lungs of small species tended to inflate with fixative to relatively larger volumes. In the mouse, for example, fixed lung volume was more than twice TLC (Lum and Mitzner, 1985). This could be due in part, to the presence of less connective tissue in smaller or younger lungs. When fixed, collagen is nonextensible so that greater amounts will restrict distention. Elastin does not fix, so that the greater elastic recoil of older or larger lungs will be retained and cause a gradual loss of volume from the value first achieved. The composition of the fixative is critical. Formaldehydes fix comparatively slowly, so inflation pressure should be maintained as long as is necessary to prevent collapse of the lung when pressure is removed. For ultrastructural morphometry, glutaraldehyde is the likely choice: this fixes rather rapidly so that its rate of flow into the lung becomes critical. In the adult rat lung, the ratio of fixed lung volume to air TLC was 0.80 when formalin was instilled and 0.60 when glutaraldehyde (Hayatdavoudi et al., 1980) was used. Normal small lungs have faster time constants and inflate more quickly than larger lungs. If glutaraldehyde is not delivered quickly enough, the larger lungs will tend not to reach maximum volume.

The buffer and final osmolality of the fixative solution has a considerable effect on the dimensions of the air–blood barrier and its cellular components (Mathieu et al., 1978). For comparative studies, the composition of the fixative solution should be consistent.

Vascular Perfusion

Vascular perfusion of fixative while lung inflation is maintained at a given transpulmonary pressure requires that one choose the perfusion *and* inflation pressures. This method has the advantage of fixing the extracapillary vessels in distention and retaining the surface lining material in the distal air spaces (Gil, 1977; Bachofen et al., 1982). To avoid edema formation, the perfusate should contain high molecular weight compounds such as dextran or polyvinylpyrrolidone that increase colloid osmotic pressure. The usual route of entry of fixative is the main, right or left pulmonary artery. In fetuses and early postnatal lungs, fixative will be shunted through the patent ductus arteriosus if it is not ligated.

Each method can be applied to lungs either in situ or excised from the chest. The behavior of the lung and its final volume are not the same under these two conditions (Bachofen et al., 1970; Wohl et al., 1968).

In lungs early in gestation or in the older fetuses of small mammals, neither instillation or perfusion may be feasible and the lungs must be fixed by simple immersion. While adequate for histological examination, this method rules out most morphometric studies.

In determining lung volume, the heart and vessels are first removed and retained for determination of ventricular weights and microscopic examination. At this stage the ductus can be examined to determine its patency.

In normal development, lung maturation may occur earlier in the apical lobe. To detect this, the volumes of individual lobes, particularly apical and diaphragmatic, should be measured. Since growth disturbances can be unilateral, the volume of each lung, left and right, may need to be measured separately.

Heart Weight

The weight of the heart is generally determined to assess right ventricular hypertrophy in developmental disturbances affecting the pulmonary vasculature. It is usually assessed for the ratio of the weight of the right ventricle to that of the left ventricle plus septum. The technique demands removal of the atria and fat (Fulton et al., 1952).

Tissue Shrinkage

Since growth is assessed from measurements on fixed tissue and tissue sections, it is crucial to recognize likely sources of error. Shrinkage is certainly one of these and occurs at three stages of processing.

The amount of shrinkage produced by fixation alone probably depends on the nature of fixative, but a rigorous comparison of the effects of various fixatives has not been performed. Adult dog lungs perfused with glutaraldehyde showed shrinkage in a linear dimension of 9% at transpulmonary pressures of 5 and 15 cm H_2O (Mazzone et al., 1980). The fact that shrinkage increased to 15% at a transpulmonary pressure of 25 cm H_2O indicated that elastic recoil continued to be effective. If the elastin content of the lung increases during development, we can expect this kind of shrinkage to be relatively greater in older lungs.

The second type of shrinkage occurs in the block cut from the lung and is generally ascribed to processing of the tissue through organic solvents and infiltrating in an embedding medium. When the block is first cut, however, retraction of tissue occurs due to its elastic content, this will be greater with increasing age of the lung. Because of practical difficulties, this is virtually never measured. The shrinkage introduced by actual processing varies with the agents used and with the fat and water content of the tissue. It is determined by measuring the dimensions of the fixed block when first removed and then after it is embedded. In adult lungs fixed in 10% formalin, the linear shrinkage due to processing for paraffin wax ranged from 17 to 30% (Lum and Mitzner, 1985). In postnatal sheep lungs fixed by instilling glutaraldehyde and embedded in methacrylate, the linear shrinkage was 11% (Davies et al., 1988). The shrinkage produced by processing is generally considered to act isotropically, so that changes in linear dimension are reflected volumetrically as the cube.

The third type of shrinkage is a compression produced by the sectioning of tissue blocks. Its greatest effect is in the direction perpendicular to the knife edge. It is determined by measuring the linear dimensions of the block face and section. In adult lungs embedded in wax, sectioning introduced a linear shrinkage rate of 5–21% (Lum and Mitzner, 1985). In postnatal sheep lungs embedded in methacrylate, no shrinkage was apparent after sectioning (Davies et al., 1988).

Tissue Sampling

At no stage of development are preacinar airways and vessels distributed uniformly through the lung; they are oriented structures and can only be sampled systematically. To determine their volume density in lung, the whole tissue must be sliced or sectioned at regular intervals. Point-counting or some other means of determining the sectional area of preacinar structures and of intra-acinar or respiratory lung is then applied to all slices: the ratio of total area of preacinar structures to total area of lung is equal to their volume density in lung. For stereological purposes the respiratory region of the normal lung can be considered nonoriented and can be sampled using either systematic or random techniques. In the developing lung certain considerations force some modification of these techniques. Differences between left and right lungs may be present as a result of unilateral hypoplasia or dysplasia, diaphragmatic eventration, or congenital diaphragmatic hernia. Within each lung regional or lobar differences occur because of differences in rate of development, and can be accentuated in disease or by the effects of treatment. Within the acinus itself certain structures show an orientation that can be important if disease is focal in distribution. Bronchopulmonary dysplasia, for example, has a centriacinar distribution that necessitates selective sampling.

B. Preacinar Airways: Quantitation

Airway branching is irregular. To measure their growth, individual branches must be classified by their level or order. There are basically three ways of doing this.

1. Counting in a centrifugal direction from the hilum, so that the mainstem or segmental bronchus is the first order and the terminal bronchiolus the last; at each branch the order is increased by one

2. Counting in a centripetal direction from the periphery, so that the terminal bronchiolus is the first order; at each branch the order is increased by one

3. The Strahler system: counts are made centripetally, but the order number of the next proximal branch is only increased when formed from two branches of the same order

When using the first two methods it is often helpful to differentiate between axial and lateral pathways, the axial being the longest pathways between the segmental hilum and the distal pleural surface (Fig. 1).

When plotted on a logarithmic scale against Strahler order, the number of branches in each order gives a straight line using regression techniques. The same is true for branch diameter and length, but the direction of change is opposite to that of the number of branches.

In studying developmental abnormalities, it is critical to determine whether the branching of preacinar airways is complete. Reduction in the number of orders indicates abnormal growth before the 16th week of intrauterine life. Because axial path lengths from bronchus to terminal bronchiolus vary with the lung segment, the number of orders should be determined in the same segment, along a long axial pathway such as that supplying the posterior basal segment. In the normal human lung, after the 16th week of gestation, the number of orders, measured according to technique (1) above, is 25 (Bucher and Reid, 1961a). The number can be determined directly by dissection under a stereoscopic microscope.

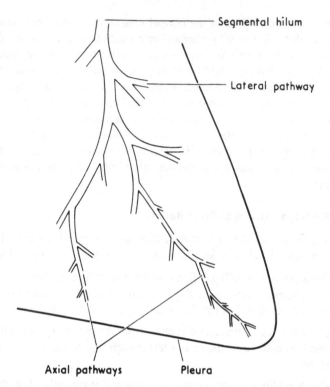

Figure 1 Human lung: diagram illustrates the location of axial and lateral airway branches.

Although this compromises the segment and possibly the lobe for other purposes, histological and morphometric examination can be carried out on the remaining lobes.

For more complete information on the branching system, an airway cast can be made. In the method of Tompsett (1970), the alveolar region is first filled by injecting a measured amount of hot gelatin into the bronchus. This is followed by a polymer injected at a standardized pressure. After polymerization, the cast is freed by digestion of the lung tissue. The airways should be filled down to the terminal bronchiolus, which can be recognized under a dissecting microscope as the last branch before a respiratory bronchiolus with alveoli in its wall.

A similar method was used by McBride and Chuang (1985), who first filled the acinar spaces with a volume of saline equal to 11% of the lung volume and then injected a silicone rubber polymer. The rubber polymerizes quickly, is not as brittle as some other materials, and can be cut with scissors into portions convenient for further study.

The diameter and length of each branch are measured under a dissecting microscope and it is assigned a level, either by giving it an order number or, in the system of McBride and Chuang (1985), by determining how much distal airway it subtends. the latter was found to be a good predictor of the cross-sectional area (Y) of an airway, with the best relationship expressed by the quadratic equation

$$\ln (Y) = A + B * \ln (X) + C (\ln[X])^2$$

where A and B are constants and X is the ratio of the sum of squared diameters of terminal bronchioli subtended relative to the sum of squared diameters of all terminal bronchioli in the lobe or lung.

C. Distal Lung: Quantitation

Distribution of the Distal Air Spaces

During the embryonic and pseudoglandular stages the airways are blind-ending tubes lined by cuboidal epithelium and surrounded by an extensive mesenchymal stroma. During the canalicular phase of fetal development, branching morphogenesis forms the acinus. This process has been studied by conventional techniques in several species. In the human (Boyden, 1974), monkey (Boyden, 1976), and dog (Boyden and Tompsett, 1961), wax reconstructions were used. Recently in the mouse it has been viewed directly in organ culture, using differential interference contrast and immunofluorescent localization of basement membrane antigens (Chen and Little, 1987). Reconstruction is now relatively easy because of the ready availability of software computer programs and megabyte storage of digitized images (Mercer and Crapo, 1987).

Stereological Assessment

The composition of the lung can be determined volumetrically by classic stereological methods, using techniques already described elsewhere (Weibel, 1979). They can be applied at several levels of resolution, at each of which increasingly detailed information is obtained. At each level, one tissue compartment serves as reference compartment for the next, higher level of resolution. The lung is, of course, the first reference and measurement of its volume enables absolute values to be obtained for all compartments. The volume density of respiratory region in lung (Vv_r) is determined by point counting lung slices or microscopic sections taken systematically through the whole organ.

The next level , usually microscopic, uses the respiratory region as reference. If an additional ultrastructural level of resolution is required, its reference compartment can continue to be the respiratory region, but it is more efficient to use the alveolar septum as reference, thus avoiding sampling fields that contain empty air space (Muller et al., 1980).

To quantify structure, a test lattice is superimposed on the fields selected. The lattices are of various kinds but each incorporates a systematic series of test points and continuous or discrete test lines with a total length that is geometrically related to the number of test points. Points and lines are contained within a test area.

The test lattice is applied to a random or systematic sample of fields. The following stereological estimates can be calculated. Volume density (V_v) represents the volume of a given compartment in a unit of the reference compartment. From the Delesse principle, it is equal to the ratio of total profile area of the compartment to the total area of reference tissue. Using a point-counting test lattice and given a sufficiently large sample, this can be estimated from the ratio of total test points falling on the compartment (P_c) and reference tissue (P_T).

$$V_v = P_c/P_T$$

Surface density (S_v) represents the surface area of a boundary per unit reference compartment. It is obtained from the ratio of the total number of intersections (I_c) made by the boundary with the test line to the total length (L_T) of the test line system.

$$S_v = 2I_c/L_T$$

Surface density is often used to determine the surface area of the alveolar epithelial or capillary endothelial surface areas. It increases with resolution, so that determinations made at the light microscopic level are appreciably less than those made at the electron microscopic level. In assessing growth, the resolution at which measurements are made should be consistent.

Length density (L_v) represents the length of a linear structure per unit reference volume. It is obtained from the ratio of the total number of transections per field (Q_c) to the total field area (A_T)

$$L_v = 2Q_c/A_T$$

It is important that the structure be thin relative to the dimensions of the section, such as elastic fibers within the alveolar septa. These indices are independent of shape.

The stereological determination of number is derived from numerical density (N_v, number per unit volume reference compartment). This is determined from the ratio of the total number (N) of profiles contained within the sampled fields to the total field area (A_T).

$$N_A = N/A_T$$

To determine N_v, this is then substituted in one of two possible equations (Weibel and Gomez, 1962):

$$N_v = (K * N_A^{3/2})/(\beta * V_v^{1/2})$$

K is a dispersion constant and has a value of 1 when the population is monodisperse. V_v is the volume density of structures in the reference compartment. β is a shape constant related to volume (v) and mean profile area (a) by the equation

$$v = \beta * a^{3/2}$$

The second equation is (De Hoff and Rhines, 1961):

$$N_v = N_A/\overline{D}$$

Where $\overline{D}$ is the mean caliper diameter, that is, the mean diameter of an infinitely large number of random cords passing through the three-dimensional structure. It can be calculated from the mean intercept made by random cords with the two-dimensional profiles in section, if one assumes a shape for the structure.

In applying stereological techniques, care must be taken to select a sufficiently large sample of fields and to choose the test lattice so that an appropriate number of points are counted for each compartment. The major difficulty is that the mean values obtained could obscure regional differences present in normal development, and the methodology does not easily permit access to sources of variability unless particular regions are sampled separately. This becomes more of a problem if disease is superimposed on the regional variability of normal development. For example, the centriacinar lesions that develop early in bronchopulmonary dysplasia will not be detected because they are swamped by the large number of fields sampled from the normal remainder of the acinus. At the least, then, the investigator needs to know something of the likely distribution of structural change before being able to demonstrate it.

Number of Saccules/Alveoli

Determining peripheral air-space number is only possible late in fetal development when saccules appear. It depends on the lung being fixed at or close to full inflation. The greater the degree of collapse the more difficult it is to resolve individual air-spaces in section. To determine the total number of saccules or alveoli, their relative number must first be estimated. This is obtained from their numerical density. The shape of air spaces is difficult to determine and both it and the size dispersion are likely to vary appreciably during development. For these reasons this remains a controversial measurement. Certainly it should not be accepted without confirmatory indices that are independent of shape and size, such as surface density or mean linear intercept, both of which can be used to give an estimate for total internal surface area.

The number of air spaces and their size are fundamental aspects of lung architecture and affect lung mechanics, perfusion, and hemodynamics. Changes in air-space number are manifestations of either primary growth disturbance of the secondary and subsequent adaptive responses to it.

An index of saccule or alveolar number within the acinus is the radial alveolar count (RAC) (Emery and Mithal, 1960). A section is scanned at low magnification and a respiratory bronchiolus identified. An imaginary line is dropped between the center of the bronchiolus and the edge of the acinus as indicated by the pleura, connective tissue septum, or vein. The number of alveoli crossed by this line is then counted. In human tissue, for the same observer, 40 counts per case gave good reproducibility; reproducibility between observers was never as good (Cooney and Thurlbeck, 1982). Greater degrees of lung inflation increase RAC. As an index of acinar development, RAC provides different information than that derived from determination of total alveolar number. Because it is measured from a respiratory bronchiolus, it is potentially more variable in those species or in later stages of alveolar development in which the generations of respiratory bronchioli increase. As a linear measurement it may not be sensitive to subtle changes that occur in three dimensions and are apparent volumetrically.

Air-Space Size

A linear measurement of the size of distal airspaces is the mean linear intercept (L_m), which is essentially the mean distance between alveolar walls (Campbell and Tomkeieff, 1952; Dunnill, 1962). It therefore includes both alveolar and alveolar duct air. As with surface density, mean linear intercept can be used to estimate total internal surface area.

The mean volume of the alveolus can be determined from the ratio of alveolar volume density to numerical density, provided that the reference compartment for both numerator and denominator is the same.

Air-Blood Barrier

This is not present before the saccular stage of development. Late in the canalicular phase, the mesenchymal tissue separating air-space epithelium and capillary endothelium condenses and at particular sites forms a thin barrier for gas exchange. The process of condensation can be demonstrated by linear measurement of airspace to capillary distance. An arithmetic thickness for the air–blood barrier is calculated from the volume to surface area ratio of the tissue sheet (Weibel and Knight, 1964).

Harmonic mean thickness is also a linear measurement made on electron micrographs, but is more relevant to gas diffusion across the barrier. It has been used to estimate so-called "morphometric diffusing capacity" (Weibel, 1971, 1979).

Possession of a thin air–blood barrier does not necessarily mean that the alveolar stage is present. Saccules are characterized by a double capillary system in which barrier thickness is quite thin.

Cellular Composition of the Distal Lung

The contribution of particular cell types to the volume of distal lung tissue can be assessed stereologically by determining their V_v. For determinations of cell number, using N_v, assumptions of shape have to be made. This can be simplified to some degree by counting the nuclei if their shape is known. For some cells, such as the type II pneumocyte, the nucleus can be assumed to be spherical. On the other hand, for the type I pneumocyte or endothelial cell, the nucleus is flattened. For rat lung, the nuclei of these cells have been reconstructed using computer software, and the value of D determined (Crapo and Greeley, 1978; Greeley et al., 1978). More recently, this has been done for the corresponding cells in adult human lungs (Crapo et al., 1982).

Autoradiography

Autoradiography has been used to evaluate cell kinetics in the developing lung (Kauffman et al., 1974; Adamson and King, 1984a).

D. Vasculature

The vasculature of the lung is comprised of a double arterial supply and a double venous drainage. The true pulmonary arterial system arises from the right side of the heart and drains to the left atrium by the pulmonary veins. It supplies the capillary bed in the respiratory region of the lung. The bronchial arterial system is a systemic supply from the left ventricle and aorta. It supplies the capillary bed of the airways down to the distal bronchioli, the venous drainage from the hilar region returns to the right side of the heart via the azygos vein. The "true" bronchial veins, from the capillary bed of the intrapulmonary airways, drain to the pulmonary veins.

Anastomoses are present between the two systems; in disease these open up.

In the pulmonary circulation there are two types of artery: conventional and supernumerary that appear at the same time in development. Conventional arteries accompany airways at all levels. The same law applies to them as to the airways: they are complete down to the terminal bronchiolus by the end of the pseudoglandular phase of lung development. Supernumerary arteries arise from conventional arteries but do not accompany airways. They provide a short-cut supply to the adjacent respiratory region. They are shorter and have smaller diameters than conventional arteries at the same level.

The dimensions of vessels within either the pulmonary or bronchial circulations are affected by the transmural distending pressure exerted by the perfusate and by the degree of smooth muscle tone. Vessels within the lung tissue are also distended by lung inflation. In comparing vessels from different regions of the lung, or in different diseases, it is important to minimize these variables.

This is best done by comparing vessels that are maximally distended. A condition close to this can be achieved by perfusing with a fixative solution at high pressure. If the pulmonary veins are ligated during this procedure, the pressure is uniform throughout the system. It can be sufficiently high to overwhelm smooth muscle tone, although perivascular edema can result. These methods can be used for examination at any level of resolution, including electron microscopy. Their major problem is that in random sections there is no way of differentiating small nonmusuclar arteries from veins. These can only be separated by microdissection or in serial sections.

A method that maintains a strict distinction between arteries and veins and that distends the vessels with reasonable uniformity is the injection of a gelatin/barium sulfate mixture at a temperature of 60 °C into the vasculature (Short, 1956). The mixture does not pass through the capillaries so any system can be selected: pulmonary or bronchial, arterial or venous. Although some vessels remain unfilled this can be an advantage, indicating routes of preferential perfusion, differentiating patent from obstructed or obliterated pathways, and ensuring that the vessels sampled are, indeed, the ones distended at known pressure because the injectate is visible in each section. Angiograms can be prepared that are useful in describing the branching system, the luminal diameters of axial pathways, and the degree of filling. The method is not suitable for electron microscopy, however. Nevertheless, at the light microscopic level it has largely been responsible for our current knowledge of the normal vascular development of the lung and the characteristics of a large number of abnormal developmental conditions (Haworth et al., 1977; Haworth and Reid, 1976, 1977a,b, 1978).

Scanning electron microscopy has proved useful in studying the three dimensional features within the acinus of the developing lung or of the residual tissue following resection (Burri et al., 1982). It has also been applied to studies of acrylic casts of the microvasculature (Caduff et al., 1986).

E. Markers of Cell Function and Differentiation

Airway Epithelium

In the airways there are two sources of mucus secretion: the goblet cells in the surface epithelium and the acinar glands that are external to the muscular layer and open onto the airway surface through an excretory duct.

In the epithelium itself nine types of cell have been identified: mucous, serous, Clara, neuroendocrine, basal, brush, ciliated, intermediate, and special (Jones and Reid, 1979; Reid and Coles, 1984). The first three secrete some of the constituents of mucus and provide material for the mucociliary escalator. Neuroendocrine cells secrete neurohormones. The others serve a variety of functions, from progenitor cells to transport and absorption, all of which have been reviewed (Reid and Coles, 1984). The relative proportions of the various cell types differ at different airway levels and the normal ratios can be altered by exogenous agents and in disease.

Various methods have been used for the morphological and histochemical study of airway epithelium and its development. Histochemical stains have been used to detect glycoproteins and glycosaminoglycans and to classify them by their possession of neutral or acidic sugar moieties (Lamb and Reid, 1972; Jones and Reid, 1973a,b; Jones et al., 1975; Mills et al., 1986) or vicinal glycols in hexoses of glycoproteins (Spicer et al., 1967). Recently, lectins conjugated to a dye have been used to localize specific sugars (Spicer et al., 1982). Techniques for the ultrastructural detection of these complex carbohydrates have been described (Spicer and Schulte, 1982).

Airway Glands

In the human fetus, the total number of mucous glands and ducts in the trachea has been studied through serial sectioning in the longitudinal and transverse plane. Transverse sections proved more suitable for the counting of ducts (Thurlbeck et al., 1961). After the 16th week of gestation, mucous gland distribution is uniform throughout most of the trachea, except in the distal one-twelfth where ducts are more numerous.

By examining glands in selected axial bronchial pathways, Bucher and Reid (1961) determined the total number, size, form, and position of the glands and their ducts, and their density relative to branch points and to cartilage plates.

The combined alcian blue and periodic acid–Schiff (AB/PAS) stain, with alcian blue used at several pH values, has been used to identify mucous glycoproteins of the bronchial glands with respect to their neutral or acidic sugar groups (Spicer and Warren, 1960; McCarthy and Reid, 1964; Lamb and Reid, 1969; 1972).

Alveolar Epithelium

In the human fetus, epithelial cell differentiation begins in the late canalicular phase, at about 20 weeks of gestational age. The presence of lamellar bodies identifies type II pneumocytes (Campiche et al., 1963; Lauweryns, 1970; Hage, 1973). When the contents of lamellar bodies are secreted, they are transformed into tubular myelin (Gil and Reiss, 1973; Paul et al., 1977). This in turn gives rise to the phospholipid monolayer at the air–fluid interface (Weibel et al., 1966; Paul et al., 1977). This monolayer represents functional surfactant. It is composed of several phospholipids, but is particularly enriched in dipalmitoyl phosphatidylcholine (King, 1974; Mason et al., 1977). In addition, three apoproteins are associated with it and these have been localized within the type II pneumocyte by the technique of in situ hybridization.

In Situ Hybridization

A cell function expressed in a protein product can now be studied dynamically by determining when the controlling gene is transcribed (deMello et al., 1988). In situ hybridization is a powerful tool that allows posttranscriptional gene expression to be studied within cells in tissue. In situ hybridization (ISH) uses nicktranslated DNA or RNA probes. To make single-stranded DNA probes complementary to specific genes, plasmid DNA containing the gene is nicked, and repaired with radiolabeled nucleotides. The plasmid is then cut and denatured to form single-stranded fragments that include fragments complementary to the gene to be studies. For RNA probes, antisense probes are generated from plasmid vectors. The relevant cDNA fragment is inserted within these vectors in reverse orientation, so that promoter sequences able to transcribe this fragment will be positioned at the far end of the gene fragment, thus transcribing an opposite or antisense strand. In medium containing radiolabeled nucleotides, a specific RNA polymerase promotes the synthesis of labeled antisense RNA. This hybridizes with complementary RNA in fixed frozen or paraffin-wax-embedded tissue sections. Often, ISH and immunohistochemistry (IH) are performed on consecutive adjacent sections to correlate posttranscriptional and posttranslational gene expression.

The criteria for tissue fixation are similar for both IH and ISH, but different from those previously given for morphometry and ultrastructure. Formaldehyde and glutaraldehyde, for example, are not considered optimal for these purposes. Paraformaldehyde is the fixative generally recommended: it gives a high signal-to-noise ratio and retains the maximal RNA. Tissue should be fixed as quickly as possible to avoid RNA degradation. However, if too prolonged, fixation can result in a poor signal because of cross-linking of proteins. Two hours at 4 °C is recommended as striking a balance between adequate preservation and excessive cross-linking.

While both IH and ISH can work with embedded tissue, frozen sections give the best results in terms of protein or mRNA retention.

F. Data Analysis

Sources of Variability

In analyzing data from developing lungs it is important to consider likely sources of variability, some of which may not be as important in normal, mature lungs. The following considerations can modify sampling strategies.

Variability can arise between fields, blocks, patients/animals, or lobes. In the sheep lung cranial lobes mature earlier than the other lobes. Usually the greatest contributor to overall variability is between individual patients or animals. Because of this, the n used in determining means and variance estimates should always reflect the number of patients or experimental animals (Gundersen and Osterby, 1981; Gupta et al., 1983). It follows from this that sampling protocols should be designed to maximize this number rather than the numbers of blocks or fields. In the study of developmental problems, however, detecting bilateral or lobar differences may make it necessary to sample these units adequately.

Differences between patients or individual animals in large part reflect intrinsic biological variability, but some other considerations should be noted:

Gender: During a critical "window" late in gestation, female lungs are more mature than those of males. During postnatal development, males tend to have larger lungs,

Position in uterus: Depending on the proximity of females to males or vice versa, lung maturation is affected. Position also may affect body weight,

Litter size: Affects body weight.

Correlating Morphometric Data

Lung weight and volume are often expressed relative to body weight to give the "specific value." Indeed, hypoplasia is best detected in this way, by lower than normal specific values. Morphometric and stereological estimates can be expressed similarly.

Over time, one structural measurement can be related to another using regression techniques. Fitting a line to data is an easy procedure if one has access to computerized analysis. In developmental studies, the data may be satisfactorily fitted with straight line, logarithmic, or exponential equations.

Allometric interpretation is a useful technique for relating the growth of two body parts, X and Y (Huxley, 1932). Its equation is

$$Y = a * X^b$$

Log conversion gives the equation for a straight line. Values for X and Y are plotted on a double log graph and a straight line added by least-squares regression. The slope is the power function, b, and the antilog of the intercept on the ordinate is a. If b is 1, the two structural compartments are growing at the same rate; if less than 1, compartment Y is growing more slowly than compartment X.

III. Interpretation

A. General Features of Normal Lung Development

The lung develops as a ventral diverticulum of the primitive foregut. In the human embryo this occurs at about the fourth week of gestation. Like the gut, the lung bud is lined by endoderm, but quickly becomes invested with mesoderm. During the first half of gestation, the lung remains small. Although considerable morphological changes are already complete by the end of the pseudoglandular stage at 16 weeks (Bucher and Reid, 1961a), only in the later part of the canalicular stage do large changes occur in the size of the lung. Between 28 weeks and term, lung volume increases exponentially (Langston et al., 1984).

Postnatally, the increase in lung volume is logarithmic when plotted against age, with the most rapid increase occurring in the first 5 years. By 2 years of age, males have a significantly greater lung volume than females, even when the values are normalized for body length (Thurlbeck, 1982).

During the first 5 years of postnatal life, the allometric relationship between lung volume (V_L) and body weight (M_B) is expressed (Zeltner et al., 1987):

$$V_L = 47.176 \ M_B^{1.051}$$

Since the power function is close to 1, lung volume is increasing at a similar rate to body weight. This is not true of all mammals. In the sheep, for example, during the first 6 months of postnatal life lung volume does not keep pace with somatic growth. The equation is (Davies et al., 1988):

$$V_L = 110 \ M_B^{0.67}$$

B. Preacinar Airways: Quantitation

During the embryonic stage, bronchi are formed by a process of irregular dichotomous branching that is stimulated by the surrounding mesoderm (Alescio and Cassini, 1962; Spooner and Wessels, 1970). In the 7-week human embryo, bronchial branching is complete to the subsegmental level (O'Rahilly, 1985). Bucher and Reid (1961a) showed that the more distal pattern of preacinar airway branches develops during the pseudoglandular phase. By 10 weeks' gestation, most pathways already have about 10 generations, shorter axial pathways are complete between 12 and 14 weeks, while longer ones are not complete until the 16th week. After this no further branches are added and, indeed, there is a slight reduction in the number of branches, presumably by alveolization of terminal bronchioli to form respiratory bronchioli. The formation of cartilage within airway walls is slower (Fig. 2). Cartilage first appears at proximal bifurcations and later appears peripherally. This process continues until 25 weeks' gestation. For any level, there is more variability in the distribution of cartilage than in the size of the airways. Generally, the longer the pathway the longer the length of airway with cartilage, so that cartilage extends to a similar relative length.

During late gestation and in the newborn, the human trachea is funnel-shaped, but during postnatal development it acquires a more cylindrical shape

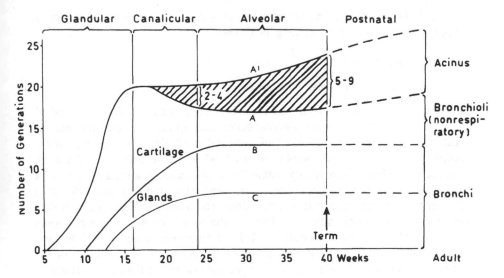

Figure 2 Human lung: the rate of development of airway generations, accompanying cartilage, and submucosal glands.

(Wailoo and Emery, 1982). An allometric analysis of the growing trachea in various mammals has shown that its length and diameter increase as $M_B^{1/3}$, as expected for a linear dimension. In the allometric double-log plot, however, the line for newborns is displaced below that of adults. Thus, in the newborn, the trachea is relatively shorter and narrower than in the adult. Predictably, tracheal volume increases as M_B^1, but once again, the line for newborns is displaced downwards and the line for tracheal volume/functional residual capacity (FRC) is appreciable lower at all body weights. Since tracheal volume is the major proportion of anatomical dead space, this may be a way of keeping dead space low, particularly when the acinar part of the lung is comparatively undeveloped (Mortola, 1983).

Postnatal growth of the airway system in humans has been studied by a number of authors whose conclusions have not always agreed. Cudmore and associates (1962) showed that the rate of increase in the length of small, peripheral airways was less than in larger ones. Hogg and his co-workers (1970) found that the diameter of peripheral airways was small in newborns and did not increase markedly. Hieronymi (1961) found that the most distal orders grow more quickly, particularly in length. Hislop and co-workers (1972) found that at all levels the diameter of airways increased proportionately, but in this study each level was defined as a fraction of the total length, which reflects the volume of lung supplied. Their result, then, indicates that at any age the diameter of an airway supplying a given volume of lung is similar.

C. Distal Lung

Formation of the Distal Air Spaces

Early in the canalicular stage, additional branching occurs at the distal end of the last generation of airways. Several generations of branches develop that are still lined by a continuous layer of cuboidal epithelium but are the future respiratory bronchioli and alveolar ducts (Boyden, 1974). From the 17th week capillaries are found increasingly within the acinus (canalization), appearing first at the periphery of the acinus. Individual capillaries gradually draw closer to the epithelium, particularly at branch points. The overlying epithelial cells flatten and become more squamous, forming the future air–blood barrier. Some flattening of the epithelium could result from migration of the cells (Chen and Little, 1987). Canalization progresses from the periphery to the center of the acinus. At the end of this series of changes, several generations of short tubes, the transitory saccules, end distally in a cluster of terminal saccules (Fig. 3). These undergo further differentiation, as described below.

1. The saccules are subdivided by secondary septa (Burri, 1974). These first appear on the primary septa as low crests containing an elastic bundle and narrow capillary. Over several days the height of each crest increases and the capillary at the tip becomes occluded and finally obliterated. The whole process has been described as uplift from the primary septum, but it could also be interpreted as growth and migration of tissue away from a tip stabilized by elastin. While the microscopic picture is the same, the two interpretations imply different mechanisms with regard, for example, to the direction from which the responsible mitogens and chemotactins will come.

2. Capillaries change their configuration: from a double system in which they are seen in section on both sides of the septum to a single one in which the capillary profiles alternate along both sides of the septum (Burri, 1974). The change appears first in the primary septa that originally separated the saccules but quickly extends into secondary septa. Its functional significance is unknown, but it may be an integral part of septal thinning. It can be detected by light microscopy, but was originally described by transmission electron microscopy (Burri, 1974; Zeltner and Burri, 1987), and, more recently, by scanning electron microscopy of microvascular casts (Caduff et al., 1986).

3. Both primary and secondary septa thin, largely due to a reduction in the interstitial volume. Because of this, intra-alveolar wall thickness decreases.

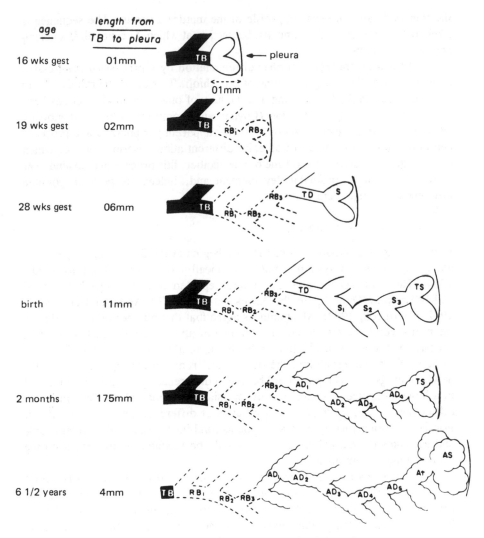

Figure 3 Human lung: diagram illustrates the growth of the acinus.

4. Individual alveoli increase in depth by growth of the intraalveolar septa running perpendicular to the axis of the alveolar duct.

There appears to be no consensus as to when, in this sequence, the air spaces can be termed alveoli. Most authors would presumably agree that steps 1 and 2 must be complete, but interpretations differ as to whether the possession of a single capillary system is enough or whether the airspace must also acquire the thin walls and cup-shaped profile of the mature alveolus. The sequence is likely to be common to all mammals, but the age at which it occurs differs among species (see below).

The acinus undergoes further differentiation by a process of progressive alveolization of the last generations of bronchioli. This almost certainly occurs to some extent in the human lung and the lung of other mammals, but has been described definitively in the dog (Boyden and Tompsett, 1961) and monkey (Boyden, 1976). Like canalization, it is first apparent peripherally and proceeds centrally. Unlike the stages of alveolar differentiation described above, which occur within a relatively short time of one another, this process may extend over a longer period of postnatal development and, indeed, is perhaps possible throughout life (see below).

Saccule/Alveolar Number

In the human, the subdivision of saccules begins at the 28th week of gestation and a single capillary system can be found locally at 32 weeks (Langston et al., 1984). The total saccular/alveolar number reported at birth ranges between 10 and 149 million (Dunnill, 1962b; Davies and Reid, 1970; Angus and Thurlbeck, 1972; Thurlbeck, 1982). Most authors agree that this number increases during the next several years to reach an adult total of up to 600 million. It seems that for any age there is considerable variability in total alveolar number. This may in part reflect real variability: the striking difference in the reported values for newborn lungs for example, could be due to differences in lung maturation and to a large incidence of undetected hypoplasia. But it may also reflect practical difficulties such as inadequate lung inflation or differences in counting criteria. For these reasons, airway generation counts combined with radial alveolar counts and estimates of internal surface area should be included in any check on the state of lung development.

In the newborn, the respiratory bronchiolus is short. Boyden and Tompsett (1965) report the appearance of new alveoli at 56 days. Zeltner and Burri (1987) report that respiratory bronchioli are first seen at the age of 6 months. This is likely to be due to the gradual alveolization of terminal bronchioli in which the lining epithelium thins, acquires a close capillary supply, and forms an alveolar-like cavity (Boyden and Tompsett, 1961). In the newborn rat, the lung is still in the saccular stage (Burri, 1974). During the first 2 weeks of postnatal life, it undergoes a rapid remodeling in which the saccules are subdivided by secondary septa (Burri, 1974). Between 3 and 8 days , the number of distal air spaces per unit sectional area triples. Alveolar development precedes that of intra-acinar arteries by as much as 7 days (Meyrick and Reid, 1982; Fig. 4).

In the sheep, saccular subdivision occurs in utero between 120 days' gestation and term (Alcorn et al., 1981). The newborn lamb has a total alveolar number

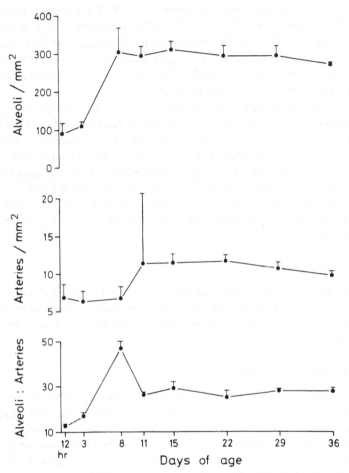

Figure 4 Postnatal lung development in rat: number of alveoli and arteries per unit area of section in rats from 12 hr to 36 days of age (from Meyrick and Reid, 1982, with permission).

of approximately 1 billion (Davies et al., 1988). Since lung volume is of the same order as in human newborns, the individual alveoli of the newborn sheep are smaller. Over the next 6 months the increase in alveolar number is logarithmic, following a curve similar to that of lung volume. Thus, numerical density of alveoli remains unchanged. As confirmation of this, alveolar volume density remains unchanged. Alveoli are being added to the lung, but the volume of the individual unit (Vv/Nv) is the same.

In the human fetus, radial alveolar count (RAC), increases exponentially when plotted against age. Values are 1–2 at 19 weeks and 4–6 at term (Cooney and Thurlbeck, 1982b). The RAC increases logarithmically after birth when plotted against age, with the period of most rapid increase occurring within the first 2 years (Cooney and Thurlbeck, 1982a). These two patterns of increase are similar to those shown by total alveolar number.

In all species there is a period following alveolar differentiation when the lung grows in volume and surface area. No study has been sufficiently exhaustive to determine how much these increases are due to the expansive growth of pre-existing alveoli or to the addition of new alveoli. Most evidence, however, suggests that alveoli are added for a prolonged period. Then the question is "By what mechanism?" One possibility is further subdivision of air spaces, but this may be genetically or mechanically limited. Another possibility is alveolization of bronchioli: these mechanisms are considered later in an analysis of adaptive lung growth.

Septal Thickness

In human fetuses, wall thickness of distal air spaces falls markedly during gestation from about 5 μm at 19 weeks to about 2 μm at term (Langston et al., 1984).

A similar change occurs in the sheep during the last 40 days of gestation (Alcorn et al., 1981; Crone et al., 1983, Fig. 5).

Internal Surface Area

In the human fetus internal surface area increases exponentially between 28 weeks and term, corresponding to a period when alveoli appear. At 28 weeks, surface area (determined by light microscopy) is 0.5 m^2: at term it is between 3 and 4 m^2. Alveolar surface area increases logarithmically after birth, with the most rapid increase occurring in the first 5 years. After two years of age, surface area in males is significantly greater than in females, even if it is normalized for height.

An ultrastructural study of human lungs in the first 5 years has calculated the allometric relationships in the growth of various lung compartments (Zeltner et al., 1987). During this time, surface area increases as lung volume to the power 1. A simple geometric increase would mean that surface area increases as the 2/3 power of volume. The power function greater than this indicates that additional surface has formed and that the surface of the lung has increased in complexity.

In the rat, between 4 and 21 of postnatal life, however, the surface area increases in relation to lung volume to a greater extent than in the human. The equation is (Burri et al., 1974):

$$S_A = 535.1 \ V_L^{1.60}$$

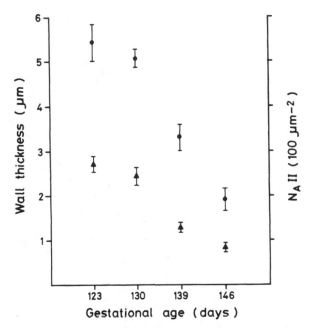

Figure 5 Fetal lung development in sheep: interalveolar wall thickness (circles) and number of type II pneumocytes per unit profile area of septal tissue (N_AII, triangles) in the lungs of four normal fetuses at different gestational ages (mean $\pm$SEM) (from Crone et al., 1983, with permission).

The power function of 1.60 clearly indicates an increase in surface complexity. On examination of microscopic sections it is evident that the original saccules are now subdivided by secondary septa (Burri, 1974).

In contrast, the lung of the newborn sheep is already alveolar and is not further subdivided. As we have seen, the postnatal increase in alveolar number parallels that of lung volume and is only a modest threefold. This is reflected in the growth of surface area, which increases at a similar rate to lung volume (Fig. 6; Davies et al., 1988).

$$S_A = 0.05 \ V_L^{1.07}$$

Capillary Surface Area

Because the capillary endothelium forms the other side of the air–blood barrier it is not surprising that its growth follows that of the alveolar surface. In the postnatal human lung, during the first 5 years, it increases at a similar rate to that of lung volume. In the rat it keeps up with alveolar surface area and increases

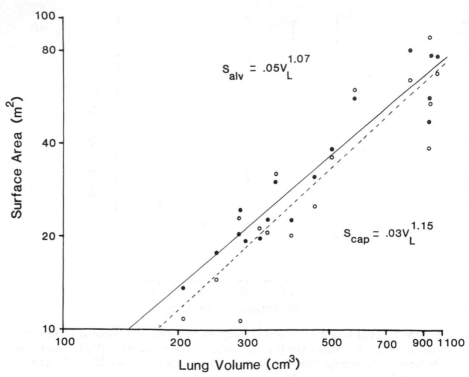

Figure 6   Postnatal lung development in sheep. Double log plots of alveolar (S_{alv}) and capillary (S_{cap}) total surface area against lung volume. The continuous line is the regression for alveolar surface area; the dotted line for capillary surface area (from Davies et al., 1983, with permission).

at a faster rate than lung volume (Burri et al., 1974). In the postnatal development of sheep (Fig. 6) capillary surface area is added at a similar rate to lung volume (Davies et al., 1988).

Although capillary surface area is clearly related to gas exchange, this is not its only function. Capillary endothelium is metabolically active and one of its functions is to process vasoactive mediators. It is responsible, for example, for converting the decapeptide angiotensin I to the octapeptide angiotensin II. This hydrolysis is catalyzed by angiotension converting enzyme (ACE) located on the luminal surface of the endothelial cell. Total enzyme activity should therefore be a reliable index of capillary surface area. In the postnatal developing lung of the sheep, this has been shown to be the case (Pitt et al., 1987). The V_{max} of ACE was measured using a tritiated molecule, BPAP, as substrate: the lungs were then fixed by instillation of fixative and processed for stereological determination of capillary surface area, S_c. Throughout the first 6 months of postnatal life, V_{max} and S_c correlate with an r value of 0.85 (Fig. 7). In the allometric relationship with lung volume, however, V_{max} increases to a significantly greater degree than capillary surface area.

$$V_{max} = 0.00003 \ V_L^{1.62}$$

$$S_c = 0.04 \ V_L^{1.08}$$

This suggests that extracapillary vasculature makes a relatively greater contribution to ACE activity during the postnatal period.

Capillary Luminal Volume

In the human during the first 5 years of postnatal life, capillary luminal volume increases to a greater degree than capillary surface. This occurs despite the fact that during the first 18 months the capillaries change from a double to a single system (Zeltner et al., 1987). This is an example of how a stereological estimate, providing a good index of the growth of a structural compartment, fails to detect a feature of structural remodeling. The sheep already has a single capillary system at birth. Throughout a 6 month period of postnatal growth, capillary volume (V_c) increases to a greater degree than lung volume, in a similar way to that of the human.

$$V_c = 0.004 \ V_L^{1.51}$$

Separate analysis of data from the first 30 days demonstrates a relatively greater increase

$$V_c = 0.004 \ V_L^{1.89}$$

During the first 30 days, the ratio of capillary luminal volume to surface increases, suggesting that the capillaries increase in diameter more than in length (Davies et al., 1988).

Interpreting Structural Changes in Relation to Body Weight

In the human lung the postnatal increase in alveolar surface area keeps pace with increasing body weight (Zeltner et al., 1987). The same is true of the rat (Burri et al., 1974). In the sheep, however, neither alveolar nor capillary surface areas increase at a similar rate to body weight, although, as we have seen, capillary luminal volume does (Davies et al., 1988). Oxygen uptake increases with body weight (as $M_B^{0.75}$; Kleiber, 1932) and activity. It is assumed that the oxygen uptake required by an active newborn lamb requires a mature lung with large surface area. Despite considerable increase in body weight, there is not a concomitant increase in

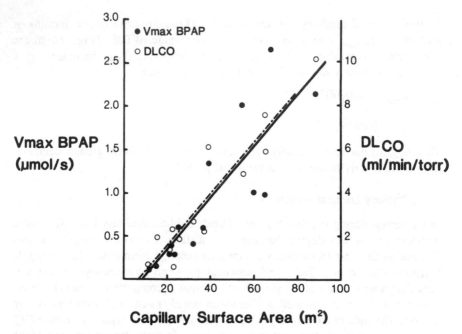

Figure 7 Postnatal lung development in sheep. Relationship of maximal velocity (Vmax) of [³H]benzoyl-phenylalanyl-proline (BPAP) hydrolysis, by angiotensin-converting enzyme (closed circles) or carbon monoxide diffusing capacity (D_{LCO}) (open circles) vs. stereologically determined capillary surface area. Solid line, regression for BPAP; dashed line, regression for D_{LCO} (from Pitt et al., 1987, with permission).

activity, so that the oxygen demand of the growing sheep can be met with a modest growth of surface area, balanced, perhaps, by the relative increase in capillary lumen, which is the only compartment in the distal lung to keep pace with body weight. Whereas

$$S_c = 7.84 \ M_B^{0.69}$$

$$V_c = 3.86 \ M_B^{1.09}$$

In the human and rat, which are helpless as newborns, the oxygen demands are accommodated by a lung that still has immature architecture. With somatic growth and an increasing level of activity, oxygen demands require extensive growth and remodeling of the surface for gas exchange.

D. Vascular Development

In the human the development of the vasculature has been studied using barium/gelatin injected material (Hislop and Reid, 1972). In the fetus the growth of the preacinar arteries follows that of the preacinar airways, so that the branching pattern of both conventional and supernumerary arteries is complete by the 16th week of gestation. After this, preacinar arteries increase in diameter and their wall structure, whether elastic or muscular, is set at this time. The medial thickness of arteries, expressed as a percentage of external diameter, is greater than postnatally (Hislop and Reid, 1972).

Immediately after birth the medial thickness of resistance arteries less than 200 μm diameter falls. With the increased arborization of acinar airways there is a corresponding increase in accompanying arteries, both conventional and supernumerary. This continues up to 8 years of age. The development of medial muscle lags behind, however, so that fewer arteries are muscular in childhood than in the adult, and medial muscle extends less far to the periphery (Hislop and Reid, 1973).

E. Markers of Cell Function

Airway Epithelium

Ciliated Cell

The ciliated cell is the predominant cell type in human airways, playing a role in mucociliary defense mechanisms as well as in the active transport of ions across the airway wall. By 13 weeks of gestation cilia are present in most major bronchi. Fetal and mature ciliated cells do not differ in their structure and organization.

Clara Cell

During the postnatal development of the rat, various changes occur in the Clara cell (Massaro et al., 1984). At 1 day of age the cell contains a large amount of glycogen and no secretory granules, whereas at 4 days this situation is reversed. In rabbits this shift occurs more slowly (Plopper et al., 1983).

Neuroendocrine Cells

In humans the first neuroendocrine cells appear at 8 weeks' gestation and by 28–32 weeks they are present at all airway levels (Cutz, 1982).

Spindel and associates (in press) have used ISH and IH to show that mRNA for gastrin-releasing peptide (GRP) first appears in proximal airways and is followed by the appearance of the peptide. This sequence later occurs in the distal airways. The peptide is apparent both proximally and distally for sometime after the "message" is no longer detectable and reaches a plateau at birth. Neuroendocrine cells are concentrated at branch points, and they may have a role in the process of branching.

In contrast to GRP, vasoactive intestinal peptide (VIP) is confined to cells in the upper respiratory tract. In infants respiratory distress syndrome (RDS), GRP content was reduced at all airway levels whereas VIP content was no different from normal (Ghatei et al., 1983).

Airway Glands

Bucher and Reid (1961b) determined that the first submucosal gland bud in the human fetus appeared in the tracheobronchial tree at 13 weeks' gestation (Fig. 8), and by 24–25 weeks the glands had a structure similar to that of the adult (Fig. 9).

Reid and de Haller (1967) found that the proportion of glandular acini staining for sialomucins was much lower than in adults. At birth any acid mucins were sulfated, sialomucins appeared at about 6 months, and increased with age relative to sulfated mucins (Lamb and Reid, 1972b). In the pig, immediately after birth, the proportion of cells in airway epithelium and submucosal glands staining positively for acidic and neutral glycoproteins increased (Mills et al., 1986). In the epithelium cells producing sulfated glycoprotein increased with age whereas

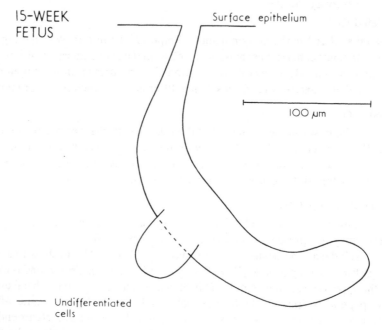

Figure 8 Human fetal lung at 15 weeks' gestation: diagram shows the appearance of a submucosal gland.

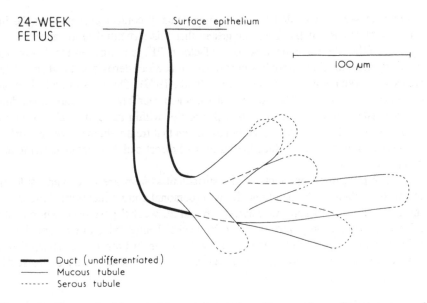

24-WEEK
FETUS

Surface epithelium

100 µm

—— Duct (undifferentiated)
—— Mucous tubule
······ Serous tubule

Figure 9 Human fetal lung at 24 weeks' gestation: diagram shows the appearance of a submucosal gland.

in the glands they were present only until day 3, when the glycoproteins became largely sialylated.

Alveolar Epithelium

In the human fetus alveolar epithelial cell differentiation begins in the late canalicular stage, at about 20 weeks' gestation, when osmiophilic lamellar bodies begin to appear in the cytoplasm (Lauweryns, 1970; Campiche et al., 1963, Hage, 1973). The number and size of the cells increase over the next few weeks. By week 25–26 some epithelial cells have become attenuated and resemble type I pneumocytes (Stahlman and Gray, 1978). In the developing lung of mice and rats autoradiographic evidence has shown that the type II cell is the precursor or stem cell of the alveolar epithelium (Adamson and Bowden, 1974; Kauffman, 1980; Kauffman et al., 1974).

The mesenchyme is important in inducing early airway branching and continues to be important at later phases of fetal growth since it is implicated in the maturation of the surfactant system. Structural studies in mice (Grant et al., 1983) and rats (Leung et al., 1980; Adamson and King, 1984b) have shown that at the time when the type II epithelial cell is differentiating (days 19 and 20), its underlying basement membrane becomes discontinuous and its basal cell membrane forms "foot processes" that often lie close to or actually contact processes of the interstitial fibroblast. It has been suggested that this structural association allows passage of the fibroblast–pneumonocyte factor (FPF), an oligopeptide that is produced by the interstitial fibroblast in response to cortiocosteroids and that stimulates surfactant production by the type II cell (Smith, 1979). The facts are a little more complicated, however. The number of basement membrane discontinuities does not correlate with the number of lamellar bodies within the epithelial cell (Grant et al., 1983) and a more recent study has shown that treatment that reduces surfactant production does not reduce the number of direct cell–cell contacts (King and Adamson, 1987).

The protein components of the surfactant system are of considerable interest. In explants of human lung, Ballard (personal communication) showed that the 6 kD apoprotein appears between 16 and 20 weeks' gestation, whereas the 35 kD apoprotein only appears after 20 weeks. Using ISH, Phelps and Floros (1988) demonstrated synthesis of the 6 kD protein in both bronchial epithelial cells and type II pneumocytes, whereas the 35 kD protein is only expressed in the type II pneumocytes (Fig. 10.)

F. Factors Influencing Lung Development

General Considerations

We will now consider specific factors that can be implicated in growth because they are absent or altered in human disease or in animal models.

In diagnosing abnormal lung development, several criteria need to be established, in particular those defining a change in mass (hypoplasia or overgrowth) and those defining inadequate differentiation.

The most frequent result of perturbed lung growth is hypoplasia (Liggins, 1984). This term is applied to a lung of small mass that has a reduced weight for age and body weight. In such cases, even if lung volume is normal due to distended, hypercompliant air spaces, the total surface area for gas exchange is reduced.

A hypoplastic lung can reproduce the normal growth pattern, exhibiting an overall reduction in the size of units at both pre- and intraacinar levels. This suggests that whatever hampered growth prevailed over the whole developmental period. Alternatively, and perhaps more usually, the pattern is disturbed, with a disproportionate reduction either in the number of preacinar airways (indicating early disruption, before the 16th week) or in the number and/or growth of intraacinar air spaces (late disruption).

Rather than being too small, lung tissue can have excessive mass due to overgrowth. If this occurs uniformly in a lobe or lung, it results in increased weight and gas exchange surface area for age and body weight. It is always accompanied by an abnormal pattern of growth. Pre- and intra-acinar structures are developmentally dissociated: there is no recorded instance of increased preacinar airway

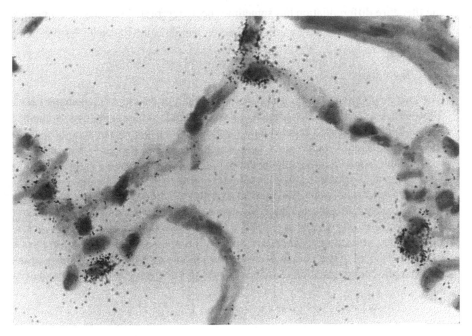

Figure 10 Human lung biopsy: in situ hybridization of 35 kD surfactant apoprotein using a [^{35}S]UTP antisense mRNA. Autoradiographic grains are localized over alveolar type II pneumocytes (220 X).

number so that overgrowth occurs largely or exclusively within the acinus. To a limited degree, this type of excessive growth is seen postnatally in normal individuals as an adaptation to reduced oxygen tension or increased oxygen demand. That it occurs in utero, when the lung is not the organ of respiration, demonstrates that other factors are at work. Overgrowth of this kind compensates for hypoplasia in the contralateral lung or ipsilateral lobe(s) and is a feature of pre- and postnatal growth of tissue remaining after lobe or lung resection.

In reviewing studies dealing with hypoplasia or overgrowth, this chapter makes no attempt to be exhaustive In many of the studies the only end-points are lung weight and volume and these are inadequate as a means of assessing how the lung has grown. In trying to interpret results, therefore, only reports that are truly morphometric or that offer interesting leads will be included.

Differentiation can be assessed in any system within the lung, but is usually judged in reference to a functional alveolar epithelium that allows independent respiration. This inevitably depends on adequate surfactant production and secretion, features that have not necessarily been evaluated by structural methods. In the literature differentiation has sometimes been interpreted as being synonymous with structural maturation. To some degree, however, the two can be dissociated (Kauffman, 1977).

Hormones

The lung developing in utero is exposed to a wealth of different hormones (Fisher, 1986). Their effects have been studied in a large number of animal models, though most of them have used biochemical measures of surfactant phospholipid as an index of maturation rather than lung structure itself (Kitterman, 1984).

Many studies have implicated the hypophyseal–pituitary axis in lung development, working through the adrenal and thyroid. Morphometry was used in a study determining the effects in fetal sheep of hypophysectomy alone or with thyroidectomy and subsequent replacement therapy with cortisol or adrenocorticotrophin (ACTH) (Crone et al., 1983). Hypophysectomy at 99–122 days' gestation resulted at term in an immature lung with more cuboidal epithelial cells per unit area (Fig. 11). The normal decrease in wall thickness that occurs late in gestation did not occur. Infusion of cortisol or ACTH for just 72–84 hr immediately

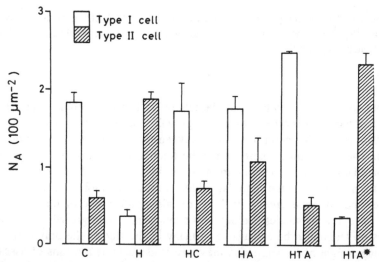

Figure 11 Fetal lung development in sheep. Effect of late gestational hormone supplementation on the number of type I and II pneumocytes per unit profile area of septal tissue (N_A) present at term. The groups are C, control; H, hypophysectomized; HC, hypophysectomized + cortisol supplementation; HA, hypophysectomized + ACTH; HTA, hypophysectomized and thyroidectomized + ACTH (mean ± SEM). Asterisk indicates the value of one animal from the HTA group (from Crone et al., 1983, with permission).

before delivery promoted lung maturation so that the number of cuboidal epithelial cells (Fig. 11) and the wall thickness both rapidly readjusted to reach control levels.

Corticosteroids stimulate production of pulmonary surfactant. The uptake of steroids has been demonstrated autoradiographically in explants of fetal rat lung (Leung et al., 1980). In the mouse lung, dexamethasone administered transplacentally between 17 days and 19 accelerated structural development. Specifically, the volume density of air spaces was increased in a dose dependent manner. This effect was achieved at a lower dosage than that required to increase the number of osmiophilic granules in type II cells, implying a dissociation beween the two processes. The synergistic action of thyroxine and hydrocortisone in promoting structural aspects of alveolar epithelial differentiation has been noted (Hitchcock, 1979).

Dexamethasone has been shown to have an effect on the major surfactantassociated protein (28–36 kD) and its mRNA. In explants of human fetal tissue at 16–23 weeks' gestation, maintained in culture for up to 5 days, dexamethasone increased the mRNA expression and protein content (Ballard et al., 1986; Liley et al., 1987).

Infants born prematurely suffer from surfactant deficiency that is manifested clinically as RDS and pathologically as hyaline membrane disease (HMD). Lungs of infants dying with HMD have been found to lack tubular myelin (Fig. 12), even though the type II pneumocytes contain abundant lamellar bodies (deMello et al., 1988; Fig. 13). A defect in the conversion of lamellar bodies to tubular myelin has been proposed as an explanation for this finding. Clinical studies utilizing surfactant preparations for replacement therapy in infants with RDS have indicated that the protein component of surfactant plays an important role (Taeusch et al., 1986; in press; Avery et al., 1986).

Whereas corticosteroids stimulate surfactant production, androgens produce the opposite effect: testosterone, for example, depressed surfactant production in rabbits (Nielsen et al., 1982) by blocking the action of fibroblast–pneumonocyte factor (FPF) (Torday, 1985). This effect provides a possible explanation for the increased incidence of RDS in male infants (Nielsen and Torday, 1987). Autoradiography in rats showed that the normal decrease in epithelial cell division seen in both sexes on days 18 and 19 of gestation (preceding the appearance of differentiated type I and II pneumocytes) was significantly greater in the female. By day 20 males have caught up. On days 18 and 19 the basement membrane underlying the type II cells was more continuous in males than females, while the number of foot processes formed by the epithelial cell was fewer. In males there were fewer contacts between epithelial and interstitial cells.

Insulin is another hormone essential for normal lung growth and maturation (Bourbon and Farrell, 1985). Infants of diabetic mothers are often born with RDS, generally thought to be due to an interference with lipid metabolism. Animal models have been created by administering pancreatic B-cell toxins to pregnant

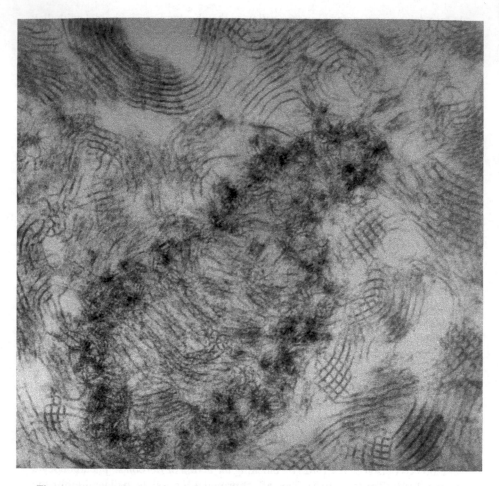

Figure 12   Electron micrograph of tubular myelin, the secreted form of pulmonary surfactant, within the alveolar space of a human newborn.

mothers, such as streptozotocin (STZ) and alloxan. STZ given to pregnant rats delayed fetal differentiation of the type II cells (Tyden et al., 1980, Gewolb et al., 1985). Fewer foot processes on the basal cell membrane that interact with underlying mesenchymal cells were formed (Grant et al., 1983). In the pregnant rabbit given alloxan, glycogen content in the epithelial cells of fetal lungs was greater than in controls. A more important finding was that fusions of the epithelial and capillary endothelial basement membranes were reduces in number so that the air–blood barrier remained thick (Sosenko et al., 1980). This indicates a profound effect on morphogenetic processes.

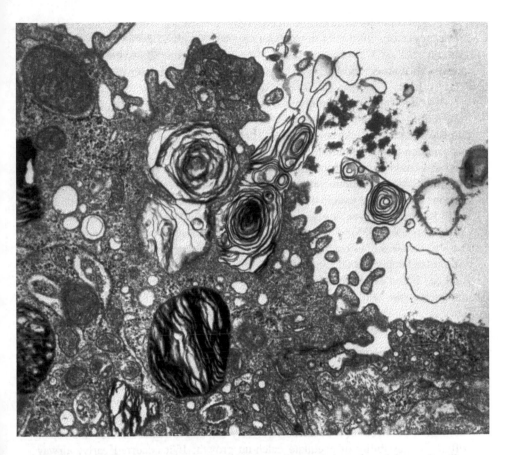

Figure 13 Electron micrograph of pulmonary surfactant being released from a lamellar body of a type II pneumocyte into the alveolar space of a human newborn.

Other hormones that appear to stimulate the maturation of the surfactant system in utero include β-adrenergic agonists, opioids, and prostaglandins (Kitterman, 1984; Smith and Bogues, 1980). They may also have effects on structure, but appropriate techniques have not been applied to establish this.

Hormones continue to be important in postnatal development. The role of the pituitary–hypophyseal axis, for example, is demonstrated in acromegaly (Brody et al., 1970). Male subjects with this condition had greater total lung capacity, FRC, and tissue volume. Female subjects, on the other hand, had normal values. Conversely, in a case of hypopituitarism, the lung was small (Jain et al., 1973).

In the newborn rat, treatment with triiodothyronine (T3) for the first 6 days of life accelerated the increase in surface area, alveolar surface to volume ratio, and reduction in mean linear intercept, despite having no effect on lung volume (Massara et al., 1986). If thyroid activity was inhibited by administering propylthiouracil (PTU) subcutaneously to the newborn pups, the normal increase in these indices was retarded: body weight and lung volume were unaffected. It is assumed that T3 increases alveolar number and complexity.

Dexamethasone, administered daily to rats from 4 to 13 days postnatally, impaired, for as long as the animals were monitored (up to 60 days), saccule subdivision that normally occurs during this period. The lungs nevertheless continued to grow and lung volume at the end was, in fact, slightly greater than normal and this could not be attributed to increased compliance or overdistention of air spaces. Surface-to-volume ratio and mean linear intercept of the air spaces, however, was larger than in untreated controls. The authors concluded that volume and, to a lesser degree, surface area increased by adding alveoli of the same size (Massaro et al., 1985). A later study showed that during this period dexamethasone had a differential effect on cell replication (Massaro and Massaro, 1986). It diminished replication of interstitial fibroblasts more than type II pneumocytes and had little or no effect on endothelial cells. These effects resulted in thinning of the alveolar wall.

Adequate Intrathoracic Space

That lung development requires sufficient intrathoracic space is suggested by hypoplastic conditions in which space has been severely compromised, such as congenital diaphragmatic hernia and thoracic dystrophy. In diaphragmatic hernia, the volumes of both lungs are often reduced but one side is usually affected more than the other. The nature of the disturbance to lung growth depends on when herniation occurred. If it occurred late, only acinar growth is affected and repair offers the possibility of adequate catch-up growth. If it occurred early, airway development is impeded, as was first shown in a case described by Kitagawa and co-workers (1971) and postrepair development does not catch up (Hislop and Reid, 1976; Thurlbeck et al., 1979).

In the sheep, diaphragmatic hernia has been created as early as 60 days, during the pseudoglandular stage. At 100 days, and at term, lung volume relative to body weight was reduced, more so on the ipsilateral side. Surgical repair done at 100 days appeared successful in one animal. At term, both lungs had volumes similar to normal.

Scoliosis mainly affects postnatal lung development. The number of alveoli has been found to be reduced (Davies and Reid, 1971; Berend and Marlin, 1979). Alveolar surface area was also reduced, but the size of distal air spaces was increased (Berend and Marlin, 1979).

Oligohydramnios is accompanied by lung hypoplasia, but whether it is due to space restriction is controversial. A simple interpretation is confused by the finding that a lung from a patient with polyhydramnios can have a reduced number of airway generations. This indicates an early abnormality, probably occurring before the volume of amniotic fluid is significant (Reid, 1984).

Respiratory Movements

The lung ventilates in utero and this is presumably one way in which available space influences the lung. Certainly in conditions associated with impaired respiratory movement, lung development is adversely affected. In most of these disorders it is difficult to differentiate the roles of respiratory movement and thoracic volume. Diaphragmatic action increases the volume of the thoracic cavity as its level descends during development.

Clinical cases of phrenic nerve agenesis are an example of a defect operating early in development. In one case studied morphometrically (Goldstein and Reid, 1980), the lung volumes were markedly reduced largely due to a reduced number of airway generations and thus of acini: acinar development, as indicated by radial alveolar count, was normal. Medial thickening of distal arteries was noted, together with extension of muscle into the normally nonmuscular precapillary artery segment. No skeletal muscle fibers were found in the diaphragm, except for a single isolated band.

Experimental studies have concentrated on later lung development. Thus, in fetal lambs, bilateral phrenectomy at 103–113 (Alcorn et al., 1980) or at 116–117 days' gestation (Fewell et al., 1981) resulted in lungs with reduced wet weight to body weight ratios and reduced radial alveolar count at term. The muscle of the diaphragm had atrophied.

Some studies have been designed to maintain diaphragmatic tone while affecting the respiratory movements proper. This was done in rabbits at 24.5 days' gestation (term=31 days) by transecting the spinal cord above the phrenic nucleus (Wigglesworth and Desai, 1979). When the fetuses were examined 3 days later, the wet weight of the lungs was less than that of controls or of animals in which the cord had been transected below the phrenic nucleus. The reduction in weight could be prevented by tracheal ligation, which suggests that the retention of fluid within the lung contributed to growth.

Another way of interfering with respiratory movements is by thoracoplasty, which increases chest compliance. This was performed on fetal sheep at 114 days' gestation. At term, wet weight of the lungs was less than normal (Liggins et al., 1981). This suggests that the active expansion of respiration, not just volume increase, is important.

Amniotic Fluid Volume and Pressure

The lung secretes liquid that augments the amniotic fluid. Respiratory movements probably encourage production of lung liquid as well as its removal from the lung.

Amniotic fluid may in turn enter the lungs, but how much is not known. Fluid flow is somewhat restricted and discontinuous so that a residual volume and slight positive pressure may always be maintained within the lung (Liggins, 1984; Vilos and Liggins, 1982). Experiments performed in sheep late in gestation have supported the idea that this fluid promotes development. Thus in fetal lambs at 105–110 days' gestation, drainage of fluid by tracheostomy resulted, at term, in reduced lung weight (absolute and normalized for body weight) and volume. Conversely, tracheal ligation at the same age produced a lung of greater weight and volume at term (Alcorn et al., 1977).

Lung development is often impaired by oligohydramnios, in which the absolute volume of fluid is reduced. One example that occurs late in gestation is when premature rupture of the amniotic membranes leads to chronic fluid loss. This is perhaps like the experimental effect of tracheostomy. In some forms of urinary obstruction and in renal agenesis and dysplasia, the lung is hypoplastic. The difficulty in interpreting data in these conditions occurs because loss of fluid inevitably results in some degree of compression to the chest or to the fetus as a whole.

In guinea pigs the effects of amniocentesis performed late in the canalicular stage (on day 45 of gestation) were examined on day 50 (term=67 days; Collins et al., 1986). Both lung volume and internal surface area relative to body weight were reduced. Total number of saccules was reduced but the high variances prevented this from being significant. Mean volume of saccules was normal. Amniotic fluid drainage seems to impair acinar development, whether by drainage of lung fluid volume or loss of a constituent needed for growth.

In rats, amniotic fluid was withdrawn at a slightly earlier stage of development, on day 16 of gestation, and the lungs examined just before term (21 days; Blachford and Thurlbeck, 1987). Lung weight/body weight and lung volume/body weight ratios were reduced, but absolute and relative internal surface area was normal. Mean linear intercept was also normal, suggesting that the saccules were of normal size. The authors presumed that total saccular number was reduced, but this was not indicated directly by the results. It was unclear how surface area remained normal, unless the number of alveoli per acinus had been increased.

In the monkey (*Macaca fascicularis*), maternal amniocentesis was performed either early in gestation (47–64 days) or later (85–95 days; term=155 days) (Hislop et al., 1984). In the early group, the animals at term had somewhat reduced body weights. When adjusted for body weight, lung volume in both groups was unaffected. In neither group was airway number abnormal. Both early and late amniocentesis, however, reduced total alveolar number, either wholly or in part through a failure to form the normal number of respiratory bronchioli. These effects occurred independently of the volume of fluid removed and, indeed, were apparent even after simple puncture of the amniotic sac. The authors speculated that puncture alone was sufficient to cause leakage and that sealing of the hole by normal repair processes occurred more slowly or less successfully than anticipated.

Interfering with the outflow of urine into the amniotic sac also produces oligohydramnios and simulates clinical conditions in which kidney function is comprised. In rabbits, the urethra was occluded on day 25 of gestation (term = 31 days). At term, lung weight to body weight ratio was reduced, but lung histological appearance was normal. In one group of animals, after the membranes were ruptured the amniotic sac was constantly infused with saline for the remainder of gestation, or the viscera were allowed to herniate through the abdominal wall, both of which procedures attempted to avoid the compression produced by abnormally large kidneys. In both groups, lung weight/body weight was higher than in the group that underwent ureteral occlusion alone, but somewhat less than normal (Nakayama et al., 1983).

The need for caution in drawing conclusions only on the basis of lung weight or volume without microscopic morphometry is illustrated by an oligohydramnios model created *early* in gestation in sheep (Hu et al., 1987). Bilateral ureteral obstruction was performed on three sheep at 60 days' gestation, during the pseudoglandular phase of lung development. At term, the animals had abnormal kidneys and lung volumes small for body weight. The numerical density of alveoli and the radial alveolar count were increased however (Fig. 14), and in two animals total alveolar number relative to body weight was slightly higher than normal. Mean linear intercept was reduced in all, indicating that the air spaces were abnormally small. Thus, although the lungs are hypoplastic by total size, acinar development as judged by alveolar multiplication was not impaired.

In a study of eight human infants with renal anomalies, lung volumes were low (Hislop et al., 1979). This was due partly to poor intraacinar development as indicated by reduced radial alveolar counts, but, unexpectedly, the number of airway generations was also reduced. Since preacinar airway generations are complete by the 16th week of gestation when renal contribution to amniotic fluid is negligible, this suggests that the development of kidney and lung is linked by a mechanism independent of the volume of amniotic fluid. One possibility is that the kidney is a major source of proline, which in its hydroxylated form is an important component of collagen. The fact that matrix components are important to development has been demonstrated in a number of experiments examined in the next section.

Metabolic Building Blocks

Collagen appears to be important to the airway branching morphogenesis that takes place in the embryonic lung (Alescio, 1973). A study performed in rats postnatally, long after airways are complete, examined its effects in intraacinar development (Das, 1980). The lathyrogen β-amino-proprionitrile (BAPN) interferes with

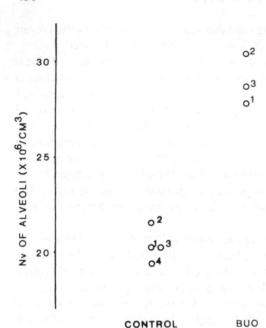

Figure 14 Bilateral ureteral obstruction performed in three fetal sheep at 60 days' gestation: numerical density (N_V) of alveoli at term compared with unoperated term controls.

collagen and elastin synthesis by binding and inactivating lysyl oxidase, which cross-links the chains as they form. BAPN was injected intraperitoneally into groups of male and female rats on days 1, 3, and 5 of postnatal life. On day 6, body weight and lung volume were reduced compared to controls that had received normal saline injection. Lung volume/body weight was slightly increased. Total length of elastin (determined from linear density, Lv) was one-fifth that of controls. Mean linear intercept was increased, but not significantly. Total alveolar number was reduced to one-third of normal.

In a more prolonged treatment, BAPN was injected into male rates every second day from day 2 to day 28 after birth (Kida and Thurlbeck, 1980). Both lung volume and lung volume/body weight were increased. Mean linear intercept was significantly increased, in large part because alveolar ducts were expanded and hypercompliant. These effects compensated volumetrically for the fact that total alveolar number was reduced, this time to about half of the control value. BAPN clearly interfered with postnatal alveolar development. This may be, in part, because during saccular subdivision elastin and, to a lesser degree, collagen are major components of the tip of the growing secondary septum (Burri, 1974).

The role of collagen in alveolar epithelial differentiation was investigated in fetal rats by injecting the proline analog *cis*-hydroxy-L-proline into the mother. While lung weights were similar in treated and control groups, the proline analog delayed morphogenetic changes and reduced the percentage of cells containing lamellar bodies, yet the number of epithelial cell–interstitial fibroblast contacts was normal (King and Adamson, 1987).

G. Factors Producing Abnormal Lung Growth

Rhesus Factor

Babies born with rhesus isoimmunization often have hypoplastic lungs. In six babies studied morphometrically, the airway generations were consistently reduced, indicating an early disturbance of growth (Chamberlain et al., 1977). Since radial alveolar count was also reduced, the effect continued into later developmental stages.

Genetic Factors

Patients with Down's syndrome and congenital heart lesions often have hypoplastic lungs. An earlier study of seven patients with or without heart disease showed that, in six, fixed lung volume was lower than predicted (Cooney and Thurlbeck, 1982c). In these same six, radial alveolar count was also reduced. The distal air spaces were larger than normal, as indicated by a higher mean linear intercept: alveolar surface area was reduced below normal. In a more recent study of 13 patients with Down's syndrome and congenital heart lesions, additional patterns of disturbed growth have been described. Ten had a reduced number of airway generations (Schloo and Reid, 1987). In the 11 patients on whose lungs radial alveolar counts could be made, 5 were within the predicted range, 1 was reduced, but the remaining 5 had counts 143–162% above predicted. This condition is called polyalveolar, a term first applied to a newborn infant with an enlarged lobe (Hislop and Reid, 1970; Fig. 15), and later to the single left lung of a 3-months-old infant with right lung aplasia (Ryland and Reid, 1971; Fig. 16). In the latter case, airway number was greatly reduced as it was in the Down's syndrome cases. The Down's syndrome patients' lungs had appreciable remodeling of the arteries with medial thickening, extension of medial muscle into normally nonmuscular arteries, and, in some, more advanced lesions.

Cystic fibrosis (CF), a common lethal genetic disease in white's, causes dysfunction of exocrine glands, notably the pancreas. This results in achylia. In patients less than 4 months old, it was found that despite obvious abnormalities in exocrine pancreatic development, the tracheal submucosal glands appeared normal (Sturgess and Imrie, 1982). These children are prone to lung infections for reasons that are not yet identified.

In the bronchial submucosal glands of young CF patients the histochemical pattern was normal, with low concentrations of sialomucins. Older patients, on the other hand, had lower concentrations than normal—a condition similar to that

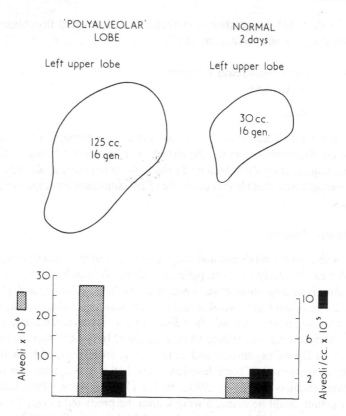

Figure 15 Human newborn diagnosed as having "lobar emphysema" in the left upper lobe. After its removal at 17 days, lobe was found to be "polyalveolar," with five times more alveoli than the normal lung shown on the right of the figure.

of chronic bronchitics (Reid and de Haller, 1967). Instead, there were higher levels of sulfomucins, as demonstrated by radiolabeled sulfate uptake studies (Reid and de Haller, 1967).

H. Adaptive Growth

Adaptation to Reduced Ambient Oxygen and Increased Oxygen Demand

Exposing an animal to high-altitude hypoxia during lung development results in larger lungs with increased internal surface area (Burri and Weibel, 1971; Lechner and Banchero, 1980) and total alveolar number (Cunningham et al., 1974). In the growing animal, increased oxygen uptake through higher levels of activity

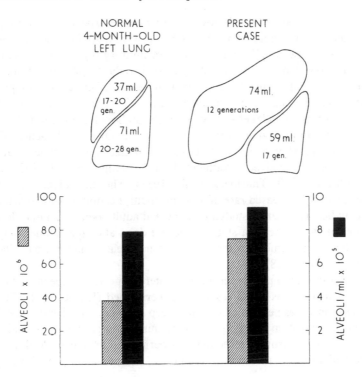

Figure 16 Human infant with right lung aplasia at 3 months. In the left lung the number of airway generations was reduced, but the lung increased its volume by increasing the total number of alveoli.

also resulted in increased surface area (Geelhaar and Weibel, 1971; Huggonaud et al., 1977; Gehr et al., 1978).

Adaptation to Increased Intrathoracic Space

In contrast to the restrictive conditions described earlier, the amount of thoracic space available to the lung may be increased. This occurs naturally in unilateral or lobar agenesis and hypoplasia, but more often is a feature of surgical intervention such as in repair of congenital diaphragmatic hernia, lobectomy, or pneumonectomy.

Removal of a lung or lobe in children because of lobar emphysema results in compensatory growth of the remaining tissue so that final volume is similar to normal for both lungs (McBride et al., 1980). Preacinar airways are incapable of increased branching, but may increase in size. Experiments in ferrets indicate that distal bronchioli increase their diameter following pneumonectomy, but this is not as much as the two-thirds power that would be expected if it were proportional to the increase in lung volume (McBride, 1985). An inadequately increased airway system now supplies an expanded respiratory region. The result is that airway function, particularly expiratory flow, is reduced (McBride et al., 1980).

Experimental studies have shown that resection stimulates compensatory growth of the alveolar surface area (Burri and Sehovic, 1979). This occurred even in adult rats (Wandel et al., 1983). Some authors have suggested that compensatory growth involving increase in alveolar number is only achieved in immature animals, whereas in adult animals increased lung volume results from growth of preexisting alveoli (Nattie et al., 1974; Langston et al., 1977; Holmes and Thurlbeck, 1979; Thurlbeck et al., 1981). The unresolved question was whether an early increased rate of alveolar multiplication was maintained. This could only be answered by studying young and adult operated animals long after resection, as was done in a study of beagle dogs after pneumonectomy. This showed that final number in the adult was no greater than normal (Davies et al., 1982; Figs.17, 18).

The response to pneumonectomy in utero appears to depend on the gestational age of the fetus when resection is performed (Hu et al., 1987). If left pneumonectomy was performed on fetal sheep at 60 days, during the pseudoglandular stage, adaptation was poor: the remaining lung grew to a normal volume for a right lung at term. If resection was performed later than 80 days, during the canalicular or later phases of development, compensatory growth was greater

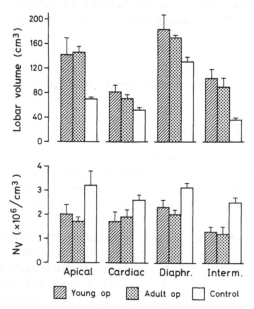

Figure 17 Left pneumonectomy in beagle dogs: compensatory growth of the remaining lung. Top: postfixation volumes of four lobes of the right lung: apical, cardiac, diaphragmatic, intermediate. Bottom: Numerical density (N_V) of alveoli in each of the four lobes (mean ± SEM). Young op, dogs operated at 6–10 weeks of age; adult op, dogs operated at 1 year (from Davies et al., 1982, with permission).

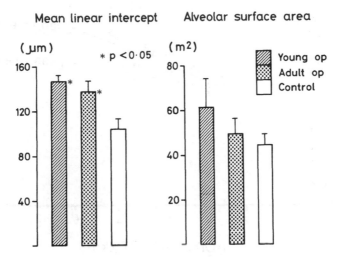

Figure 18 Left pneumonectomy in beagle dogs: compensatory growth of the remaining lung. Mean linear intercept and alveolar surface area. Young op, dogs operated at 6–10 weeks of age; adult op, dogs operated at 1 year.

and, although variable, could be complete. Animals operated upon at 100 or 120 days had lung volumes and surface areas equal to those of two normal lungs. The allometric equation relating alveolar surface area to lung volume

$$S_A = 0.074 \ V_L^{1.07}$$

shows that the surface increased in complexity (Fig. 19). The numerical density of alveoli did not change, nor did their volume density or mean linear intercept. Thus alveoli *of the same size* (volume and linear dimensions) were being added. This means that the remaining lung that compensated successfully had, at term, a total alveolar number twice normal. It was thus polyalveolar.

Mechanisms of Lung Adaptation

In its response to reduced ambient oxygen or increased oxygen demand, localized hypoplasia, or resection, the lung demonstrates a remarkable ability to adapt its structure to functional needs. Because these conditions can all be simulated experimentally, the structural biologist has an excellent opportunity to study the adaptive repertoire of the lung.

The growth of pre- and intraacinar structures is dissociated in time and biological control. The number of preacinar airways does not increase after the pseudoglandular phase: an increase in lung volume and gas exchange surface area must be achieved within existing acini.

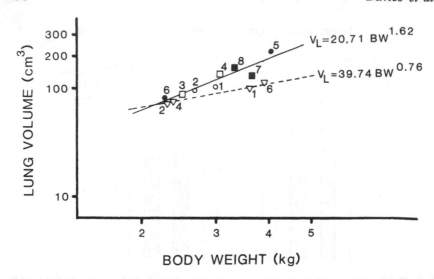

Figure 19 Left pneumonectomy in fetal sheep: volume of the remaining (right) lung (V_L) at term; double log plot against body weight (BW), with the least-squares regression line and derived allometric equation. Inverted triangles and dotted line, unoperated controls: open circles, operated at 60 days; open squares, at 80 days; closed circles, at 100 days; closed squares, at 120 days (from Hu et al., 1987, with permission).

The increase in the volume of the acinar or respiratory region can be produced in several ways: (1) by increase in size of the distal air spaces with little or no growth, which includes distention of alveoli and alveolar ducts but fails to compensate for loss of gas-exchange surface area; (2) increase in size of distal air spaces and therefore of surface area through lengthening of the alveolar walls; or (3) increase in the number of alveoli. Combinations of two or more of these mechanisms are possible and, as we have seen in experimental pneumonectomy, appear to depend on whether the animal is still in a stage of active morphogenesis. Thus, mechanisms (1) and (2) may occur at any age, but mechanism (3) is confined to immature stages and leads, at least temporarily, to a polyalveolar condition.

As has been mentioned, the polyalveolar condition was first defined in lung specimens from patients. It occurred in a localized idiopathic form (Hislop and Reid, 1970). As compensation it was found in residual lung and also when an earlier perturbation of growth of a lung or lobe had caused failure to form the normal number of preacinar airway generations, which inevitably reduces the number of acini (Ryland and Reid, 1971).

The mechanisms by which extra alveoli are added to the lung in adaptive growth have not been demonstrated, but several possibilities can be suggested on the basis of studies in the normal developing lung:

1. Additional subdivision of saccular spaces by secondary septa. This may only be possible during a limited developmental "window" when the process of subdivision normally occurs. The result is to increase the number of alveoli along each generation of alveolar duct. If the alveoli grow to normal size, the duct will be longer than normal. Mature airspaces seem incapable of undergoing further subdivision.

2. Increased generations of intraacinar airways. During the canalicular stage, additional canaliculi (putative respiratory bronchioli and ducts) could form. This has not yet been demonstrated, however. Once the airways are lined by airspaces it is difficult to imagine how de novo formation of new branches would occur.

3. At later stages of development the number of generations could be increased by further stimulating the normal process by which the centri- and preacinar airways form alveoli in their walls. Thus, respiratory bronchioli become alveolar ducts and terminal bronchioli become respiratory.

These are not trivial considerations in terms of acinar ventilation and blood supply and drainage. Mechanisms (2) and (3) could produce an abnormally long or wide acinus, perhaps more prone to collapse. Its distal air spaces could contain oxygen at lower partial pressure than in the normal-sized acinus because of uptake by capillaries arranged in parallel along the whole length of the acinus and/or a greater acinar length, which may reduce partial pressure, even allowing for the rapid rate of oxygen diffusion. Since arteries accompany airways, the blood volume and pressure distributions resulting from the altered configuration of the small arteries will be different from normal and from each other.

The effects of such growth on vascular structure are already apparent. The study of pneumonectomy in fetal sheep showed a clear relationship between the degree of compensatory growth and arterial remodeling that was demonstrated by an inverse straight line regression relating the percentage of arteries muscularized to lung volume (Fig. 20). This relationship has already been noted in clinical cases of lung hypoplasia, and perhaps reflects the effect of relative increased blood flow to a capillary bed reduced in luminal volume and surface area. That it occurred in utero, when the absolute flow to the distal lung is less than 10% of postnatal values, demonstrates the sensitivity of the precapillary segment to relative change in hemodynamic conditions.

The association of lung hypoplasia and vascular adaptation is directly pertinent to patient survival, as demonstrated in a study of seven infants who underwent surgical repair of congenital diaphragmatic hernia (Geggel et al., 1985). They were classified into two groups: those in which PaO_2 in the descending aorta was transiently greater than 150 mmHg (the so-called "honeymoon" group) and those in which PaO_2 was never more than 85 mmHg ("no honeymoon"). All lungs

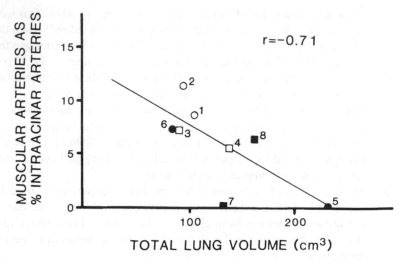

Figure 20 Left pneumonectomy in fetal sheep: relationship between the proportion of intraacinar muscular arteries and total right lung volume at term in operated animals. Symbols as in Figure 19 (from Hu et al., 1987, with permission).

were hypoplastic judged by weight and volume, the ipsilateral lung being the more severely affected. In three of the four patients in the no honeymoon group, ipsilateral lung volume was very low. As compared with the honeymoon group, lungs in the no honeymoon group demonstrated more remodeling of the distal arteries, external diameter of arteries at all levels was even more reduced and their medial thickness increased (Fig. 21), while medial muscle extended as far as the alveolar wall level (Fig. 22). Thus, cross-sectional luminal area available for perfusion was restricted. In the no honeymoon group, the external diameter of intraacinar arteries was reduced, to a greater extent, in the contralateral lung than the ipsilateral. Since it seems intuitively that the contralateral lung is the one that will change most in the early stages of correction, this could be the critical factor determining survival.

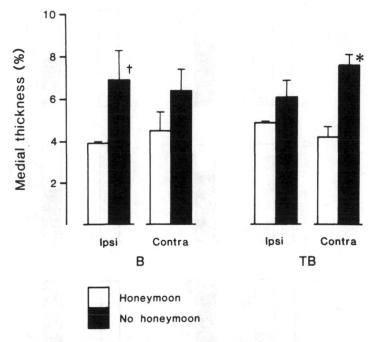

Figure 21 Human infants after repair of congenital diaphragmatic hernia. Arterial medial thickness (as % of external diameter) of barium/gelatin-filled arteries accompanying bronchioli (B) and terminal bronchioli (TB) in ipsi- and contralateral lungs from infants in honeymoon and no honeymoon groups (mean ± SEM) (from Geggel et al., 1985, with permission).

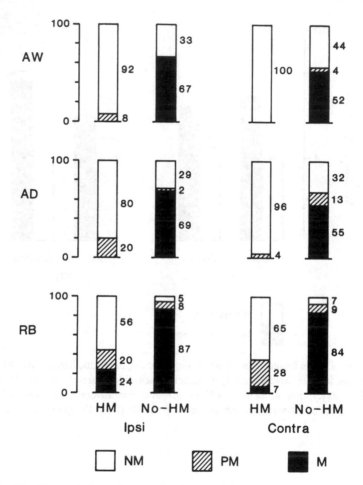

Figure 22 Human infants after repair of congenital diaphragmatic hernia. Histogram displays percentage arteries accompanying alveolar walls (AW), alveolar ducts (AD), or respiratory bronchioli (RB) in ipsi- and contralateral lungs of infants in honeymoon (HM) and no-honeymoon (no-HM) groups that were fully muscular (M), partially muscular (PM), or nonmuscular (NM) (from Geggel et al., 1985, with permission).

IV. Perspectives

The new methods that exploit molecular biology bring us to a new phase when the signals responsible for lung growth and maturation can be identified. But the lessons we have already learned must be remembered for maximum return from these new methods. Signal is not enough: the cell must accept it. Timing of change is still a critical factor.

A plethora of chemical compounds that stimulate cells and tissue in vitro (mediators, growth factors, or hormones) are now available, but because something happens in a test tube does not necessarily mean it is important in life. In vivo studies must be included: new methods are available to help here. Using the methods of in situ hybridization, the strength and surge of signal can be assessed, as well as the amount of mediator present. In fact, production and effect of signal can be at their highest when tissue content is at its lowest. Again, timing is important. In vitro and in vivo studies are complementary and both are essential to detect significant bioactive agents.

This is true for the analysis of normal growth. Analyzing the abnormal lung makes several additional demands. Lung growth in animals, especially mammals, is sufficiently similar to human to permit useful experiments, but analysis of clinical cases and material is still the touchstone. Abnormality, either in the animal model or in spontaneous disease, represents a basic injury as well as adaptation to it. The tissue does not necessarily behave as would normal: In vitro studies should include tissues or cells from the abnormal as well as the normal organ. The abnormal event is imposed on the normal framework, with timing being a critical factor. A given challenge or perturbation does not produce the same effect at each stage of development. Signal delivered and receptivity of tissue must both be critically analyzed, with due regard to the fourth dimension of time.

References

Adamson, I. Y. R., and Bowden, D. H. (1974). The type II cell as progenitor of alveolar epithelial regeneration. A cytodynamic study in mice after exposure to oxygen. *Lab Invest.* **32**:736–745.

Adamson, I. Y. R., and King, G. M. (1984a). Sex differences in development of fetal rat lung. I. Autoradiographic and biochemical studies. *Lab. Invest.* **50**:456–460.

Adamson, I. Y. R., and King, G. M. (1984b). Sex differences in development of fetal rat lung. II. Quantitative morphology of epithelial–mesenchymal interactions. *Lab. Invest.* **50**:461–468.

Alcorn, D., Adamson, T. M., Lambert, T. E., Maloney, J. E., Ritchie, B. C., and Robinson, P. M. (1977). Morphological effects of chronic tracheal ligation and drainage in the fetal lamb lung. *J. Anat.* **123**:649–660.

Alcorn, D., Adamson, T. M., Maloney, J. E., and Robinson, P. M. (1980). Morphological effects of chronic bilateral phrenectomy or vagotomy in the fetal lamb lung. *J. Anat.* **130**:683–695.

Alcorn, D., Adamson, T. M., Maloney, J. E., and Robins, P. M. (1981). A morphologic and morphometric analysis of fetal lung development in the sheep. *Anat. Rec.* **201**:655–667.

Alescio, T., and Cassini, A. (1962). Induction *in vitro* of tracheal buds by pulmonary mesenchyme grafted on tracheal epithelium. *J. Exp. Zool.* **50**:83–94.

Angus, G. E., and Thurlbeck, W. M. (1972). Number of alveoli in the human lung. *J. Appl. Physiol.* **32**:483–485.

Avery, M. E., Taeusch, H. W., and Floros, J. (1986). Surfactant replacement. *N. Engl. J. Med.* **315**:825–826.

Bachofen, H., Hildebrandt, J., and Bachofen, M. (1970). Pressure–volume curves of air- and liquid-filled excised lungs: surface tension *in situ. J. Appl. Physiol.* **29**:422–431.

Bachofen, H., Ammann, A., Wagensteen, D., and Weibel, E. R. (1982). Perfusion fixation of lungs for structure–function analysis: credits and limitations. *J. Appl. Physiol.* **53**:528–533.

Ballard, P. L., Hawgood, S., Liley, Hl., Wellenstein, G., Gonzales, L. W., Benson, B., Cordell, B., and White, R. T. (1986). Regulation of pulmonary surfactant apoprotein SP 28–36 gene in fetal human lung. *Proc. Natl. Acad. Sci. USA* **83**:9527–9531.

Berend, N., and Marlin, G. E. (1979). Arrest of alveolar multiplication in kyphoscoliosis. *Pathology* **11**:485–491.

Berend, N. Skoog, C., Waszkiewicz, L., and Thurlbeck, W. M. (1980). Maximum volumes in excised human lungs: effects of age, emphysema, and formalin inflation. *Thorax* **35**:859–864.

Blachford, K. G., and Thurlbeck, W. M. (1987). Lung growth and maturation in experimental oligohydramnios in the rat. *Pediatr. Pulmonol.* **3**:328–333.

Bourbon, J. R., and Farrell, P. M. (1985). Fetal lung development in the diabetic pregnancy. *Pediatr. Res.* **19**:253–267.

Boyden, E. A. (1974). The mode of origin of pulmonary acini and respiratory bronchioles in the fetal lung. *Am. J. Anat.* **141**:317–328.

Boyden, E. A. (1976). The development of the lung in the pigtail monkey (*Macaca nemestrina L.*). *Anat. Rec.* **186**:15–37.

Boyden, E. A., and Tompsett, D. H. (1961). The postnatal growth of the lung in the dog. *Acta Anat.* **47**:185–215.

Boyden, E. A., and Tompsett, D. H. (1965). The changing patterns in the developing lungs of infants. *Acta Anat.* **61**:164–192.

Brody, J. S., Fisher, A. B., Gocmen, A., and Dubois, A. B. (1970). Acromegalic pneumonomegaly: lung growth in the adult. *J. Clin. Invest.* **49**:1051–1060.

Bucher, U. G., and Reid L. (1961a). Development of the intrasegmental bronchial tree: the pattern of branching and development of cartilage at various stages of intra-uterine life. *Thorax* **16**:207–218.

Bucher, U. G., and Reid, L. (1961b). Development of the mucus-secreting elements in human lung. *Thorax* **16**:219–225.

Burri, P. H. (1974). The postnatal growth of the rat lung. III. Morphology. *Anat. Rec.* **180**:77–98.

Burri, P. H., Dbaly, J., and Weibel, E. R. (1974). The postnatal growth of the rat lung. I. Morphometry. *Anat. Rec.* **178**:711–730.

Burri, P. H., Pfrunder, H. B., and Berger, L. C. (1982). Reactive changes in pulmonary parenchyma after bilobectomy: a scanning electron microscopic investigation. *Exp. Lung Res.* **4**:11–28.

Burri, P. H., and Sehovic, S. (1979). The adaptive response of the rat lung after bilobectomy. *Am. Rev. Respir. Dis.* **119**:769–777.

Burri, P. H., and Weibel, E. R. (1971). Morphometric estimation of pulmonary diffusion capacity. II. Effect of PO_2 on the growing lung, adaptation of the growing rat lung to hypoxia and hyperoxia. *Respir. Physiol.* **11**:247–264.

Campiche, M. A., Gautier, A., Hernandez, E. L., and Reymond, A. (1963). An electron microscope study of the fetal development of human lung. *Pediatrics* **32**:976–994.

Caduff, J. H., Fischer, L. C., and Burri, P. H. (1986). Scanning electron microscopic study of the developing microvasculature in the postnatal rat lung. *Anat. Rec.* **216**:154–164.

Cambell, H., and Tomkeieff, S. A. (1952). Calculation of the internal surface of the lung. *Nature* **170**:116–117.

Chamberlain, D., Hislop, A., Hey, E., and Reid, L. (1977). Pulmonary hypoplasia in babies with severe rhesus isoimmunization: a quantitative study. *J. Pathol.* **122**:43–52.

Chen, J.- M., and Little, C. D. (1987). Cellular events associated with lung branching morphogenesis including the deposition of collagen type IV. *Dev. Biol.* **120**:311–321.

Collins, M. H., Moessinger, A. C., Kleinerman, J., James, L. S., and Blanc, W. A. (1986). Morphometry of hypoplastic fetal guinea pig lungs following amniotic fluid leak. *Pediatr. Res.* **20**:955–960.

Cooney, T. P., and Thurlbeck, W. M. (1982a). The radial alveolar count of Emery and Mithal: a reappraisal. I. Postnatal lung growth. *Thorax* **37**:572–579.

Cooney, T. P., and Thurlbeck, W. M. (1982b). The radial alveolar count of Emery and Mithal: a reappraisal. 2. Intrauterine and early postnatal growth. *Thorax* **37**:580–583.

Cooney, T. P., and Thurlbeck, W. M. (1982c). Pulmonary hypoplasia in Down's syndrome. *N. Engl. J. Med.* **307**:1170–1173.

Crapo, J. D., Barry, B. E., Gehr, P., Bachofen M., and Weibel, E. R. (1982). Cell number and cell characteristics of the normal human lung. *Am. Rev. Respir. Dis.* **125**:332–337.

Crapo, J. D., and Greeley, D. A. (1978). Estimation of the mean caliper diameter of cell nuclei. II. Various cell types in rat lung. *J. Microsc.* **114**:41–48.

Crone, R., Davies, P., Liggins, G., and Reid, L. (1983). The effects of hypophysectomy, thyroidectomy, and postoperative infusion of cortisol or adrenocorticotrophin on the structure of the ovine fetal lung. *J. Dev. Physiol.* **5**:281–288.

Cudmore, R. E., Emery, J. L., and Mithal, A. (1962). Postnatal growth of the bronchi and bronchioles. *Arch. Dis. Child.* **37**:481–484.

Cunningham, E. I., Brody, J. S., and Jain, B. P. (1974). Lung growth induced by hypoxia. *J. Appl. Physiol.* **37**:362–366.

Cutz, E. (1982). Neuroendocrine cells of the lung. An overview of morphological characteristics and development. *Exp. Lung. Res.* **3**:185–208.

Das, R. M. (1980). The effect of α-aminoproprionitrile on lung development in the rat. *Am. J. Pathol.* **101**:711–722.

Davies, G., and Reid, L. (1970). Growth of the alveoli and pulmonary arteries in childhood. *Thorax* **25**:669–681.

Davies, G., and Reid, L. (1971). Effect of scoliosis on growth of alveoli and pulmonary arteries and on right ventricle. *Arch. Dis. Child.* **46**:623–632.

Davies, P., McBride, J., Murray, G., Wilcox, B., Shallal, J., and Reid, L. (1982). Structural changes in the canine lung and pulmonary arteries after pneumonectomy. *J. Appl. Physiol.* **53**:859–864.

Davies, P., Reid, L., Lister, G., and Pitt, B. (1988). Postnatal growth of the sheep lung: a morphometric study. *Anat. Rec.* **220**:281–286.

De Hoff, R. T., and Rhines F. M. (1961). Determination of number of particles per unit volume from measurements made on random plane sections: the general cylinder and the ellipsoid. *Trans. AIME* **221**:975.

deMello, D., Phelps, D., Floros, J. and Lagunoff, D. (1988). Expression of 35 kDA (SP-A) and low molecular weight surfactant proteins (SP-B,C) in lungs of infants with respiratory distress syndrome. *Pediatr. Res.*, **23**: (abstract).

Dunnill, M. S. (1962a). Quantitative methods in the study of pulmonary pathology *Thorax* **17**:320–328.

Dunnill, M. S. (1962b). Postnatal growth of the lung. *Thorax* **17**:329–333.

Emery, J. L., and Mithal, A. (1960). The number of alveoli in the terminal respiratory unit of man during late intrauterine life and childhood. *Arch. Dis. Child.* **35**:544–547.

Fewell, J. E., Lee, C., and Kitterman, K. A. (19810). Effects of phrenic nerve section on the respiratory system of fetal lambs. *J. Appl. Physiol.* **51**:293–297.

Fisher, D. A. (1986). The unique endocrine milieu of the fetus. *J. Clin. Invest.* **78**:603–611.

Fulton, R. M., Hutchinson, E. C., and Jones, A. M. (1952). Ventricular weight in cardiac hypertrophy. *Br. Heart J.* **14**:413–420.

Geelhaar, A., and Weibel, E. R. (1971). Morphometric estimation of pulmonary diffusion capacity. III. The effect of increased oxygen consumption in Japanese waltzing mice. *Respir. Physiol.* **11**:354–366.

Geggel, R., Murphy, J., Langleben, D., Crone, R., Vacanti, J., and Reid, L. (1985). Congenital diaphragmatic hernia: arterial structural changes and persistent pulmonary hypertension after surgical repair. *J. Pediatr.* **107**:457–464.

Gehr, P, Hugonnaud, C., Burri, P. H., Bachofen H., and Weibel, E. R. (1978). Adaptation of the growing lung to increased VO₂. III. The effect of exposure to cold environment in rats. *Respir. Physiol.* **32**:345–353.

Gewolb, I. H., Rooney, S. A., Barrett, C., Ingleson, L. D., Light, D., Wilson, C. M., Smith, G, J., Gross, I., and Warshaw, J. B. (1985). Delayed pulmonary maturation in the fetus of the streptozotocin-diabetic rat. *Exp. Lung Res.* **8**:141–151.

Ghatei, M. A., Sheppard, M. N., Henzen-Logman, S., Blank, M. A., Polak, J. M., and Bloom, S. R. (1983). Bombesin and vasoactive intestinal polypeptide in the developing lung: marked changes in acute respiratory distress syndrome. *J. Clin. Endocrin. Metabol.* **57**:1226–1232.

Gil, J. (1977). Preservation of tissues for electron microscopy under physiological criteria. In *Techniques of Biochemical and Biophysical Morphology.* Vol. 3. Edited by D. Glick and R. M Rosenbaum. New York, John Wiley, pp. 19–44.

Gil, J., Bachofen, H., Gehr, P., and Weibel, E. R. (1979). Alveolar volume-surface area relation in air- and saline-filled lungs fixed by vascular perfusion. *J. Appl. Physiol.* **47**:990–1001.

Gil, J., and Reiss, O. K. (1973). Isolation and characterization of lamellar bodies and tubular myelin from rat lung homogenates. *J. Cell. Biol.* **58**:152–171.

Goldstein, J. D., and Reid, L. (1980). Pulmonary hypoplasia resulting from phrenic nerve agenesis and diaphragmatic amyoplasia. *J. Pediatr.* **97**:282–287.

Grant, N. M., Cutts,N. R., and Brody, J. S. (1983). Alterations in lung basement membrane during fetal growth and type II cell development. *Dev. Biol.* **97**:173–183.

Grant, N. M., Cutts, N. R., and Brody, J. S. (1983). Influence of maternal diabetes on basement membranes, type 2 cells, and capillaries in the developing rat lung. *Dev. Biol.* **104**:469–476.

Greeley, D., Crapo, J. D. and Vollmer, R. T. (1978). Estimation of the mean caliper diameter of cell nuclei. I. Serial section reconstriction method and endothelial nuclei from human lung. *J. Microsc.* **114**:31–39.

Gundersen, H. J. G., and Osterby, R. (1981). Optimizing sampling efficiency of stereological studies in biology: or "Do more less well." *J. Microsc.* **121**:65–73.

Gupta, M., Mayhew, T. M. Bedi, K. S. Sharma, A. K., and White, F. H. (1983). Inter-animal variation and its influence on the overeall precision of morphometric estimates based on nested sampling designs. *J. Microsc.* **131**:147–154.

Hage, E. (1973). The morphological development of the pulmonary epithelium of human fetuses studied by light and electron microscopy. *Z. Anat. Entwickl.-Gesch.* **140**:271–279.

Haworth, S. G., and Reid, L. (1976). Persistent fetal circulation: newly recognized structural features. *J. Pediatr.* **88**:614–620

Haworth, S. G., and Reid, L. (1977a). Structural study of pulmonary circulation and of heart in total anomalous pulmonary venous return in early infancy. *Br. Heart. J.* **39**:80–92.

Haworth, S. G., and Reid, L. (1977b). Quantitative structural study of pulmonary circulation in the newborn with aortic atresia, stenosis or coarctation. *Thorax* **32**:121–128.

Haworth, S. G., and Reid, L. (1978). A morphometric study of regional variation in lung structure in infants with pulmonary hypertension and congenital cardiac defect. A justification of lung biopsy. *Br. Heart J.* **40**:825–831.

Haworth, S. G., Sauer, U., Buhlmeyer, K., and Reid, L. (1977). Development of the pulmonary circulation in ventricular septal defect: a quantitative structural study. *Am. J. Cardiol.* **40**:781–788.

Hayatdavoudi, G., Crapo, J. D., Miller, F. J., and O'Neil, J. J. (1980). Factors determining degree of inflation in intratracheally fixed rat lungs. *J. Appl Physiol.* **48**:389–393.

Hieronymi, G. (1961). Uber den durch das alter bedingten formwandel menschlicher lungen. *Ergeb. Allg. Pathol. Anat.* **41**:1–62.

Hislop, A., and Reid, L. (1970). New pathological findings in emphysema of childhood. I. Polyalveolar lobe with emphysema. *Thorax* **25**:682–690.

Hislop, A., and Reid, L. (1972). Intra-pulmonary arterial development during fetal life branching pattern and structure. *J. Anat.* **113**:35–48.

Hislop, A., and Reid, L. (1973). Pulmonary arterial development during childhood: branching pattern and structure. *Thorax* **28**:129–135.

Hislop, A., and Reid, L. (1976). Persistent hypoplasia of the lung after repair of congenital diaphragmatic hernia. *Thorax* **31**:450–455.

Hislop, A., Muir, D. C., Jacobsen, M., Simon G., and Reid L. (1972). Postnatal growth and function of the pre-acinar airways. *Thorax* **27**:265–274.

Hislop, A., Hey, E., and Reid, L. (1979). The lungs in congenital bilateral renal agenesis and dysplasia. *Arch. Dis. Child.,* **54**:32–38.

Hislop, A., Fairweather, D. V. I., Blackwell, R. J., and Howard, S. (1984). The effect of amniocentesis and drainage of amniotic fluid on lung development in *Macaca fascicularis. Br. J. Obst. Gyn.,* **91**:835–842.

Hitchcock, K. R. (1979). Hormones and the lung. I. Thyroid hormones and glucocorticoids in lung development. *Anat. Rec.* **194**:15–40.

Hogg, J.C., Williams, J., Richardson, J. B., Macklem, P. T., and Thurlbeck, W. M. (1970). Age as a factor in the distribution of lower-airway conductance and in the pathologic anatomy of obstructive lung disease. *N. Engl. J. Med.* **282**:1283–1287.

Holmes, C., and Thurlbeck, W. M. (1979). Normal lung growth and response after pneumonectomy in the rat at various ages. *Am. Rev. Respir. Dis.* **120**:1125–1136.

Hu, L.-M., Davies, P., Adzick, N. S., Harrison, M. R., and Reid, L. (1987a). Lung growth after intrauterine bilateral ureteral obstruction. *Fed. Proc.* **46**:1151a.

Hu, L.-M., Davies, P., Adzick, N. S., Harrison, M. R., and Reid, L. (1987b). The effects of intrauterine pneumonectomy in lambs: a morphometric study of the remaining lung at term. *Am. Rev. Respir. Dis.* **135**:607-612.

Hugonnaud, C. P., Gehr, P., Weibel, E. R., and Burri, P. H. (1977). Adaptation of the growing lung to increased oxygen consumption. II. Morphometric analysis. *Respir. Physiol.* **29**:1-10.

Huxley, J. S. (1932). *Problems of Relative Growth*. London, Methuen.

Jain, B. P., Brody, J. S., and Fisher, A. B. (1973). The small lung of hypopituitarism. *Am. Rev. Respir. Dis.* **108**:49-55.

Jones, R., Baskerville, A., and Reid, L. (1975). Histochemical identification of glycoproteins in pig bronchial epithelium: (a) normal and (b) hypertrophied from enzootic pneumonia. *J. Pathol.* **116**:1-11.

Jones, R., and Reid, L. (1973a). The effect of pH on Alcian Blue staining of epithelial acid glycoproteins. I. Sialomucins and sulphomucins (singly or in simple combinations). *Histochem. J.* **5**:9-18.

Jones, R., and Reid, L. (1973b). The effect of pH on Alcian Blue staining of epithelial acid glycoproteins. II. Human bronchial submucosal gland. *Histochem. J.* **5**:19-27.

Kauffman, S. L. (1977). Acceleration of canalicular development in lungs of fetal mice exposed transplacentally to dexamethasone. *Lab. Invest.* **36**:395-401.

Kauffman, S. L. (1980). Cell proliferation in the mammalian lung. *Int. Rev. Exp. Pathol.* **22**:131-191.

Kauffman, S. L., Burri, P. H., and Weibel, E. R. (1974). The postnatal growth of the rat lung. II. Autoradiography. *Anat. Rec.* **180**:63-76.

Kida, K., and Thurlbeck, W. M. (1980). The effects of α-aminoproprionitrile on the growing rat lung. *Am. J. Pathol.* **101**:693-670.

King, G. M., and Adamson, I. Y. R. (1987). Effects of *cis*-hydroxyproline on type II cell development in fetal rat lung. *Exp. Lung. Res.* **12**:347-362.

King, R. J. (1974). The surfactant system of the lung. *Fed. Proc.* **33**:2238-2247.

Kitterman, J. A. (1984). Fetal lung development. *J. Dev. Physiol.* **6**:67-82.

Kleiber, M. (1932). Body size and metabolism. *Hilgardia* **6**:315-353.

Lamb, D., and Reid, L. (1969). Histochemical types of acidic glycoprotein produced by mucous cells of the tracheobronchial glands in man. *J. Pathol.* **98**:213-229.

Lamb, D., and Reid, L. (1972a). Quantitative distribution of various types of acid glycoprotein in mucous cells of human bronchi. *Histochem. J.* **4**:91-102.

Lamb, D., and Reid, L. (1972b) Acidic glycoproteins produced by the mucous cells of the bronchial submucosal glands in the fetus and child: a histochemical autoradiographic study. *Br. J. Dis. Chest* **66**:248-253.

Langston, C., Sachdeva, P., Cowan, M. J., Haines, J., Crystal, R. G., and Thurlbeck, W. M. (1977). Alveolar multiplication in the contralateral lung after unilateral pneumonectomy in the rabbit. *Am. Rev. Respir. Dis.* **115**:7–13.

Langston, C., Kida, K, Reed, M., and Thurlbeck, W. M. (1984). Human lung growth in late gestation and in the neonate. *Am. Rev. Respir. Dis.* **129**:607–613.

Lauweryns, J. M. (1970). Hyaline membrane disease. *Human Pathol.* **1**:175.

Lechner, A. J., and Banchero, N. (1980). Lung morphometry in guinea pigs acclimated to hypoxia during growth. *Respir. Physiol.* **42**:155–169.

Leung, C. K., Adamson, I. Y., and Bowden, D. H. (1980). Uptake of ^{3}H prednisolone by fetal lung explants: role of intercellular contacts in epithelial maturation. *Exp. Lung. Res.* **1**:111–120.

Liggins, G. C. (1984). Growth of the fetal lung. *J. Dev. Physiol.* **6**:237–248.

Liggins, G. C., Vilos, G. A., Campos, G. A. Kitterman, J. A., and Lee, C. H. (1981). The effect of bilateral thoracoplasty on lung development in fetal sheep. *J. Dev. Physiol.* **3**:275–282.

Liley, H. G., Hawgood, S., Wellenstein, G. A., Benson, B., Tyler-White, R., and Ballard, P. (1987). Surfactant protein of molecular weight 28,000–36,000 in cultured human fetal lung: cellular localization and effect of dexamethasone. *Mol. Endocrinol.* **1**:205–215.

Lum, H., and Mitzner, W. (1985). Effects of 10% formalin fixation on fixed lung volume and lung tissue shrinkage. *Am. Rev. Respir. Dis.* **132**:1078–1083.

Mason, R. J., Dobbs, L. G., Greenleaf, R. D., and Williams, M. C. (1977). Alveolar type II cells. *Fed. Proc.* **36**:2697–2702.

Massaro, D., and Massaro, G. D. (1986). Dexamethasone accelerates postnatal alveolar wall thinning and alters wall composition. *Am. J. Physiol.* **251**:R218–224.

Massaro, D., Teich, N., and Massaro, G. D. (1986). Postnatal development of pulmonary alveoli: modulation in rats by thyroid hormones. *Am. J. Physiol.* **250**:R51–55.

Massaro, D., Teich, N., Maxwell, S., Massaro, G. D., and Whitney, P. (1985). Postnatal development of alveoli. Regulation and evidence for a critical period in rats. *J. Clin. Invest.* **76**:1297–1305.

Massaro, G. D., Davis, Loranine, and Massaro, D. (1984). Postnatal development of the bronchiolar Clara cell in rats. *Am. J. Physiol.* **16**:C197–C203.

Mathieu, O., Claassen, H., and Weibel, E. R. (1978). Differential effect of glutaraldehyde and buffer osmolarity on cell dimensions: a study on lung tissue. *J. Ultrastruct. Res.* **63**:20–34.

Mazzone, R. W., Kornblau, S., and Durand, C. M. (1980). Shrinkage of lung after chemical fixation for analysis of pulmonary structure–function relations. *J. Appl. Physiol.* **48**:382–385.

McBride, J. T. (1985). Postpneumonectomy airway growth in the ferret. *J. Appl. Physiol.* **58**:1010–1014.

McBride, J. T., and Chuang, C. (1985). A technique for quantitating airway size from bronchial casts. *J. Appl. Physiol.* **58**:1015–1022.

McBride, J. T., Wohl, M. E. B., Strieder, D. J., Jackson, A. C., Morton,, J. R., Zwerdling, R. G., Griscom, N. T., Treves, S., Williams, A. J., and Schuster, S. (1980). Lung growth and airway function after lobectomy in infancy for congenital lobar emphysema. *J. Clin. Invest.* **66**:962–970.

McCarthy, C., and Reid, L. (1964). Acid mucopolysaccharide in the bronchial tree in the mouse and rat (sialomucin and sulphate). *Q. J. Exp. Physiol.* **49**:81–84.

Meyrick, B., and Reid, L. (1982). Pulmonary arterial and alveolar development in normal postnatal rat lung. *Am. Rev. Respir. Dis.* **125**:468–473.

Mills, A. N., Lopez-Vidriero, M. T., Haworth, S. G. (1986). Development of the airway epithelium and submucosal glands in the pig lung: changes in epithelial glycoprotein profiles. *Br. J. Exp. Pathol.* **67**:821–829.

Mortola, J. P. (1983). Dysanaptic lung growth: an experimental and allometric approach. *J. Appl. Physiol.* **54**:1236–1241.

Muller, A. E., Cruz-Orive, L. M., Gehr, P., and Weibel, E. R. (1980). Comparison of two subsampling methods for electron microscopic morphometry. *J. Microsc.* **123**:35–49.

Nakayama, Glick, Harrison, Villa, and Noall (1983). Experimental pulmonary hypoplasia due to oligohydramnios and its reversal by relieving thoracic compression. *J. Pediatr. Surg.* **18**:347–353.

Nielsen, H. C., and Torday, J. S. (1987). The sex difference in fetal lung surfactant production. *Exp. Lung Res.* **12**:1–20.

Nielsen, H. C., Zinman, H. M., and Torday, J. S. (1982). Dihydrotestosterone inhibits fetal rabbit pulmonary surfactant productions. *J. Clin. Invest.* **69**:611–66.

O'Rahilly, R. (1985). The early prenatal development of the human respiratory system. In *Pulmonary Development. Transition From Intrauterine to Extrauterine Life*. Edited by G. H. Nelson. New York, Marcel Dekker, pp. 3–18.

Paul, G. W., Hassett, R. H., Reiss, O. K. (1977). Formation of lung surfactant films from intact lamellar bodies. *Proc. Natl. Acad. Sci. USA* **74**:3617–3620.

Phelps, D. S., and Floros, J. (1988). Localization of surfactant protein synthesis in human lung by *in situ* hybridization. *Am. Rev. Respir. Dis.* **137**:939–942.

Plopper, C. G., Alley, J. L., Serabjit-Singh, C. J., and Phelpot, R. M. (1983). Cytodifferentiation of the non ciliated bronchiolar epithelial (Clara) cell during rabbit lung maturation: an ultrastructural and morphometric study. *Am. J. Anat.* **167**:329–357.

Pitt, B. R., Lister, G., Davies, P., and Reid, L. (1987). Correlation of pulmonary ACE activity and capillary surface area during postnatal development. *J. Appl. Physiol.* **62**:2031–2041.

Reid, L. (1984). Lung growth in health and disease (Tudor Edwards Lecture). *Br. J. Dis. Chest.* **78**:113–134.

Reid, L., and Coles, S. J. (1984). The bronchial epithelium of humans: cytology, innervation and function. In *The Endocrine Lung in Health and Disease.* Edited by K. L. Becker and A. F. Gazdar. Philadelphia, W. B. Saunders, pp. 56–78.

Reid, L., and de Haller, R. (1967). The bronchial mucous glands—their hypertrophy and change in intracellular mucus. *Mod. Probl. Pediatr.* **10**:195–199.

Ryland, D., and Reid, L. (1971). Pulmonary aplasia—a quantitative analysis of the development of the single lung. *Thorax* **26**:602–609.

Scherle, W. F. (1970). A simple method for volumetry of organs in quantitative sterology. *Mikroskopie* **26**:57–60.

Schloo, B. L., and Reid L. (1987). Intrasegmental airway generations and radial alveolar counts in Down's syndrome—patterns of disturbed lung growth. *Am. Rev. Respir. Dis.* **135**:A24.

Short, D. S. (1956). Post-mortem pulmonary arteriography with special reference to the study of pulmonary hypertension. *J. Fac. Radiol.* **7**:118–131.

Smith, B. T. (1979). Lung maturation in the fetal rat: acceleration by injection of fibroblast-pneumocyte factor. *Science* **204**:1094–1095.

Smith, B. T., and Bogues, W. G. (1980). Effects of drugs and hormones on lung maturation in experimental animal and man. *Pharmacol. Ther.* **9**:51–74.

Sosenko, I. R. S., Frantz, I. D., Roberts, R. J., and Meyrick, B. (1980). Morphologic disturbance of lung maturation in fetuses of alloxan diabetic rabbits. *Am. Rev. Respir. Dis.* **122**:687–696.

Spicer, S. S., Horn, R. G., and Leppi, T. J. (1967). Histochemistry of connective tissue mucopolysaccharides. In *The Connective Tissue.* Edited by B. M. Wagner and D. E. Smith. Baltimore, Williams & Wilkins, pp, 251–303.

Spicer. S. S., and Schulte, B. A. (1982). Ultrastructural methods for localizing complex carbohydrates. *Hum. Pathol.* **13**:343–354.

Spicer, S. S., and Warren, L. (1960). The histochemistry of sialomucins. *J. Histochem. Cytochem.* **8**:135–137.

Spicer, S. S., Schulte, B. A., Thomopoulos, G. F. N., Parmley, R. T., and Takagi, M. (1982). Cytochemistry of complex carbohydrates by light and electron microscopy: available methods and their application. In *Connective Tissue and Diseases of Connective Tissue.* Edited by B. M. Wagner and R. Fleismajer. Baltimore, Williams & Wilkins.

Spindel, E. R., Sunday, M. E., Hofler, H., Wolfe, J. H., Habener, J. F., and Chin, W. W. 1987. Transient elevation of mRNAs encoding gastrin-releasing peptide (GRP), a putative pulmonary growth factor, in human fetal lung. *J. Clin. Invest.* **80**:1172–1179.

Spooner, B. S., and Wessells, N. K. (1970). Mammalian lung development: interactions in primordium formation and bronchial morphogenesis. *J. Exp. Zool.* **175**:445–454.

Stahlman, M. T., and Gray, M. E. (1978). Anatomical development and maturation of the lungs. *Clin. Perinatol.* **5**:181–196.

Sturgess, J., and Imrie, J. (1982). Quantitative evaluation of the development of tracheal submucosal glands in infants with cystic fibrosis and control infants. *Am. J. Pathol.* **106**:303–311.

Taeusch, H. W., Keough, K. M. W., Williams, M., Slavin, R., Lee, A. S., Phelps, D. S., Kariel, N., Floros, J., and Avery, M. E. (1986). Characterization of an exogenous bovine surfactant for infants with respiratory distress syndrome. *Pediatrics* **77**:572–581.

Taeusch, H. W., Phelps, D. S., Floros, J. (in press). Surfactant treatment of respiratory distress syndrome and the relevance of surfactant-associated proteins. *J. Austr. Perinat. Soc.*

Thurlbeck, W. M. (1982). Postnatal human lung growth. *Thorax* **37**:564–571.

Thurlbeck, W. M., Benjamin, B., and Reid, L. (1961). A sampling method for estimating the number of mucous glands in the fetal human trachea. *Br. J. Dis. Chest.* **55**:49–53.

Thurlbeck, W. M., Galaugher, W., and Mathers, J. (1981). Adaptive response to pneumonectomy in puppies. *Thorax* 424–427.

Tompsett, D. H. (1970). *Anatomical Techniques*. Edinburgh, Livingstone, pp. 123–129.

Torday, J. S. (1985). Dihydrotestosterone inhibits fibroblast–pneumonocyte factor mediated synthesis of saturated phosphatidylcholine by fetal rat lung cells. *Biochem. Biophys. Acta* **835**:23–28.

Tyden, O., Berne, C., and Eriksson, U. (1980). Lung maturation in fetuses of diabetic rats. *Pediatr. Res.* **14**:1192–1195.

Usher, R., McLean, F., and Scott, K. E. (1966). Judgment of fetal age. II. Clinical significance of gestational age and an objective method for its assessment. *Pediatr. Clin. North Am.* **13**:835–862.

Vilos, G., and Liggins, G. C. (1982). Intrathoracic pressure in fetal sheep. *J. Dev. Physiol.* **4**:247–256.

Wailoo, M. P., and Emery, J. L. (1982). Normal growth and development of the trachea. *Thorax* **37**:584–587.

Wandel, G., Berger, L. C., and Burri, P. H. (1983). Quantitative changes in the remaining lung after bilobectomy in the adult rat. *Am. Rev. Respir. Dis.* **128**:968–972.

Weibel, E. R. (1971). Morphometric estimation of pulmonary diffusion capacity. I. Model and method. *Respir. Physiol.* **11**:54–75.

Weibel, E. R. (1979). *Stereologic Methods*. New York, Academic Press.

Weibel, E. R., and Gomez, D. (1962). A principle for counting tissue structures on random sections. *App. Physiol.* **17**:343–348.

Weibel, E. R., and Knight, B. W. (1964). A morphometric study on the thickness of the pulmonary air-blood barrier. *J. Cell Biol.* **21**:367–384.

Weibel, E. R., Kistler, J. S. and Tondury, G. (1966). A stereologic electron microscope study of "tubular myelin figures" in alveolar fluids of rat lungs. *F. Zellforsch. Microsk.* **69**:418–427.

Wigglesworth, J. S., and Desai, R. (1979). Effects on lung growth of cervical cord section in the rabbit fetus. *Early Hum. Dev.* **3**:51–65.

Wohl, M. E. B., Turner, J., and Mead, J. (1968). Static volume–pressure curves of dog lungs *in vivo* and *in vitro*. *J. Appl. Physiol.* **24**:348–354.

Yen, R. T., Zhuang, F. Y., Fung, Y. C., Ho, H. H., Tremer, H., and Sobin, S. S. (1984). Morphometry of cat pulmonary arterial tree. *J. Biochem. Eng.* **106**:131–136.

Zeltner, T. B., and Burri, P. H. (1987). The postnatal development and growth of the human lung. II. Morphology. *Respir. Physiol.* **67**:269–282.

Zeltner, T. S., Caduff, J. H., Gehr, P., Pfenninger, J., and Burri, P. H. (1987). The postnatal development and growth of the human lung. I. Morphometry. *Respir. Physiol.* **67**:247–267.

15

Purification and Primary Culture of Type II Pneumocytes and Their Application in the Study of Pulmonary Metabolism

YUTAKA KIKKAWA and NEAL METTLER

New York Medical College
Valhalla, New York

I. Introduction

The investigation of any metabolic function in the lung is complicated by the unique anatomy and physiology of this organ. These complications are further compounded by the extensive cellular heterogeneity found in the lung, making the assignment of a particular biochemical event to one specific cell type very tenuous.

A case in point is the role of the type II alveolar epithelial cell. This cell makes up only 14.2% of the lung cell population (Crapo et al., 1980) including the alveolar macrophage, endothelial cell, interstitial cell, and other types of epithelial cells. Therefore, results obtained from whole lung tissue for the study of the metabolism of pulmonary surfactant would be difficult to interpret with respect to any specific cell type. Considerable improvements in our understanding of pulmonary surfactant metabolism would be expected if type II cells could be isolated, free from the other cell types in the lungs. This was accomplished in 1974 when type II cells were isolated from the lung with high purity (Kikkawa and Yoneda, 1974) and were shown to possess the ability to synthesize some of the major components of surfactant lipids (Kikkawa et al., 1975).

The isolation of the type II cell offered a new dimension to the study of lung biology. Until this time, the alveolar macrophage had been the only cell from the lung to be isolated and studied in detail (Tierney, 1974) and the only methods available for studying the functions of other cells; for example, the type II cell, involved using perfused lungs, lung minces, or lung slices. Each of these techniques, as mentioned earlier, had the common disadvantage of not being specific for any one cell type as well as possessing certain experimental disadvantages (Tierney, 1974). With the advent of procedures for isolating type II cells, our understanding of the biochemical events (Kikkawa et al., 1975; Batenburg et al., 1978; Smith and Kikkawa, 1979; Smith et al., 1980), neurohormonal mechanisms (Ballard et al., 1978, Dobbs and Mason, 1979; Post et al., 1980; Brown and Longmore, 1981; Mettler et al., 1981), and the role of a cellular matrix (Leslie et al., 1985) in controlling the synthesis and secretion of pulmonary surfactant has increased substantially.

The various methods used for the isolation and primary culture of type II cells, as well as some of the applications and experimental results obtained using isolated type II cells, are discussed in this chapter.

II. Type II Cell Isolation

The first successful isolation of type II pneumocytes was described from this laboratory in 1974 (Kikkawa and Yoneda, 1974) and was soon followed by a number of other reports (Mason et al., 1975; King, 1977; Pfleger, 1977; Batenburg et al., 1978; Brown and Longmore, 1981; Mettler et al., 1981) employing similar procedures. All of these methods share the same sequence of steps.

1 . Mechanical removal of free lung cells, primarily alveolar macrophages and blood cells

2 . Enzymatic dispersion of lung parenchymal cells

3 . Isolation of type II cells by centrifugation, cell culture, or both

Removal of alveolar macrophages is accomplished through pulmonary lavage using calcium free Joklik's minimum essential medium or phosphate-buffered saline containing EDTA. This step, however, removes only a portion of macrophages residing in the alveolar space; removal of interstitial and remaining alveolar macrophages is carried out by mincing the lung parenchyma as described in detail below. Blood cells are removed by vascular perfusion of the lung with concomitant ventilation with room air.

Trypsin appears to be the enzyme of choice, with the grade of this enzyme ranging from fairly pure preparations (Mason et al., 1975; Mettler et al., 1981) to relatively impure material (Kikkawa and Yoneda, 1974; Pfleger, 1977; King, 1977). Successful use of elastase has also been reported (Dobbs and Mason, 1979; Dobbs et al., 1980), while collagenase was used for the isolation of crude isolates of type II cells (Wolf et al., 1968; Ayuso et al., 1973). More recently protease, a mixture of several proteolytic enzymes, was also used with success (Devereux and Fouts, 1981). Although there are no systematic studies available, evidence indicates that high concentrations of trypsin used for dissociation may be detrimental to phospholipid synthesis in type II cells. The rates of choline, glucose, and palmitate incorporation into freshly isolated type II cells with 1% trypsin are 2–10 times lower than those reported by Finkelstein et al. (1983) with the use of mixtures of 0.025 mg/ml trypsin, 1.3 units/ml elastase, and 10 µg/ml of DNAase. A mixture of trypsin/EDTA/DNAase of 0.05/0.02/0.001% appeared to incorporate 10 times more choline into type II cell phosphatidylcholine (Scott et al., 1983) than that reported earlier (Smith and Kikkawa, 1979). Our current procedure utilizes 0.1% trypsin with DNAase (Kikkawa et al., 1983). Use of a mixture of enzymes—0.008% elastase, 0.2% collagenase, 0.005% DNAase, and 0.05% trypsin (Weller and Karnovsky, 1986)—is apparently successful in dissociating type II cells but this effect is not understood. Since 0.05% trypsin is singularly successful in a dissociation process, an addition of others may be harmful to the cells. However, it should be noted that with the use of any of these proteolytic enzymes, there is a potential for cell damage to occur with subsequent metabolic changes and loss of hormone sensitivity (Wolf et al., 1968; Kono and Barham, 1971; Mango et al., 1975; Mason et al., 1976). This is probably not the case in recent years with the use of lower concentrations of enzymes.

There are several procedures available for the enzymatic dissociation of type II cells from the lung. Intratracheal instillation of the enzyme solution is used by several laboratories (Mason et al., 1975; Pfleger, 1977; Mettler et al., 1981). This is followed by a timed incubation and mincing of the lung parenchyma. King (1977), on the other hand, expands the lung with his enzyme solution and then partially minces it prior to incubation, while Kikkawa and Yoneda (1974) mince the lung tissue before exposing it to the enzyme solution. Regardless of the method of enzyme digestion or choice of enzyme used, all incubations are carried out for 10–30 min. When the tissue is minced, there is an inevitable consequence of free radical injury to the tissue. We have, therefore, used some antioxidants, such as ascorbate and the constituent amino acids of glutathione, in isolation solution. The results indicate that the level of glutathione in isolated type II cells is 100% higher and plating efficiency increased fivefold (Mettler et al., 1984).

In those procedures involving trypsin, the proteolytic action of the enzyme is stopped with either soybean trypsin inhibitor (Mason et al., 1975; Pfleger, 1977, Mettler et al., 1981) or calf serum (Kikkawa and Yoneda, 1974), while Dobbs and Mason (1979) use fetal calf serum to inhibit elastase.

To facilitate the removal of any remaining alveolar macrophages from type II cells, investigators have used colloidal barium sulfate (Kikkawa and Yoneda, 1974; Mettler et al., 1981), fluorocarbon–albumin mixture (Mason et al., 1975;

Pfleger, 1977, Dobbs and Mason, 1979), or mineral oil (King, 1977). These substances are phagocytosed by the macrophage, altering its density and allowing it to be centrifuged into the heavier zones of the density gradients (Kikkawa and Yoneda, 1974; King, 1979).

Current methods use either Ficoll, a neutral, highly branched polymer of sucrose (Kikkawa and Yoneda, 1974; King, 1977; Pfleger, 1977; Mettler et al., 1981); albumin (Mason et al., 1975); Metrizamide (Devereux and Fouts, 1981; Geppert et al., 1980); or Percoll (Weller and Karnovsky 1986) in forming density gradients. These materials share the property of producing relatively low osmotic pressures on the type II cell. There are no studies available to evaluate the advantages or disadvantages, if any, of these gradients on the type II cell. All methods employ at least a gradient in the 1.040–1.050g/ml range and several also use an additional heavier gradient between 1.060 and 1.080 g/ml, to cover the range of type II cell densities, which are between 1.045 and 1.085 g/ml (Kikkawa and Yoneda, 1974; Mason et al., 1975). The inclusion of this second heavier gradient, although isolating greater numbers of type II cells, overlaps with the average density of the alveolar macrophage (1.074 g/ml) (Kikkawa and Yoneda, 1974) and increases the amount of contamination by macrophages. However, regardless of the number of density gradients used, all are centrifuged at relatively low gravity spins for periods ranging from 10 to 60 min.

More recently, centrifugal elutriation, (Greenleaf et al., 1979; Devereux and Fouts, 1981) has been used in an attempt to obtain relatively pure populations of type II cells with reduced time and handling. This technique has not been entirely successful, however, in that discontinuous density centrifugation had to be used before (Greenleaf et al., 1979) or after (Devereux and Fouts, 1981) elutriation centrifugation to achieve fractions of type II cells with a purity of 80% or greater.

Perhaps the most sophisticated method to date used for the isolation of almost pure fractions of type II cells has been the application of laser flow cytometry (Leary et al., 1982) in lieu of any preparative centrifugation. This technique involves measuring the high fluorescence of the lipophilic stain, Phosphine 3R, in the lamellar bodies of type II cells, simultaneously with another parameter that involves low-angle light scatter. This latter determination actually measures the volume of cytoplasmic structures (organelles) within the cell. With laser cytometry and using these parameters, populations of type II cells with greater than 98% purity can be isolated from crude lung cell populations. The major drawback of this system, however, is the length of time required to obtain a large quantity of type II cells. Only $5-10 \times 10^5$ type II cells can be isolated per hour as compared with the $5 \times 10^7-1 \times 10^8$ type II cells that can be obtained using bulk methods (Leary et al., 1982).

Recently, a new method was developed for removing non-type-II cell contaminants from a heterogeneous cell mixture that was enriched in type II cells by a prior density gradient centrifugation using Percoll. To remove contaminating cells, primarily macrophages and lymphocytes from this mixture, these authors have exploited the fact that these contaminating cells have leukocyte common antigen (LC) on their surface whereas type II cells do not. This procedure yielded approximately 2×10^7 cells per rat lung. Since type II cells were recovered from a broad range of Percoll gradient densities, the authors estimated that they have collected 90% of the dissociated type II cells, while in most other procedures the recoveries are in the range of 10–20%. Unfortunately, there are no functional studies comparable to others to assess the merit of this procedure (Weller and Karnovsky, 1986). After the cells are isolated from the gradients, cell counts and purity checks are made.

A comparison of many of the methods for isolating type II cells is shown in Table 1. It can be seen that the only common parameters that allow the comparison between these procedures are cell yield and purity. These values range from 65 to 95%, with approximately 20×10^6 cells being isolated (purity 85%) per rabbit (Kikkawa et al., 1975) to 15–20 $\times 10^6$ (purity 65–80%) being isolated from a single rat (Mason et al., 1975; King, 1977; Mettler et al., 1981) to only 5×10^6 cells being obtained from the Syrian hamster (Pfleger, 1977). Purity is assessed using either a modified Papanicolaou stain (Kikkawa and Yoneda, 1974) or a fluorescent dye, Phosphine 3R (Mason et al., 1975). With use of the former stain, the type II cells are easily identified by their bright blue inclusions, each surrounded by a clear halo, while in the latter, the intracellular inclusions fluoresce when excited at 466 nm.

It is difficult to evaluate the metabolic integrity of type II cells obtained using these different procedures. Oxygen consumption is the most frequently reported with values ranging from 46 to 200 nmoles/10^6cell/hr. However, in light of the discussions presented so far, the most rational approach for obtaining the best functioning type II cells would involve (1) the use of an improved isolation vehicle containing the necessary nutrients for type II cells and antioxidants to minimize oxidant damage, (2) the use of low concentrations of proteolytic enzymes and continuous removal of cells as they become detached during digestion, and (3) avoidance of centrifugal damage to the cells with the removal of macrophages and lymphocytes through cell surface leukocyte common antigen or other lectins and differential plating.

III. Primary Culture of Type II Cells

The first successful primary culture of type II cells was reported from our laboratory (Diglio and Kikkawa, 1977). Successful primary cultures were obtained only after utilizing high-density cell plating (3×10^5 viable cells/cm²) and allowing an attachment time of 48 hr. Attachment efficiency of the isolated cell

Table 1 Comparison of Type II Cell Isolation Methods

Study	Year	Enzyme	Isolation Medium	Inhibitor	Purification	Animal	Cell Yield	Purity (%)	O_2 (nm/10⁶/hr)	Protein (mg/10⁶)	Lipid (mg/10⁶) P	Results
Kikkawa and Yoneda Kikkawa et al.	1974 1975	Trypsin 1% 1:250 Difco	Joklik	25% FCS	Ficoll 1.047	Rat Rabbit	4×10^5 $20\text{--}30 \times 10^6$	95%	—	57	1.10	PC DPPC PG[a] 63.3 49.3 4.3
King	1977	Trypsin type 2 cryst. 5 mg/ml	MEM	10% FCS	Ficoll 1.040 1.100	Rat	$2\text{--}3 \times 10^6$	70–90	70–90	—	—	—
Mason et al.	1977	Trypsin cryst. 3 mg/ml instilled	PBS	Soybean inhibitor	Albumin 1.040 1.080	Rat	$20\text{--}30 \times 10^6$	60	76	113	1.18	PC DPPC PG[a] 71.8 31.4 10.3
Fisher and Furia	1977	Trypsin 1% Cryst.	Krebs Ringer	Soybean inhibitor	Ficoll 1.058	Rat	4×10^6	73	46.9	—	—	CO_2-17[b] lactate-17[b] pyruvate-8.7[b]
Pfleger	1977	Trypsin 0.3% (1:250) instilled	Joklik	Soybean inhibitor	Ficoll 1.041	Hamster	6×10^6	79	—	—	—	—
Greenleaf et al.	1979	Cryst. Trypsin 3 mg/ml instilled	PBS	Soybean inhibitor	Elutriation Albumin 1.040 −1.080	Rat	2×10^6	86	101 w/succinate	—	1.1	PC PG[a] 71.8 10.3
Fisher et al.	1980	Trypsin 0.25%, 10 µg DNAase, 1% chicken sera	MEM	1:7 FBS	3 hr culture followed by 22 hr culture	Rat	3.5×10^6	82–90	202	140	—	Lactate 58[b] Pyruvate 25[b] CO_2 56[b] Glucose 104[b]
Dobbs et al.	1980	Porcine elastase, 40 units/ml, instilled	PBS	25% FCS	Albumin 1.040 1.080	Rat	25×10^6	80	75	—	—	Palmitic acid oxidation 4[b]

Reference	Year	Enzyme	Medium	Additives	Separation	Species	Yield	Viability				Markers
Devereau and Fouts	1981	Type I protease 0.1%, instilled	Krebs, Ringer	0.5% BSA 0.5% DNA	Elutriation Fraction II Metrizamide 1.060	Rabbit	25×10^6	80	—	—	—	NADPH cyt[c] red[d] 49[e] / P450 6.5 0.07[c] .080[e]
Finkelstein et al.	1983	Elastase 0.3 mg/ml Trypsin 0.025 mg/ml, instilled	Joklik	20% FCS	Ficoll 1.058	Rabbit	100×10^6	80–90	158	—	—	PC 66.7 / PG[a] 3.5 NADPH cyt.c red[d] / cyt.c 161.4[e]
Robinson et al.	1984	Elastin 4.3U/ml	DMEM	DNAase	Metrizamide 1.040–1.090	Human	1.3–4.8	20	—	—	—	PC 63.4 / PG[a] 8.9
Metler et al.	1984	Trypsin 0.1%	Joklik MEM, ascorbate, etc.	25% FCS DNAase	Ficoll 1.047	Rabbit	20×10^6	87	—	—	—	GSH 11.7[f]
Weller and Karnovsky	1986	Elastase 0.008% Collagenase 0.2% DNAase 0.005% Trypan 0.05% Instilled	PBS	Trypsin inhibitor 4% FCS	Percoll Antirat LC ascites fluid (30 min.) Petri dish coated with IgG fraction of goat anti-mouse IgG 2 hr	Rats	$10–20 \times 10^6$	95	—	—	—	

[a] Percentage liquid phosphorus.
[b] nmoles/10^6 cells/hr.
[c] nmoles/mg protein.

[d] NADPH cytochrome c reductase
[e] nmoles cytochrome c reduced/mg protein/min
[f] μg GSH/10^8 cells

preparations was highest when medium was supplemented with 10% fetal calf serum. Light, phase, and electron microscopic examination demonstrated that these primary cultures were indeed type II cells. The principal morphological feature was the presence of dense lamellar granules in these cells. Primary cultures retained the characteristic type II features for 3–5 days in vitro, after which cultures exhibited a progressive deterioration and loss of their phenotypic properties. This behavioral pattern of type II cells in culture may represent both accelerated proliferation and transformation of these cells into type I epithelial cells. A requirement for high-density plating has been experienced by others (Mettler et al., 1981; Dobbs et al., 1986). The large number of cells required for the establishment of primary cultures is compounded by the low plating efficiency seen with this cell type, usually between 5 and 15% (Diglio and Kikkawa, 1977; Dobbs and Mason, 1979; Mettler et al., 1981). Supplementation of the isolation medium with ascorbate increased plating efficiency several-fold (Mettler et al., 1984).

Geppert and co-workers (1980) have developed a primary culture of rat type II cells that uses floating collagen membranes. This procedure was an attempt to prolong the biochemical and morphological features of type II cells for periods greater than 3–5 days. This system shows some promise in retaining morphological features of type II cells. Cells cultured on a matrix derived from corneal endothelial cells contain lamellar bodies for up to 7–10 days in culture (Mason et al., 1982). Type II cells cultured on human amniotic basement membrane retained their cuboidal shape, lamellar bodies and, surface microvilli for up to 8 days (Lwebuga-Mukasa et al., 1984). Acellular lung matrix has also been used as the substrate to which type II cells were seeded and cultured (Lwebuga-Mukasa, 1986). In one study, cells retained a cuboidal shape as well as lamellar bodies, while in other studies cells behaved similarly to those cultured under plastic dish culture conditions.

While cells cultured on a matrix generally retained phenotypic characteristics of type II cells better than those on the plastic dishes, the phospholipid composition of the two conditions was similar. Table 2 shows the percentage change in phosphatidyl choline and phosphatidyl glycerol composition in type II cells under various culture conditions. Despite the fact that the morphological features of type II cells are better maintained on a matrix, the content of phosphatidyl choline and phosphatidyl glycerol with time in culture does not differ from cells cultured on plastic (Dobbs et al., 1985; Table 2). Table 2 illustrates that the percentage of phosphatidyl choline does not change markedly with time in culture but that of phosphatidyl glycerol decreases markedly by 48 hr. Since phosphatidyl glycerol is more specific for surfactant lipid than all classes of phosphatidyl choline, these studies suggest that there may be a significant decline of "surfactant lipids" in type II cells by 48 hr. Very few studies have partitioned phosphatidyl choline into disaturated species, which is more specific for "surfactant lipid " than all classes of phosphatidyl choline. Smith et al., (1980) have shown that the content

Table 2 Percentage Change in Phosphatidylcholine and Phosphatidylglycerol Composition

		Days in Culture									Author	Animal	Label	Incubation Time with Label (hr)
		0	1	2	3	4	5	6	7	8				
Plastic	PC	—	73.0	53.3	38.2	39.6	—	—	54.2	57.7	Dobbs et al., 1985	Rats	Acetate	24
	PG	—	6.0	2.0	0.7	0.7	—	—	0.6	0.6				
Matrix	PC	—	71.2	51.4	40.1	46.2	—	—	54.4	56.3				
	PG	—	6.1	1.9	0.8	0.7	—	—	0.7	0.8				
Plastic	PC	44.5[a]	—	55.6	58.6	57.4	—	—	—	—	Smith et al. 1980	Rabbit	Palmitate	1
	PG	6.3	—	4.8	1.4	1.4	—	—	—	—				
Plastic	PC	—	65.6 (60%)[b]	—	56.0	—	—	—	—	—	Lwebuga-Mukasa et al., 1986	Rat	Acetate	22
	PG	—	8.6	—	3.1	—	—	—	—	—				
Matrix	PC	—	—	—	72.0	—	—	—	—	—				
	PG	—	—	—	3.1	—	—	—	—	—				
Matrix	PC	—	56.6[a]	—	49.7	—	54.7	—	—	—	Geppert et al., 1980	Rat	Glycerol	24
	PG	—	16.8	—	4.2	—	2.4	—	—	—				
Matrix	PC	—	67.6	57.0	56.4	60.8	—	—	—	—	Saito et al., 1986	Rat Newborn	Glycerol	24
	PG	—	12.3	2.9	1.5	1.2	—	—	—	—				

[a]For all lipids.
[b]Disaturated.

of DSPC in phosphatidyl choline is decreased from 50% to 25% by 24 hr. These data on both DSPC and PG seem to indicate that surfactant lipid synthesis rapidly decreases even if the morphology of type II cells may resemble in vivo type II cells. While type II cells in culture may be merely dedifferentiating, the possibility also exists that type II cells in culture behave similarly to type II cells in vivo by transforming into type I cells (Evans et al., 1973), as suggested at the time of initial successful primary culture of type II cells (Diglio and Kikkawa, 1977). Brandt (1982) reported that the alveolar surface of type I and type II cells have different lectin-binding properties. *Maclura pomifera*, a lectin binding to alpha-galactose residues, binds to the apical surfaces of type II but not type I cells, while *Ricinus communis I*, a lectin specific for beta—galactose residues, binds to the surface of type I but not type II cells in. Utilizing these lectins, Dobbs et al., (1985) found evidence that type II cells culture acquire characteristics of type I pneumocytes. The ultimate fate, dedifferentiation to type I cells, cannot be avoided, short of transformation of type II cells, for example, with SV_{40} virus. The fetal pretype II cells can be cultured and maintained as type II cells for some period of time (Scott et al., 1983). These cells appear to retain a reasonably high proportion of disaturated phosphatidyl choline and phosphatidyl glycerol (38% DSPC, 5, 2% PG) after 2 weeks in culture. Since there are no differences in the surfactant machinery in adult and newborn lungs, this cell culture system appears to be the most promising model for study at present.

IV. Applications

Despite some drawbacks in using freshly isolated and cultured type II cells, numerous studies have been reported in the past decade that have given us new insight into some of the biochemical, hormonal, and intracellular mechanisms operating within this cell, in relation to surfactant production and secretion, electrolyte and glucose uptake, and drug metabolism.

From available reports, it appears that type II cells prefer glucose, acetate, and palmitate as substrates for the synthesis of surfactant phospholipids (Kikkawa et al., 1975; Batenburg et al., 1978; Smith and Kikkawa, 1979). From studies of isolated type II cells, evidence has accumulated indicating that the CDP-choline pathway is the major, if not the only, route for the de novo synthesis of phosphatidyl choline. Experiments with isolated type II cells indicate that dipalmitoylphosphatidyl choline can be produced directly by the CDP-choline pathway (Smith and Kikkawa, 1978; Post et al., 1983). However, a significant portion of dipalmitoylphosphatidyl choline is probably produced by a remodeling mechanism (Smith and Kikkawa, 1978). The major mechanisms that have been proposed for this remodeling process are a deacylation–reacylation cycle catalyzed by phospholipase A_2 and lysophosphatidyl choline acyltransferase, respectively; and a deacylation–transacylation process involving the sequential action of phospholipase A_2 and lysophosphatidyl-choline/lysophosphatidyl choline acyltransferase (Batenburg and Van Golde, 1979). In both adult and fetal type II cells the conversion of the intermediate 1-palmitoyl-sn-glycerol-3-phosphocholine proceeds by reacylation rather than by transacylation (Batenburg et al., 1979). Studies with type II cells isolated from adult and fetal rat lung (Post et al., 1984; Van Golde et al., 1985) led to the conclusion that choline phosphate cytidyltransferase catalyses an important regulatory step in the formation of surfactant phosphatidylcholine. Moreover, a ratio of 8:1 is seen for dipalmitoylphosphatidyl choline and phosphatidyl glycerol in both type II cell and lung lavage, which suggests that the type II cell is the only source for pulmonary surfactant lipids (Kikkawa and Smith, 1983).

The study of the surfactant-associated proteins, which are 35kDA and 6–10 Kda, has been aided by the development of CDNA probes for these proteins. This technology enabled preparations of these proteins to be obtained in pure form. With these, isolated type II cells have been used to elucidate the role of the 35,000 dalton protein (SAP 35). This protein was shown to inhibit surfactant secretion (Rice et al., 1987), while Crouch et al., (1987) have used isolated type II cells to study the posttranslational modifications of SAP-35). Phelps and Taeusch (1987) used indirect immunoperoxidase staining to localize this apoprotein to type II cells. This protein has been shown to be important in aiding the spreading of surfactant.

Dobbs and Mason (1979) reported that the secretion of phosphatidyl choline was stimulated by beta-adrenergic agonists. This was confirmed by others (Brown and Longmore, 1981; Mettler et al., 1981). These workers have shown that intracellular cyclic AMP content in type II cell is increased with beta adrenergic agonists. The calcium ionophore A23187 (Mason et al., 1977) and 12-0-tretradecanoyl phorbol 13-acetate also stimulate surfactant secretion in isolated type II cells by manipulating intracellular calcium levels. It is interesting that beta-adrenergic agonists A23187 and tetradecanoyl phorbol acetate have additive effect on secretion; it has, therefore, been suggested that cyclic AMP and calcium act as intracellular "second messengers" in the coupling of stimuli to the secretion of surfactant (Dobbs et al., 1986). The study by Brown et al., (1985) seems to indicate the involvement of the microtubule–tubulin system in the secretion of surfactant by beta-adrenergic agonist, which appears to be operating as a receptor-independent mechanism (Fabisiak et al., 1986).

In addition, using freshly isolated rat type II cells, Ballard and his co-workers (1978) demonstrated that these cells possessed specific glucocorticoid-binding sites. Others (Post et al., 1980) have supported this observation by demonstrating that the glucocorticoid cortisol stimulates the formation of phosphatidylglycerol and dipalmitoyl phosphatidyl choline. Another investigation (Geppert et al., 1980) has revealed that type II cells, in the presence of another glucocorticoid, dexamethasone, have an improved surfactant synthesis and plating efficiency onto a collagen matrix.

It is of interest to note that isolated type II cells have been shown to secrete substances other than surfactant-associated substances. Sage et al. (1983) have shown that type II cells produced components of extracellular matrix. Fractionation by ion-exchange chromatography of radiolabeled protein secreted into the culture medium resulted in the partial purification of two of these components: fibronectin and type IV procollagen. The levels of these secreted proteins were measured by radioimmune precipitation. Of the total radiolabeled culture medium protein secreted during a 24 hr period by the granular pneumocytes, fibronectin, type IV procollagen, and thrombospondin represented 3–15%, 2%, and 3%, respectively. The biosynthesis by alveolar epithelial cells of proteins that constitute or are closely associated with the alveolar basement membrane implies that this structure is at least partially derived from the cells themselves. Furthermore, it suggests that the type II epithelial cell is involved in pulmonary cytodifferentiation, in lung morphogenesis and repair, and in certain interstitial lung disorders in which derangement of the extracellular matrix occurs.

The alveolar air space of the adult mammalian lung is maintained virtually free of fluid with only a thin film, the alveolar lining layer, covering the epithelial surface. The factors responsible for maintaining this balance between the alveolar fluid and extracellular fluid of the interstitium are complex and most likely involve a number of interrelated passive and active forces. One important factor may be a process of active ion transport across the alveolar epithelium, which results in fluid movement from the alveolus to the interstitium. Recently, however, a study on the transport properties of primary cultured monolayers of type II alveolar cells from adult rat lung, which showed that the epithelium absorbs sodium chloride and water, has opened up new possibilities for the epithelium to be studied (Goodman et al., 1983). Utilizing mouse type II cell monolayer culture, Schneider et al. (1985) investigated membrane channels of these cells. In excised plasma membrane patches in the "outside-out" configuration, they observed anion-selective channels with a conductance of 350–400 pS, and burst lengths lasting seconds. When patches were bathed in solutions with equal chloride concentrations, channels opened and closed spontaneously at membrane voltages close to zero, but tended to close when the potential was shifted by $\pm$ 10 mV, particularly in the negative direction. Other anions, for example, iodine and bromine, could pass through these channels. In contrast, there was a very low permeability for the sodium cation. On the other hand, calcium ions reduced channel conductance. These authors suggested that the channels have a role in salt absorption. Furthermore, Cott et al. (1986) have shown that physiological and pharmacological agents that alter intracellular cyclic AMP appear to play an important role in regulating active ion transport across the alveolar epithelium. While these studies are useful it would be necessary to determine that this property of "type II cells" may indeed by a property of type I cells. Such markers are available today.

Type II pneumocytes have been studied in some pathological conditions. After acute exposure to bleomycin, the isolated type II cells were hypertrophic and possessed an increased capacity to synthesize surfactant lipids (Kikkawa et al., 1976). Wright et al. (1982) studied the changes of type II cell lipid metabolism after exposure to NO_2. A general increase in cell content of biosynthetic enzyme activities (units/mg DNA) was observed in type II cells from NO_2-exposed rats, but no change was detected in the activity of the microsomal marker enzyme, NADPH cytochrome c reductase. Glycerolphosphate acyltransferase and choline phosphotransferase increased 171 and 168%, respectively, and phosphatidate phosphohydrolase increased 69%. Glycerolphosphate phosphatidyltransferase increased 143% and succinate cytochrome c reductase, the mitochondrial marker enzyme, increased 111%. The increases in protein content and activity of phospholipid biosynthetic enzymes in type II cells are consistent with a general hypertrophy of type II cells, which includes stimulation of surfactant phospholipid biosynthesis 2 days after exposure to NO_2 when type II cell proliferation is occurring.

Skillrud and Martin (1984) studied the direct effect of paraquat on isolated type II cells. Paraquat, a widely used herbicide, causes severe, often fatal, lung damage. In vivo studies suggest that the alveolar epithelial cells (types I and II) are specific targets of paraquat toxicity. This study used ^{51}Cr-labeled type II cells to demonstrate that paraquat ($10^{-5}M$) resulted in type II cell injury in vitro, independent of interacting immune effector agents. With ^{51}Cr release expressed as the cytotoxic index (CI), type II cell injury was found to accelerate with increasing paraquat concentrations ($10^{-5}M$, $10^{-4}M$, and $10^{-3}M$, resulting in a CI of 12.5 $\pm$ 2.2, 22.8 $\pm$ 1.8, and 35.1 $\pm$ 1.9, respectively). Paraquat-induced cytotoxicity ($10^{-4}M$, with a CI of 22.8 $\pm$ 1.8) was effectively reduced by catalase, 1,000 U/ml (CI 8.0 $\pm$ 3.2, p <0.001), superoxide dismutase, 300 U/ml (CI 17.4 $\pm$ 1.7, p <0.05), and alpha-tocopherol 10 μg/ml (CI 17.8 $\pm$ 1.6, p <0.05). Paraquat toxicity ($10^{-3}M$) was potentiated in the presence of 95% 0_2 with an increase in CI from 31.1 $\pm$ 1.7 to 36.4 $\pm$ 2.3 (p <0.05). Paraquat-induced type II cell injury was noted as early as 4 hr after incubation by electron microscopy, which showed swelling of mitochondrial cristae and dispersion of nuclear chromatin. Thus, this in vitro model indicates that paraquat-induced type II cell injury can be quantitated, confirmed by morphological ultrastructural changes, significantly reduced by antioxidants, and potentiated by hyperoxia.

Freeman et al. (1986) measured the changes in the antioxidant enzymes CuZn superoxide dismutase, Mn superoxide dismutase, catalase, glutathione peroxidase, and glucoce-6-phosphate dehydrogenase from both air control rats and those exposed to 85% oxygen. The oxygen-exposed group had a significantly higher level of these enzymes than did the air-exposed group. Moreover, when the antioxidant enzyme levels in isolated type II cells were known, combined with morphometric data, the total enzyme activity in the lung accounted for by type

II cells could be calculated. In both the air- and oxygen-exposed groups, type II cells accounted for over 50% of glucose-6-phosphate dehydrogenase activity and less than 10% of the Mn superoxide dismutase. On the other hand, the level of CuZn superoxide dismutase increased by 10% from 8 to 18%.

Several laboratories have started to use isolated type II cells to investigate the role of this cell in the process of metabolizing pulmonary toxins and carcinogens.

Devereux and her colleagues (1981, 1982) studied drug-metabolizing enzyme system in type II cells. These studies indicate that type II cells are enriched with cytochromes P450 and b5 and NADPH cytochrome c reductase. Ipomeanol, a lung toxin, has selective binding to type II cells and Clara cells. The binding of Ipomeanol is prevented by piperonyl butoxide, an inhibitor of Iponeanol-dependent cytochrome P450 metabolism.

Others (Jones et al., 1982) have examined the ability of type II cells and Clara cells to metabolize potentially carcinogenic compounds such as beta-naphthoflavone. Their data suggest the importance of type II cells and Clara cells in the metabolism of drugs as well as chemical carcinogens.

References

Ayuso, M. S., Fisher, A. B., Parilla, R., and Williamson, J. R. (1973). Glucose metabolism by isolated rat lung cells. *Am. J. Physiol.* **225**:1153–1160.

Ballard, P. L., Mason, R. J., and Douglas, W. H. (1978). Glucocorticoid binding by isolated lung cells. *Endocrinology* **102**:1570–1575.

Batenburg, J. J., and Van Golde, L. M. G. (1979). Formation of pulmonary surfactant in whole lung and in isolated type II alveolar cells. In *Reviews in Perinatal Medicine* Vol. 3. Edited by E. M. Scarpelli and E. V. Cosmi. New York, Raven Press, pp. 73–114.

Batenburg, J. J., Longmore, W. J., and Van Golde, L. M. G. (1978). The synthesis of phosphatidylcholine by adult rat lung alveolar type II epithelial cells in primary culture. *Biochim. Biophys. Acta.* **529**:160–170.

Batenburg, J. J., Longmore, W. J., Klazinga, W., and Van Golde, L. M. G. (1979). Lysolecithin acyltransferase and lysolecithin: lycolecithin acyltransferase in adult rat lung alveolar type II epithelial cells. *Biochim. Biophys. Acta.* **573**:136–144.

Brandt, A. E. (1982). Cell surface saccharides of rat lung alveolar type 1 and type 2 cells. *Fed. Proc.* **42**:755A. 2830.

Brown, L. A. S., and Longmore, W. J. (1981). Adrenergic and cholinergic regulation of lung surfactant secretion in the isolated perfused rat lung and in the alveolar type II cell in culture. *J. Biol. Chem.* **256**:66–72.

Brown, L. A. S., Pasquale, S. M., and Longmore, W. J. (1985). Role of microtubules in surfactant secretion. *J. Appl. Physiol.* **58**:1866–1873.

Cott, G. R., Sugahara, K., and Mason, R. J. (1986). Stimulation of net active ion transport across alveolar type II cells monolayers. *Am. J. Physiol.* **250**: (2 Pt 1) 222–227.

Crapo, J. D., Barry, B. E., Foscue, H. A., and Shelburne, J. (1980). Structural and biochemical changes in rat lungs occurring during exposures to lethal and adaptive doses of oxygen. *Am. Rev. Respir. Dis.* **122**:123–143.

Crouch, E., Rust, L., Moxley, M., and Longmore, W. (1987). 2'-dipyridyl inhibits the secretion of surfactant apoprotein by type II pneumocytes. Evidence for a role of the collagenous domain in apoprotein secretion. *Am. Rev. Respir. Dis.* **135**:A62.

Devereux, T. R., and Fouts, J. R. (1981). Xenobiotic metabolism by alveolar type II cells isolated from rabbit lung. *Biochem. Pharmacol.* **30**:1231–1237.

Devereux, T. R., Jones, K. G., Bend, J. R., Fouts, J. R., Stratham, C. N., and Boyd, M. R. (1982). *In vitro* metabolic activation of the pulmonary toxin, 4-ipomeanol, in nonciliated bronchilar epithelial (Clara) and alveolar type II cells isolated from rabbit lung. *J. Pharmacol. Exp. Ther.* **220**:223–227.

Diglio, C. A., and Kikkawa, Y. (1977). The type II epithelial cells of the lung IV. Adaption and behavior of isolated cells in culture *Lab. Invest.* **37**:622–630.

Dobbs, L. G., and Mason, R. J. (1979). Pulmonary alveolar type II cells isolated from rats. Release of phosphatidylcholine in response to beta-adrenergic stimulation *J. Clin. Invest.* **63**:378–387.

Dobbs, L. G., Geppert, E. F., Williams, M. C., Greenleaf, R. D., and Mason, R. J. (1980). Metabolic properties and ultrastructure of alveolar type II cells isolated with elastase. *Biochim. Biophys. Acta* **618**:510–523.

Dobbs, L. G., Williams, M. G., and Mason, R. J. (1985). Changes in biochemical characteristics and pattern of lectin binding of alveolar type II cells with time in culture. *Biochim. Biophys. Acta* **846**:155–166.

Dobbs, L. G., Gonzalez, R. F., Meriuari, L. A., Mescher, E. J., and Hawgood, S. (1986). The role of calcium in the secretion of surfactant by rat alveolar type II cells. *Biochim. Biophys. Acta* **877**:305–313.

Evans, M. J., Cabral, L. J., Stephens, R. J., and Freeman, G. (1973). Renewal of alveolar epithelium in the rat following exposure to NO_2. *Am. J. Pathol.* **70**:175–198.

Fabisiak, J. P., Rannels, S. R., Vesell, E. S., and Rannels, D. E. (1986). Receptor-independent sequestration of beta-adrenergic ligands by alveolar type II cells. *Am. J. Physiol.* **250**:871–879.

Finkelstein, J. N., Maniscalco, W. M., and Shapiro, D. S. (1983). Properties of freshly isolated type II alveolar epithelial cells. *Biochim. Biophys. Acta* **762**:398–404.

Fisher, A. B., and Furia, L. (1977). Isolation and metabolism of granular pneumocytes from rat lungs. *Lung* **154**:155–165.

Fisher, A. B., Furia, L., and Berman, H. (1980). Metabolism of rat granular pneumocytes isolated in primary culture. *J. Appl. Physiol.* **49**:743–750.

Freeman, B. A., Mason, R. J., Williams, M. C., and Crapo, J. D. (1986). Antioxidant enzyme activity in alveolar type II cells after exposure of rats to hyperoxia. *Exp. Lung. Res.* **10**:203–222.

Geppert, E. F., Williams, M. C., and Mason, R. J. (1980). Primary culture of alveolar type II cells on floating collagen membranes. *Exp. Cell Res.* **128**:363–374.

Goodman, F. E., Fleischer, R. W., and Crandall, E. D. (1983). Evidence for active Na^{2+} transport by cultured monolayers of pulmonary alveolar epithelial cells. *Am. J. Physiol.* **245**: (Cell physiol. 14): C78-C83.

Greenleaf, R. D., Mason, R. J., and Williams, M. C. (1979). Isolation of alveolar type II cells by centrifugal elutriation. *In Vitro* **15**:673–684.

Jones, K. G., Holland, J. F., and Fouts, J. R. (1982). Benzo (a)-pyrene hydroxylase activity in enriched populations of Clara cells and alveolar type II cells from control and beta-naphthoflavone pretreated rats. *Cancer Res.* **42**:4658–4663.

Kikkawa, Y., and Smith, F. (1983). Cellular and biochemical aspects of pulmonary surfactant in health and disease. *Lab. Invest.* **49**:122–139.

Kikkawa, Y., and Yoneda (1974). The type II epithelial cells of the I. Method of isolation. *Lab. Invest.* **30**:76–84.

Kikkawa, Y., Yoneda, K., Smith, F., Packard, B., and Syzuki, K. (1975). The type II epithelial cells of the lung. II. Chemical composition and phospholipid synthesis. *Lab Invest.* **32**:295–301.

Kikkawa, Y., Aso, Y., Yoneda, K., and Smith, F. (1976). Secithin synthesis by normal and bleomycin-treated type II cells. In: *Lung Cells and Disease*. Edited by A. Bouhys. Amsterdam/New York/Oxford, North Holland Publishing Co., pp. 139–146.

King, R. J. (1977). Metabolic fate of the apoproteins of pulmonary surfactant. *Am. Rev. Respir. Dis.* **115**:73–79.

King, R. J. (1979). Utilization of alveolar epithelial type II cells for the study of pulmonary surfactant. *Fed. Proc.* **38**:2637–2643.

Kono, T., and Barham, F. W. (1971). The relationship between the insulin-binding capacity of fat cells and the cellular response to insulin. *J. Biol. Chem.* **246**:6210–6216.

Leary, J. F., Finkelstein, J. N., Notter, R. H., and Shapiro, D. L. (1982). Isolation of type II pneumocytes by laser flow cytometry. *Am. Rev. Respir. Dis.* **125**:326–330.

Leslie, C. C., McCormick-Shannon, K., Robinson, P. C., and Mason, R. J. (1985). Stimulation of DNA synthesis in cultured rat alveolar type II cells. *Exp. Lung Res.* **8**:53–66.

Lwebuga-Mukasa, J. S., Thulin, G., Madri, J. A., Barrett, C., and Warshaw, J. (1984). An acellular human amnionic membrane model for in vitro culture of type II pneumocytes. The role of the basement membrane in cell morphology and function. *J. Cell Physiol.* **121**:215-225.

Lwebuga-Mukasa, J. S., Lugbar, D. H., and Madri, J. A. (1986). Repopulation of a human alveolar matrix by adult rat type II pneumocytes *in vitro*. *Exp. Cell Res.* **162**:423-435.

Mango, J. A., McSherry, N. R., Butscher, F., Irwin, K., and Barber, T. (1975). Dispersed rat paratid acinar cells. I. Morphological and functional characterization. *Am. J. Physiol.* **229**:553-559.

Mason, R. J., Williams, M. C., and Clements, J. A. (1975). Isolation and identification of type 2 alveolar epithelial cells. *Chest* **67** (supplement):365.

Mason, R. J., Williams, M. C., and Greenleaf, R. D. (1976). Isolation of lung cells. In *Lung Cells in Disease*. Edited by A. Bouhuys. Amsterdam/New York/Oxford, North Holland Publishing Co., pp. 39-52.

Mason, R. J., Williams, M. C., and Dobbs, L. G. (1977). Secretion of disaturated phosphatidyl choline by primary cultures of type II alveolar cells. In *Pulmonary Macrophage and Epithelial Cells*. Edited by C. L. Sanders, R. P. Schneider, G. E. Dagle, and H. A. Ragar. 16th Annual Hanford Biology Symposium. Springfield, VA, Energy Research and Development Administration, pp. 280-297.

Mason, R. J., Williams, M. C., Widdicombe, J. M., Sauders, M. J., Misfeld, D. S., and Berry, L. C. Jr. (1982). Transepithelial transport by pulmonary alveolar type II cells in primary culture. *Proc. Natl. Acad. Sci. USA* **79**:6033-6037.

Mettler, N. R., Gray, M. E., Schuffman, S., and LeQuire, V. S. (1981). Beta-adrenergic induced synthesis and secretion of phosphatidylcholine by isolated pulmonary alveolar type II cells. *Lab. Invest.* **45**:575-586.

Mettler, N. R., Yano, S., Kikkawa, Y., and Ivasauskas, E. (1984). Type II epithelial cells of the lung. VII. The effect of ascorbic acid and glutathione. *Lab Invest.* **51**:441-448.

Pfleger, R. C. (1977). Type II epithelial cells from the lung of syrian hamsters: isolation and metabolism. *Exp. Mol. Pathol.* **27**:152-166.

Phelps, D. S., and Taeusch, H. W. (1987). Characterization, N-terminal sequence and immunolocalization of a bouine low molecular weight surfactant protein. *Am. Rev. Respir. Dis.* **135**:A378.

Post, M., Batenberg, J. J., and Van Golde, L. M. G. (1980). Effects of cortisol and thyroxine on phosphatidylcholine and phosphatidylglycerol synthesis by adult rat alveolar type II cells in primary culture. *Biochim. Biophys. Acta.* **618**:308-317.

Post, M., Schuurmans, A. J. M., Batenberg, J. J., and Van Golde, L. M. G. (1983). Mechanisms involved in the synthesis of disaturated phosphatidylcholine by alveolar type II cell isolated from adult rat lung. *Biochim. Biophys. Acta.* **750**:68-77.

Post, M., Torday, J. S., and Smith, B. T. (1984). Alveolar type II cells isolated from fetal rat lung organotypiccultures synthesize and secrete surfactant associated phospholipids and response to fibroblast–pneumocyte factor. *Exp. Lung. Res.* **7**:53–65.

Rice, W. R., Singleton, F. M., Tannenbaum, D. A., and Ross, G. F. (1987). Regulation of surfactant secretion by the major surfactant associated protein (SAP-35) and lectins. *Am. Rev. Respir. Dis.* **135**:A377.

Robinson, P. C., Voilker, D. R., and Mason, R. J. (1984). Isolation and culture of human alveolar type II epithelial cells. *Am. Rev. Respir. Dis.* **130**:1156–1160.

Sage, H., Forin, F. M., Striker, G. E., and Fisher, A. B. (1983). Granular pneumocytes in primary culture secrete several major components of the extracellular matrix. *Biochemistry* **22**:2148–2155.

Saito, K. J. S., Lwebuga-Mukasa, J. S., Barrett, C., Light, D., and Warshaw, J. B. (1985). Characteristics of primary isolates of alveolar type II cells from neonatal rats. *Exp. Lung. Res.* **8**:213–225.

Schneider, G. T., Cook, D. T., Gage, P. W., and Young, J. A. (1985). Voltage sensitive, high conductance chloride channels in the luminal membranes of cultured pulmonary alveolar (type II) cells. *Pflugers Arch.* **404**:354–357.

Scott, J. E., Possmayer, G., and Harding, P. G. R. (1983). Alveolar pre-type II cells from the fetal rabbit lung. *Biochim. Biophys. Acta.* **753**:195–204.

Skillrud, D. M., and Martin, W. J. II (1984). Paraquat induced injury of type II alveolar cells. *Am. Rev. Respir. Dis.* **129**:995–999.

Smith, F. B., and Kikkawa, Y. (1978). The type II epithelial cells of the lung. III. Lecithin synthesis: a comparison with pulmonary macrophages. *Lab. Invest.* **38**:45–51.

Smith, F. B., and Kikkawa, Y. (1979). The type II epithelial cells of the lung. V. Synthesis of phosphatidyl glycerol in isolated type II cells and pulmonary alveolar macrophages. *Lab Invest.* **40**:172–177.

Smith, F. B. Kikkawa, Y., Diglio, C. A., and Dalen, R. C. (1980). The type II epithelial cells of the lung. VI. Incorporation of ^{3}H-choline and ^{3}H-palmitate into lipids of cultured type II cells. *Lab. Invest.* **42**:296–301.

Tierney, D. F. (1974). Lung metabolism and biochemistry. *Annu. Rev. Physiol.* **36**:209–231.

Van Golde, L. M. G., Post, M., Batenburg, J. J., DeVries, A. C. J., and Smith, B. T. (1985). Synthesis of surfactant lipids in developing rat lung: studies with isolated alveolar type II cells. *Biochem. Soc. Trans.* **13**:86–89.

Weller, N. K., and Karnovsky, M. J. (1986). Improved isolation of rat alveolar type II cells. More representative recovery and retention of cell polarity. *Am. J. Pathol.* **122**:92–100

Wolf, B. M. J., Rubinstein, D., and Beck, J. C. (1968). The metabolism of isolated pneumocytes from rabbit lung. *Can. J. Biochem.* **46**:151–154.

Wright, E. S., Vang, M. J., Finkelstein, J. N., and Mavis, R. D. (1982). Changes in phospholipid biosynthetic enzymes in type II cells and alveolar macrophages isolated from rat lungs after NO_2 exposure. *Toxicol. Appl. Pharmacol.* **66**:305–311.

16

Endothelial Cells of the Lung

EDWARD J. MACARAK

Connective Tissue Research Institute
University of Pennsylvania
Philadelphia, Pennsylvania

I. Introduction

A. Historical Background of Endothelial Cell Culture and Lung Endothelial Cell Culture

Techniques for the isolation and in vitro propagation of vascular endothelial cells were perfected in the early 1970s (Jaffe et al., 1973a; Gimbrone et al., 1974; Lewis et al., 1973). While a number of approaches have been used, the most useful ones for obtaining large numbers of cells involve either collagenase digestion (Jaffe et al., 1973a; Gimbrone et al., 1974) or scraping of the intima (Lewis et al., 1973; Ryan et al., 1978). It had been shown previously that the use of proteolytic enzymes such as trypsin did not produce viable, dividing endothelial cultures while both collagenase digestion and intimal scraping did (Maruyama, 1973). These two approaches have been used by many laboratories and are still the techniques of choice for the isolation of endothelial cells from large blood vessels, such as aorta. The initial reports on endothelial culture established that endothelium from large vessels such as the aorta, vena cava, and pulmonary artery could be maintained in vitro for relatively long periods of time (reviewed in Jaffe, 1984). Subsequently, other procedures have been developed for the isolation and culture of endothelial cells from microvascular beds, including those in the lung (Folkman et al., 1979; Davidson et al., 1980; Wagner and Mathews, 1975; Habliston et al., 1979), although problems still exist with the in vitro characterization and long-term maintenance of these cells.

As with the culture of any type of differentiated cell, it is necessary to characterize its phenotype in vitro. For endothelial cells this is relatively simple since factor VIII/von Willebrand's protein is unique to the endothelial layer within the vessel wall (Jaffe et al., 1973b). Because factor VIII/von Willebrand's protein is not produced by other tissue components of the vessel wall, its presence can be used to characterize cells in vitro as bona fide endothelium. Commercial antibodies raised against this protein are available, which make it relatively easy for investigators to verify the presence of this antigen in their cultured cells. All endothelial cells, including those from large vessels and microvascular sources, appear to produce this antigen (Wagner et al., 1982). The ability of endothelial cells to express factor VIII/von Willebrand's protein also seems to be a relatively stable phenotypic trait, since it appears to be continuously expressed throughout the in vitro lifespan of endothelium (Rosen et al., 1981).

B. Statement of Goals

The lung contains an extensive vascular bed that has previously been shown to be extremely dynamic in terms of its metabolic activities (Bakhle and Vane, 1977). Because the pulmonary circulation is thought to be the site of action for many physiological processes, there is strong interest in learning more about its properties. Cell culture lends itself well to such studies, since one can isolate and maintain individual cell types to characterize their properties. For example, vascular endothelial cells (Ryan et al., 1978; Johnson, 1980), smooth muscle cells (Dunn and Franzblau, 1982), fibroblasts (Bruel et al., 1980), and alveolar epithelial cells (Kikkawa and Yoneda, 1974) from the lung have been studied. Of major importance is determining if segments of the vascular tree within the lung have different properties. For example, it is not known whether the physiological properties and activities of pulmonary artery endothelium differ from those of alveolar capillary endothelium. The goal of this chapter is to describe state-of-the-art approaches to the culture of vascular cells from the lung that have been successfully applied in the author's and other investigators' laboratories. The chapter will not attempt to be an exhaustive review of all available methods, but will rely on procedures with which the author has firsthand experience. For the most part, these methods have been published and are standard techniques used in a number of laboratories. In addition, comments have been provided to help individuals not familiar with these particular cell culture techniques. Since this chapter must be selective, other effective methods may have been excluded. For this, only the author can be faulted.

It is assumed that the readers have some expertise with standard techniques of cell culture; those unfamiliar with them should refer to reference books such as Paul's *Cell and Tissue Culture* (1972) and Kruse and Patterson's *Tissue Culture: Methods and Applications* (1973) or other similar literature.

II. Method for Culture of Lung Endothelial Cells (Macarak et al., 1976)

A. Reagents

1. Collagenase (GIBCO, Grand Island, NY, cat. No. 840-701811) prepared from *Cl. histolyticum* 1 mg/ml in Medium 199

2. Medium 199/Earles salts (KC Biological, Lenexa, KS, cat. No. DM-310) with L-glutamine, without sodium bicarbonate

3. Fetal bovine serum (Hazelton Research Products, Denver, PA) 20% (v/v)

4. Amphotericin-B (GIBCO, cat. No. 600-5295) 2.5 μg/ml

5. Gentamicin sulfate (GIBCO, cat. No. 600-5710) 50 μg/ml

6. HEPES (N-2-hydroxyethyl piperazine-N'-2-ethane sulfonic acid; Research Organics, Inc., Cleveland, OH, cat No. 6003H), 15 mM

7. Bacto-Peptone (Difco Laboratories, Detroit, MI) 0.5 g/L

8. Glucose, 3 g/L

9. BME amino acids (GIBCO Laboratories, Grand Island, NY)

10. BME vitamins (GIBCO Laboratories, Grand Island, NY)

B. Types of Endothelium Obtainable From the Lung

Human, bovine, and porcine tissues are the most widely used sources of lung endothelial cells. In the author's laboratory, bovine cells are used almost exclusively because the tissue can be obtained easily. Although human cells appear to be more fastidious to maintain for long periods of time in the laboratory, they may have advantages for experimental purposes, for example, if an investigator has developed assays or reagents that will only work with human tissues. While the methods described here use bovine blood vessels as a source for endothelium, similar approaches have also worked using human tissues (Thornton et al., 1983).

At most cattle abattoirs, blood vessels from adults or fetuses can be obtained for a nominal fee. It is good practice to obtain the vessels as soon as possible after the death of the animal and to immerse them in a chilled (4 °C) saline solution. Cell viability appears directly related to the speed with which these initial steps are performed. A buffered saline solution can be used to transport the abattoir to the laboratory; however, serum-free culture medium can also be used as a "transport medium."

As mentioned above, blood vessels can be obtained from fetuses. Our laboratory has routinely used fetal blood vessels as a source of cells for the past 6 years (Macarak, 1984). One problem that arose from our prior use of calf blood vessels was that the animals from which the vessels were taken were raised as sources of premium veal and, as such, were fed a diet deficient in vitamins and iron (ostensibly to improve the quality of the veal by making the meat more tender). In reality, these animals were unhealthy and the endothelial cells derived from them were difficult to maintain. It has been our experience that cells isolated from such animals showed a high incidence of bacterial and perhaps viral infection. As a result of our difficulties with cells derived from animals raised on such diets, we began to use fetal vessels as sources for our endothelium. To date, our success has been very good with the isolation and long-term culture of both endothelial and smooth muscle cells from fetal vessels.

We routinely choose a good-sized fetus from which to remove the blood vessels (crown–rump length over 50 cm). The heart is dissected out together with the pulmonary artery, ductus arteriosus, thoracic aorta, and vena cava. Dissection of the fetal heart together with the connecting blood vessels is performed as aseptically as possible by the operator wearing sterile gloves. Sterile scissors are used to open the chest cavity of the animal prior to removal of the heart. Before the chest of the fetus is opened, the axillary muscles are cut to permit the forelegs to lie flat. To remove the heart, a ventral incision is made at the lower margin of the right rib cage and extended toward the sternum along the lower edges of the ribs. The sternum is cut and two other incisions are made between the ribs, 3–4 inches apart, and extended toward the spinal cord of the animal. The ribs are either cut or broken at the spinal column and the flap of ribs bent back to expose the pleural cavity. The diaphragm is cut along the right dorsal margin to separate it from the body wall and is also cut around the vena cava. The vena cava is freed from the liver by cutting around it.

Anterior to the heart and ascending aorta, the trachea and esophagus are cut and the dorsal aorta exposed by pulling the heart, lungs, and associated tissue away from the spinal column. The aorta lies against the dorsal wall of the thoracic cavity and can be removed by careful cutting of each of the paired intercostal arteries as they branch off of the aorta. When this is accomplished, the heart, lungs, and associated blood vessels can be removed from the chest cavity. The pulmonary arteries are cut distal to the bifurcation into the right and left trunks, the remaining lung tissue is cut away, and the tissue is placed in chilled medium on ice. Alternatively, the lungs can be left intact, and transported back to the laboratory together with the heart and associated blood vessels.

In the laboratory, the sample container is placed in a laminar flow hood after the external surfaces have been wiped down with 70% alcohol. The working area within the hood is arranged so that a sterile towel can be placed in the center of the hood to serve as a sterile field. Items to be placed in the hood include the following: a sterile instrument tray containing large and small scissors, several pairs of forceps, several small hemostats, a pair of sterile gloves, a sterile towel and sterile 15 ml beakers (one beaker for each vessel to be processed). Sterile technique should be followed when processing the vessels.

Using the hemostat, the tissue is removed from the transport container and placed on the sterile towel. The pulmonary artery, aorta, and vena cava are individually dissected free of the heart and placed in the small beaker containing medium and antibiotics (50 μg/ml gentamicin sulfate and 2.5 μg/ml amphotericin B).

The transport container and towel are removed from the hood, the hood surface wiped with 70% alcohol, and a new sterile towel laid out. The investigator dons new gloves and removes a blood vessel from its sterile beaker. If collagenase digestion is to be performed, the lower end of the vessel is clamped with a small hemostat, the vessel is filled with a solution of collagenase (1 mg/ml in medium 199), and the top is closed with another hemostat. The vessel is then set aside for approximately 15 min to permit the collagenase to digest the intima. The other vessels are treated sequentially. At the end of the digestion period, the vessel contents are emptied, the vessel is filled three times with medium, and the washes are collected and saved for each vessel. The total effluent from each vessel is centrifuged and the the pellets containing endothelial cells are resuspended in medium supplemented with 20% fetal bovine serum and antibiotics (50 μg/ml gentamicin sulfate and 2.5 μg/ml amphotericin B), inoculated into culture flasks (typically two T-25 flasks/vessel) and incubated at 37 °C in an atmosphere of 5% CO_2 in air.

Alternatively, each vessel can be opened and laid out, endothelial side up on the sterile towel. A sterile scalpel blade is mounted on a handle and gently dragged across an area of the vessel surface. A residue of the endothelial cell layer should be visible on the edge of the scalpel blade. The cells on the edge of the blade are released by swirling it in a centrifuge tube containing medium.

Bovine pulmonary endothelial cells are maintained in medium 199 (M199) with Earle's salts supplemented as described by Lewis et al. (1973). These supplements include Bacto-Peptone (0.5 g/L), glucose (3 g/L), 10 ml of 100 × concentrated BME amino acids, 10 ml of 100 × BME vitamins (GIBCO Laboratories, Grand Island, NY). In addition, we routinely add (HEPES 15 mM). It should be noted that other investigators have cultured bovine endothelial cells in other culture media (Ryan et al., 1978; Gajdusek and Schwartz, 1983). We routinely add 20% fetal bovine serum to the medium as well as antibiotics (gentamicin sulfate, 50 μg/ml, and amphotericin B, 2.5 μg/ml).

C. Cloning of Endothelium

The following reagents are used:

1. Sterile-filtered conditioned medium from smooth muscle cells

2. Nalgene disposable filter (115 mL), 0.20 μm (Nalge Company, Rochester, NY)

3. Costar 96-well tissue culture clusters with flat bottom wells (catalogue No. 3596, Costar, Cambridge, MA)

4. Trypsin solution: 0.025% trypsin (GIBCO, cat. No. 610-5095)–0.05% disodium ethylenediamine tetracetate (EDTA, Fisher Scientific, cat. No. S-311, King of Prussia, PA) in a Ca^{2+}- and Mg^{2+}- free saline.

Regardless of the care with which primary cultures are initiated, a percentage of cultures will always contain smooth muscle cells. In some preparations, contamination is negligible, while in others smooth muscle cells begin to overgrow the endothelial cells. One way to alleviate this problem is to clone the cells present in the primary cultures (Rosen et al., 1981; Gajdusek and Schwartz, 1983). A clone, as defined in the *Tissue Culture Association Manual* (Schaeffer, 1978), is "a population of cells derived from a single cell by mitosis. A clone is not necessarily homogeneous and, therefore, the terms *clone* and *cloned* must not be used to indicate homogeneity in a cell population, genetic or otherwise." To prepare endothelial clones, one must use "conditioned" medium to promote cell division in low-density cultures. Conditioned medium is prepared by removing medium from a culture of smooth muscle cells (Ross, 1971) and replacing it with serumless medium. The serum-free medium is incubated with the cells for 72 hr, after which it is removed and filtered through a 0.20 μm Nalgene filter, aliquoted into sterile containers, and stored at 4 °C until used. To prepare growth medium, modified M199 is mixed 1:1 (v/v) with conditioned medium. To this mixture is added fetal bovine serum such that the final concentration of added serum is 20% v/v. For cloning, a subconfluent primary culture of endothelial cells in a T-25 flask is used. The medium is removed from this culture and the cells washed once with serumless M199. The cell layer is trypsinized using 1 ml trypsin solution (0.025% trypsin–0.05% EDTA in a Ca^{2+}- and Mg^{2+}-free saline) and diluted to 10 ml with medium 199 modified and supplemented with serum and antibiotics as described above (20% FBS and antibiotics). An aliquot of this cell suspension is counted and diluted with medium such that the final cell concentration is approximately 100 cells/ml.

Using a micropipette, 10 μl of this suspension is added to each well of a 96-well plate and the plates incubated in a humidified atmosphere of 5% CO_2 in air for several hours to permit cell attachment. (Fill the outer rows of the well plate with sterile distilled water to prevent dessication.) The wells are then screened using a phase-contrast inverted phase microscope to determine which wells contain a single endothelial cell. Such wells are marked while wells containing either no cells or more than one cell are filled with sterile water. Additional medium is then added to wells containing a single cell. After 3–4 days of incubation, the marked wells are rescreened to determine wells in which proliferation has occurred. Subsequently, the growth medium is changed every 2 days until confluent monolayers are obtained. These are expanded by subcultivation until sufficient numbers of cells are present.

III. Interpretation of Results

A. Endothelial Cell Morphology

Endothelial cells in vitro show a distinctly squamous epithelial morphology. Squamous epithelial cells exhibit the following histological characteristics: the cells form a contiguous sheet; the upper surface of the sheet is "free" (i.e., it does not share a border with other tissues); the epithelial sheet is underlain by a basement membrane; and the cells within the sheet have little or no intercellular space (i.e., the cells are closely associated with one another via specialized cell–cell junctions). When freshly isolated endothelial cells are first placed in culture, they initially are associated with one another in the form of small clumps. These quickly attach to the culture dish and spread to form island-like clusters, which eventually merge to form a complete, contiguous monolayer of polygonal endothelial cells (Fig. 1). Both smooth muscle and fibroblast cells have distinct morphologies at confluency, which permit an experienced investigator to identify them if they contaminate endothelial cell cultures. However, since experimental conditions can promote alterations in cell and culture morphology, the optimal endothelial morphology, as described above, may not be apparent, and other approaches should be used to verify rigorously the identity of cells.

An approach used in many laboratories is to determine if factor VIII/von Willebrand's protein is present in the cultured cells (Jaffe et al., 1973b). Its presence unequivocally identifies the cells as endothelium.

B. Use of Factor VIII/von Willebrand's Staining

The following reagents are used:

1. Lab-Tek tissue culture chamber/slides, no. 4804 with removable gasket (Miles Scientific, Naperville, IL)

2. Triton X-100, product No. CS282-100 (Fisher Scientific, Orangeburg, NY)

3. Phosphate-buffered saline

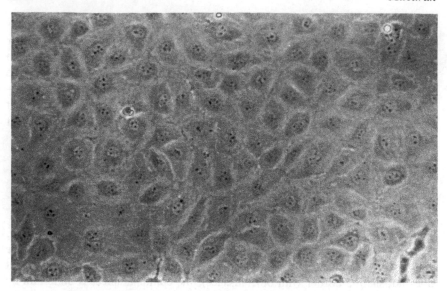

Figure 1   Phase photograph of a confluent culture of cloned fetal bovine pulmonary artery endothelial cells (×225).

4.   Antiserum to human factor VIII/von Willebrand's protein (rabbit), product No. 782301 (Calbiochem-Behring Corp., La Jolla, CA)

5.   Rhodamine-conjugated goat antirabbit IgG (Capell, Cochranville, PA)

Staining of cells with antifactor VIII/von Willebrand's protein is performed following the method of Wagner et al. (1982). Cells are grown in Lab-Tek tissue culture chamber/slides, fixed in 3.7% formaldehyde in phosphate-buffered saline (PBS) for 20 min, and subsequently permeabilized with 0.5% Triton X-100 in PBS for 15 min at room temperature. The cells are incubated first with antiserum to factor VIII/von Willebrand's protein for 30 min at 37°C, followed by a second rhodamine-conjugated antibody. After incubation with both reagents, slides should be washed several times with PBS. After the final washing, the coverslips should be mounted in glycerol. While dilutions for each immune reagent should be empirically determined by "checkerboard titration," good starting dilutions are 1:50 for the primary antiserum and 1:100 for the secondary antibody. The actual dilution used should be the highest that gives a positive signal relative to controls used at the same dilution. Prior to their use, both the primary and secondary antisera should be centrifuged in a microfuge (Eppendorf) to remove aggregates. The slides are examined using a Zeiss microscope equipped with epifluorescence optics. We routinely use a water–glycerol–oil-immersible Zeiss 60x lens. This permits the cells to be examined and photographed without a coverslip. For photography, Kodak Tri-X film is used followed by development in Diafine, which increases the effective film speed to 1600 (the normal ASA of Tri-X film is 400).

Cells that stain positively with the antibody exhibit a perinuclear granular fluorescence with little or no direct nuclear staining (Fig. 2). Controls that can be stained with normal rabbit serum or saline do not exhibit this unique granular staining. If the granular staining is not apparent, a lens with a higher numerical aperature should be used to obtain greater resolution. In our hands, endothelial cells always exhibit, to varying degrees, a granular fluorescence when stained with antiserum to factor VIII/von Willebrand's protein. In some instances, cells artifactually stain positive with the normal rabbit serum; however, this staining has a more uniform, "painted" appearance, with the nucleus as well as the cytoplasm showing uniform staining. Controls of this type do not exhibit the granular cytoplasmic staining characteristic of factor VIII/von Willebrand's staining. Another control to help in this regard is to incubate replicate wells of endothelial cells with PBS instead of normal rabbit serum (NRS). When cells are so treated there is little or no fluorescence present. If difficulties arise with respect to identification of positive cells, additional purification of the antisera may be necessary. In general, there will be less nonspecific fluorescence if purified IgG rather than antiserum is used.

One must always be cautious in interpreting immunocytochemical data. Fluorescence should always be interpreted relative to controls, with special attention paid to the pattern of the "specific" fluorescence. Intensity of fluorescence is not a useful measure since it cannot be proven that all available epitopes within cells or tissues are accessible to the antibody. We have observed that all cells within an examined area do not stain equally with the antibody. This suggests that the concentration of epitope may be different from cell to cell and that in some cells the assay may not be sensitive enough to visualize all available epitopes. Because of these problems and the others stated above, immunocytochemistry cannot be quantitative; however, when performed carefully, this approach can be used successfully to characterize endothelial cells in culture.

C. Value of Method

There is obvious value in being able to isolate and maintain differentiated lung endothelial cells in culture that can be used for in vitro studies. While cell culture has some obvious advantages over whole animal studies, one must always be concerned about the artifactual nature of cells maintained in vitro. In most instances, cells are grown on a plastic substrate that is distinctly different from what they normally experience in vivo. A number of approaches can be used to mimic more faithfully the extracellular environment that cells normally interact with: (1) maintaining cells on purified matrix proteins that have been coated on the cell culture dish, (2) using a three-dimensional "native" collagen matrix, and (3) growing

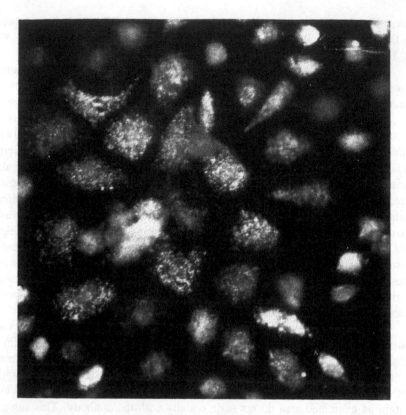

Figure 2 Immunofluorescence photograph of pulmonary artery endothelial cells incubated with a 1:50 dilution of rabbit antifactor VIII/von Willebrand's protein antiserum and a 1:100 dilution of rhodamine-conjugated goat antirabbit IgG (×400).

endothelial cells on a matrix previously deposited by endothelial cells. With the use of such approaches it has been demonstrated, in a variety of cell systems, that the extracellular matrix environment of cells plays an important role in defining their phenotypic expression (Macarak and Howard, 1983; Canfield et al., 1986; Grinnell, 1982; Bissell et al., 1982).

Another advantage of the cell culture approach is that pure populations of cells can be isolated and their individual cell products purified. These substances can then be used with other cell types found in the same organ to determine if one cell's products affect the behavior of another different cell. For example, it is now known that there is signaling between smooth muscle and endothelial cells within the vessel wall. By isolating the products of one cell type and adding them to the culture medium of th other, considerable new information has been obtained about the interactions between these cells (Mecham et al., 1987). It is also possible to isolate large quantities of endothelial cell products by using innovative culture techniques. For example, substantial increases in the surface area available for cell growth can be provided by microcarrier beads (Davies, 1981), which make it possible to purity substantial quantities of cell products.

IV. Perspectives for Future Development and Study

A. Isolation of Endothelium from Intermediate-Sized Blood Vessels

There is no information on the properties of endothelium isolated from the pulmonary arterioles. It should be possible to dissect out some of the smaller vessels of the pulmonary arterial vascular "tree" and to isolate small quantities of endothelium from them using methods developed for the culture of capillary endothelium (Zetter, 1984). With the use of conditioned medium to promote growth at low seed densities, these initial endothelial cells should be able to be expanded to produce useful numbers of cells for experimental purposes.

B. Capillary or Microvascular Cells from the Lung

Isolation and culture of lung microvascular endothelial cells have been reported by Habliston et al. (1979): however, they caution that it is unlikely that all the cells isolated are from the pulmonary microcirculation. If this is true, it would be hard to draw conclusions about the unique properties of the microvascular cells, since their origin cannot be rigorously substantiated. A major problem with the isolation and culture of lung microvascular endothelium is that, at the present time, there are no markers for microvascular or capillary cells. If such markers were available, it would make it possible to characterize such cells rigorously after their isolation from the lung. Until such markers are identified, it seems unlikely that pure populations of lung microvascular cells can be studied easily and unambiguously.

Acknowledgments

This work was supported by NIH grants AM20553, 41882 and HL34005. It is a pleasure to acknowledge the collaboration of Pamela S. Howard, Stephen Gorfien and Jeanne C. Myers and the technical assistance of Josephine DeFonso and Lynda Muirhead.

References

Bakhle, Y. S., and Vane, J. R. (1977). *Metabolic Functions of the Lung*. New York, Marcel Dekker.

Bossell, M., Hall, G., and Parry, G. (1982). How does the extracellular matrix direct gene expression? *J. Theor. Biol.* **99**:31–68.

Bruel, S., Bradley, K., Hance, A., Schafer, M., Berg, R., and Crystal, R. (1980). Control of collagen production by human diploid fibroblasts. *J. Biol. Chem.* **255**:5250–5260.

Canfield, A., Schor, A., Schor, S., and Grant, M. (1986). The biosynthesis of extracellular-matrix components by bovine retinal endothelial cells displaying distinct morphological phenotypes. *Biochem. J.* **235**:375–383.

Davidson, P. M., Bensch, K., and Karasek, M. (1980). Isolation and growth of endothelial cells from the microvessels of the newborn human foreskin in cell culture. *J. Invest. Dermatol.* **75**:316–321.

Davies, P. F.(1981). Microcarrier culture of vascular endothelial cells on lid plastic beads. *Exp. Cell Res.* **134**:367–376.

Dunn, D. M., and Franzblau, C. (1982). Effects of ascorbate on insoluble elastin accumulation and cross-link formation in rabbit pulmonary artery. *Biochemistry* **21**:4195–4202.

Folkman, J., Haudenschild, C. C., and Zetter, B. R. (1979). Long term culture of capillary endothelial cells. *Proc. Natl. Acad. Sci. USA* **76**:5217–5221.

Gajdusek, C., and Schwartz S. (1983). Technique for cloning bovine aortic endothelial cells. *In Vitro* **19**:394–402.

Gimbrone, M. A., Jr., Cotran, R. S., and Folkman, J. (1974). Human vascular endothelial cells in culture. Growth and DNA synthesis. *J. Cell Biol.* **60**:673–684.

Grinnell, F. (1982). Cell-collagen interactions: Overview. In: *Methods in Enzymology*, Vol. 82. Edited by L. Cunningham and D. Frederiksen. New York, Academic Press.

Habliston, D. L., Whitaker, C., Hart, M. A., Ryan, U. S., and Ryan, J. W. (1979). Isolation and culture of endothelial cells from the lungs of small animals. *Am. Rev. Respir. Dis.* **119**:853–868.

Jaffe, E. A., Hover, L. W., and Nachman, R. (1973a) Synthesis of antihemophilic factor antigen by cultured human endothelial cells. *J. Clin. Invest.* **52**:2757–2764.

Jaffe, E. A., Nachman, R. L., and Becker, C. G. (1973b). Culture of human endothelial cells derived from umbilical veins. Identification by morphological and immunological criteria. *J. Clin. Invest.* **52**:2745–2756.

Jaffe, E. A. (1984). Culture and identification of large vessel endothelial cells. In: *Biology of Endothelial Cells*. Edited by E. Jaffe. Hingham, MA. Martinus Nijhoff.

Johnson, A. R. (1980). Human pulmonary endothelial cells in culture, activities of cells from arteries and cells from veins. *J. Clin. Invest.* **65**:841–850.

Kikkawa, Y., and Yoneda, K. (1974). The type II epithelial cell of the lung. I Method of Isolation. *Lab. Invest.* **30**:76–84.

Kruse, Paul F., and Patterson, M. K., Jr. (1973). *Tissue Culture: Methods and Applications.* New York, Academic Press.

Lewis, L. J., Hoak, J. C., Maca, R. D., and Fry G. (1973). Replication of human endothelial cells in culture. *Science* **181**:453–454.

Macarak, E. (1984). Collagen synthesis by cloned pulmonary artery endothelial cells. *J. Cell. Physiol.* **119**:175–182.

Macarak, E., and Howard, P. (1983). Adhesion of endothelial cells to extracellular matrix proteins. *J. Cell. Physiol.* **116**:76–86.

Macarak, E. J., Howard, B., and Kefalides, N. (1976). Properties of calf endothelial cells in culture. *Lab. Invest.* **36**:62–67.

Maruyama, Y. (1973). The human endothelial cell in tissue culture. *Z. Zellforsch. Mikrosk. Anat.* **60**:69–79.

Mecham, R., Whitehouse, L., Wrenn, D., Parks, W., Griffin, G., Senior, R., Crouch, E., Stenmark, K., and Voelkel, N. (1987). Smooth muscle-mediated connective tissue remodeling in pulmonary hypertension. *Science* **237**:333–464.

Paul, J. (1972). *Cell and Tissue Culture.* Baltimore, Williams & Wilkins.

Rosen, E. M., Mueller, S., Noueral, J., and Levine, E. (1981). Proliferative characteristics of clonal endothelial cell strains. *J. Cell. Physiol.* **107**:123–137.

Ross, R. (1971) The smooth muscle cell II. Growth of smooth muscle in culture and formation of elastic fibers. *J. Cell Biol.* **50**:172–186.

Ryan, U. S., Clements, E., Habliston, D., and Ryan, J. W. (1978). Isolation and culture of pulmonary artery endothelial cells. *Tissue Cell* **10**:535.

Sheaffer, W. J. (1978). Proposed usage of animal tissue culture terms. *Tissue Culture Association, Inc. Manual* **4**:(1) 779–782.

Thornton, S. C., Mueller, S. N., and Levine, E. (1983). Human endothelial cells: Use of heparin in cloning and long-term serial cultivation. *Science* **222**:623–625.

Wagner, R. C., and Mathews, M. A. (1975). The isolation and culture of capillary endothelium from the epidymal fat pad. *Microvasc. Res.* **10**:286–297.

Wagner, D., Olmstead, J., and Marder, V. (1982). Immunolocalization of von Willebrand protein in Weibel-Palade bodies of human endothelial cells. *J. Cell Biol.* **95**:355–360.

Zetter, B. (1984). Culture of capillary endothelium from neural capillaries. In: *Biology of Endothelial Cells.* Edited by E. Jaffe. Hingham, MA, Martinus Nijhoff.

17

Autoradiography

RONALD E. GORDON

Mount Sinai Medical Center
New York, New York

I. Introduction

Autoradiography is the method of recording a picture or pattern from a radiographic source in which the source of the radioactivity is contained within the tissue or specimen. Autoradiography is an old technique that preceded and ultimately contributed to the discovery of radioactivity by the combined efforts of Niepce de St. Victor in 1867, Henry Becquerel in 1896, (1896a,b,c,d); and the Curies in 1898. It was not until the 1940s that autoradiography was transformed from a curiosity to a legitimate scientific technology, when Leblond (1943) demonstrated the distribution of radioactive iodine in thyroid tissue.

The technique itself is extremely simple. All that is required is that the tissue be treated in some way with a radioactive material and an appropriate photographic emulsion to record the location of the radioactive emissions. The photographic emulsion consists of large numbers of silver halide crystals suspended in a solid phase, usually gelatin. In the past, photographic emulsions have been used mounted on glass, plastic, or directly on the tissue as a fluid or removed from the glass or plastic support and placed directly over the tissue. Recently, investigators have less often used emulsions removed from glass or plastic and more often use a liquid dipping type emulsion. Glass plates are almost never used anymore. For

507

a more detailed description of the techniques and materials necessary for such procedures, consult the list of references for "Preparation, Exposure, and Processing."

A wide variety of photographic emulsions are now available to detect many if not all of the radioisotopes used in biological and medical research, and in diagnosis. The three leading manufacturers of photographic film and liquid emulsions are Kodak, Ilford, and Agfa Gevaert. The radioactivity released from the isotopes can be detected at a macro-, micro-, or ultrastructural level.

It is important to understand that autoradiographic techniques are normally only 10% efficient compared to other modes of detecting radioactivity, such as scintillation and Geiger counting. However, these other methods do not identify the specific location of the source of the radiation within the specimen, but only confirm its presence or absence in relative amounts. Autoradiography is the only technique that allows for the localization of radioactivity in situ. Since efficiency of detection and recording can be a major drawback, a variety of procedures have been developed to improve the efficiency of detection by autoradiography. These procedures will be described later in the discussion of their applications.

The principal focus of this discussion will be on the application of autoradiographic technology to the study of the mechanisms involved in the cause, development, detection, diagnosis, and treatment of lung disease. Of particular importance is the early detection of neoplastic and non-neoplastic diseases and the detection of the effectiveness of treatment protocols used.

General theory, methods, and applications of autoradiography are detailed in publications by Rogers (1979a,b); Williams (1977); Gahan (1972); Roth and Stumpf (1969); Baserga and Malamud (1969).

II. Designing Experimental Protocols

The first step for any investigator in selecting the appropriate autoradiographic technique is to determine at what level the data must be exhibited to answer the questions posed. The initial step in planning autoradiographic studies is to evaluate the parameters carefully so that the data recorded will be optimal. Some of the important parameters include (1) the level at which the autoradiographic image will provide the optimal information (i.e., macro-, micro-, ultrastructural); (2) the type of radioactive label; (3) the amount of radioactivity labeled material given; (4) the type of photographic emulsion; (5) the length of exposure; and (6) whether or not enhancers are appropriate. For instance, electron microscopic autoradiography would not be appropriate when determining the regions of drug or hormone binding in the body of an experimental animal or human. However, macroscopic autoradiography with or without computer-assisted morphometric analyses would be most appropriate. Electron microscopic analysis would be the technique of choice, if it is necessary to determine the target for these drugs or hormones. It could identify a specific cell type in a particular organ or tissue, and these specific cells could be identified on the basis of ultrastructure. Once the level (macro or micro) at which autoradiographic data will be collected is established, it will then be much easier to decide on the radionuclide to be used. In many instances molecules and probes have limited availability and cannot be incorporated or bound to certain probes or molecular species of interest. Some may also be prohibitive in cost. Another consideration in the choice of nuclides used in autoradiographic technology is choosing the appropriate energy level, particle emission, and half-life. The theoretical considerations of radionuclide energy levels, particle types, and half-life will not be discussed here. The reader can find additional information in any available handbook of chemistry and physics.

We will discuss here the practical aspects of these important considerations. For example, the use of indium or ^{32}P label may not be appropriate for electron microscopic autoradiographs, but are appropriate labels for macroautoradiographs and human tissue labeling. They emit radioactive particles with adequate energy to reach a photographic film or detector. They have a very short half-life, making the overall period of exposure limited. They are not generally applicable to electron microscopic autoradiographs because adequate exposure periods, in excess of 1 months, are required and, these radionuclides are totally dissipated within 1–4 weeks. Many of these radionuclides emit more than one type of radioactive particle and, therefore, possess variable energies that can be applied to autoradiographic technology. For instance, indium 111, and iodine 125 are primarily gamma emitters. Gamma emissions are much too energetic to record in a photographic emulsion, particularly those very thin layers over ultrathin sections for ultrastructural analyses. However, both have a lower-energy auger electron with approximately the same energy as tritium and are therefore compatible with macro-, micro-, and ultrastructural autoradiography. As a result of the multiple emissions, all of which are not equal in terms of detection, techniques have evolved to enhance the detection of those emissions. These improve the efficiency of recording and overcome the problem of a short half-life and dissipation of radioactive emissions. The second major consideration is, therefore, selecting the appropriate radioactive label to provide the predicted data needed to achieve the research goals.

III. Application

This chapter will describe applications of autoradiographic technology to experimental lung research. Most autoradiographic technologies and applications have been developed in other systems, organs, and tissues. Many, if not all, of the examples we will use have been applied in other than the lung but are applicable to the lung.

We will also identify gaps in the lung literature that autoradiography may help to fill by contributing pertinent data on the elucidation of molecular mechanisms of normal and diseased lung function. These methods can also be applied to the study of the alterations that may lead to the development of diseases, particularly chronic diseases such as chronic obstructive pulmonary disease (COPD) and fibrosis. This discussion will be divided into topics of experimental study ranging from determination of the in situ distribution of a molecular species throughout the body and lung to their identification in biochemical isolations. Each will be considered according to macro-, micro-, and ultrastructural autoradiographic technologies when applicable. Where applicable, new developments in the technology are described.

IV. Determination of Distribution

Radioactively tagged molecules such as drugs, toxins, and specific antibodies can be used effectively as tracers to determine distribution within the lung or throughout the entire body. It is possible to determine the most effective mode of administration and time course by sequential sacrifice of animals. Dependent upon the animal or lung size, the solubility of the tagged molecule, and the energy of the radioactive tag, the animal or organ can be evaluated first in situ. A specimen can then be fixed and embedded in paraffin, plastic, or frozen, depending on the sectioning equipment available. Sections can be cut and mounted on wax paper, glass, or grids. These sections are overlaid with either a single or double layer of emulsion, based on the energy levels of the radioactive tags and the efficiency of detection. The double-sided x-rays are most efficient with higher-energy radioactive tags, especially when it is anticipated that the concentration may be low. One rule to keep in mind, however, is that the higher the energy, the higher the potential for improving the efficiency of detection. The converse is that there will be reduced resolution. Although detection efficiency is the most important factor in macroautoradiographs, the degree of resolution should be taken into account. One argument against the use of high-energy radioactive tags is that light histological or ultrastructural localization with the same tissue is difficult, if not impossible, unless the radioactive component has a secondary emission of intermediate or low energy. Energies of radionuclides such as ^{14}C, ^{35}S, and ^{45}Ca are appropriate for light microscopic autoradiographs. They provide a significant degree of resolution and efficiency of detection by light microscopy, but are virtually useless for electron microscopy.

In many cases it may be feasible to select a radioactive emitter of lower energy such as ^{3}H or radionuclides with multiple emissions (^{111}In or ^{125}I), and sacrifice any efficiency at the macro level for resolution at the light histological and ultrastructural levels using ^{3}H, ^{111}In, and ^{125}I.

Autoradiographic methods at the molecular level are applicable in the lung for studies of drug localization and distribution by identification of specific binding sites. These are used in the treatment of various diseases such as asthma, bronchitis, bronchiolitis, emphysema, fibrosis, bacterial infections, and neoplasms. They are useful in diagnostic procedures to identify tumors, especially if these are small and otherwise undetectable; sites of inflammation; airway obstruction; and defective clearance mechanisms. These methods are as important, if not more important, when studying the molecular mechanisms of most diseases of the lung. These methods are alternatives to immunocytochemical procedures. Immunocytochemical analyses are generally not applicable in tissues in situ, since invasive techniques such as biopsy must be used. In the lung, where antibodies to specific proteins or enzymes are frequently difficult to prepare, autoradiography may be the next method of choice. Autoradiography affords the clinician and/or research investigator the ability to identify the specific localizations in organisms, organs, tissues, and cells and, by alternative methods, facilitates identification of specific molecular binding components in biochemical preparations.

In the case of indium and iodine, the gamma emissions are excellent for macrovisualization by computer-assisted scintillation camera analyses of distribution. The soft auger electrons are excellent for light histological and electron microscopic autoradiographic analyses. An example of ^{111}In localization in red cell fragments is illustrated in the spleen (Fig. 1a) and in the liver (Fig. 1b). The label is identified in a macrophage population. When the amount of concentration of radioactivity is high, even brief exposures due to very short half-lives are effective, as seen in Figure 2 of a lung macrophage that has incorporated an ^{125}I-labeled antibody.

A. Use of Enhancers

When the levels of detection are low, their efficiency of detection could be enhanced by the use of a fluorescent molecule (Przybylski, 1969). This is theoretically similar to the mechanisms by which scintillation counters record radioactive emissions in a solubilized form. Scintillation theory and methods are available in many publications, one of which discusses the issues at length (Bransome, 1970). The theory is that the energy of the radioactive emission is absorbed by a fluorescent molecule added to the emulsion, thereby energizing an electron and causing it to jump to an outer atomic electron shell. As the electron drops back to its orginal electron shell, energy is released as light. The light then produces the latent image within the ionic silver emulsion, much like standard photographic technology, rather than a direct transfer of energy from the radioactive emission particle imparted to the ionic silver. This technology is available for macroautoradiography in the form of intensifying phosphor screens mounted adjacent to the film during exposure. Scintillant solutions are applicable for use in macro-, micro-, and

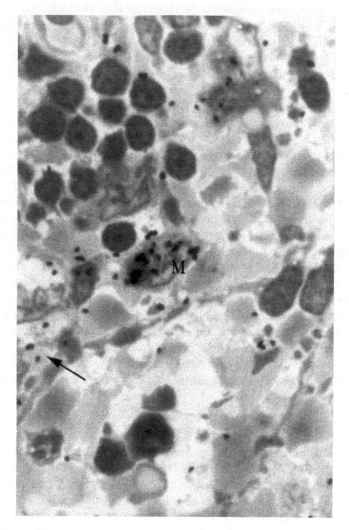

Figure 1a Autoradiograph of a 1 μm Epon section of rabbit splenic tissue fixed with glutaraldehyde 24 hr after injection with indium 111-treated erythrocyte fragments. It is possible to identify labeled erythrocyte fragments in the sinusoids of the spleen (arrow) and in macrophages (M) that ingested the radiolabeled fragments.

ultrastructural autoradiography by submerging the x-ray film, or dried emulsions on x-ray films, light microscope slides, and electron microscope grids into the appropriate scintillant solution and allowing them to dry before storing them for exposure. Radioactive molecules such as ^{125}I that primarily emit high-energy

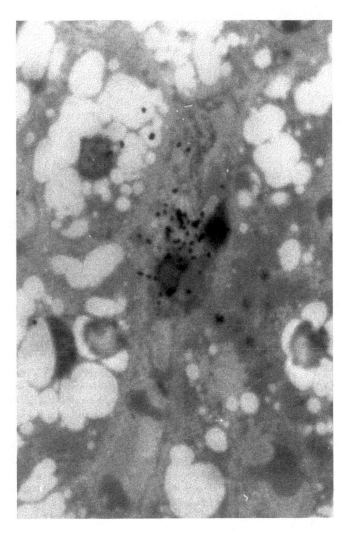

Figure 1b   A section of liver removed and fixed from the same animal. A labeled macrophage (Kupffer's cell) can be seen within a sinusoid. Autoradiographs were prepared by coating the 1 μm sections mounted on glass slides with 1:1 dilution of Kodak NTB2 photographic emulsion, exposed for 2 weeks in the dark, developed, fixed, and stained with methylene blue and azure II ($\times$1600).

gamma particles also emit secondary low-energy auger electrons that have approximately the same energy as the soft, low-energy, B emissions of tritium. These low-energy particles make it possible to detect and record the location of these

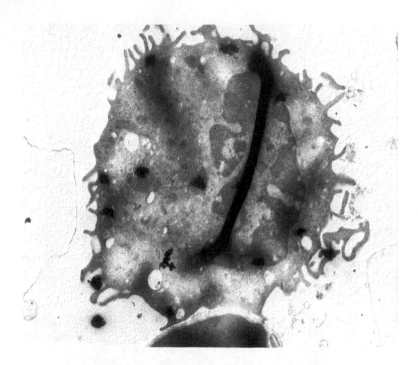

Figure 2 Transmission electron microscopic autoradiograph of a rat lung macrophage that had been incubated in situ with ^{125}I-labeled antibody to surface antigen Ia. The tissue was fixed with glutaradehyde, embedded in Epon, sectioned for electron microscopy, and prepared for autoradiography. It is possible to identify the specific sites of labeling at the membrane surface and in lysosomes. Autoradiographs were prepared by mounting thin sections on copper grids. The grids were anchored to glass slides with double-sided cellophane tape. With use of a double dip of formvar and then emulsion (Ilford L4) on a glass slide, the double layer was stripped from the slide in a water bath and the grids overlaid with the emulsion. After 6 weeks of exposure the grids were developed, fixed, and stained with uranyl acetate and lead citrate (×9900).

radioactive isotopes with these autoradiographic techniques. An example of the use of scintillant within the emulsion of light histology seen in Figure 3 can be compared with those examined without the use of scintillant seen in Figure 1. The number of silver grains over the macrophages is significantly increased. An autoradiographic electron micrograph (Fig. 4) of a lung macrophage illustrates silver grains in the emulsion, which had been treated with scintillant by the technique of Durie and Salmon (1975) to enhance detection. Enhancement was necessary since the ^{111}In label had a short half-life and there was not as much label given to this animal as to the animal whose tissue is shown in Figure 2. Additional references for the techniques of image intensification are in the Appendix section "Light Microscopic Intensification Techniques."

B. Use of Tritium

It is also possible to use a tritium label for some types of macroautoradiographs. Tritium is appropriate with large slices of tissue or directly on tissues, organs, or whole animals, when it can be assumed that only the label in tissues at the surface of contact will be detected and recorded. Using the appropriately fixed material, it will be possible to identify not only sites of labeling but also the cell types that could be identified by light microscopy and, if necessary, by electron microscopy (Fig. 5), to reveal organelle(s) and molecular binding sites or sites of incorporation. In Figure 5 one can identify the binding sites of taurine in the ciliated cells of hamster tracheal epithelium. In Figure 6, it is possible to identify a ciliated cell in regenerating tracheal epithelium labeled with tritiated thymidine while in the DNA synthetic phase. It would be possible to determine from each such labeling if there were any secretory granules along with basal bodies in these cells. Basal bodies and sometimes even small secretory granules cannot be identified at the light histological level (Gordon and Lane, 1984). Furthermore, with the aid of computerized image analysis systems, it is possible to evaluate the areas, cells, and subcellular sites of preferential localization. If the amount of label bound or incorporated is not adequate, intensification or enhancement techniques can be applied at the electron microscopic level (see references for "Electron Microscopic Intensification").

If human tissues are being studied in situ or if the animal tissues to be studied are sensitive to radiation injury and the exposures would not be considered acute (i.e., lymphocytes, proliferative cells), it would be appropriate to use a labeled probe with a very short half-life, such as indium. If counts or detection efficiency are low, enhancement techniques could then be used to improve the efficiency of detection of the secondary, auger electrons. This would be most important if the tissue was to be analyzed morphometrically and adequate amounts of radioactivity had to be detected and recorded for significance in comparisons, particularly at the macro- and microautoradiographic levels.

In addition to the localization of specific molecules, binding, or tracking synthetic function by radioactively labeled precursor molecules, cell and tissue proliferation can easily be detected and quantified. It has been shown that proliferation is a secondary response and a measure of injury in many tissues, especially the lung. Autoradiographic methodology is invaluable when making these proliferative measurements, since they relate to injury or methods of treatment or prophylaxis. Because this technique allows one to identify and differentiate tissues and cells histologically, it is that much easier to elucidate specific mechanisms

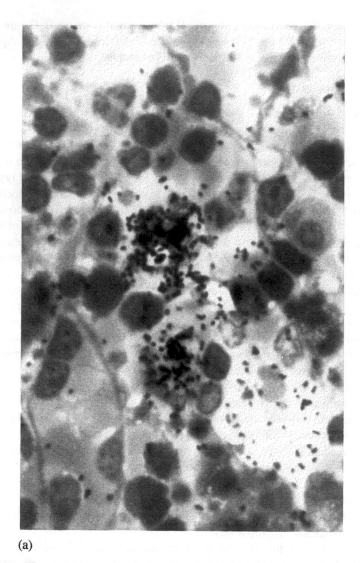

(a)

Figure 3 These light autoradiographs (a, spleen; b, liver) are additional sections taken from the same blocks described in Figure 1. The only difference in autoradiographic preparation was the addition of a scintillant solution to the dried emulsion before storage in the dark. The number of grains over the cells are significantly greater compared to those in Figure 1. The scintillant clearly enhanced the efficiency of detection and recording (×1600).

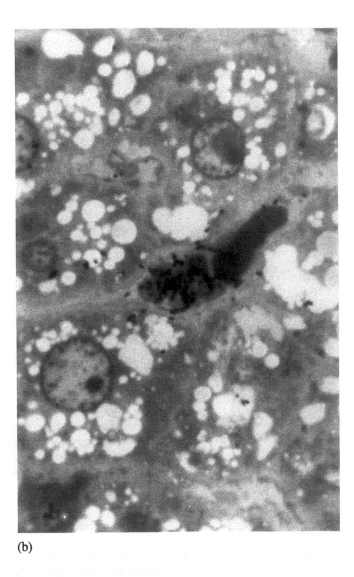

(b)

and processes contributing to normal lung development and function compared to altered conditions. In these cases, the choice is a DNA precursor, thymidine, labeled with tritium. The radioactively labeled thymidine is incorporated into DNA during the DNA synthetic phase of the cell cycle. If given as a flash label at a single time in situ with a cold chase of excess unlabeled thymidine, it is possible to identify the cells involved in DNA synthesis at the time the labeled precursor was available for incorporation. The number of labeled cells compared to the

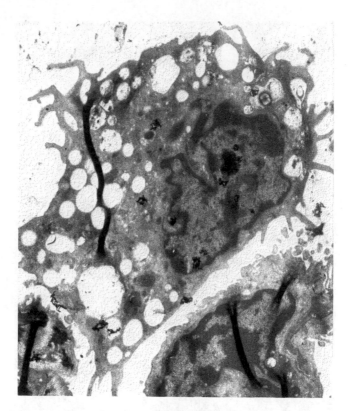

Figure 4 Transmission electron microscopic autoradiograph of a hamster lung macrophage treated with indium 111. When the grids were prepared as described in Figure 2 for autoradiography, no label was detected. However, when scintillant was added to the emulsion layer, the efficiency of detection is significantly improved. This illustrates that radionuclides with a lower concentration and very short half-life can be detected at the ultrastructural level. Sections were stained with uranyl acetate and lead citrate (×7500).

entire population provides a percentage termed the *labeling index*. This index allows the investigator to compare proliferative rates at various times or under various conditions. If the investigator wants to be sure that the cells continued to cycle through to mitosis, the percentage of labeled mitoses can also be calculated. Because mitosis is generally a very brief period in the cell cycle (1 hr or less), the efficiency of detection of the mitotic cells can be improved by using an appropriate statmokinetic agent such a colchicine, colcemid, vinblastin, or others. These agents disrupt spindle formation and therefore arrest the cells in mitotic metaphase. This increases the number of cells seen in metaphase, which

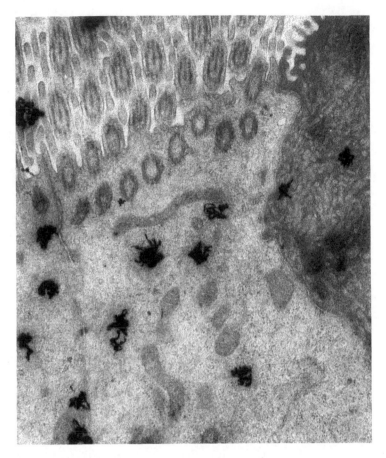

Figure 5 Electron microscopic autoradiograph of tracheal epithelium from a hamster treated with [³H]taurine. The tissue had been fixed with glutaraldehyde, embedded in Epon, thin sectioned, and prepared as described in Figure 2 for autoradiography. It is possible to identify the ultrastructural sites of binding of taurine in the ciliated and adjacent secretory cells of the tracheal epithelium. Sections were stained with uranyl acetate and lead citrate (×20,0000).

improves the efficiency of mitotic cell counts. A labeled mitotic cell can be identified in Figure 7 in regenerating tracheal epithelium treated with tritiated thymidine followed by colchicine treatment. It is possible to detect and record this label at all autoradiographic levels.

A variety of autoradiographic techniques are available to determine if cells synthesize DNA, divide and go into a resting G_1 or differentiating stage G_0 or

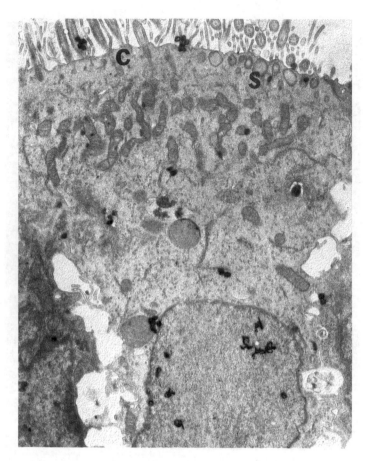

Figure 6 Electron microscopic autoradiograph of a transitional cell in a regenerative lesion fixed 72 hr after injury. The tissue was labeled with [³H]thymidine 39 hr after injury. After preparation for autoradiography, thin sections were exposed for 4 weeks and developed. The transition cell exhibits silver grains over its nucleus. This cell is relatively electron translucent. Cilia (C) and secretory (S) cells can be seen at the apical surface of the cell. Sections were stained with uranyl acetate and lead citrate (×8600).

G_2, or continue to cycle. Investigators frequently need to know if the cells of interest proliferate and continue to cycle and what, if any, is the fate of their progeny. This is of particular importance when studying chemotherapeutic modalities, and their effectiveness in influencing cycling and killing tumor cells. Autoradiographic techniques are appropriate when one is studying the fate of epithelial and interstitial cells under various normal and altered conditions and

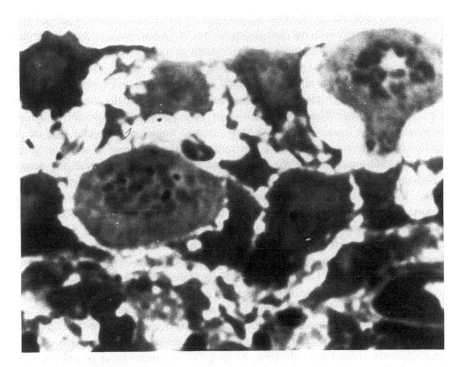

Figure 7 A light autoradiograph of a 1μm plastic section of tracheal epithelium from a rat pulse labeled by intraperitoneal injection of [³H]thymidine 14 hr after trauma. The trachea was immersed in fixative immediately upon excision, 13 hr later. One of the two mitotic cells exhibit silver grains over the chromosomes, demonstrating incorporation of the labeled thymidine during DNA synthesis. Section was stained with methylene blue and azure II (×3000).

how treatment or proposed treatment will modify the conditions. This type of autoradiographic methodology allows one to measure deviation from the norm. The critical issue is whether the stem or progenitor population can be obliterated or sterilized so that it can no longer divide. Most, if not all, of these questions can be answered by autoradiographic techniques. One important technique that has been used but has come under attack is the morphometric analysis involved in mean half-grain counting (Quastler and Sherman, 1959). The theory is, that cells going through the first DNA synthetic phase with the radioactively labeled precursor available will incorporate an average amount of radioactive precursor and will be detected in autoradiographs as X number of grains per nucleus. These cells would have been fixed immediately after labeling, before it is possible for them to divide. Based on the synthesis and separation of duplicate DNA strands into separate cells during the first cell division, there should be a theoretic halving of incorporated radioactivity and therefore 1/2 X grains/nucleus. Following the same reasoning, a second division would exhibit 1/4 X grains/nucleus and so on. However, a significant percentage of cells always either do not cycle or drop out of the cycle. These cells alter the statistical analysis, especially if a significant part of the population goes through only one cell cycle. It is usually not known what percentage drop out. This technique is based on a statistical average for the entire population. One cannot know the history for any single cell. Another method of statistically analyzing grain counts per nucleus is the technique of absolute grain counting (Sklarew et al., 1976). This makes it possible to determine how many divisions any individual cell has undergone in its life since being labeled and identified. Cells that were not cycling within a mixed population fixed some time after incorporation of radioactive precursor will also be identified by the absence of label. However, this technique, which is also statistical, has been criticized. One technique that allows a direct analysis for each cell and avoids the above criticisms is double-label autoradiography. This involves the use of the same thymidine precursor labeled with ^{3}H or ^{14}C. Use of the two different labels given to animals or cultures at different times allows the identification of individual cells involved in DNA synthesis at both times the labeled precursors were available for synthesis (Gordon and Lane, 1977). Double-label autoradiography technology allows for detection of the ^{3}H label in the first layer of photographic emulsion due to its low energy and relatively shorter half-life (Fig. 8) than ^{14}C. A spacer layer of gelatin, colloidin, or Formvar is used to separate the second layer of emulsion from the first, and to distance the second layer far enough from the tissue so that it is not affected by the low-energy β emissions from the tritium. The exposure time of the first layer must be relatively brief (not longer than 1 week), so that the emissions from the ^{14}C would be minimal and statistically insignificant. The second layer can be exposed for whatever period is necessary for detection. There could be no interference from the ^{3}H, since it cannot reach the second layer of photographic emulsions (Fig. 9). It is also important to have a higher concentration of [^{3}H]thymidine available to the cells to incorporate and optimize detection during the short exposure period.

The technique of double-label autoradiography can be performed two ways. The first is to develop the first layer and count the cells and grains by phase microscopy (Fig. 8), put on the spacer layer and a second emulsion layer, develop the layers, and count cells labeled with ^{14}C (Fig. 9) with or without ^{3}H. The presence of ^{14}C is identified by tracking in the uppermost layer of emulsion. Cells exhibiting label only in the first layer of emulsion, representing ^{3}H-labeled cells, are counted. Subtracting the second ^{3}H count from the first gives the number of double-labeled cells. With those two calculations, it is possible to calculate immediately the number of cells labeled with only ^{14}C. A second method involves to following the above procedure without counting between processes,

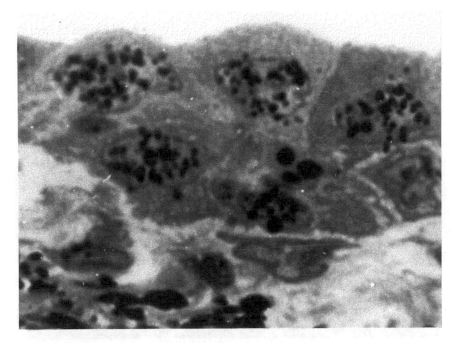

Figure 8 This light autoradiograph of a 1 μm plastic section of regenerating tracheal epithelium shows only the tritium label as grains in the first layer of emulsion directly over the nucleus. The tissue was labeled with [^{3}H]thymidine during the first DNA synthetic phase and with ^{14}C during a second DNA synthetic phase. Sections were stained with methylene blue and azure II (×3000).

and then counting as the final step by focusing each layer individually. However, as seen in Figure 10, this method could be confusing and time-consuming. Therefore, I recommend the first procedure. The confusion arises in double-labeled cells when overlying silver grains can obscure those beneath. References to other techniques of this type are available in the Appendix section "Double-Emulsion Techniques."

At the other end of the spectrum, macroautoradiographs are very useful in combination with the blotting techniques associated with biochemical isolations of proteins, DNA, and RNA. These include single- and double-dimension chromatography, electrophoresis, isoelectric focusing and others. This technology enables the investigator to use a radioactively labeled precursor molecule that can be bound or incorporated into specific molecular species such as DNA, RNA, protein, enzyme, lipids, or carbohydrate in situ. This label can be recovered by various biochemical separations to identify specific molecules or fractions it has

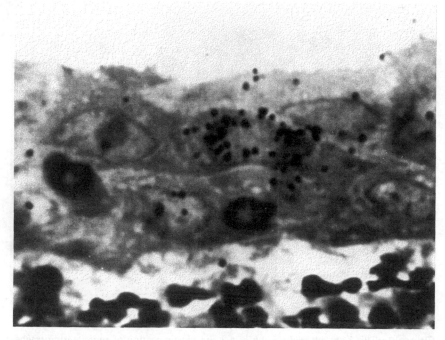

Figure 9 This autoradiograph of a 1 μm plastic section of regenerative tracheal epithelium exhibits only the incorporation of the ^{14}C label as a halo of silver grains tracking into the cytoplasm in the second of two layers of emulsion. This is a section of the same trachea described in Figure 8. Sections were stained with methylene blue and azure II ($\times$3000).

been bound to or incorporated. An alternative technique would be to perform separations using the various biochemical techniques and then use radioactively labeled antibodies or DNA or RNA hybridization techniques to identify specific molecules, molecular components, and DNA or RNA sequences or loci.

These techniques require only that the radioactive label be within the sample and that the x-ray film be applied directly adjacent to these labeled materials. Of course, the appropriate film and enhancement techniques should be considered, dependent upon the radioactive species used. These technologies are extremely important to lung research, particularly in the isolation of degradative enzymes; collagen and elastin components; protein factors stimulating cell proliferation and differentiation; and binding sites for drugs, toxic materials, and stimulatory factors. References to publications on these techniques can be found in the Appendix.

Other applications of autoradiographic technology are generally special applications for electron microscopy. One of these is the autoradiography of freeze-

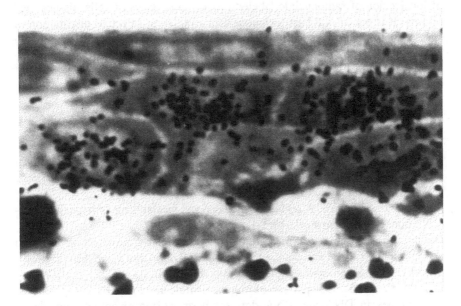

Figure 10 Light autoradiograph of a 1 μm plastic section of tracheal epithelium of a rat in which [^{3}H]thymidine was injected intraperitoneally 16 hr after mild surface abrasion and sacrificed 28 hr later. The excised trachea was pulse labeled for 30 min in medium containing [^{14}C]thymidine and fixed. The sections were coated with two layers of NTB2 nuclear emulsion separated by a thin layer of gelatin. These double-labeled cells have silver grains in two planes. The ^{14}C appears as a spray of silver grains in both planes around the perimeter of the nucleus, while the ^{3}H only exhibits grains in one plane directly over the nucleus. Sections were stained with methylene blue and azure II ($\times$3000).

fractured specimens. This technique is especially useful when determining localization of binding sites, particularly membrane-binding sites (Schiller et al., 1978). It is also extremely useful in localizing water-soluble or diffusible molecules such as amino acids, drugs, and some hormones (Rix et al., 1976). With this technology the tissues can be rapidly frozen after treatment with the radioactive molecules and localized to the structural components of the cell. Since the tissue is frozen, there is little or no chance for molecules to change position. If the molecule in question binds to membranes, freeze fracture can determine which side and in association with the ultrastructural component. The most appropriate radionuclide for these procedures is either tritium or iodine.

Another application of autoradiography is to scanning electron microscopy. This procedure allows for a wide range of localizations with increased resolution over light microscopy. It becomes possible with this technology to observe sites of localization of radionuclide-labeled probes on surfaces of cells, tissues, and organs. It is one of the best techniques for evaluating selective distribution of labeled molecules. References with precise descriptions of the the methodology can be found in the Appendix section "Scanning Electron Microscope Autoradiography."

Appendix

Techniques of Macroscopic Autoradiography

Anderson, N. L., and Anderson, N. G. (1977). High resolution two-dimensional electrophoresis of human plasma proteins. *Proc. Natl. Acad. Sci. U.S.A.* **74**:5421–5425.

Babcock, T. A. (1976). A review of methods and mechanisms of hypersensitization. *AAS Photo-Bull.* **13**:3–8

Bartnik, E., Borsuk, P., and Pieniazek, N. J. (1981). Sensitive detection of tritium in southern blot and plaque hybridizations. *Anal. Biochem.* **116**:237–240.

Berent, S. L., Mahmoudi, M., Torczynski, E. M., Bragg, P. W., and Bollow, A. P. (1985). Comparison of oligonucleotide and long DNA fragments as probes in DNA and RNA dot, southern, northern, colony and plaque hybridizations. *Biotechniques* May/June, 208–220.

Berlin, M., and Ullberg, S. (1963). Accumulation and retention of mercury in the mouse. *Arch. Exp. Health* **6**:27–38.

Bochner, B. R., and Ames, B. N. (1983). Sensitive fluorographic detection of H-3 and C-14 on chromatograms using methyl anthranilate as a scintillant. *Anal. Biochem.* **131**:510–515.

Bonner, V. M., and Laskey, R. A. (1974). A film detection method for tritium-labelled proteins and nucleic acids in polyacrylamide gels. *Eur. J. Biochem.* **46**:83–88.

Bonner, W. M., and Stedman, J. D. (1978). Efficient fluorography of H-3 and C-14 on thin layers. *Anal. Biochem.* **89**:247–256.

Chamberlain, J. P. (1979). Flurographic detection of radioactivity in polyacrylamide gels with the water-soluble fluor, sodium salicylate. *Anal. Biochem.* **98**:132–135.

Choo, K. H., Cotton, R. G. H., and Danks, D. M. (1980). Double-labeling and high precision comparison of complex protein patterns on two-dimensional polyacrylamide gels. *Anal. Biochem.* **103**:33–38.

Cooper, P. C., and Burgess, A. W. (1982). Simultaneous detection of S-35- and P-32-labeled proteins on electrophoretic gels. *Anal. Biochem.* **126**:301–305.

Cross, S. A. M., Groves, A. D., and Hesselbo, T. (1974). A quantitative method for measuring radioactivity in tissues sectioned for whole body autoradiography. *Int. J. Applied Radiat. Isot.* **25**:381–386.

Curtis, C. G., Cross, S. A. M., McCulloch, R. J., and Powell, G. M. (1981). *Whole Body Autoradiography*. New York, Academic Press.

d'Argy, R., Ullberg, S., Stalwacke, C. G., and Langstom, B. (1984). Whole body autoradiography using C-11 with double-tracer applications. *Int. J. Appl. Radiat. Isotopes* 35(2):120–134.

Fairbanks, G., Jr., Levinthal, C., and Reeder, R. H. (1965). Analysis of C-14-labeled proteins by disc electrophoresis. *Biochem. Biophys. Res. Commun.* 20:393–399.

Franklin, E. R. (1985). The use of measurements of radiographic response of x-ray films in quantitative and semiquantitative whole body autoradiography. *Int. J. Appl. Radiat. Isot.* 36:193–196.

Garrels, J. I. (1979). Two dimensional gel electrophoresis and computer analysis of proteins synthesized by clonal cell lines. *J. Biol. Chem.* 254:7961–7977.

Gibson, W. (1983). Replica images of silver-stained gels using direct duplicating film. *Anal. Biochem.* 132:171–173.

Gruenstein, E. I., and Pollard, A. L. (1976). Doubled-label autoradiography on polyacrylamide gels with H-3 and C-14. *Anal. Biochem.* 76:452–457.

Iwao, H., Nakamura, N., Ikemoto, F., and Yamamoto, K. (1983). Whole body autoradiographic distribution of exogenously administered renin in mice. *J. Histochem. Cytochem.* 31:776–782.

Keller, F., and Wasser, P. G. (1982). Quantification in macroscopic autoradiography with carbon-14—an evaluation of the method. *Int. J. Appl. Radiat. Isot.* 33:1427–1432.

Kronenberg, L. H. (1979). Radioautography of multiple isotopes using color films. *Anal. Biochem.* 93:189–195.

Laskey, R. A. (1980). The use of intensifying screens or organic scintillators for visualizing radioactive molecules resolved by gel electrophoresis. *Method Enzymol.* 65:363–371.

Laskey, R. A., and Mills, A. D. (1977). Enhanced autoradiographic detection of P-32 and I-125 using intensifying screens and hypersensitized film. *FEBS Lett.* 82:314–316.

LeCocq, R. E., Hepburn, A., and Lamy, F. (1982). The use of L-S-35-methionine and L-SE-75-selenomethionine for double-label autoradiography of complex protein patterns on two-dimensional polyacrylamide gels: a drastic shortening of the exposure time. *Anal. Biochem.* 127:293–299.

Liss, R. H., and Kensler, C. J. (1976). Radioautographic methods for physiologic disposition and toxicology studies. In *Advances in Modern Toxicology* Edited by M. A. Mehlman, R. E. Shapiro, and H. Blumenthal, pp. 273–305.

Longshaw, S., and Fowler, J. S. L. (1978). A poly (Methyl-C-14) methacrylate source for use in whole-body autoradiography and beta-radiography. *Xenobiotica* 8:289–295.

Maxam, A. M., and Gilbert, W. (1977). A new method for sequencing DNA. *Proc. Natl. Acad. Sci.* 74:560–564.

McConkey, E. H. (1979). Double-label autoradiography for comparison of complex protein mixtures after gel electrophoresis. *Anal. Biochem.* **96**:39–44.

Noren, O., and Sjostrom, H. (1979). Fluorography of tritium-labelled proteins in immunoelectrophoresis. *J. Biochem. Biophys. Methods.* **1**:59–64.

Ornstein, D. L., and Kashdan, M. A. (1985). Sequencing DNA using S-35-labeling: a troubleshooting guide. *Biotechniques* **3**:476–483.

Phillips, C. A., Smith, A. G., and Hahn, E. J. (1986). Recent developments in autoradiography. In *Proceedings from the Second International Symposium on the Synthesis and Application of Isotopically Labeled Compounds.* Edited by R. R. Muccino. New York, Elsevier, pp. 189–194.

Randerath, E. (1970). An evaluation of film detection methods for weak-emitters, particularly tritium. *Anal. Biochem.* **43**:188–205.

Reivich, M., Jehle, J., Sokoloff, L., and Kety, S. S. (1969). Measurement of regional cerebral blood flow with antipyrine-C-14 in awake cats. *J. Appl. Physiol.* **43**:296–300.

Schibeci, A., and Martonosi, A. (1980). Detection of Ca^{+2} binding proteins on polyacrylamide gels by ^{45}Ca autoradiography. *Anal. Biochem.* **104**:335–342.

Som, P., Yonekura, Y., Oster, Z. H., Meyer, M. A., Pelletteri, M. L., Fowler, J. S., MacGregor, R. R., Russell, J. A. G., Wolf, A. P., Fand, I., McNally, W. P., and Brill, A. B. (1983). Quantitative autoradiography with radiopharmaceuticals, part 2: applications in radiopharmaceutical research: concise communication. *J. Nucl. Med.* **24**:238–244.

Stevens, G. W. W. (1976). Production of multi-level radioactive source for autoradiographic calibration. *J. Microsc.* **106**:285–289.

Swanstrom, R., and Shank, P. R. (1978). X-ray intensifying screens greatly enhance the detection by autoradiography of the radioisotopes P-32 and I-125. *Anal. Biochem.* **86**:184–192.

Tomoike, H., Ogata, I., Maruoka, Y., Sakai, K. Kurozumi, T., and Nakamura, M. (1983). Differential registration of two types of radionuclides on macroautoradiograms for studying coronary circulation: concise communication. *J. Nucl. Med.* **24**:693–699.

Ullberg, S. (1977). The technique of whole-body autoradiography: Cryosectioning of large specimens. In: *Science Tools,* Special Issue. Edited by O. Alvfeldt. Sweden, L. K. B. Producter A. B.

Waddell, W. J., and Marlowe, C. (1977). Method and techniques. In *Autoradiography In Drug Fate and Metabolism.* Edited by E. R. Garrett and J. L. Hirtz. New York, Marcel Dekker, pp. 1–25.

Yonejura, Y., Brill, A. B., Som, P., Bennett, G. W., and Fand, I. (1983). Quantitative autoradiography with radiopharmaceuticals, part 1: digital film-analysis system by videodensitometry: concise communication. *J. Nucl. Med.* **234**:231–237.

Preparation, Exposure, and Processings

Azmitia, E. C., and Gannon, P. J. (1982). A light and electron microscope analysis of the selective retrograde transport of H3-5-hydroxytryptamine by serotonergic neurons. *J. Histochem. Cytochem.* **30**:799–804.

Bader, A. V., and Steinberg, R. L. (1972). A photographic developing unit for use in autoradiographic electron microscopy. *Stain Technol.* **47**:213–214.

Barnawell, E. B., Barnerjee, M. R., and Rogers, F. M. (1970). A compact apparatus for liquid emulsion autoradiography in total darkness. *Stain Technol.* **45**:40–41.

Baserga, R., and Nemeroff, K. (1962). Factors which affect efficiency of autoradiography with tritiated thymidine. *Stain Technol.* **37**:21–26.

Brock, M. L., and Brock, T. D. (1968). The application of microautoradiographic techniques to ecological studies. *Int. Assoc. Theor. Appl. Limnol.* **15**:1–29.

Budd, G. C. (1971). Recent developments in light and electron microscope radioautography. *Int. Rev. Cytol.* **31**:21–56.

Coleman, E. J. (1965). A simplifdied autoradiographic dipping procedure slides handled in groups of five. *Stain Technol.* **40**:240–241.

Davis, T. L., Spencer, R. F., and Sterling, P. (1979). Preparing autoradiograms of serial sections for electron microscopy. *J. Neurosci. Methods.* **1**:179–183.

Descarries, L. (1975). High resolution radioautography of noradrenergic axon terminals in the neocortex. Appendix II: A technique for high resolution radioautography. In *Vision in Fishes—New Approaches in Research*. Edited by M. A. Ali. New York, Plenum Press, pp. 224–232.

Edwards, S. B., and Hendrickson, A. (1981). The autoradiographic tracing of axonal connections in the central nervous system. In *Neuroanatomical Tract-Tracing Methods*. Edited by L. Heimer and M. J. Robards. New York, Plenum Press. pp. 171–205.

Elias, J. M. (1964). A desiccant-holding slide box for radioautographic exposure. *Stain Technol.* **39**:235–236.

Fitzgerald, P. J., Ord, M. G., and Stocken, L. A. (1961). A dry mounting autoradiographic technique for the localization of water soluble compounds. *Nature* **189**:55–56.

Haissig, B. E. (1969). A two-level autoradiographic emulsion tank for easy filling and dipping in total darkness. *Stain Technol.* **44**:253–255.

Hammarstrom, L., Appelgren, L. E., and Ullberg, S. (1965). Improved method for light microscopy autoradiography with isotopes in water-soluble form. *Exp. Cell Res.* **37**:608–613.

Hartzell, H. C. (1980). Distribution of muscarinic acetylcholine receptors and presynaptic nerve terminals in amphibian heart. *J. Cell Biol.* **86**:6–20.

Heywood, P., Hodge, L., Davis, F., and Simmons, T. (1977). A simple method for holding electron microscope grids during autoradiography of serial sections. *J. Microsc.* **110**:167–169.

Hillemann, H. H., and Ritschard, R. L. (1964). A light-tight, dust-free drying box for autoradiography. *Stain Technol.* **39**:327–328.

Hiraoka, J. (1972). A holder for mass treatment of grids, adapted especially to electron staining and autoradiography. *Stain Technol.* **47**:297–301.

Ison, E. J., and Seridan, P. J. (1981). Autoradiography of diffusible substances–a practical approach. *Am. J. Med. Technol.* **47**(1):38–42.

Jenkins, E. C. (1972). Wire-loop application of liquid emulsion to slides for autoradiography in light microscopy. *Stain Technol.* **47**(1):23–26.

Jerry, N. L. (1967). Quantity dipping and processing of radioautographic slides. *J. Biol. Photogr. Assoc.* **35**:73–82.

Kennedy, A. R., and Little, J. B. (1974). Autoradiography using dry mounted freeze-dried sections for localization of carcinogens in the lung. *J. Histochem. Cytochem.* **22**:361–367.

Kopriwa, B. (1967). A semiautomatic instrument for the radioautographic coating technique. *J. Histochem. Cytochem.* **14**:923–928.

Kopriwa, B. M. (1963). A model dark room unit for radioautography. *J. Histochem. Cytochem.* **11**:553–555.

Kopriwa, B. M. (1973). A reliable standardized method for ultrastructural electron microscope radioautography. *Histochemie* **37**:1–17.

Kopriwa, B. M., and Leblond, C. P. (1962). Improvements in the coating technique of radioautography. *J. Histochem. Cytochem.* **10**:269–284.

Markov, D. V. (1984). A simple apparatus for radioautographic coating. *J. Microsc.* **133**:281–283.

McGuffee, L. J., Hurwitz, L., and Little, S. A. (1977). A method for coating multiple slides for light microscopic autoradiography. *J. Histochem. Cytochem.* **25**:1107–1108.

Pickering, E. R. (1966). Autoradiography of mobile C^{14}-labeled herbicides in sections of leaf tissue. *Stain Technol.* **41**:131–137.

Rechenmann, R. V. (1967). Autoradiography by electron microscopy. *J. Nucl. Biol. Med.* **11**:111–131.

Sanderson, J. (1975). New techniques for the preparation of uniform layers of nuclear emulsions for use in microautoradiography. *J. Microsc.* **104**:179–185.

Sikov, M. R. (1965). A commercial encapsulated desiccant for use in slide containers during autoradiographic exposure. *Stain Technol.* **40**:239.

Stumpf, W.. E. (1976). Techniques for the autoradiography of diffusible compounds. *Methods Cell Biol.* **13**:171–193.

Stumpf, W. E. (1971). Autoradiographic techniques for the localization of hormones and drugs at the cellular and subcellular level. *Karolinska Symp. Res. Methods Reprod. Endocrinol.* 205–222.

Stumpf, W. E., and Roth, L. J. (1966). High resolution autoradiography with dry mounted, freeze-dried frozen sections: comparative study of six methods using two diffusible compounds ³H-estradiol and ³H-mesobilrubinogen. *J. Histochem. Cytochem.* **14**:274–287.

Stumpf, W. E. (1968). High resolution autoradiography and its application to in vitro experiments: Subcellular localization of 3H-estradial in rat uterus. In *Radioisotopes in Medicine—In vitro Studies*: Proceedings. Edited by R. L. Hayes, F. A. Goswitz, and B. E. P. Murphy. Tennessee, U.S. Atomic Energy Commission, pp. 633–660.

Telford, J. N., and Matsumura, F. (1969). The expandable loop: an improved wire-loop device for producing thin photographic films suited to auto-radiographic electron microscopy. *Stain Technol.* **44**:259–260.

Traurig, H. H. (1967). Equipment for simultaneous processing of groups of slides for radioautography. *Stain Technol.* **42**(2):97–100.

Williamson, J. R. and Bosch, H. Van Den. (1971). High resolution autoradiography with stripping film. *J. Histochem. Cytochem.* **19**:304–309.

Young, W. S. III, and Kuhar, M. J. (1979). A new method for receptor autoradiography: H3-opioid receptors in rat brain. *Brain Res.* **179**:255–270.

Double-Emulsion Techniques

Baserga, R., and Nemeroff, K. (1962). Two-emulsion radioautography. *J. Histochem. Cytochem.* **10**:628–635.

Dawson, K. B., Field, E. O., and Stevens, G. W. (1962). Differential autoradiography of tritium another B-emitter by a double stripping film technique. *Nature* **195**:510–511.

Field, E. O., Dawson, K. B., and Gibbs, J. E. (1965). Autoradiographic differentiation of tritium and another B-emitter by a combined colour-coupling and double stripping-film technique. *Stain Technol.* **40**:295.

Han, S. S., and Kim, M. K. (1972). An improved method for double-isotope and double-emulsion radioautography using epoxy resin sections. *Stain Technol.* **47**:291–296.

Pickworth, J. W., Cotton, K., and Skyring, A. P. (1963). Double emulsion autoradiography for identifying tritium-labelled cells in sections. *Stain Technol.* **38**:237.

Ruter, A., Aus, H. M., Harms, H., Haucke, M. ter Meulen, V., Maurer-Schultze, B., Korr, H., and Kellerer A. (1979). Automated quantitative analysis of single and doubled label autoradiography. *J. Histochem. Cytochem.* **27**:217–224.

Schultze, B., Maurer, W., and Hagen Busch, H. (1976). A two-emulsion autoradiographic technique and the discrimination of the three different types of labelling after double labelling with H3- and C14-thymidine. *Cell Tissue Kinet.* **9**:245–255.

Vincent, P. C., Borner, G., Chanana, A. D., Conkite, E. P., Greenberg, M. L., Joel, D. D., Schiffer, L. M., and Stryckmans, P. A. (1969). Studies on lymphocytes: XIV. Measurement of DNA synthesis time in bovine thoracic duct lymphocytes by analysis of labeled mitoses and by double labeling, before and after extracorporeal irradiation of the lymph. *Cell Tissue Kinet.* **2**:235–247.

Light Microscopic Intensification Techniques

Durie, B. G. M., and Salmon, S. E. (1977). High speed scintillation autoradiography. *Science* **190**:1093–1095, 1975. (see also Goldgefter, I., Toder V., Scintillator distribution in high-speed autoradiography) and response by Durie, B. G. M. *Science* **195**:208.

Kopriwa, B. M. (1980). Quantitative investigation of scintillator intensification for light and electron microscope radioautography. *Histochemistry* **68**:265–279.

Kopriwa, B. M. (1979). Examination of various methods for the intensification of radioautography by scintillators. *J. Histochem. Cytochem.* **27**:1524–1526.

Mourier, J. P. (1979). Incorporation of H^3-thymidine in the nephron of gasterosteus aculeatus L. and its stimulation methyltestosterone. A high-speed scintillation autoradiographic study. *Cell Tissue Res.* **201**:249–262.

Panayi, G. S., and Neill, W. A. (1972). Scintillation autoradiography—a rapid technique. *J. Immunol. Methods* **2**:115–117.

Przybylski, R. J. (1969) Scintillation radioautography: a new technique designed to augment silver grain number in radioautographs. *J. Cell Biol.* **43**:108a.

Smith, D. M., Shelley, S. A., and Balis, J. U. (1979). Reduced cell proliferation in fetal lung after maternal administration of pilocarpine. a scintillation autoradiographic study. *Am. J. Anat.* **155**:131–137.

Stanilus, B. M., Sheldon, S., Grove, G. L., and Cristofalo, V. J. (1979). Scintillation fluid shortens exposure times in autoradiography. *J. Histochem. Cytochem.* **27**:1303–1307.

Woodcock, C. L. F., D'Amico-Martel, A., McInnis, C. J., and Annunziato, A. T. (1979). How effective is "high-speed" autoradiography? *J. Microsc.* **117**:417–423.

Electron Microscopy Intensification

Buchel, L. A., Delain, E., and Bouteille, M. (1978). Electron microscope fluoroautoradiography: improvement of efficiency. *J. Microsc.* **112**:223–229.

Fischer, H. A. (1979). High resolution scintiautoradiography. *J. Histochem. Cytochem.* **27**:1527–1528.

Fisher, H. A., Korr, H., Thiele, H., and Werner, G. (1971). Shorter radioautographic exposure of elctron microscopical preparations by means of scintillators. *Naturwissenschaften* **58**:101–102.

Kopriwa, B. M. (1979). Examination of various methods for the intensification of radioautographs by scintillators. *J. Histochem. Cytochem.* **27**:1524–1526.

Kopriwa, B. M. (1980). Quantitative investigation of scintillator intensification for light and electron microscope radioautography. *Histochemistry* **68**:265–279.

Mizuhira, V., Shiihashi, M., and Futaesaku, Y. (1981). High-speed electron microscope autoradiographic studies of diffusible compounds. *J. Histochem. Cytochem.* **29**:143–160.

Scanning Electron Microscope Autoradiography

Darley, P. J., and MacFarlane, B. J. (1977). Scanning electron microscope autoradiography. A technique for the location and study of microscope B-active particles. *Health Physics* **32**:259–270.

Downs, G. L. (1981). A new technique for microautoradiography and tritium profiling. *Proc. Environ. Degrad. Eng. Mater. Hydrogen Conf.* 425–435.

Hodges, G. M., and Muir, M. D. (1976). Scanning electron microscope autoradiography. In *Principles and Techniques of Scanning Electron Microscopy*, volume 5. Edited by M. A. Hayat. New York, Van Nostrand Reinhold Company, pp. 78–93.

Hodges, G. M., and Muir, M. D. (1974). Autoradiography of biological tissues in the scanning electron microscope. *Nature* **247**:383–385.

Junger, E., and Bachmann, L. (1980). Methodological basis for an autoradiographic demonstration of insulin receptor sites on the surfaces of whole cells: a study using light and scanning electron microscopy. *J. Micros.* **119**:199–211.

Le, T. D., and Wilde, B. E. (1983). An autoradiographic technique for studying the segregation of hydrogen absorbed into carbon and low alloy steels. *Corrosion N.A.C.E.* **39**:258–265.

Paul, D., Grove, A., and Zimmer, F (1970). Autoradiography in the scanning electron microscope. *Nature* **227**:488–489.

Schutten, W. H., Van Horn, D. L., Bade, B. J., and Faculjak, M. L. (1980). Corneal endothelial autoradiography with the scanning electron microscope. *Invest. Opthalmol. Vis. Sci.* **19**:417–420.

Sudar, F., and Csaba, G. (1978). Scanning-autoradiography—a new method for the demonstration of membrane-surface-associated structures. *Experientia* **34**:416–417.

Weiss, R. L. (1980). Scanning electron microscope autoradiography of critical point dried biological samples. *Scanning Electron Microsc.* **IV**:123–132.

References

Baserga, R., and Malamud, D. (1969). *Autoradiography Techniques and Applications*. New York, Harper and Row.

Becquerel, H. (1896a). Sur les radiations emises par phosphorescence. *C. R. Seances Acad. Sci.* **122**:420–421.

Becquerel, H. (1896b). Sur les radiations invisibles emises par les corps phosphorescents. *C. R. Seances Acad. Sci.* **122**:501.

Becquerel, H. (1896c). Sur les radiations invisibles emises par les sels d'uranium. *C. B. Seances Acad. Sci.* **122**:689.

Becquerel,H. (1896d). Emission de radiations nouvelles par l'uranium metallique. *C. R. Seances Acad. Sci.* **122**:1086.

Bransome, E. D. (1970). *The Current Status of Liquid Scintillation Counting.* New York and London, Grune & Stratton.

Curie, S. *C. R. Seances Acad.Sci.* **126**:1101.

Durie, B. G. M., and Salmon, S. E. (1975). High speed scintillation autography. *Science* **190**:1093–1095.

Gahan, P. B. (1972). *Autoradiography for Biologists.* New York, Academic Press.

Gordon, R. E., and Lane, B. P. (1977). Cytokinetics of rat tracheal epithelium stimulated by mechanical trauma. *Cell Tissue Kinet.* **10**:171–181.

Gordon, R. E., and Lane, B. P. (1984). Ciliated cell differentiation in degenerating rat tracheal epithelium. *Lung* **162**:233–243.

Leblond, C. P. (1943). Localization of newly administered Iodine in the thyroid gland as indicated by radio-iodine. *J. Anat.* **77**:149–152.

Neipce de St. Victor. (1867). Sur une nouvelle action de la lumier. *C. R. Seances Acad. Sci.* **65**:505.

Przybylski, R. J. (1969). Scintillation radioautography: a new technique designed to augment silver grain number in radioautographs. *J. Cell Biol.* **43**:108a.

Quastler, H., and Sherman, F. G. (1959). Cell population kinetics in the intestinal epithelium of the mouse. *Exp. Cell Res.* **17**:420–438.

Rix, E., Schiller, A., and Taugner, R. (1976). Freeze-fracture-autoradiography. *Histochemistry* **50**:91–101.

Rogers, A. W. (1979a). *Techniques of Autoradiography.* New York, Elsevier.

Rogers, A. W. (1979b). *Practical Autoradiography.* Review No. 20, Arlington Heights, IL, Amersham Corp.

Roth, L. J., and Stumpf, W. E. (1969). *Autoradiography of Diffusible Substances.* New York, Academic Press.

Schiller, A., Rix, E., and Taugner, R. (1978). Freeze-fracture autoradiography: the in-vacuo coating technique. *Histochemistry* **59**:9–16.

Sklarew, R. J., Hoffman, J., and Post, J. (1976). Analysis of grain count decay patterns in human breast cancer labeled with tritiated thymidine in vivo. *J. Cell Biol.* **70**:239a

Williams, M. A. (1977). *Autoradiography and Immunocytochemistry.* Amsterdam, Elsevier Press North-Holland.

Yagoda, H. (1949). *Radiactive Measurements with Nuclear Emulsions.* John Wiley and Sons, New York.

18

Structural Methods for Studying Bronchiolar Epithelial Cells

CHARLES G. PLOPPER

School of Veterinary Medicine
University of California
Davis, California

I. Introduction

From a cell biological perspective, the respiratory system is highly complex. Of the more than 40 different cell types that have been identified within the respiratory system, at least 12 are found in the epithelial population that lines the conducting air passages (Breeze and Wheeldon, 1977; Jefferey, 1983; Plopper et al., 1983c, 1987; Widdicombe and Pack, 1982). Many of these cell types are restricted to specific zones within the airways. The methods included in this chapter resulted from recognition that these cell populations have specific anatomical distributions and that an adequate understanding of their biology can be derived only from studies in which the microenvironment of the populations involved is carefully identified. This applies not only to evaluations of specific cell types but also to the utility of those studies for extrapolating information from other animal species to humans. This chapter focuses on methodologies currently available for evaluating one particular group of cells: those associated with the distal conducting airways (or bronchioles) and the junction between the conducting airways and the gas exchange area, that is, the centriacinar region.

A. Definitions

Bronchioles

Bronchioles can be defined as more distal intrapulmonary branches of the conducting airway tree. The luminal surface is lined by simple cuboidal epithelium containing a mixture of nonciliated secretory cells, which are not mucous goblet cells, and ciliated cells. The peribronchiolar connective tissue elements associated with this epithelial population consist of a variety of fibroblastic cells, collagen and elastic fibers, and smooth muscle cells. The abluminal surface of the walls of these airways are lined by type 1 and type 2 alveolar epithelium and their associated capillary bed. Lymphatics are also found within the matrix.

Bronchi

Bronchi are more proximal in the conducting airway tree than are bronchioles, but distal to the trachea. The characteristics that differentiate bronchi from bronchioles include increased complexity of the epithelial lining and the content of the peribronchial interstitial matrix. The epithelium has more than two cell types: basal cells, ciliated cells, mucous goblet cells, and a variety of intermediate forms. In addition, the epithelial population has invaginations into the surrounding extracellular matrix to form submucosal glands. The matrix components surrounding bronchi include cartilage plates, smooth muscle cells, fibroblasts, and collagen and elastic fibers. Lymphatics and the branches of the bronchial artery are also found in the walls of these airways.

Centriacinar Region

The centriacinar region encompasses the junction between the most distal branches of the conducting airways and the proximal portions of the gas exchange area. This produces an admixture of both interstitial and epithelial components that is characteristic of both conducting airways and gas exchange area. It has been established that this region was markedly different in architectural configuration in different species (Tyler, 1983). In the majority of smaller mammals (laboratory rodents, guinea pigs, and rabbits) and in larger domesticated species (horse and cow) this junctional complex shows very little interspersion of gas exchange area and conducting airway epithelium. Conducting airways intermixed with alveolar gas exchange tissue are therefore minimal. In a variety of mid-sized species (including cat, dog, and primates), this intermixing is extensive.

Respiratory Bronchioles

The result of this intermixing, respiratory bronchioles, have bronchiolar epithelium intermixed with alveoli of several generations in cat, dog, ferret, and primates.

B. Rationale

It is assumed that there is wide variability between species and within regions of the same species, not only in the composition of the epithelial populations but also in their interactions with toxins and their relationship to extracellular matrix and other interstitial elements. This assumption was the stimulus for developing approaches to obtain specifically identified and characterized samples for the study of bronchiolar cell biology. Two important parameters must be established before meaningful information on the biology of bronchiolar epithelial cells can be obtained. The first is the location of the epithelial population by region within the respiratory system. This includes identification of the length and number of generations of airway branching to the site being studied and the position of the site with relation to gravitational forces. The second is the location of the epithelial population in relation to the gas exchange region (the alveolarized area of the central acinus). This second factor includes two considerations: (1) the relationship of the population to both alveolarized and nonalveolarized areas and (2) the position proximal to distal within the central acinus (most distal bronchiole, first generation respiratory bronchiole, second generation respiratory bronchiole, first alveolar duct bifurcation). It is now apparent that these factors are not translatable from species to species and must be determined for each species. Epithelial and interstitial cellular populations in specifically defined microenvironments, even if they are the same cell types, may react very differently to the same types of stimuli.

II. Methodology

Three approaches have been used to obtain defined bronchiolar cell populations for biological characterization. Each has its own strengths and weaknesses, and each has proven useful for obtaining different types of information on the biological character of bronchiolar cell populations. The first approach relies on the use of fixed lung for the selection of specimens. The second approach uses fixed lung specimens that have been processed for examination by microscopic techniques. The third approach uses specimens from unfixed lung tissue.

A. Microdissection of Fixed Lung

Procedures

The first requirement for microdissection of airways is fixed lung tissue in which the air spaces are inflated with either fixative or air. Tissue can be fixed for microdissection either by installation via the trachea or perfusion via the pulmonary arterial system. The advantages and disadvantages of various types of fixatives and their method of application to the respiratory system have been discussed

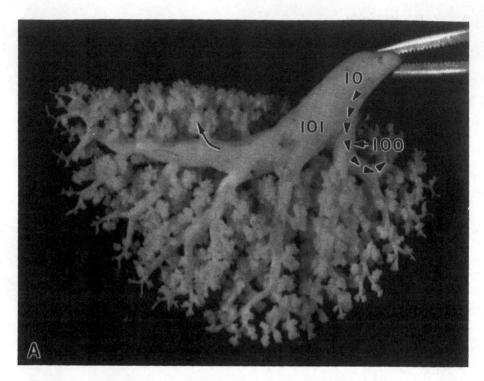

Figure 1 Comparison of (A) an airway cast and (B) microdissected airways in the left lobe of adult rat lung. (A) Silastic cast of airway tree. Numbering scheme indicates lobar bronchus (10), major daughter (axial continuation) (100), and minor daughter (100). Arrowheads trace the shortest pathway distally to a pulmonary acinus. Curved arrow indicates minor daughter from axial pathway. (B) Microdissection of fixed left lobe of adult rat lung exposing axial pathway and minor daughter branches in the plane of the dissection. Arrowheads and curved arrows show the same airways on the cast.

in detail previously (Dungworth et al., 1985; Tyler et al., 1985). Once the lung has been adequately fixed, preferably using aldehydes, the next phase is the exposure of the tracheobronchial airway tree by microdissection. The equipment necessary to microdissect lungs successfully includes a high-resolution dissecting microscope (we use a Wild M-8 dual viewing microscope, which has the added advantage of allowing one person to microdissect the airways while a second diagrams the microdissection and records the position of the specimens taken), and fine scissors and forceps (we obtain all of our equipment from Roboz Surgical Instruments Company, Washington, D.C.). A wide variety of scissors are useful for the dissection. For removal of tissue specimens once the airways

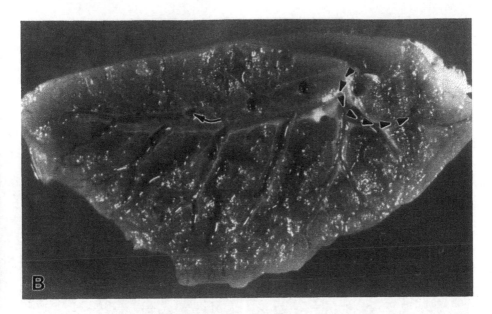

are dissected, we use portions of thin, double-edged razor blades that are cut into the shape of scalpel blades with tin snips and held in curved mosquito hemostats. Because of the toxicity of the fumes generated by fixed tissue, we use an exhaust system at the level of the specimen to pull air and fumes away from the dissector.

Before beginning dissection of the airways, we generate airway casts such as those described in Chapter 9. Figure 1 compares a silicon rubber cast of the left lobe of the rat with a microdissection following the major pathways. The lung in Figure 1b was fixed by airway infusion. The pathways of the principal side branches can be followed with the dissecting microscope during the process of the dissection. Based on the question being asked, we define specific regions and the types of airways to be sampled before beginning the dissection. The casting methods can be relied on for general airway patterns and mirror what you can expect to find during dissection (Plopper et al., 1983b). Before one begins a dissection, the lobe is removed from the trachea at the hilum. Using a pair of forceps in one hand and relatively large scissors in the other, we begin by look-ing directly into the lobar bronchus. If the decision is to expose as many of the airway paths as possible, as illustrated in Figure 1b, the initial cuts are made in the plane that will expose the majority of the airways. This plane is determin-ed from the cast. In most species, including mouse, rat, hamster, rabbit, cat, bonnet monkey, rhesus monkey, sheep, goat, horse, and human, the majority of the major pathways are in one plane. As illustrated in Figure 2, it is possible

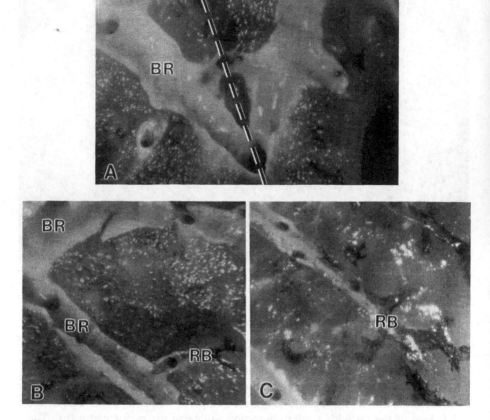

Figure 2 Views of fixed cat lung through dissecting microscope during dissection of airways. (A) Major bronchus (BR) dissected near the lobar bronchus showing both halves. The separation of airways is at the plane of dissection shown by the dashed line. (B) One-half of microdissected intrapulmonary airways shows various generations of branching from a large bronchus (BR) to a respiratory bronchiole (RB). (C) Higher magnification of respiratory bronchiole (RB) shows multiple generations of branching. The dark pigmented material is carbon deposited in the interstitium surrounding proximal respiratory bronchioles.

to look directly into the airways with the dissecting microscope. Figure 2a shows the dissection of a large bronchus in the right apical lobe of the cat.

All the airways are cut so that the entire face of the cut surface is exposed at the same time. Figure 2b shows the appearance of a number of airway generations, varying in size from a large bronchus to a respiratory bronchiole, as they appear during the dissection. Figure 2c is a high magnification view of the

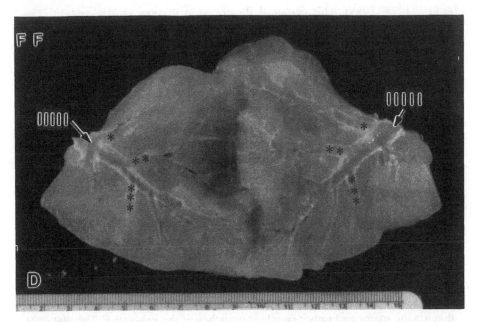

Figure 2 (continued) (D) Both halves of a microdissected right diaphragmatic lobe from a bonnet monkey. The lobar bronchus of this lobe is numbered IIIII. Complementary halves of dissected airways are indicated by asterisks (*, **, ***). From this type of microdissected specimen every airway generation through the axial pathway can be selected down to the respiratory bronchiole. One side can be used for scanning electron microscopy or histochemical preparations and the other side for transmission electron microscopy.

multiple generations of respiratory bronchiole from the same lung. The cat lung used in this illustration was exposed to diesel exhaust for 2 years. The dark pigment is carbon particles deposited in macrophages associated with the proximal generations of respiratory bronchioles. As the dissection progresses towards more distal airway generations, the entire lung will be exposed. Figure 2d shows an example of a microdissected lung from a bonnet monkey. This is a right diaphragmatic lobe and shows the major and minor daughter pathways exposed through a plane parallel to the costal surface. Once the dissection has been carried nearly to the pleural surface on all sides except for the reflections of the costodiaphragmatic area, the lungs are opened and photographed using Polaroid film. Prints are immediately availabale on which to record airway generations while specimens are removed for tissue processing.

The most important aspect of the dissection from this point on is the numbering of each airway generation. We use the binary system developed by Phelan et al., (1978, 1983). This system is illustrated in Figures 1a, 1b, and 2d. The trachea is numbered "I", and the larger of the two branches, or daughters, that arise from it is "II". In most species, the right mainstem bronchus is the larger daughter branch. The smaller daughter is numbered "I0." In the left lobe of the rat (Fig. 1b) at the first bifurcation of the lobar bronchus (I0), the smaller daugher (I00) carries air to the cranial quadrant of the lung. With each succeeding generation of branching, either "I" or "0" is added to the number for the next airway. By contrast, the extrapulmonary airway leading to the right diaphragmatic lobe always has the largest branches. Since three other lobes branch from the airway prior to the diaphragmatic lobe, in the bonnet monkey, the lobar bronchus for that lobe has the number IIIII (Fig. 2d). This number indicates that there are four branch points before this airway is formed and that it is the largest daughter branch at each point. This system allows one to assign each airway branch, or generation, its own number, which also gives its branching history. For ease of reporting, all the airways that are the same number of branch points away from the trachea are given the same generation number. The trachea is considered generation 0 and both primary bronchi, each one branch point from the trachea are considered generation 1. Four airways (III, II0, I0I, and I00) are two branch points from the trachea and are considered generation 2. The eight airway generations that are the major and minor daughter branches of the generation 2 airways are considered generation 3, and so on. The advantage of this type of numbering system for morphological and cell biological studies is that all of the history for each airway branch, including its regional position in the lung, is defined by the specimen number. A further advantage is that this can be tied to information developed from casts, as discussed in Chapter 9.

Once generations are recorded, the specimens can be selected for further processing for any of a wide variety of morphological techniques, including immunocytochemistry, carbohydrate cytochemistry, and scanning and transmission electron microscopy. Tissue samples are excised from the lungs for processing. The areas from which specimens were taken and the numbers of the airways included in the specimen are recorded. For light and transmission electron microscopy, we use large epoxy blocks as discussed below. We take specimens as large as 10×30 mm. Specimens for scanning electron microscopy, which can be the complementary halves for those used in transmission electron microscopy, are processed by conventional methods, as discussed in Chapter 3.

If the area of interest is the centriacinar region of the lung and the branching history of the airways is unimportant, the methodology proposed by Barry et al. (1985) offers a quick alternative for fixed lung tissue that is unprocessed. The lungs are sliced into slabs 1 cm or smaller in thickness. The slabs are laid under a dissecting microscope and terminal bronchiole alveolar duct junctions are isolated by punching them out of the slabs with a sharpened cannula. In most cases, cross-sections of terminal bronchioles are isolated. The terminal bronchioles are then embedded in BEEM capsules oriented so that the terminal bronchiole is sectioned first and a cross-section of the airway is obtained. The type of information generated by this approach is discussion below.

Observations

The airway microdissection approach to lung tissue selection has altered the general definitions of the airway tree as they apply to the tracheobronchial epithelial populations. Adults of three species have been studied in detail: sheep (Mariassy and Plopper, 1983c, 1984), rabbit (Plopper et al., 1983c, 1984, 1987), and rhesus monkey (Tyler and Plopper, 1985; Plopper et al., 1989). These airway generation studies have demonstrated that the composition of the tracheobronchial epithelium in adult mammals is very species-specific. For instance, mucous goblet cells, which are thought to be the predominant secretory cell type in tracheobronchial airways, are restricted to only the most proximal five intrapulmonary airway generations in the rabbit, are found in approximately two-thirds of the intrapulmonary airway generations in the sheep, and are in all airway generations, including respiratory bronchioles, in the rhesus monkey. Clara cells are the predominant nonciliated cell type throughout the airway tree in the rabbit, are restricted to the most distal 10 generations in the sheep, and are found only as a small population on one side of respiratory bronchiolar walls in the rhesus monkey. The relative proportions of these cell types within the population also vary in different airway generations in the same species and between species for any given airway generation. The carbohydrate composition of the secretory product stored in cells throughout the tracheobronchial airways also varies not only between species at any given airway generation, but between different airway generations in the same species. In the sheep, the majority of proximal secretory cells on the surface produce sulfated and acidic glycoconjugate (Mariassy et al., 1988a,b), as is the case for the rhesus monkey (Plopper et al., 1989). However, in the rabbit (Plopper et al., 1984) the predominant secretory product is a neutral glycoconjugate that in many cases in nonreactive with any of the conventional carbohydrate cytochemical methods. Quantitation of stored glycoconjugate has shown regional variation within the trachea (Heidsek et al., 1987) and between airway generations (Plopper et al., 1989) of macaques. Airway microdissection has also demonstrated the heterogeneity of epithelial populations in the respiratory bronchioles of macaque monkeys (Tyler and Plopper, 1985). When the location of tissue samples within the airways is known, the variability of the peribronchial connective tissue elements (i.e., the smooth muscle, cartilage, and submucosal glands) between species also becomes evident. In the rabbit, cartilage extends into only three generations of intrapulmonary airway and glands are not found distal to the trachea. In sheep, cartilage, glands, and surface mucous goblet cells extend the same number of generations distally. In the rhesus monkey, cartilage plates are found in the proximal two or three generations of respiratory bronchioles. An overview of these comparative studies emphasizes the high degree of diversity in different species of mammals. Epithelial populations of the tracheobronchial tree vary in the same airway generation in different species, in different airway generations in the same species, in types of cells present in all airway generations, in the overall abundance of secretory product and number of cells, in carbohydrate content of secretory granules, in the amount of stored secretory product, and in the ultrastructural composition of the same cell type in different species.

This approach has also proven extremely useful in characterizing the pathological response of the lung to injurants that cause their principal damage in the centriacinar region. The deposition and resulting fibrosis associated with long-term inhalation of diesel exhaust are focused in the proximal two generations of respiratory bronchioles and the distal portion of the terminal bronchiolar region in the cat (Plopper et al., 1983c; Hyde et al., 1985). The fibrosis and epithelial metaplasia and hyperplasia associated with long-term exposure to ambient concentrations of ozone produce a similar focal change in the first and second orders of respiratory bronchioles in the bonnet monkey (Fujinaka et al., 1985). The formation of extensive respiratory bronchioles by remodeling of the centriacinar region and invasion of bronchiolar epithelium into gas exchange area has been demonstrated in rats chronically exposed to high ambient levels of ozone (Barr et al., 1988). Samples from microdissected airways have also been used to demonstrate the influence of airway patterns on the deposition of aerosolized crysotile asbestos at the terminal bronchiole–alveolar duct junction and the associated fibrotic response (Pinkerton et al., 1986). Regional differences were demonstrated in centriacinar asbestos fiber size, number, and mass. All three were inversely related to pathway length and to the number of bifurcations along each airway path. Fiber burden within each region was found to be proportional to the relative degree of tissue injury in the terminal bronchiole and proximal alveolar ducts. These findings suggest that differences in centriacinar tissue injury from region to region in lungs exposed to asbestos result from regional differences in the deposition and retention of these substances in the lungs. Airway characteristics, such as branch angle, number of branch points, and total airway path length to the centriacinar site of deposition, play important roles in the deposition and subsequent bronchiolar injury caused by particulates and environmental pollutants.

Value and Future Prospects

The use of carefully defined specimens seems almost limitless for the precise characterization of pathological responses in the respiratory system. Use of this technique has defined the need to know the tracheobronchial airway pattern through which air and associated toxins pass to a site of injury. It has also been used to establish the need to define carefully the airway branching history of an epithelial or interstitial population within the tracheobronchial tree that is used for a study of alterations in epithelial populations. This is especially emphasized when subtle alterations are being assessed, such as those associated with changes in the amount or chemical composition of secretory product, the abundance of specific epithelial cell types or interstitial components, and the presence and abundance of metabolically active enzyme systems. The prospects for use of this technique are high, because it is straightforward and simple in execution and yields large amounts of information in relation to the effort required.

B. Processed Specimens

This approach relies on the selection of specimens for further study from tissue that has been partially or completely processed for embedment in various media. It has been used successfully to characterize the regional variability within the centriacinus of pathological processes and to define the species differences in epithelial populations in the region of the terminal bronchiole–alveolar duct junction.

Procedures

Key features necessary for successful use of processed specimens include osmication of the tissue to provide contrast and dehydration to clear the lung tissue so it can be viewed. It is essential for this type of embedding that all the processing steps be increased in time to ensure that the large blocks of lung tissue are completely dehydrated and infiltrated. This involves either replacement with fresh solutions of the same concentration during each processing step, doubling of the time for each processing step, or both. Most commonly, specimens have been carried all the way to embedding resin that has been polymerized. This involves large epoxy resin blocks and the selection of specimens from those blocks (Dungworth et al., 1985). Once the epoxy block is hardened, a number of alternatives are available. The most widely used approach is to embed the specimens in molds used for the Sorvall JB4 microtome as illustrated in Figure 3A. The large blocks of tissue are embedded in plastic and 1 μm sections made from the blocks (Fig. 4A). Relevant areas are selected using the stained 1 μm sections for orientation. One approach is to pyramid the blocks while they are still on the large aluminum chucks (Lowrie and Tyler, 1973). An alternative method is to remove the large blocks from the chucks and mount them on cardboard with double-sided cellophane tape (Fig. 3B). A razor blade (Fig. 3C) is used to cut the specimens from the plastic chips. The plastic pieces are then mounted on BEEM capsules using epoxy resin (Fig. 3D). Thin sections are made from pyramided specimens. Another approach is to cut the large epoxy blocks into slices and identify terminal

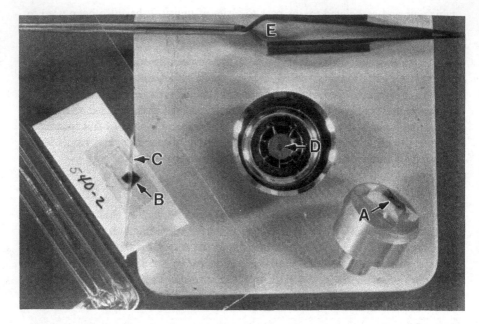

Figure 3 Steps in tissue selection from large blocks embedded in epoxy resin for transmission electron microscopy. (A) Large block of lung tissue embedded in a JB4 mold and mounted on a chuck. (B) A piece of embedded lung tissue removed from a JB4 block and mounted on double-sided-tape-backed cardboard. (C) Position of razor blade placed through the block to cut out specific areas for remounting on blank BEEM capsules. (D) Small piece of fixed and embedded lung tissue removed from large block and remounted on BEEM capsule. (E) Tweezers used to hold pieces of tissue removed from large blocks prior to mounting on BEEM capsule.

bronchioles, alveolar ducts, or the bifurcations on the slices held to cardboard with double-sided tape (Chang et al., 1986a). Figure 4 illustrates the degree of specificity available with methods that rely on 1 μm sections of large blocks for orientation followed by removal and remounting of specific pieces.

Another method is to use partially processed specimens that have been through a number of the steps but are not yet in polymerized plastic. The earliest of these used large blocks of lung tissue that were osmicated, dehydrated, and infiltrated with unpolymerized resin (Stephens and Evans, 1973). These cleared specimens contain clearly outlined centriacinar regions that can be cut from unpolymerized specimens. The final embedding is done in the caps from BEEM capsules. These specimens are then sectioned for light and electron microscopy. All of the above methods rely on 1 μm sections of large pieces of tissue, whether

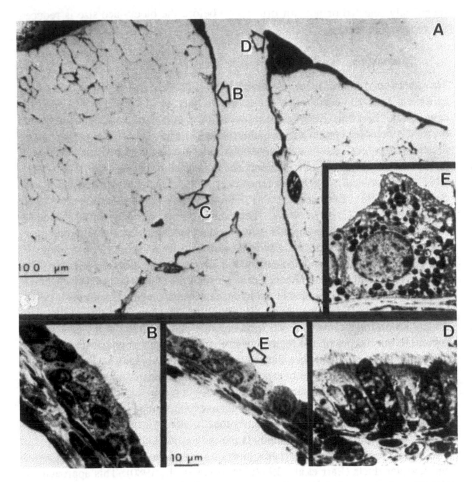

Figure 4 Selection of specific areas of respiratory bronchiole from a cat lung for light and transmission electron microscopy. (A) Photomicrograph of portion of 1 μm section from large epoxy-embedded block of cat lung tissue. (B) High-resolution light micrograph of area marked in (A). (C) High-resolution light micrograph of epithelium in area marked by (C) in (A). Both B and C contain Clara cells associated with different extracellular matrix constituents and different numbers of ciliated cells. (D) High magnification of area D marked in (A), which contains a different population composition than the epithelium slightly more distal at position B. This epithelium contains mucous cells, basal cells, and ciliated cells and the matrix behind it contains a small cartilage plate. (E) Low-magnification electron micrograph of the cell in (C).

specifically selected or not, for orientation before the block face size is reduced for ultrathin sectioning.

Observations

The selection of areas from processed lung specimens was designed to obtain specimens from the centriacinar region of the lung. As a consequence, a significant body of literature has been generated on the effects of a variety of toxicants on the bronchiolar epithelium and interstitium in this region. The earliest work illustrated the focal nature of the epithelial and interstitial injury associated with the inhalation of oxidant air pollutants (ozone and nitrogen dioxide) (Barr et al., 1988; Boorman et al., 1980; Castleman et al., 1985; Chang et al., 1986a,b; Chow et al., 1979, 1981; Eustis et al., 1981; Fujinaka et al., 1985; Moffat et al., 1987; Plopper et al., 1973; 1978, 1979). These showed that the majority of the injury occurred at the terminal bronchiole–alveolar duct junction, or terminal bronchiole–respiratory bronchiole junction, and that more distal air spaces in the acinus or more proximal epithelium in the conducting airway tree had less or no injury. This approach was also used to define the critical role of the Clara cell as the progenitor for renewal of bronchiolar epithelium in the centriacinar region (Evans et al., 1978; Lum et al., 1978; Castleman et al., 1980). It has also been established that with a focal lesion the elevation in cell turnover rate is augmented in the centriacinar region, but not in peripheral regions where cellular injury is undetected. The use of precision tissue sampling has allowed the selection of subregions within the acinus for defining effects of long-term exposure to oxidant pollutants on bronchiolar cells. In the rat there is a greater than threefold increase in respiratory bronchiolar volume in the centriacinar region, in addition to changes in terminal bronchiole diameter, and formation of fused basement membranes in newly formed respiratory bronchioles. In chronically exposed animals the only epithelial damage is observed in a very small region of the acinus at the junction of respiratory bronchiolar epithelium and alveolar ducts (Barr et al., 1988; Boorman et al., 1980). This approach has also been used to demonstrate the presence of inhaled particles at the first alveolar duct junction in the centriacinar region (Brody et al., 1981). The majority of the particles were deposited there initially and retained there. Because precision selection of tissue was used, it was also possible to demonstrate the incorporation of asbestos fiber particles into alveolar epithelium and their passage into the interstitial space at the first alveolar duct bifurcation. This method has also been successfully used to define the effects of subchronic inhalation of low concentrations of nitrogen dioxide in the proximal alveolar region of adult rats and to compare that with juveniles (Chang et al., 1986a). It has also been used for defining the response of respiratory bronchiolar epithelium to chronic oxidant injury (Fujinaka et al., 1985; Moffat et al., 1987). Hypertrophy and hyperplasia of these cells occur only in the most proximal respiratory bronchioles.

This approach has also been used for defining the cytodifferentiation of the Clara cell in the centriacinar region (Plopper et al., 1983a; Hyde et al., 1983; Massaro et al., 1984). Most of the hallmarks of a differentiated Clara cell in the rabbit (Plopper et al., 1983a, Hyde et al., 1983) and in the rat (Massaro et al., 1984) appear postnatally in terminal bronchiolar regions of this species. The pattern of differentiation is loss of glycogen, increase in agranular and granular endoplasmic reticulum, and the appearance of secretory granules. These changes occur postnatally and lag well behind differentiation of type 2 alveolar epithelium and the blood–air barrier in the same region. Factors that affect postnatal differentiation of Clara cells include protein deprivation to the mother (Massaro et al., 1984) and hyperoxia in the newborn (Massaro et al., 1986).

Careful localization of centriacinar areas has also established that there are marked ultrastructural differences in Clara cells in adults of different species (Plopper et al., 1980a,b,c). In most species, the predominant organelles are agranular endoplasmic reticulum, secretory granules, and granular endoplasmic reticulum. In a small number of species, the predominant cellular constituent is not an organelle, but cytoplasmic glycogen. In nonhuman primates and humans the predominant organelles are mitochondria and granular endoplasmic reticulum. Comparative studies of airway generations in the rabbit (Plopper et al., 1983c) show that these characteristics tend to vary with airway level. These studies reemphasize the importance of using defined samples that are comparable from animal to animal when comparing alterations in Clara cell biology.

Value and Prospects for Future Use

For studies of bronchiolar epithelium and its response to lung injurants whose focus of damage is the centriacinar region, and where the branching history of the conducting airways proximal to the sites of injury can be shown to be irrelevant, the approaches using large pieces of lung tissue embedded in resin have a number of advantages over the microdissection methodology. These include the speed with which specimens can be translated from the fixed state into usable blocks and the freedom from careful recording of branching history. The large block method and the microdissected centriacinus method share several advantages. Both allow selection of bronchiolar cells in close approximation to the gas exchange area. The samples are carefully defined so that reproducibility is maximized. It is also possible to define better the microecology of bronchiolar epithelial interaction with tissue compartments throughout the junction between conducting airways and gas exchange area. The disadvantage of the large block method is the inability to define the airway generation. Most current literature on the biology of bronchiolar epithelial cells in situ uses the selection of bronchioles in processed specimens, whether these have been microdissected before embedding or not. This suggests that this is now the method of choice and that future studies designed to provide detailed understanding of the response of bronchiolar epithelium to mediators and injurants must rely on this approach.

C. Microdissection of Unfixed Specimens

The most exciting new procedure for studying the biology of Clara cells from distal bronchioles is the use of selected airway specimens microdissected from fresh lung tissue. This procedure allows definition of the biology of these cells with a minimum of enzymatic disruption. Methodologies have been developed for studying Clara cells isolated from lung tissue by protease digestion (Devreaux et al., 1979, 1980, 1981). The principal drawbacks of this approach are that the microenvironment from which the cells were isolated is unknown because the isolated population consists of cells from the entire airway tree the cells and that the effects of protease on the metabolic activity of these cells is unknown.

Procedures

The two features of this procedure that appear to be critical for maintaining viable and active cells in dissected airways are use of a balanced salt solution (such as Waymouth's), that contains no nutrients, and maintenance of the medium and the lungs ice-cold once they are removed from the animal. In other respects the approach is very similar to that used with fixed tissue. The airways are dissected along the peribronchiolar connective tissue and not through the airway lumen. The entire procedure is by blunt dissection. The airway entering a specific lobe and the pulmonary artery and vein leading into that lobe are stabilized with either ligatures or hemostats. With use of a high-resolution dissecting microscope, the connective tissue is then dissected away from the airway and each branch is followed down as close to the parenchyma as possible. If the lung is filled with fluid without the animal being exsanguinated and the animal is held in the cold for at least a half an hour, the blood will coagulate in the lungs. This makes the microdissection procedure much more rapid and allows easy identification of the blood-filled parenchyma and major blood vessels, greatly facilitating their separation from airway tissue. Figure 5 illustrates the airway path in the right diaphragmatic lobe of an 80-day fetal rhesus monkey. This shows that even small branches of these airways can be dissected free. While not all of the branches can be identified, a significant number of individual acini can be isolated by careful dissection. In our laboratory we have applied this methodology to animals as small as mice and fetal rhesus. The larger the animal, the easier and more rapid the dissections. The microdissected airways have been used in two different ways. The first has been to cannulate them and apply standard digestion procedures involving either elastase or protease to isolate airway cells from known distal airway generations (St. George et al., 1986, 1987). Isolated airways have also been used as explants to evaluate the metabolic activity of Clara

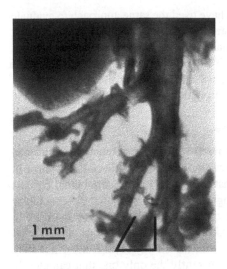

Figure 5 Microdissection of conducting airway tree in the right diaphragmatic lobe of a fetal monkey of 80 days' gestational age. This tissue is unfixed. The triangle at the base of the photograph outlines an individual acinus.

cells in their normal microenvironment (Bond et al., 1988; Suverkropp et al., 1988).

Observations

We have used this approach for isolating cells from proximal and distal airways in the same animal to demonstrate the culture requirements for differentiation of fetal airway epithelium cells (St. George et al., 1986, 1987). The primary conclusions are that fetal airway epithelial cells of rhesus monkey will maintain their differentiated status *in vitro*; the degree of differentiation found *in vitro* reflects that found *in vivo*; cells grown on collagen gel substrate were the most differentiated regardless of their status in proximal or distal airways; and in the adult rhesus monkey, where the epithelial population is the same in both proximal and distal airways, factors necessary to maintain a differentiated state are the same for cells isolated from both proximal and distal airways. Using microdissected explants from the mouse, we have demonstrated that the metabolic activity of the epithelial cells in the isolated airways is up to 10-fold greater per milligram of protein than the activity of microsomal fractions isolated from the whole lung (Suverkropp et al., 1988). We have also shown that the cytotoxic response to a bioactivated xenobiotic compound specific for Clara cell toxicity in vivo (naphthalene) occurs in a time- and dose-dependent fashion in isolated airway explants. There is less metabolic activity per milligram of protein in proximal airway generations than in distal airways. The most distal airway generations have the highest metabolic activity per milligram protein. To date, we have been able to maintain the metabolic activity of isolated explants for 6 hr with no degradation in function or signs of cytotoxicity. Others have used the same approach to evaluate the xenobiotic metabolizing enzyme systems in the airways of the dog (Bond et al., 1988). They have shown that the metabolic activity for a variety of substrates varies both by intensity and substrate specificity in an airway-generation-dependent fashion. In the dog, the most activity is not necessarily in the most distal airways, but is found in varying degrees throughout the airway tree.

Value and Prospects for Future Use

The value of using microdissected airways from fresh specimens has not been fully evaluated to date. However, the few studies that have been done, most in preliminary or abstract form, suggest that there is tremendous potential for this method in evaluating the variety of biological functions attributed to bronchiolar epithelium. It is now apparent that if the epithelial population varies in different airway generations in the species being used, the culture conditions under which they will remain differentiated may vary with the airway level from which they are isolated. This particular approach is currently the only one that can clearly define these conditions. Metabolic studies suggest the need to perform primary functional studies using microdissected airways. At the very least, metabolic observations of epithelial cells isolated from the entire airway tree should be confirmed by comparing their function with that of Clara cells still in the intact state. The prospects for future use of this approach seem almost unlimited. Use of defined airway populations will be the only way to identify the metabolic byproducts that produce such phenomena as Clara cell degranulation and cytotoxicity.

III. Conclusions

It is arguable that the biology of bronchiolar epithelial cells can be adequately studied with methodological approaches that do not have the precision described here. For instance, it is possible that a Clara cell is a Clara cell anywhere within the airway tree. A Clara cell's interactions with neighboring cells and extracellular matrix and its response to biological stimulants and toxicants are the same regardless of where it is and which species is studied. While this may be true for some aspects of bronchiolar epithelial cell biology, most of the biological functions that have been studied in carefully defined populations have shown that this is not the case. Responses of bronchiolar epithelial cells to both bloodborne and airborne toxicants have the highest intensity of response in the epithelial cells closest to the junction with the gas exchange area. This is especially true when inhaled toxic particles or gases are studied in a species that has minimal respiratory bronchioles, such as the rat. In this case, the most distal bronchiolar epithelium has the strongest response, while bronchiolar epithelium in more proximal airway generations is essentially unaffected. In species with extensive respiratory bronchioles, the area of major effect is the proximal one to three generations of respiratory bronchiole. The bronchiolar epithelium of more proximal generations and in more distal generations of the respiratory bronchiole appear to be minimally affected, if at all. The same is true for differentiation of bronchiolar epithelium in species with respiratory bronchioles, such as the rhesus monkey. The bronchiolar epithelium not only differentiates later than in more proximal generations, but the two epithelial populations of the primate respiratory bronchiole also differentiate at different times in relation to the pulmonary arteriole. Metabolic studies of the ability of different bronchiolar epithelial populations throughout the tree to metabolize xenobiotics indicate that there is an airway-generation-specific difference in species with an extensive population of Clara cells throughout the airways, such as the mouse, as well as for species with Clara cells restricted to very distal airways, such as the dog. This also appears to be true for the types of secretory products synthesized in different airway levels in different species. Only a few species have acidic and sulfated material throughout the airway tree. The methodology required to produce a sample of airway epithelium that has higher information content than that of randomly selected samples is not that complex. As outlined above, direct approaches can be used to define the exact position of the epithelium in relation to other portions of the conducting airway tree. These regions represent different microenvironments, both in terms of the cell populations composing the epithelium and in terms of the extracellular matrix and its cellular composition. Defining the biological activity of bronchiolar epithelium in relation to microenvironment is now a reality, making the types of information that can be derived from studies of these biological populations far more useful than was previously possible.

References

Barr, B. C., Hyde, D. M., Plopper, C. G., and Dungworth, D. L. (1988). Distal airway remodeling in rats chronically exposed to ozone. *Am. Rev. Respir. Dis.* **137**:924–938.

Barry, B. E., Miller, F. J., and Crapo, J.D. (1985). Effects of inhalation of 0.12 and 0.25 parts per million ozone on the proximal alveolar region of juvenile and adult rats. *Lab. Invest.* **53**:692–704.

Bond, J. A., Harkema, J. R., and Russel, V. (1988). Regional distribution of xenobiotic metabolizing enzymes in respiratory airways of dogs. *Drug Metab. Dispos.* **16**:116–124.

Boorman, G. A., Schwartz, L. W., and Dungworth, D. L. (1980). Pulmonary effects of prolonged ozone insult in rats: Morphometric evaluation of the central acinus. *Lab. Invest.* **43**:108–115.

Breeze, R. G., and Wheeldon, E. B. (1977). The cells of the pulmonary airways. *Am. Rev. Respir. Dis.* **116**:705–777.

Brody, A. R., Hill, L. H., Adkins, B., and O'Conner, R. W. (1981). Chrysotile asbestos inhalation in rats. Deposition pattern and reaction of alveolar epithelium and pulmonary macrophages. *Am. Rev. Respir. Dis.* **123**:670–679.

Castleman, W. L., Dungworth, D. L., and Tyler, W. S. (1975). Intrapulmonary airway morphology in three species of monkeys: A correlated scanning and transmission electron microscopic study. *Am. J. Anat.* **152**:107–122.

Castleman, W. L., Dungworth, D. L., Schwartz, L. W., and Tyler, W. S. (1980). Acute respiratory bronchiolitis: an ultrastructural and autoradiographical study of epithelial cell injury and renewal in Rhesus monkeys exposed to ozone. *Am. J. Pathol.* **98**:811–827.

Chang, L-Y., Graham, J. A., Miller, F. J., Ospital, J. J., and Crapo, J.D. (1986a). Effects of subchronic inhalation of low concentrations of nitrogen dioxide: I. The proximal alveolar region of juvenile and adult rats. *Toxicol. Appl. Pharmacol.* **83**:46–61.

Chang, L-Y., Mercer, R., and Crapo, J. D. (1986b). Differential distribution of brush cells in rat lung. *Anat. Rec.* **216**:49–54.

Chow, C. K., Plopper, C. G., and Dungworth, D. L. (1979). Influence of dietary vitamin E on the lungs of ozone-exposed rats: a correlated biochemical and histological study. *Environ. Res.* **20**:309–317.

Chow, C. K., Plopper, C. G., Chiu, M., and Dungworth, D. L. (1981). Dietary vitamin E and pulmonary biochemical and morphological alterations of rats exposed to 0.1 ppm ozone. *Environ. Res.* **24**:315–324.

Dungworth, D. L., Tyler, W. S., and Plopper, C. G. (1985). Morphologic methods for gross and microscopic pathology. In *Handbook of Experimental Pharmacology*. Edited by J. P. Witschi, and J. D. Brain. New York, Springer-Verlag, **75**:229–258.

Eustis, S. L., Schwartz, L. W., Kosch, P. C., and Dungworth, D. L. (1981). Chronic bronchiolitis in nonhuman primates after prolonged ozone exposure. *Am. J. Pathol.* **105**:121–137.

Evans, M. J., Cabral-Anderson, L. J., and Freeman, G. (1978). Role of the Clara cell in renewal of the bronchiolar epithelium. *Lab Invest.* **38**:648–655.

Fujinaka, L. E., Hyde, D. M., Plopper, C. G., Tyler, W. S., Dungworth, D. L., and Lollini, L. O. (1985). Respiratory bronchiolitis following long-term ozone exposure in bonnet monkeys: a morphometric study. *Exp. Lung Res.* **8**:167–190.

Heidsiek, J. G., Hyde, D. M., Plopper, C. G. and St. George, S. A. (1987). Quantitiative histochemistry of mucosubstances in the tracheal epithelium of the macaque monkey. *J. Histochem. Cytochem.* **35**:435–442.

Hyde, D. M., Plopper, C. G., Kass, P. H., and Alley, J. L. (1983). Estimation of cell numbers and volumes of bronchiolar epithelium during rabbit lung maturation. *Am. J. Anat.* **167**:359–370.

Hyde, D. M., Plopper, C. G., Weir, A. J., Murnane, R. D., Warren, D. L., and Last, J. A. (1985). Periabronchiolar fibrosis in lungs of cats chronically exposed to diesel exhaust. *Lab. Invest.* **52**(2):195–206.

Jeffery, P. K. (1983). Morphologic features of airway surface epithelial cells and glands. *Am. Rev. Respir. Dis.* **128**:S14–S20.

Lowrie, P. M., and Tyler, W. S. (1973). Selection and preparation of specific tissue regions for TEM using large epoxy-embedded blocks. *Proc. Annu. Meet. Electron Microsc. Soc. Am.* **148**:324–325.

Lum, H., Schwartz, L. W., Dungworth, D. L., and Tyler, W. S. (1978). A comparative study of all renewal after exposure to ozone or oxygen. Response of terminal bronchiolar epithelium in the rat. *Am. Rev. Respir. Dis.* **118**:335–345.

Mariassy, A. T., and Plopper, C. G. (1983). Tracheobronchial epithelium of the sheep: I. Quantitative light microscopic study of epithelial cell abundance and distribution. *Anat. Rec.* **205**:263–275.

Mariassy, A. T., and Plopper, C. G. (1984). Tracheobronchial epithelium of the sheep: II. Ultrastructural and morphometric analysis of the epithelial secretory cell types. *Anat. Rec.* **209**:523–534.

Mariassy, A. T., St. George, J. A., Nishio, S. J., and Plopper, C. G. (1988a). Tracheobronchial epithelium of the sheep: III. Carbohydrate histochemical and cytochemical characterization of secretory epithelial cells. *Anat. Rec.* **221**:540–549.

Mariassy, A. T., Plopper, C. G., St. George, J. A., and Wilson, D. W. (1988b). Tracheobronchial epithelium of the sheep: IV. Lectin histochemical characterization of secretory epithelial cells. *Anat. Rec.* **222**:49–59.

Moffatt, R. K., Hyde, D. M., Plopper, C. G., Tyler, W. S., and Putney, L. F. (1987). Ozone-induced adaptive and reactive cellular changes in respiratory bronchioles of bonnet monkeys. *Exp. Lung Res.* **12**:57–74.

Phalen, R. F., and Oldham, M. J. (1983). Tracheobronchial airway structure as revealed by casting techniques. *Am. Rev. Respir. Dis.* **128**:51–53.

Phalen, R. F., Yeh, H. D., Schum, G. M., and Raabe, O. G. (1978). Application of an idealized model to morphometry of the mammalian tracheobronchial tree. *Anat. Rec.* **190**:167–176.

Pinkerton, K. E., Plopper, C. G., Mercer, R. R., Roggli, V. L., Patra, A. L., Brody, A. R., and Crapo, J. D. (1986). Airway branching patterns influence asbestos fiber location and the extent of tissue injury in the pulmonary parenchyma. *Lab. Invest.* **55**(6):688–695.

Plopper, C. G., Dungworth, D. L., and Tyler, W. S. (1973). Pulmonary lesions in rats exposed to ozone: A correlated light and electron microscopic study. *Am. J. Pathol.* **71**:375–394.

Plopper, C. G., Chow, C. K., Dungworth, D. L., Brummer, M. G., and Nemeth, T. J. (1978). Effect of low level of ozone on rat lungs. 2. Morphological responses during recovery and re-exposure. *Exp. Mol. Pathol.* **29**:400–411.

Plopper, C. G., Dungworth, D. L., Tyler, W. S., and Chow, C. K. (1979). Pulmonary alterations in rats exposed to 0.2 and 0.1 ppm ozone: A correlated morphological and biochemical study. *Arch. Environ. Health* **34**(6):390–395.

Plopper, C. G., Mariassy, A. T., and Hill, L. H. (1980a). Ultrastructure of the nonciliated bronchiolar epithelial (Clara) cell of mammalian lung: I. A comparison of rabbit, guinea pig, rat, hamster and mouse. *Exp. Lung. Res.* **1**:139–154.

Plopper, C. G., Mariassy, A. T., and Hill, L. H. (1980b). Ultrastructure of the nonciliated bronchiolar epithelial (Clara) cell of mammalian lung: II. A comparison of horse, steer, sheep, dog, and cat. *Exp. Lung Res.* **1**:155–170.

Plopper, C. G., Hill, L. H., and Mariassy, A. T. (1980c). Ultrastructure of the nonciliated bronchiolar epithelial (Clara) cell of mammalian lung: III. A study of man with comparison of 15 mammalian species. *Exp. Lung Res.* **1**:171–180.

Plopper, C. G., Alley, J. L., Serabjit-Singh, C. J., and Philpot, R. M. (1983a). Cytodifferentiation of the nonciliated bronchiolar epithelial (Clara) cell during rabbit lung maturation: an ultrastructural and morphometric study. *Am. J. Anat.* **167**:329–357.

Plopper, C. G., Mariassy, A. T., and Lollini, L. O. (1983b). Structure as revealed by airway dissection: A comparison of mammalian lungs. *Am. Rev. Respir. Dis.* **128**:S4–S7.

Plopper, C. G., Halsebo, J. E., Berger, W. J., Sonstegard, K. S., and Nettesheim, P. (1983c). Distribution of nonciliated bronchiolar epitheliam (Clara) cells in intra- and extrapulmonary airways of the rabbit. *Exp. Lung Res.* **5**:79–98.

Plopper, C. G., Hyde, D. M., and Weir, A. J. (1983d). Centriacinar alterations in lungs of cats chronically exposed to diesel exhaust. *Lab. Invest.* **49**:391–399.

Plopper, C. G., St. George, J. A., Nishio, S. J., Etchison, J. R., and Nettesheim, P. (1984). Carbohydrate cytochemistry of tracheobronchial airway epithelium of the rabbit. *J. Histochem. Cytochem.* **32**:209–218.

Plopper, C. G., Cranz, D. L., Kemp, L., Serabjit-Singh, C. J., and Philpot, R. M. (1987). Immunohistochemical demonstration of cytochrome P-450 monooxygenase in Clara cells throughout the tracheobronchial airways of the rabbit. *Exp. Lung Res.* **13**:59–68.

Plopper, C. G., Heidsiek, J. G., Weir, A. J., St. George, J. A., and Hyde, D. M. (1989). Tracheobronchial epithelium in the adult rhesus monkey: a quantitative histochemical and ultrastructural study. *Am. J. Anat.* **184**:31–40.

St. George, J., Edmonson, S., Wong, V., Cranz, D., Weir, A., Wu, R., and Plopper, C. (1986). Control of differentiation of mucous cells from nonhuman primate respiratory airways. *J. Cell Biol.* **103**:200a

St. George, J. A., Wu, R., Neu, K., and Plopper, C. G. (1987). Growth and differentiation of fetal rhesus airway epithelium *in vitro. Am. Rev. Respir. Dis.* **135**:A364.

Stephens, R. J., and Evans, M. J. (1973). Selection and orientation of lung tissue for electron microscopy. *Environ. Res.* **6**:52–59.

Suverkropp, C., Plopper, C., and Buckpitt, A. (1988). Characterization of airway explants as a model for assessing pulmonary metabolism and toxicity of environmental chemicals. FASEB J. **2**(4):2935.

Tyler, N. K., and Plopper, C. G. (1985). Morphology of the distal conducting airways in rhesus monkey lungs. *Anat. Rec.* **211**:295–303.

Tyler, W. S. (1983). Comparative subgross anatomy of lungs: pleuras, interlobular septa, and distal airways. *Am. Rev. Respir. Dis.* **128**:S32–S36.

Tyler, W. S., Dungworth, D. L., Plopper, C. G., Hyde, D. M., and Tyler, N. K. (1985). Structural evaluation of the respiratory system. *Fundam. Appl. Toxicol.* **5**:405–422.

Tyler, N. K., Hyde, D. M., Hendrickx, A. G., and Plopper, C. G. (1988). Morphogenesis of the respiratory bronchiole in rhesus monkey lungs. *Am. J. Anat.* **182**:215–223.

Widdicombe, J. G., and Pack, R. J. (1982). The Clara cell. *Eur. J. Respir. Dis.* **63**:202–220.

Part Three

STRUCTURAL TECHNIQUE IN EXPERIMENTAL PATHOLOGY

19

Experimental Pulmonary Microsurgery

PETER H. BURRI

University of Berne
Berne, Switzerland

Only a few microsurgical interventions can be performed on the lungs of small animals. In this chapter we discuss three types of operations that have been used frequently in various species, but particularly in smaller mammals, among which the rat was the most commonly used. The three topics are the resection of lung tissue, the occlusion of the pulmonary artery, and the artificial creation of pulmonary atelectasis. Other surgical interventions that require the use of larger animals are not treated.

Although the small size of rats is technically a limiting factor, the investigation of small lungs also presents great advantages. The structure of the lung parenchyma of small mammals is very homogeneous. It is not complicated by the presence of marked gradients in alveolar or capillary size. This fact can greatly simplify the sampling problem in all investigations in which a statistical approach is needed, and may therefore improve the reliability of the findings.

I. Experimental Lung Tissue Resection

A. Historical Background

Attempts to remove parts of the lung and investigation of the consequences of such operations have a long history. According to a quotation by Murphy (1898), the first pneumonectomy was tried in dogs as far back as the year 1492 by Rolandus. Starting in the 1880s, a larger number of reports are available about successful and unsuccessful outcomes of pneumonectomies in various experimental animals, particularly dogs and rabbits (Block, 1881; Gluck, 1881, Schmid, 1881, Haasler, 1892). Although some authors were aware of Lister's work and operated using antiseptic precautions, the rate of success was usually low; many animals did not survive the operation or died within days or a few weeks due to infections or other complications. Therefore, as reported by Haasler (1892), these early experiments mainly addressed the feasibility of pulmonary surgery and usually no comments were made about any observed changes in the remaining lung tissue. After the turn of the century the techniques improved and the survival of operated animals began to reach promising rates. Simultaneously, interest began to focus on whether the lung was capable of any kind of compensatory growth, as was already known to occur in the kidney. Good reviews of the early experimental developments are given in articles by Kawamura (1914) and of Heuer and Dunn (1920; see footnotes).

B. Surgical Technique

General Description

The technique presented here is that used in rats of all ages in our own experiments on lung regeneration. We could not pretend that other approaches are less adequate. Usually, once a technique yields satisfactory results, further technical experimentation is avoided and one remains with what has given satisfaction. The operation consists of the so-called bilobectomy, that is, the removal of the upper and middle lobes of the right lung. It can easily be adapted to the resection of other lung portions, such as the whole left lung, which is the simpler and most frequently chosen approach in this type of experiments.

The animal in deep anesthesia is fixed in supine position onto the surgical board by means of rubber bands passed around its pasterns. The head is pulled upwards and held down in a similar way by passing a rubber band behind the incisors. After moistening the skin of the whole thorax with a commercially available saponaceous antiseptic solution, the area where the incision is to be made is shaved. The cleanest shave is achieved with a large scalpel blade. This method is by far superior to the use of any electric shaver. For adequate ventilation of the animal during surgery, the trachea is blindly cannulated through the mouth with a transparent and flexible polyethylene tubing of appropriate size (internal diameter approximately 0.5 mm) reinforced by a metal wire functioning as a mandrel. The latter gives the tube the necessary stiffness so that it can be adequately guided through the pharynx and larynx. Care must be taken that the tip of the tubing has no sharp edges and that the wire does not protrude. The wire should be rapidly removed, as soon as the tube is properly positioned, and the animal is allowed to breath through the cannula until the chest is opened. The skin is incised along the fifth rib and freed from the underlying muscle layer. Then the intercostal muscles of the fifth intercostal space are separated with forceps and the slit is kept open by a small retractor or by silk threads passed around the upper and lower ribs. As soon as the pleural cavity is entered, the intratracheal tube is connected to an animal respirator, usually of the small rodent type from Harvard Apparatus. The gas used can be medicinal compressed air or O_2 with 5% CO_2.

The respirator is turned on and the animal is moved under an operating microscope. A loop of silk thread (3/0) is then positioned on the thorax opening. One end is held by a clamp fixed and positioned on a large lump of plasticine. The other end remains free, so that it can be rapidly pulled. After respiration is stopped with the lung in a deflated position, the right upper lobe is pulled out. The best technique is to take a pair of fine but blunt and possibly rubber-padded forceps in each hand, and grasp the collapsed lung lobe at an edge, pull it delicately through the opening, alternating with the left and right forceps, and eviscerate it completely. The operating microscope with its direct illumination provides a good view of the operating site, even deep inside the thorax. The loop of silk thread automatically comes to lie around the hilum of the eviscerated lung portion. After the hilum is tied with a double knot, the lung tissue is cut off, the threads are severed, and the hilar stump is allowed to slip back into the thorax. Ventilation is restarted immediately, the intercostal and thorax muscles are carefully sutured with three to five stitches of 3/0 silk. The skin is then closed with either clips or discontinuous silk stitches. As soon as the chest wall is airtight the cannula is disconnected from the ventilator, and if the animal breathes normally, it is extubated.

Critical Steps

The Question of Sterility

The rat has an astonishing resistance to infection. Sterility, as needed for human and large animal surgery, is not necessary, and the wearing of gloves and masks is indicated only for the investigator's protection if the animals have been treated with infectious or radioactive agents. It is clear, however, that hands and instruments should be washed, preferably with a commercially available antiseptic solution. During the operation, the tips of the instruments are usually kept in one of these solutions.

Artificial Ventilation and Intubation

Artificial ventilation is a must for this operation. It can, however, be carried out with more or less sophistication. In our most recent experiments we have worked with the small rodent respirator from Harvard Apparatus. In the beginning, however, we used a very simple home-built device illustrated in Figure 1. The

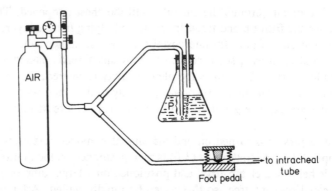

Figure 1 Schematic drawing of simple animal respirator. Maximal applied pressure P (cmH$_2$O) is controlled by water level in the Erlenmeyer flask. Pedal allows the operator to stop inflation. Deflation occurs by elastic recoil of lung and thorax because the intratracheal tube is never airtight.

Erlenmeyer flask allows one to adjust and monitor precisely the pressure applied to the lungs. With the pedal the respiration can be controlled by squeezing the tubing rhythmically. Gas mixtures of either controlled (medicinal) compressed air or 95% oxygen with 5% carbon dioxide were satisfactory. With pur O$_2$ we observed frequent focal atelectases, which could hardly be overcome by increasing the ventilatory pressure.

Simpler procedures for intubation are also possible. Rannels and co-workers (1979) simply inserted the tube delivering oxygen into the animal's mouth and rhythmically sealed the nose and mouth with thumb and forefinger. With intratracheal intubation, however, it is advisable before the start of the experiments to sacrifice a rat of similar size, expose its trachea, and select the tube size in accordance with the tracheal diameter. The intratracheal intubation has been described by Brody and co-workers (1977), who routinely exposed the trachea to control the positioning of the cannula. We found that it was possible, after some training, to guide the cannula into the trachea blindly (i.e., without tracheal exposure). Therefore, the tube should be shaped into a slight bend with a wire and the tongue pulled out gently with a pair of padded forceps. Whether the intubation is successful or not is indicated by a feeling of slight roughness (trachea) or smoothness (esophagus). The blind intubation is at least as difficult as the pneumonectomy or lobectomy itself. To minimize the extracorporeal dead space, the cannula must be kept as short as possible.

During the intubation, passage of the tubing through the larynx can be difficult and depends significantly on the depth of the anesthesia (reflex closure of the glottis). Pushing too hard at any time during the cannulation must be strictly avoided; lesions of pharynx, larynx, or trachea could ensue. There is also a danger of suffocating the animal if the tip of the cannula is pressed relentlessly against the closed glottic rim.

At any time during intubation, the formation of a meniscus of fluid or secretion in the cannula must be avoided. Premedication by subcutaneous injection of atropine sulfate 20–30 min before surgery can be used to decrease secretory activity.

Evisceration

Pulling out the lungs is a delicate procedure. Sqeezing or pulling the tissue too hard will lead to disruption of the lung accompanied by massive bleeding. Even though the ruptured lobe will ultimately be removed, rupture of the tissue usually leads to death of the animal. The airway stump is much more resistent to rupture. If it is pulled too strongly, however, the mediastinal structures will be distorted, particularly the thin walled veins, thus preventing influx of the blood into the heart. In several series of experiments (over 150 operations), this proved to be fatal twice, probably due to a reflex arrest of the heart.

Lung Reexpansion

To obtain reproducible conditions at the end of the surgery, it is advisable to inflate the lungs carefully before closing the chest wall. This is particularly important in experimental designs involving morphometric analyses performed soon after surgery.

Postoperative Treatment

The removal of one lung or two lobes is a short (15–20 min) but serious surgical intervention, and the question whether the animal should receive postoperative pain relief needs to be addressed. All available data show that rats tolerate the operation well, as long as the amount of lung tissue resected is less than 45–50% of the total lung volume. The limit of tolerance seems to be set not by the respiratory functions, but by the capacity of the pulmonary circulation to accommodate the surplus flow. The animals start to move and feed rapidly after surgery. After a period as short as 24 hr their body weights can match those of controls (Berger and Burri, 1985) or are only slightly lowered. Therefore, in striking the balance between the benefits of analgesic treatment and the need to handle the animals during the critical hours after surgery, most researchers consider that it is preferable to leave the animals undisturbed: At least no mention is made of analgesic treatment in most papers. In larger experimental animals medication can be given more easily without the need to handle the animal.

Since the rat's body temperature drops during anesthesia, keeping the animals warm after surgery is recommended. This can be performed very simply by lighting a bulb about 50 cm above the cage. During pneumonectomy of rabbits, Boatman and co-workers (1983) used heating pads throughout the surgical intervention.

Which Controls?

It is evident that the injuries created on the chest wall will affect thorax mechanics, particularly in the immediate postoperative period. To assess the impact of such disturbances on the lung, so-called sham-operated animals are needed as controls in addition to normal, unoperated animals. Sham surgery should include all the steps needed to gain access to the lungs from the skin incision and the opening of the chest wall to the ventilation of the lungs. It has been argued that sham opera-tion should only include skin and thorax incision, without entering the pleural cavity (Thurlbeck et al., 1981), because collapse of one lung was found to induce cellular multiplication (Simnett, 1974; Inselman et al., 1977) and weight increase (Tartter and Goss, 1973) in the contralateral lung. In our view, however, to differentiate the effects of thoracotomy, lung ventilation, and collapse from the effects of lung tissue resection alone, it appears essential that the sham-operated animals undergo all the procedures except tissue resection. Hence, two groups of controls should be provided: normal unoperated animals and sham-operated ones.

C. Alterations in the Remaining Lung

Introductory Remarks

It is well established that resection of pulmonary tissue, either by lobectomy or pneumonectomy, is followed by a volume increase in the remaining lung. It is also accepted that this volume gain corresponds to a real compensatory growth involving a burst of cell proliferation followed by an increase in weight of the remaining lung. Controversy remains, however, as to the extent of the "recovery" and the structural and functional results of the compensatory growth process. The mechanisms regulating the "regenerative" events are also still com-pletely obscure.

In the literature, the analysis of the sequelae of lung tissue resection in the remaining lung can be grouped into three types of approaches.

The biochemical approach usually measures the changes in the synthesis of DNA, RNA, and protein. In some investigations the biochemical analyses are combined with DNA labeling by tritiated thymidine.

The morphometric approach provides quantitative structural data. In our view it allows the best assessment of the final result of the regenerative process.

The morphological approach describes the structural alterations observed with light (LM), or transmission (TEM) and scanning electron microscopy (SEM).

The best understanding of the events would obviously be gained from an investigation combining all three approaches in a single experiment. This would guarantee the homogeneity of the experimental conditions and the comparability of the data. Until such a study is available, we must derive our understanding from data obtained under differing experimental conditions: strain of species, age and body weight of animals, surgical approach, anesthesia, localization and amount of lung tissue removed. This article cannot review all the experiments and do justice to all the data available on this topic. In an attempt to present a coherent synopsis of the postpneumonectomy tissue reaction in the remaining lung, we chose to consider only results obtained in rats. They represent by far the most comprehensive set of data available. Furthermore, data obtained in other species do not contradict the rat data; they merely deviate in timing and in the magnitude of the reaction.

Biochemical Events: Changes in DNA, RNA, and Protein Content

Following a left pneumonectomy in young rats (body weight 85 g), right lung dry weight remained constant during the first 2 days. On day 3 after the operation right lung dry weight started to increase and equalled the weight of both lungs of nonoperated animals on the 7th day after the operation (Rannels et al., 1984). The accumulation of DNA, RNA, and protein in the right lung matched closely the increase in lung weight. In an earlier study, Brody and co-workers (1978) found that DNA synthesis started to increase on postpneumonectomy day 3 and peaked on day 6. By means of autoradiography (Fig. 2) these authors found DNA to be labeled first in pleural mesothelial cells on the first day after pneumonectomy. The endothelial and interstitial cell labeling was highest on days 3 and 4, respectively. Finally, labeling of cells of the alveolar wall (mainly type 2 pneumocytes) was different in peripheral (peak on day 3) and more central alveoli (peak on day 6). Nijjar and Thurlbeck (1980) were able to demonstrate that prior to the enhanced cell proliferation, the activity of various enzymes was elevated in the remaining lung (Fig. 3). They found that adenylate cyclase, cyclic AMP phosphodiesterase, cyclic-AMP-dependent protein kinase, phosphorylase, and glucose-6-phosphatase activity was maximally elevated during the first 3 days after left pneumonectomy. The authors interpreted their findings as an indication of enhanced catabolism of glycogen to produce the glucose needed after pneumonectomy.

Morphometric Analysis

In a combined morphological and morphometric study of the time course of recovery, we were able to establish the sequence of histological events in the remaining lung after a right bilobectomy (resection of right upper and middle

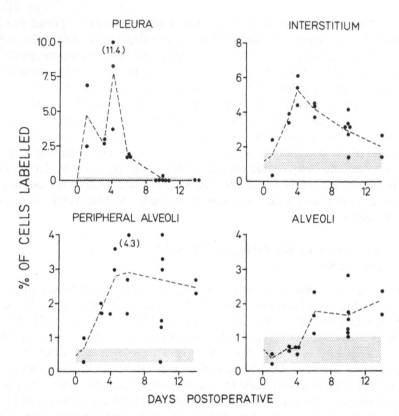

Figure 2 Time course of lung cell labeling with tritiated thymidine following left-sided pneumonectomy in rats aged 8–10 weeks. Values are expressed as percentage of cells in pleura (mesothelium only), interstitium, peripheral alveoli (adjacent to pleura), and alveolar walls containing four or more grains of [^{3}H]thymidine. Zero time is the control value, the shaded area represents the mean $\pm$ SE for the control value. Points represent values of individual animals (redrawn with permission from Brody et al., 1978).

lobes) in young rats (age 23 days). The bilobectomy represents a loss of 25% of total lung volume, unlike left pneumonectomy, which corresponds to a 35–37% volume reduction. The investigation consisted of an assessment of the morphological changes on LM, TEM, and SEM and a quantitative analysis at the LM and EM level of the volumetric alterations observed on days 1, 4, 6, 9, 12, 18, and 30 after surgery (Burri et al., 1982; Berger and Burri, 1985). From an earlier experiment performed on rats of the same age and strain and with identical techniques, we knew that 45 days after bilobectomy all investigated lung parameters were indiscernible from control and sham values (Burri and Sehovic,

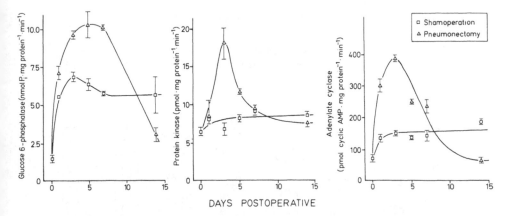

DAYS POSTOPERATIVE

Figure 3 Activity of various enzymes in the remaining lung of 6-weeks-old rats after left-sided pneumonectomy. The results represent means ± SE from three separate experiments (redrawn with permission from Nijjar and Thurlbeck, 1980).

1979). The main quantitative findings are presented in Figures 4 and 5 and can be summarized as follows: On the first day after surgery, the remaining lung had expanded to about seven-eights of the original total lung size. This was achieved mainly by overinflation of the remaining lung and, to a smaller extent, by an increase in capillary volume. While the tissue mass was unchanged, the airspace compartment was enlarged by 24% over that of the "reduced" lungs, which are defined as control lungs after subtraction of their upper and middle lobes. Three days later (day 4), this trend was even more pronounced. Due to dilatation of the air spaces, lung volume was back to control levels with no measurable change in tissue volume. On day 6, however, tissue volume was increased together with a massive augmentation in capillary volume. While relative air-space volume had regresssed, the total septal volume closely matched the control values. We were surprised to find, on days 6 and 9, that the volume densities of epithelial and endothelial cells and of the interstitial compartment were not significantly different from control lungs (Burri, unpublished observations). This means that 1 week after surgery tissue composition reflected a compensatory growth that was well balanced between compartments. In a quantitative ultrastructural study of pneumonectomized adult rats, Thet and Law (1984) observed also an equilibrated increase in the various lung compartments on postpneumonectomy day 7. Except for the type 2 pneumocytes, the volume increase amounted to between 22 and 27% (interstitial cells and matrix, 22%; endothelial cells, 24%, air spaces, 26%, type 1 pneumocytes, 27%).

Concomitantly with the tissue proliferation, the gas exchange surface areas returned to control level (Fig. 5). From day 6 onward, there were no significant

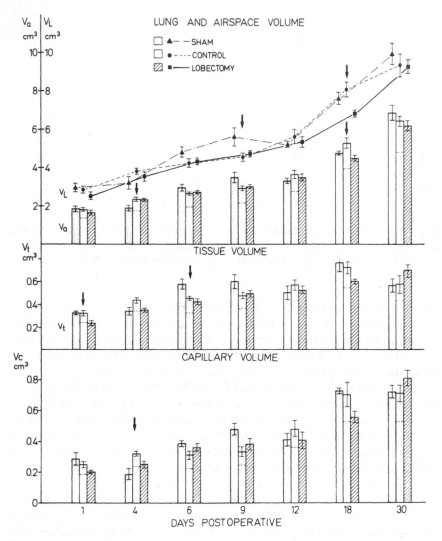

Figure 4 Time course of volume changes in various parameters in the remaining lung of 3-week-old rats after a bilobectomy. VL, lung volume; Va, volume of parenchymal airspaces; Vt, volume of parenchymal tissue; Vc, volume of capillaries. Columns or points represent group means, brackets include two standard errors. Arrows point to significant differences between groups in the Kruskal-Wallis test: $0.05 > p > 0.01$. The dotted line in columns of control groups indicates the mean for so-called "reduced lungs," that is, control lungs minus their right upper and middle lobes. The difference between lobectomy and reduced lung values documents the extent of compensatory growth. Volume recovery in lobectomy group is due to an increase in air-space volume on days 1 and 4. Tissue volume is augmented from day 6 onwards. Sham-operated rats show a depression of lung growth on day 4 followed by a rebound phenomenon on days 6 and 9 (redrawn with permission from Berger and Burri, 1985).

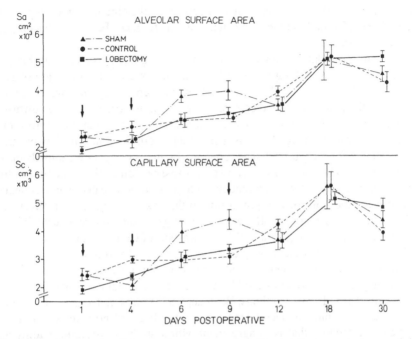

Figure 5 Time course of changes in alveolar and capillary surface area in the remaining lung of 3-week-old rats following a right-sided bilobectomy. Points are group means; brackets include two standard errors. Arrows point to significant differences between groups. In lobectomy-group, "recovery" is completed from day 6 onwards. Sham-group shows a growth depression on day 4 and a rebound effect on days 6 and 9 (redrawn with permission from Berger and Burri, 1985).

differences between operated and control animals except for day 18, when operated animals again had smaller lungs than controls (Fig. 4). We proposed that this deficit in lung size was due to an interference between the regenerative effort just produced and the normal growth of the lung. The control data indeed showed a growth spurt between days 9 and 18. It seemed plausible that the catch-up growth effort produced by the tissues of operated lungs could disrupt the timing of the normal growth phases and blunt the peak of an expected growth spurt. The slower growth obviously did not jeopardize the final size of the regenerated lung (at least for the period investigated), since results obtained on postoperative days 30 and 45 failed to reveal any significant differences between lobectomized and control animals.

From these data we conclude that the rat lung has a tremendous ability to make up for a loss of lung tissue. Within about 1 week after resection of 25% of the total lung volume, the dimensions of the remaining lung were restored to close to those of unoperated controls. The morphometrically derived diffusion capacity of the pulmonary parenchyma of operated rats was at the control level from day 6 onwards (Fig. 6). This means that, from the point of view of structure available for the gas exchange function, the lungs of operated animals match normal lungs. There is, however, a serious reservation to this statement. Morphometric analysis of the nonparenchymal lung portion revealed that the conductive airways did not grow in proportion to the gas exchange region (Burri and Sehovic, 1979). Six and a half weeks after surgery, the volume density of conducting airways in "regenerated" lungs was significantly smaller than in controls. This means that the airway tree did not match the volume increase of the whole lung,* which is a behavior already observed in the adaptive response of lung to hypoxia (Brody et al., 1977) and in the postpneumonectomy growth of ferrets (McBride, 1985). In rabbits, Boatman (1977) measured longer airway branches in casts of pneumonectomized lungs. He observed no changes in airway diameters, a finding that is in agreement with a study on pneumonectomy and lung mechanics by Yee and Hyatt (1983), but that contradicts results McBride (1985) obtained in pneumonectomized ferrets. The latter found that central airways showed increased cross-sectional areas. These different observations may reflect interspecies differences, but they make clear that for a realistic evaluation of pulmonary function of "regenerated" lungs the occurrence of ventilation-perfusion inequalities must be considered.

Morphology of Compensatory Growth

Observations in Scanning Electron Microscopy

The morphological observations made in SEM match closely the morphometric findings (Burri et al., 1982). On day 1 after surgery, the remaining lung appeared dilated (Fig. 7), particularly the alveolar ducts. Three days later, the dilatation had spread to the alveoli: all airspaces were clearly widened when compared with their normal counterparts. On day 6, the dilatation had regressed and on day 9

---

*In 1974 Green and co-workers coined the adjective "dysanaptic" to describe this unequilibrated growth. In fact, this term duplicates the expression of negative allometry or negative allometric growth, which perfectly describes the behavior of the bronchial tree with respect to lung volume.

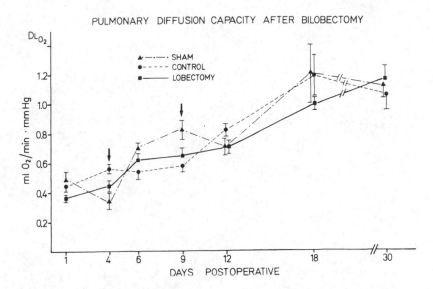

PULMONARY DIFFUSION CAPACITY AFTER BILOBECTOMY

Figure 6 "Structural" pulmonary diffusion capacity in 3-week-old bilobectomized rats calculated from the morphometric data using the model described by Weibel (1970/71). Symbols are as in Figure 5. Notice rapid recovery in lobectomy group and low and high values in sham-operated animals on days 4 and 9, respectively (reproduced with permission from Berger and Burri, 1985).

operated lungs could no longer be differentiated from controls in SEM pictures (Fig. 8).

Alveolization: Yes or No?

The above observations immediately raise the question of whether the "morphological recovery" is associated with the formation of new alveoli. Indeed, if the evident dilatation of air spaces disappears in a lung that has increased in volume (Fig 8), one is tempted to assume that formation of additional units must occur. This is all the more tempting if morphometry indicates a full compensation of the lost gas exchange surface area (Thurlbeck, 1975). There is, however, an alternative mode of compensatory growth that satisfactorily explains the observed morphological and quantitative structural changes. In a simplified two-dimensional lung model, we tried to demonstrate how, by dilatation of the airways followed by growth of the existing interalveolar walls, the morphology could apparently return to normal (Burri et al., 1982). Even quantitative measurements performed on the model failed to detect significant differences. It is obvious that the statistical assessment of relatively small changes in alveolar size will be more difficult on lung sections than in the model case, because of the irregularity of air-space size and the corrugation of the interalveolar septa.

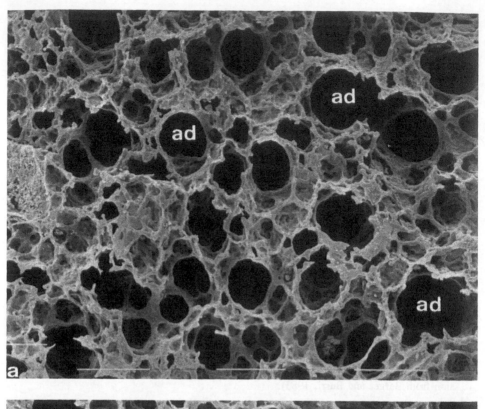

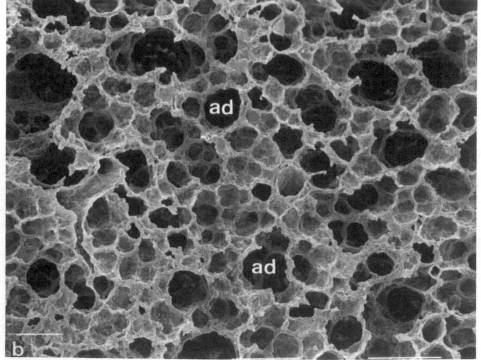

The question of whether the lung recovers by what was called "hyperplasia" (an inadequate term meaning formation of new alveoli) or by "hypertrophy" (enlargement of existing units) is almost as old as the problem of lung regeneration iself. Somehow, the opinion always prevailed that a recovery based on new units would be functionally more beneficial than one resulting from simple enlargement. To clarify this problem, alveolar counting has often been performed in the last two decades, but has failed to answer the question definitively. The application of seemingly adequate morphometric techniques yielded contradictory results (Sery et al., 1969; Buhain and Brody, 1973; Boatman, 1977; Langston et al., 1977). The reasons for these discrepant findings are multiple. The various reports dealt with different species and animals of different ages. The equations used to calculate alveolar number per unit lung volume call for knowledge of alveolar shape, which was often assumed to be constant; this is certainly an erroneous assumption for developing alveoli. The issue may further be complicated by the finding of Nattie and co-workers (1974) that subpleural alveoli may behave differently from deeper ones. The former were found to have proliferated but the latter simply to have dilated after a left-sided pneumonectomy. The three-dimensional reconstruction studies by Hansen and Ampaya (1974) finally demonstrated competently that large errors are made when ducts and alveoli have to be defined in planar sections. The introduction of improved sterological counting methods, such as the disector (Sterio, 1984) or the selector method (Cruz-Orive, 1987), may soon contribute to solving the problem. These techniques are basically bias-free. They rely upon the analysis of two parallel sections apart from each other. The application of these methods to this specific problem is highly welcome.

The problem of "regenerative" alveolar formation is also closely related to the mode of alveolization during lung development. We know that the prerequisite for the formation of alveoli is the presence of primitive (or immature) septa containing a double capillary network, as are found in the alveolar stage of lung development (Burri, 1974; Zeltner and Burri, 1987). Unless alveoli are formed by an alternative mode during "regeneration" (a very unlikely eventuality), the lung, in order to perform new alveoli, must still contain primitive septa. At 3 weeks of age, the lungs of rats look mostly mature: the originally thick septa appear slender and the double capillary system seems to have disappeared. In a study of the postnatal transformation of the capillary bed, we were able to show that during the third week the double capillary network has reduced to a single one over wide areas by a complex process involving both capillary fusions and

Figure 7 (a) Scanning electron micrograph of remaining lung of 3-week-old rat, 1 day after bilobectomy. Alveolar ducts are dilated and show rounded contours (190×; reproduced with permission from Burri et al., 1982). (b) Scanning electron micrograph of control lung at same magnification.

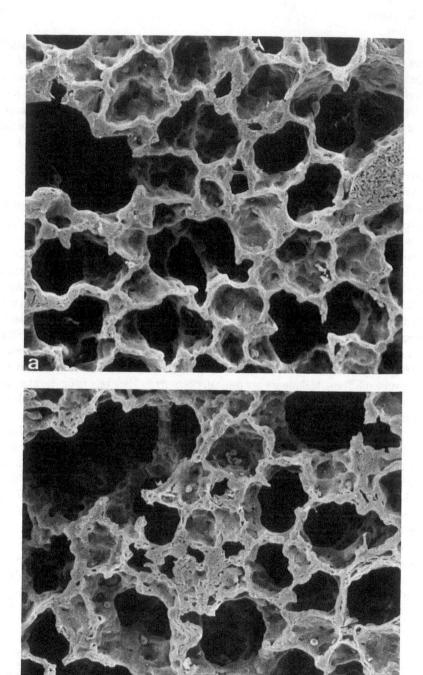

preferential growth (Caduff et al., 1986). This transformation to a single flat network occupying the whole width of the interalveolar septum is, however, never absolutely complete. Even in adult rat lungs, sites can be found where the septa or portions of a septum appear immature. This means that the morphological substrate for a minimal increase in alveolar number is still present in the adult rat. Hence, the idea that septal morphology could provide the answer to the problem fades away. A new concept was presented by Davies and co-workers (1982), who interpreted the lack of change in alveolar number 5 years after surgery in young and old pneumonectomized dogs: alveolar proliferation reported shortly after pneumonectomy could simply represent a speeding up of normal developmental patterns.

Lung Tissue Resection in Young Versus Adult Animals

On the grounds of general biological principles, one expects the catch-up growth to be more complete in young than in older animals. Indeed, there are numerous reports that the regenerative response is limited in adult animals (Bremer, 1936/37; Longacre and Johansmann, 1940; Buhain and Brody, 1973; Nattie et al., 1974; Wilcox et al., 1979; Holmes and Thurlbeck, 1979). Often the observed differences hinted at alveolar formation in the young and simple air-space dilatation in the old animals. An example of the biochemical recovery response in old (330 g) rats is given in Figure 9. The data are from the work of Rannels and co-workers (1979) and show a marked compensatory growth, which, within a week, however, does not reach the values of both sham lungs. In another study this group could show that restoration seems to be delayed progressively as the animals grow older. In animals of 193 g and 322 g average body weight, it took up to 14 days for full recovery. However, the rate of net gain in protein per day was similar in young and old rats. Thus, the longer period needed for recovery reflects the fact that more tissue was resected in larger rats. When net gain in protein was expressed per mg lung protein and hour, however, young rats proved to be twice as efficient as old animals. This is in accord with the decrease in lung growth rate described as occurring with age (Burri et al., 1974; Zeltner et al., 1987).

In a morphometric analysis of adult female rat lungs 45 days after bilobectomy, we observed a compensatory growth closely resembling that found in 3-week-old rats (Wandel et al., 1983). The absolute volumes of air spaces and capillary surface areas and the arithmetic and harmonic mean thicknesses of the air–blood barrier were identical in lung-resected and control animals. This result was very surprising. Between birth and three weeks rats are quadrupling their body weight and almost tripling their lung volume. Twenty-five percent of the lung volume (VL) at 3 weeks (VL $\sim$ 0.5 cm^3) represents a mere 8% of the lung volume 45 days later (VL $\sim$ 6.2 cm^3). In female adult rats, the situation is very different. They

Figure 8 Comparison of scanning electron micrographs of remaining lung of 3-week-old rat (a) 9 days after bilobectomy and (b) nonoperated control. No difference in airspace dimensions can be detected (350×).

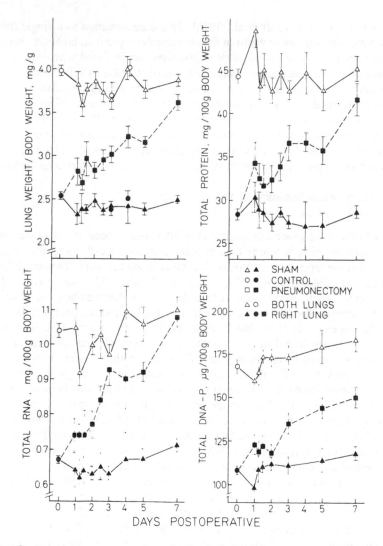

Figure 9 Incomplete recovery 1 week after left-sided pneumonectomy in older rats (320–330 g). Total lung RNA per 100 g body weight in pneumonectomized animals matches the value for both lungs in sham-operated animals; the other parameters do not yet reach control values (redrawn with permission from Rannels et al., 1979).

have only a relatively limited growth potential beyond 300 g body weight, despite the fact that rats are known to grow for almost as long as they live (Donaldson, 1924). In our study, the adult female rats gained only between 60 and 70 g during the experiment, that is, about 25–30% of their initial weight. From allometric growth curves for lung and body weight (Burri et al., 1974), it is possible to infer that during this time their lungs increased by about 20%. The lungs of the lobectomized animals, however, grew by more than 70%. This finding opposes the argument that compensatory growth is partly explained by the fact that the absolute amount of resected lung tissue in a young animal is negligible when compared to the adult lung volume. The similarity in the morphometric compensatory growth response in young and adult rats observed by Wandel and co-workers (1983) is backed by the findings of Watkins and co-workers (1985), who observed practically identical pattern in the changes of total protein, RNA, and DNA between younger (4 weeks, 82 g) and older rats (320 g).

In summary, it seems that, at least in the rat, compensatory lung growth is manifest even in the adult. Rodents may, however, represent a special case, because of the continuous growth of their pleural cavity (Bremer, 1936/37).

Sham Findings

In a large number of publications sham surgery was found to produce no response in the lung tissue. More recently, however, several authors reported changes induced by the preparative surgical manipulations. Brody and coauthors (1978) showed that entering the pleural cavity increased the labeling index of pleural cells to values at least as high as pneumonectomy on day 1 after surgery, but not later on. Labeling of lung cells proper, however, was not increased. The authors comment that pleural irritation was unlikely to be the cause, since, in a wax plombage experiment, no pleural labeling was observed. They suggested that the collapsing of the lung was likely to be responsible for the observation. It was reported that mere collapsing of one lung elicited a response on the contralateral side (Simnett, 1974; Inselman et al., 1977). Nijjar and Thurlbeck (1980) reported an increase in the activity of various enzymes, including adenylate cyclase, cyclic AMP, phosphodiesterase, and glucose-6-phosphatase in sham-operated rats (Fig. 3). Sham operation consisted only of skin incision with no entering of the pleural space. As the authors discussed, the observed augmentations lasted for the duration of the experiment and might therefore be attributed more to continued normal growth in the 6-week-old animals than to operative stress. No unoperated controls were used.

In our study on the time course of the compensatory growth (Berger and Burri, 1985), in which sham operation consisted of cannulation of the trachea, chest surgery, and collapse and ventilation of the lung, we observed significant differences in various morphometric parameters between sham and unoperated control animals. Air-space, tissue, and capillary volumes, as well as alveolar and capillary surface areas, were smaller on day 4 and larger on days 6 and 9 than those of control animals (Figs. 4, 5). On day 1 and after 12 days, sham results were within control range. These findings were interpreted as a brief and transitory depression of normal growth during the first few days afer surgery, followed by a rebound phenomenon. The latter propelled the morphometric diffusion capacity to values over 40% above those of control and bilobectomized animals for a brief period of time around day 9 (Fig. 6). One must therefore keep in mind that pneumonectomy or lobectomy findings are likely to be superimposed on a background of sham effects.

Control of Compensatory Response

A wide variety of hypotheses have been put forward to explain how the lung remnants could be triggered to compensate for a loss of lung mass. They have been reviewed by Thurlbeck (1975), Burri and Sehovic (1979), Davies and Reid (1982), and, more recently, by Rannels and Rannels (1988b). It would be beyond the scope of this article to discuss all the pros and cons for each of the proposed mechanisms. In the following, therefore, we simply list the factors mentioned and propose a selection of pertinent references that comment on or argue either in favor or against the corresponding factor: hypoxia/hypoxemia (Phillips et al., 1941; Nattie et al., 1974); functional load (Romanova et al., 1971); mechanical factors: intrathoracic space or lung distention (Addis, 1928; Hilber, 1934; Cohn, 1939; Charbon and Adams, 1952; Fisher and Simnett, 1973; Cowan and Crystal, 1975; Brody et al., 1978; Pecora and Hohenberger, 1979; Thurlbeck et al., 1981; Arnup et al., 1984); ventilation (Bartlett, 1972); increased blood flow through the remaining lung (Romanowa et al., 1971; Tartter and Goss, 1973); chalones (Bullough, 1965; Simnett et al., 1969; Fisher and Simnett, 1973); hormones (Brody and Buhain, 1973; Jain et al., 1973; Bennett et al., 1985, 1987; Rannels et al., 1986, 1987); somatomedins (Smith et al., 1980d; Sosenko et al., 1982; Stiles et al., 1983a; Thurlbeck et al., 1984; Faridy et al., 1988); various serum factors (Romanova et al., 1971; Romanova and Zhikhareva, 1972; Orlova et al., 1978; Smith et al., 1980a; Thurlbeck et al., 1981; Faridy et al., 1988); polyamines and the enzymes of polyamine metabolism (Pegg et al., 1970; Wiegand and Pegg, 1978; Tabor and Tabor 1984; Tatar-Kiss et al., 1984; Rannels et al., 1986; Bardocz et al., 1986); cyclic AMP (Nijjar and Thurlbeck, 1980).

We may summarize here that this highly complex problem of compensatory growth initiation and regulation is far from being solved and that most likely, as with the causes of cancer, a multitude of factors may be involved and interplay with the goal of maintaining lung tissue homeostasis.

D. Physiological Consequences of Lung Tissue Resection

It is clear that for practical technical reasons physiological studies on pneumonectomy's effects are mostly performed in dogs or other larger animal species.

Circulatory Adaptations

The first serious investigations of the pathophysiological effects of pneumonectomy on the circulation and respiration date back to the 1920s. Heuer and Dunn (1920) noted heart action while performing a left pneumonectomy in dogs. Ligation of pulmonary arteries and veins, division of the bronchus, and removal of the entire left lung had little or no effect on cardiac pulse and blood pressure. Cardiac rhythm was, however, very sensitive to traction on the lungs and bronchus and to the degree of lung expansion. Somewhat different observations were made by Andrus (1923). Within a few hours after surgery, total blood flow through the remaining lung was found to be greater than the total amount flowing through both lungs before the operation. Within 3 days blood flow began to decrease and on the 10th day reached the level flowing through both lungs before surgery. The pulse rate similarly increased rapidly after surgery to fall back to preoperative levels within 10 days. Since pulse rate decreased more than blood flow, Andrus deduced that blood output per beat increased by about 11 %, which is in accord with some observations of a right ventricular hypertrophy. In 1939 Edwards observed immediate changes in size of the pulmonary arteries and veins following either left or right pneumonectomies in rabbits. These vessels were found to dilate to practically double their size in the course of a few minutes and to remain in this state for the rest of the animal's life. Bronchial arteries also underwent a considerable increase in dimensions.

Since these early studies the acute effects of lung tissue resection on blood pressure and flow have been more clearly elucidated. As well as the traditional measurements of pulmonary vascular resistance, the changes in vascular impedance have received particular attention in more recent work (Pouleur et al., 1978; Lucas et al., 1983; Crouch et al., 1987). Vascular impedance is a function of frequency and is expressed as an impedance spectrum computed by Fourier analysis from pressure and flow data. It has the advantage over the traditional resistance measurements of taking into account the pulsatile flow characteristic of the pulmonary circulation. Although it must be expected that a loss of about 35–45 % of the lung tissue will seriously affect the hemodynamics of the pulmonary vascular bed, it is remarkable that in most experiments the pneumonectomized animals showed only minor changes in heart rate, left atrial pressure, aortic pressure, and cardiac output, while pulmonary arterial pressure and pulmonary vascular resistance were usually increased. In the long run, the higher workload on the right heart must be expected to show its effects. Right ventricular hypertrophy was found 5 years after a left-sided pneumonectomy performed in dogs at 6–10 weeks or at 1 year of age. In the older group, media hypertrophy was present mainly in the preacinar pulmonary arteries, while the young operated group paradoxically lacked structural changes in the pulmonary arteries (Davies et al., 1982). The authors have no explanation for this finding. The young operated animals were also found surprisingly to be more reactive to hypoxia than the older ones (Murray et al., 1986). The latter authors suggest that future studies of compensatory lung growth after pneumonectomy should also investigate the effects on right ventricular growth.

Changes in Respiratory Parameters

Heuer and Andrus (1922) found that after removal of one lung in dogs there was a temporary increase in alveolar CO_2, a fall in alveolar O_2, a slight increase in

CO_2 of the blood, and marked decrease of blood oxygenation for a period of about 30 days. More recently in the rat, however, the hypoxemia was absent or of such short duration that it was unlikely to be responsible for the regenerative response of the remaining lung (Nattie et al., 1974).

Changes in lunch mechanics and lung volumes are expected. It is generally found that total lung capacity (TLC) is almost completely restored, while the ratios of functional and residual capacities to TLC are increased following a left pneumonec- tomy (Ford et al., 1981). Lung compliance is found to be decreased and elastic recoil to be increased (Wilcox et al., 1979; Ford et al., 1981). The physiological measure- ments of pulmonary diffusion capacity show that the degree of compensatory adap- tation again depends on the age of the animals at surgery (Pimmel et al., 1981).

In several extensive clinical studies of patients undergoing lung tissue resec- tion, it was concluded that the operation is relatively well tolerated, particularly when it is performed in young patients (Schilling, 1965; Stiles et al., 1969; McBride et al., 1980; Laros, 1982; Laros and Westerman, 1987). Other studies support this statement (Eigen et al., 1976; Frenckner and Freyschuss, 1982). In one case a pa- tient who underwent left pneumonectomy at the age of 23 years had the right upper and middle lobes removed at the age of 54; he survived and lived an active life with just his right lower lobe (Judd et al., 1985).

II. Other Pulmonary Microsurgical Techniques

Very few other surgical interventions can be performed on the lungs of small mam- mals. Occlusion of a pulmonary artery and experimental atelectasis of a single lobe or of one lung have not infrequently been performed on rats. These will be discuss- ed here briefly.

The experimental creation (and repair) of diaphragmatic hernia and its clinical implications, however, represents a broad subject that is beyond the scope of this chapter. The operation is usually performed in fetal lambs with the ∩im of in- vestigating the role of intrathoracic space for lung development. The different surgical techniques are described in several articles (de Lorimier et al., 1967; Harrison et al., 1980; Soper et al., 1984; Pringle, 1984) and the clinical implications have been reviewed recently in detail by Cullen and co-workers (1985).

A. Occlusion of Pulmonary Artery

One of the primary reasons to experiment with pulmonary artery occlusion was to model the postpneumonectomy state in patients before surgery (Carlens et al., 1951; Brofman et al., 1957; Jezek, 1970; Olsen et al., 1975) or to simulate the clinical symptoms of pulmonary embolism and study its effect on the circulatory system and/or on the mechanical, ventilatory, and metabolic lung functions (Morgan and Edmunds, 1967; Massaro et al., 1971). In other cases, the operation was perform- ed to determine the influence of blood flow alterations on lung development (Pickard et al., 1979; Thomasson et al., 1979; Haworth et al., 1981a,b) or compensatory growth processes (Romanova et al., 1971; Tartter and Goss, 1973).

In principle, the pulmonary artery can be occluded using two very different approaches: intraarterial occlusion by a balloon catheter or ligation of the pulmonary artery.

The former procedure was advocated many years ago to investigate preoperatively the pulmonary function in patients undergoing lung tissue resection. It was first tested in dogs and successfully applied in three awake patients who felt no discomfort during the procedure (Carlens et al., 1951). The practicability of this technique was confirmed later in several studies of more than 100 patients (Uggla, 1956; Brofman et al., 1957). This technique obviously can only be used in larger animals and does not directly involve microsurgery. In patients, it is difficult and time-consuming and has therefore fallen into disuse (Olsen, 1988).

Ligation of the pulmonary artery is the alternative approach. In small animals, it is a difficult operation, however, and Romanova and co-workers (1971) define it as "technically more complex than left-sided pneumonectomy." If it must be performed in rats, they recommend animals weighing at least 250–300 g. In rats of this size, the surgical technique for ligating the right pulmonary artery is briefly described by Tartter and Goss. With the animal receiving assisted ventilation, the pleural cavity is entered between the fourth and fifth ribs adjacent to the sternum. After exposure of the junction between the pulmonary vein, the superior vena cava, and the main bronchus, curved forceps are passed under the artery to grasp a 3/0 silk thread. The vessel is then tied and the chest wall closed. Most frequently, however, the operation was performed in larger species: rabbits, dogs, and cattle.

It is worth mentioning here that pulmonary arterial ligation can successfully be performed in sheep and even in rabbit (!) fetuses. In sheep, the technique consisted of "marsupializing" the fetal thorax skin to the edges of the hysterotomy (Pickard et al., 1979), and in rabbits of cauterizing the amniotic membranes to the fetal skin after hysterotomy (Thomasson and Ravitch, 1969; Thomasson et al., 1979). In both cases the goal of this particular procedure was to avoid leakage of amniotic fluid.

B. Experimental Pulmonary Atelectasis

As with pulmonary artery occlusion, there are two different approaches to collapsing a lung. Atelectasis can be achieved by so-called plombage of the pleural cavity with some inert material such as oil, paraffin, wax, plastic, silicon rubber, or even sponge (Addis, 1928, Cohn 1939; Charbon and Adams, 1952; Fisher and Simnett, 1973; Cowan and Crystal, 1975). The second method is by obstruction of a major airway, by ligature (Tartter and Goss, 1973), by injection of a solidifying plastic into the airways (Inselman et al., 1977), or in larger animals, by bronchoscopically painting the inside of the bronchus with some caustic agent, such as a silver nitrate solution (Loosli et al., 1949).

The plombage technique was once popular for the management of tuberculosis (Olsen, 1988). Experimentally, it has often been used either alone or in combination with pneumonectomy in attempts to investigate the role of intrathoracic space in postpneumonectomy compensatory growth (Addis, 1928; Cohn, 1939; Tartter and Goss, 1973; Cowan and Crystal, 1975). The technique for introducing the material into the pleural cavity can be more or less sophisticated. The material is either simply injected through the thorax wall or introduced through a thoracotomy opening. The choice of the method is determined by the material and also by the desire to have visual control of the operation.

Airway obstruction by ligature of the bronchus necessitates a surgical approach comparable to that for pneumonectomy or pulmonary artery ligation.

The injection of a liquid plastic into a stem bronchus of young rats is described by Inselman and co-workers (1977). It is probably the method most likely to result in postoperative infection. The authors therefore routinely treated the animals with procaine penicillin.

III. Outlook

From the selection of microsurgial techniques presented in this chapter, the resection of lung tissue is by far the most popular operation. In comparison, the other interventions are of minor importance. All the procedures can be used to investigate what factors are controlling the compensatory growth.

Thanks to the large number of careful biochemical or quantitative structural studies performed in the last 10–20 years, the timetable of the compensatory growth response in the remaining lung is now fairly well established. From the wealth of data we can crystallize the following facts:

The organism senses that lung tissue has been removed.

Blood flow through the remaining lung is increased.

Synthesis of RNA, DNA, and protein are activated in the remaining lung tissue.

Mitotic activity is turned on in the various cell types. It peaks in the following order: pleural mesothelial cells on day 1, interstitial and endothelial cells on day 3, and type 2 pneumocytes on day 6 after surgery.

The additional tissue mass is arranged in an orderly manner into normal lung structures.

The activated catch-up growth is rapid and continues until lung replacement is complete. Then it is shut down.

With a little speculation, we can deduce that the only way to control organ or tissue homeostasis so tightly in a paired organ is initially by humoral factors.

Nervous control can hardly play a role, since nerves are practically nonexistent at the alveolar level. This means that either directly or indirectly, the link to the cells of the remaining lung is humoral. This understanding clearly shows the route for further experimentation, which is already being used by a number of investigators. In 1972 Romanova and Zhikhareva found that after left pneumonectomy in pregnant rats the lungs of the fetuses grew at a faster rate and showed increased cell labeling. They suggested that compensatory growth was associated with the release into the circulation of organ-specific factors stimulating cell proliferation. In 1980 Smith and co-workers postulated the presence of some kind of growth-initiating factor(s) in the serum of pneumonectomized animals and tested serum of operated animals on alveolar type 2 cells in vitro. They found that this serum, obtained 9 and 21 days after surgery, enhanced the incorporation of tritiated thymidine into the DNA of the cultured pneumocytes, but not into the DNA of skin fibroblasts. Late after the operation (140 days) the serum no longer elicited the response. The authors first proposed the hypothetical factor to be a somatomedin or somatomedinlike peptide. On the basis of additional data, they later revised their statement. The observed postpneumonectomy factor was unlikely to correspond to somatomedin-C (Thurlbeck et al., 1984). It remains to be identified.

The experiments of Rannel's group are interesting as well. They demonstrated that type 2 pneumocytes prepared on the sixth day after surgery from pneumonectomized rats exhibited elevated rates of polyamine uptake and, after 24 hr in culture, an increased DNA and protein content. The cells were also larger than those obtained form nonoperated controls (Rannel and Rannels, 1988a). Furthermore, under the influence of serum from pneumonectomized rats they increased the uptake of exogenous spermidine while cells from control animals did not (Rannels and Rannels, 1987).

The interlude with the hypothetical somatomedin illustrates some of the difficulties encountered in this type of experimentation. The testing of full serum is still relatively crude, but the direction is clear. It is a matter of time until the use of modern methods in biochemistry and cellular and molecular biology will provide new insights into the mechanisms of compensatory growth. The lung is certainly an excellent model for such studies.

References

Addis, T. (1928). Compensatory hypertrophy of the lung after unilateral pneumectomy. *J. Exp. Med.* **47**:51–56.

Andrus, W. D. W. (1923). Observations on the total lung volume and blood flow following lobectomy or pneumonectomy. *Bull. Johns Hopkins Hosp.* **34**:119–121.

Arnup, M. E., Greville, H. W., Oppenheimer, L., Mink, S.N., and Antonisen, N. R. (1984). Dynamic lung function in dogs with compensatory lung growth. *J. Appl. Physiol.* **57**:1569–1576.

Bardocz, S., Tatar-Kiss, S., and Kertai, P. (1986). The effect of alpha-difluoromethylornithine on ornithine decarboxylase activity in compensatory growth of mouse lung. *Acta Biochim. Biophys. Hung.* **21**:59–65.

Bartlett, D. Jr. (1972). Postnatal development of the mammalian lung. In *Regulation of Organ and Tissue Growth*. Edited by R. J. Goss. New York, Academic Press, pp. 197–209.

Bennett, R. A., Colony, P. C., Addison, J. L., and Rannels, D. E. (1985). Effects of prior adrenalectomy on postpneumonectomy lung growth in the rat. *Am. J. Physiol.* **248**:E70–E74.

Bennett, R. A., Addison, J. L. and Rannels, D. E. (1987). Static mechanical properties of lungs from adrenalectomized pneumonectomized rats. *Am. J. Physiol.* **253**:E6–11.

Berger, L. C., and Burri, P. H. (1985). Timing of the quantitative recovery in the regenerating rat lung. *Am. Rev. Respir. Dis.* **132**:777–783.

Block, (1881). Experimentelles zur Lungenresection. *Dtsch. Med. Wochenschr.* **VII**:634–636.

Boatman, E. S. (1977). A morphometric and morphological study of the lungs of rabbits after unilateral pneumonectomy. *Thorax* **32**:406–417.

Boatman, E. S., Ward, G., and Martin, C. J. (1983). Morphometric changes in rabbit lungs before and after pneumonectomy and exposure to ozone. *J.Appl Physiol.* **54**:778–784.

Bremer, J. L. (1936). The fate of the remaining lung tissue after lobectomy or pneumonectomy. *J. Thorac. Surg.* **6**:336–343.

Brody, J. S., and Buhain, W. J. (1973). Hormonal influence on post-pneumonectomy lung growth in the rat. *Respir. Physiol.* **19**:344–355.

Brody, J., Lahiri, S. Simpser, M., Motoyama, E., and Velasquez, T. (1977). Lung elasticity and airway dynamics in Peruvian natives to high altitude. *J. Appl. Physiol.* **42**:245–251.

Brody, J. S., Bürki, R., and Kaplan, N. (1978). Deoxyribonucleic acid synthesis in lung cells during compensatory lung growth after pneumonectomy. *Am. Rev. Respir. Dis.* **117**:307–316.

Brofman, B. L., Charms, B. L., Kohn, P. M., Elder, J., Newman, R., and Rizika, M. (1957). Unilateral pulmonary artery occlusion in man. Control studies. *J. Thorac. Surg.* **34**:206–227.

Buhain, W. J., and Brody, J. S. (1973). Compensatory growth of the lung following pneumonectomy. *J. Appl. Physiol.* **35**:898–902.

Bullough, W. S. (1965). Mitotic and functional homeostasis: A speculative review. *Cancer Res.* **25**:1683–1727.

Burri, P. H. (1974). The postnatal growth of the rat lung. III. Morphology. *Anat. Rec.* **180**:77–98.

Burri, P. H., and Sehovic, S. (1979). The adaptive response of the rat lung after bilobectomy. *Am. Rev. Respir. Dis.* **119**:769–777.

Burri, P. H., Dbaly, J., and Weibel, E. R. (1974). The postnatal growth of the rat lung. I. Morphometry. *Anat. Rec.* **178**:711–730.

Burri, P. H., Pfrunder, B., and Berger, L. C. (1982). Reactive changes in pulmonary parenchyma after bilobectomy: a scanning electron microscopic investigation. *Exp. Lung Res.* **4**:11–28.

Caduff, J. H., Fischer, L. C., and Burri, P. H. (1986). Scanning electron microscopic study of the developing microvasculature in the postnatal rat lung. *Anat. Rec.* **216**:154–164.

Carlens, E., Hanson, E., Nordenstroem, H., and Nordenstroem, B. (1951). Temporary unilateral occlusion of the pulmonary artery. A new method of determining separate lung function and of radiologic examinations. *J. Thorac. Surg.* **22**:527–536.

Charbon, B. C., and Adams, W. E. (1952). A study to determine the effect of prevention of overdistention of the remaining lung tissue on the elevated right ventricular pressures, following the resection of lung tissue in dogs. *J. Thorac. Surg.* **23**:341–347.

Cohn, R. (1939). Factors affecting the postnatal growth of the lung. *Anat. Rec.* **75**:195–205.

Cowan, M. J., and Crystal, R. G. (1975). Lung growth after unilateral pneumonectomy: quantitation of collagen synthesis and content. *Am. Rev. Respir. Dis.* **111**:267–277.

Crouch, J. D., Lucas, C. L., Keagy, B. A., Wilcox, B. R., and Ha, B. (1987). The acute effects of pneumonectomy on pulmonary vascular impedance in the dog. *Ann. Thorac. Surg.* **43**:613–616.

Cruz-Orive, L. M. (1987). Arbitrary particles can be counted using a disector of unknown thickness: the selector. *J.Microsc.* **145**:121–142.

Cullen, M. L., Klein, M. D., and Philippart, A. I. (1985). Congenital diaphragmatic hernia. *Surg. Clin. North. Am.* **65**:1115–1138.

Davies, P., and Reid, L. (1982). Developmental constraints in compensatory postnatal growth of the lung. *J. Dev. Physiol.* **4**:265–272.

Davies, P., McBride, J., Murray, G. F., Wilcox, B. R., Shallal, J. A., and Reid, L. (1982). Structural changes in the canine lung and pulmonary arteries after pneumonectomy. *J. Appl. Physiol.* **53**:859–864.

DeLorimier, A. A., Tierney, D. F., and Parker, H. R. (1967). Hypoplastic lungs in fetal lambs with surgically produced congenital diaphragmatic hernia. *Surgery* **62**:12–17.

Donaldson, H. H. (1924). *The Rat*, 2nd ed. Memoirs of the Wistar Institute, No. 6, Philadelphia.

Edwards, F. R. (1939). Studies in pneumonectomy and the development of a two-stage operation for the removal of a whole lung. *Br. J. Surg.* **27**:392–413.

Eigen, H., Lemen, R. J., and Waring, W. W. (1976). Congenital lobar emphysema: long-term evaluation of surgically and conservatively treated children. *Am. Rev. Respir. Dis.* **113**:823.

Faridy, E. E., Sanij, M. R., and Thliveris, J. A. (1988). Influence of maternal pneumonectomy on fetal lung growth. *Respir. Physiol.* **72**:195–210.

Fisher, J. M., and Simnett, J. D. (1973). Morphogenetic and proliferative changes in the regenerating lung of the rat. *Anat. Rec.* **176**:389–396.

Ford, G. T., Galaugher, W., Forkert, L., Fleetham, J. A., Thurlbeck, W. M., and Anthonisen, N. R. (1981). Static lung function in puppies after pneumonectomy. *J. Appl. Physiol.* **50**:1146–1150.

Frenckner, B., and Freyschuss, U. (1982). Pulmonary function after lobectomy for congenital lobar emphysema and congenital cystic adenomatoid malformation. A follow-up study. *Scand. J. Thorac. Cardiovasc. Surg.* **16**:293–298.

Gluck, T. (1881). Experimenteller Beitrag zur Frage der Lungenexstirpation. *Berl. Klin. Wochenschr.* **18**:645–648.

Green, M., Mead, J., and Turner, J. M. (1974). Variability of maximum expiratory flow–volume curves. *J. Appl. Physiol.* **37**:67–74.

Haasler, F. (1892). Ueber compensatorische Hypertrophie der Lunge. *Virchows Arch. Pathol. Anat. Physiol.* **128**:527–536.

Hansen, J. E., and Ampaya, E. P. (1974). Lung morphometry: a fallacy in the use of the counting principle. *J. Physiol.* **37**:951–954.

Harrison, M. R., Jester, J. A., and Ross, N. A. (1980). Correction of congenital diaphragmatic hernia in utero. I. The model: intrathoracic balloon produces fatal pulmonary hypoplasia. *Surgery* **88**:174–182.

Haworth, S. G., de Leval, M., and Macartney, F. J. (1981a). Hypoperfusion and hyperperfusion in the immature lung. Pulmonary arterial development following ligation of the left pulmonary artery in the newborn pig. *J. Thorac. Cardiovasc. Surg.* **82**:281–292.

Haworth, S. G., McKenzie, S. A., and Fitzpatrick, M. L. (1981b). Alveolar development after ligation of left pulmonary artery in newborn pig: clinical relevance to unilateral pulmonary artery. *Thorax* **36**:938–943.

Heuer, G. J., and Dunn, G. R. (1920). Experimental pneumectomy. *Bull. Johns Hopkins Hosp.* **31**:31–42.

Heuer, G. J., and Andrus, W. D. W. (1922). The alveolar and blood gas changes following pneumectomy. *Bull. Johns Hopkins Hosp.* **33**:130–134.

Hilber, H. (1934). Experimenteller Nachweis des formativen Einflusses der Atemluft auf regenerierende Rattenlungen. *Morphol. Jahrb.* **74**:171–220.

Holmes, C., and Thurlbeck, W. M. (1979). Normal lung growth and response after pneumonectomy in rats at various ages. *Am. Rev. Respir. Dis.* **120**:1125–1136.

Inselman, L. S., Mellins, R. B., and Brasel, J. A. (1977). Effect of lung collapse on compensatory lung growth. *J. Appl. Physiol. Exercise Physiol.* **43**(1):27–31.

Jain, B. P., Brody, J.S., and Fisher, A. B. (1973). The small lung of hypopituitarism. *Am. Rev. Respir. Dis.* **108**:49–55.

Jezek, V. (1970). Pulmonary haemodynamics and blood gases during unilateral pulmonary artery occlusion and after lung resection. *Bull. Physiopathol. Respir.* **6**:255–264.

Judd, D. R., Vincent, K. S., Kinsella, P. W., and Gardner, M. (1985). Long-term survival with the right lower lobe as the only lung tissue. *Ann. Thorac. Surg.* **40**:623–624.

Kawamura, K. (1914). Experimentelle Studien über die Lungenexstirpation. *Dtsch. Z. Chir.* **131**:189–222.

Langston, C., Sachdeva, P., Cowan, M. J., Haines, J., Crystal, R. G., and Thurlbeck, W. M. (1977). Alveolar multiplication in the contralateral lung after unilateral pneumonectomy in the rabbit. *Am. Rev. Respir. Dis.* **115**:7–13.

Laros, C. D. (1982). Lung function data on 123 persons followed up for 20 years after total pneumonectomy. *Respiration* **43**:81–87.

Laros, C. D., and Westermann, C. J. (1987). Dilatation, compensatory growth, or both after pneumonectomy during childhood and adolescence. A thirty-year follow-up study. *J. Thorac. Cardiovasc. Surg.* **93**:570–576.

Longacre, J. J., and Johansmann, R. (1940). An experimental study of the fate of the remaining lung following total pneumonectomy. *J. Thorac. Surg.* **10**:131–149.

Loosli, C. G., Adams, W. E., and Thornton, T. M. Jr. (1949) The histology of the dog's lung following experimental collapse. *Anat. Rec.* **105**:697–721.

Lucas, C. L., Murray, G. F., Wilcox, B. R., and Shallal, J.A. (1983). Effects of pneumonectomy on pulmonary input impedance. *Surgery* **94**:807–816.

Massaro, D., Weiss, H., and White, G. (1971). Protein synthesis by lung following pulmonary artery ligation. *J. Appl. Physiol.* **31**:8–14.

McBride, J. T. (1985). Postpneumonectomy airway growth in the ferret. *J. Appl. Physiol.* **58**(3):1010–1014.

McBride, J. T., Wohl, M. E., Strieder, D. J., Jackson, A. C., Morton, J. R., Zwerdling, R. G., Griscom, N. T., Treves, S., Williams, A. J., and Schuster, S. (1980). Lung growth and airway function after lobectomy in infancy for congenital lobar emphysema. *J. Clin. Invest.* **66**:962–970.

Morgan, T. E., and Edmunds, L. H. (1967). Pulmonary artery occlusion. III. Biochemical alterations. *J. Appl. Physiol.* **22**:1012–1016.

Murphy, J. B. (1898). Surgery of the lung. Five of the nine animals operated upon died of sepsis. *J. A. M. A.* **31**:151.

Murray, G. F., Lucas, C. L., Wilcox, B. R., and Shallal, J. A. (1986). Cardiopulmonary hypoxic response 5 years postpneumonectomy in beagles. *J. Surg. Res.* **41**:236–244.

Nattie, E. E., Wiley, C. W., and Bartlett, D. Jr. (1974). Adaptive growth of the lung following pneumonectomy in rats. *J. Appl. Physiol.* **37**:491–495.

Nijjar, M. S., and Thurlbeck, W. M. (1980). Alterations in enzymes related to adenosine 3', 5'-monophosphate during compensatory growth of rat lung. *Eur. J. Biochem.* **105**:403–407.

Olsen, G. N. (1988). Pre- and postoperative evaluation and management of the thoracic surgical patient. In *Pulmonary Diseases and Disorders*. Edited by A. P. Fishman, 2nd ed., vol. 3. New York, McGraw-Hill, pp. 2413–2432.

Olsen, G. N., Block, A. J., Swenson, E. W., Castle, J. R., and Wynne, J. W. (1975). Pulmonary function evaluation of the lung resection candidate: a prospective study. *Am. Rev. Respir. Dis.* **111**:379–387.

Orlova, I. I., Lisatova, N. G., and Mikhal'chenko, S. D. (1978). Action of allogeneic serum from a pregnant pneumonectomized rat on embryonic lung tissue. *Biull. Eksp. Biol. Med.* **86**:84–87.

Pecora, D. V., and Hohenberger, M. (1979). Effects of postpneumonectomy distention on pulmonary compliance and vascular resistance. *Am. Surg.* **45**:797–801.

Pegg, A. E., Lockwood, D. H., and Williams-Ashman, H. G. (1970). Concentrations of putrescine and polyamines and their enzymic synthesis during androgen-induced prostatic growth. *Biochem. J.* **117**:17–31.

Phillips, F. J., Adams, W. E., and Hrdina, L. S. (1941). Physiologic adjustment in sublethal reduction of lung capacity in dogs. *Surgery* **9**:25–39.

Pickard, L. R., Tepas, J. J. III, Inon, A., Hutchins, G. M., Shermeta, D. W., and Haller, J. A. Jr. (1979). Effect of pulmonary artery ligation on the developing fetal lung. *Am. Surg.* **45**:793–796.

Pimmel, R. L., Friedman, M., Murray, G. F., Wilcox, B. R., and Bromberg, P. A. (1981). Forced oscillatory resistance and compliance parameters following pneumonectomy in beagle dogs. *Respiration* **41**:17–24.

Pouleur, H., Lefevre, J., van Eyll, Ch., Jaumin, P. M., and Charlier, A. A. (1978). Significance of pulmonary input impedance in right ventricular performance. *Cardiovasc. Res.* **12**:617–629.

Pringle, K. C. (1984). Fetal lamb and fetal lamb lung growth following creation and repair of a diaphragmatic hernia. In *Animal Models in Fetal Medicine*. Edited by P. W. Nathanielsz. Perinatology Press, Chapter 6, pp. 109–148.

Rannels, D. E., and Rannels, S. R. (1987). Acute changes in pulmonary uptake of polyamines following partial pneumonectomy. *Chest* **91**:25S–26S.

Rannels, S. R., and Rannels, D. E. (1980a). Alterations in type II pneumocytes cultures after partical pneumonectomy. *Am. J. Physiol.* **254**:C684–C690.

Rannels, D. E., and Rannels, S. R. (1988b). Compensatory growth of the lung following partial pneumonectomy. Minireview. *Exp. Lung Res.* **14**:157–182.

Rannels, D. E., White, D. M., and Watkins, C. A. (1979). Rapidity of compensatory lung growth following pneumonectomy in adult rats. *J. Appl. Physiol.* **46**(2):326–333.

Rannels, D. E., Burkhart, L. R., and Watkins, C. A. (1984). Effect of age on the accumulation of lung protein following unilateral pneumonectomy in rats. *Growth* **48**:297–308.

Rannels, D. E., Addison, J. L., and Bennett, R. A. (1986). Increased pulmonary uptake of exogenous polyamines after unilateral pneumonectomy. *Am. J. Physiol.* **250**:E435–E440.

Rannels, D. E., Karl, H. W., and Bennett, R. A. (1987). Control of compensatory lung growth by adrenal hormones. *Am. J. Physiol.* **253**:E343–E348.

Romanova, L. K., and Zhikhareva, I. A. (1972). On the humoral regulation of restorative growth in the lungs, kidneys, and liver. *Biull. Eksp. Biol.* **73**:84–87.

Romanova, L. K., Leikina, E. M., Antipova, K. K., and Sokolova, T. N. (1971). The role of function in the restoration of damaged viscera. *Ontogenez* **2**:479–486.

Schilling, J. A. (1965). Pulmonary resection and sequelae of thoracic surgery. In *Handbook of Physiology* , Sect. 3, Vol II, *Respiration*. Washington, D.C., American Physiological Society, pp. 1531–1552.

Schmid, H. (1881). Experimentelle Studien über partielle Lungenresektion. *Berl. Klin. Wochenschr.* **18**:757–759.

Sery, Z., Ressl, J., and Vyhnalek, J. (1969). Some late sequels of childhood pneumonectomy. *Surgery* **65**:343–351.

Simnett, J. D. (1974). Stimulation of cell division following unilateral collapse of the lung. *Anat. Rec.* **180**:681–686.

Simnett, J. D., Fisher, J. M., and Heppleston, A. G. (1969). Tissue-specific inhibition of lung alveolar cell mitosis in organ culture. *Nature* **233**:944–946.

Smith, B. T., Galaugher, W., and Thurlbeck, W. M. (1980a). Serum from pneumonectomized rabbits stimulates alveolar type II cell proliferation in vitro. *Am. Rev. Respir. Dis.* **121**:701–707.

Smith, L. J., Kaplan, N. B., and Brody, J. (1980b). Response of normal and beige mouse alveolar type 2 cells to lung injury. *Am. Rev. Respir. Dis.* **122**:947–957.

Soper, R. T., Pringle, K. C., and Schofield, J.C. (1984). Creation and repair on diaphragmatic hernia in the fetal lamb. *J. Pediatr. Surg.* **19**:33–40.

Sosenko, I., Stiles, A., Boyer, A., D'Ercole, A. J., and Smith, B. T. (1982). Somatomedin-C and compensatory organ growth: possible paracrine mechanism. *Pediatr. Res.* **16**:117.

Sterio, D. C. (1984). The unbiased estimation of number and sizes of arbitrary particles using the disector. *J. Microsc.* **134**:127–136.

Stiles, Q. R., Meyer, B. W., Lindesmith, G. G., and Jones, J. C. (1969). The effects of pneumonectomy in children. *J. Thorac. Cardiovasc. Surg.* **58**:394–400.

Stiles, A. D., Sosenko, I. R. S., Smith, B. T., and D'Ercole, A. J. (1983). Rapid rise in pulmonary somatomedin-C levels in the regenerating lung. *Pediatr. Res.* **17** Suppl.:391A.

Tabor, C. W., and Tabor, H. (1984). Polyamines. *Annu. Rev. Biochem.* **53**:749–790.

Tartter, P. I., and Goss, R. J. (1973). Compensatory pulmonary hypertrophy after incapacitation of one lung in the rat. *J. Thorac. Cardiovasc. Surg.* **66**:147–152.

Tatar-Kiss, S., Bardocz, S., and Kertai, P. (1984). Changes in L-ornithine decarboxylase activity in regenerating lung lobes. *FEBS Lett.* **175**:131–134.

Thet, L. A., and Law, D. J. (1984). Changes in cell number and lung morphology during early postpneumonectomy lung growth. *J. Appl. Physiol.* **56**:975–978.

Thomasson, B., and Ravitch, M. M. (1969). Fetal surgery in the rabbit. *Surgery* **66**:1092–1102.

Thomasson, B., Kero, P., Laensimies, E., Toikkanen, S., and Vaelimaeki, I. (1979). Pulmonary aeration after unilateral fetal ligation of pulmonary artery in neonatal rabbits. *Biol. Neonate* **36**:92–98.

Thurlbeck, W. M. (1975). Postnatal growth and development of the lung. *Am. Rev. Respir. Dis.* **111**:803–844.

Thurlbeck, W. M., Galaugher, W., and Mathers, J. (1981). Adaptive response to pneumonectomy in puppies. *Thorax* **36**:424–427.

Thurlbeck, W. M., D'Ercole, A. J., and Smith, B .T. (1984). Serum somatomedin-C concentrations following pneumonectomy. *Am. Rev. Respir. Dis.* **130**:499–500.

Uggla, L. G. (1956). Indications for and results of thoracic surgery with regard to respiratory and circulatory function tests. *Acta Clin. Scand.* **111**:197–213.

Wandel, G., Berger, L. C., Burri, P. H. (1983). Morphometric analysis of adult rat lung after bilobectomy. *Am. Rev. Respir. Dis.* **128**:968–972.

Watkins, C. A., Burkhart, L. R., and Rannels, D. E. (1985). Lung growth in response to unilateral pneumonectomy in rapidly growing rats. *Am. J. Physiol.* **248**:E162–E169.

Weibel, E. R. (1970/71). Morphometric estimation of pulmonary diffusion capacity. I. Model and method. *Respir. Physiol.* **11**:54–75.

Wiegand, L., and Pegg, A. E. (1978). Effects of inhibitors of S-adenosylmethionine decarboxylase and ornithine decarboxylase on DNA synthesis in rat liver after partical hepatectomy. *Biochim. Biophys. Acta* **517**:169–180.

Wilcox, B. R., Murray, G. F., Friedmann, M., and Pimmel, R. L. (1979). The effects of early pneumonectomy on the remaining pulmonary parenchyma. *Surgery* **86**:294–300.

Yee, N. M., and Hyatt, R. E. (1983). Effect of left pneumonectomy on lung mechanics in rabbits. *J. Appl. Physiol.* **54**:1612–1617.

Zeltner, T. B., and Burri, P. H. (1987). The postnatal development and growth of the human lung. II. Morphology. *Respir. Physiol.* **67**:269–282.

Zeltner, T. B., Caduff, J. H., Gehr, P., Pfenninger, J., and Burri, P. H. (1987). The postnatal development and growth of the human lung. I. Morphometry. *Respir. Physiol.* **67**:247–267.

20

Quantitative Evaluation of Minimal Injuries

LING-YI CHANG and JAMES D. CRAPO

Duke University Medical Center
Durham, North Carolina

I. Introduction

There is no specific level of pathologic change that defines a minimal injury. It is safe to state that a minimal injury is one in which the lesion is hard to detect with standard methods. Why, then, is it important to study minimal injuries? One reason is that the reagent that causes injury may be a common health hazard and as such may have adverse effects on a large segment of the general public. Two examples are the oxidant pollutants ozone and nitrogen dioxide. Ozone at a concentration of 1 ppm and NO_2 at 30 ppm have been demonstrated to induce pulmonary lesions in rats, dogs, and monkeys (Stokinger et al., 1957; Freeman and Haydon, 1964; Freeman et al., 1972; Schwartz et al., 1976; Mellick et al., 1977; Hyde et al., 1978; Dawson and Schenker, 1979, Plopper et al., 1979). Ambient levels of O_3, or NO_2, in major metropolitan areas in the United States are far below 1 ppm and 30 ppm, respectively, but frequently exceed the current air quality standards (NASC, 1977; CARB, 1983). For example, O_3 levels reaching 0.25 ppm have been recorded at Los Angeles for the summer months, while the national air quality standard for O_3 is 0.12 ppm. Whether subchronic or chronic exposures to ambient levels of oxidants causes injury is, therefore, an important health concern. Another reason why the study of minimal injury

597

is important lies in the fact that many toxicants induce a minimal amount of damage at the beginning of the injury. This site of early injury is a key to the pathogenesis of the lesion and a study of the early process of injury evolution is fundamental to understanding disease processes.

During the past few years, our laboratory has been engaged in the study of injuries caused by inhalation of a variety of environmental toxicants (Barry et al., 1982, 1985b, 1987; Crapo et al., 1984, 1987; Chang and Crapo, 1985; Chang et al., 1986, 1987). The magnitude of these injuries is small because the studies involve either a brief exposure period or low concentrations of reagents. Before presenting the animal models we used and summarizing some of the experimental results, we will review the approaches one may take when evaluating an injury that is hard to detect.

For the quantitation of minimal injuries, an overriding consideration is to increase the sensitivity of measurement procedures. For instance, electron microscopy (EM) increases resolution and consequently increases the sensitivity of detection. However, the effectiveness of EM morphometry on random tissue samples as a tool for quantiation of minimal injury depends on two factors. First, injury must be homogeneously distributed. The effects of a highly localized injury generally are lost when random samples for the whole organ are used. Second, the investigator must identify the correct indicator(s) of injury. This is particularly true for the lung, which contains a substantial variety of tissue types and cell populations. In a study aimed at defining early structural changes caused by 100% O_2, Barry et al. (1985a) found that the earliest quantitative changes in the pulmonary capillary bed during exposure were increases in platelet volume in the lung microvasculature and in the endothelial surface they cover. This was followed by an influx of neutrophils. These events took place before statistically significant changes in the shapes, volumes, or numbers of either epithelial or endothelial cells could be demonstrated (Crapo et al., 1978, 1980). If appropriate sensitive histological indicators of injury are available, quantitation of homogeneous minimal injuries can be carried out according to methods outlined previously (Weibel, 1963, 1980a,b; DeHoff and Rhines, 1968; Underwood, 1970; Elias and Hyde, 1983; Pinkerton and Crapo, 1985; Barry and Crapo, 1985). These procedures will not be discussed here.

An accurate quantitation of localized injury requires that the investigation be limited to the particular regions where changes occur. The two-animal models of minimal injuries presented in the following discussion are both models involving localized injury. Ambient levels of O_3 or NO_2 and a 1 hr exposure to chrysotile asbestos fibers cause tissue reactions in the area of the bronchiolar-alveolar duct junctions (BAD junction). Figure 1 illustrates one scheme to explain how these injuries may arise. Due to factors such as dilution of the oxidant gas concentration as it moves distally, and protection of the proximal airways by a thick mucous layer, O_3 and NO_2 cause an injury that is localized to the

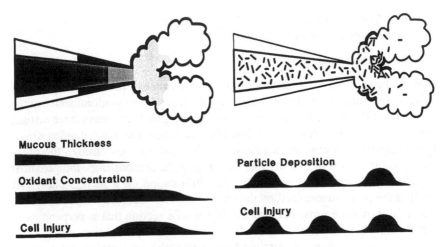

Figure 1 Schematic diagram of lung injury patterns by reactive oxidant gases in low concentrations and particles. Inhaled oxidants cause a focal injury in the region of the BAD junction. Their rapid consumption at the airway surfaces dilutes the concentration of the reactive species available to attack the more distal tissue. Oxidant gases affect epithelial tissues in the area of the BAD junction uniformly since these gases are diffusible. Particles, on the other hand, deposit primarily at branch points. In this simplified illustration of the lung, no airway branch point is shown. In the gas-exchange region, the most severe tissue response is found on the first alveolar duct bifurcation and on the adjacent portions of the alveolar duct walls.

tissues in the area of the BAD-junction (Freeman et al., 1968; Stephen et al., 1972, 1973; Parkinson and Stephens, 1973; Dungworth et al., 1975; Schwartz et al., 1976; Mellick et al., 1977; Boorman et al., 1980; Eustis et al., 1981). In the gas exchange region, asbestos fibers are preferentially deposited on the most proximal alveolar duct bifurcations (Brody et al., 1981, 1984; Brody and Roe, 1983) and have been shown to cause enlargement of the first bifurcation (Warheit et al., 1984). In both of these injury models, when whole lung sections or random samples of lung are examined, the injuries are undetectable. In our studies, we used microdissection to isolate the BAD junctions for quantitative histological evaluations. Three types of samples can be isolated from the region of the BAD junction: the terminal bronchiole (TB) (Barry and Crapo, 1985; Barry et al., 1987), the first alveolar duct bifurcation (Warheit et al., 1984), and the proximal alveolar region (PAR) (Barry and Crapo, 1985; Barry et al., 1985b; Chang et al., 1986), depending on how the section plane transects the BAD junction.

II. Special Methods for Morphometric Analysis of Focal Lesions

A. Microdissection

A key to the study of focal lesions is to be able to isolate reproducibly the region of the injury. Since many inhaled toxic agents cause injury predominantly in the region of the BAD junction, an approach to isolating and sectioning three distinct areas in this region is described below (Fig. 2). These include obtaining cross-sections through terminal bronchioles (Fig. 2b) and sections through the proximal alveolar region in two orientations. Figure 2a demonstrates the proximal alveolar region and the first alveolar duct bifurcation on a section plane that parallels the plane determined by the axes of the two secondary alveolar ducts. Figure 2c shows the proximal alveolar region on a section that is perpendicular to the axial plane at the site of the first alveolar duct bifurcation.

To obtain sections of terminal bronchioles, the proximal alveolar region or the first alveolar duct bifurcation, pieces of the left lung (approximately 2 mm × 4 mm × 6 mm) are embedded in Epon 812. The blocks are softened with mild heat and cut with a razor blade into slices less than 0.5 mm thick. The slices are then examined serially with a dissecting microscope. The small branches of a conducting airway are followed until one terminates at a BAD junction. If a slice contains a segment of the terminal bronchiole that is cut in cross-section by the slice, a small block containing the terminal bronchiolar segment is cut out and glued to a blank block for ultrathin sectioning. Sections are then picked up on grids containing a single round hole that has been coated with a supportive film. Uranyl acetate- and lead-citrate-stained ultrathin sections are examined and photographed with an electron microscope.

In similar fashion, when the cutting surface of a slice is found to transect the BAD junction in a way that allows the PAR to be sampled as illustrated in Figures 2c and 3, a square containing the BAD junction (and PAR) is cut out and glued to a blank block (Fig. 3). Most of the excess tissue above the bifurcation should be trimmed away with a razor blade before the block is mounted on the ultramicrotome. The block face is then smoothed and 1 μm sections are cut until the first bifurcation of the primary alveolar duct is reached and the profiles of two separate secondary alveolar ducts are formed. At that point the block face is trimmed to the edge of the duct wall and ultrathin sections of 70–90 nm thickness are obtained. Sections are individually picked up on 200 mesh grids, stained, and examined.

To obtain longitudinal sections through first alveolar duct bifurcations, lung slices that reflected the outlines of the respiratory bronchiole and the two secondary alveolar ducts already transected (Fig. 2a) on their cut surface are selected. A square enclosing the longitudinally sliced BAD junction is cut out and glued on a blank block. The block is then faced until an intact section demonstrating

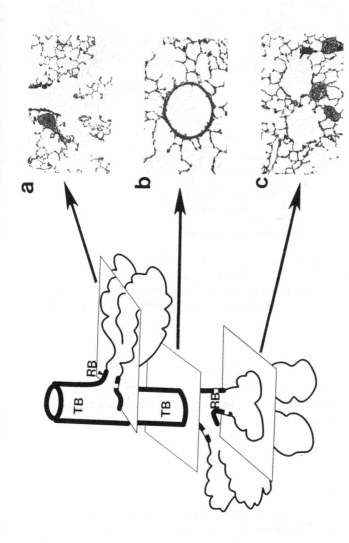

Figure 2 Three distinct anatomical regions used in analysis of localized injuries. In the rat, the acinus (thin line) is frequently joined to the terminal bronchiole (TB, thick line) with a very short segment of respiratory bronchiole (RB, broken thick line). The proximal portion of the acinus can be sampled by a plane that is either parallel or perpendicular to the axes of the secondary alveolar ducts. When the transecting plane is parallel, micrographs showing a longitudinal view of the first bifurcation and the two ducts can be obtained (a, first bifurcation). When the plane is perpendicular to the axes, a cross-sectioned view of the bifurcation and its adjacent ducts is obtained (c, PAR). The terminal bronchiole (TB) is most easily sampled by a plane that is perpendicular to the direction of the bronchiole. (b) A perpendicular cross-section of the terminal bronchiole is shown.

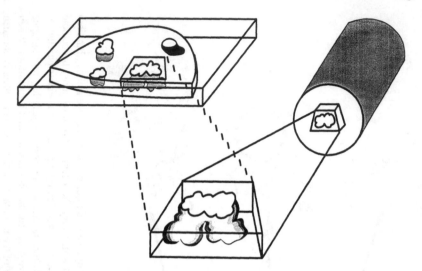

Figure 3 Microdissection of samples of the proximal alveolar region. Serial slices of lung tissue in Epon 812 are examined to identify PARs transected as illustrated in Figure 2c. The identified sample is cut out and glued to a cylindrical blank block for ultrathin sectioning. Terminal bronchioles and first alveolar duct bifurcations are isolated in the same manner.

The respiratory bronchiole and the two alveolar ducts separated by the first bifurcation is obtained. Ultrathin sections are then cut and picked up on coated grids that contained a single 1 mm hole.

B. Morphometric Analysis

General Methods

Microscopy and Counting

Tissue is sampled from the sections of PARs by taking two pictures per grid square: one at the upper left corner and another at the lower right corner in a manner described by Weibel (1963, 1980a). A photograph of a calibration grid is taken at the same magnification. Pictures are printed on 11 × 14 inch photographic paper at a final magnification of 8500X. A multipurpose overlay consisting of 112 lines, each 2 cm long, is used to perform point and intercept counting (Weibel, 1980a). Volume and surface area measurements are expressed as a ratio of the volume or surface area to the surface area of epithelial basement membrane (Barry and Crapo, 1985; Barry et al., 1985; Chang et al., 1986). The reference space is the proximal alveolar region that contains both tissue and air space.

The first alveolar duct bifurcation is sampled by taking overlapping photographs of all tissue on the bifurcation at 2000X. The boundaries of the first bifurcation are defined as the air margins on the top and both sides of the bifurcation and as the midpoint of the bifurcation stem at the bottom. The bifurcation stem is defined as the septum(s) separating the alveolar ducts and the first alveolus just below the branching point (Warheit et al., 1984). Micrographs are printed on 11 × 14 inch photographic paper at 8500X. The area for point counting on each micrograph is marked by dividing the overlapping areas between adjacent pictures. A 448 line overlay with 1.37 cm long lines is used for point and intercept counting. Volume and surface area measurements are expressed in a manner similar to that used for the proximal alveolar region. However, in this case the reference space is the first alveolar duct bifurcation and involves only tissue.

Terminal bronchioles are sampled by taking overlapping photographs that cover the entire circumference of the bronchiole (Barry and Crapo, 1985; Barry et al., 1987). Pictures are printed at 8500X on 11 × 14 inch papers. After the epithelial area is marked for morphometeric analysis on each micrography, the micrograph is placed under a Merz curvilinear overlay (Merz, 1967) consisting of 224 points. The surface area of the bronchiolar basement membrane is used to normalize the data. The reference space is the tissue area of bronchiolar epithelium.

Morphometric Analysis of Volume/Surface Ratios and Surface/Surface Ratios (Barry and Crapo, 1985)

For a standard morphometric study, the absolute volumes of alveolar tissues are calculated by multiplying their respective volume densities first with the volume fraction of parenchymal tissue in the lung and then with the lung volume. To calculate the absolute volumes of tissues in the PAR, TB, or on the first alveolar duct bifurcation, the volume fraction of each of these three anatomical regions in the whole lung would need to be known. Because it is often difficult to identify PAR, TB, or first alveolar duct bifurcations properly in a random lung section, the volume fraction of each cannot be accurately measured. Simply to report volume densities of the lung tissues would be misleading, since tissue volume and reference volume may change proportionally, leaving no detectable change in volume density. An alternative is to express volume as a ratio to the surface area of the epithelial basement membrane. The surface area of the basement membrane is a collagenous structure that undergoes minimal stretching at physiological levels of inflation and is not significantly affected by low levels of toxicants (Barry et al., 1985b; Chang et al., 1986). It can, therefore, serve as a stable reference point. Table 1 lists the formula used to convert volume density and surface density to volume/surface and surface/surface ratios.

Table 1 Formulas for Analyzing Morphometric Data from Localized Anatomical Regions

Normalizing volume (V) and surface (SA) to the surface area of epithelial and basement membrane (bm).

Volume/surface ratio

$$V_{SA_{bm}} = \frac{V}{SA_{bm}} = \frac{V_V}{S_{V_{bm}}}$$

Surface/surface ratio

$$S_{SA_{bm}} = \frac{SA}{SA_{bm}} = \frac{S_V}{S_{V_{bm}}}$$

Cell characteristics

Cell number

$$N_{SA_{bm}} = \frac{N}{SA_{bm}} = \frac{N_V}{S_{V_{bm}}}$$

Mean cell size

$$\bar{V} = \frac{V_{SA_{bm}}}{N_{SA_{bm}}}$$

Mean cell surface

$$\overline{SA} = \frac{S_{SA_{bm}}}{N_{SA_{bm}}}$$

$V_{SA_{bm}}$, volume per unit area of epithelial basement membrane; SA_{bm}, surface area of the epithelial basement membrane; $S_{V_{bm}}$, surface density of the epithelial basement bembrane; $S_{SA_{bm}}$, surface area per unit area of epithelial basement membrane; $N_{SA_{bm}}$, number of cells per unit area of epithelial basement bembrane.

C. Cell Characteristics

Specific cell characteristics can be derived to assess the response of individual cell types to minimal injuries. While changes in tissue volumes and thicknesses may indicate the extent of an injury, they reveal little about the mechanism of injury. Changes in cell characteristics such as cell number, cell size, and cell shape can reveal individual cell susceptibility, cell ability to differentiate or regenerate, and relationships between cell populations as well as between cells and matrix components.

In analysis of specific regions such as the PAR, TB, or the first alveolar duct bifurcation, cell number must be expressed in relation to the surface area of basement membrane, that is, the number of cells per unit area (N_{SA}) (Table 1). This is determined by dividing the numerical density (N_V) of a cell type by the surface density of the basement membrane. Mean cell volume and mean cell surface area can be derived by dividing the volume density or the surface density of that cell type by its numerical density. It can be seen that the measurement of numerical density is central to the determination of cell characteristics. The determination of numerical density or the number of particles per unit volume is one of the more difficult morphometric problems. Both the frequency and general appearance of particle profiles on two-dimensional sections are determined to a large extent by the size and shape of the particle in three-dimensional space. Larger particles have a greater probability of presenting a profile on a section because they are more likely to be sectioned. Likewise, a group of ellipsoids may present more profiles on sections than a group of spheres that are equal in number and volume. A number of methods have been proposed for estimating numerical density. The most easily applied of these techniques will be described here.

Determination of Numerical Density by N_A, Profiles per Unit Area, and $\overline{D}$, the Mean Caliper Diameter

The morphometric formula, $N_V = N_A/\overline{D}$, was first described by Abercombie in 1946. This formula is simplified from its more general form and assumes that section thickness is sufficiently thin and resolution is sufficiently high that errors caused by section thickness and by missing cap sections are negligible. The mean caliper diameter ($\overline{D}$) is defined as the average distance across the particle averaged over all possible orientations of the object. The incorporation of $\overline{D}$ in the denominator corrects for the fact that the probability of a particle being sectioned by a random plane is proportional to both the size and shape of the particle.

The number of profiles per unit area (N_A) is determined by counting the number of profiles on a section and measuring the actual area over which all the profiles counted were located. The unbiased counting rule (Gundersen, 1977) that uses a "forbidden line" to exclude particles that intersect with two continuous sides and the three corners of the test area in contact with the two sides should be adapted. For most cell counts, it is easier to count nuclear profiles because they are more easily recognized than cell borders. Each cell is commonly assumed to have one nucleus.

To determine N_A on sections of the proximal alveolar region, the number of nuclei from each class of alveolar cells in grid squares completely covered by the section is recorded. The sides of the grid square serve as the unbiased counting frame (Barry and Crapo, 1985). The area on which the cell number is obtained is calculated by multiplying the number of grid squares used with the area occupied by a single grid square. The latter is determined by digitizing photographs of at least one representative grid square taken from the same grids on which the nuclear counts were performed. The N_A for cells on the first alveolar duct bifurcations and terminal bronchioles is determined by first counting nuclei from a montage of the entire bifurcation or bronchiole and then summing the areas obtained by digitizing each micrograph of the montage.

The mean caliper diameter of the particle (nucleus in the case of cell counting) can be determined by a variety of methods (Abercrombie, 1946; Greeley and Crapo, 1978; Greeley et al., 1978; Loud et al., 1978; Woody et al., 1979; Cruz-Orive, 1980). If one can safely assume that the particles conform to a certain shape, such as spherical, oblate ellipsoids, or prolate ellipsoids, $\overline{D}$ can be calculated from established formulae using data obtained from measuring profiles on sections (Kauffman et al., 1974; Haies et al., 1981). $\overline{D}$ can also be derived from histograms of random profile diameter by the mean profile diameter method (Weibel, 1973, 1980b) or the size class analysis method (Saltykov, 1967; Greely and Crapo, 1978). There are also methods for $\overline{D}$ determination that are independent of shape assumptions. We have determined nuclear $\overline{D}$ for a variety of pulmonary cells by three-dimensional reconstructions of their nuclei (Greeley et al., 1978; Woody et al., 1979). Figure 4 shows two nuclei reconstructed from serial EM sections of an alveolar type II cell and a fibroblast, respectively. In practice, we reconstruct cell nuclei from serial 0.5 μm thick plastic sections. Section thickness is determined by measuring the size of a tissue block before and after sectioning using a Vernier micrometer or by using a dissecting microscope to measure the change in depth of a notch placed on the cutting surface. The nucleus to be reconstructed may be selected by identifying the center of the nucleus closest to a random point on the slide. This technique introduces a known bias into the selection in that nuclei will be chosen in proportion to their $\overline{D}$. This bias can be removed by calculating a weighted mean to obtain information about the general population of nuclei (Greeley et al., 1978). A more unbiased technique to select the nuclei for reconstruction is to use the "selector," as described later in this chapter. This technique identifies nuclei without bias for shape or size, but may increase the workload substantially without improving accuracy significantly over the simple selection method described above.

The reconstruction is accomplished by tracing profiles of a serially sectioned nucleus into the computer memory. The stack of nuclear profiles must be carefully aligned for correct orientation. In computer memory the three-dimensionally reconstructed nucleus is rotated randomly. The position of a pair of parallel planes perpendicular to the x axis and in contact with the outermost surface of the nucleus are found. The distance between the two planes is determined. The mean of a large number of these measurements for each nucleus is an unbiased estimate of the mean caliper diameter of the nucleus. In the case

Epithelial Type II Cell   Fibroblast

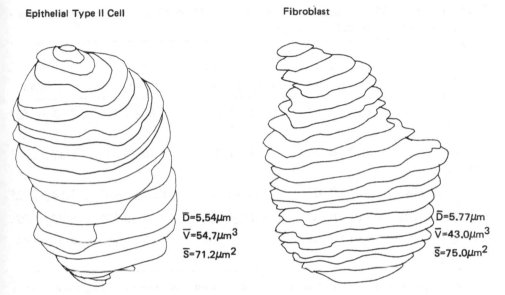

$\bar{D}$=5.54μm

$\bar{V}$=54.7μm^3

$\bar{S}$=71.2μm^2

$\bar{D}$=5.77μm

$\bar{V}$=43.0μm^3

$\bar{S}$=75.0μm^2

Figure 4 Three-dimensional reconstruction of lung cell nuclei of an epithelial type II cell and a fibroblast. The average section thickness is 100 nm. Only the tracings from every fourth section are reproduced.

of terminal bronchioles, mean caliper height ($\bar{H}$) instead of $\bar{D}$ is used for the determination of N_V. This is because the terminal bronchiole is sampled with a preferred orientation. Serial sections through perpendicular cross-sectioned terminal bronchioles were used. For a series of cells the number of serial sections needed to transect each nucleus completely is multiplied with the average section thickness to obtain $\bar{H}$.

Table 2 lists the mean caliper diameters of nuclei of major classes of cells found in the alveolar region of mammalian lungs (Greeley et al., 1978; Pinkerton and Crapo, 1985). For each value, weighted means were determined to correct for the known sampling bias: that larger nuclei were more likely to be selected for three-dimensional modeling (Greeley and Crapo, 1978). Other potential errors include overestimating $\bar{D}$ due to the Holmes effect (Holmes, 1927) and underestimation of $\bar{D}$ caused by an inability to recognize "cap" sections. Evaluation of the methods used to determine these values have shown that these errors are negligible and can be ignored (Greeley and Crapo, 1978).

The disadvantage of the method described above is the relatively large workload involved in three-dimensional reconstructions. However, we have found that $\bar{D}$ of various lung cells do not change significantly over a wide range of

Table 2 Mean Caliper Diamaeter of Nuclei from Major Classes of Cells Found in the Alveolar Region of Mammalian Lungs

Type of Nucleus	Rat		Dog	Human
	Fischer 344	Sprague-Dawley		
Alveolar type I cell	8.29 ± 1.05^a	7.97 ± 0.66^a	7.69 ± 1.33^a	7.85 ± 0.52^a
Alveolar type II cell	7.23 ± 0.65^a	6.92 ± 0.57^a	7.10 ± 0.53^a	7.57 ± 0.44^a
Interstitial cells	7.53 ± 0.92^a	7.64 ± 1.10^a	8.07 ± 0.93^a	7.54 ± 0.72^a
Endothelial cells	7.01 ± 0.64^a	8.00 ± 0.76	7.07 ± 0.37^a	8.77 ± 1.09
Alveolar macrophages	7.53 ± 0.79^a	7.65 ± 0.54^a	6.39 ± 0.27	8.33 ± 0.86

All data are given in μm and are mean $\pm$ standard deviation. The means are weighted to account for and reduce sampling bias (Greeley et al., 1978).
[a]For each type of cell nucleus, all data marked by the same letter subscript are not statistically different from each other.

pathological states (Crapo et al., 1982; Pinkerton et al., 1984). Consequently, in studies of minimal injury patterns, these measurements can be used reliably to determine cell number. In our experience, this method has been proven to be both effective and time efficient.

Determination of N_V by Use of the Disector or Selector (Sterio, 1983; Cruz-Orive, 1980, 1987)

The disector is a three-dimensional probe for obtaining unbiased estimates of the number of particles per unit volume. It consists of a pair of parallel planes, a known distance (h) apart. One of the planes contains an unbiased counting frame (Gundersen, 1977) of a known area, a. The particles that fall within the counting frame are first identified by following the unbiased counting rule. The number (Q) of particles identified on planes that do not transect the second parallel plane is then determined. Numerical density is calculated by dividing Q by the volume bounded by the counting frame and the parallel planes (axh).

$$N_V = \frac{Q}{a \times h}$$

The disector method in effect counts only particles whose top is located within a given reference volume (axh). This reduces every particle to a single point regardless of its size. Consequently, the bias of preferentially counting large particles is removed. A practical example of the disector is illustrated in Figure 5, in which three new profiles are identified on plane 4 that appear not to present on plane 1.

There are two limitations for the disector method: (1) the distance h must be shorter than the height of any of the particles in the direction perpendicular to the planes of the disector; (2) the relationship between the sets of profiles on the two planes of the disector (i.e., whether two profiles on separate planes actually belong to the same particles) has to be identifiable. The reason for the first requirement is to prevent particles from being missed because they are fully contained between the two planes. The importance of the second condition can be readily recognized by comparing the two planes of the disector to the actual block illustrated in the middle of Figure 5. Note that of the three profiles regarded as new starts on section 4, two belong to the same particle. A fourth profile was not recognized as a new start because it happened to superimpose on another profile in Section 1. In this example, the two mistakes cancelled each other, but one can easily see how this type of mistake can create serious measurement errors. More than two planes, (i.e., serial sections) appear to be needed to facilitate the recognition of particles when the dissector method is used. Moreover, it can be difficult to measure accurately the distance between two specific section planes because h must be small to satisfy requirement. As a result, the average distance between sections measured from a longer serial sectioning run may be needed to determine h with sufficient accuracy.

The selector is derived from the disector with some major differences (Gundersen and Jensen, 1985; Cruz-Orive, 1987). First, a series of parallel planes is required. Second, the distance between the planes need not be known. Third, numerical density is determined by dividing total volume density of the particles by the mean particle volume ($N_v = V_v/\overline{V}$) (Gundersen and Jensen, 1983, 1985). The selector works by unbiasedly selecting the particle for study using the principles described for the disector, namely identifying particles that start between the two planes of the selector. Figure 5 shows how this selection can be done. Once a particle is selected, all the sections in the series containing the particle are picked out. An array of randomly oriented lines is overlaid on top of the particle profiles on each of the sections. When the line intercepts the particle, the line length between the two intercept points is measured. The volume of the particle can be estimated by the mean length ($\overline{l}$) of the intercepts.

$$V = (\pi/3)\overline{l}^3$$

The average or mean particle volume ($\overline{V}$) is obtained after the volumes of a number of particles are measured. Because the volume density (V_v) of a class

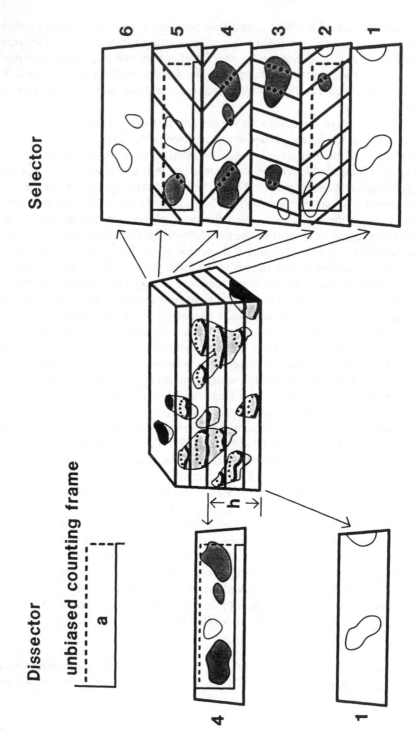

Selector

Dissector

unbiased counting frame

(i) of particles is equal to the total volume of the particles, $i(V_i)$ divided by the total volume of the reference space (V_T) and V_i is equal in turn to the product of mean particle size $(\overline{V})$ and particle number (N_i), it follows that:

$$V_v = \frac{V_i}{V_T} = \frac{N_i \times \overline{V}}{V_T} \text{ or } \frac{V_v}{\overline{V}} = \frac{N_i}{V_T}$$

and,

$$N_V = \frac{N_i}{V_T} = \frac{V_v}{\overline{V}}$$

The volume density (V_v) is determined easily by point counting. The advantage of the selector method over the disector is that the distance between the planes need not be known. In addition, since the selector merely selects and N_v is not determined by counting, the effect of identifiability (requirement 2) is not as significant. However, serial sections are actually used to analyze mean intercept length so that identifiability is enhanced. It is evident that for both the disector or the selector to function properly, serial sections are required. This increases workload. For studies of limited scale, these methods can be very effective.

Another method similar to the disector was reported by Howard et al. (1985). A three-dimensional probe, or a brick, is used to count particle number directly with a tandem scanning reflected light microscope (TSRLM). This method could be efficient, but the TSRLM is not widely available.

Figure 5 Schematic illustration of the disector and the selector. The disector consists of two parallel planes separated by a known distance, (h), and an unbiased counting frame (of area a) superimposed on one of the planes. Profiles within the counting frame not identified on the other plane are counted. The selector consists of a series of sections. Particles fully contained within the block are selected by a pair of parallel planes at one end of the series of sections. These two planes are called the selector and can be either planes 1 and 2 or planes 6 and 5 in this figure. Because the selected particles must be completely contained within the block, the top (plane 6) or the bottom (plane 1) section is the reference plane and the other section in the pair is the "look-up" plane. A new start of a particle is identified on the look-up plane; then the other sections that contain profiles of the particle are picked out for point-intercept analysis. Sets of randomly oriented lines are superimposed on the sections as shown. The length of the intercept (dotted portion of the line) can be measured.

III. Examples of Experimental Findings

With the use of microdissection and morphometric analysis on electron micrographs, the localized injuries caused by a brief exposure or a very low dose of toxic agents can be defined clearly.

A. Oxidants: O_3 and NO_2

Both low O_3 concentrations (0.12 and 0.25 ppm) and low NO_2 concentrations (0.5 and 2.0 ppm) have been found to cause lung structural modifications. The major histological responses in the proximal alveolar region to O_3, and NO_2 are hypertrophy and hyperplasia of the alveolar epithelium (Barry et al., 1985; Chang et al., 1986). Six-week-old male Fisher 344 rats were exposed to either room air, O_3 or NO_2 for 6 weeks. The O_3 exposures were for 12 hr/day, 7 days/week and the NO_2 exposures were for 23 hr/day, 7 days/week. The animals were housed in 0.323 m^3 stainless steel Rochester-type chambers. Air flow in the chamber, methods to generate O_3, and NO_2, and maintenance of oxidant concentrations were as reported by Barry et al. (1985b) and Chang et al. (1986). At the end of the 6-week exposure, the lungs were first deflated by puncturing the hemidiaphragm through a cut in the abdomen and then fixed by installation of 2% glutaraldehyde through an intratracheal cannula at a pressure of 20 cmH_2O (Pinkerton and Crapo, 1985). Lung volume was determined by fluid displacement (Scherle, 1970) after a minimal fixation time of 24 hr.

Ozone

Figure 6 demonstrates a thickened type I epithelium after exposure to 0.25 ppm O_3 for 6 weeks. At the same time, regions of type II cell hyperplasia were found. The volume of type I epithelium increased 23% while that of the type II epithelium increased by 33%. The number and volume of alveolar macrophages in the proximal alveolar region doubled. Changes also occurred in the interstitium, including an increase in the volumes of both cells and matrix. The reactions of alveolar tissues to 0.25 ppm O_3 are summarized in Figure 7. Morphological evidence of matrix accumulation is shown in Figure 8. These observations agree with previous description of the O_3-induced injuries found at the BAD junction (Schwartz et al., 1976; Plopper et al., 1979; Boorman et al., 1980).

Exposure to 0.12 ppm O_3 caused a significant increase in the volume of type I epithelium in the proximal alveolar region. The use of microdissection and morphometric analysis of selective lung regions is sufficiently sensitive that changes can be detected at the current air quality standard for ozone. The effect of the two ozone exposures on the volume of type I epithelium at the level of individual sites was compared by plotting the type I epithelial volume per square millimeter basement membrane from individual proximal alveolar regions (Fig.

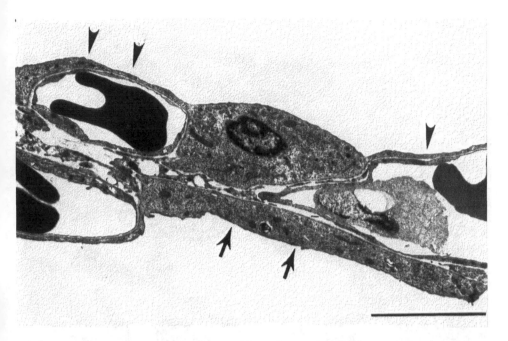

Figure 6 Alveolar septum from a young adult rat exposed to 0.25 ppm O_3 for 6 weeks since 6 weeks of age. An area of thickened type I epithelial cytoplasm is shown (arrows). An adjacent area of type I epithelium appears to be normal (arrowhead). Bar = 5 μm.

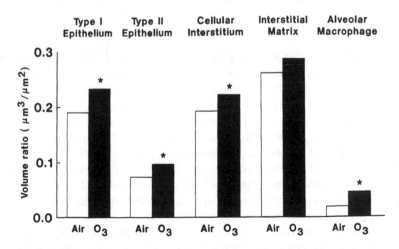

Figure 7 Patterns of tissue reaction in the proximal alveolar region of the young adult rats exposed to 0.25 ppm O_3 for 6 weeks. Volumes of the various tissues are normalized to the surface area of the epithelial basement membrane. *p < 0.05.

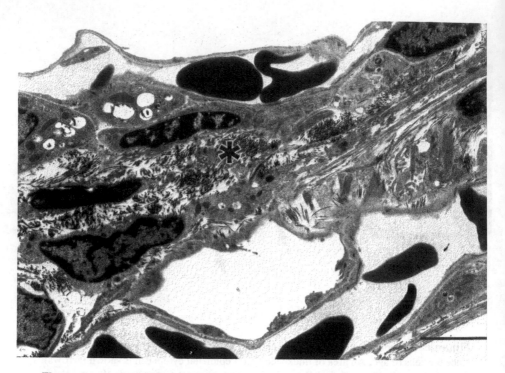

Figure 8   An area of thickened interstitium from a young adult rat exposed to 0.25 ppm O_3 for 6 weeks. There are more collagen fibrils (*) in the interstitium. Bar = μm.

9). These data illustrate that biological variability existed both within and between animals. Two important conclusions can be drawn from the data shown in Figure 9. The first is that there appears to be a linear relationship between O_3 concentration and type I epithelial reaction. The second is that the response of type I epithelium occurred in virtually all of the proximal alveolar regions throughout the lungs of exposed animals because the whole group of data appears to be displaced toward a higher volume of type I epithelium. Despite the significant overlap of values found in control animals and values found in exposed animals, two-thirds of the sites from control animals had values below 0.175 while 90% of the sites from 0.12-ppm-exposed animals and virtually all sites from the 0.25-ppm-exposed animals were above 0.175.

   Studies of cell characteristics indicated how the epithelial changes evolved. Figure 10 shows the changes in characteristics of type I and type II epithelial cells after exposure to 0.25 ppm O_3. It can be seen that type I cells were increased in number, became smaller in size, covered less basement membrane surface

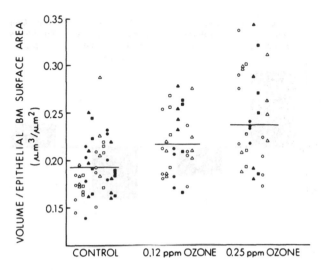

Figure 9 Volume density of type I epithelium in the proximal alveolar regions of control and O_3 exposed rats. Each column of four identical symbols represents the four data values calculated for this parameter for each animal. The horizontal bar in each cluster of symbols represents the mean for the animals in that group. This figure illustrates the linear relationship between O_3 concentration and increases in the volume density of type I epithelium after 6 weeks of exposure.

area, and were significantly thicker than type I cells from control animals. The number of type II epithelial cells increased, but mean cell volume, mean cell surface area, and mean cell thickness of the type II cells were not significantly altered. Similar findings were observed for epithelial cells exposed to 0.12 ppm O_3. The changes described for type I epithelial cells suggest that an increased portion of these cells were at an intermediate stage of cell differentiation between type I and type II cells (Evans et al., 1975, 1976). The combined findings of qualitative evidence of type I cell injury, increased numbers of type II cells and morphometric data documenting changes in type I cell structure indicate that inhalation of 0.12 or 0.25 ppm O_2 cause injury to the epithelium and an increase in the rate of epithelial cell proliferation (Evans et al., 1976; Lum et al., 1978; Stephen et al., 1978).

Nitrogen Dioxide

Many of the changes found in animals exposed to O_3, have also been observed in animals exposed to NO_2 (Freeman et al., 1968; Stephens et al., 1972). Thickened type I epithelium, an elevated type I cell number, an increased

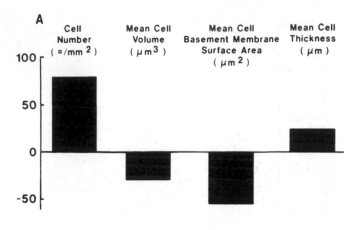

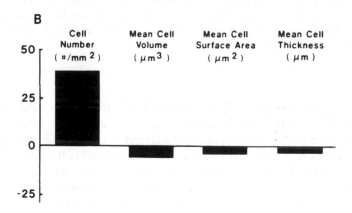

Figure 10   Changes in characteristics of type I and type II epithelial cells in the proximal alveolar regions after a 6 week exposure to 0.25 ppm O_3.

interstitial volume, and an increase in alveolar macrophage volume were found after 6 weeks of exposure to 2.0 ppm NO_2. The severity of the injuries at this level of NO_2 was less intense than that found with 0.25 ppm O_3, and the concentration–response relationship was not as evident as it was with O_3. The lower concentration of NO_2 (0.5 ppm) caused a greater increase in the total epithelial cell volume (+ 24%) than did the higher concentration (+ 17%). The mean thickness of type II cells was increased 25% by exposure to 0.5 ppm NO_2, but was increased only 10% by exposure to 2.0 ppm NO_2. The 0.5 ppm NO_2 response consisted mainly of a dramatic increase in the volume of type II epithelium (+ 110%) while the 2.0 ppm exposure caused increases in the volumes of both the type I epithelium (+ 22%) and the type II epithelium (+62%). At first glance, these results appeared confusing. Analysis of cell characteristics again provided insight to understanding these tissue reactions.

Epithelial type II cell reactions were studied in three groups of NO_2 exposed animals. In addition to the 6-week-old rats exposed to 0.5 or 2.0 ppm NO_2, a group of juvenile rats exposed to 0.5 ppm NO_2 for 6 weeks since 1 day of age was also studied together with matched control animals. No major tissue reactions were found in the proximal alveolar region of 0.5-ppm NO_2-exposed juvenile rats. The thickness of the type II epithelial cells of these animals, however, was significantly reduced (Fig. 11a). Examination of the trend of type II epithelial cell responses in the three groups of animals suggests that type II cells can respond to oxidant challenge in a number of ways, depending on the nature or severity of the injury to the alveolar epithelium (Chang et al., 1986). Figure 12 uses cell spreading and cell hypertrophy as indices of type II cell injury. We postulate that when loss of type I epithelial cells is minimal, the first reaction of the type II cells is to undergo mild hypertrophy and to spread and cover the vacated alveolar surface. There is minimal type II cell proliferation and differentiation found in juvenile rats. Because the degree of type II cell spreading exceeds the degree of cell hypertrophy, the net effect was a reduced arithmetic mean thickness of type II cells.

As the injury progresses, greater stress was placed on the type II cells. The type II cells initially underwent extensive hypertrophy to prepare for a four- to fivefold increase in volume and more than 50-fold increase in surface area needed to undergo a transition to a mature type I cell. As a result, the thickness of type II cells increased (Fig. 11b). Adult animals exposed to 0.5 ppm NO_2 showed this type of response.

At a still higher level of injury, type II cell proliferation was induced and an increased number of type I cells can be observed, as is the case in animals exposed to 2.0 ppm NO_2. Due to the low rate of cell proliferation even at the 2.0 ppm NO_2 injury level, the degree of cell proliferation is not reflected directly in type II cell numbers. The need for cell differentiation to replace type I cells kept the type II cell population at a seemingly normal level. Therefore, the rate of type II cell proliferation is expressed by the number of type I cells, the product of type II cell proliferation, and differentiation. In addition to cell proliferation, spreading and hypertrophy still occurred. The average type II cell thickness, at this time, returns toward the normal value, possibly because the hypertrophy response was being counterbalanced by a higher rate of cell division. The type II cell characteristics found with O_3 exposure are also consistent with these

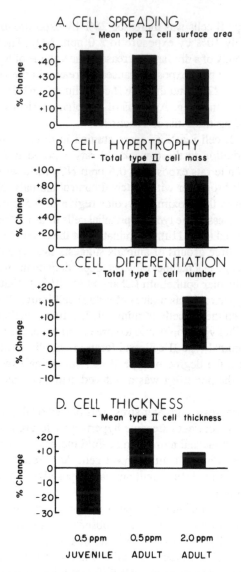

Figure 11 Suggested pattern for the reactions of the alveolar epithelium to increasing degree of injury caused by NO_2. The primary types of type II epithelial cell reaction are illustrated as: (A) cell spreading, expressed as mean type II cell surface area; (B) cell hypertrophy, expressed as mean type II cell volume; and (C) cell division and differentiation, indicated by changes in the population of product cells (type I); (D) arithmetic mean cell thickness, which is the ratio of cell volume (hypertrophy) to cell surface area (spreading). The widely different responses in cell thickness can be explained by following the patterns of reactions in (A) to (C).

hypotheses. Ozone at both dosages described in the previous section induced greater injuries that appear to have further stimulated type II cell division. The type II cell spreading and hypertrophy responses were then not recognizable because there were more type II cells as well as a greater population of transitional type I cells.

B. Asbestos

The hazards to human beings of inhaled asbestos are well known (Selikoff and Lee, 1978). The pathological characteristics of asbestosis in chronically exposed humans and animals have been studied in detail. Brody and co-workers (Brody et al., 1981, 1984; Brody and Row, 1983; Warheit et al., 1984, 1985, 1986) have established that inhaled chrysotile asbestos fibers accumulate preferentially on the more proximal alveolar duct junctions. This coincides with earlier findings showing that the fibrotic changes of asbestosis occur predominantly around the BAD junctions (Begin et al., 1981, 1982, 1983; Craighead et al., 1982). Subsequent studies by Warheit et al. (1984) demonstrated that significant numbers of pulmonary macrophages accumulate at sites of asbestos deposition within 48 hr after a 1 hr exposure. This influx of macrophages was associated with a significant increase of the cross-sectional tissue area of the alveolar duct bifurcations (Warheit et al., 1984). These changes appeared to be the earliest detectable lesions of asbestosis. The specific cell and tissue responses involved in this acute reaction to chrysotile asbestos and its progression into localized fibrosis can be demonstrated by performing morphometric analysis on isolated first alveolar duct bifurcations.

Animals used in the study summarized below were 7-week-old CD(SD)BR rats (Charles River Laboratories). The exposure chambers and engineering aspects of the asbestos exposure (chamber construction, dose calibration, etc.) have been reported in detail previously (Brody et al., 1981; Brody and Roe, 1983; Timbrell et al., 1968, 1970). Chrysotile asbestos fibers, whose characteristics have been carefully studied, were used (Pinkerton et al., 1983). The respiratory mass of the aerosolized chrysotile asbestos was 13 mg/m^3 and the exposure duration was 1 hr. Rats were held in Plexiglas nose-only holders during the exposure. Sham (air)-exposed rats were used as controls. After the exposure, rats were kept in room air for either 2 days or 1 month. At the end of the recovery period, rats were sacrificed and their lungs fixed by intratracheal instillation of 2% buffered glutaraldehyde.

Figure 13 shows a series of low-magnification electron micrographs of first alveolar duct bifurcations from control and exposed rats. The bifurcations isolated from asbestos-exposed animals 2 days after the exposure exhibited greater cellularity (Fig. 13b). Morphometric analysis revealed that the total tissue volume (not including capillary lumen or any blood elements) increased by 67%. The

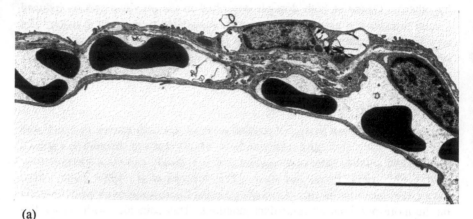

(a)

Figure 12 Difference of type II cell morphology in relation to its response to NO_2. (a) A spreading type II cell from a 6-week old juvenile rat exposed to 0.5 ppm NO_2 for 6 weeks. (b) A hypertrophic type II cell from a young adult rat exposed to 0.5 ppm NO_2 for 6 weeks. Bar = 5μm.

epithelium covering the first bifurcation underwent hypertrophy and hyperplasia. The volumes of the type I epithelium and type II epithelium increased 43% and 126%, respectively, while the number of type I cells on the bifurcation increased 82% and the number of type II cells increased nearly 38%. These hyperplastic epithelial cells demonstrate cell characteristics similar to those found after oxidant injuries. The mean cell volume and mean cell surface areas of type I epithelial cells decreased approximately 25%, while their thickness increased. Type II epithelial cells became larger and covered a larger average surface area. The epithelium on the first bifurcation was therefore made up of a greater population of intermediate or transitional cells.

Alveolar macrophages accumulated on the first bifurcation, resulting in a 10-fold increase in their volume and a 12-fold increase in their number locally. The interstitial fibroblasts were not changed at this time, and there was no evidence of edema or fibrosis. However, the volume and number of interstitial macrophages were increased by over 200%.

Figure 14 summarizes the changes induced by chrysotile asbestos. An acute imflammatory phase and a progressive fibrotic phase can be demonstrated. First alveolar duct bifurcations isolated from exposed animals that were kept in room air for 1 month after the brief exposure continued to show an increased tissue volume (Fig. 13c). However, the major site of tissue reaction at the first alveolar duct bifurcation at this time was different. The intense alveolar macrophage response had subsided. While the epithelial reactions were not fully resolved, epithelial cell volumes were close to the control values and there was no evidence

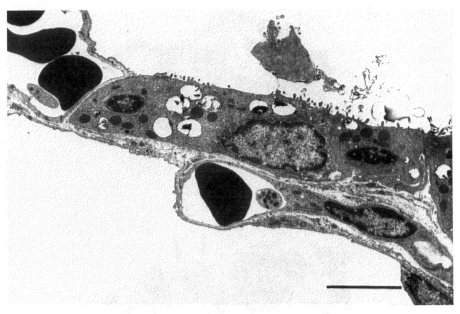

(b)

of epithelial hyperplasia. The volume and number of interstitial macrophages persisted at elevated levels, and the resident interstitial cell population, comprised of a mixture of fibroblasts, myofibroblasts, and smooth muscle cells, increased 25%. This was accompanied by a 53% increase in the volume of interstitial matrix in the first bifurcation. Qualitative morphological examination of the first bifurcation indicated that during the acute phase, asbestos fibers were found only in the air space or within cells such as alveolar macrophages or epithelial cells. The distribution of fibers in the first bifurcation shifted after a month, with the fibers being found in the interstitial matrix or in interstitial cells. Some fibers were found embedded in the center of microcalcifications, as described by Brody and Hill (1982).

   The "minimal injury" described in this model suggests that a brief inhalation of asbestos leads to fiber deposition on the first alveolar duct bifurcations followed by a local inflammatory response characterized by the influx of macrophages. The accumulation of alveolar macrophages is related to a complement-activated chemotactic activity (Warheit et al., 1985, 1986). Epithelial cell hyperplasia occurs either through direct epithelial interaction with asbestos or as a consequence of stimulation of injury by alveolar-macrophage-derived substances, or both (Davies et al., 1974; Craighead and Mossman, 1982; Kouzan et al., 1985; Leslie et al., 1986). As the chrysotile fibers translocate into the

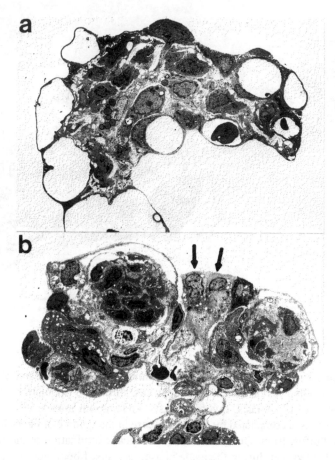

Figure 13 Low-magnification electron micrographs of first alveolar duct bifurcations isolated from sham- or asbestos-exposed rats. (a) Sham-exposed control bifurcation. (b) Asbestos-exposed bifurcation after 2 days of "recovery." Alveolar macrophages are found on the bifurcation (arrows). (c) Asbestos-exposed bifurcation after 1 month of "recovery." The bifurcation is still visibly enlarged. Bar = 15 μm.

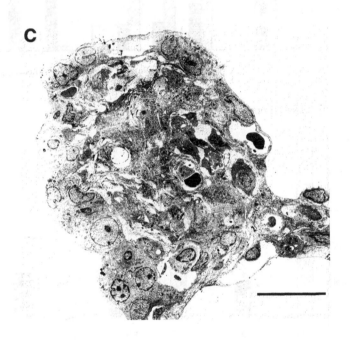

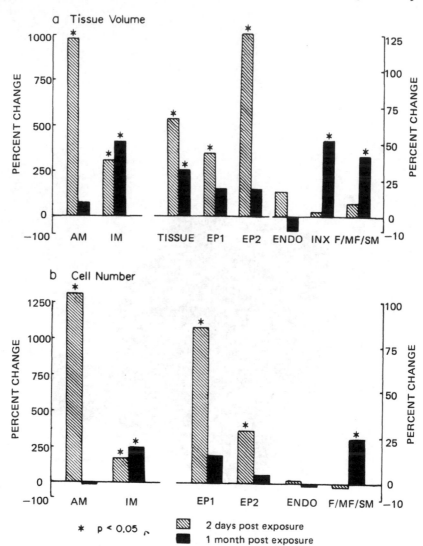

Figure 14 The pattern of tissue and cell reactions in first alveolar duct bifurcation 2 dyas or 1 month after rats were exposed briefly to chrysotile asbestos fibers. (a) Changes of volumes of alveolar macrophages (AM), interstitial macrophages (IM), type I epithelium (EP1), type II epithelium (EP2), fibroblast/myofibroblast/smooth muscle cells (F/MF/SM), interstitial matrix (IN-X), endothelium (ENDO), and total tissue. (b) Changes in the numbers of cells in the first bifurcations. Asterisks indicate statistically significant ($p < 0.05$) changes compared to their respective control values.

interstitium, the epithelial reaction subsides, but interstitial hyperplasia occurs and results in localized interstitial fibrosis.

IV. Conclusions

A. General Discussion

The examples described above illustrate how animal models can be established for experimental pathological examination of localized injuries, how morphometric analysis can be carried out to evaluate the injuries, and how results can be intepreted to define the types and pathogenesis of injuries. All the structural changes that occurred in these models can be considered minimal injuries because they are hard to detect qualitatively on whole lung sections or random samples of the lung. The quantitation of statistically significant changes in the case of oxidant- or asbestos-induced injuries required the application of two techniques: microdissection and morphometry.

It is apparent that restricting the study to the site of specific injury greatly enhanced the ability to detect subtle changes. Another less obvious advantage offered by studying a defined anatomical site is that due to the structural uniformity of a single defined anatomical region, the variance resulting from morphometric measurements of the tissue is reduced. Figure 15 illustrates the variances of type I epithelial cell volume between animals. Alveolar tissues were samples from normal unexposed rat lungs either randomly or specifically from the proximal alveolar region. As predicted (Snedecor and Cochran, 1980), variances between animals decreased as more animals were studied. The variance in type I epithelial cell volume from random samples of the alveolar region is consistently higher than that found in the proximal alveolar region. Reduced variance has the effect of making small changes more easily detected as statistically significant and, therefore, makes the test more sensitive.

The importance of studying localized injuries, however, is not limited to detecting small structural modifications. Because of the unique structure of the lung, with its branching airways leading into a vast gas-exchange area, injury can be different at various parts of the lung. This has been demonstrated by Pinkerton et al. (1986) in a study of rats exposed to asbestos for 1 year. It related the number of airway branch points preceding a specifically selected tissue block to the asbestos fiber content in that block (Roggli and Brody, 1984; Roggli et al., 1986). Microdissection of fixed but unembedded lung was carried out according to the method of Plopper et al. (1983). Tissue blocks distal to the terminal bronchioles in the dorsal, lateral, cranial, costolateral, and caudal regions of the middle right lobe were taken. Figure 16a plots the number of preceding branch points against the number of fibers found in these tissue blocks and shows that fiber load was diminished as the inhaled fibers passed more branch points. Figure 16b

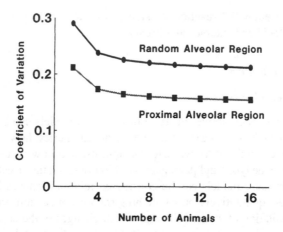

Figure 15 Sample variances of the volume of type I epithelium between animals. Volume of type I epithelium was measured from sections of tissue blocks randomly selected from the alveolar region of the lung (●) or selected from the proximal alveolar region by microdissection (■). Coefficient of variation is the square root of sample variance divided by the group mean. Sample variance between animals can be reduced by increasing the number of animals studied, but variance for the alveolar samples is consistently greater than the variance for the proximal alveolar samples.

compares the severity of tissue reactions in three of the five regions studied. It is clear that the toxic reaction to asbestos fiber inhalation was influenced by airway branching pattern. Similar situations may exist for a great variety of pollutants. This means that analysis of random tissue samples may not be adequate for the accurate interpretation of the injury pattern in the lung, even at high pollutant dosage that create apparently homogeneous lesions. Microdissection appears to be uniquely suited to studying the specific relationships between dosage and injury.

The high resolution offered by electron microscopy makes it possible to identify specific tissue or cell changes. In the asbestos study, for example, the association between fiber location, alveolar macrophage influx, and epithelial cell hyperplasia in the acute phase could be demonstrated and the relationship between fibroblast hypertrophy and hyperplasia, matrix accumulation, and the persistent presence of more interstitial macrophages at the later fibrotic phase could also be demonstrated. Analysis of individual cell characteristics, as demonstrated by the oxidant studies described here, illustrates how each type of cell in the epithelium reacts to the oxidant challenges in the process of injury and repair.

The importance of studying these localized injuries is twofold. For environmental toxicologists, these models and techniques offer a highly sensitive tool for evaluating the effects of pollutants. In the case of O_3, the data presented

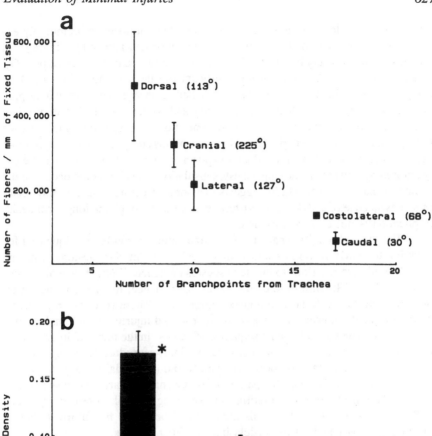

Figure 16 The relationship of airway branching pattern (from the trachea) to alveolar tissue density. (a) Fiber numbers found for parenchymal tissues in the dorsal, cranial, lateral, costolateral, and caudal region. The cumulative branch angle relative to the trachea for each airway is given in parentheses. (b) Alveolar tissue density in the cranial, costolateral, and caudal regions of control (CONT) and asbestos-exposed (Exp) animals. Alveolar tissue density represents the relative fraction of the alveolar region occupied by septal tissue. All values are expressed as mean ± S.D. *p<0.05 for comparison to control.

represent injury detected at ambient levels to which humans are commonly exposed. The sensitivity of the technique is demonstrated amply by Figure 9, in which structural remodeling of the lung is shown after exposure to 0.12 ppm O_3. In addition, there appears to be no significant treshhold for the effect of O_3 in causing a change in the type I epithelium. Whether dosages lower than 0.12 ppm O_3 can cause structural changes in the lung and whether a cummulative effect is caused by the product of time and concentration are interesting questions for further studies. The results presented by the asbestos study raised questions about the potential toxic effect of accidental exposures, such as those that may be encountered by workers removing asbestos insulation when breakage occurs in the protective barriers. This study suggests that some structural changes may result as a consequence of accidental short-term exposure, although the long-term health implication is not clear at this time.

For the pathologist, analysis of localized injury provides a unique tool for studying the early responses and pathogenesis of pulmonary lesions. Chronic models of in vivo reactions to inhaled asbestos (Wagner, 1965; Holt et al., 1966; Davis et al., 1978; Craighead et al., 1982), for instance, involve complex reactions of numerous cellular and matrix components. These reactions represent a mixture of acute injuries and progressions from old injuries, which make it difficult to elucidate the pathogenic sequence of events. In the minimal injury model illustrated, the dosage of asbestos delivered is highly reproducible and specifically located. The tissue response occurred rapidly and interstitial fibrosis developed within a short time. Because the pattern of tissue changes parallels those described for generalized diffused interstitial fibrosis (Wagner et al., 1980; Barry et al., 1983; Pinkerton et al., 1984), this model could be useful for future studies of the mechanisms that lead to particle-induced lung fibrosis.

B. Future Directions

We have identified three areas involving histological analyses in which advances can be made to widen the scope of our knowledge of structural–function relationships, cellular mechanisms, and possibly regulation of cell response to injury.

Cytodynamics

Although morphometric studies can provide evidence of population shifts, they measure only static characteristics of the cell population. No kinetic data about cell death, cell proliferation, or cell differentiation can be detailed by morphometric studies alone. Cytodynamic studies, however, can be a highly sensitive index of cell injury. The sample of NO_2-induced epithelial changes demonstrated that hypertrophy can be offset by cell spreading and cell proliferation. Cell proliferation can in turn be offset by cell differentiation. In mild injuries, the rate of cell proliferation and differentiation (repair) may keep pace with the rate of cell injury so that no overt injury can be detected, unless the rate of cell death is measured.

A model for normal cell turnover in mild injury is presented in Figure 17. This model was devised according to work by Cleaver (1967) and Hollis and Paradise (1978). In mild injuries, the rate of cell turnover reaches a new but elevated level of steady state after an initial phase of change in the rate of cell proliferation. The model, which mimicks alveolar epithelium, includes a progenitor cell (II) and a differentiated cell (I).

The rate constant K_B represents the rate of daughter cell II generation (cell birth). The constants K_I and K_E are rates of cell influx and efflux for the cell II population and K_i and K_e are the corresponding rate constants for cell population I. Because only half of the daughter cell II population represents true influx, $K_B = 2K_I$. Also, since cell populations are in equilibrium, $K_I = K_E = K_i = K_e = K_B/2$.

To measure the cell flux constants, the rate of cell birth had to be known. This can be accomplished by continuously labeling the dividing cells with [³H]thymidine and sampling the cell population for the number of labeled cells at regular intervals (Getreke et al., 1976). There are two method of labeling cell populations continuously: (1) with frequent injections (Getreke et al., 1976) of [³H]thymidine, and (2) with [³H]thymidine-filled miniosmotic pumps implanted subcutaneously or in the peritoneum (Gould et al., 1982; Theeuwes, 1980). The later method has been tried in our laboratory, to label cell populations during exposures to O_3 or NO_2, with satisfactory results.

The rate of cell fluxes for cell populations not in a steady state can also be calculated if the size of the population at each time is known. If the numbers of cell I at two points A and B are N_A and N_B, respectively, and the time lapse is Δt, the rate of cell I death is:

$$K_e = \frac{N_A - N_B}{\Delta t} + K_i$$

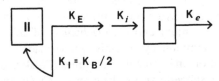

Figure 17 A diagram of a cell turnover model. The system consists of two cell populations in equilibrium, one (cell II) is the progenitor and the other (cell I) is the product cell. K_B, rate of cell birth; K_I, rate of cell influx of cell II population; K_E, rate of cell efflux of cell II population; K_i, rate of cell influx of cell I population; K_e, rate of efflux or rate of cell death of cell I population.

Labeling index studies have been performed with oxidant-exposed lung tissue. A higher turnover rate was found in the epithelia of the BAD junction (Evans et al., 1972, 1975, 1976; Stephens et al., 1978). Relatively high dosages of oxidants were used in these studies and cell turnover was studied by measuring labeling indices after a pulse injection of [^{3}H]thymidine. Unfortunately, rate constants cannot be calculated using this approach.

The major limitation of a cytodynamic study using continuous labeling is that it is time consuming. The morphometric and autoradiographic procedures are laborious. The turnover rate of epithelial cells in the lung, even in mild injury, is very low (Bertalanffy and Leblond, 1953; Leblond and Walker, 1956; Kleinerman, 1970; Evans et al., 1971) compared to other types of epithelium. This requires the use of large sample sizes. We believe, however, that cytodynamic studies will provide a fuller understanding of lung cell turnover in general and cell response and interaction in injury specifically.

Immunocytochemistry

Castleman et al. (1973a,b) have shown, using light microscopic histochemical studies, that the densities of staining for acid phosphatase and β-glucuronidase were present in the epithelium of terminal bronchioles and adjacent alveoli of O_3-exposed rats. They suggested that the enzymes might have leaked into the cytoplasm of epithelial cells due to O_3 damage to lysosomal membranes. Ozone inhalation was also reported to cause an increase in the amount of succinate oxidase in lung tissue and reductions in the amount of NADP cytochrome c reductase and glycose-6-phophate dehydrogenase (Mustafa et al., 1977). Changes in enzyme localization or concentration may be either a direct result of injury or a result of altered metabolism subsequent to injury (Mustafa and Tierney, 1978). These changes may take place before structure remodeling occurs. Biochemical alterations, therefore, are another sensitive indicator for early injury. Moreover, cell function is related to its profile of enzyme content and enzyme concentrations. For example, an increased concentration of CuZn superoxide dismutase in endothelial cells after exposure to 85% O_2 is believed to make endothelial cells more resistant to hyperoxia (Crapo et al., 1980; Chang et al., 1987). Studies of enzyme localization and enzyme concentration may provide valuable information on the nature of cell injury and the mechanism of cell responses.

Recent advances in immunocytochemistry have made it possible to examine the subcellular distribution and content of enzymes (Polak and Varndell, 1984). We have reported the subcellular localization of CuZn and Mn superoxide dismutases (SODs) in rat hepatocytes on ultrathin cryosections (Slot et al., 1986). Induction of both SODs in lung cells after exposure to 85% O_2 was also observed on ultrathin cryosections labeled with the SOD antibodies (Chang et al., 1987). Figure 18 shows Mn SOD localized in the mitochondria of a type II cell after

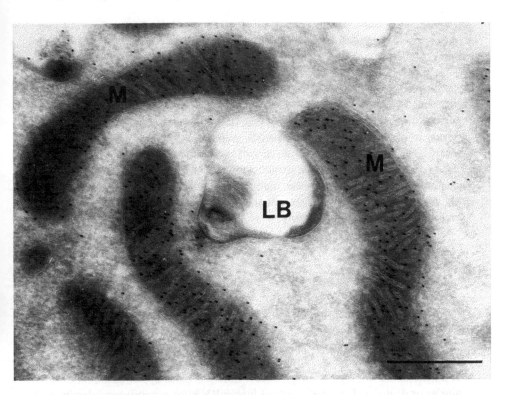

Figure 18 Localization of Mn SOD in the mitochondria of alveolar type II epithelial cells exposed to 85% O_2 for 14 days. Mn SOD is localized on ultrathin cryosections of the lung with 9 nm gold particles complexed with protein A. Lb, lamellar body; M, mitochondria. Bar = 0.5 μm.

85% O_2 exposure. Posthuma et al. (1987) have devised an ingenious method of quantitating the absolute enzyme concentration in cell compartments. The technique incorporates an en bloc enzyme standard of known concentration mounted alongside the tissue block. Labeling of antibody and protein A–gold on ultrathin cryosections can be limited to the surface of the section by embedding the composite block in polyacrylamide. This procedure makes it possible to compare labeling density among different organelles as well as to calculate enzyme concentrations from the ratio of labeling densities on an organelle and on the enzyme standard. This technique has been used to quantitate the concentration of CuZn superoxide dismutase in hepatocyte organelles (Slot et al., 1987). In addition, the total content of CuZn superoxide dismutase has been determined in liver by integrating morphometry and quantitative immunocytochemistry. The result obtained by immunocytochemical methods (0.39 mg CuZn SOD/g liver) closely approximates the result obtained by biochemical assay (0.38 mg CuZn SOD/g liver).

In the case of localized injuries, immunocytochemistry is the only practical method that can be used to study cellular or subcellular biochemical alterations. Just as EM morphometry dissects the analysis of tissue reactions to specific cell responses, immunocytochemistry may further define cell responses to specific enzyme changes. If appropriate probes (antibodies) are available, a host of metabolic processes and their role in cell response to injury can be examined. This has the potential to advance greatly our understanding of the most fundamental aspects of injury and cell response; namely, the interactions of reagent to specific subcellular components and intracellular protein modifications that lead to alterations in cell structure or function.

Molecular Biology

Cell functional and structural changes are ultimately regulated by changes in gene expression. In this sense, modification in the cell's gene expression is one of the earliest indicators of cell injury and an important component of the mechanism of cell response. New techniques in molecular biology have become available, and it is now possible to study the molecular basis for regulation of gene expressions of specific cellular enzymes. Although a body of literature already exists on the molecular biology of lung cancer, little work has appeared on other lung injuries or diseases. Elevated levels of lung mRNA for procollagen, elastin, and fibronectin have been found following acute lung injury induced by bleomycin (Kelley et al., 1985; Raghow et al., 1985). Mueller et al. (1987) recently demonstrated paraquat-induced mRNA staining on paraffin sections by in situ hybridization. In situ hybridization can be readily applied to the study of regulation of gene expression in localized injury.

By studying the cytodynamics of cell populations with autoradiography, quantitating enzyme concentrations with immunocytochemistry, and studying gene regulation using molecular biological techniques, one should be able to explore the mechanisms of cell responses to injury at every level of cell function.

References

Abercrombie, M. (1946). Estimation of nuclear population from microtome sections. *Anat. Rec.* **94**:239–247.

Barry, B. E., and Crapo, J. D. (1985). Application of morphometric methods to study diffuse and focal injury in the lung caused by toxic agents. *CRC Crit. Rev. Toxicol.* **14**:1–32.

Barry, B. E., Miller, F. J., and Crapo, J. D. 1982). Alveolar epithelial injury caused by inhalation of 0.25 ppm of ozone. *Adv. Mod. Environ. Toxicol.* 3:299–309.

Barry, B. E., Wong, K. C., Brody, A. R., and Crapo, J. D. (1983). Reaction of rat lungs to inhaled chrysotile asbestos following acute and subchronic exposures. *Exp. Lung Res.* 5:1–22.

Barry, B. E., Freeman, B. A., and Crapo, J. D. (1985a). Patterns of accumulation of platelets and polymorphonuclear leukocytes in rat lungs during exposure to 100% and 85% oxygen. *Am. Rev. Respir. Dis.* 132:548–555.

Barry, B. E., Miller, F. J., and Crapo, J.D. (1985b). Effects of inhalation of 0.12 and 0.25 ppm ozone on the proximal alveolar region of juvenile and adults rats. *Lab Invest.* 53:692–704.

Barry, B. E., Mercer, R. R., Miller, F. J., and Crapo, J. D. (1987). Effects of inhalation of 0.25 ppm ozone on the terminal bronchioles of juvenile and adults rats. *Exp. Lung Res.* in press.

Begin, R., Rola-Pleszczynski, M., Sirois, P., Lemaire, I., Nadeau, D., Bureau, M. A., and Masse, S. (1981). Early lung events following low-dose asbestos exposure. *Environ. Res.* 26:391–401.

Begin, R., Masse, S., and Bureau, M. A. (1982). Morphologic features and functions of the airways in early asbestosis in the sheep model. *Am. Rev. Respir. Dis.* 126:870–876.

Begin, R., Rola-Pleszczynski, M., Masse, S., Lemaire, I., Sirois, P., Boctor, M., Nadeau, D., Drapeau, G., and Bureau, M. A. (1983). Asbestos-induced injury in the sheep model: the initial alveolitis. *Environ. Res.* 30:195–210.

Bertalanffy, F. D., and Leblond, C. P. (1953). The continuous renewal of the two types of alveolar cells in the lung of the rat. *Anat. Rec.* 115:515–541.

Boorman, G. A., Schwartz, L. W., and Dungworth, D. L. (1980). Pulmonary effects of prolonged ozone insult in rats. *Lab Invest.* 43:108–115.

Brody, A. R., and Hill, L. H. (1982). Interstitial accumulation of inhaled chrysotile asbestos fibers and consequent formation of microcalcifications. *Am. J. Pathol.* 109:107–114.

Brody, A. R., and Roe, M. W. (1983). Deposition pattern of inorganic particles at the alveolar level in the lungs of rats and mice. *Am. Rev. Respir. Dis.* 128:724–729.

Brody, A. R., Hill, L. H., Adkins, B. A., and O'Connor, R. W. (1981). Chrysotile asbestos inhalation in rats: deposition pattern and reaction of alveolar epithelium and pulmonary macrophages. *Am. Rev. Respir. Dis.* 123:670–679.

Brody, A. R., Warheit, D. B., Chang, L., Roe, M. W., George, G., and Hill, L. H. (1984). Initial deposition in pattern of inhaled minerals and consequent pathogenic events at the alveolar level. *Ann. N.Y. Acad. Sci.* 428:108–120.

California Air Resources Board (CARB) (1983). California Air Quality Data, Vol. 15. Annual Summary.

Castleman, W. L., Dungworth, D. L., and Tyler W. S. (1973a). Cytochemically detected alterations of lung acid phosphatase reactivity following ozone exposure. *Lab. Invest.* **29**:310–319.

Castleman, W. L., Dungworth, D. L., and Tyler, W. S. (1973b). Histochmically detected enzymatic alterations in rat lungs exposed to ozone. *Exp. Mol. Pathol.* **19**:402–421.

Chang, L. and Crapo, J. D. (1985). Morphometric analysis of focal lung injury caused by low levels of nitrogen dixoide. *Acta Stereol.* **4**:201–206.

Chang, L., Graham, J. A., Miller, F. J., Ospital, J. J., and Crapo, J. D. (1986). Effect of subchronic inhalation of low concentrations of nitrogen dioxide. The proximal alveolar region of juvenile and adult rats. *Toxicol. Appl. Pharmacol.* **83**:46–61.

Chang, L., Slot, J. W., and Crapo, J. D. (1987). Immunocytochemical demonstration of superoxide dismutase induction in alveolar cells during oxygen adaptation. *Am. Rev. Respir. Dis.* **135**:A16.

Cleaver, J. E. (1967). *Thymidine Metabolism and Cell Kinetics*. Amsterdam, North Holland Publishers.

Craighead, J. E., and Mossman, B. T. (1982). The pathogeneis of asbestos-associated disease. *N. Engl. J. Med.* **306**:1446–1455.

Craighead, J. E., Abraham, J. L., Churg, A., Green, T-H.Y., Kleinerman, J., Pratt, P. C., Seemayer, T. A., Vallyathan, V., and Weill, H. (1982). The pathology of asbestos-associated disease of the lung and pleural cavities: diagnostic criteria and proposed grading schema. *Arch. Pathol. Lab. Med.* **106**:543–596.

Crapo, J. D., Peters-Golden, M., Marsh-Salin, J., and Shelburne, J. D. (1978). Pathologic changes in the lungs of oxygen-adapted rats. A morphometric analysis. *Lab. Invest.* **39**:640–653.

Crapo, J. D., Barry, B. E., Foscue, H. A., and Shelburne, J. (1980). Structural and biochemical changes in rat lungs occurring during exposures to lethal and adaptive doses of oxygen. *Am. Rev. Respir. Dis.* **122**:123–143.

Crapo, J. D., Barry, B. E., Gehr, P., Bachofen, M., and Weibel, E. R. (1982). Cell number and cell characteristics of the normal human lung. *Am. Rev. Respir. Dis.* **125**:740.

Crapo, J. D., Barry, B. E., Chang, L., and Mercer, R. R. (1984). Alterations in lung structure caused by inhalation of oxidants. *J. Toxicol. Environ. Health* **13**:301–321.

Crapo, J. D., Huang, Y., Chang, L., and Mercer, R. R. (1987). Assessment of lung injury caused by oxidant air pollutants using electron microscopic morphometric techniques. *Proceedings of the Air Pollution Control Association*, 80th Meeting.

Cruz-Orive, L. M. (1980). On the estimation of particle number. *J. Microsc.* **120**:15–27.

Cruz-Orive, L. M. (1987). Particle number can be estimated using a disector of unknown thickness: the selector. *J. Microsc.* **145**:121–142.

Davies, P., Allison, A. C., Ackerman, J., Butterfield, A., and Williams, S. (1974). Asbestos-induced selective release of lysosomal enzyme from mononuclear phagocytes. *Nature* **251**:423–425.

Davis, J. M. G., Beckett, S. T., Bolton, R. E., Collings, P., and Middleton, A. P. (1978). Mass and number of fiber in the pathogenesis of asbestos-related lung disease in rats. *Br. J. Cancer* **37**:673–688.

Dawson, S. V., and Schenker, M. B. (1979). Health effects of inhalation of ambient concentrations of nitrogen dioxide. *Am. Rev. Respir. Dis.* **120**:281–292.

DeHoff, H., and Rhines, F. N. (1968). *Quantitative Microscopy*. New York, McGraw-Hill.

Dungworth, D. L., Castleman, W. L., Chow, C. K., Mellick, P. W., Mustafa, M. G., Tarkington, B., and Tyler, W. S. (1975). Effect of ambient levels of ozone on monkeys. *Fed. Proc.* **34**:1670–1674.

Elias, H., and Hyde, D. (1983). *A Guide to Practical Stereology*. New York, S. Karger.

Eustis, S. L., Schwartz, L. W., Kosch, P. C., and Dungworth, D. L. (1981). Chronic bronchiolitis in nonhuman primates after prolonged ozone exposure. *Am. J. Pathol.* **105**:121–137.

Evans, M. J., Bils, R. F., and Clayton, C. T. (1971). Effects of ozone on cell renewal in pulmonary alveoli of aging mice. *Arch. Environ. Health* **22**:450–453.

Evans, M. J., Stephens, R. J., Cabral, L. J., and Freeman, G. (1972). Cell renewal in the lungs of rats exposed to low levels of NO_2. *Arch. Environ. Health* **24**:180–188.

Evans, M. J., Cabral, L. J., Stephens, R. J., and Freeman, G. (1975). Transformation of alveolar type II cells to type I cells following exposure to NO_2. *Exp. Mol. Pathol.* **22**:142–150.

Evans, M. J., Johnson, L. V., Stephens, R. J., and Freeman, G. (1976). Cell renewal in the lungs of rats exposed to low levels of ozone. *Exp. Mol. Pathol.* **24**:70–83.

Freeman, G., and Haydon, G. B. (1964). Emphysema after low-level exposure to NO_2. *Arch. Environ. Health* **8**:125–128.

Freeman, G., Stephens, R. J., Crane, S. C., and Furiosi, N. J. (1968). Lesion of the lung in rats continuously exposed to two parts per million of nitrogen dioxide. *Arch. Environ. Health* **17**:181–192.

Freeman, G., Crane, S.C., Furiosi, M. J., Stephens, R. J., Evans, M. J., and Moore, W. D. (1972). Covert reduction in ventilatory surface in rats during prolonged exposure to subacute nitrogen dioxide. *Am. Rev. Respir. Dis.* **106**:563–579.

Getreke, D., Jegsen, A., and Gross, R. (1976). Continuous labeling method for autoradiographic analysis of cell cycle parameters in steady state systems. *Experimentia* **32**:1088–1090.

Gould, T. R. L., Brunmette, D. M., and Dorey, J. (1982). Cell turnover in the periodontal ligament determined by continuous infusion of ^{3}H-thymidine using osmotic minipumps. *J. Periodont. Res.* **17**:662–668.

Greeley, D., and Crapo, J. D. (1978). Practical approach to the estimation of the overall mean caliper diameter of a population of spheres. *J. Micros.* **114**:261–269.

Greeley, D., Crapo, J. D., and Vollmer, R. T. (1978). Estimation of the mean caliper diameter of cell nuclei. I. Serial section reconstruction method and endothelial nuclei from human lung. *J. Microsc.* **114**:31–39.

Gundersen, H. J. G. (1977). Notes on the estimation of the numerical density of arbitrary profiles: the edge effect. *J. Microsc.* **111**:219–224.

Gundersen, H. J. G., and Jensen, E. B. (1983). Particle sizes and their distributions estimated from line-and-point sampled intercepts. Including graphic unfolding. *J. Microsc.* **131**:291–310.

Gundersen, H. J. G., and Jensen, E. B. (1985). Stereological estimation of the column-weighted mean volume of arbitrary particles observed on random sections. *J. Microsc.* **138**:127–142.

Haies, D. M., Gil, J., and Weibel, E. R. (1981). Morphometric study of rat lung cells. I. Numerical and dimensional characteristics of parenchymal cell pupulation. *Am. Rev. Respir. Dis.* **123**:533–541.

Hollis, G. T. B., and Paradise, L. J. (1978). Cytokinetics of lung. In *Pathogenesis and Therapy of Lung Cancer*. Edited by C. C. Harris. New York, Marcel Dekker, pp. 369–418.

Holmes, A. H. (1927). Petrographic Methods and Calculations. London, Murby and Company.

Holt, P. F., Mills, J., and Young, D. K. (1966). Experimental asbestosis in the guinea pig. *J. Pathol.* **92**:185–195.

Howard, V., Reid, S., Baddeley, A., and Boyde, A. (1985). Unbiased estimation of particle density in the tandem scanning reflected light microscope. *J. Microsc.* **138**:203–212.

Hyde, D., Orthoefer, J., Dungworth, D., Tyler, W., Carrter, R., and Lum, H. (1978). Morphometric and morphologic evaluation of pulmonary lesions in beagle dogs chronically exposed to high ambient levels of air pollutants. *Lab Invest.* **38**:455–469.

Kauffman, S. L., Burri, P. H., and Weibel, E. R. (1974). The postnatal growth of the rat lung. II. Autoradiography. *Anat. Rec.* **180**:63–76.

Kelley, J., Chrin, L., Shull, S., Rowe, D. W., and Cutroneo, Q. K. R. (1985). Bleomycin selectively elevates mRNA levels for procollagen and fibronectin following acute lung injury. *Biochem. Biophys. Res. Commun.* **131**:838–843.

Kleinerman, J. (1970). Effects of NO_2 in hamsters: Autoradiographic and electron microscopic aspects. In *Inhalation Carcinogenesis*. Edited by M. G. Hana, Jr., P. Nettesheim, and R. Gilbert. Oakridge, TN, US Atomic Energy Commission, pp. 271–279.

Kouzan, S., Brody, A. R., Nettesheim, P., and Eling, T. (1985). Production of arachidonic acid metabolites by macrophages exposed to *in vitro* to asbestos, carbonyl iron particles, or calcium ionophore. *Am. Rev. Respir. Dis.* **131**:624–632.

Leblond, C. P., and Walker, B. E. (1956). Renewal of cell populations. *Physiol. Rev.* **36**:255–276.

Leslie, C. L., McCormick-Shannon, K., Cook, J. L., and Mason, R. J. (1986). Macrophages stimulate DNA synthesis in rat alveolar type II cells. *Am. Rev. Respir. Dis.* **132**:1246–1252.

Loud, A. V. Peiro, A., Giacomelli, F., and Weiner, J. (1978). Absolute morphometric study of myocardial hypertrophy in experimental hypertension. I. Determination of myocyte size. *Lab. Invest.* **38**:586–596.

Lum, H., Schwartz, L. W., Dungworth, D. L., and Tyler, W. S. (1978). A comparative study of cell renewal of exposure to ozone or oxygen. Response of terminal bronchiolar epithelium in the rat. *Am. Rev. Respir. Dis.* **118**:335–345.

Mellick, P. W., Dungworth, D. L., Schwartz, L. W., and Tyler, W. S. (1977). Short-term morphologic effects of high ambient levels of ozone on lungs of rhesus monkeys. *Lab. Invest.* **36**:82–90.

Merz, W. A. (1967). Die streckenmessung an gerichteten strukturen im mikroskop and ihre an endung zur bestimmung von oberflachen-volume-relationen im knochengewebe. *Mikroskopie* **22**:132–140.

Mueller, M. P., Dubaybo, B. A., Allen, P. L., and Thet, L. A. (1987). Patterns of extracellular matrix gene expression during the genesis of pulmonary fibrosis. *Am. Rev. Respir. Dis.* **135**:A251.

Mustafa, M. G., and Tierney, D. F. (1978). Biochemical and metabolic changes in the lung with oxygen, ozone and nitrogen dioxide toxicity. *Am. Rev. Respir. Dis.* **118**:1061–1090.

Mustafa, M. G., Hacker, A. D., Ospital, J. J., Hussain, Z. M., and Lee, S. D. (1977). Biochemical effects of environmental oxidant pollutants in animal lungs. In *Biochemical Effects of Environmental Pollutants*. Edited by S. D. Lee and B Peirano. Ann Arbor, MI, Ann Arbor Science Publishers, pp. 59–96.

National Academy of Sciences Committee (NASC) on the Medical and Biological Effects of Environmental Pollutants (1977). Atmospheric concentrations of photochemical oxidants. In *Ozone and Other Photochemical Oxidants*. Washington, D.C., National Academy of Sciences, pp. 126–194.

Parkinson, D. R., and Stephens, R. J. (1973). Morphological surface changes in the terminal bronchiolar region of NO_2-exposed rat lung. *Environ. Res.* 6:37–51.

Pinkerton, K. E., and Crapo, J. D. (1985). Morphometry of the alveolar region of the lung. In *Handbook of Experimental Pharmacology*, Vol. 75. Edited by H. P. Witschi and J. D. Brain. Berlin, Springer-Verlag, pp. 259–285.

Pinkerton, K. E., Brody, A. R., McLaurin, D. A., Adkins, B. Jr., O'Connor, R. W., Pratt, P. C., and Crapo, J. D. (1983). Characterization of three types of chrysotile asbestos after aerosolization. *Environ. Res.* 31:32–53.

Pinkerton, K. E., Pratt, P. C., Brody, A. R., and Crapo, J. D. (1984). Fiber localization and its relationship to lung reaction in rats after chronic inhalation of chrysotile asbestos. *Am. J. Pathol.* 117:142–156.

Pinkerton, K. E., Plopper, C. G., Mercer, R. R., Roggli, V. L., Patra, A. L., Brody, A. K., and Crapo, J. D. (1986). Airway branching patterns influence asbestos fiber location and the extent of tissue injury in the pulmonary parenchyma. *Lab. Invest.* 55:688–695.

Plopper, C. G., Dungworth, D. L., Tyler, W. S., and Chow, C. K. (1979). Pulmonary alterations in rats exposed to 0.2 and 0.1 ppm ozone: a correlated morphological and biochemical study. *Arch. Environ. Health* 34:390–395.

Plopper, C. G., Mariassy, A. T., and Lollini, L. O. (1983). Structures as revealed by airway disection: a comparison of mammalian lungs. *Am. Rev. Respir. Dis.* 128:S4–S7.

Polak, J. M., and Varndell, I. M. (1984). *Immunolabelling for Electron Microscopy*. New York, Elsevier.

Posthuma, G., Slot, J. W., and Genze, H.J. (1987). Usefulness of the immunogold technique in quantitation of a soluble protein in ultrathin sections. *J. Histochem Cytochem.* 35:405–410.

Raghow, R., Luvie, S., Seyer, J. M., and Kang, A. H. (1985). Profiles of steady state levels of messenger RNSs coding for type I procollagen elastin, and fibronectin in hamster lungs undergoing bleomycin-induced interstitial pulmonary fibrosis. *J. Clin. Invest.* 76:1733–1739.

Roggli, V. L., and Brody, A. R. (1984). Changes in numbers and dimensions of chrysotile asbestos fibers in lungs of rats following short-term exposure. *Exp. Lung Res.* 7:133–147.

Roggli, V. L., Pratt, P. C., and Brody, A. R. (1986). Asbestos content of lung tissue in asbestos associated disease: a study of 110 cases. *Br. J. Ind. Med.* 43:18–28.

Saltykov, S. V. (1967). The determination of the size distribution of particles in an opaque material from measurement of the size distribution of their sections. In *Stereology*. Edited by H. Elias. New York, Springer, pp. 163–173.

Scherle, W. (1970). A simple method for volumetry of organs in quantitative stereology. *Mikroskopie* **26**:57–60.

Schwartz, L. W., Dungworth, D. L., Mustafa, M.G., Takington, B. K., and Tyler, W. S. (1976). Pulmonary responses of rats to ambient levels of ozone. Effects of 7-day intermittent or continuous exposure. *Lab. Invest.* **34**:565–578.

Selikoff, I. J., and Lee, D. H. K. (1978). *Asbestos and Disease*. New York, Academic Press.

Slot, J. W., and Geuze, H. J. (1982). Ultracryotomy of polyacrylamide embedded tissue for immunoelectron microscopy. *Biol. Cell.* **44**:325–328.

Slot, J. W., Geuze, H. J., Freeman, B. A. and Crapo, J. D. (1986). Intracellular localization of the copper-zinc and manganese superoxide dismutases in rat liver parenchymals cells. *Lab Invest.* **55**:363–371.

Slot, J. W., Posthuma, G., Chang, L., Crapo, J. D., and Geuze, H. J. (1987). Quantitative assessment of immuno-gold labeling in cryosections. In *Immuno-Gold Probes in Cell Biology*. Edited by A. J. Verkleij and J. L. M. Leunissen. Boca Raton, FL, CRC Press.

Snedecor, G. W., and Cochran, W. G. (1980). *Statistical Methods*. Ames, IA, Iowa State University Press.

Stephens, R. J., Freeman, G., and Evans, M. J. (1972). Early response of lungs to low levels of nitrogen dioxide. *Arch. Environ. Health* **24**:160–179.

Stephens, R. J., Sloan, M. F., Evans, M. J., and Freeman, G. (1973). Early response of lung to low levels of ozone. *Am. J. Pathol.* **74**:31–38.

Stephens, R. J., Sloan, M. F., Groth, D. G., Negi, D. S., and Lunan, K. D. (1978). Cytologic response of postnatal rat lungs to O_3 and NO_2 exposure. *Am. J. Pathol.* **93**:183–199.

Sterio, D. C. (1983). The unbiased estimation of number and sizes of arbitraray particles using the disector. *J. Microsc.* **134**:127–136.

Stokinger, H. E., Wagner, W. D., and Dobrogorski, O. J. (1957). Ozone toxicity studies. III. Chronic injury to lungs of animals following exposure to a low level. *Arch. Ind. Health* **16**:514–522.

Theuwes, F. (1980). Delivery of active agents by osmosis. In *Controlled Release Technologies: Methods, Theory, and Applications*, Vol. II. Edited by A. F. Kydonieus. Boca Raton, FL, CRC Press, pp. 195–205.

Timbrell, V., Hyett, A. W., and Skidmore, J. W. (1968). A simple dispenser for generating dust clouds from standard reference samples of asbestos. *Ann. Occup. Hyg.* **11**:273–281.

Timbrell, V., Skidmore, J. W., Hyett, A. W., Wagner, J.C. (1970). Exposure chambers for inhalation experiments with standard reference samples of asbestos of the International Union Against Cancer. *Aerosol. Sci.* 1:215–223.

Underwood, E. E. (1970). *Quantitative Stereology*. Reading, MA, Addison-Wesley.

Wagner, J. C., Barry, B., Skidmore, J. W., and Pooley, F. C. (1980). The comparative effects of three chrysotiles by injection and inhalation in rats. In *Biological Effects of Mineral Fibers*. Edited by J. C. Wagner. Lyon, IARC Scientific Publications, pp. 363–372.

Warheit, D. B., Chang, L., Hill, L. H., Hook, G. E. R., Crapo, J. D., and Brody, A. R. (1984). Pulmonary macrophage accumulation and asbestos-induced lesions at sites of fiber deposition. *Am. Rev. Respir. Dis.* 129:301–310.

Warheit, D. B., George, G., Hill, L. H., Snyderman, R., and Brody, A. R. (1985). Inhaled asbestos activates a complement-dependent chemoattractant for macrophages. *Lab. Invest.* 52:505–514.

Warheit, D. B., Hill, L. H., George, G., and Brody, A. R. (1986). Time course of chemotactic factor generation and the corresponding macrophage response to asbestos inhalation. *Am. Rev. Respir. Dis.* 134:128–133.

Weibel, E. R. (1963). *Morphometry of the Human Lung*. New York, Springer-Verlag.

Weibel, E. R. (1973). Stereological techniques for electron microscopic morphometry. In *Principle and Techniques of Electron Microscopy*, Vol. 3. Edited by M. Hayat. New York, Von Nostrand, Reinhold, p. 237.

Weibel, E. R. (1980a). *Stereological Methods*, Vol. 1. *Practical Methods for Biological Morphometry*. London, Academic Press.

Weibel, E. R. (1980b). *Stereological Methods*, Vol. 2. *Theoretical Foundations*. London, Academic Press.

Woody, D., Woody, E., and Crapo, J. D. (1979). Determination of the means caliper diameter of lung nuclei by a method which is independent of shape assumptions. *J. Microsc.* 118:421–477.

21

Acute Alveolar Injury: Experimental Models

STEPHEN F. RYAN

St. Luke's-Roosevelt Hospital
New York, New York

I. Introduction

When Ashbaugh and co-workers (1967) first called attention to the adult respiratory distress syndrome (ARDS) they described a fairly consistent clinical-pathologic picture. The patients were previously healthy, young to middle-aged adults, most of whom suffered a clearly recognized catastrophic event, usually trauma, and soon thereafter developed dyspnea, refractory hypoxemia, decreased lung compliance, and diffuse infiltrates on chest x-ray. Microscopically, the lungs of these patients showed capillary congestion, diffuse microatelectasis, and hyaline membranes. The authors noted the similarity of these findings to those in the respiratory distress syndrome of the newborn. Since then, the number of conditions associated with this histopathologic picture, which has come to be known as acute alveolar injury (AAI) (Bowden, 1981) or diffuse alveolar damage (Katzenstein et al., 1976), has steadily increased (Ryan, 1983) (Table 1). What is most notable about these conditions or antecedent events is their variety and the apparent lack of a common mechanism by which this quite stereotyped pattern of lung injury and repair might be precipitated.

Attempts to reproduce acute lung injury in experimental animals began in the 1960s. Most employed hemorrhagic hypotension, tissue injury, or sepsis in

Table 1 Causes or Antecedent Events in 53 Cases of Histologically Proven AAI (Open Biopsy-29, Autopsy-16, Both Open Biopsy and Autopsy-8)

Unknown spontaneous onset	Infectious-like prodrome	17
Uncertain; multiple possible causes		
Sepsis vs. shock	5	
Sepsis vs. chemotherapy	2	
Sepsis vs. aspiration	1 ⟹	11
Shock vs. chemotherapy	1	
Aspiration vs. shock	1	
Post surgical	1	
Infections		
Bacterial pneumonia		6
Pneumocystis pneumonia		3
Miliary tuberculosis		3
Toxoplasma pneumonia		1
Aspiration		3
Sepsis		3
Trauma		2
Drug overdose		2
Lupus		1
Hemorrhagic shock		1

AAI was strictly defined (see text) and judged to be the major cause of respiratory failure. In the cases associated with pulmonary infections (bacterial pneumonia, miliary tuberculosis, toxoplasmosis, *pneumocystis carinii*) AAI coexisted with and clearly affected many areas of lung not involved by the infectious lesion. Description of the ultrastructural features is based on examination of 24 cases (12 open biopsies, 6 autopsies, 6 open biopsies folowed by autopsy). In the autopsies, the lungs were removed within two hours postmortem and fixed by tracheal inflation with cold formol-gluteraldehyde at a pressure of 40 cm H_2O.

efforts to recapitulate the antecedents most frequently recognized in humans (Willwerth et al., 1967; Pomerantz and Eisman, 1968; Sealy, 1968; Clowes et al., 1968; Cook, 1968). While various physiologic abnormalities and even abnormalities of surfactant were induced (Henry, 1968), structural changes in the lung were either incompletely described or not described at all. When given, they usually included patchy congestion, edema, and atelectasis without hyaline membranes, although in one early study in dogs 24 hours after a 4-hour period of hemorrhagic shock, electron microscopic examination revealed apparently selective injury to alveolar type II cells (Barkett et al., 1969).

A 1976 conference on acute respiratory failure sponsored by the National Institutes of Health (Murray, 1977) reviewed possible mechanisms which might cause acute alveolar injury and the known structural and functional responses to such injury. Noting the lack of a valid experimental model as a major impediment to progress, the conference included among its recommendations for future efforts the development of such models. Since then information derived from several models has provided critical insights into possible mechanisms of human acute lung injury and the structural and metabolic consequences thereof.

II. Morphologic Features of AAI

Because AAI is defined morphologically, a clear understanding of its structural evolution is critical in evaluating the relevance of any experimental model to the condition seen in humans. Several excellent descriptions of the histologic and electron microscopic features of AAI are available and they are in general agreement on the more important features (Katzenstein et al., 1976; Pratt, 1978; Bachofen and Weibel, 1977, 1982, and Nash et al., 1974). The following description is based upon the authors' light microscopic examination of 53 cases, 24 of which were also studied by electron microscopy (Ryan 1983). The clinical features of these cases are presented in Table 1.

While AAI is often presented as a two-stage process, an early exudative phase and a later proliferative phase (Bachofen and Weibel, 1977), it is more accurate to think of it as a continually evolving process resulting from an acute episode of cellular injury and requiring roughly two weeks to complete. This consistent picture is often partly obscured or complicated by repeated episodes of injury or by therapeutic hyperoxia.

The earliest histologic changes recognizable as AAI are seen 24 to 48 h after the onset of respiratory symptoms. At that time the lung architecture is normal but the peribronchiolar and perivascular connective tissue is edematous and patches of alveoli are often fluid filled. Small airspaces are lined by hyaline membranes which, at first, are more prominent in alveolar ducts where they often selectively cover the inner edges of the alveolar septa (the alveolar entrance rings)

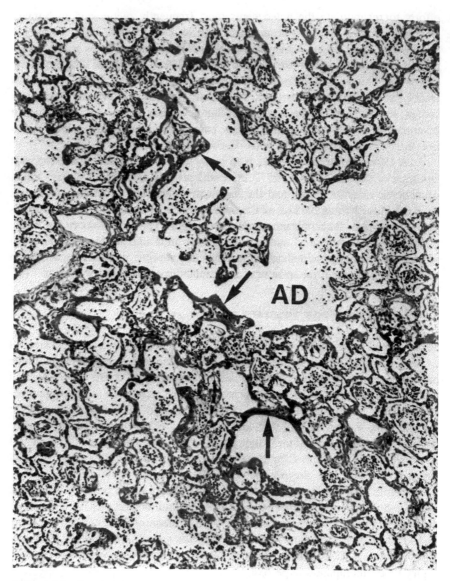

Figure 1 Microscopic section of lung from a patient 5 days after onset of respiratory failure following massive trauma with shock. Lung architecture is only minimally disturbed by slight enlargement of alveolar ducts and shrinkage of alveoli. The airspaces are lined by hyaline membranes (arrows) (hematoxylin-eosin × 87). Reprinted from Ryan, 1983.

(Figs. 1, 2). Over the next 3-5 days, the hyaline membranes thicken and become more extensive. Typically they are waxy and acidophilic and they sometimes contain fragmented and pyknotic nuclei (Fig. 3). Electron microscopy during this phase shows extensive injury of alveolar epithelium. Type I cells are swollen, fragmented, or completely absent (Fig. 4) and similar changes are often seen in the type II cells (Fig. 5). Hyaline membranes invariably lie on basement membranes whose epithelium is destroyed. They are composed of cell membranes, degenerated organelles, and nuclei admixed with granular debris, indicating that they are, at least in part, the remains of destroyed epithelial cells (Fig. 6). Often, capillary lumens contain neutrophils, but fibrin or platelet thrombi are seen only infrequently. Occasionally capillary endothelial cells are swollen or lifted from their basement membranes but extensive injury or separation of cells at junction complexes are seldom seen.

By 5 to 7 days after the onset of symptoms, hyaline membranes are extensive and thick and regeneration of alveolar epithelium, which usually begins 2-3 days after onset becomes prominent (Fig. 7). The regenerating epithelial cells first appear in alveoli near small blood vessels and airways in the centriacinar area. They are large and round or elliptical and contain vacuolated cytoplasm and large nuclei with prominent nucleoli and occasional mitoses. Between 5-7 and 10-12 days after onset, these regenerating cells increase in number, until many alveoli are completely lined by them. During this time they steadily become more uniform and cuboidal (Fig. 8). Electron microscopy reveals that many of the regenerating cells during this 3-7 day period closely resemble the epithelial cells of the fetal lung and are devoid of lamellar bodies (Fig. 9). By 7 days, some alveolar lining cells are recognizable as mature type II cells by the presence of typical cytoplasmic lamellar bodies (Fig. 10). The number of these apparently mature type II cells increases until the 10th to the 12th day, when they comprise the entire population of regenerating cells (Fig. 11). Particularly in the early part of this regenerative period, some of the epithelial cells have bizarre elliptical shapes and many irregular surface microvilli. These may be regenerating cells which themselves are being injured, perhaps by the same mechanism which initiated the original injury, or by therapeutic hyperoxia. In milder cases of AAI the process may progress to recovery with little or no residual disturbance of lung architecture; however, in the most severe cases, architectural disorganization becomes marked and irreversible. This disorganization is caused principally by two processes, namely irreversible closure of alveoli and organization of hyaline membranes (Ryan, 1983). Irreversible alveolar closure begins as early as 5-7 days after onset. During this period, clusters of shrunken or closed alveoli can be seen interspersed among enlarged small airspaces, most of which appear to be alveolar ducts or respiratory bronchioles (Fig. 12). In the ensuing days, alveolar closure becomes more extensive and the remaining open spaces more enlarged (Fig. 13). Ultrastructurally, the closed alveoli are devoid of epithelium and the denuded

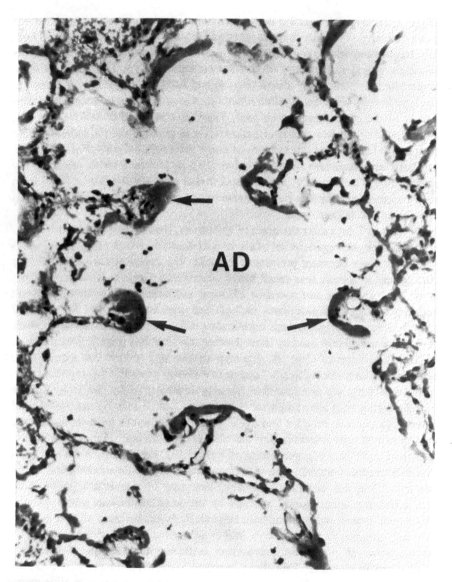

Figure 2   Alveolar duct from the same section depicted in Figure 1. Hyaline membranes (arrows) selectively involve the inner edges of the alveolar septa which form the alveolar entrance rings. (Hematoxylin-eosin × 200.) Reprinted from Ryan, 1983.

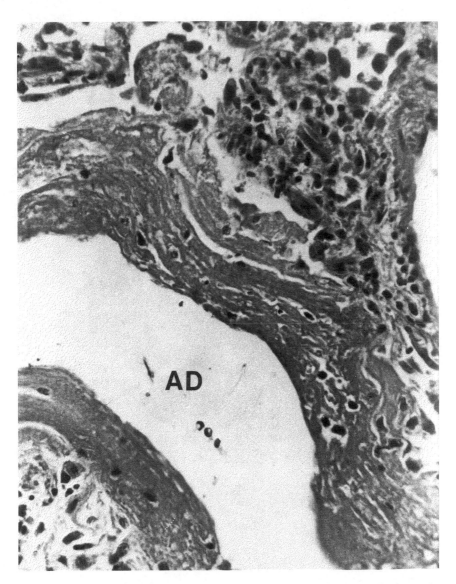

Figure 3 A typical, well developed hyaline membrane is thick and waxy and contains pyknotic and fragmented nuclei. (Hematoxylin and eosin × 366.) Reprinted from Ryan, 1983.

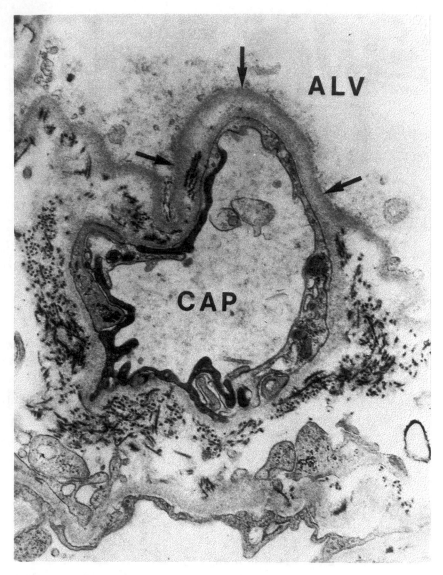

Figure 4   Electron micrograph of an alveolar septum from a patient 5 days after onset of respiratory failure following trauma. Histologic sections showed typical AAI. The alveolar epithelium is destroyed leaving a denuded basement membrane (arrows) while the capillary endothelium is intact (×16,200). Reprinted from Ryan, 1983.

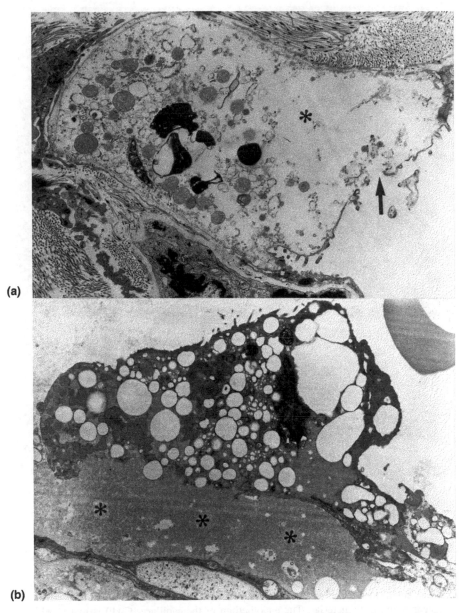

Figure 5 Electron micrographs of type II alveolar epithelial cells from 2 cases of early AAI. In (a) the cell membrane is disrupted (arrow) and the cytosol is rarified (asterisk) (×16,000). In (b) the cell, recognizable as a type II cell by its content of 1 or 2 lamellar bodies, is lifted from the basement membrane by fluid (asterisks) and the cytosol is vacuolated. (×16,000) Reprinted from Ryan, 1983.

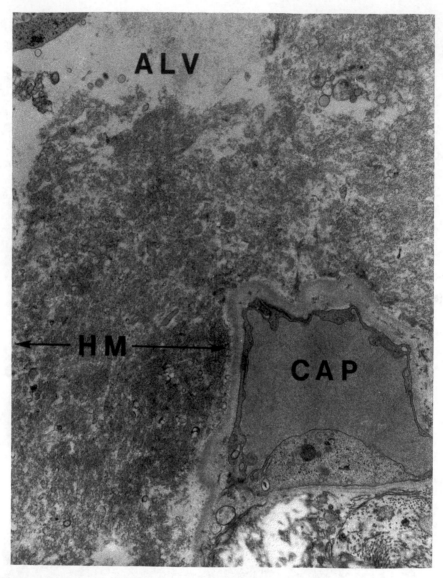

Figure 6 Electron micrograph of a typical hyaline membrane (HM). It lies on a denuded basement membrane and is composed of remnants of organelles admixed with flocculent or granular material. The endothelium of the capillary (CAP) is intact. ALV = alveolus (×10,200) Reprinted from Ryan, 1983.

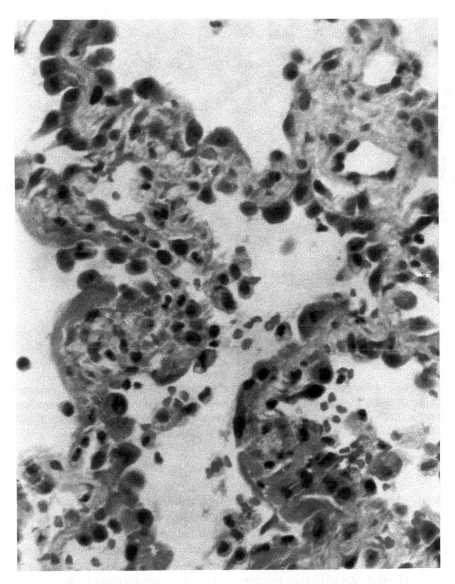

Figure 7 Early epithelial regeneration in a patient with typical AAI of unknown cause 6 days after onset of respiratory symptoms. Alveoli are partly lined by irregular epithelial cells with large, hyperchromatic nuclei. (Hematoxylin-eosin × 400.) Reprinted from Ryan 1983.

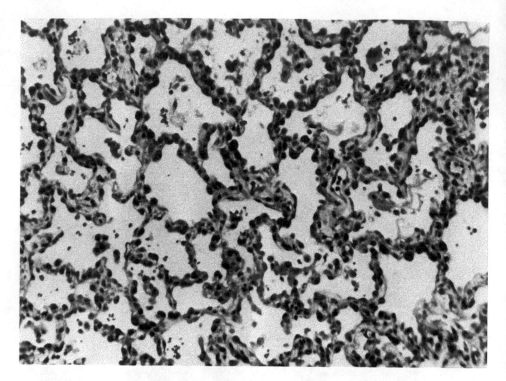

Figure 8 Advanced epithelial regeneration in a patient with typical AAI of unknown cause 14 days after onset of respiratory symptoms. In this area of the biopsy the architecture has not been disturbed. Airspaces are almost completely lined by fairly regular cuboidal cells. (Hematoxylin-eosin × 210.) Reprinted from Ryan 1983.

basal laminae of opposite sides of the alveolus lie close to or in direct contact with one another (Figs. 14 and 15). We have referred to this closure as irreversible because the spaces cannot be opened by inflating the lung with liquid at high inflation pressures (Ryan, 1983). The process is accompanied by significant volume loss of the lung.

Organization of hyaline membranes usually begins between 10 and 14 days after onset of symptoms with the growth of fibroblasts from the alveolar septum into the hyaline membrane, and ends with complete fibrous replacement of the membrane (Figs. 16 and 17). The resulting thick layers or plugs of fibrous tissue, like their predecessors, the hyaline membranes, are more prominent in alveolar ducts. As they mature, their shrinkage further distorts the lung architecture. Late in the process, their origin is often no longer recognizable and the tissue they

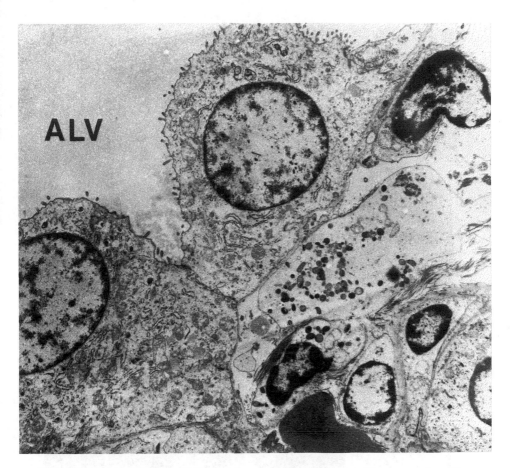

Figure 9 Electron micrograph of alveolar epithelial cells during early regeneration. The nuclei are large and the cytosol contains abundant rough surfaced endoplasmic reticulum but no lamellar bodies (×4300). Reprinted from Ryan 1983.

occupy can only be characterized as showing fibrosis without a clear statement as to whether its distribution is interstitial or alveolar.

From very early in the course of AAI, the volume of the interstitium of the lung is increased, partly due to edema and infiltration of neutrophils and partly because of proliferation of resident cells. Many of these cells are fibroblastlike cells which show marked increase in rough endoplasmic reticulum with extensive dilatation of cisternae. The function of these cells during AAI has received little attention, but there is no evidence that they are functioning as fibroblasts and little evidence that true interstitial fibrosis occurs after acute alveolar injury.

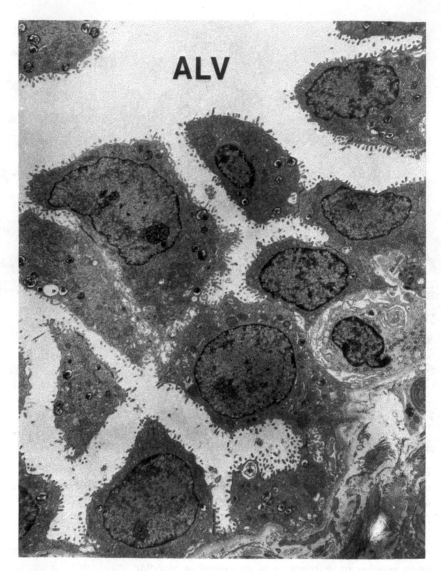

Figure 10   Electron micrograph of alveolar epithelial cells during early regeneration. At this stage, some cells contain some lamellar bodies while others contain none. (×3200). Reprinted from Ryan 1983.

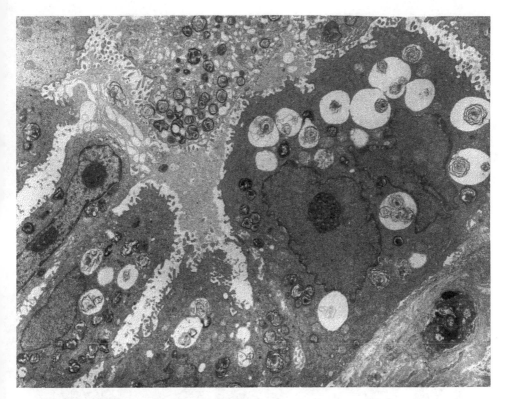

Figure 11 Electron micrograph of alveolar epithelial cells during late regeneration. The alveolus is entirely lined by type II cells containing many lamellar bodies. (×5900). Reprinted from Ryan 1983.

It seems clear from these studies that necrosis of alveolar epithelium is the major determinant of most of the important structural features of AAI. It is this lesion that results in the formation of hyaline membranes with their eventual organization in some cases, as well as to irreversible alveolar closure, both of which lead to the eventual architectural revision of the lung. These observations also suggest that the regenerating alveolar epithelial cells, believed to be derived from type II cells, are dedifferentiated and closely resemble those of the developing fetal lung. They are devoid of lamellar bodies, suggesting that they, like other types of regenerating cells, undergo obligate loss of cytoplasmic function as they rapidly divide to reline the alveoli. After repair is accomplished, cytoplasmic lamellar bodies, and presumably surfactant synthesis, reappear. It follows from these observations that injury to type II cells may not be a necessary prerequisite

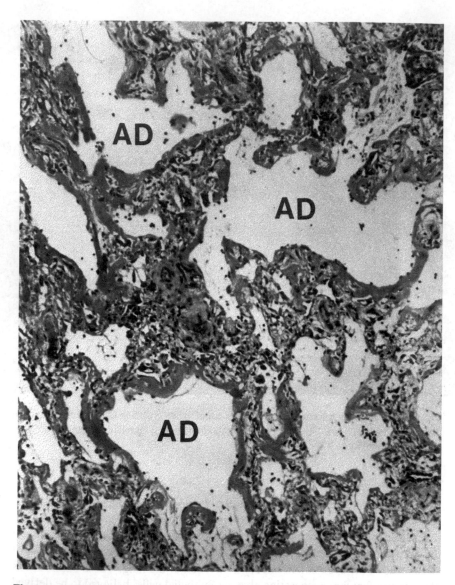

Figure 12 Moderate architectural revision in a patient with typical AAI of unknown cause 9 days after onset of respiratory symptoms. Hyaline membranes line enlarged alveolar ducts (AD) which are separated by clusters of closed alveoli. (Hematoxylin-eosin ×110.) Reprinted from Ryan 1983.

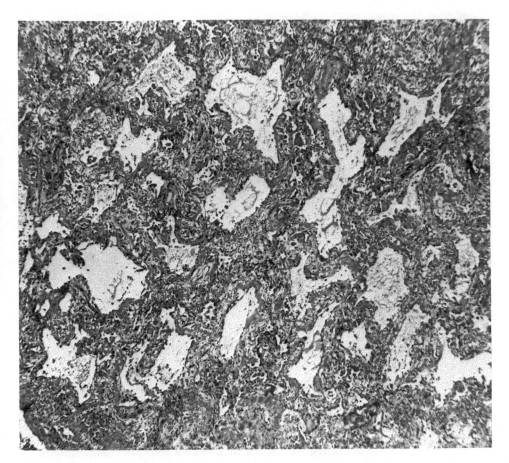

Figure 13 Severe architectural revision in a patient with AAI of unknown cause, 14 days after onset of respiratory symptoms. Enlarged and contorted airspaces are lined by hyaline membranes and separated by large clusters of closed or shrunken alveoli. (Hematoxylin-eosin × 75.) Reprinted from Ryan 1983.

to decreased surfactant synthesis by the injured lung. Necrosis, primarily or exclusively of type I cells, may lead to decreased production of surfactant as the type II cells dedifferentiate and temporarily lose their surfactant-producing capacity as they divide to reestablish the integrity of the alveolar epithelium.

It is important to point out that most electron microscopic studies of AAI have not shown extensive injury of capillary endothelium (Bachofen and Weibel, 1977, 1982; Nash et al., 1974; Ryan, 1983). Some have suggested that endothelial

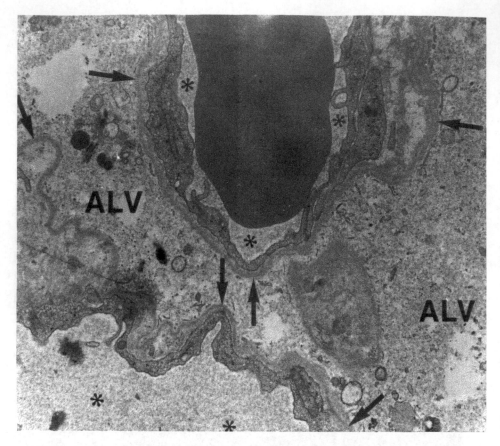

Figure 14   Electron micrograph of a closed alveolus which is entirely devoid of epithelium (arrows indicate the denuded basement membranes). The residual tiny lumen is filled with granular material and cell debris. Two capillaries (asterisks) on opposite sides of the alveolus and formerly widely separated are now almost in contact with each other. Their endothelial cells are intact (×14,000). Reprinted from Ryan 1983.

injury may occur early in the course of AAI and that its rapid repair reestablishes integrity by the time epithelial necrosis and hyaline membranes are seen. Very few electron micrographs of lung tissue from patients during the first 24 hours of symptoms, who were doubtless in the process of developing AAI, are available. The few studies that have been made have not demonstrated significant endothelial abnormalities. For example, Bachofen and Weibel (1982) studied lung tissue from a 16-year-old male who died with pulmonary edema 7 hours after massive trauma with shock sustained in an auto accident. While epithelial necrosis consistent with

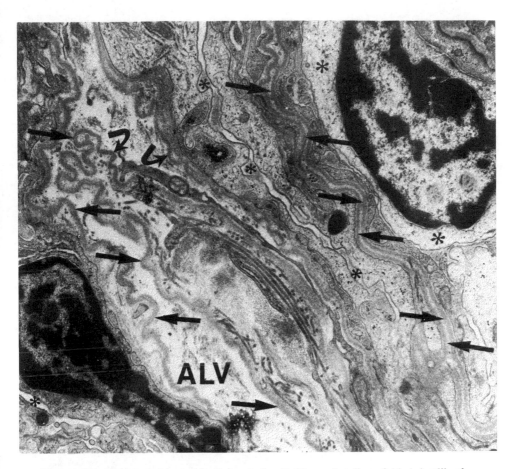

Figure 15 Electron micrograph of closed alveoli. Three alveoli are folded shut like the pleats of an accordion. The alveolus on the left (ALV) is completely devoid of epithelium and its denuded basement membrane is indicated by arrows. One of its capillaries, marked by asterisks, lies at the extreme left of the micrograph. The alveolus in the center, marked by curved arrows, is lined on its left side by an intact type I cell and devoid of epithelium on its right side. The alveolus on the right side is devoid of epithelium and its basement membranes, marked by arrows, lie directly apposed to one another. Intact capillaries on either side of this closed alveolus are marked by asterisks (×10,000). Reprinted from Ryan 1983.

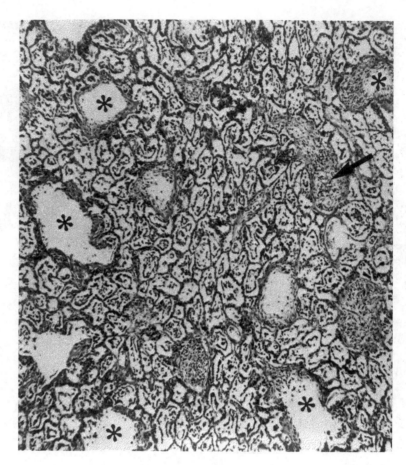

Figure 16 Organization of hyaline membranes in a patient who died 18 days after spontaneous onset of respiratory symptoms. Open lung biopsy 6 days after onset showed typical AAI. Alveolar ducts (asterisks) are lined by layers of avascular fibrous tissue. Some of the smaller airspaces (arrow) are filled with fibrous tissue and the process here resembles organizing pneumonia. (Hematoxylin-eosin ×150.) Reprinted from Ryan 1983.

early AAI was found, no abnormality of capillary endothelium was seen. Thus electron microscopy has not provided an explanation for the usually marked increase in extravascular lung water which is so consistently seen that increased permeability of the capillary endothelium is considered by many investigators to be the primary abnormality in AAI.

Many other details of altered lung structure in human AAI also remain to be investigated. The types of inflammatory cells, their sequence of accumulation

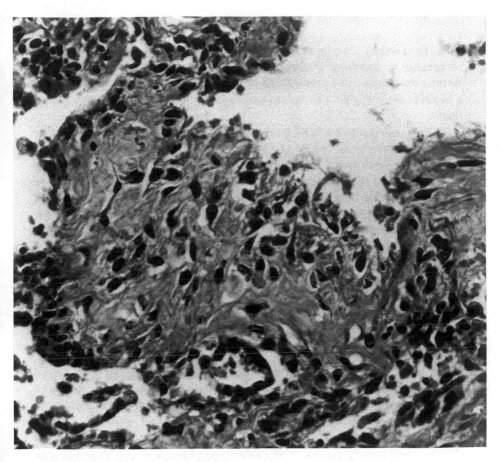

Figure 17 Section of an organized hyaline membrane showing vertically oriented fibroblasts. (Hematoxylin-eosin ×310.) Reprinted from Ryan 1983.

in the lung, and the time relationships of their recruitment to the onset of cellular injury are poorly understood. The mechanism by which fairly selective injury can be inflicted upon epithelial cells by inflammatory cells presumably recruited from the blood remains to be explained. The types and functions of the proliferating resident cells in the interstitium are also poorly understood. Certain of these cells may be important to the repair of alveolar injury by modulating mesenchymal-epithelial interactions.

III. Experimental Models

To be most clearly relevant to human AAI, an experimental model must not only recapitulate its histologic evolution; it also must share a common pathogenesis. Because the etiology and pathogenesis of AAI are not yet clearly understood such a model is not available. Current investigations are proceeding along two general lines.

First, assuming that basic mechanisms of acute inflammation and tissue injury are common to all organs and tissues, investigators have used information derived from experiments with other organs and tissues or from in vitro systems to design studies of acute lung injury. In some of these studies of mediators of acute inflammation or injury in vivo, features other than histology (e.g., changes in lung water) have been used as indices of injury. In others, histologic examination of tissue from the experimentally injured lung has shown acute injury which does not closely conform to that seen in human AAI. All studies of experimental lung injury which purport to relate to AAI should include histologic and electron microscopic examination of lung tissue, but because of possible variability among species, differences in intensity and time course of injury between experimental and human injury and possibly other as yet unknown variables, these models should not necessarily be expected to reproduce the histologic changes and evolution of the human injury in every detail. Much of the growing body of information concerning mediators of acute lung injury in which the induced injury is not closely similar to human AAI may eventually be shown to be highly relevant to the human situation. Nevertheless, the minimum requirements for such a model should be that the induced injury is acute and diffuse, is confined to the *cells* of the interalveolar septum without destruction of connective tissue, involves primarily alveolar epithelium, capillary endothelium, or both, and is not associated with a significant alveolar inflammatory exudate or any other type of reaction such as granulomas or vasculitis.

The second line of investigation utilizes empirical models to study structural or biological *consequences or correlates* of AAI. Because their only index of relevance to the human injury is their structural similarity, such models must closely reproduce the human injury at least in its major characteristics, that is, they must be acute and diffuse, feature necrosis and regeneration of alveolar epithelial cells, and follow the temporal evolution of the human process.

In each model discussed below, the volume of evidence and the number of conclusions derived from morphologic observations is far outweighed by those derived from other kinds of experiments. In fact, in certain models (e.g., neutrophil-induced lung injury) morphologic data are quite limited and entirely nonquantitative. Even if they were more complete, morphologic findings could not be adequately considered without the context of data from other disciplines. For this reason the following discussions include but are not confined to structural data.

IV. Neutrophil-Mediated Oxidant Lung Injury

The major links in the chain of events which leads to the protective action of neutrophils are now understood. When this sequence takes place locally, for example, at a site of tissue invasion by bacteria, the action is contained and the results tend to be beneficial or protective to the host. When it occurs massively and systemically in the vascular compartment, the results can be catastrophic. It has recently become apparent that damage can be inflicted upon the lung by just this chain of events which culminates with activated neutrophils sequestered in the pulmonary capillaries. While many questions remain unanswered, evidence is accumulating that the cascade of events culminating in this damage may play a major role in acute lung injury of diverse etiologies.

A. Neutrophil-Mediated Inflammation

Several recent articles review this topic in detail (Fantone and Ward, 1982, 1985; Fridovich 1986; Palmblad 1984; Harlan 1985; Henson and Johnson, 1987; Freeman and Crapo, 1981; Weiss, 1989). Briefly, activation of complement by antigen antibody complexes (e.g., during opsonization), via the alternative pathway (e.g., by endotoxin), or by exposure of serum to certain foreign surfaces generates the complement fragment C5A and its proteolysis product C5A-des arg. These fragments induce a set of changes in neutrophils collectively referred to as activation. These changes comprise increased adherence to endothelial cells and to other neutrophils, chemotaxis, and the generation and release of toxic oxygen radicals and proteases. The neutrophil possesses a specific membrane receptor for C5A and C5A des arg called the MO-1 receptor, which is a surface lipoprotein heterodimer. The binding of C5A or C5A des arg to these receptors initiates the process of activation (Chenoweth and Hugli, 1978). Increased adhesiveness of the cell involves increased expression of MO-1 through recruitment of additional molecules into the membrane from an intracellular compartment (Arnaout et al., 1984; Todd III et al., 1984). This increased expression is associated with a decrease in the net negative charge of the cell (Todd et al., 1981; Arnaout et al., 1983, 1985). The decreased charge appears to be responsible for increased adhesiveness and the resultant leukoaggregation as well as adherence to endothelial cells (Gallin et al., 1975). It has also been shown (Ryan et al., 1981; Tate and Repine 1983) that exposure of cultured bovine pulmonary artery endothelial cells to lysates of neutrophils causes unmasking of latent receptors for the complement fragment C3b and for the Fc portion of IgG. This unmasking is presumably due to the activity of proteolytic enzymes such as elastase produced by the activated neutrophils. It is possibly abetted by toxic oxygen radicals and myeloperoxidase also produced by the activated neutrophils which interfere with plasma antiproteases (Corp and Janoff, 1979; Matheson et al., 1979). Because these antiproteases normally inactivate proteases, the latter are unchecked in these circumstances and membrane damage may be augmented. Exposure of endothelial C3b and F_c receptors may act to fix immune complexes or complement to the cell surface and increase neutrophil adherence through similar receptors present on them. Neutrophil proteases have also been shown to promote neutrophil adherence to cultured monolayers of endothelial cells by degrading endothelial cell surface fibronectin (Vercelloti et al., 1983). Another and probably very important augmenting factor is the cleavage of membrane lipids of activated neutrophils by phospholipases to form arachidonic acid, the precursor of prostaglandins, thromboxanes, and leukotrienes. The subsequently formed lipoxygenase products include two compounds, leukotriene B4 and 5-hydroxyeicosatetraenoic acid which are powerfully chemotaxic for neutrophils (Ford-Hutchinson et al., 1980).

In most cells, the superoxide radical ($O_2^{\cdot-}$) is a quantitatively minor product of various oxidation reactions which occur by the single electron reduction of molecular oxygen. The great majority of molecular oxygen is utilized in mitochondrial oxidative phosphorylation coupled with electron transport in which 2-electron reduction of oxygen, driven by cytochrome oxidase occurs.

Neutrophils appear to have appropriated the production of the superoxide radical ($O_2^{\cdot-}$) into their armamentarium for killing phagocytized organisms. During activation, the neutrophil exhibits a burst of oxygen comsumption driven by an unusual cell membrane-associated enzyme, NADPH oxidase. This respiratory burst results in the production of $O_2^{\cdot-}$ which is rapidly dismuted to hydrogen peroxide (H_2O_2) (Fantone and Ward, 1982; Weiss, 1989). The highly reactive hydroxyl radical ($OH\cdot$) is believed to be produced from the reaction of $O_2^{\cdot-}$ and H_2O_2 in the presence of transition metal ions, especially Fe^{2+}-Fe^{3+} (the Fenton reaction, Fig. 18), although this has not been proven to occur in neutrophils. Other reactive products are formed from the metabolism of H_2O_2. The most important of these is the reaction of myeloperoxidase with H_2O_2 in the presence of chloride to form hypochlorous acid. The neutrophils use these radicals and compounds as well as the proteases contained in the cytoplasmic granules, to kill ingested organisms in the relatively contained phagolysosome. Unfortunately, these toxic products are also released into the extracellular milieu.

Cells contain antioxidant enzymes such as the superoxide dismutases, which catalyze the reduction of $O_2^{\cdot-}$ to H_2O_2 and catalase and glutathione peroxidase which, by different mechanisms, convert H_2O_2 to water. By scavenging $O_2^{\cdot-}$ and H_2O_2 , these enzymes protect the cell, probably by preventing the formation by Fenton chemistry of $OH\cdot$. Because they are present only in low concentrations in the extracellular environment, the potential of toxic oxygen radicals released from activated neutrophils to injure other cells is virtually unopposed. In addition

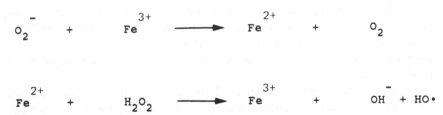

Figure 18 The iron catalyzed Fenton reaction. Ferrous ion (Fe + +) is generated by oxidation of O_2^-. The ferrous ion so generated, catalyzes the generation of hydroxyl radical (HO·) by acting as an electron donor in the reduction of hydrogen peroxide.

to this potential, O_2^- appears capable of augmenting recruitment of neutrophils by reacting with an extracellular precursor, probably arachidonic acid, to generate a derivative which is chemotactic for neutrophils (Perez and Goldstein, 1980).

Among the several chemical mechanisms operating to kill or inactivate ingested microorganisms, the one with the greatest potential for incidental injury to host cells is peroxidation of cell membrane lipids by toxic oxygen radicals. The fatty acid radicals formed by proton extraction from polyunsaturated fatty acids react with O_2 to form fatty acid peroxy radicals which, in a chain reaction, can lead to oxidation of other lipids and proteins (Frankel, 1985; Mead, 1976).

B. Neutrophils in Acute Lung Injury

The possibility that neutrophils are involved in the pathogenesis of AAI began to attract serious attention with the observation that transient marked neutropenia consistently occurs in patients during hemodialysis and that decreased pulmonary diffusing capacity and mild hypoxemia occur in some of them. This effect was shown to depend upon activation of complement via the alternative pathway when plasma comes into contact with the cellophane dialyzer membrane. Animal experiments in which complement was activated by exposure of plasma to dialyzer membranes demonstrated that the neutropenia which developed was due to sequestration of neutrophils in small vessels of the lung (Craddock et al., 1977a, 1977b). A similar chain of events also occurs in patients during nylon fiber leukaphoresis, a method used to separate neutrophils from blood for donation to neutropenic patients (Hammerschmidt et al., 1978). In this procedure, as in hemodialysis, plasma is exposed to a foreign polymeric surface and activation of complement, as well as sequestration of neutrophils in the vessels of the lung, occurs (Nusbacher et al., 1978). Some patients experience transient mild pulmonary dysfunction during leukaphoresis. Gel filtration, ultrafiltration, and antiserum inhibition studies have demonstrated that C5A is the complement fragment responsible for the leukostasis in the lung under these circumstances (Nusbacher et al., 1978). These observations in humans, together with the animal experiments to which they led, have suggested not only that neutrophils may cause acute lung dysfunction but that systemic activation of complement may initiate the pulmonary leukoaggregation.

Several subsequent clinical studies have clearly demonstrated that the plasma of most patients following massive trauma, major abdominal surgery, acute pancreatitis, disseminated intravascular coagulation, burns or septic shock contains elevated levels of C5a (Duchateau et al., 1984; Weinberg et al., 1984). Only one group has reported a sufficiently high correlation between elevated plasma C5a and the development of acute lung injury in these patients to suggest a predictive value in the plasma C5a level (Hammerschmidt et al., 1980). Others have concluded that serum complement activation is probably a necessary but not sufficient condition for the development of acute lung injury (Duchateau et al., 1984) while still others have questioned whether intravascular complement activation plays any significant role in the pathogenesis of ARDS. These doubts have been based upon the finding of substances chemotactic for neutrophils in alveolar lavage fluid but not in the plasma of patients with ARDS (Fowler et al., 1987).

Increased numbers of marginated or aggregated neutrophils have been found in the pulmonary capillaries during AAI by several investigators (Pratt, 1978; Bachoven and Weibel, 1977) and others have reported that they compose a large proportion of the cells recovered by bronchoalveolar lavage from these patients (Lee et al., 1981; McGuire et al., 1982; Fowler et al., 1982). Furthermore, there is good correlation between the severity of the respiratory failure and the numbers of neutrophils recovered by lavage (Fowler et al., 1987; Weiland et al., 1986).

A recent study has presented indirect but convincing evidence that oxidants are generated in the lungs of patients with ARDS by demonstrating that a fraction of the alpha-$_1$ protease inhibitor recovered in bronchoalveolar lavage from these patients had been inactivated by oxidation, presumably of the methionyl residue at the reaction site of the enzyme (Cochrane et al., 1983). Serine elastase (Lee et al., 1981; McGuire et al., 1982; Idell et al., 1985) and collagenase (Christner et al., 1985) have also been identified in bronchoalveolar lavage fluid from patients with ARDS although recent studies have demonstrated increases in antiproteases, both alpha$_1$-protease inhibitor and alpha$_2$-macroglobulin, which appear in some patients to shift the protease-antiprotease balance toward antiprotease (Wewers et al., 1988).

Other toxic products of neutrophils (e.g., hypochlorous acid) have not yet been reported in bronchoalveolar lavage fluid from these patients.

Reports of AAI in profoundly neutropenic patients have provided evidence that neutrophils are not critical in the pathogenesis of injury at least in some patients (Braude et al., 1985; Maunder et al., 1986). In view of the fact that toxic oxygen radicals are produced by alveolar macrophages as well as by neutrophils, it seems reasonable to suppose that these cells may contribute to the injury in all patients and that they may be sufficient in the occasional neutropenic patient to produce oxidant-mediated lung injury in the near absence of neutrophils.

C. In Vivo Experimental Models and In Vitro Systems (in Lung Cells)

Experimental models have provided critical insights into the mechanism by which neutrophils are recruited to the lung, the mediators of the injury they cause, and the structural features of this injury and its repair. A large amount of recent data on every aspect of neutrophil-mediated lung injury is available in several recent reviews (Brigham and Meyrick, 1984; Tate and Repine 1983; Till and Ward, 1986; Ward et al., 1986). What follows is a brief outline of that evidence with emphasis on that derived from structural models and commentary on the relevance of this injury to human acute alveolar injury.

When pneumococci are injected intravenously into guinea pigs, complement activation through C5a, neutropenia, and localization of a significant fraction of the injected bacteria in the lung rapidly follow (Hosea et al., 1980). This process is accompanied by increased permeability of pulmonary capillaries to albumin. Localization of bacteria in the lung does not occur in C5-deficient animals and the infusion of purified C5a without bacteremia increases pulmonary vascular permeability only in animals that are not neutropenic. These observations have been interpreted to suggest that in bacteremia, lung injury with increased vascular permeability is mediated by neutrophils and that activation of complement through C5a is necessary for sequestration of neutrophils in the lung. Microscopic or electron microscopic examination of lung tissue from these experimental animals was not performed by these investigators.

Others have induced acute lung injury by the intravascular activation of complement. Intravenous injection in rats (Till et al., 1982) or mice (Tvedten et al., 1985) of a bolus of purified cobra venom factor, a potent activator of complement, causes rapid, profound neutropenia and transitory (3-4 h) increase in pulmonary vascular permeability the degree of which is markedly reduced if the blood is depleted of neutrophils or of complement prior to injection of cobra venom factor. Electron microscopic examination during the period of increased permeability shows plugging of capillaries by neutrophils, discontinuity, and even destruction of capillary endothelial cells and fibrin and erythrocytes in alveolar spaces. Amazingly, by 4 h after injection of cobra venom factor, capillary permeability returns to normal and injury to capillary endothelium is no longer detectable by electron microscopy. The mechanism of this extremely rapid reestablishment of endothelial integrity is not clear but, repair of small circumferential areas of mechanically denuded endothelium in the aorta has been shown to occur by spreading of cytoplasmic extensions of adjacent cells not requiring cell division

(Reidy and Schwartz, 1981). By this process, repair of an area one or two endothelial cells in width was completed within 8 h. A similar process may accomplish rapid repair of the very small foci of endothelial necrosis in lung microvessels described in these reports.

Activation of complement with generation of C5a induced by cutaneous thermal burn in rats (Till et al., 1983) and by infusion of complement-activated plasma in sheep (Meyrick and Brigham, 1984) have also been found to induce sequestration of neutrophils in the lung with injury morphologically similar to or identical with that following injection of cobra venom factor. The time course of increased vascular permeability varies considerably with the method used to activate complement.

Not all investigators, however, have been able to demonstrate lung injury with complement activation. Using the rabbit, one group was able to demonstrate lung injury following injection of cobra venom factor only when the infusion was accompanied by hypoxia or by the infusion of prostaglandin E2 (Henson et al., 1982). Others have demonstrated lung injury in the rabbit after prolonged infusion (4 h) of zymosan-activated plasma (Hohn et al., 1980). It appears then that complement-induced pulmonary leukostasis can cause lung injury in all species studied but that species may differ in their susceptibility to the injury. The role of cyclooxygenase products of arachidonic acid is unclear but most of the evidence available suggests that these compounds alter only vascular and airway resistance but play little part in vascular permeability.

Depletion of platelets with antiplatelet serum in mice has been shown to distinctly but incompletely limit the increased pulmonary vascular permeability caused by cobra venom factor (Tvedten et al., 1985). The mechanism by which platelets potentiate complement-mediated acute lung injury is not clear, but it may be related to augmented production of lipoxygenase products of arachadonic acid by neutrophils caused by transfer of arachadonate intermediates into them from activated platelets (Marcus et al., 1982).

D. Mediators of Lung Injury

Following systemic complement activation by cobra venom factor in rats and in mice, and following cutaneous burns in rats, treatment with antioxidants decreases the severity of the resultant lung injury (Till et al., 1982, 1983; Tvedten et al., 1985). This has provided convincing evidence that oxidants play a major part in this injury. Because catalase (Till et al., 1982; Tvedten et al., 1985), iron chelators, and scavengers of hydroxyl radical such as dimethylsulfoxide (Ward et al., 1983) confer greater protection than superoxide dismutase, it appears that the major toxic oxidant is hydroxyl radical generated from H_2O_2 and O_2^- by the Fenton reaction. Killing of endothelial cells from both human umbilical vein and bovine elastic artery when they are exposed to activated neutrophils in vitro has been shown to be mediated by a hydrogen peroxide-dependent mechanism (Sacks et al., 1978; Martin, 1984). The myeloperoxidase-dependent hypohalite ion is also capable of killing bovine endothelial cells in vitro (Martin, 1984) and may do so in complement-triggered experimental acute lung injury, although this effect has not yet been demonstrated in vivo.

Another major group of mediators of neutrophil-induced tissue injury, the proteases, appears to play a lesser role in cellular injury in complement-triggered acute lung injury at least during the very early phase. The major evidence for this is that the severity of early lung injury (30 minutes) induced by complement activation is no different in beige mice whose neutrophils contain almost no neutral proteases than in mice with normal levels of neutrophil neutral proteases (Tvedten et al., 1985). However, proteases are released from activated neutrophils during this type of lung injury (Schraufstatter et al., 1984) and there is evidence that neutrophil elastase may be cytotoxic to microvascular endothelial cells in vitro under certain conditions (Smedly et al., 1986). Recent evidence (Fligiel et al., 1984) also indicates that the presence of small amounts of H_2O_2 enhances the ability of neutrophil serine proteases to degrade hemoglobin, fibronectin and proteins of glomerular basement membrane. This enhancement apparently resulted from a direct effect of H_2O_2 on the substrate protein. Protease-mediated cytotoxicity during neutrophil-induced acute lung injury may also be enhanced by the apparent inhibitory effect of oxidant radicals on serum antiproteases (Corp and Janoff, 1979).

It appears that alveolar epithelial cells are more resistant than endothelial cells to oxidant-induced injury. However, epithelial cells have been shown to be injured by activated neutrophils by an oxygen metabolite-independent mechanism which requires intact MO-1 surface receptors and close adhesion of the neutrophils to the target epithelial cell (Simon et al., 1986). The mechanism of this injury remains to be clarified.

Peroxidation of membrane lipids is believed to be the major mechanism of lung cell injury by toxic oxygen radicals. That lipid peroxidation occurs is indicated by the finding in the lung and in the plasma of products of lipid peroxidation during neutrophil-mediated acute lung injury (Till et al., 1985). The quantities of these products correlate directly with the severity of the induced lung injury (Ward et al., 1985).

The above evidence, taken together, provides strong support for the hypothesis that complement-activated neutrophils can cause injury to pulmonary endothelium. The relationship of this type of injury to acute alveolar injury in the human is less clear. The severity of the experimental process does not approach that of the human lesion and injury to alveolar epithelium has not been described. Species differences, the very short duration of the experiments reported to date, or factors which intensify the acute inflammatory response in the human condition may, singly or in some combination, explain the morphologic differences.

Nevertheless, until the histologic picture of human AAI is more closely reproduced, conclusions about the role of complement activation and neutrophil sequestration in human AAI must be tentative.

V. Hyperoxic Lung Injury

Experimental hyperoxic lung injury has been studied more intensively than any other lung injury model and the effort has provided valuable insights into mechanisms of oxidant injury and its morphologic consequences. At least in humans, hyperoxia of sufficient degree and duration causes AAI, morphologically indistinguishable from that of any other cause (Gould et al., 1972; Nash et al., 1967; Katzenstein et al., 1976). Its etiology is thus potentially relevant to that of AAI in general.

The evidence is strong that hyperoxic injury is caused, at least principally, by toxic oxidant radicals apparently generated by two closely interrelated mechanisms. The first and probably the initiating mechanism involves production of toxic oxidant radicals by lung cells while in the second they are generated by inflammatory cells recruited into the lung. The lung is the major target organ in hyperoxic injury not because lung cells are more susceptible than those of other organs, but because they are directly exposed to the increased oxygen tension whereas all other cells are protected to a very large degree by the limited oxygen carrying capacity of blood after hemoglobin saturation is reached.

The initial phase of hyperoxic injury occurs because lung cells exposed to increased oxygen tension produce toxic oxygen radicals in excess of the ability of the cells antioxidant enzymes to quench them. Under normoxic conditions more than 90% of the molecular oxygen utilized by lung cells is reduced to water by transfer of pairs of electrons at the end of the electron transport chain in the mitochondria during oxidative phosphorylation (Freeman et al., 1981, 1982). This divalent reduction must be achieved enzymatically, by enzymes such as cytochrome oxidase because the natural tendency of molecular oxygen, due to an electron spin restriction, is toward single electron or univalent reduction (Fridovich, 1978). The oxygen which leaks into this univalent pathway is reduced by a variety of reductant compounds which undergo autooxidation including components of the respiratory chain such as pyridine-linked dehydrogenases, ubiquinone, other cytochromes and several other compounds related only by their relatively high reduction potential with respect to oxygen (Freeman et al., 1985; Turrens et al., 1982a). The quantity of superoxide produced via this pathway even at normal oxygen tension apparently poses a threat to the cell because antioxidant enzymes (superoxide dismutases, catalase, and peroxidase) have evolved in virtually all aerobic organisms. As the oxygen tension of inspired gas increases so does the production of O_2^-. Because small amounts of cyanide inhibit components of the cytochrome system, thereby blocking divalent reduction of oxygen, the level of cyanide-resistant respiration has been employed as an indirect measure of O_2^- production in cells. It allows only a rough estimation of the O_2^- production which would occur at a given oxygen tension in the absence of cyanide because of other effects of cyanide upon mitochondrial respiration, some of which tend to increase and others to decrease O_2^- production (Freeman and Crapo, 1981). In lung slices and homogenates, the fraction of total oxygen consumption which is cyanide resistant increases from 7–9% under normoxic conditions to 17–18% with exposure to 80–85% oxygen, indicating a significant increase in O_2^- production under these conditions (Freeman and Crapo, 1981; Freeman et al., 1982). Hyperoxia also increases H_2O_2 production by lung mitochondria and lung microsomes. H_2O_2 production by mitochondria rose linearly with respect to oxygen concentration up to 60% O_2, but then rose more sharply at higher concentrations. It was suggested that this precipitate rise may occur at an O_2 concentration which begins to overwhelm the intramitochondrial antioxidant enzymes (Turrens et al., 1982b). As the ability of the antioxidant enzymes to quench these radicals is overcome, both O_2^- and H_2O_2 become available and the presence of iron salts allows the iron catalyzed Fenton reaction to produce the highly reactive hydroxyl radical and possibly singlet oxygen. It is its ability to take part in the Fenton reaction with H_2O_2 rather than any major potential to cause injury by itself that makes O_2^- dangerous; however, there is evidence that H_2O_2 generated by neutrophils can by itself cause call damage in vitro (Nathan et al., 1979; Weiss et al., 1981). Several lines of evidence suggest that these oxidants, generated by lung cells, are capable of damaging the lung, independent of recruited inflammatory cells. In one study, exposure of lung homogenates to 85% O_2 for 7 days resulted in marked increases in O_2^- production (as inferred from cyanide-resistant O_2 consumption) and in products of lipid peroxidation (Freeman et al., 1982). When lung homogenates from normal animals were exposed to 80% O_2 for only 75 minutes, lipid peroxidation products increased significantly (these products were present in measurable amounts even in homogenates maintained under normoxic conditions) and the increase was suppressed by superoxide dismutase, by catalase, and to a greater degree by superoxide dismutase and catalase together than by either alone. In another study, lung explants whose blood vessels were perfused free of neutrophils showed marked increases in cell lysis (as reflected by release of ^{51}Cr) when exposed to 95% O_2 for 18 hours (Martin et al., 1981). This cytotoxicity was reduced by superoxide dismutase, catalase, and the antioxidants, ascorbate and alpha-tocopherol. In addition, hyperoxia has been shown to directly injure cultured alveolar epithelial cells (Simon et al., 1979), lung fibroblasts (Balin et al., 1976), and pulmonary artery endothelial cells (Suttorp and Simon, 1982). Moreover, electron microscopic studies in rats have shown significant cellular injury in the lungs of animals exposed to hyperoxia well before significant influx of neutrophils (Barry and Crapo, 1985).

Recruitment of neutrophils from the circulation into the lung begins at intervals after onset of hyperoxic exposure which appear to vary with the concentration of inspired oxygen and probably with species. In rats the number of neutrophils in the capillaries of the lung markedly increases between 40 and 60 h after the start of exposure to 100% oxygen, whereas after exposure to 85% oxygen the increase is less marked, begins at 72 h, peaks at 120 h, and returns toward normal thereafter (Barry and Crapo, 1985). Intervals exceeding 117 h in mice (Parrish et al., 1984) and 66 h in rats (Fox et al., 1981) between the onset of exposure to 95% O_2 and a significant increase in neutrophils in alveolar lavage fluid suggests considerable species differences in rates of neutrophil recruitment. The mechanism of recruitment of neutrophils into the lung is incompletely understood, apparently complex, and probably multifactorial. One possible initiating or participating mechanism in this recruitment is hyperoxic injury to capillary endothelial cells resulting in increased adherence of neutrophils to them. A recent study (Bowman et al., 1983) demonstrated that exposure of cultured bovine pulmonary artery endothelial cells to 95% oxygen caused injury indicated by release of cytoplasmic lactic dehydrogenase and depressed growth rate. Nonactivated neutrophils adhered to these oxygen damaged cells in significantly greater numbers than to control endothelial cells exposed to normoxia for the same period of time. The mechanism of this increased adherence is unknown but it may be mediated by arachidonic acid metabolites produced by the injured cells or by unmasking of C3 receptors on them. If the same injury-induced adherence of neutrophils occurs in vivo in capillary endothelium it may be an important initiating event in neutrophil recruitment.

One well established source of chemotaxins that might contribute to neutrophil recruitment during hyperoxia is the alveolar macrophage. These cells are known to produce one or more factors chemotactic for neutrophils upon exposure to a variety of substances including bacteria and immune complexes (Hunninghake et al., 1980). Hyperoxia has been shown to injure alveolar macrophages in vivo and to stimulate them to release factors which are chemotactic for neutrophils (Fox et al., 1981). After exposure of rats to 95% oxygen for 66 h, alveolear lavage fluid was found to contain neutrophil chemotaxins the activity of which was apparently not reduced by prior depletion of complement. The appearance of these chemotactic factors coincided with a sharp increase in numbers of neutrophils in alveolar lavage fluid (Fox et al., 1981).

There is also some evidence that complement fragments contribute to neutrophil recruitment during hyperoxia. Mice that are congenitally deficient in C5 have been shown to recruit fewer neutrophils into the lung, to have a delayed and less intense inflammatory reaction, and a lower mortality rate (25 vs. 95%) after exposure to hyperoxia than non-C5-deficient mice (Parrish et al., 1984). These differences between the two strains of mice are abolished by transfusing the C5-deficient mice with plasma containing C5 prior to exposure to hyperoxia.

That significant numbers of neutrophils are recruited to the lung even in C5-deficient mice suggests redundancy in the recruitment mechanisms. The manner in which complement might be activated during hyperoxia is unknown but it has been suggested that it might occur locally in the small pulmonary vessels without triggering of the entire complement cascade by direct cleavage of C5 to C5a and C5a des arg by proteases from neutrophils or macrophages (Ward and Hill, 1970; Synderman et al., 1972).

That platelet-derived chemotactic factors may play a part in the process of recruitment is suggested by the observation that an increased number of platelets in lung capillaries coincides with the earliest endothelial injury at about 40 hours of 100% oxygen exposure in rats, well before a significant increase in neutrophils is observed in the lung (Barry and Crapo, 1985). Activated platelets adherent to injured endothelial cells or extracellular matrix are known to release a variety of mediators, several of which are chemotactic for neutrophils (Deuel et al., 1981, 1982; Weksler and Coupal, 1975).

The role of neutrophils in hyperoxic lung injury is at present not entirely clear, but considerable evidence suggests that they contribute to the injury probably by releasing toxic oxidants and possibly proteases. Mortality in rats exposed to hyperoxia follows closely upon the period of rapid increase in neutrophils in the lung and is lower in mice that recruit reduced numbers of neutrophils because of C5 deficiency (Parrish et al., 1984). Furthermore, the number of neutrophils recovered by lung lavage after hyperoxia correlates with the severity of lung injury as reflected by the quantity of albumin recovered in the lavage and by an increase in extravascular lung water (Shasby et al., 1982). On the other hand, experiments in which neutropenic animals are exposed to hyperoxia have produced conflicting results. In one study rabbits depleted of neutrophils by a single injection of nitrogen mustard showed less severe lung injury (reflected by an increase in extravascular lung water and in albumin recovered by alveolar lavage) after 72 hours of hyperoxia than rabbits replete with neutrophils (Shasby et al., 1982). However, in a similar study, rabbits made and kept severely neutropenic by repeated injections of nitrogen mustard survived no longer in 100% oxygen than control animals not injected with nitrogen mustard and sustained similar degrees of lung injury as reflected by increased extravascular lung water (Raj et al., 1985). The reason for the conflicting results between these studies remains unexplained. While the preponderance of evidence suggests that neutrophils take part at least in the later phase of hyperoxic injury, available information does not yet allow quantitative estimation of the contribution of endogenous and of neutrophil-derived sources of injury.

To complicate matters, there is evidence that lung cells previously exposed to hyperoxia are more susceptible to injury by activated neutrophils than those not previously exposed. One group demonstrated that cultured lung epithelial cells, lung fibroblasts and bovine pulmonary artery endothelial cells, after exposure to prolonged hyperoxia are more susceptible to injury by activated neutrophils than are control cells after exposure to air for the same period of time (Suttorp and Simon, 1982). Release of ^{51}Cr was the index of cell injury used and durations of hyperoxic exposure for each of the three cell types were chosen so that quantities of ^{51}Cr released from posthyperoxic cells prior to neutrophil exposure were no different from those released from post normoxic cells. This time period was shorter for endothelial cells (24 h) than for the other two cell types (48 h) because the longer period used for the latter caused overt injury to the endothelial cells with marked increase in ^{51}Cr release. All three cell types preexposed to hyperoxia (630 torr) released significantly more ^{51}Cr during 4 h incubation in air with activated neutrophils than control cells preexposed to normoxia (140 torr). Incubation with activated neutrophils after exposure to normoxia also caused some degree of injury and endothelial cells were more susceptible than the other two types. Thus endothelial cells appear to be more susceptible to hyperoxic exposure alone and to injury by activated neutrophils. The cell injury during exposure to activated neutrophils after hyperoxia was attributed to H_2O_2 because it was largely prevented by catalase. The most likely mechanism for this increased susceptibility to neutrophil injury by hyperoxia is simply the additive effect upon the cell of the burdens of endogenous oxidants induced by hyperoxia and of exogenous oxidants introduced by neutrophils. Experimental conditions were chosen such that the period of hyperoxia was not sufficient to cause ^{51}Cr release but was probably sufficient to have consumed the cells' reserves of antioxidants so that subsequent exposure to neutrophil-derived oxidants led to overt injury.

In experiments using isolated perfused rabbit lung, another group demonstrated that infusions of nonactivated neutrophils and prior exposure of the lung to hyperoxia act synergistically to produce high permeability lung edema (Kreiger et al., 1985). Isolated lungs perfused with Hanks' balanced salt solution containing 2% albumin and ventilated for 2 hours with 95% oxygen were then ventilated with 15% oxygen for 2 hours. At the beginning of the second 2-h period, unactivated neutrophils were injected into the pulmonary artery perfusate. During this second period, lung weight was recorded and, at the end, neutrophil counts, measurements of protein content, and assays for chemotactic activity, prostacyclin, and thromboxane A2 were done on alveolar lavage and perfusates. Malondialdehyde was quantified in lung homogenates as an index of lipid peroxidation. The same procedure was carried out using lungs ventilated with 15% oxygen for both 2 h periods and with 15% oxygen and 95% oxygen for the first and second 2-h periods, respectively. In addition, lungs ventilated with 95% oxygen and others with 15% oxygen for the entire 4-h period were perfused with saline rather than with neutrophils during the second 2-h period. Only in the lungs receiving 95% oxygen prior to perfusion with neutrophils (95–15%-NEUT) did significant weight gain and increase in protein content of alveolar lavage occur. Neither those ventilated with 95% oxygen only during the second 2-h while being perfused with neutrophils (15–95%-NEUT) nor those ventilated for 4 hours with 95% oxygen but not infused with neutrophils (95–95% no NEUT) showed significant increases. Alveolar lavage fluid from the 95–15% NEUT lungs also contained more neutrophils and more of both prostaglandins assayed than that from the 15%–15%-NEUT lungs. Perfusates, but not lavage from the 95–15%-NEUT lungs contained significantly more chemotactic activity and the tissue homogenates contained significantly greater amounts of malondialdehyde than those from the 15–15%-NEUT lungs. These data indicate that after hyperoxia, neutrophils can be recruited and activated to injure the lung by mechanisms which reside in the lung and are not the result of exposure of neutrophils to hyperoxia. If lung edema and evidence of lipid peroxidation were the results of direct activation of neutrophils by hyperoxia these changes should have been most severe in the 15–95%-NEUT lungs. In fact they were most severe in the 95–15%-NEUT lungs. If injury were due solely to addition of neutrophils, then the 15–15%-NEUT lungs should have shown significant edema which was not the case. The mechanism of neutrophil recruitment and activation in these experiments is uncertain, but it occurred in a system in which complement was not available and it seems unlikely that it was due to chemotactic factors from alveolar macrophages because chemotactic activity increased only in perfusate and not in alveolar lavage. The most likely mechanism for neutrophil recruitment as well as for the synergism between hyperoxia- and neutrophil-induced injury is hyperoxic injury to endothelial cells with activation of neutrophils by the injured cells, possibly mediated by chemotaxins such as arachidonate products.

Together, the evidence from experiments using cultured cells (Bowman et al., 1983; Suttorp and Simon, 1982) and these data using the isolated perfused rabbit lung (Krieger et al., 1985) suggest that increased adherence of neutrophils, enhanced susceptibility of hyperoxia preexposed endothelial cells to injury by activated neutrophils, and at least the early phase of increased permeability of the alveolar capillary wall are related phenomena all based on injury to capillary endothelial cells by hyperoxia. Later in the course of hyperoxia, lung injury may be augmented by redundancy in the recruiting-activating mechanisms introduced by complement activation, alveolar macrophage chemotactic factors, and possibly by platelet-derived chemotactic factors.

A. Morphologic Changes

The morphologic changes in the lung after exposure to hyperoxia vary with the species and age of the animal, the concentration of inspired oxygen, and the duration of exposure. Oxygen at concentrations of 95% or higher at 760 torr for a sufficient period of time causes fatal lung injury in the majority of adults of all species so far studied. (Young animals, at least of some species, are more resistant to hyperoxia than adults.) Injury or necrosis of capillary endothelial cells and interstitial edema are structural features common to all species under these conditions, but there are significant differences among species in the rate of occurrence and severity of injury to the alveolar epithelium. In rats, extensive injury of capillary endothelial cells and interstitial edema are seen after 60 hours of exposure to 100% oxygen, but the alveolar epithelium shows little or no injury even in animals exposed until death (Crapo et al., 1980). In mice (Adamson et al., 1970) and monkeys (Kapanci et al., 1969), on the other hand, extensive necrosis of type I epithelial cells accompanies the endothelial damage. In all of these species, proliferation of alveolar epithelial cells follows the phase of cellular injury regardless of whether it is preceded by electron microscopically apparent epithelial injury.

Morphometric studies (Crapo et al., 1978, 1980, Thet et al., 1984, 1986) in rats have quantified changes in several parameters of lung structure after exposure to 100% oxygen. Of these, the estimates of cell numbers are among the most informative. The stereologic method used in these studies to estimate cell numbers per unit volume of specimen (the numerical density, Nv) involves: (1) Counting nuclei of the cell type in question per unit area (Na) on electron microscopic sections from tissue blocks obtained by appropriate stratified random sampling of the lung. (2) Measuring the mean nuclear diameter (D) of the cell type in question. It has been shown (Dehoff and Rhine, 1961) that:

$$Nv = Na/D$$

The volume of the fixed lung is measured by water displacement and the fraction of the total volume which comprises alveolar region is estimated by point-counting on light microscopic sections. Total cells per lung can then be calculated by multiplying the volume in which Nv was measured by the total volume of alveolar region. The data can be expressed as cells per total volume of alveolar region or, if the tissue component of this volume has been measured, as cells per volume of tissue in alveolar region. Details of this method for estimating Nv are given by Haes et al. (1981) and by Crapo et al. (1978) and those of an alternative method by Weibel and Gomez (1962) and by Haes et al. (1981). A review of this and the stereologic methods used to obtain other data reviewed here is given by Weibel (1973).

Endothelial Cells

After 60 hours exposure of rats to 100% oxygen, 30% of the endothelial cells in alveolar region had been destroyed and, after fatal exposure (mean time in 02–64 h), 44% had been destroyed (Crapo et al., 1980). The total capillary surface area and blood volume and the total mass of endothelial cells decreased in proportion to the reduction in cell numbers. By 3 days after removal from oxygen, after 60 h exposure, the numbers of endothelial cells had returned to slightly above those of normal controls and by 7 days after removal they had increased even further to values almost double those of controls (Thet et al., 1984). The endothelial cells during recovery were significantly larger than controls and rich in organelles. The finding by electron microscopy of avascular areas in interalveolar septa immediately after 60 hours exposure to 100% oxygen led to the speculation that some capillaries were destroyed (Thet, 1986). Electron microscopic images of avascular septa or of intermediate stages in this destruction were not presented but it seems likely that extensive loss of endothelium along an entire capillary segment rather than true destruction implying not only complete loss of endothelium but of basal lamina and of basement membrane as well was occurring. To the author's knowledge, destruction of connective tissue matrix implying proteolysis has not been reported in hyperoxic injury.

Cell kinetic studies which reflect mitotic activity in lung cells during repair of hyperoxic injury have also been done (Adamson and Bowden, 1974). In these studies mice were exposed to 90% O_2 for 6 days and the 20% that survived were studied at 12-h intervals up to 3 days after return to air. Each animal was injected with [³H]thymidine and colchicine prior to sacrifice and thin (0.5-0.75 μm) sections of glycol-methacrylate embedded tissue were studied with the light microscope after developing as autoradiographs or after staining with hematoxyln-eosin. The proportions of cells labelled with [³H]thymidine or in metaphase were expressed in 2 ways: (1) total labelled or metaphase cells (without regard to type) as percent of total lung cells (the labelling index and the mitotic index, respectively), (2) the percent of total labelled or of total metaphase cells comprising labelled or metaphase cells of the type in question (i.e., as differential counts).

$$\frac{\text{No. labelled endothelial cells}}{\text{Total labelled cells of all types}} \times 100$$

For 24 hours after removal from O_2, the labeling and mitotic indices differed little from those of unexposed controls, probably because of suppression of mitotic activity by 90% oxygen (Hackney et al., 1981). Both the labelling and the mitotic indices increased sharply after 24 h, peaked at 2.5 days and returned to near baseline at 3 days (Fig. 19).

For 24 hours, the percent of labelled endothelial cells in the differential was no different from controls but it was increased sharply at 36 hours and had returned to control levels at 3 days (Fig. 20). The percents of metaphase endothelial cells in the differentials showed a corresponding rise, peaked at 2 days and returned to control levels at 3 days. Increased mitoses and presumably endothelial damage were limited to capillaries and to small pre- and postcapillary vessels up to 200 μm in diameter (Bowden and Adamson, 1974).

Assuming no large decrease in pool size or mitotic rate in other cell populations, the data indicate that endothelial mitoses increased sharply at 36 and 48 h, but we must remember that the numbers of endothelial mitoses and of

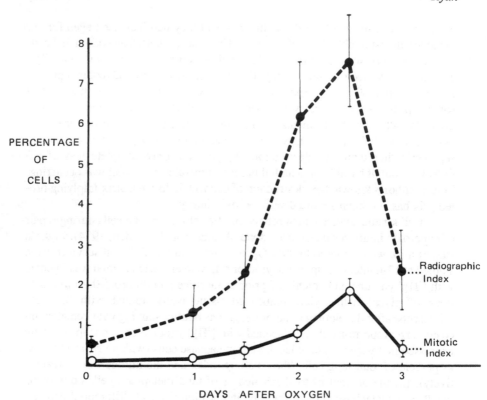

Figure 19 Labelling and mitotic indices at intervals after removal of mice from 100% oxygen after 6 days exposure. Reprinted from Adamson and Bowden 1974.

[3H]-labelled endothelial cells are expressed as percentages of all alveolar cell typesin mitosis or labelled and not as percent of total endothelial cells. Because of this it is not possible to conclude that the peak of mitotic activity in endothelial cells occurs at 2 days. A relatively greater increase in mitotic rate in one or more other cell types would lead to a decrease in mitotic index and labelling index in endothelial cells even if the mitotic rate in the endothelial population were increasing. In order to reach conclusions about change in mitotic rate in the endothelium or in any other cell population, mitotic counts or counts of cells incorporating [3H]thymidine must be expressed as percent of the cell type in question.

It is also useful to know the size of each population of cells at each time point at which mitotic rate is measured in order to account for attrition rates. Ideally, therefore, studies of this kind should include morphometric estimates of the size of each cell population in question and counts of mitoses expressed as percent of that population. Such data are not yet available.

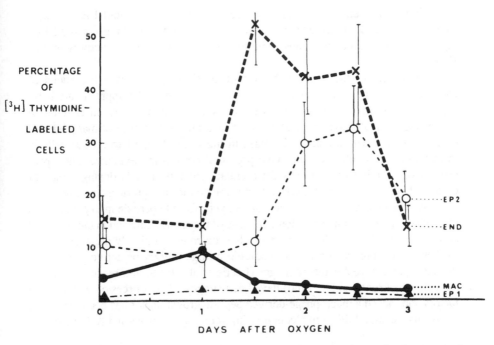

Figure 20 Percentages of labeled cells in differential counts at intervals after removal of mice from 100% oxygen after 6 days exposure. EP2 = type II epithelial cells, End = endothelial cells, Mac = macrophages, EP1 = type I epithelial cells. Reprinted from Adamson and Bowden 1974.

B. Alveolar Epithelial Cells

Immediately after 60 hours exposure to 100% O_2 and after lethal exposure the alveolar epithelial cells of the rat lung appeared normal by electron microscopy and their numbers, mean cell volume and surface area were not changed from control values (Crapo et al., 1980). Three days after 60 hours exposure, the number of type II cells was increased by about 150% and by 7 days it had decreased to about double that of control (Thet et al., 1984).

In the kinetic studies in mice described above (Adamson and Bowden, 1974) extensive necrosis of type I epithelial cells and only minimal injury to type II cells was seen by electron microscopy immediately after 6 days in 90% O_2. The percents of labeled type II cells in the differentials were significantly increased two days after removal from oxygen and peaked at 2.5 days (Fig. 20). The percent of type II cells in mitosis in the differentials showed a similar pattern, increasing sharply at 2 days and peaking at 2.5 days. The percent of labeled type

I cells in the differentials decreased from about 2 at 1 day to about 1 at 2.5 days. The considerations discussed above with respect to endothelial mitotic and labeling counts also limit conclusions about the time course of mitotic activity in type II cells.

In the same study, the authors counted silver grains over the nuclei of [^{3}H]thymidine-labeled epithelial cells in autoradiographs. An approximate halving of grain counts over nuclei of type II cells occurred three days after removal of animals from 90% oxygen and coincided with the appearance of many labeled type I cells and with a marked increase in the ratio of type I to type II cells in the alveoli. These observations have provided convincing evidence that a proportion of dividing type II cells differentiates to type I cells during repair of epithelial injury and that type I cells do not divide in significant numbers.

It is interesting to note that during recovery from hyperoxic injury, increased mitotic activity occurs in the endothelial cells of the mouse lung and in the epithelial cells of the rat lung in the absence of electron microscopically apparent injury of these cells. This suggests that a soluble mitogen might be active in the lung during repair. Alveolar macrophages (Leslie et al., 1985) and recruited inflammatory cells (Shami et al., 1986) have been suggested as sources of such growth factors and indeed a macrophage-derived peptide growth factor has been recovered in alveolar lavage fluid from bleomycin-injured lungs (Kovacs and Kelley, 1986).

C. Interstitial Cells

Morphometric studies in the rat after 60 hours in 100% O_2 have shown marked increases in total interstitial cells, but because of the large variety of cells in this compartment both resident and transient, the dynamics of each population is unknown (Crapo et al., 1980). It is clear, however, that neutrophils increased markedly from practically none in controls to about 10% of the population. The resident cells in the interstitium are the least understood population of cells in the lung. Only recently have subsets of these cells such as myofibroblasts and contractile interstitial cells been separated on structural grounds from the general population of interstitial fibroblast-like cells (Evans et al., 1982; Kapanci et al., 1974). It appears that one subset of these cells, at least in the fetal lung, synthesizes and secretes an oligopeptide growth factor which promotes maturation of type II cells and stimulates the production of surfactant phospholipids (Smith, 1979). Whether these cells subserve a similar function during repair of injury has not been investigated.

While a detailed discussion is beyond the scope of this chapter, we cannot fail to mention the fascinating phenomenon of adaptation in which animals of some species, after prolonged exposure to sublethal hyperoxia are able to survive exposure to concentrations of inspired O_2 which would be lethal to previously unexposed animals (Rosenbaum et al., 1969). Morphometric studies

(Crapo et al., 1980) have shown that the sublethal or adaptive exposure (85% O_2 in rats) causes injury to capillary endothelium that is similar in extent but more gradual than that caused by 100% O_2 and is followed after 7 days by enlargement of the remaining cells whose numbers remain stable at about 60% of normal thereafter. Type II cells enlarge and proliferate during the adaptive exposure (Crapo et al., 1980) and dilation of cisternae of endoplasmic reticulum and numbers of ribosomes as well as characteristic mitochondrial changes occur. These consist of increased size and volume, increased density of matrix and changes in the arrangement of cristae (Rosenbaum et al., 1969; Yamamoto et al., 1970). Homogenates of the lungs of rats after adaptive exposure to hyperoxia contain increased activity of both manganese and of copper-zinc superoxide dismutases and species such as guinea pigs, hamsters, and mice that do not develop O_2 tolerance show much smaller increases in superoxide dismutase activity after exposure (Crapo et al., 1980; Crapo and Tierney, 1974). The induction of superoxide dismutase is currently held to be the most likely explanation for the phenomenon of adaptation. Because superoxide dismutase is an intracellular enzyme one can infer that the protection afforded by its increased activity is against oxidant radicals generated within the lung cells and not against those generated by inflammatory cells. It would be of interest to know whether O_2-tolerant lungs recruit and/or activate neutrophils from the circulation upon reexposure to oxygen or if activated neutrophils perfused into the lung inflict the same injury upon O_2-tolerant lungs that they do on non-O_2-tolerant lungs. Such experiments might provide further insight into mechanisms of recruitment of neutrophils during hyperoxia and of the relative importance of intrinsic and of neutrophil-derived damage in the overall picture of hyperoxic injury.

The morphologic, morphometric, and kinetic data briefly reviewed above are very useful because, together, they present a coherent if not entirely complete picture of the structural consequences in the lung of an injury almost certainly mediated predominantly by toxic oxidant radicals. They thus provide a frame of reference which broadly defines oxidant injury in general. They demonstrate significant species differences in which of the two cell populations principally involved, the capillary endothelial cells and the type I alveolar epithelial cells sustains the major injury. These observations have implications which bear on experimental models of AAI. They indicate that in studies of etiology it is important to recognize that one fundamental etiology may lead to a significantly different distribution of injury and repair in different species and that overly rigid morphologic definitions in this type of model should be avoided. These species' differences also dictate caution in extrapolating pathophysiologic consequences of oxygen toxicity from experimental animals to man. For example, abnormal lung mechanics associated with dysfunction of the surfactant system may result from a variety of causes including synthesis of intrinsically abnormal surfactant by injured or dedifferentiated type II cells and contamination of intrinsically normal surfactant by plasma components. Because available observations suggest that in humans, hyperoxia causes necrosis of alveolar epithelium (Gould et al., 1972; Nash and Pontopidan,1967; Katzenstein et al., 1976), the monkey, in which necrosis of alveolar epithelium is a prominent result of hyperoxic injury would be an ideal species in which to study this problem.

Finally, the morphometric and kinetic data provide a quantitative basis for comparison with results of therapeutic intervention such as treatment with antioxidant compounds in various forms.

VI. Acute Alveolar Injury Caused by *N*-Nitroso-*N*-Methylurethane

Interest in *N*-nitroso-*N*-methylurethane (nitrosourethane) as a cause of lung injury began in 1967 when it was noted that this potent carcinogen caused interstitial injury to the lungs of hamsters when injected subcutaneously (Herrold, 1967). Subsequent studies showed that repeated weekly injections of small doses led, after several weeks, to respiratory insufficiency and to histologic changes in the lungs which closely resembled those in human fibrosing alveolitis (Ryan, 1972). Larger doses led to more rapid onset of respiratory failure and to more florid lung injury, often with hyaline membranes. It has since been shown that a single subcutaneous injection of 5-8 mg/kg body weight of nitrosourethane in dogs causes AAI practically indistinguishable by light and by electron microscopy from that seen in humans (Ryan et al., 1976). Not only the evolving morphologic features but also the time course are similar. Because of this structural similarity, because the severity of the physiologic and structural changes induced by nitrosourethane are dose-related and because large animals can be used, this empirical lung injury has provided an ideal model in which to study mechanical, physiologic, and metabolic consequences of AAI.

After injection of 5 mg of nitrosourethane per kilogram body weight, about a third of the animals die of respiratory failure, often complicated by pneumothorax, most between 5 and 18 days postinjection (Barrett et al., 1979).

Morphologically, perivascular edema and cytoplasmic vacuolation of type II alveolar epithelial cells are seen at 24 hours and by two days scattered epithelial cells are necrotic. Numbers of necrotic epithelial cells of both types increase until 5 or 6 days postinjection when many are seen and large areas of basement membrane are denuded (Figs. 21, 22). Hyaline membranes, while not as numerous as in the most severely injured human lungs, are readily identified and identical with them, resting on denuded basement membranes and composed of granular material enmeshing fragmented membranes and disrupted organelles (Fig. 23). During this early period alveolar septa are infiltrated by neutrophils and atelectasis and patchy alveolar edema are marked (Fig. 24). As early as two days

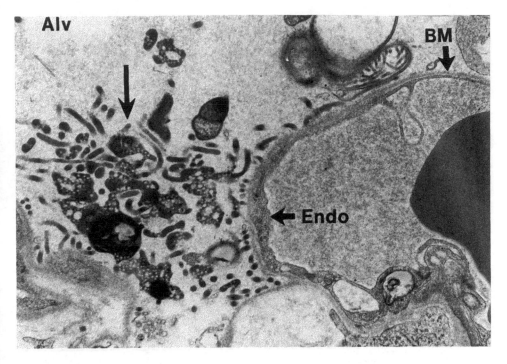

Figure 21 Electron micrograph of lung from a dog 7 days after injection of nitrosoure-thane. An alveolar epithelial cell, probably a type I cell (arrow), is necrotic and partly sloughed leaving an exposed basement membrane. Capillary endothelium (endo) is intact. BM = basement membrane, ALV = alveolus ($\times$17,500). Reprinted from Ryan et al., 1976.

postinjection, a few alveoli, almost always centriacinar, are partly lined by regenerating epithelial cells (Fig. 25). These cells, which steadily increase in number, are at first irregularly shaped with swollen mitochondria and only rare lamellar bodies (Fig. 26). Nuclei and nucleoli are large. As their numbers increase, the new cells become more uniform and cuboidal, and by 9-10 days postinjection, when they have come to occupy a large proportion of the alveolar surface, lamellar bodies are numerous (Figs. 27, 28). This process of maturation continues until by 20 days almost all the cells closely resemble mature type II cells. During the entire process, capillary endothelium remains essentially intact with only scattered subendothelial vesicles and a rare swollen cell. Because the lungs in these studies were examined after inflation fixation at a constantly maintained translung pressure of 25 cm of water, architectural revision was readily recognized. As early as 5 days postinjection, groups of alveoli, mostly in centriacinar zones showed loss of volume or collapse. In the most severely injured

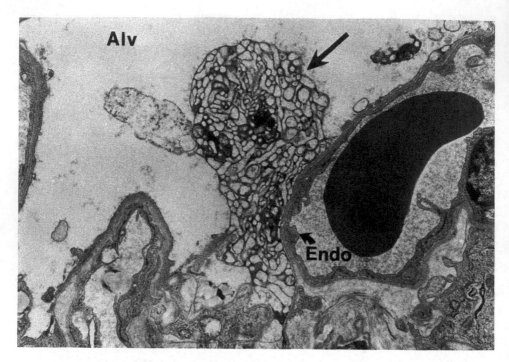

Figure 22 Electron micrograph of lung from a dog, 7 days after injection of nitrosourethane. An alveolar epithelial cell, probably a type II cell (arrow) is necrotic while the adjacent capillary endothelium is intact (×9600). Reprinted from Ryan et al., 1976.

lungs, the volume loss was extensive and random and not confined to centriacinar zones. The extent of patchy alveolar closure increased with time and was accompanied by enlargement of intervening small airspaces, mostly alveolar ducts, which gave a coarse appearance to the parenchyma even on gross examination. This close similarity of the evolving morphologic picture with that in the human, dominated by epithelial injury and repair with early edema and late irreversible architectural revision is the basis for the utility of nitrosourethane injury as a model of human disease.

A. Lung Mechanics

Within 48 hours after injection of nitrosourethane, lung compliance decreases and elastic recoil increases. These alterations increase in severity until between 8 and 12 days after injection following which they improve toward control values (Fig. 29). Late in the recovery phase (later than 21 days post injection) elastic

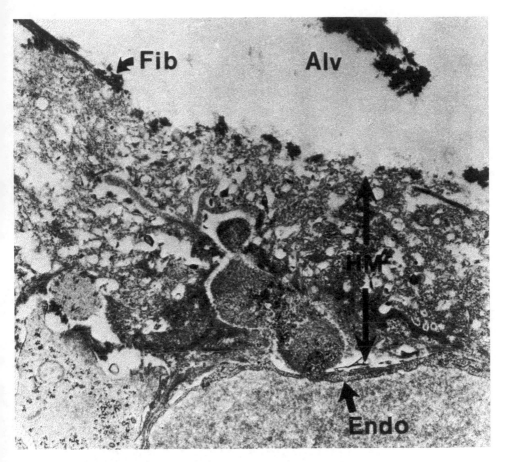

Figure 23 Electron micrograph of lung from a dog 5 days after nitrosourethane injection. A hyaline membrane (HM) covers a denuded basement membrane and is composed of fragments of necrotic cells, organelles and granular material. A few strands of fibrin (Fib) lie on its surface (×1450). Reprinted from Ryan et al., 1976.

lung recoil near the resting expiratory level is not significantly different from control values while that observed at maximum lung inflation remains significantly abnormal (Fig. 29).

B. Volume-Pressure and Morphometric Observations

To further examine the causes of decreased lung compliance during nitrosourethane-induced AAI, volume-pressure diagrams using excised lungs and data

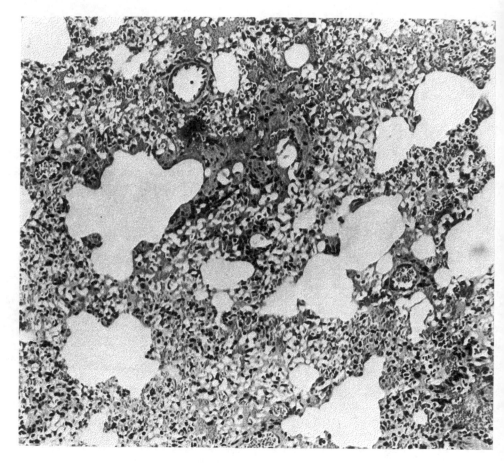

Figure 24 Microscopic section of lung from a dog 7 days after injection of nitrosourethane. Extensive microatelectasis involves small airspaces and those remaining patent are enlarged. (Hematoxylin-eosin ×120.) Reprinted from Ryan et al., 1976.

derived from them were compared with morphologic observations and with the extent of irreversible alveolar closure (Ryan et al., 1978).

In these studies, one lung of each dog was used for volume-pressure measurements and the other was prepared for morphologic and morphometric studies by fixing it via a bronchial cannula with 10% buffered formalin at a constant inflation pressure of 40 cm of water. Airspaces that remained closed after liquid inflation fixation at this pressure were defined as irreversibly closed (Figs. 30, 31). Airspace sizes in microscopic sections from tissue blocks selected by stratified random sampling were measured by the mean linear intercept (Lm)

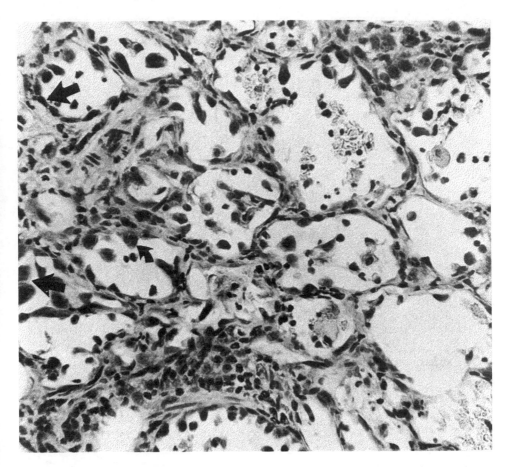

Figure 25 Microscopic section of lung from a dog 5 days after injection of nitrosourethane. Alveoli are partly lined by irregular epithelial cells with large nuclei (arrows). (Hematoxylin-eosin ×450). Reprinted from Ryan et al., 1976.

method. The Lm is the average distance between alveolar surfaces and is directly proportional to the average size of the small airspaces. Only the surfaces of open airspaces were counted as intercepts and thus the Lm increased in direct proportion to the number of irreversibly closed alveoli.

Comparisons of the volume-pressure characteristics of the lung during inflation-deflation with air and with saline allow estimation of the contributions of surface forces and of tissue forces to lung recoil because inflation-deflation with saline largely eliminates surface tension in the lung. The normal lung inflated with saline typically reaches total volume at lower pressure and shows less

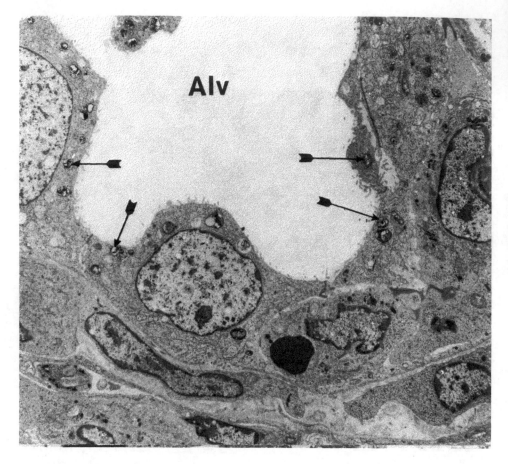

Figure 26 Electron micrograph of lung from a dog 3 days after injection of nitrosourethane. An alveolus (ALV) is partly lined by regenerating cells with large nuclei and a few cytoplasmic lamellar bodies (×4100). Reprinted from Ryan et al., 1976.

hysteresis than when inflated with air (Fig. 32). Figure 33 shows volume-pressure diagrams of dog lungs during deflation after inflation with air at intervals after injection of nitrosourethane. In these curves, volumes are expressed as percent of predicted total lung capacity to normalize for differences in lung size among the animals. There is a progressive downward shift in the entire curve from the control to the 3-4 day period and a further significant downward shift at 5-7 days. In contrast, the deflation curve after inflation with saline is not significantly different from control during the 3-4 day period but shows a marked downward shift at 5-7 days (Fig. 33).

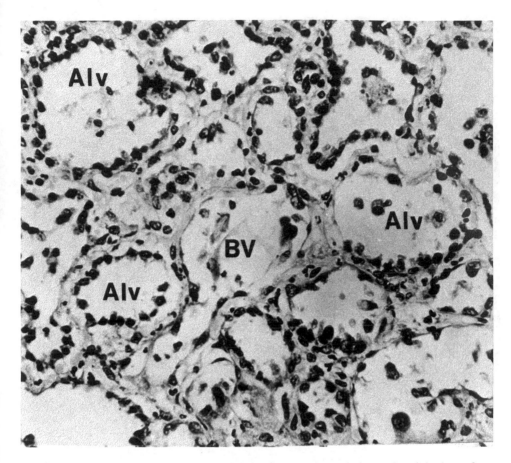

Figure 27 Microscopic section of lung from a dog 16 days after injection of nitrosourethane. Interalveolar septa are uniformly thickened and alveoli are almost completely lined by uniform cuboidal epithelial cells. BV = blood vessel. (Hematoxylin-eosin ×420.) Reprinted from Ryan et al., 1976.

When the volume pressure diagrams are corrected for loss of lung volume by expressing lung volume as a percent of observed total lung capacity, only the volume pressure behavior of open or recruitable airspaces is measured and the effect of closed airspaces is excluded. Such curves are shown in Figure 34. The diagrams on air deflation show a downward shift at 3-4 days with no further reduction, while the diagrams on saline deflation show no downward shift from controls. Lm was slightly increased at 3-4 days and steadily increased thereafter (Fig. 35, top). In addition, Lm showed significant inverse linear correlation with total

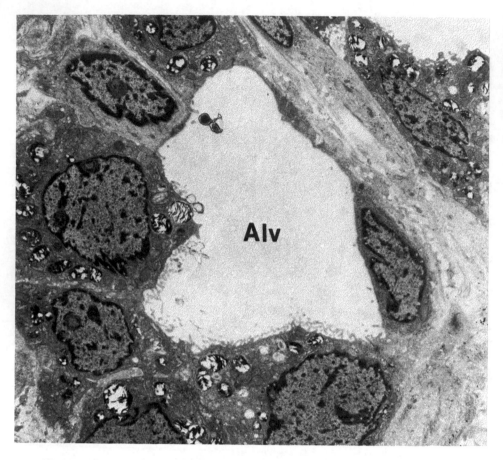

Figure 28 Electron micrograph of lung from a dog 20 days after injection of nitrosourethane. An alveolus is almost completely lined by mature appearing type II cells containing many lamellar bodies (×4800). Reprinted from Ryan et al., 1976.

lung capacity (the volume of the excised lung measured at a translung pressure of 30 cm H_2O) and with compliance measured during deflation with saline (Fig 35 middle and bottom). These correlations as well as morphologic examination eliminate emphysema as a cause of increased Lm and support the conclusion that it is due entirely to irreversible alveolar closure. Morphologic examination ruled out obstruction of small airways and decreased tissue distensibility due to fibrosis. The remaining possible causes of loss of lung volume and compliance are therefore three: (1) increased surface tension; (2) decreased distensibility of recruitable airspaces due to vascular congestion and interstitial edema; (3) irreversible alveolar closure.

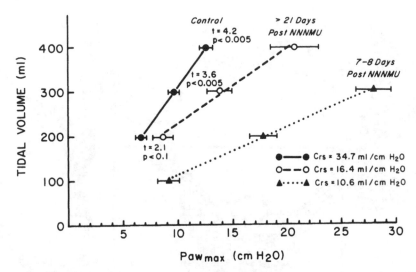

Figure 29 Volume-pressure diagrams during peak illness and recovery. Horizontal bars represent 1 SE. Eleven dogs were studied 7-8 days after injection of nitrosourethane and 5 of these were again studied 21-42 days after injection. Crs = compliance of respiratory system (slope of best fitting straight line drawn through points by the method of least squares). Reprinted from Barrett et al., 1979.

Analysis of the volume-pressure, morphologic, and morphometric changes together permit estimation of the relative contributions of each of these three possible causes of decreased lung compliance during the evolution of lung injury and repair. During early injury (3-4 days) loss of lung volume and distensibility are due almost entirely to increased surface tension because the volume-pressure curves on saline deflation are entirely normal while those on air deflation show a marked downward shift. At this time the extent of irreversible alveolar closure is small and edema (lung weight-body weight ratio) is not yet marked. When the altered volume-pressure behavior during air deflation is corrected for loss of volume by expressing volume as percent of observed total lung capacity, abnormal compliance is still present. Thus the abnormal compliance on air deflation is due partly to closure of airspaces which cannot be reversed with air inflation to 30 cm H_2O (but can be reversed by saline inflation) and partly to increased surface tension in recruitable alveoli.

Another picture emerges by 5-7 days when loss of lung volume and distensibility can no longer be explained by the effects of increased surface tension alone. Now the lungs are even less distensible when inflated with air and they also exhibit decreased distensibility when inflated with saline. Clusters of irreversibly closed alveoli are present in larger numbers and Lm is significantly increased.

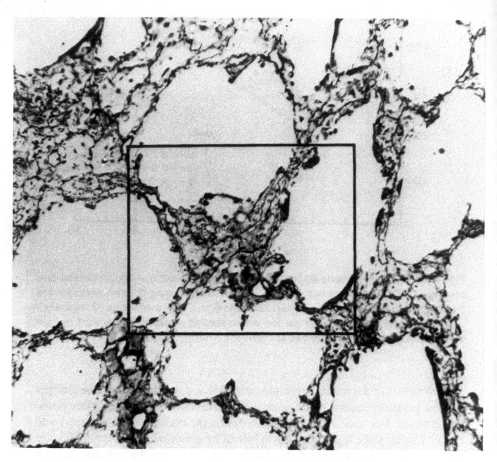

Figure 30 Microscopic section of lung from a dog 11 days after injection of nitrosourethane. The lung was fixed by inflation with 10% buffered formalin at a pressure of 40 cm H_2O. Clusters of closed alveoli are separated by patent airspaces which are markedly enlarged. (Gomori's silver reticulin × 140.) Reprinted from Ryan et al., 1978.

At 9-14 days, volume-pressure diagrams with air and with saline are not significantly different from those at 5-7 days. Morphologically there are more clusters of irreversibly closed alveoli and Lm is even larger. Inverse correlations between Lm and total lung volume and between Lm and compliance on saline deflation support the concept that irreversible alveolar closure is the principal factor decreasing lung distensibility during the second week of nitrosourethane injury. However, a persistent contribution by increased surface tension is demonstrated by the persistent abnormality in the volume-corrected air volume-

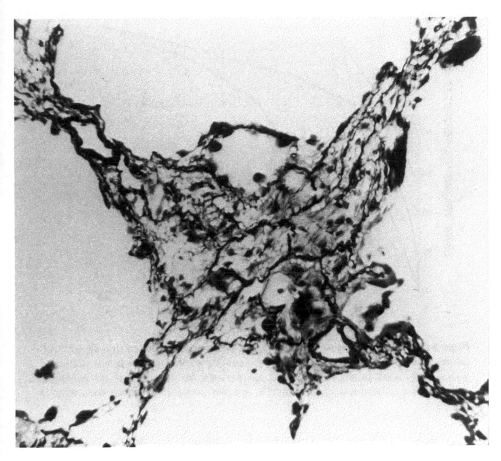

Figure 31 Detailed view of the field enclosed by the square in Fig. 29. The reticulin framework of the closed alveoli is clearly seen but there is no evidence of interstitial fibrosis. (Gomori's silver reticulin ×360.) Reprinted from Ryan et al., 1978.

pressure curve (Fig. 34). The normal volume-pressure characteristics of the volume-corrected saline volume-pressure diagrams at all phases of injury argues against abnormal tissue distensibility as a significant cause of abnormal compliance even though lung weights steadily increase and interstitial edema is prominent. These latter findings are in agreement with other studies showing that interstitial edema decreases lung compliance only slightly and that the major effect of pulmonary edema is mediated through the increase in surface tension caused by alveolar edema (Nobel et al., 1975; Cook et al., 1959; Johnson et al., 1964).

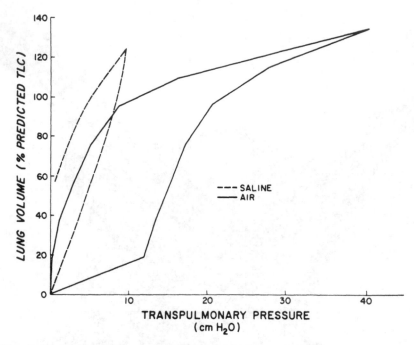

Figure 32 Volume-pressure diagrams during inflation and deflation with air and with saline in the excised lung of a normal dog. Because the air-surface interface is eliminated with saline inflation, surface tension and its contribution to static pressure are also eliminated and the saline inflated lung inflates to TLC at lower pressure. Reprinted from Ryan et al., 1978.

The above observations support the following hypothesis. In the earliest phase of AAI, decreased lung compliance and volume are primarily due to alveolar collapse caused by increased surface tension. The contribution of increased surface tension is maximized by 5-7 days and remains constant through the following week. Beginning at 5-7 days and increasingly thereafter, irreversible closure of alveoli effects lung distensibility. As the number of irreversibly closed alveoli increases, the distensibility of the lung progressively decreases. The number of irreversibly closed alveoli must be important in determining the degree to which lung compliance remains abnormal after recovery.

These observations are reviewed in some detail because it seems highly probable that they pertain as well to human AAI. They support the observations subsequently made in human lungs (Ryan, 1983) that irreversible closure of airspaces contributes significantly to volume loss and to architectural revision. They also suggest that increased surface tension is a major cause of mechanical

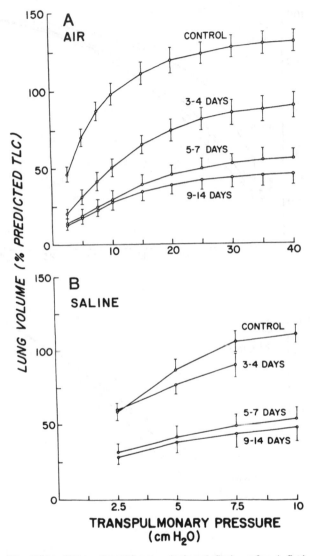

Figure 33 Volume-pressure diagrams during deflation after inflation with air (A) or saline (B) at intervals after injection with nitrosourethane. The points and bars represent means ± 1 SE (22 experimental, 8 control dogs). The volume axis is expressed as percent of predicted total lung capacity (TLC) to normalize for difference in size among the animals. Reprinted from Ryan et al., 1978.

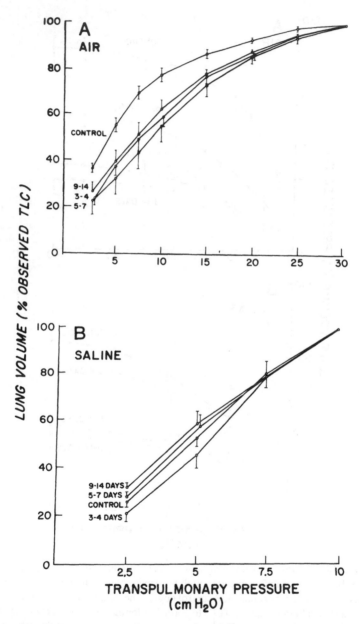

Figure 34 Volume-pressure diagrams during deflation after inflation with air (A) or saline (B) at intervals after injection with nitrosourethane. Symbols are the same as in Fig. 33. The volume axis is expressed as percent of observed total lung capacity (TLC) to correct for loss of lung volume thus measuring volume pressure characteristics only of recruitable airspaces. Reprinted from Ryan et al., 1978.

abnormality during early to peak injury and they call into question the assumption that these abnormalities are due solely to high permeability pulmonary edema.

C. Surfactant Alterations

Surface tension might be increased in the lung during injury by reduction in the quantity of surfactant, by intrinsic abnormality of surfactant or by changes induced by components of edema fluid. During acute nitrosourethane injury both profound reductions in the quantity and intrinsic abnormality of the remaining surfactant occur. Changes in quantities of major surfactant phospholipids in alveolar lavage after injury are shown in Table 2 (Liau et al., 1984). Exhaustive lavage was carried out with large volumes of normal saline (10 ml/g normal lung weight) to assure quantitative recovery of phospholipids possibly sequestered in reversibly closed alveoli. Disaturated phosphatidylcholine (DSPC), the major surface active phospholipid decreases to between 20 and 35% of control levels (depending on the dose of nitrosurethane) at peak injury (6-8 days) and returns to about 80% of control levels late in recovery (15-20 days). Total phospholipids, phosphatidylcholine (PC), and phosphatidylethanolamine follow the same pattern but the changes are somewhat smaller. Phosphatidylglycerol (PG) decreases strikingly to about 20% of control at peak injury but, in contrast to PC and DSPC, it remains at this low level during recovery. The quantity of phosphatidylinositol is significantly higher during late recovery (days 15-20) than in controls or at peak injury.

Lung tissue DSPC also decreases during early and peak injury and the reduction in the quantity estimated to be derived from type II cells (as opposed to that from all other lung cell sources which was assumed to remain constant during injury) was about equivalent to the reduction in alveolar DSPC. This equivalent reduction indicates that impaired lipid synthesis rather than increased degradation is the main cause of the deficiency of alveolar surfactant. During early and peak injury (days 2-7) lung compliance shows a significant linear correlation (p = 0.02) with quantities of DSPC in alveolar lavage (Ryan et al., 1981).

D. Qualitative Alterations in Surfactant

To examine its function and composition in the injured lung, surfactant was purified by centrifugation on a NaBr density gradient (Liau et al., 1987) using a modification of a previously described method (Shelley et al., 1977). This method has been shown to remove a large proportion of contaminating plasma phospholipids and proteins from the surfactant.

Purified surfactants from lungs during early (2-4 day) and peak (6-8 day) injury failed to lower surface tension below 20 dynes/cm when spread as films and compressed on a modified Wilhelmy balance even when excess material was applied to the surface, whereas those from lungs after recovery were not different

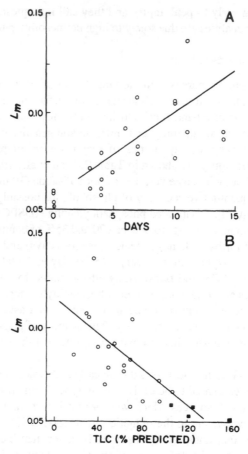

Figure 35   Relations between mean linear intercept (Lm) and day of disease (A), total lung capacity (B) and compliance on saline deflation (C). In (B) and (C) data from control animals are denoted by squares and data from nitrosourethane injected animals are denoted by circles. Correlation coefficients and P values are (A) r = 0.614, P < 0.01; (B) r = 0.611, P < 0.01; (C) r = 0.571, P < 0.02. Reprinted from Ryan et al., 1978.

from those from control animals (Fig. 36). During early recovery (day 10-12), surface tensions were lowered to levels intermediate between those at peak injury and those after recovery. In addition, the rate of adsorption of surfactant into the surface film from the hypophase was decreased during early and peak injury and returned to normal during recovery. Analysis of phospholipids and neutral lipids of the purified surfactant (Table 3) revealed that the percentages of all the major phospholipids except PG in all phases of injury and recovery

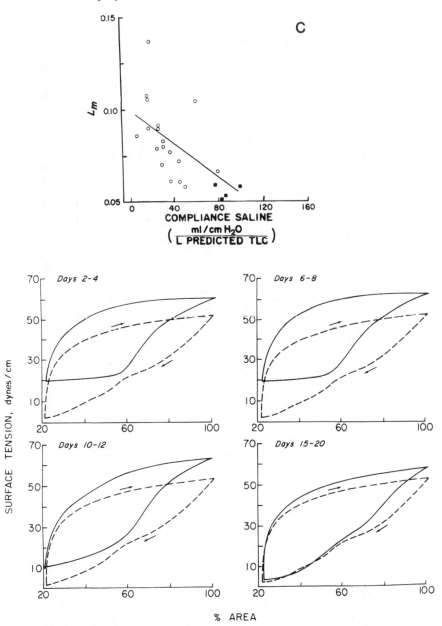

Figure 36 Representative surface tension versus area diagrams generated on a Wilhelmy balance by purified surfactants from lungs of dogs at intervals after injection of nitrosourethane. A diagram generated by surfactant from the lung of a normal dog (dashed line) is included for comparison in the diagrams of surfactant from each phase of injury. Reprinted from Liau et al., 1987.

Table 2 Quantities of Phospholipids in Alveolar Lavage During Acute Alveolar Injury Induced by Nitrosourethane[a]

| | Day post-NNNMU | | | | | Days 6-7 vs. 0 | Days 15-20 vs. 0 | Days 15-20 vs. 6-8 |
	0 (8)	2-4 (5)	6-8 (5)	10-12 (5)	15-20 (7)		P	
Total phospholipids	2.06±0.41	1.27±0.27	0.97±0.52	1.06±0.71	1.78±0.65	<0.005	NS	<0.05
Phosphatidylcholine	1.56±0.31	0.94±0.20	0.70±0.26	0.80±0.54	1.37±0.63	<0.005	NS	<0.05
Disaturated phosphatidylcholine	0.76±0.20	0.35±0.05	0.27±0.15	0.35±0.27	0.61±0.21	<0.001	NS	<0.01
Phosphatidylglycerol	0.13±0.04	0.04±0.01	0.03±0.02	0.02±0.02	0.02±0.02	<0.001	<0.001	NS
Phosphatidylinositol	0.06±0.02	0.06±0.01	0.06±0.02	0.06±0.02	0.11±0.03	NS	<0.005	<0.005
Phosphatidylethanolamine	0.06±0.02	0.04±0.02	0.03±0.01	0.03±0.02	0.05±0.02	<0.005	NS	NS
Sphingomyelin	0.07±0.02	0.11±0.03	0.08±0.02	0.05±0.02	0.08±0.03	NS	NS	NS
Lysophosphatidylcholine	0.06±0.02	0.05±0.01	0.04±0.02	0.04±0.02	0.07±0.03	NS	NS	NS
Dyphosphatidylglycerol	nd	nd	nd	nd	nd			

Abbreviations: NS = not significantly different; nd = not detectable.
[a]Values are means ± SD and are given as milligrams per gram predicted normal lung weight. Figures in parentheses are number of dogs in each group.

Table 3 Composition of Phospholipids, Neutral Lipids, and Phospholipid/Protein Ratio of Purified Surfactant During Acute Alveolar Injury Induced by Nitrosourethane[a]

	Day post-NNMU					P		
	0 (8)	2–4 (5)	6–8 (5)	10–12 (5)	15–20 (7)	Days 6–8 vs. 0	Days 15–20 vs. 0	Days 15–20 vs. 6–8
Phospholipids								
Phosphatidylcholine	73.5±2.5	72.5±5.9	74.2±2.8	75.3±3.7	77.3±3.3	NS	NS	NS
Disaturated phosphatidylcholine	43.6±2.8	43.7±3.7	44.1±1.6	45.2±2.4	46.1±2.4	NS	NS	NS
Phosphatidylglycerol	12.8±1.3	4.4±0.6	4.2±0.7	4.0±0.9	3.0±0.7	<0.001	<0.001	NS
Phosphatidylinositol	3.3±1.1	5.1±0.8	6.1±1.4	5.7±2.1	6.3±1.6	<0.005	<0.001	NS
Phosphatidylethanolamine	1.9±0.6	2.2±0.6	2.0±0.5	1.5±0.4	1.7±0.6	NS	NS	NS
Phosphatidylserine	nd	nd	nd	nd	nd			
Sphingomyelin	4.6±1.3	10.7±4.8	7.8±1.6	4.4±0.8	4.5±1.1	<0.005	NS	<0.005
Lysophosphatidylcholine	3.9±1.2	4.6±1.5	5.9±2.1	8.7±1.5	7.2±2.0	<0.005	<0.005	NS
Neutral lipids								
Cholesterol	10.4±1.4	11.6±1.9	12.3±2.4	11.6±2.1	10.9±1.8	NS	NS	NS
Cholesteryl ester	1.4±0.3	2.8±1.0	3.1±0.5	1.7±0.4	1.8±0.8	<0.001	NS	<0.001
Triglyceride	8.6±1.2	3.9±0.5	3.6±0.6	4.9±0.5	5.6±0.8	<0.001	<0.001	<0.001
Phospholipid/protein (mg/mg)	7.8±1.2	6.2±1.2	6.2±1.8	6.8±1.4	8.3±1.5	NS	NS	NS

Abbreviations: *see* Table 2.

[a]Values are means ± SD and are given as percentages of total phospholipids for each phospholipid and as percentages of total lipids (phospholipids plus neutral lipids) for each neutral lipid. Figures in parentheses are the number of dogs in each group. Protein in this table is total surfactant-associated protein, not surfactant apoprotein.

were not different from those of control animals. PG decreased markedly during injury and remained at minimal levels throughout recovery while both PI and lysophosphatidylcholine behaved reciprocally with PG, increasing and remaining increased during injury and recovery. An increase in sphingomyelin (SPH) led to a reduced PC/SPH (L/S) ratio.

When subjected to continuous sucrose density gradient centrifugation (0.1-1.0 M) partially purified surfactant (ppt from lavage after centrifugation at 27,000 x g 2 h) showed marked changes in density during injury and recovery. Normal surfactant yielded two distinct bands with isopycnic densities of 1.05 and 1.09 while surfactant from dogs at peak injury yielded only 1 band with density of 1.02. During early injury and early recovery, bands with densities intermediate between these extremes (d = 1.04-1.05) were separated while at recovery the bands were identical with those from normal surfactant.

The major surfactant apoprotein (mw-38,000) in this fraction as determined by SDS-PAGE began to decrease during early injury (2-4 days), reached a nadir at peak injury and increased during recovery approaching normal during late recovery.

Decreases of surface activity and changes in phospholipid composition very similar to those found in surfactant from the nitrosourethane injured lung have recently been described in patients with ARDS (Hallman et al., 1982). Similar patterns in surfactant density on continuous sucrose density gradient centrifugation have also been found in surfactant from such patients (Petty et al., 1977, 1979). The compositional abnormalities responsible for impaired surface function in the surfactant from the nitrosourethane-injured lung have not yet been identified but the most dramatic change, the marked and sustained decrease in PG has been ruled out because during late recovery when surfactant surface activity has returned to normal, PG remains maximally reduced (Liau et al., 1985). Deficiency of PG has been found in the lungs of prematurely born rabbits and humans and of patients with ARDS (Hallman, et al., 1982, 1977; Hallman and Gluck, 1976) and has been considered to be essential for surfactant function (Hallman and Gluck, 1976). However, recent data (Beppu et al., 1983; Hallman et al., 1985) supported by the finding in the nitrosourethane-injured lung (Liau et al., 1985) suggests that PG is not an essential component for normal surface function as measured by commonly used parameters. It appears that deficient PG secretion is a characteristic of immature type II cells whether they be fetal or regenerating.

The possibility that deficiency of apoprotein is at least partly responsible for surface abnormalities is more attractive because an increased phospholipid-apoprotein ratio during injury coincides with impaired function of purified surfactant and restoration of normal phospholipid apoprotein ratio is accompanied by restoration to normal of surfactant function.

Indeed, reconstitution experiments in which apoprotein is added incrementally to phospholipids extracted from purified surfactant and surface active liposomes are generated from this phospholipid-apoprotein mixture, have shown that the apoprotein is essential for normal adsorption (Liau, unpublished observations). However, there is as yet no direct evidence that the surfactant apoprotein is essential for normal surface tension lowering. On the contrary, there is considerable evidence that various mixtures of phospholipids without protein are capable in vitro and probably in vivo of reducing surface tension to levels equally as low as those achieved by natural surfactant (Liau et al., 1984b; Tanaka et al., 1986; Yu et al., 1984). One remarkable finding (Liau, unpublished observation) is that phospholipid extracts of purified normal surfactant, when tested in the form of liposomes, lower surface tension normally while the same preparation of surfactant from the nitrosourethane lung does not. However, when the cold acetone precipitate from this extracted lipid is used, the liposomes lower surface tension normally. Thus a cold acetone extractable compound may be responsible for the abnormal surface activity.

Whatever its cause it is unlikely that abnormal surface activity of the surfactant remaining in the lung during acute injury contributes significantly to abnormal lung mechanics. The observed quantitative deficits in the order of 70-80% during peak injury must far outweigh the relatively mild impediment of function in the remaining material. The possibility remains however that the abnormal surfactant represents the only recoverable remnant of a much larger surfactant fraction which has been degraded beyond recoverability by the massive oxidative injury to which all the other lipids and proteins in the lung are subjected.

VII. Acute Lung Injury Induced by Oleic Acid

Probably the most extensively studied empirical model of acute lung injury is that induced by the intravascular injection of oleic acid (OA). The original impetus for the investigation of the injurious effects of this free fatty acid upon the lung was provided by clinical and experimental studies of pulmonary fat embolism. Clinicians observed that patients with posttraumatic fat embolism usually developed respiratory symptoms only after a latent period of 24-72 h (Scully, 1956; Sprougle et al., 1964; Peltier and Wertzberger, 1968; Peltier, 1967). At autopsy the lungs of some of these patients showed not the acute cor pulmonale of massive small artery occlusion, but hemorrhagic pulmonary edema (Wyatt and Khoo 1950; Robb-Smith, 1941). One hypothesis to which these observations have led is that fat emboli cause respiratory impairment not by mechanically occluding small vessels but by releasing fatty acids into the lung as a result of hydrolysis by lipase which is abundant in the lung (Harris et al., 1939; Peltier, 1957). Unhydrolyzed neutral lipids have been shown to produce minimal reaction when introduced into tissue whereas after hydrolysis of these lipids severe tissue injury, caused by free fatty acids, occurs (Hirsch, 1941). The immediate effect upon the lung of intravenous injection of oleic acid is quite different from that of injection of neutral fat. Animals tolerate nearly four times as much neutral lipid as oleic acid and its injection in lethal amounts causes rapid death with pallor of the lungs and dilation of the right ventricle whereas OA in lethal doses causes hemorrhagic pulmonary edema (Newman, 1948; Peltier, 1956). In one study, large but sublethal doses of neutral lipid in the rat led to pulmonary edema which was present on the second day postinjection and maximal on the fourth. The appearance of edema coincided with an increase in quantity of fatty acids and of lipase in lung tissue (Fonte and Hausberger, 1971). These animal studies have provided some support for the hypothesis that lung injury in fat embolism is caused by free fatty acid but it will be important to reproduce these findings in other laboratories, to learn whether the free fatty acids are derived from lipids of the emboli or from some other source and to establish that the relatively small quantity of free fatty acids found are capable of causing significant lung injury.

After the original studies aimed at clarifying the mechanism of lung injury in fat embolism, the OA model became popular as a model of diffuse acute lung injury without reference to fat embolism because it reproduced many of the clinical, physiologic, and pathologic features of ARDS.

The morphologic consequences of intravascular administration of OA depend to a large degree upon the dose. Large doses (those above the LD_{50}) cause massive hemorrhagic pulmonary edema and death within several hours (Peltier, 1956; Uzawa and Ashbaugh, 1969). The ultrastructural features of this fulminant injury have not been studied.

The structural changes resulting from a large but sublethal dose of OA have been fairly well studied and it is this sort of injury which is used as a model of ARDS. Because of its insolubility in water, OA, after intravenous injection, arrives in the pulmonary arterioles and capillaries largely as droplet emboli. This has been clearly shown both in histologic sections (Derks and Jacobovitz-Derks, 1977; Motohiro et al., 1986) and in vivo using a thoracic window (King et al., 1971). In the latter study, in which small airspaces and their blood vessels were observed through the pleura during the intracardiac infusion of OA, some arterioles (about 40-50 μm diameter) became completely occluded by OA emboli and remained so with arrested blood flow through them and into their capillaries throughout the observation period of several hours. Other emboli transiently occluded or slowly passed through arterioles. Passage of the embolus through the arteriole was followed within 8 minutes by a periarteriolar red blush which expanded into the subtending alveoli until by 3 hours, many of them were overrun by it and almost invisible. This blush was attributed to leakage of red blood cells and of hemoglobin from lysed red blood cells into the alveoli following direct endothelial damage by the OA. It seems likely that other droplets were small enough to pass arterioles and arrest in capillaries or to pass the pulmonary circulation altogether. Unfortunately, the zones of lung tissue involved by the process were not examined by light and electron microscopy. This would have provided a unique opportunity to study the structural features of early OA injury of precisely known duration.

This largely embolic distribution with arrest in precapillary vessels appears to be a principal determinant of a consistent feature of single dose OA injury, namely its patchy or multifocal and peripheral distribution which has been pointed out by several authors (Derks and Jacobovitz-Derks, 1977; Schoene et al., 1984; Hedlund et al., 1985). The distribution of the early lesions of OA injury has been graphically demonstrated in a recent study in which computed tomograms of the lower thorax were made at intervals of 10-30 minutes for up to 4 hours after the injection of OA (45 mg/kg) in dogs (Hedlund et al., 1985). The animals were killed at 10, 30, 60, and 240 minutes after injection of OA, frozen in dry ice and the thorax cut with a bandsaw into 1 cm sections at the same level as that of the CT scan. The CT scan showed lobular patches of increased density averaging about 1 cm in diameter, mostly in the subpleural zone. These lobular patches were first visible 30 minutes after injection of OA and steadily increased in density during the ensuing 3.5 hours. In contact radiographs of the frozen lung sections, these patches, which were apparent as early as 10 minutes after injection of OA, were surrounded and separated by radiographically normal lung and appeared to surround or emanate from small arteries (Figs. 37 , 38). Histologic sections of these areas, prepared after freeze substitution, showed perivascular and interstitial edema at 30 minutes postinjection and alveolar edema and neutrophil infiltration of alveolar septa after 1 h. These observations again suggest that OA arrives in the lung as arteriolar microemboli, apparently of relatively uniform size and that the tissue subtending the embolized vessels is injured.

Labeled OA injected intravenously also shows a preferential distribution in the subpleural zone of the lung (Tarver et al., 1986). The reason for this distribution is not clear but it has been suggested that it is related to the size of the OA droplets because labelled microspheres of 137 μm diameter arrest in larger numbers in the subpleural lung zone than in the deeper zones whereas microspheres 15 μm in diameter are uniformly distributed through the lung.

Several microscopic studies of lung tissue during the first 24 hours after OA injection have described vascular congestion and interstitial and alveolar edema beginning as early as a few minutes after injection and followed within 26 hours by an acute inflammatory infiltrate (Dickey et al., 1981; Gemer et al., 1975; Motohiro et al., 1986; Hedlund et al., 1985; Slotman et al., 1982; Ashbaugh and Uzawa, 1968). Most of these reports have not provided details or photographic documentation of such key features as specific type of alveolar cell sustaining injury, types of inflammatory cell involved or reparative changes in the epithelium. Three studies have provided reasonably complete light and electron microscopic descriptions in animals sacrificed at several intervals after OA injection. While

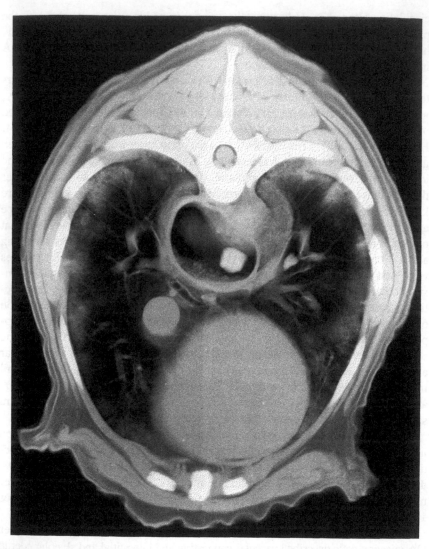

Figure 37 Contact radiograph of a frozen section of thorax from a dog 240 minutes after injection of oleic acid. Patchy areas of increased density separated by normally transradiant lung occupy the peripheries of both lungs. Small bronchovascular shadows can be seen entering the apices of several of the triangular subpleural densities. Reprinted from Hedlund et al., 1985.

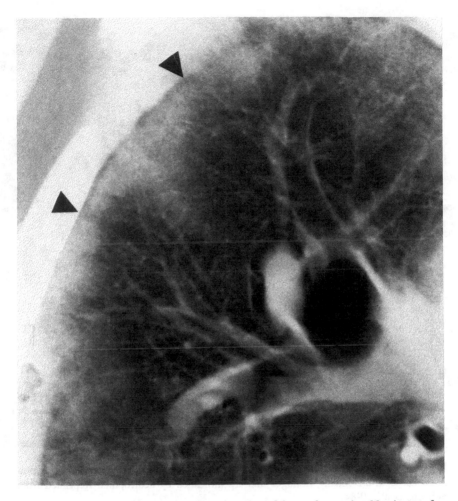

Figure 38 Contact radiograph of a frozen section of thorax from a dog 30 minutes after injection of oleic acid. The size and shape of the subpleural densities (one delimited by arrow points) and their relation to bronchovascular structures suggest that they are terminal respiratory units or aggregates thereof e.g. secondary labules. Reprinted from Hedlund et al., 1985.

these studies were done in different species, and with different doses of OA, [dog 45 and 90 mg/kg (Derks and Jacobovitz-Derks, 1977), dog 90 mg/kg (Schoene et al., 1984), baboon 35 to 125 mg/kg (Johanson et al., 1982)], they do provide the most detailed picture of the morphologic evolution of OA injury available.

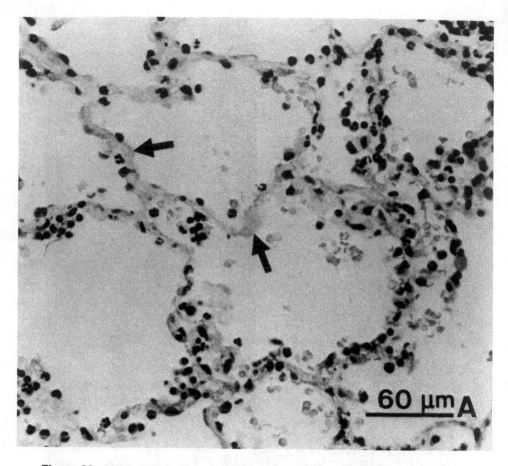

Figure 39 Microscopic section of lung from a dog 24 hours after injection of oleic acid. Alveolar septa are focally smudged and are infiltrated by neutrophils. Reprinted from Schoene et al., 1984.

The lungs of dogs 1 hour after OA showed capillary congestion and patchy alveolar edema and by electron microscopy capillary endothelial cells and type I epithelial cells were necrotic (not illustrated) and some capillaries were occluded by fibrin and platelet thrombi (not illustrated) (Derks and Jacobovitz-Derks, 1977). At 2 hours after injection, neutrophils infiltrated interalveolar septa and alveolar and interstitial edema, which began at 1 hour, increased to peak between 6 and 12 hours. The edema was decreased at 24 hours and by 48 hours only perivascular edema remained. In the other dog study (Schoene et al., 1984), alveolar and interstitial edema and neutrophil infiltrate were present at 24 hours (the first interval studied) (Fig. 39) and electron microscopic examination showed

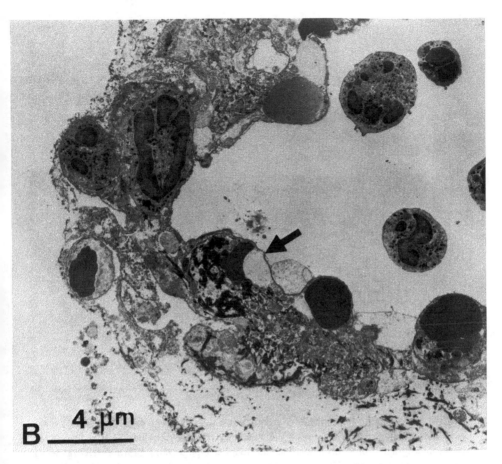

Figure 40 Electron micrograph of lung from a dog 24 hours after injection of oleic acid. Alveolar epithelial cells are extensively necrotic but basal laminae (arrow) remain intact. Endothelial cells are swollen or necrotic and a capillary (at arrow) contains fibrin. Reprinted from Schoene et al., 1984.

necrosis of both endothelium and alveolar epithelium (Figs. 40 and 41). One image shows a sloughed and probably necrotic type II cell (Fig. 41). Connective tissue and basal laminae remained intact. By four days after injection (90 mg/kg) (Schoene et al., 1984) architectural revision was severe and disorderly cellular hyperplasia was apparent. The hyperplastic cells could not be easily identified because of collapse of airspaces and architectural revision but were thought to include both endothelial and epithelial cells. At 7 days after the lower dose of OA [0.45 mg/kg (Derks and Jacobovitz-Derks, 1977)] the acute edema had resolved and the lung architecture was normal except for scattered foci of probable

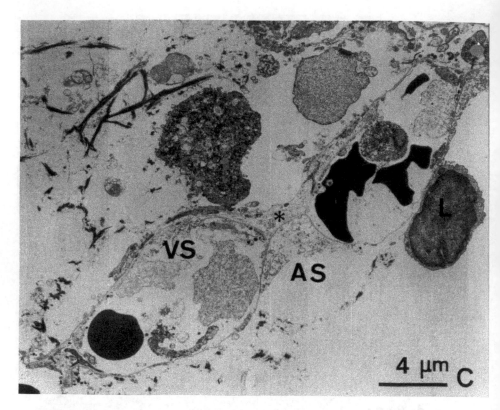

Figure 41   Electron micrograph of lung from a dog 24 hours after injection of oleic acid. Endothelium and epithelium are extensively necrotic. The dark cell at the center of the micrograph is probably a necrotic and sloughed type II cell. Fibrin and cell organelles lie in the alveolar space. VS = vascular space, AS = airspace, asterisk = interstitium. Reprinted from Schoene et al., 1984.

irreversible alveolar closure. Some alveoli were lined by increased numbers of type II epithelial cells which persisted for several weeks (Fig. 42). Following the larger dose (90 mg/kg) of OA the lung architecture at 7 days was more extensively revised and regenerating type II cells were numerous (Fig. 43).

In the study using baboons, doses of OA were such that ventilatory support was necessary for survival of the animals. Post OA intervals are not clearly stated but appear to range from 2 to 10 days or more. During the early phase, edema and hyaline membranes were prominent and extensive microatelectasis was demonstrated by transvascular fixation with the lung inflated with air at a transpulmonary pressure of 20 cm H_2O. Electron microscopic examination

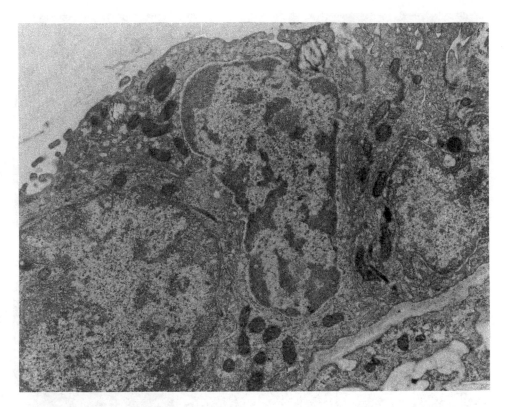

Figure 42 Electron micrograph of lung from a dog 14 days after injection of oleic acid. An alveolus is lined by cuboidal epithelial cells. (×16,000). Reprinted from Derks and Jacobovitz-Derks, 1977.

showed loss of both types of alveolar epithelial cells. Some capillaries were occluded by masses of cell organelles or possibly thrombi of degranulated platelets but overt injury of capillary endothelium was not described. The reparative phase was characterized by increased cellularity of the interstitium which was said to include fibroblasts (not illustrated) and hyperplasia of type II epithelial cells (not illustrated). Residual hyaline membranes were seen to undergo fibrous organization (not convincingly illustrated).

This evolving process of injury to cells of the alveolar wall including epithelial cells followed by their regeneration qualifies the process as acute alveolar injury *in the loci in which it occurs*. In the author's experience, AAI in the human is not always diffuse and may involve the lung in a patchy distribution but the distinctly focal and peripheral distribution of OA injury has no counterpart in

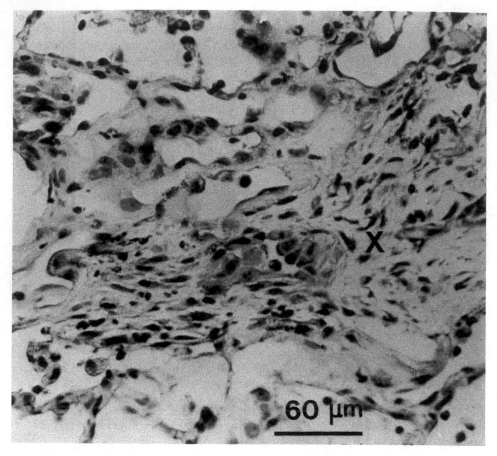

Figure 43 Microscopic section of lung from a dog 168 hours after injection of oleic acid. Lung architecture is severely disturbed with extensive closure (X). Remaining open airspaces are incompletely lined by regenerating cells. Reprinted from Schoene et al., 1984.

the human process. This distribution must be taken into account not only in considerations of pathogenesis but even in the evaluation of the injury as a valid model of human AAI.

Many studies have shown that the alterations in hemodynamics, gas exchange, and lung mechanics after OA injection are similar to those seen in human ARDS (Gemer et al., 1975; Julien et al., 1986; Uzawa and Ashbaugh, 1969; Motohiro et al., 1986; Schoene et al., 1984; Slotman et al., 1982; Johanson et al., 1982; Olanoff et al., 1984; Ashbaugh and Uzawa, 1968; Grossman et al.,

1980; Schuster and Trulock, 1984; Boiteau et al., 1986). Hypoxemia associated with an increased shunt fraction, decreased lung compliance and increased extravascular lung water (EVLW), all occurring within as little as 15-30 minutes after injection are universal findings. In two studies, one in the dog (Olanoff et al., 1984) and the other in the sheep (Julien et al., 1986) the increase in EVLW was found to be dose related (in the same animal but not necessarily among different animals). In studies that are sufficiently prolonged (Schoene et al., 1984) and in which the injury is severe, the decrease in oxygenation peaked in about a day and returned nearly to normal within about one week. One group using dogs (Schuster and Trulock, 1984) found a strong correlation between the initial fall in PaO_2 and the EVLW 90 minutes after OA but unless EVLW was increased by at least 150% (reflecting severe lung injury) some dogs showed rapid improvement in PaO_2 and decrease in venous admixture during the several hours after peak hypoxia without any measurable decrease in EVLW. The authors attributed this rapid improvement to compensatory hemodynamic mechanisms such as redistribution of pulmonary blood flow away from injured areas to more normal ones.

Pulmonary vascular resistance has also been increased in every study in which it was measured, but cardiac output (CO) and consequently pulmonary artery pressure have been variable. Most investigators have found significantly decreased CO but in some studies (Johanson et al., 1982; Ashbaugh and Uzawa, 1968) the decrease has been transitory (5 min, 1 h) while in others it has been more persistent (Motohiro et al., 1986; Slotman et al., 1982; Schuster and Trulock, 1984; Boiteau et al., 1986), and in still others no change was recorded (Gemer et al., 1975; Schoene et al., 1984). Some of the conflicting data (Motohiro et al., 1986; Schoene et al., 1984; Slotman et al., 1982; Schuster and Trulock, 1984) are from studies using the same species (dog) and the study that reported no change in CO used the highest dose of OA. The reasons for these discrepancies are unclear. Decreased CO during OA-induced injury is believed to be due either to depressed myocardial function caused by OA or some mediator released from injured lung or to increased systemic vascular resistance and afterloading of the left ventricle (Prewitt and Wood, 1981).

The cause of decreased lung compliance during OA injury has not been extensively investigated. One study using rabbits 30 minutes after OA injection concluded that the decrease is caused mostly by alveolar edema because the observed decrease in functional residual capacity (FRC) was almost entirely accounted for by the measured increase in EVLW in the excised lung. This decrease in FRC, in turn, was nearly sufficient to have caused the observed decrease in lung distensibility. Foaming of edema fluid was excluded as a contributory cause by the near identity of deflation volume-pressure curves before and after elimination of foam by degassing the lung. When the volume axis of the volume-pressure curves was expressed as observed total lung capacity, those from control and from OA-injured lungs were nearly identical. This finding suggests that recruitable alveoli were normally compliant and militated against increased surface tension as a significant contributor to abnormal compliance. Similar studies have not been done during the later phases of OA-induced injury, but on the basis of morphologic findings it is likely that other forces such as irreversible volume loss and decreased surfactant may supervene to cause decreased lung compliance if the injury is sufficiently severe.

It is generally agreed that the increase in EVLW during OA-induced injury is due principally to increased permeability of the capillary-alveolar barrier. The finding of normal or decreased pulmonary artery wedge pressures supports this view. Strong support is also provided by the observation of marked increase in lymph flow from the lung with persistence of the normal concentration of protein in the lymph during OA-induced injury (Julien et al., 1986). This proportionate increase in water and protein in lung lymph is characteristic of edema caused by increased barrier permeability. Because OA arrests in pulmonary microvessels as emboli, it is possible that the increased transvascular filtration of fluid which occurs is due to increased microvascular pressure or flow in nonoccluded microvessels as it appears to be in some other types of experimental microembolism (Ohkuda et al., 1978). This possibility has been examined in the isolated, blood-perfused dog lung injured by OA (Ehrhart and Hofman, 1981). After introduction of OA into the perfusing blood, the investigators found a steady, dose-related increase in lung weight even though perfusion pressure was kept constant as pulmonary vascular resistance increased. This weight gain was accompanied by a transfer of labeled albumin from blood to airway fluid in high concentrations. The authors concluded that these findings were indicative of increased microvascular permeability and that high vascular pressure and increased flow velocities in patent vessels were not essential factors in causing edema in this preparation. However, in a subsequent study using the same preparation with constant perfusion pressure, these investigators demonstrated an increase in venous resistance and in capillary pressure after OA (Hofman and Ehrhart, 1983). Lobes that had been made edematous by elevating venous pressure did not show this increase in venous resistance, thus eliminating edema itself as a cause of the increase. They suggested that the increased capillary pressure may lead to increased fluid filtration and contribute to OA-induced edema.

The distribution of increased EVLW between interstitial-perivascular connective tissue and alveoli in OA-induced edema appears to differ from that in hydrostatic edema as it does in other types of permeability edema (Vreim and Staub, 1976). In a recent study, physiologic changes, microscopic distribution of edema fluid, permeability of the alveolar-capillary barrier to dextran tracers and ultrastructure of the barrier in hydrostatic edema and in that induced by OA were compared using dogs (Montaner et al., 1986). Hydrostatic edema was induced by intravenous fluid overload and simultaneous inflation of an aortic balloon catheter manipulated so that the quantities of EVLW in the two types of edema were very similar. The dogs were sacrificed two hours after injection of OA or induction of hydrostatic edema. Physiologically the PaO_2 was lower and the shunt fraction higher in OA-injected dogs than in those with hydrostatic edema. In the OA-injured lungs there was significantly more alveolar edema and less perivascular edema suggesting that alveolar flooding occurs at a lower level of total EVLW than in hydrostatic edema. In addition, significantly more intravenously administered dextran (150,000 mw) appeared in the alveoli during OA induced edema than during hydrostatic edema. Dextran (40,0000 mw) instilled into the lung prior to OA administration showed a sudden large increase in concentration in the blood immediately after OA whereas that instilled into normal animals showed a slow steady rise in blood concentration. The authors attributed this sudden rise to disruption of alveolar epithelium, a conclusion supported by the electron microscopic finding of disruption of epithelium in OA-induced but not in hydrostatic edema.

Another group measured the clearance rate of ^{99m}Tc DTPA (diethylene triamine pentacetate, mw 492, a small hydrophilic tracer) from alveoli to blood in rabbits after injection of either triolein (neutral lipid), OA or saline (Jones et al., 1982). In the animals injected with OA there was an almost immediate increase in permeability of the alveolar-capillary barrier to the tracer, in alveolar-arterial O_2 difference (A-aPO$_2$) and in EVLW and a decrease in dynamic lung compliance. Animals injected with saline or with triolein showed no significant change in any of these parameters. This study provides supporting evidence that permeability of alveolar epithelium increases rapidly after injection of OA. The rapidity of this increase suggest that cellular changes other than the necrosis and disruption seen electron microscopically after 2 h are responsible for the increased permeability during the early phase. The nature of these changes remains to be studied.

The cause of OA-induced increase in permeability of the capillary-alveolar barrier has been investigated in several ways. In light of the evidence reviewed above that blood cells mediate lung injury after complement activation, their role in OA injury has been investigated using the isolated ventilated dog lung depleted of blood and perfused at constant pressure with a blood free dextran-ion solution (Hofman and Ehrhart, 1984). Under these conditions, addition of OA to the perfusate led to a very rapid (15 min) and marked gain in lobe weight and in wet to dry weight ratio while control lobes without OA showed a much smaller and slower gain. In fact, the gain in EVLW was greater in lobes perfused with dextran than in those perfused with blood at similar pressures and given the same dose of OA. Oncotic pressures of perfusate and of fluid collected from airways of OA lobes were nearly equivalent, indicating increased barrier permeability. The authors concluded that OA injures lung cells either directly or by the release of chemical mediators from pulmonary endothelium and that blood components are not essential to the process of injury. Indeed they suggested that some component of the blood may partially protect against this injury.

Several studies have documented a marked increase in neutrophils in alveolar lavage fluid or lung lymph which peaks during the first 24 h after OA (Julien et al., 1986; Schoene et al., 1984; Eiermann et al., 1983), but the two that have attempted to analyze their participation in the lung injury have reached opposite conclusions. One, using sheep, found that depletion of blood leukocytes (to about 5% of normal) with a nitrogen mustard had no effect on the severity of OA induced lung injury as measured by pulmonary vascular resistance, A-aPo$_2$ and lung lymph flow and protein concentration (Julien et al., 1985). The other, using rats, found that depletion of neutrophils by rabbit antineutrophil serum (to about 5% of normal) reduced but did not prevent OA induced edema as reflected in the lung "permeability index" (estimated by comparing quantity of intraperitoneally administered labeled albumin in blood-free lung with that in blood) (Eierman et al., 1983). It has been pointed out (Julien et al., 1986) that vascular pressures were not measured in the experiments using rats (Eierman et al., 1983) and that protein flux across the capillary-alveolar barrier in which endothelium is injured is extremely sensitive to small changes in this pressure (Huchon et al., 1981). On the basis of results in the blood-free perfused lung and in the sheep model (Julien et al., 1986) it seems unlikely that neutrophils are the prime effector of lung injury after OA but it remains possible that they play a secondary role and this may differ with species.

Thrombocytopenia and deposition of platelets in the lung have also been shown to occur during OA injury (Spragg et al., 1983; King et al., 1971) but depletion of blood platelets with antiplatelet serum (Julien et al., 1986) or inhibition of platelet aggregation with indomethacin (Dickey et al., 1981) have not affected the severity of the injury.

The role of arachidonate metabolites in OA-induced lung injury has not been extensively investigated but the available data indicate that the cyclooxygenase products thromboxane A2 and prostacyclin are released from the lung into the circulation during OA injury. Whether these compounds contribute to edema by increasing vascular permeability or by affecting vascular tone or are merely released from injured lung cells without contributing to edema is uncertain (Olanoff et al., 1984; Katz et al., 1987).

Finally, the isolated guinea pig lung perfused with buffer containing bovine serum albumin has been shown to release histamine into the effluent vascular perfusate during oleic acid injury if histamine catabolism is blocked by aminoguanidine (Selig et al., 1986). Blocking by aminoguanidine alone without administering OA did not cause this histamine release. Pretreatment with antagonists of H$_1$-histamine receptors significantly reduced lung weight gain and capillary alveolar loss of protein induced by OA. Degranulation of lung mast cells with compound 48/80 in the absence of OA led to a transitory weight increase which was much smaller than that induced by OA and was not associated with increased loss of protein from capillaries to alveoli. Because histamine is known to cause postcapillary vasoconstriction in the lung (Hakim et al., 1979; Linehan et al., 1982) the authors suggested that the increased postcapillary resistance demonstrated during OA injury (Hofman and Ehrhart, 1983) is mediated by this vasoactive amine released by pulmonary mast cells and that the resulting increase in capillary hydrostatic pressure augments fluid and protein loss across the capillary-alveolar barrier directly injured by OA.

It thus appears that OA injures the lung directly and that recruitment of in-flammatory cells and the release of arachidonate metabolites and vasoactive amines are secondary phenomena which may in some species contribute to capillary loss of water and protein by augmenting vascular permeability or by increasing microvascular pressure. The precise mechanism of OA injury remains unknown.

The OA-injured lung has provided investigators with a tool through which valuable insights into various aspects of permeability pulmonary edema have been gained. However, it is important to recognize that single-dose OA injury differs from human AAI in the following respects: (1) the injury is distinctly multifocal and not diffuse as is the usual human AAI. (2) With the doses and methods of administration most commonly employed, the onset of physiologic abnormalities and of pulmonary edema and their resolution are much more rapid than in human AAI. (3) Morphologically there appears to be more endothelial damage and a larger component of alveolar edema than in the human process. Thus the single-dose OA model as presently used closely reproduces neither the structural features not the temporal evolution of the human process and should not be considered an acceptable empirical model of human AAI. Nevertheless, OA does selective-ly injure the cells of the alveolar wall and it should be possible to manipulate dosage, method of administration, and physical state of the OA to produce a more diffuse and less fulminant lung injury which more closely resembles the human injury. It is interesting to note how little attention has been paid to the physical state of administered OA when this is apparently such an important determinant of the resulting injury. Prior to administration, OA is usually diluted in normal saline in which it forms a droplet suspension because of its near complete in-solubility in water. The size of these droplets appears to be a major determinant of distribution and possibly of severity of the induced lung injury. Droplet size is determined by the vigor and duration with which the suspension is shaken and upon the concentration of OA. After shaking, the droplets gradually coalesce un-til the OA is completely separated from the aqueous phase. Thus concentration, shaking and interval between shaking and administration probably all affect droplet size and the size range in such a suspension is probably quite large. The in-travascular droplet size of OA administered in the usual way is not known but it too is probably quite variable. The arterioles in which many droplets arrest are about 50 microns in diameter in the dog lung. A frequently stated but

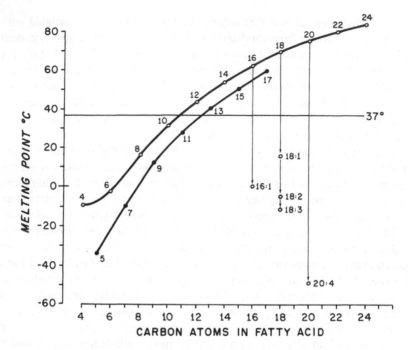

Figure 44 Melting point versus chain length of fatty acids. The effect of unsaturation is illustrated for C16, C18 and C20 fatty acids. Courtesy of S. A. Hashim, unpublished.

erroneous concept is that injected OA is bound to albumin in the circulation and that the capacity of albumin to bind it must be exceeded before the fatty acid appears free in the serum. Fatty acids mobilized from fat stores are bound to albumin but only the infinitesimally small fraction of exogenous OA that is solubilized is so bound. The remainder circulates as droplets of free fatty acid and it apparently is contact between these droplets and endothelial cells, possibly prolonged by the occlusive or near occlusive size of the droplets, that initiates OA injury. Free fatty acids might be prepared in such a way that droplets are smaller and more uniform. For example, the sodium salt of OA has also been shown to cause lung injury (Selig et al., 1986) but because the carboxyl group of the salt is polar and the hydrocarbon chain is intrinsically insoluble in water, it readily disperses in water to form micelles, a finer and more uniform dispersion. It is conceivable that this size difference may determine a more diffuse lung injury than that induced by the acid. Stabilized emulsions of free fatty acids might also induce more diffuse lung injury although it is possible that the emulsifying agent might interfere with the injurious effect of the acid. Another approach would involve the use of shorter chain fatty acids. Figure 44 shows the melting points (inversely related to solubility in water) of various fatty acids. The melting point and water solubility are related not only to chain length but to degree of unsaturation. It would be of interest to investigate the effects upon the lung of free fatty acids of various degrees of water solubility. A final variation would be the repeated administration of the free fatty acid in smaller doses so that the resulting injury is less fulminant. A more diffuse and less fulminant free fatty acid injury would be more relevant to human AAI.

References

Adamson, I. Y. R., Bowden, D. H., and Wyatt, J. P. (1970). Oxygen poisoning in mice-ultrastructural studies during exposure and recovery. *Arch. Pathol.* **90**:463–472.

Adamson, I. Y. R., and Bowden, D. H. (1974). The type II cell as progenitor of alveolar epithelial regeneration. A cytodynamic study in mice after exposure to oxygen. *Lab Invest.*, **30**(1):35–42.

Arnaout, M. A., Dana, N., Pitt, J., and Todd, R. F. III, (1985). Deficiency of two human leukocyte surface membrane glycoproteins (MOI and LFA-1). *Fed Proc.* **44**(10):2664–70.

Arnaout, M. A., Todd, R. F. III, Dana, N., Melamed, J., Schlossman, S. F., and Colten, H. R. (1983). Inhibition of phagocytosis of complement C3 or immunoglobulin G coated particles and of C3Bi by monoclonal antibodies to a monocyte-granulocyte membrane glyco-protein (MOI). *J. Clin. Invest.* **72**:171–179.

Arnaout, M. A., Spits, H., Terhorst, C., Pitt, J., and Todd, R. F. (1984). Deficiency of a leukocyte surface glycoprotein (LFA-1) in two patients with MO-1 deficiency: Effects of cell activation on MO-1/LFA-1 surface expression in normal and deficient leukocytes. *J. Clin. Invest.* **74**:1291–1300.

Ashbaugh, D. G., Bigelow, D. B., Petty, T. L., and Levine, B. E. (1967). Acute respiratory distress in adults. *Lancet* **2**:319–323.

Ashbaugh, D. G., and Uzawa, T. (1968). Respiratory and hemodynamic changes after injection of free fatty acids. *J. Surg. Res.* **8**:417–23.

Babior, B. M., Kipnes, R. S., and Curnutte J. T. (1973). The production by leukocytes of superoxide, a potential bactericidal agent. *J. Clin. Invest.* **52**:741–44.

Bachofen, M., and Weibel, E. R. (1977). Alterations of the gas-exchange apparatus in adult respiratory insufficiency associated with septicemia. *Am. Rev. Resp. Dis.* **116**:589–615.

Bachofen, M., and Weibel, E. R. (1982). Structural alterations of lung parenchyma in the adult respiratory distress syndrome. In *Clinics in Chest Medicine* **3**(1):35–36.

Balin, A. T. D., Goodman, B. P., Rassmusen, H., and Christofalo, V. J. (1976). The effect of oxygen tension on the growth and metabolism of WT-38 cells. *J. Cell Physiol.* **89**:235-250.

Barkett, V. M., Coalson, J. J., and Greenfield, L. (1969). Early effects of hemorrhagic shock on surface tension properties and ultrastructure of canine lungs. *Bull. Johns Hopkins* **124**:87-94.

Barrett, C. R., Bell, A. L. L., and Ryan, S. F. (1979). Alveolar epithelial injury causing respiratory distress in dogs. Physiologic and electron-microscopic correlations. *Chest* **75**:705-711.

Barrett, C. R., Bell, A. L. L., and Ryan, S. F. (1981). Effects of positive end-expiratory pressure on lung compliance in dogs after acute alveolar injury. *Am. Rev. Respir. Dis.* **124**:705-708.

Barry, B. E., and Crapo, J. D. (1985). Patterns of accumulation of platelets and neutrophils in rat lung during exposure to 100% and 85% oxygen. *Am. Rev. Resp. Dis.* **132**:548-555.

Beppo, O. S., Clements, J. A., and Goerke, J. (1983). Phosphatidyl-glycerol-deficient lung surfactant has normal properties. *J. Appl. Physiol.* **55**:496-502.

Boiteau, P., Ducas, J., Schick, U., Girling, L., and Prewett, R. (1986). Pulmonary vascular pressure-flow relationship in canine oleic acid pulmonary edema. *Am. J. Physiol.* **251** *(Heart Circ. Physiol.* **20**):H1163-1170.

Bowden, D. H., and Adamson, I. Y. R. (1974). Endothelial regeneration as a marker of the differential vascular responses in oxygen induced pulmonary edema. *Lab. Invest.* **30**:350-357.

Bowden, D. H., (1981). Alveolar response to injury. *Thorax* **36**(11):801-804.

Bowman, C. M., Butler, E. N., and Repine, J. E. (1983). Hyperoxia damages cultured endothelial cells causing increased neutrophil adherence. *Am. Rev. Respir. Dis.* **128**:469-472.

Braude, S., Apperley, J., Krausz, T., Goldman, J. M., and Royston, D. (1985). Adult respiratory distress syndrome after allogeneic bone-marrow transplantation: Evidence for a neutrophil independent mechanism. *Lancet* **1**(8440):1239-42.

Brigham, K. L., and Meyrick, B. (1984). Interaction of granulocytes with the lungs. *Circ. Res.* **54**:623-635.

Cassidy, S. S., Robertson, C. H. Jr., Pierce, A. K., and Johnson, R. L. (1978). Cardiovascular effects of positive end-expiratory pressure in dogs. *J. Appl. Physiol. Respir. Environ. Exercise Physiol.* **44**:743-750.

Chenowith, D. E., and Hugli, T. E. (1978). Demonstration of a specific C5A receptor on intact human polymorphonuclear leukocytes. *Proc. Natl. Acad. Sci. (U.S.A.)* **75**:3393-3947.

Christner, P., Fein, A., Goldberg, S., Lippmann, M., Abrams, W., and Weinbaum, G. (1985). Collagenase in the lower respiratory tract and patients with adult respiratory distress syndrome. *Am. Rev. Resp. Dis.* **131**:690-695.

Clowes, G. H. A. Jr., Zuschneid, W., Dragacevic, S., and Turner, M. (1968). The nonspecific pulmonary inflammatory reactions leading to respiratory failure after shock, gangrene, and sepsis. *J. Trauma* **8**:899–914.

Cochrane, C. G., Spragg, R., Revak, S. D., (1983). Pathogenesis of the adult respiratory distress syndrome. *J. Clin. Invest.* **71**:754–761.

Cook, C. D., Mead, J., Schreiner, G. L., Frank, W. R., and Craig, J. M. (1959). Pulmonary mechanics during induced pulmonary edema in anesthetized dogs. *J. Appl. Physiol.* **14**:177–186.

Cook, W. A. (1968). Experimental shock lung model. *J. Trauma* **8**:793–796.

Corp, H., and Janoff, A. (1979). In vitro suppression of serum elastase-inhibitory capacity by reactive oxygen species generated by phagocytosing polymorphonuclear leukocytes. *J. Clin. Invest.* **63**:793–797.

Craddock, P. R., Fehr, J., Brigham, K. L., Kronenberg, R. S., and Jacob, H. S. (1977a). Compliment and leukocyte-mediated pulmonary dysfunction in hemodialysis. *N. Engl. J. Med.* **296**:769–774.

Craddock, P. R., Fehr, J., Dalmasso, A. P., Brigham, K. L., and Jacob, H. S., (1977b). Hemodialysis leukopenia. Pulmonary vascular leukostasis resulting from complement activation by dialyzer cellophane membranes. *J. Clin. Invest.* **59**:879–888.

Crapo, J. D., and Tierney, D. F. (1974). Superoxide dismutase and pulmonary oxygen toxicity. *Am. J. Physiol.* **226**:1401–1407.

Crapo, J. D., Peters-Golden, M., Marsh-Salin, J., and Shelburne, J. S. (1978). Pathologic changes in the lungs of oxygen-adapted rats. *Lab. Invest.* **39**:640–653.

Crapo, J. D., Barry, B. E., Foscue, H. A., and Shelburne, J. (1980). Structural and biochemical changes in rat lung occurring during exposure to lethal and adaptive doses of oxygen. *Am. Rev. Resp. Dis.* **122**:123–43.

Davis, W. B., Rennard, S. I., Bitterman, P. B., and Crystal, R. G. (1983). Pulmonary oxygen toxicity: early reversible changes in human alveolar structures induced by hyperoxia. *N. Engl. J. Med.* **309**:878–83.

DeHoff, R. T., and Rhines, F. N. (1961). Determination of the number of particles per unit volume from measurements made on random plane sections: The general cylinder and the ellipsoid. *Trans. Am. Inst. Mining Met. Engrs.* **224**:474.

Derks, C. M., and Jacobovitz-Derks, D. (1977). Embolic pneumopathy induced by oleic acid. *Am. J. Pathol.* **87**:143–158.

Deuel, T. F., Senior, R. M., Chang, D., Griffin, G. L., Heinrikson, R. L., and Kaiser, E. T. (1981). Platelet factor 4 is chemotactic for neutrophils and monocytes. *Proc. Natl. Acad. Sci.* (U.S.A.) **78**:4584–4587.

Deuel, T. F., Senior, R. M., Huang, J. S., and Griffin, G. L. (1982). Chemotaxis of monocytes and neutrophils to platelet derived growth factor. *J. Clin. Invest.* **69**:1046–1049.

Dickey, B. F., Thrall, R. S., McCormick, J. R., and Ward, P. A. (1981). Oleic-acid-induced lung injury in the rat. Failure of indomethacin treatment or complement depletion to ablate lung injury. *Am. J. Pathol.* **103**:376–383.

Duchateau, J., Haas, M., Schreyen, H., Radoux, L., Sprangers, I., Noel, F. X., Braun, M., and Lamy, M. (1984). Complement activation in patients at risk of developing the adult respiratory distress syndrome. *Am. Rev. Resp. Dis.* **130**:1058–1064.

Dunegan, L. J., Knight, D. C., Harken, A., O'Connor, N., and Morgan, A. (1975). Lung thermal volume in pulmonary edema. *Ann. Surg.* **181**:809–812.

Ehrhart, I. C., and Hofman, W. F. (1981). Oleic acid dose-related edema in isolated canine lung perfused at constant pressure. *J. Appl. Physiol.* **54**(4):926–933.

Eiermann, G. J., Dickey, B. F., and Thrall, R. S. (1983). Polymorphonuclear leukocytes participation in acute oleic-acid induced lung injury. *Am. Rev. Respir. Dis.* **128**:845–850.

Elkes, J. (1949). Studies on the etiology of fat embolism and its relation to the use of finely dispersed oil in water emulsions in intravenous alimentation. Thesis: University of Birmingham.

Evans, J. N., Kelley, J., Low, R. B., and Adler, K. B. (1982). Increased contractility of isolated lung parenchyma in an animal model of pulmonary fibrosis induced by bleomycin. *Am. Rev. Respir. Dis.* **125**:89–94.

Fantone, J. C., and Ward, P. A. (1982). Role of oxygen derived free radicals and metabolites in leukocyte dependent inflammatory reactions. *Am. J. Pathol.* **107**:397–418.

Fantone, J. C., and Ward, P. A. (1985). Polymorphonuclear leukocyte-mediated cell and tissue injury: oxygen metabolites and their relations to human disease. *Hum. Pathol.,* **16**:973–978.

Fligiel, S. E., Lee, E. C., McCoy, J. P., Johnson, K. J., and Varani, J. (1984). Protein degradation following treatment with hydrogen peroxide. *Am. J. Pathol.* **115**:418–425.

Fonte, D. A., and Hausberger, F. X. (1971). Pulmonary free fatty acids in experimental fat embolism. *J. Trauma* **11**(8):668–672.

Ford-Hutchinson, A. W., Bray, M. A., Doig, M. V., Shipley, M. E., and Smith, M. J. H. (1980). Leukotriene B, a potent chemokinetic and aggregating substance released from polymorphonuclear leukocytes. *Nature* **286**:264–265.

Fowler, A. A., Walchak, S., Gielas, P. C., Henson, P. H., and Hyers, T. M. (1982). Characterization of antiproteinase activity in the adult respiratory distress syndrome. *Chest* **81**(Suppl):505–15.

Fowler, A. A., Hyers, T. A., Fisher, B. J., Bechard, D. E., Centor, R. M., and Webster, R. O. (1987). The adult respiratory distress syndrome-cell populations and soluble mediators in the air spaces of patients at high risk. *Am. Rev. Resp. Dis.* **136**:1225-1231.

Fox, R. B., Hoidal, J. R., Brown, D. M., and Repine, J. E. (1981). Pulmonary inflammation due to oxygen toxicity. Involvement of chemotactic factors and polymorphonuclear leukocytes. *Am. Rev. Resp. Dis.* **123**:521-523.

Frankel, E. N. (1985). Chemistry of free radical and singlet oxidation of lipids. *Prog. Lipid Res.* **23**:197-221.

Freeman, B. A., and Crapo, J. D. (1981). Hyperoxia increases oxygen radical production in rat lung mitochondria. *J. Biol. Chem.* **256**:10986-10992.

Freeman, B. A., Toplosky, M. K., and Crapo, J. D. (1982). Hyperoxia increases oxygen radical production in rat lung homogenates. *Arch. Biochem. Biophys.* **216**:477-484.

Freeman, B. A., Mason, R. J., Williams, M. C., and Crapo, J. D. (1985). Antioxidant enzyem activity in alveolar type II cells after exposure of rats to hyperoxia. *Exp. Lung Res.* **10**(2):203-22.

Fridovich, I. (1978). The biology of oxygen radicals. *Science* **201**:875-79.

Fridovich, I. (1986). Biological effects of the superoxide radical. *Arch. Biochem. Biophys.* **247**:1-11.

Fridovich, I., and Freeman, B. (1986). Antioxidant defenses in the lung. *Ann. Rev. Physiol.* **48**:693-702.

Gallin, J. I., and Durocher, J. R., Kaplan, A. P. (1975). Interaction of leukocyte chemotactic factors with the cell surface I chemotactic factor-induced changes in human granulocyte surface charge. *J. Clin. Invest.* **55**:967-974.

Gemer, M., Dunegan, L. J., Lehr, J. L., Bruner, J. D., Stetz, C. W., Don, H. F., Hayes, J. A., and Drinker, P. A. (1975). Pulmonary insufficiency induced by oleic acid in the sheep-a model for investigation of extracorporeal oxygenation. *J. Thorac. Cardiovasc Surg.* **69**(5):793-798.

Gould, V. E., Tosco, R., Wheelis, R. F., Gould, N. S., and Kapanci, Y. (1972). Oxygen pneumonitis in man: ultrastructural observations on the development of alveolar lesions. *Lab. Invest* **26**(5):499-508.

Grossman, R. F., Jones, J. G., and Murray, J. F. (1980). Effects of oleic acid-induced pulmonary edema on lung mechanics. *J. Appl. Physiol.* **48**(6):1045-1051.

Hackney, J. D., Evans, M. J., Spier, C. E., Anzar, U. T., and Clark, K. W. (1981). Effect of high concentrations of oxygen on reparative regeneration of damaged alveolar epithelium in mice. *Exper. Mol. Pathol.* **34**:338-344.

Haies, D. M., Gil, J., and Weibel, E. R. (1981). Morphometric studies of rat lung cells: numerical and dimensional characteristics of parenchymal cell population. *Am. Rev. Resp. Dis.* **123**:533-541.

Hakim, T. S., Dawson, C. A., and Linehan, J. H. (1979). Hemodynamic response of dog lung lobe to lobar venous occlusion. *J. Appl. Physiol.* **47**:145-152.

Hallman, M., and Gluck, L. (1976). Phosphatidylglycerol in lung surfactant III. Possible modifier of surfactant function. *J. Lipid Res.* **17**:257–262.

Hallman, M., Feldman, H. B., Kirkpatrick, E., and Gluck, L. (1977). Absence of phosphatidylglycerol in respiratory distress syndrome of the newborn. *Pediatr. Res.* **11**:714–720.

Hallman, M., Spragg, R., Harrell, J. H., Mose, K. M., and Gluck, L. (1982). Evidence of lung surfactant abnormality in respiratory failure. *J. Clin. Invest.* **70**:673–683.

Hallman, M., Enhorning, G., and Possmayer, F. (1985). Composition and surface activity of normal and phosphatidylglycerol deficient lung surfactant. *Pediatr. Res.* **19**:286–292.

Hammerschmidt, D. E., Craddock, P. R., McCullogh, F. Kronenberg, R. S., Delmasso, A. P., and Jacob, H. S. (1978). Complement activation and pulmonary leukostasis during nylon fiber filtration leukopheresis. *Blood* **51**:721–730.

Hammerschmidt, D. E., Weaver, L. J., Hudson, L., Craddock, P. R., and Jacobs, H. (1980). Association of complement activation and elevated plasma-C5a with adult respiratory distress syndrome. *Lancet* **1**:947–949.

Harlan, J. M. (1985). Leukocyte-endothelial interactions. *Blood* **65**:513–525.

Harris, R. I., Perrett, T. S., and MacLachlin, A. (1939). Fat embolism. *Ann. Surg.* **110**:1095–1114.

Hedlund, L. W., Bates, E., Beck, J., Goulding, P., and Putman, C. (1982). Pulmonary edema: a CT study of regional changes in lung density following oleic acid injury. *J. Comput. Assist. Tomogr.* **6**:939–946.

Hedlund, L. W., Vock, P., Effman, E. L., and Putman, C. E. (1985). Morphology of oleic acid induced lung injury. Observations from computed tomography, specimen radiology, and histology. *Invest. Radiol.* **20**:2–8.

Henry, J. N. (1968). The effect of shock on pulmonary alveolar surfactant., *J. Trauma* **8**:756–769.

Henson, P. M., Larsen, G. L., Webster, R. O., Mitchell, B. C., Goins, A. J., and Henson, J. E. (1982). Pulmonary microvascular alteration and injury induced by complement fragments; synergistic effect of complement activation, neutrophil sequestration and prostaglandins. *Ann. N. Y. Acad. Sci.* **384**:287–300.

Henson, P. M., and Johnston, R. B. Jr. (1987). Tissue injury in inflammation: oxidants, proteinases and cationic proteins. *J. Clin. Invest.* **79**:669–674.

Herrold, K. M. (1967). Fibrosing alveolitis and atypical proliferative lesions of the lung: an experimental study in Syrian hamsters. *Am. J. Pathol.* **50**:639–651.

Hirsch, E. F. (1941). Relation of the chemical composition of lipids to characteristic tissue lesions. *Arch. Pathol.* **31**:516–527.

Hofman, W. F., and Ehrhart, I. C. (1983). Methylprednisolone prevents venous resistance increase in oleic acid lung injury. *J. Appl. Physiol.* **54**(4):926–933.

Hofman, W. F., and Ehrhart, I. C. (1984). Permeability edema in dog lung depleted of blood components. *J. Appl. Physiol.* **57**(1):147–153.

Hohn, D. C., Meyers, A. J., Gherini, S. T., Beckmann, A., Markison, R. E., and Churg, A. M. (1980). Production of acute pulmonary injury by leukocytes and activated complement. *Surgery* **88**(1):48–58.

Hosea, S., Brown, E., Hammer, C., and Frank, M. (1980). Role of complement activation in a model of adult respiratory distress syndrome. *J. Clin. Invest.* **66**:375–382.

Huchon, G. J., Hopewell, P. C., and Murray, J. F. (1981). Interactions between permeability and hydrostatic pressure in perfused dog's lung. *J. Appl. Physiol.* **50**:905–911.

Hunninghake, G. W., Gadek, J. E., Fales, H. M., and Crystal, R. G. (1980). Human alveolar macrophage-derived chemotactic factor for neutrophils. Stimuli and partial characterization. *J. Clin. Invest.* **66**:473–483.

Idell, S., Kucich, U., Fein, A., Kueppers, F., James, H. L., Walsh, P. N., Weinbaum, G., Colman, R. W., and Cohen, A. B. (1985). Neutrophil elastase releasing factors in bronchoalveolar lavage from patients with adult respiratory distress syndrome. *Am. Rev. Resp. Dis.* **132**:1098–1105.

Jamieson, D., Chance, B., Cadenas, E., and Boveris, A. (1986). The relation of free radical production to hyperoxia. *Ann. Rev. Physiol.* **48**:703–19.

Johanson, W. G. Jr., Holcomb, J. R., and Coalson, J. (1982). Experimental diffuse alveolar damage in baboons. *Am. Rev. Respir. Dis.* **126**:142–151.

Johnson, J. W. C., Permutt, S., Sipple, J. H., and Salem, E. S. (1964). Effect of intra-alveolar fluid on pulmonary surface tension properties. *J. Appl. Physiol.* **19**:769.

Jones, J. G., Minty, B. D., Beeley, J. M., Royston, D., Crow, J., and Grossman, R. F. (1982). Pulmonary endothelial permeability increased after embolization with oleic acid but not with neutral fat. *Thorax* **37**:169–174.

Julien, M., Hoeffel, J. M., and Flick, M. R. (1986). Oleic acid lung injury in sheep. *J. Appl. Physiol.* **60**(2):433–440.

Kapanci, Y., Kaplan, H. P., Weibel, E. R., and Robinson, F. R. (1969). Pathogenesis and reversibility of the pulmonary lesions of oxygen toxicity in monkeys. *Lab Invest.* **20**(1):101–18.

Kapanci, Y., Assimacopoulos, A., Irle, C., Zwahlen, A., and Gabriani, G., (1974). "Contractile interstitial cells" in pulmonary alveolar septa: a possible regulation of ventilation-perfusion ratio? *J. Cell. Biol.* **60**:375–392.

Kaplan, H. P., Kapanci, Y., Robinson, F. R., and Weibel, E. R. (1969). Pathogenesis and reversibility of the pulmonary lesions of oxygen toxicity in monkeys. I-Clinical and light microscopic studies. *Lab. Invest.* **20**(1):94–100.

Katz, S. A., Halushka, P. V., Wise, W. C., and Cook, J. A. (1987). Oleic acid induces pulmonary injury independant of eicosanoids in the isolated, perfused rabbit lung. *Circ. Shock.* **22**(3):221–30.

Katzenstein, A. A., Bloor, C. M., and Liebow, A. A. (1976). Diffuse alveolar damage-the role of oxygen, shock and related factors. *Am. J. Pathol.* **85**(1):210–228.

King, E. G., Wagner, W. W., Ashbaugh, D. G., Latham, L. P., and Halsey, D. R. (1971). Alterations in pulmonary micro-anatomy after fat embolism. In vivo observations via thoracic window of the oleic acid embolized canine lung. *Chest* **59**(5):524–530.

King, E. G., Weily, H. S., Genton, E., and Ashbaugh, D. G. (1971). Consumption coagulopathy in the canine oleic acid model of fat embolism. *Surgery* **69**:533–541.

Kistler, G. S., Caldwell, P. R. B., and Weibel, E. R. (1967). Development of fine structural damage to alveolar and capillary lining cells in oxygen poisoned rat lungs. *J. Cell. Biol.* **32**(3):605–28.

Kovacs, E. J., and Kelley, J. (1986). Intra-alveolar release of a competence-type growth factor after lung injury. *Am. Rev. Respir. Dis.* **133**:68–72.

Krieger, B. P., Loomis, W. H., Czer, G. T., and Spragg, R. (1985). Mechanisms of interaction between oxygen and granulocytes in hyperoxic lung injury. *J. Appl. Physiol.* **58**(4):1326–1330.

Lee, C. T., Fein, A. M., Lippman, M., Kimbel, P., and Weinbaum, G. (1981). Elastolytic activity in pulmonary lavage fluid from patients with adult respiratory distress syndrome. *N. Engl. J. Med.* **304**:192–196.

Leslie, C. C., McCormick-Shannon, K., Cook. J. L., and Mason, R. J. (1985). Macrophages stimulate DNA synthesis in rat alveolar type II cells. *Am. Rev. Respir. Dis.* **132**:1246–1252.

Liau, D. F., Barrett, C. R., Bell, A. L. L., Cernansky, G., and Ryan, S. F. (1984a). Diphosphatidylglycerol in experimental acute alveolar injury in the dog. *J. Lipid Res.* **25**:678–683.

Liau, D. F., Barrett, C. R., Bell, A. L. L., and Ryan, S. F. (1984b). Artificial surfactant prepared from a mixture of dipalmitoylphosphatidylcholine and dipalmitoylphosphatidyl ethanolamine. *Amer. Rev. Respir. Dis.* **129**:224. (Abstract)

Liau, D. F., Barrett, C. R., Bell, A. L. L., and Ryan, S. F. (1985). Normal surface properties of phosphatidylglycerol deficient surfactant from dogs after acute lung injury. *J. Lipid Res.* **26**:1338–1344.

Liau, D. F., Barrett, C. R., Bell, A. L. L., and Ryan, S. F. (1987). Functional abnormalities of lung surfactant in experimental acute alveolar injury in the dog. *Am. Rev. Respir. Dis.* **136**:395–401.

Linehan, J. H., Dawson, C. A., and Rickaby, D. A. (1982). Distribution of vascular resistance and compliance in a dog lung lobe. *J. Appl. Physiol.* **53**:158–168.

Marcus, A. J., Boekman, M. J., Ullman, H. L., and Islam, N., (1982). Formation of leukotrienes and other hydroxy acids during platelet-neutrophil interactions in vitro. *Biochem. Biophys. Res. Commun.* **109**:130–137.

Martin, W. J. II, Gadek, J. E., Hunninghake, G. W., Gary, W., and Crystal, R. G. (1981). Oxidant injury of lung parenchymal cells. *J. Clin. Invest.* **68**:1277–1288.

Martin, W. J. (1984). Neutrophils kill pulmonary endothelial cells by a hydrogen peroxide dependent pathway. *Am. Rev. Respir. Dis.* **130**:209–213.

Matheson, N. R., Wong, P. S., and Travis, J. (1979). Enzymatic inactivation of human alpha-1-proteinase inhibitors by neutrophil myeloperoxidase. *Biochem. Biophys. Res. Commun.* **88**:402–409.

Maunder, R. J., Hackman, R. C., Riff, E., Albert, R. K., and Springmeyer, S. C. (1986). Occurrence of the adult respiratory distress syndrome in neutropenic patients. *Am. Rev. Respir. Dis.* **133**:313–316.

McGuire, W. W., Spragg, R. D., Cohen, A. M., and Cochrane, C. G. (1982). Studies on the pathogenesis of the adult respiratory distress syndrome. *J. Clin. Invest.* **69**:543–553.

Mead, J. F. (1976). Free radical mechanism of lipid damage and consequences for cellular membranes. In *Free Radicals in Biology*. Vol I. Edited by W. A. Pryor, New York Academic Press, pp. 51–68.

Meyrick, B. O., and Brigham, K. L. (1984). The effect of a single infusion of Zymosan-Activated plasma on the pulmonary microcirculation of sheep. *Am. J. Pathol.* **114**:32–45.

Motohiro, A., Furukawa, T., Yasumoto, K., and Inokuchi, K. (1986). Mechanisms involved in acute lung edema induced in dogs by oleic acid. *Eur. Surg. Res.* **18**:50–57.

Montaner, J. S. G., Tsang, J., Evans, K. G., Mullen, J. B. M., Burns, A. R., Walker, D. C., Wiggs, B., and Hocc, J. C. (1986). Alveolar epithelial damage, a critical difference between high pressure and oleic acid induced low pressure pulmonary edema. *J. Clin. Invest.* **77**(6):1786–1796.

Murray, J. F., and Staff of Division of Lung Diseases, National Heart, Lung, and Blood Institute. (1977). Conference Report-Workshop on mechanisms of acute respiratory failure. *Am. Rev. Respir. Dis.* **115**:1071–1078.

Nash, G., Blennerhassett, J. B., and Pontoppidan, H. (1967). Pulmonary lesions associated with oxygen therapy and artificial ventilation. *N. Engl. J. Med.* **276**:367–74.

Nash, G., Foley, F. D., and Langlinais, P. C. (1974). Pulmonary interstitial edema and hyaline membranes in adult burn patients-electron microscopic observations. *Hum. Pathol.* **5**:149–160.

Nathan, C. F., Silverstein, S. C., Brukner, L. H., and Cohn, Z. A. (1979). Extracellular cytolysis by activated macrophage and granulocytes II hydrogen peroxidase as a mediator of cytotoxicity. *J. Exp. Med.* **149**:100–113.

Newman, P. H. (1948). The clinical diagnosis of fat embolism. *J. Bone Joint Surg.* **30B**:290–297.

Noble, W. H., Kay, J. C., and Obdrzalek, J. (1975). Lung mechanics in hypervolemic pulmonary edema. *J. Appl. Physiol.* **38**:681–687.

Nusbacher, J., Rosenfeld, S. I., MacPherson, J. L., Thiem, P. A., and Leddy, J. P. (1978). Nylon fiber leukopheresis-associated complement component changes and granulocytopenia. *Blood* **51**:359–365.

Ohkuda, K., Nakahara, K., Weidner, W. J., Binder, A. and Staub, N. C. (1978). Lung fluid exchange after uneven pulmonary artery obstruction in sheep. *Circ. Res.* **43**:152:161.

Olanoff, , L. S., Reines, D. H., Spicer, K. M. and Halushka, P. V. (1984). Effects of oleic acid on pulmonary capillary leak and thromboxanes. *J. Surg. Res.* **36**:597–605.

Palmblad, J. (1984). The role of granulocytes in inflammation. *Scand. J. Rheumatol* **13**:163–172.

Parrish, D. A., Mitchell, B. C., Henson, P. M., and Larsen, G. L., (1984). Pulmonary response of fifth complement-sufficient and complement-deficient mice to hyperoxia. *J. Clin. Invest.* **74**:956–65.

Peltier, L. F. (1956). Fat embolism III The toxic properties of neutral fat and free fatty acids. *Surgery* **40**:665–670.

Peltier, L. F. (1957). Appraisal of the problem of fat embolism. *Surg. Gynecol. Obstet.* (Int. Abst. of Surg.) **104**:313–324.

Peltier, L. F. (1967). Fat embolism: a pulmonary disease. *Surgery* **62**:756–758.

Peltier, L. F., and Wertzberger, J. J. (1968). Fat embolism: the importance of arterial hypoxia. *Surgery* **63**:626–629.

Perez, H. D. and Goldstein, J. M. (1980). Generation of a chemotactic lipid from arachadonic acid by exposure to a superoxide generating system. *Inflammation* **4**(3):313–328.

Petty, T. L., Reiss, O. K., Paul, G. W., Silvers, G. W., and Elkins, N. D. (1977). Characteristics of pulmonary surfactant in adult respiratory distress syndrome associated with trauma and shock. *Am. Rev. Resp. Dis.* **115**:531–536.

Petty, T. L., Silvers, G. W., and Paul, G. W. (1979). Abnormalities in lung elastic properties and surfactant function in adult respiratory distress syndrome. *Chest* **75**:571–574.

Pomerantz, M., and Eisman, B. (1968). Experimental shock lung model. *J. Trauma* **8**:782–787.

Pratt, P. C., (1978). Pathology of adult respiratory distress syndrome. In *The Lung, Structure, Function and Disease.* Edited by W. Thurlbeck and M. R. Abell, Internat. Acad. of Path. Monograph No 19: pp. 43–57.

Prewett, R. M., and Wood, L. D. H. (1981). Effect of sodium nitroprusside on cardiovascular function and pulmonary shunt in canine oleic acid pulmonary edema. *Anesthesiology* **55**:537–541.

Raj, J. U., Hazinski, T. A., and Bland, R. D. (1985). Oxygen induced lung microvascular injury in neutropenic rabbits and lambs. *J. Appl. Physiol.* **58**(3):921-927.

Reidy, M. A., and Schwartz, S. M. (1981). Endothelial regeneration III. Time course of intimal changes after small defined injury to rat aortic endothelium. *Lab. Invest.* **44**:301-308.

Robb-Smith, A. H. T. (1941). Pulmonary fat embolism. *Lancet* **1**:135-141.

Rosenbaum, R. M., Wittner, M., and Lenger, M. (1969). Mitochondrial and other ultrastructural changes in great alveolar cells of oxygen adapted and poisoned rats. *Lab. Invest.* **20**:516-528.

Ryan, S. F. (1972). Experimental fibrosing alveolitis. *Am. Rev. Respir. Dis.* **105**:776-791.

Ryan, S. F., Bell, A. L. L., and Barrett, C. R. (1976). Experimental acute alveolar injury in the dog-morphological-mechanical correlations. *Am. J. Pathol.* **82**:353-372.

Ryan, S. F., Barrett, C. R., Lavietes, M. H., Bell, A. L. L. and Rochester, D. F. (1978). Volume-pressure and morphometric observations after acute alveolar injury in the dog from N-nitroso-N-methly-urethane. *Am Rev. Respir. Dis.* **118**:735-745.

Ryan, S. F., Liau, D. F., Bell, A. L. L., Hashim, S. A., and Barrett, C. R., (1981). Correlation of lung compliance and quantitation of surfactant phospholipids after acute alveolar injury from N-nitroso-N-methylurethane in the dog. *Am. Rev. Respir. Dis.* **123**:200-204.

Ryan, S. F., (1983). The acute diffuse infiltrative lung diseases and the role of the open lung biopsy in their diagnosis. *Prog. in Surg. Pathol.* **5**:209-242.

Ryan, U. S., Schultz, D. R., and Ryan, J. W. (1981). Fc and C3b receptors on pulmonary endothelial cells: induction by injury. *Science* **214**:557-558.

Sacks, T., Moldow, C. F., Craddock, P. R., Bowers, T. K., and Jacob, H. S. (1978). Oxygen radicals mediate endothelial damage by compliment-stimulated granulocytes. *J. Clin. Invest.* **61**:1161-1167.

Schoene, R. B., Robertson, H. T., Thorning, D. R., Springmeyer, S. C., Hlastala, M. P., and Cheney, F. W. (1984). Pathophysiological patterns of resolution from acute oleic acid lung injury in the dog. *J. Appl. Physiol.* **56**(2):472-481.

Schraufstatter, I. U., Revak, S. D., and Cochrane, C. G., (1984). Proteinases and oxidants in experimental pulmonary inflammatory injury. *Clin. Invest.* **73**:1175-1184.

Schuster, D. P., and Trulock, E. P. (1984). Correlation of changes in oxygenation, lung water and hemodynamics after oleic acid-induced acute lung injury in dogs. *Crit. Care Med.* **12**:1044-1048.

Scully, R. E. (1956). Fat embolism in Korean battle casualties. *Am. J. Pathol.* **32**:379.

Sealy, W. C. (1968). The lung in hemorrhagic shock. *J. Trauma* **8**:774-781.

Selig, W. M., Patterson, C. E., Henry, D. P. and Rhoades, R. A. (1986). Role of histamine in acute oleic acid lung injury. *J. Appl. Physiol.* **61**(1):233–239.

Shami, S. G., Evans, M. J., and Martinez, L. A. (1986). Type II cell proliferation related to migration of inflammatory cells into the lung. *Exper. Mol. Pathol.* **44**:344–352.

Shasby, D. M., Fox, R. B., Harada, R. N., and Repine, J. E. (1982). Reduction of the edema of acute hyperoxia lung injury by granulocyte depletion. *J. Appl. Physiol. Respirat. Eviron. Exercise Physiol.* **52**(5):1237–1244.

Shelly, S. A., Paciga, J. E., and Balis, J. U. (1977). Purification of surfactant from lung washings and washings contaminated with blood constituents. *Lipids* **12**:505–510.

Simon, L. M., Raffin, T. A., Douglas, W. H. J., Theodore, J. and Robin, E. D. (1979). Effects of high oxygen exposure on bioenergetics in isolated type II pneumocytes. *J. Appl. Physiol.* **47**(1):98–103.

Simon, R. H., DeHart, P. D., and Todd, R. F. III (1986). Neutrophil induced injury of rat pulmonary alveolar epithelial cells. *J. Clin. Invest.* **78**:1375–1386.

Slotman, G. J., Machiedo, G. W., Casey, K. F., and Lyons, M. J. (1982). Histologic and hemodynamic effects of prostacyclin and prostaglandin E, following oleic acid infusion. *Surgery* **92**:93–100.

Slutsky, A. S., Scharf, S. M., Brown, R., and Ingram, R. H. (1980). The effect of oleic acid induced pulmonary edema on pulmonary and chest wall mechanics in dogs. *Am. Rev. Respir. Dis.* **121**:91–96.

Smedly, L. A., Tonnesen, M. G., Sandhaus, R. A., Haslett, C., Guthrie, L. A., and Johnston, R. B. Jr. (1986). Neutrophil mediated injury to endothelial cells. Enhancement by endotoxin and essential role of neutrophil elastase. *J. Clin. Invest.* **77**:1233–1243.

Smith, B. T. (1979). Lung maturation in the fetal rat: Acceleration by injection of fibroblast-pneumocyte factor. *Science* **204**:1094–1095.

Spragg, R. G., Abraham, J. L., and Loomis, W. H. (1982). Pulmonary platelet deposition accompanying acute oleic-acid-induced pulmonary injury. *Am. Rev. Respir. Dis.* **126**:553–557.

Sproule, B. J., Brady, J. L., and Gilbert, J. A. L. (1964). Studies of the syndrome of fat embolism. *Can. Med. Assoc J.* **90**:1243.

Suttorp, N. and Simon, L. M. (1982). Lung cell oxidant injury enhancement of polymorphonuclear leukocyte-mediated cytotoxicity in lung cells exposed to sustained in vitro hyperoxia. *J. Clin. Invest.* **70**:342–350.

Synderman, R., Shin, H. S., and Dannenberg, A. M. (1972). Macrophage proteinase and inflammation: the production of chemotactic activity from the fifth component of complement by macrophage proteinase. *J. Immunol.* **109**:896–898.

Tanaka, Y., Takei, T., Aiba, T., Masuda, K., Kiuchi, A., and Fugiwara, T. (1986). Development of synthetic lung surfactant. *J. Lipid. Res.* **27**:475–485.

Tarver, O. A., Tsai, J., and Hedlund, L. W. (1986). Regional pulmonary distribution of Iodine-125 labeled oleic acid: Its relationship to the pattern of oleic acid edema and pulmonary blood flow. *Invest Radiol.* **21**:102–107.

Tate, R. M., and Repine, J. E. (1983). Neutrophils and the adult respiratory distress syndrome. *Amer. Rev. Respir. Dis.* **128**(3):552–559.

Thet, L. A., Parra, S. C., and Shelburne, J. D. (1984). Repair of induced lung injury in adult rats-the role of ornithene decarboxylase and poly-amines. *Am. Rev. Respir. Dis.* **129**:174–181.

Thet, L. A. (1986). Repair of oxygen induced lung injury. In *Physiology of Oxygen Radicals*, Clinical Physiology Series, Ed. by A. E. Taylor, S. Matalon, and P. A. Ward, Bethesda, Maryland, American Physiologic Society, p. 93.

Till, G. O., Johnson, K. J., Kunkel, R., and Ward, P. A. (1982). Intravascular activation of complement and acute lung injury-dependency on neutrophils and toxic oxygen metabolites. *J. Clin. Invest.* **69**:1126–1135.

Till, G. O., Beauchamp, C., Menapace, D., Tourtellote, W., Kunkel, R., Johnson, K. J., and Ward, P. A. (1983). Oxygen radical dependent lung damage following thermal injury to rat skin. *J. Trauma* **28**:269–277.

Till, G. O., Hatherhill, J. R., Tourtellotte, W. W., Lutz, M. J., and Ward, P. A. (1985). Lipid peroxidation and acute lung injury after thermal trauma to skin. Evidence of a role for hydroxyl radical. *Am. J. Pathol.* **119**:376–384.

Till, G. O., and Ward, P. A. (1986). Systemic complement activation and acute lung injury. *Fed. Proc.* **45**:13–18.

Todd, R. F. III, Nadler, L. M., and Schlossman, S. F. (1981). Antigens on human monocytes identified by monoclonal antibodies. *J. Immunol.* **126**:1435–1442.

Todd, R. F. III, Arnaout, M. A., Rosin, R. E., Crowley, C. A., Peters, W. A., and Babior, B. M. (1984). Subcellular localization of the large subunit MO1 (MOIa, formerly g.p.110), A surface glycoprotein associated with neutrophil adhesion. *J. Clin. Invest.* **74**:1280–1290.

Turrens, J. F., Freeman, B. A., Levitt, J. G., and Crapo, J. D. (1982a). The effect of hyperoxia on superoxide production by lung submitochondrial particles. *Arch. Biochem. Biophys.* **217**:401–410.

Turrens, J. F., Freeman, B. A., and Crapo, J. D. (1982b). Hyperoxia increases H2O2 release by lung mitochondria and microsomes. *Arch. Biochem. Biophys.* **217**:411–421.

Tvedten, H. W., Till, G. O., and Ward, P. A. (1985). Mediators of lung injury in mice following systemic activation of complement. *Am. J. Pathol.* **119**:92–100.

Uzawa, T. and Ashbaugh, D. G. (1969). Continuous positive pressure breathing in acute hemorrhagic pulmonary edema. *J. Appl. Physiol.* **26**(4):427–432.

Vercellotti, G., McCarthy, J., Furcht, L., Jacob, H., and Moldow, C. (1983). Inflamed fibronectin: an altered fibronectin enhances neutrophil adhesion. *Blood* **62**:1063–1067.

Vreim, C. E., and Staub, N. C. (1976). Protein composition of lung fluids in acute alloxan edema in dogs. *Am. J. Physiol.* **230**:376–379.

Ward, P. A., and Hill, J. H. (1970). C5 Chemotactic fragments produced by an enzyme in lysosomal granules of neutrophils. *J. Immunol.* **104**:535–543.

Ward, P. A., Till, G. O., Kunkel, R., and Beauchamp, C. (1983). Evidence for role of hydroxyl radical in complement and neutrophil dependent tissue injury. *J. Clin. Invest.* **72**:798–801.

Ward, P. A., Till, G. O., Hatherhill, J. R., Annesley, T. M., and Kunkel, R. G. (1985). Systemic complement activation, lung injury and products of lipid peroxidation. *J. Clin. Invest.* **76**:517–527.

Ward, P. A., Johnson, K. J., and Till, G. O. (1986). Animal models of oxidant lung injury. *Respirat.* **50**(suppl):5–12.

Weibel, E. R., and Gomez, D. M. (1962). A principle for counting tissue structures on random sections. *J. Appl. Physiol.* **17**:343–348.

Weibel, E. R. (1973). Steriological techniques for microscopic morphometry. In *Principles and Techniques of Electron Microscopy* Vol. 3. Biological Applications. Edited by M. A. Hyat. Van Nostrand Reinhold Co. New York. pp. 239–296.

Weiland, J. E., Davis, W. B., Holter, J. F., Mohammed, J. R., Dorinsky, P. M., and Gadek, J. E. (1986). Lung neutrophils in the adult respiratory distress syndrome-clinical and pathologic significance. *Am. Rev. Respir. Dis.* **133**:218–225.

Weinberg, P. F., Matthay, M. A., Webster, R. O., Roskos, K. V., Goldstein, I. M., and Murray, J. F. (1984). Biologically active products of complement and acute lung injury in patients with the sepsis syndrome. *Am. Rev. Respir. Dis.* **130**:791–796.

Weiss, S. J., Young, J., LoBuglio, A. F., Slivka, A. and Nimeh, N. F. (1981). Role of hydrogen peroxide in neutrophil-mediated destruction of cultured endothelial cells. *J. Clin. Invest* **68**:714–721.

Weiss, S. J., (1989). Tissue destruction by neutrophils. *N. Engl. J. Med.* **320**:365–376.

Weksler, B. B., and Coupal, C. E. (1975). Platelet-dependent generation of chemotactic activity in serum. *J. Exp. Med.* **137**:1419–1430.

Wewers, M. D., Herzyk, D. J., and Gadek, J. E. (1988). Alveolar fluid neutrophil elastase activity in the adult respiratory distress syndrome is complexed to alpha-2-macroglobulin. *J. Clin. Invest.* **82**:1260–1267.

Williwerth, B. M., Crawford, F. A., Young, W. G. Jr., and Sealy, W. C. (1967). The role of functional demand on the development of pulmonary lesions during hemorrhagic shock. *J. Thorac. Cardiovasc. Surgery* **54**:658–65.

Wyatt, J. P., and Khoo, P. (1950). Fat embolism in trauma. *Amer. J. Clin. Pathol.* **20**:637–640.

Yamamoto, E., Wittner, M. and Rosenbaum, R. M. (1970). Resistance and susceptibility to oxygen toxicity by cell types of the gas-blood barrier of the rat lung. *Am. J. Pathol.* **59**:409–35.

Yu, S. H., Harding, P. G. R., and Possmayer, F. (1984). Artificial pulmonary surfactant. Potential role for hexagonal H II phase in the formation of a surface active monolayer. *Biochem. Biophys. Acta* **776**:37–47.

22

Experimental Induction of Pulmonary Fibrosis

DRUMMOND H. BOWDEN

University of Manitoba
Winnipeg, Manitoba, Canada

Pulmonary fibrosis is a disabling disease with a poor prognosis. Aside from the recognized causes of chronic inflammatory diseases of the lung, such as the infectious and industrial granulomas, in a large group of patients the cause is unknown. Initial injury to the alveolar lining cells is common to most forms of pulmonary fibrosis. The term *fibrosing alveolitis*, introduced by Scadding (1964), is useful because it stresses the role of the inflammatory response in the genesis of inappropriate and disorderly repair in the lung (Keogh and Crystal, 1982).

Elucidation of the sequential events that determine whether the lung will heal normally or heal with fibrosis requires an animal model. Textbooks on pulmonary disease usually provide long lists of drugs and other therapeutic or environmentally derived agents known to have induced pulmonary fibrosis in susceptible persons. In the search for satisfactory experimental models of this disease, many of these agents have been tested in animals. Most of them do not meet the exacting requirements of a reproducible experimental model and they will not be included in the descriptions that follow here.

Two notable exclusions require comment: fibrosis induced by toxic particles and by immunogenic agents. Pulmonary fibrosis related to occupational exposure to silicious particles and fibers has been studied extensively in experimental animals. In addition to a wealth of information about the pathogenesis of these diseases, these experiments provide the foundation for more recent work on the interplay of macrophages, fibroblasts, and their products in the induction of fibrosis. These topics are described elsewhere in this volume by Kleinerman and Wagner. Pulmonary injury induced by immunological maneuvers may lead to granulomatous disease and fibrosis. Although such methods are useful for the study of immunological events, they have not been used frequently as models for the production of fibrosis (Johnson et al., 1979).

I. Criteria Used in the Selection of Experimental Methods

In selecting a particular method for the induction of pulmonary fibrosis, the following criteria are considered:

1. The simplicity and reproducibility of the method
2. The need to identify the specific cell types involved in all phases of injury and repair
3. The need to quantitate the injury and the resulting fibrosis
4. The appropriateness of the species and strain of animal to the experiment

Two considerations require particular attention: the susceptibility of a particular animal to the injurious agent, and the presence of overt or hidden infection in the stock to be used. The latter factor is of particular importance in mice and rats: Murine chronic respiratory disease caused by *Mycoplasma pulmonis* is so common in rats that many investigators are not entirely familiar with the normal histological appearance of the lungs in these animals. Excellent accounts of this and other endemic infections in rats and mice are given by Brownstein (1985) and by Schoeb and Lindsay (1985).

II. Experimental Methods

The methods to be described are culled from a miscellany of chemicals and other agents that, at one time or another, have been reported to induce pulmonary fibrosis in animals. Chemical injury of the lung is the most popular method in current use, and of the many potentially fibrogenic chemicals, bleomycin, paraquat, and butylated hydroxytoluene (BHT) are the most reliable. The fibrotic response to paraquat and BHT may be potentiated by the administration of high concentrations of oxygen. Irradiation, administered locally or to the whole body, has also been used to induce fibrosis, and, since this method is suitable for the study of mechanisms of injury and repair, it is described in some detail. Finally, the paradoxical response of neonatal animals to the inhalation of high concentrations

Table 1 Experimental Pulmonary Fibrosis

Mode of Injury	Species	References
Chemicals		
Bleomycin	Mouse	Adamson and Bowden (1974, 1979)
	Hamster	Snider et al. (1978)
		Starcher et al. (1978)
	Rat	Selman et al. (1985)
	Dog	Fleischman et al. (1971)
	Pheasant	Bedrossian et al. (1977)
Paraquat	Rat	Smith et al. (1974)
Paraquat and Oxygen	Rat	Selman et al. (1985)
BHT	Mouse	Adamson et al. (1977)
BHT and Oxygen	Mouse	Kehrer and Witschi (1980)
Irradiation		
	Mouse	Adamson et al. (1970a,b),
		Adamson and Bowden (1983)
	Rat	Adamson et al. (1970a)
	Hamster	Pickrell et al. (1976)
	Dog	Pickrell et al. (1976, 1978)
Oxidant gases		
Oxygen	Newborn Mouse	Bonikos et al. (1976)

of oxygen is used as a model for the production of bronchopulmonary dysplasia in the newborn.

All of the methods to be presented here have been adequately tested in several laboratories, usually in more than one species (Table 1).

A. Fibrosis Induced by Chemicals

Bleomycin

Bleomycin, an antibiotic derived from *Streptomyces verticillatus*, is used as a cytotoxic agent in the treatment of a variety of tumors including squamous cell carcinomas and lymphomas. Introduced largely because it has no major effects on the kidneys or bone marrow, it soon became apparent that some patients receiving the drug became dyspneic and developed pulmonary fibrosis. The cytotoxicity of bleomycin is inversely related to the activity of a cellular hydroxylase that degrades the drug. The level of activity of this enzyme is particularly low in epidermal cells, so that bleomycin is able to reach the nucleus where it induces fragmentation of DNA with subsequent derangement in the synthesis of DNA, RNA, and protein.

In the lung, bleomycin is preferentially bound by endothelial and epithelial cells. The attenuated lining cells on both sides of the air–blood barrier are damaged by the drug, whereas the type 2 cells, although they take up bleomycin, are not obviously affected. As these cells divide to replace the injured type 1 epithelium, the presence of bleomycin makes them particularly susceptible to derangements of DNA. Delayed and disorderly epithelial repair is a common sequel, and this may be an important factor in the development of fibrosis (Bowden, 1985).

Species Variation

The pulmonary toxicity of bleomycin varies widely in different species of animal. As little as 0.4 mg/kg induces pulmonary fibrosis in dogs, whereas the corresponding dosage in mice is 20 mg/kg by repeated intraperitoneal injection or 120 mg/kg given as a single intravenous injection (Table 2). In selecting an appropriate animal for the induction of pulmonary fibrosis, the variation of response in different strains of mice is also an important consideration; collagen production is high in C57B1/6 mice, intermediate in DBA/2 and Swiss mice, and low in the BALB strain (Schrier et al., 1983).

Experimental Methods

Three routes of administration have been used: intravenous, intraperitoneal, and endotracheal. The sequence of events induced by each method is similar. A single intravenous injection to a mouse leads to pulmonary fibrosis in 2 weeks (Adamson and Bowden, 1979), intraperitoneal injections produce fibrosis after 4–8 weeks (Adamson and Bowden, 1974), and administration through the trachea induces fibrosis within 30 days (Starcher et al., 1978).

Intraperitoneal Injection (Mouse). In the Swiss albino mouse, the administration of 1 mg of bleomycin twice weekly is uniformly fatal within 5 weeks (Adamson and Bowden, 1974). The response to lower dosages is illustrated in Table 3. The earliest lesions, at 2 weeks, are in the endothelial cells. Necrosis of type 1 epithelial cells, observed at 4 weeks, is associated with intra-alveolar aggregates of fibrin. Intra-alveolar and interstitial fibrosis develops from the fourth week on and, by 8–12 weeks, the fibrosis, predominantly subpleural in distribution, is well established. The results of this dosage experiment indicate that 0.5 mg, given twice weekly to mice weighing 25 g (20 mg/kg), induces pulmonary fibrosis in a significant number of animals.

Table 2 Pulmonary Fibrosis Induced by Bleomycin

Animal	Route of Administration	Dosage[a]	References
Swiss Albino Mouse	Intraperitoneal	20 mg/kg twice/week	Adamson and Bowden (1974,1979)
	Intravenous	120 mg/kg	Adamson and Bowden (1974,1979)
Rat	Intratracheal	20 units/kg	Selman et al. (1985)
Hamster			
	Intratracheal	5 units/kg	Snider et al. (1978)
	Intratracheal	10 units/kg	Starcher et al. (1978)
Dog	Intravenous	0.4 mg/kg	Fleischman et al. (1971)
Baboon	Intramuscular	1.5 units/kg twice/week	McCullough et al. (1978)
Pheasant	Intravenous	4.12 mg/kg twice/week	Bedrossian et al. (1977)

[a]1 mg is approximately 1 unit of bleomycin sulfate.

Table 3 Response of Mice (Groups of 10) to Various Dosages of Bleomycin

	Bleomycin (mg, twice per week intraperitoneally)					
	0.01	0.02	0.04	0.1	0.5	1.0
Number of injections	16	16	16	16	16	10
Time of death or sacrifice (weeks)	8–20	8–20	8–20	8–20	8–20	5
Mortality	0/10	1/10	1/10	6/10	3/10	10/10
Pulmonary lesion	0/10	3/10	3/10	7/10	7/10	10/10
Fibrosis	No	No	No	Yes(2)	Yes(2)	Yes(2)

Source: Adamson and Bowden (1974).

Intravenous Injection (Mouse). The introduction of this method (Adamson and Bowden, 1974) eliminates the need for repeated injection. A single injection of bleomycin, 120 mg/kg to Swiss Albino mice, results in a mortality rate of 50%; fibrosis is found in the majority of the survivors. The method offers a distinct advantage to the investigator who requires a precise time for the start of kinetic studies of cells and collagen. The methods for labeling cells with [^{3}H]thymidine, and for determining the kinetics and the distribution of the drug following the injection of tritiated bleomycin, are described by Adamson and Bowden (1979).

The earliest detectable alteration is demonstrated by bronchoalveolar lavage. The number of cells in control animals usually is 20×10^4, almost all of them macrophages. Between 1 and 3 days after bleomycin administration, the yield is more than doubled and up to 10% are polymorphonuclear leukocytes. In sections of the lung, perivascular edema and endothelial swelling with vacuolation are observed at 5 days. In some animals, injury does not progress beyond endothelial swelling with concomitant interstitial edema and egress of inflammatory cells. Such lesions are reversible: the endothelial cells regenerate with complete restitution of the vascular lining.

The most critical cellular event in the genesis of pulmonary injury is destruction of the thin type 1 epithelium. Damage to the epithelial barrier, although not directly visible by light microscopy, may be inferred by the exudation of fibrin into the air sacs. If this is massive, the animals die; if limited or multifocal, the animals survive but the repair process usually involves fibrosis. The evolution of fibrosis is observed in two compartments of the lung: in fibrin-filled alveoli and in the interstitium (Fig. 1). Proliferation of fibroblasts is well established between 10 and 14 days, and the laying down of collagen is progressive thereafter. Fibrosis, which is predominantly subpleural and most marked in the perivascular and peribronchiolar spaces, is well established by 2–4 weeks after a single injection of bleomycin.

The development of fibrosis is invariably accompanied by distorted regeneration of alveolar epithelium. Early necrosis of type 1 cells is followed by division of type 2 cells, but instead of differentiating to reconstitute a thin layer of type 1 cells, the proliferated cuboidal cells persist, creating tubule-like alveoli. The cuboidal cells may exhibit a variety of metaplastic changes: giant forms almost filling the lumen are observed together with ciliated alveolar cells and squamous differentiation with keratin production (Adamson and Bowden, 1979).

Endotracheal Injection (Hamster). Administration of a single endotracheal dose of bleomycin provides a simple method of inducing pulmonary fibrosis in animals. Snider et al. (1978) described the reaction in hamsters; a single instillation of 5 units/kg induces focal fibrosis by day 8, and fibrosis with "honeycombing" by day 60. This report provides details of pulmonary function studies. In the same year, Starcher et al. reported the morphological changes in hamsters after the instillation of 10 units/kg; they correlate the morphological findings with

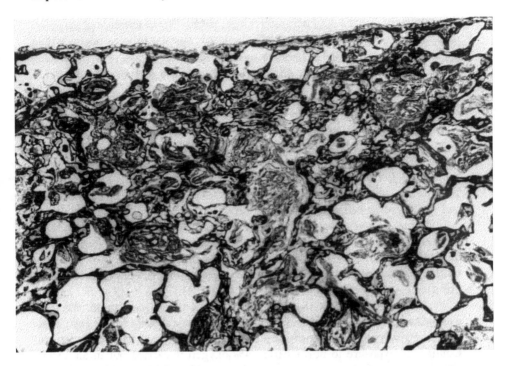

Figure 1 Mouse lung, 2 weeks after a single intravenous injection of bleomycin. Interstitial and intra-alveolar fibrosis is predominantly subpleural (silver methenamine).

quantitative measurements of collagen and elastin. After 30 days the lungs show fibrosis involving the air spaces, terminal bronchioles and pleura; collagen and elastin levels are more than doubled. These changes are readily induced by a single injection of 10 units/kg; doubling the dosage results in a high mortality rate.

Paraquat

Accidental or deliberate ingestion of the weed killer paraquat has been responsible for many deaths. Severe injury to the lungs with fulminating pulmonary fibrosis leads to respiratory failure. The severity of the initial injury and the high mortality rate following the administration of a single large dose of paraquat have limited its use as an experimental model for pulmonary fibrosis. With smaller dosages, administered over a prolonged period, pulmonary fibrosis is readily induced. In contrast to the results with bleomycin, this reaction is produced only when the drug is administered systemically; inhalation of paraquat aerosols produces more damage to the alveolar epithelium than doses systemic administration but pulmonary fibrosis does not result (Gage, 1968). The description of paraquat-induced fibrosis by Smith et al. (1973) indicates that the process begins within the alveoli rather than the pulmonary interstitium. The process is rapid, leading to obliteration of alveoli with dilatation of bronchioles and alveolar ducts.

Experimental Methods
Paraquat Alone. The paper by Smith et al. (1973) describes two sets of experiments, acute and chronic. In the acute experiment the mortality rate is very high. Rats given more than 60 mg paraquat/kg body weight by intraperitoneal injection die within 40 hr; when given 40 mg/kg the animals survive long enough to develop a florid fibroblastic picture. The chronic experiment is more successful. The majority of rats given 10 mg/kg at intervals between 1 and 106 days survive, and they develop diffuse pulmonary fibrosis.

    Paraquat and Oxygen. This method, described by Selman et al. (1985), is a modification of the paraquat model, in which repeated low dosages of paraquat are given to animals exposed to high concentrations of oxygen. The procedure is characterized by a low mortality rate and the development of severe diffuse pulmonary fibrosis in 3–4 weeks (Table 4).

    The method is described for male rats of the Wistar strain. Paraquat dichloride is dissolved in saline at 0.5% concentration freshly before each injection, and given intraperitoneally at a dosage of 2.5–5.0 mg/kg of body weight. Continuous exposure to normobaric 74% oxygen is carried out in an airtight chamber made of sheets of acrylic, with a capacity of 160 L. Air and oxygen are combined in an oxygen mixer, with a flow of 8–9 L/min, and humidified with a nebulizer prior to delivery to the chamber. The animals are removed from the chamber only at the time of the intraperitoneal injection of paraquat.

BHT

Butylated hydroxytoluene, a common food additive, has been used to induce acute injury to alveolar endothelial and epithelial cells. The repair process is accompanied by proliferation of interstitial cells, with the production of multifocal areas of interstitial fibrosis. By morphological criteria fibrosis is not severe, there is no great distortion of pulmonary architecture, and the animals appear to be well.

Experimental Methods
BHT Alone. The method is described by Adamson et al. (1977) for Swiss Webster mice; the dosage of BHT required to induce interstitial fibrosis is 400 mg/kg of body weight, dissolved in corn oil, and given by a single intraperitoneal injection. Proliferation of fibroblasts with excess deposition of collagen is seen as early as 9 days after the injection.

    BHT and Oxygen. Oxygen potentiates the fibrogenic effect of BHT (Kehrer and Witschi, 1980). Mice injected with BHT, as described above, are exposed

Table 4 Pulmonary Fibrosis Induced in Rats by Paraquat and Normobaric 74% Oxygen

Dosage (mg/kg Body Weight)	Number of Injections	Interval between Injections (hr)	Duration of Exposure to 74% Oxygen (weeks)	Number of Animals Surviving	Collagen Content (mg/Lung)
Control	—	—	—	20|20	11.1 ± 2.6
4.0	10	72	3	12&12	17.8 ± 2.9^a
3.5	10	72	5	4&8	28.6 ± 7.5^a
2.5	20	72	9_710	5&8	17.1 ± 3.4^a

^{a}Significant at $p < 0.01$.
Source: Selman et al. (1985).

to normobaric 70% oxygen for 6 days. They are subsequently kept in room air for the remainder of the experiment. When the animals are removed from the oxygen chamber at 6 days, the hydroxyproline content of the lung is not significantly different from that observed in mice given only BHT. Thereafter, a steady increase is observed, and at 15 days, the following figures are reported: control 200 μg; BHT alone 300 μg; and oxygen 600 μg.

B. Fibrosis Induced by Irradiation

Although diffuse pulmonary fibrosis is readily induced by thoracic irradiation, its use in experimental animals is not without problems; there is a high mortality rate, and the effects of irradiation may be difficult to differentiate from those induced by bacterial pneumonia (Smith, 1963). The complications arising from postirradiation infection are ameliorated by using young, healthy, preferably cesarean-born barrier-sustained animals, and by the addition of chlortetracycline (2 g/L) to the drinking water before irradiation and up to the time of sacrifice. With this treatment, the survivors show no evidence of pneumonia and the pulmonary changes are considered to be a direct result of injury induced by irradiation (Fig. 2) (Adamson et al., 1970b).

Three methods for the production of pulmonary fibrosis are available: thoracic irradiation, whole body irradiation, and the inhalation of radioactive particles. The sequence of events from the initial injury to the development of fibrosis is qualitatively similar whichever method is used. The capillary endothelium is particularly sensitive to irradiation and the attenuated type 1 cells of the alveolar epithelium exhibit focal necrosis; the type 2 cells and the cells of the bronchiolar epithelium are less affected by the dosages of irradiation required to induce fibrosis. Regeneration of alveolar epithelial cells is accomplished with little subepithelial

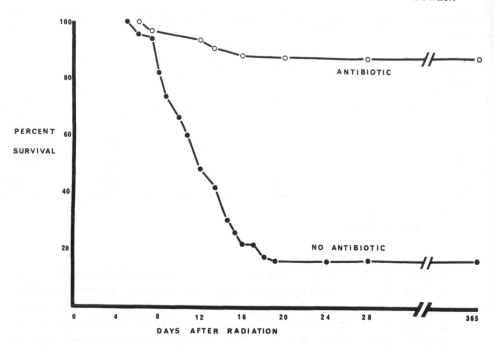

Figure 2   Percentage survival of mice after 1,100 rads whole body irradiation, with and without the antibiotic chlortetracycline (from Adamson et al., 1970b).

scarring, and fibrosis is related almost exclusively to small vessels, including capillaries (Adamson et al., 1970a; Adamson and Bowden, 1983).

Experimental Methods

Thoracic Irradiation

The method is suitable for rats and larger animals; it is more difficult to control in small animals such as mice. The following description pertaining to rats is derived from papers by Smith (1963), Adamson et al. (1970a,b), and Naimark et al. (1970).

The animals are anesthetized with thiopental sodium (25 mg/kg) and fastened to a board in the supine position by means of rubber bands about the extremities. Two animals are placed side by side on the board and are covered by a lead shield 2.54 cm thick with a central slot measuring 4.5 cm in width, in which a movable lead insert of the same thickness is set (Fig. 3). The lower border of the slot is placed over the costal margin at the midclavicular line, and the upper margin should extend to the axilla. By adjusting the position of the slot and the insert, it is possible to protect the spleen and to irradiate the right hemithorax of both

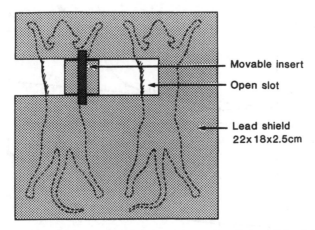

Figure 3 Shielding apparatus allowing simultaneous irradiation of the right lung of two rats (after Smith, 1963).

animals simultaneously. This allows the left lung to be used as a control. The roentgen ray unit is operated at 250 KV and 15 mA, with no filter. The target to skin distance is 42 cm, and a dosage of 3,000 rads is administered in 16 min. The administration of such a large dosage of irradiation to the chest results in severe esophagitis and consequent inanition. Mortality from this complication may be reduced from 40% to 20% by providing a liquid diet for 2 weeks following irradiation (Naimark et al., 1970).

Whole Body Irradiation

This method, which is suitable for a variety of experimental animals, is particularly useful for the induction of fibrosis in mice. Following irradiation of the whole body, the number of circulating leukocytes falls rapidly and remains low for 2 weeks. By the end of the third week, the hematological values are within the normal range (Adamson and Bowden, 1983). The amount of cellular injury and fibrosis is dependent on the size of the dosage administered to the animals. After 650 rads, focal fibrosis only is observed with no increase in the content of hydroxyproline in the lung. More extensive fibrosis with significant elevation of the hydroxyproline level is induced by 1,000 rads (Fig. 4).

The following description of the method is taken from papers by Adamson et al. (1970b) and Adamson and Bowden (1983).

Mice in sets of 10 are irradiated in individual sections of a lucite box 20 × 20 × 5 cm. The animals are equidistant from a ^{60}Co source of irradiation. Aside from treatment with chlortetracycline, as described above, no special measures are required to ensure a high rate of survival (Fig. 2). Following

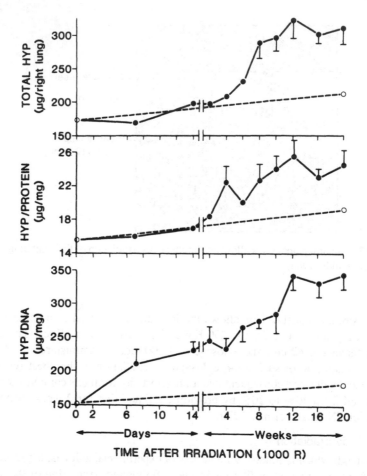

Figure 4 Mean values of total hydroxyproline, the hydroxyproline/protein ratio, and the hydroxyproline/DNA ratio at intervals after 1,000 rads (solid circles). Standard deviation is shown where values are significantly above control (broken line) ($p < 0.05$; (from Adamson and Bowden, 1983; reprinted with permission).

irradiation, serial studies may be made for periods up to 1 year (Adamson et al., 1970b).

Inhalation of Radioactive Particles

This method was developed by investigators concerned primarily with health hazards associated with human exposure to radioactive aerosols (Pickrell et al., 1976, 1978). Hamsters and dogs exposed to alpha and beta gamma radiation develop pulmonary fibrosis, the severity of which is dependent upon the initial dosage rate and the total amount of irradiation. The need for special radioprotective facilities precludes the use of this method in the standard laboratory.

C. Fibrosis Induced by Oxidant Gases

The inhalation of oxidant gases such as oxygen, ozone, and nitrogen dioxide is associated with acute injury to alveolar and bronchiolar cells. Although collagen deposition is observed during repair, this is not a particularly good method for the induction of diffuse pulmonary fibrosis. While this is true of adult animals, the paradoxical response of neonatal animals to breathing pure oxygen provides an interesting and potentially useful model for studying the special problems of lung injury and repair in the newborn. Whereas adult animals exposed to 100% oxygen die from respiratory failure within a few days, newborn animals are unusually resistant, exhibiting a 60% survival rate at 7 days (Bonikos et al., 1976). Their resistance to the acute toxic effects of oxygen is not without cost, because the surviving animals develop widespread pulmonary fibrosis with bronchiolitis and emphysema. The morphological appearance is not unlike that seen in infants with bronchopulmonary dysplasia.

Experimental Method: 100% Oxygen to Newborn Mice

The method is described by Bonikos et al. (1976) for the C 57 BL/Ka mouse. Naturally born animals are assigned to "standard" litters, so that each litter contains only one newborn from the "natural" litter. Each "standard" litter, with a foster mother, is placed in a cage in a controlled environment chamber through which 100% oxygen is passed at a rate of 6 L/min. Glove ports and a lock system on the chamber allow for the interchange of mothers, food renewal, and cleaning. The mothers are exchanged between control and experimental litters every 24 h to prevent oxygen poisoning of the adult animals. The survival rates over a period of 6 weeks are shown in Table 5.

Table 5 Survival of Newborn Mice Exposed to 100% Oxygen

Duration of Exposure (weeks)	% Survival
1	60
2	57
3	59
4	48
6	18

Source: Bonikos et al. (1976).

After 4 weeks exposure to 100% oxygen, there is continuing evidence of epithelial and vascular injury, with widespread intra-alveolar and interstitial fibrosis. Biochemical measurements of the amount of collagen are not reported.

III. Evaluation of Methods

A. Chemical Methods

The administration of toxic chemicals is the most widely used means of inducing pulmonary fibrosis in animals and, of the three agents described (bleomycin, BHT, and paraquat), bleomycin is undoubtedly the most popular method as judged by the number of publications in the last 10 years. The reasons for this preference are clear. Over a wide range of species this drug induces fibrosis within a short period, provided care is taken to use a dosage that is appropriate to both the species and the strain. The three routes of administration (intravenous, intraperitoneal, and intratracheal) are simple to use, and the results with each method are similar. For most studies, particularly those involving collagen metabolism, the intratracheal method is the easiest. When one is studying the precise sequence of cellular events, however, a single intravenous injection may be the method of choice.

The popularity of bleomycin as a fibrogenic agent may also be related to the clinical importance of pulmonary fibrosis as a complication of bleomycin therapy. Furthermore, in most institutions, the ready availability of the agent and the possibility of using up outdated supplies are added attractions.

As with most experimental models, bleomycin has its drawbacks. Although it is excellent for studying the pathogenesis of pulmonary fibrosis, the degree of fibrosis, and its distribution, make it less than ideal for physiological studies. Severe restrictive disease is not usual following bleomycin because the lesions are predominantly subpleural and in the peribronchiolar and perivascular spaces. Paraquat, which is very toxic to the alveolar cells, rapidly induces a fibroblastic response, but its high mortality rate has limited its usefuless for the study of chronic pulmonary fibrosis. The recent combination of lower dosages of paraquat and high concentrations of oxygen appears to offer a more manageable model. Augmentation of the toxic effect of BHT with oxygen also offers a reliable method for the induction of fibrosis.

The overall experience with oxidant gases as fibrogenic agents has not been good. Maintaining the required conditions of exposure is not easy, and the degree of fibrosis induced is seldom impressive. However, the response of newborn animals to hyperoxia offers an excellent opportunity to investigate injury and repair in the developing lung.

B. Irradiation

Local irradiation of the thorax with lead shielding has the distinct advantage of delivering a high dose of rays to the right lung while shielding the left lung, the spleen, and whole of the hematopoietic system. Given good care in the postir-radiation period, the animals survive the hazards of radiation-induced esophagitis, and infections are limited by antibiotics. The ability to compare an injured lung with a control lung in the same animal is a clear advantage to the method.

Whole-body irradiation is more easily administered and, when prophylactic antibiotics are given, the mortality rate is very low. Both lungs are affected equally, although less severely than after a high dose to one lung, and the administration of a single dose facilitates sequential cellular studies. The depression of leukocytes for 2 weeks may be considered a disadvantage, but it does enable the investigator to study cellular changes without an overlay of inflammatory cells.

IV. Prospects for Development

Two streams relating to experimental models of pulmonary fibrosis are identified in the literature. In one, the search for animal counterparts to human diseases is dominant; the other represents a seemingly endless hunt for a model for idiopathic pulmonary fibrosis. In clinical terms, the identification of the known causes of pulmonary fibrosis, such as infections, immunological reactions, the inhalation of fibrogenic dusts, and the iatrogenic induction of pulmonary disease by therapeutic agents, leaves a majority of causes unexplained. Since it is not possible to develop a model for a disease or a group of diseases of unknown cause, there is an increasing realization that the hunt for a model of idiopathic pulmonary fibrosis may be without end.

The majority of investigators in this field now concentrate their efforts on mechanisms rather than cause and, for this purpose, the experimental methods described in this chapter are entirely suitable. With the diminishing need for additional methods, a more focused approach to specific areas of concern is required. Two of the methods described use oxygen as a supplementary agent to chemically induced pulmonary injury. The possible synergism between chemicals and oxidant gases has broad implications. The additive effects of various environmental agents is a matter of some concern, and experimental protocols need to be developed for the investigation of these particular problems. The possibilities are almost unlimited, but a few examples will suffice:

1. Combination of antineoplastic drugs.

2. Antineoplastic drugs and irradiation.

3. Viral infections and inhaled materials, such as tobacco smoke, oxidant gases, and fibrogenic particles. Studies of the interaction of these agents may be particularly useful in the elucidation of bronchiolar injury and fibrosis, which, increasingly, is thought to play a role in the pathogenesis of chronic obstructive pulmonary disease. Examples of rodent infections

suitable for this type of experiment are described by Brownstein (1985) and by Schoeb and Lindsay (1985).

4. The interplay of genes and environment. The variability of human reactions to drugs such as bleomycin suggests that individual susceptibility may be important in determining the outcome of a particular injury. Bleomycin-induced lung injury may be ideal for the study of the interplay of genes and environment. We have seen how the toxicity of drugs varies from species to species and the variability of response in different strains of mice is truly remarkable (Schrier et al., 1983). The reasons for such varied responses are not known. There may be a relationship between the major histocompatibility complex and the ability of cells to repair DNA (Walford and Bergmann, 1979). With the availability of the techniques of molecular genetics, the investigation of genetic susceptibility or resistance to pulmonary injury and fibrosis is now possible.

References

Adamson, I. Y. R., and Bowden, D. H. (1974). The pathogenesis of bleomycin-induced pulmonary fibrosis in mice. *Am. J. Pathol.* **77**:185–198.

Adamson, I. Y. R., and Bowden, D. H. (1979). Bleomycin-induced injury and metaplasia of alveolar type 2 cells. *Am. J. Pathol.* **96**:531–544.

Adamson, I. Y. R., and Bowden, D. H. (1983). Endothelial injury and repair in radiation-induced pulmonary fibrosis. *Am. J. Pathol.* **112**:224–230.

Adamson, I. Y. R., Bowden, D. H., and Wyatt, J. P. (1970a). A pathway to pulmonary fibrosis: An ultrastructural study of mouse and rat following radiation to the whole body. *Am. J. Pathol.* **58**:481–498.

Adamson, I. Y. R., Bowden, D. H., and Wyatt, J. P. (1970b). Radiation survival: improved rates following chlortetracycline administration. *Radiat. Res.* **44**:478–483.

Adamson, I. Y. R., Bowden, D. H., Cote, M. G., and Witschi, H. (1977). Lung injury induced by butylated hydroxy toluene: cytodynamic and biochemical studies in mice. *Lab. Invest.* **36**:26–32.

Bedrossian, C. W. M., Greenberg, S. D., Yawn, D. H., and O'Neal, R. M. (1977). Experimentally induced bleomycin sulfate pulmonary toxicity. Histopathologic and ultrastructural study in the pheasant. *Arch. Pathol. Lab. Med.* **101**:248–254.

Bonikos, D. S., Bensch, K. G., and Northway, W. H. Jr. (1976). Oxygen toxicity in the newborn: the effect of chronic continuous 100 percent oxygen exposure on the lungs of newborn mice. *Am. J. Pathol.* **85**:623–650.

Bowden, D. H. (1985). Bleomycin-induced injury, mouse: a model for pulmonary fibrosis. In *Monographs on Pathology of Laboratory Animals: Respiratory*

System. Edited by T. C. Jones, A. Mohr, and R. D. Hunt. Berlin, Springer-Verlag, pp. 160–166.

Brownstein, D. G. (1985). Sendai virus infection, lung, mouse and rat. In *Monographs on Pathology of Laboratory Animals: Respiratory System*. Edited by T. C. Jones, A. Mohr, and R. D. Hunt. Berlin, Springer-Verlag, pp. 195–203.

Fleischman, R. W., Baker, J. R., Thompson, G. R., Schaeppi, U. H., Ilievsky, V. R., Cooney, D. A., and Davis, R. D. (1971). Bleomycin induced interstitial pneumonia in dogs. *Thorax* **26**:675–682.

Gage, J. C. (1968). Toxicity of paraquat and diquat aerosols generated by a size-selective cyclone effect of particle size distribution. *Br. J. Ind. Med.* **25**:304–314.

Johnson, K. J., Chapman, W. E., and Ward, P. A. (1979). Immunopathology of the lung: a review. *Am. J. Pathol.* **95**:795–844.

Kehrer, J. P., and Witschi, H. (1980). In vivo collagen accumulation in an experimental model of pulmonary fibrosis. Exp. Lung Res. **1**:259–270.

Keogh, B. A., and Crystal, R. G. (1982). Alveolitis: the key to the interstitial lung disorders. *Thorax* **37**:1–10.

McCullough, B., Schneider, S., Greene, N. D., and Johanson, Jr. W. G., (1978). Bleomycin-induced lung injury in baboons: alteration of cells and immunoglobulins recoverable by bronchoalveolar lavage. *Lung* **155**:337–358.

Naimark, A., Newman, D., and Bowden, D. H. (1970). Effect of radiation on lecithin metabolism, surface tension and compliance of rat lung. *Can. J. Physiol. Pharmacol.* **48**:685–694.

Pickrell, J. A., Harris, D. V., Mauderley, J. L., and Hahn, F. F. (1976). Altered collagen metabolism in radiation-induced interstitial pulmonary fibrosis. *Chest* **69**:311–316 (Supp.)

Pickrell, J. A., Schnizlein, C. T., Hahn, F. F., Snipes, M. B., and Jones, R. K. (1978). Radiation-induced pulmonary fibrosis: study of changes in collagen constituents in different lung regions of beagle dogs after inhalation of beta-emitting radionuclides. *Radiat. Res.* **74**:363–377.

Scadding, J. G. (1964). Fibrosing alveolitis. Br. Med. J. **2**:686.

Schoeb, T. R., and Lindsay, R. J. (1985). Murine respiratory mycoplasmosis, lung rat. In *Monographs on Pathology of Laboratory Animals: Respiratory System*. Edited by T. C. Jones, U. Mohr, and R. D. Hunt. Berlin, Springer-Verlag, pp. 213–218.

Schrier, D. J., Kunkel, R. G., and Phan, S. H. (1983). The role of strain variation in murine bleomycin-induced pulmonary fibrosis. *Am. Rev. Respir. Dis.* **127**:63–66.

Selman, M., Montano, M., Montfort, I., and Perez-Tomayo, R. (1985). *Exp. Mol. Pathol.* **43**:375–387.

Smith, J. C. (1963). Experimental radiation pneumonitis. *Am. Rev. Respir. Dis.* **87**:656–665.

Smith, P., Heath, D., and Kay, J. M. (1974). The pathogenesis and structure of paraquat-induced pulmonary fibrosis in rats. *J. Pathol.* **144**:57–67.

Snider, G. L., Celli, B. R., Goldstein, R. H., O'Brien, J. H., and Lucey, E. C. (1978). Chronic interstitial pulmonary fibrosis produced in hamsters by endotracheal bleomycin. *Am. Rev. Respir. Dis.* **117**:289–297.

Starcher, B. C., Kuhn, C., and Overton, J. E. (1978). Increased elastin and collagen content of the lungs of hamsters receiving an intra-tracheal injection of bleomycin. *Am. Rev. Respir. Dis.* **117**:299–305.

Walford, R. L., and Berkmann, K. (1979). Influence of genes associated with the main histo-compatibility complex on deoxyribonucleic acid excision repair capacity and bleomycin sensitivity in mouse lymphocytes. *Tissue Antigens* **14**:336–342.

23

Overview of Experimental Pneumoconiosis

J. CHRISTOPHER WAGNER*

Weymouth, Dorset, England

In 1913 Mavrogordato undertook some basic experiments in Haldane's private laboratory in Oxford. Early in these studies he realized two essential facts; to obtain satisfactory results the animals must inhale from a dust cloud, and unplanned pulmonary infection must be prevented. He continued the work at the South African Institute of Medical Research in Johannesburg, South Africa, where he explored the effects of exposing the animals to a variety of dusts, including freshly fractured quartz.

Meanwhile, E. H. Kettle had become interested in the association of the tubercle bacillus and quartz. It is rumored that this interest stemmed from work being undertaken by the Medical Research Council, which had been established in Britain in 1916. This early investigation by the Medical Research Committee suggested that gas gangrene contamination of wounds was more severe if the trauma occurred in Flanders rather than France, because in Flanders battle injuries were sustained over well-manured farm land and there was more free quartz in the soil there than from samples taken from the battlefields in France.

Leroy Gardner, while recovering from tuberculosis at the Trudeaux Sanatorium at Saranac Lake, once again emphasized that the only manner in which experimental

*Medical Research Council, Retired.

studies could help to elucidate the dust in humans was by inhalation. He also undertook a series of experiments with asbestos as well as with silica. He most often used Canadian chrysotile dust and was able to establish a "subtile" (sic) difference in the response of different species of animals to a similar dust exposure. Arthur Vorwald joined Gardner's team and experimented using the inhalation, subcutaneous, and intratracheal routes, but relied mainly on inhalation. His work was continued by Gerrit Scheepers, who wrote up and published all the previous findings. Scheepers also produced tumors by inoculating beryllium into the bones of rabbits.

E. J. King, who had been in Best's laboratory in Toronto, accepted the post of Professor of Biochemistry at the Post Graduate Medical School in London. King had shown some interest in the pneumoconioses; when established in London, he returned to his experimental studies using the intratracheal method of exposure, but he also realized that inhalation studies were needed.

By the 1950s inhalation studies were being undertaken in the United States, Britain, France, Germany, Italy, and South Africa. The means of producing the dust clouds varied from laboratory to laboratory, and some were bizarre.

Following the appointment of Martin Wright as experimental pathologist at the Medical Research Council's Pneumoconiosis Unit in Wales in 1947, the research became more organized. Martin developed and calibrated a scientifically successful dust dispenser that could be produced commercially.

Wright's studies were mainly on coal dust, but he also did dusting for Heppelston, who exposed animals to quartz and mixed quartz and hematite. Having established the methods, Wright returned to his role as an instrument maker. In this he made major contributions to instruments required in pulmonary physiology. Unfortunately, he did not pursue this work long enough to tackle the problems due to exposure of animals to mineral fibers. This was investigated by his successor. With this apparatus we were able to carry out our series of studies on the various forms of asbestos and their substitutes, both natural and synthetic.

I have not mentioned here the numerous in vitro investigations that have been undertaken. These are well covered in the proceedings of three conferences: Brown, R. C., Gormley, I. P., Chamberlain, M., and Davies, R. (eds.) (1980). *The in Vitro Effects of Mineral Dusts*. London, Academic Press; Beck, E. G., and Bignon, J. (eds.) (1985). *In Vitro Effects of Mineral Dusts*, Berlin, Springer-Verlag; *Tissue Culture in Medical Research, Proceedings of the Symposium* (1973); London, Heinemann and a fourth conference is being undertaken at this writing, at Sherbrooke, Canada, under the chairmanship of Dr. Brook Mossman.

I. Overview

A. Reasons for Undertaking Animal Experiments with Mineral Dusts

To state that you are undertaking experiments to increase scientific knowledge is both too bland and most unlikely. Experiments are undertaken to establish the biological effects of exposure to certain dusts in which there is interest. Unlike the majority of experimental medical studies, those in which mineral dust are involved always have political implications. At some stage of the investigation, the interests of trades union, industry, or government departments is inevitable. In recent years this has become complicated by legal constraints. Today, workers involved in this type of study must report their results with the knowledge that their publication or presentation will be studied from both the scientific and the legal point of view. However, the studies should be initiated by the observations or surmise that exposure to a certain mineral dust may have had a deleterious result.

B. Financial Support

This type of experiment is expensive to undertake and the current studies have taken at least 3 years to complete. It is therefore unlikely that future experiments will be economically viable. Further studies must be either undertaken in or be supported by government agencies. The possibility of obtaining financial support from either industry or trades union may be considered, but this must be the concern of the government agency and not reseachers. Representation of both industry and unions on the committee supervising the investigation will help to make the final results and conclusions acceptable to all parties.

C. Planning

These studies must be multidisciplinary in concept. Experimental pathologists, veterinarians, dust physicists, biochemists, mineralogists, and statisticians are required. None of these scientists will be fully employed in any one project, but all should be available, particularly during the initial planning. It is most important that statisticians should be involved in the study from the start. They will be responsible for the final assessment of the results, which will thus depend upon their devising an investigation that is acceptable for any planned analyses.

D. Materials

Dusts

The dust for experimental use should ideally consist of a respirable fraction of a dust to be studied. In addition, both negative and positive controls are required and sufficient material is required for inhalation and implantation and in vitro studies. This would mean that at least 3.0 kg of dust should be available and sufficient bulk material to allow dust to be prepared by other workers undertaking confirmatory studies, if necessary. A standard reference dust available in large quantities and fully characterized would be most valuable. The reference dust should be used as a basis for comparison with the dust or dusts under investigation. The standard reference samples would be available to all research workers undertaking experiments involving the biological effects of inhaling mineral dusts.

Preparation of these dusts requires the skills of dust physicists and mineralogists, with support and expert advice from industry.

These investigations are time consuming and expensive, but the investigator must make certain that the dust being studied is the exact material that requires investigation. If a number of samples are available from industry, work should concentrate on those samples in current production or that will be used in the future. It is pointless to study a material to which workers will never be exposed.

We have some knowledge of the size of the respirable mineral dust samples. These can be classified into the particulate and fiber particles. The particulate samples should be in the vicinity of a 5.0 μm density sphere. The fibers to be respired should be less than 3.0 μm in diameter. Further information is available concerning the fibers. A diameter of less than 0.25 μm is significant for the production of diffuse mesotheliomas. It is generally accepted that essential length is between 5.0 and 30.0 μm. However, fibers longer than 300 μm have been observed in the human lung.

Controls

The most suitable positive control for particulate dusts is quartz dust (alkaline washed after milling). The most satisfactory dust of the fibrous dusts is Oregon erionite. This latter material, when milled to respirable size, is extremely dangerous and precautions must be enforced when it is used. For the negative control, particulates to be used are titanium dioxide. The fibrous control dust is a more difficult problem. The best suggestion is to use a glass wool more than 1.0 μm in diameter and particles 10–20 μm long. Advice should be sought from industry for preparation of the sample.

Standard reference samples, as mentioned previously, should be prepared in the same way, in consultation with industry. An example of how a standard reference sample or samples can be obtained with industrial collaboration comes from our experiments with asbestos. The industry agreed to provide 1 ton of each of the following materials: anthophyllite from Finland, amosite and crocidolite from South Africa, and chrysotile from Canada and Zimbabwe. There was a complication with the Canadian chrysotile, since it was decided that the sample should come from nine separate mines, with the amount submitted related to annual production. The industry in South Africa provided shredding and milling facilities for production of the samples, and for the packaging for distribution. The preparation was supervised by Dr. V. Timbrell with the assistance of Mr. Rendall from the South African Research Institute. These samples were designed to be 90% respirable. The mineral content of the samples was assessed by Professor F. D. Pooley and others. Distribution was from the Johannesburg and Cardiff Units. These samples have been criticized by people who have not understood the concept behind their creation: to form a basis both for comparison of techniques between various laboratories and the biological effects of a specific dust against the sample. For these purposes, the samples were ideal. A more practical method of obtaining a standard sample of quartz was successfully undertaken by Professor E. J. King. Walking in London one day, he espied a huge crystal in an optician's window. This proved to be pure Brazilian quartz. King's powers of persuasion were such that after 20 minutes of discussion, the optician parted with the crystal in the "interests of medical research." This material was used as the British quartz standard for many years.

Animals

Studies require an animal that is known to be susceptible to pulmonary fibrosis and tumor of the lung and pleura following treatment with noxious dusts. The animal must have clean lungs, that is, be free of endemic pulmonary disease. It should be easy to keep, live for a suitably long time, and be small enough that a statistically significant number of animals would be available in any one experiment. Numerous animals have been used, varying in size from mules and pit ponies to tadpoles.

Primates or other simians might give a closer comparison to humans than other species. However, these animals' environmental requirements would be too expensive in temperate climates. In South Africa we had the opportunity of using Vervet monkeys and baboons. These animals were trapped in the "bush," isolated for a period of time, and tested for the presence of tuberculosis. The problems were that it was difficult to treat a statistically significant sample of animals, and because they had come from the wild, many were shown to have chronic lung disease, including mites living in the major airways.

The problem was solved by using a much more mundane animal, the Norwegian albino rat. These animals had been used for many years, but the results of studies of dust on the lungs were always suspect because of their high natural rate of pulmonary infection, frequently leading to bronchiectasis and subsequent pulmonary fibrosis.

This problem was overcome by Dr. P. C. Elmes, who in the later 1950s started to use so-called specific pathogen-free rats, which he obtained from the ICI's Pharmaceutical Division. Recognizing that all laboratory rats have pulmonary infection, which they obtained during weaning, ICI and other establishments investigated the problem. The answer was to develop colonies of rats obtained by cesarean operation and kept completely isolated from other rats (so-called barrier protected). Animals of this type can be obtained from major breeders in various countries. Despite guarantees from industry, it is the responsibility of each research group to check that the lungs of the animals to be used in the studies are free of pulmonary disease. This can be done by culling stock on arrival by killing an accepted number of randomly selected animals. It is essential to know of any other disease in the animal stocks. This information can be obtained from long-term studies, for example, from the Wistar Breeding Foundation; a number of colonies of Fischer 344 rats have been studied over their full lifetime.

Methods of Exposure

The following methods of exposure have been used in the study of experimental pneumoconioses: subcutaneous inoculation; intravenous inoculation; gavage and ingestion; intratracheal administration; implantation in coelomic cavities: intrapleural and ingestion.

Subcutaneous Inoculation

This was used in the early studies by Kettle for producing granulomata with quartz that could then be exacerbated by intravenous inoculation with tubercle bacilli. Fallon used the subcutaneous route to show the similarity of the lipoid content in tuberculous and silica-induced granulomata. This method of treatment was used by Rivers and others to compare the effect of different types of dusts and mixtures of minerals by comparing the size of profusions of granulomata in the liver.

A problem occurred following the intravenous injection of quartz; rabbits treated died of what appeared to be acute shock. Only in the 1950s did Jack Harington demonstrate that the quartz formed minute emboli that blocked the cerebral capillaries.

Gavage and Ingestion

This method of treatment became popular in the attempt to produce peritoneal mesotheliomas following exposure to various types of mineral fiber. Large dosages of the various types of asbestos were used, which wasted a considerable amount of the standard reference samples. These studies were exactly the same as putting the animals on a high-fiber diet: Everything passed right through and the animals were probably fitter for the experience.

Intratracheal Administration

This method had been used by several of the earlier workers, including King and Gardner. It was used most extensively by King and his colleagues. Their results were unfortunately invalidated by the fact that the rats that were treated had rat bronchitis. This led to bronchiectasis and the reactive production of fibrous tissue, making the interpretation difficult. the two major objections to this route of exposure are that the dust is unevenly distributed and in many cases it completely blocks smaller airways in a completely arbitrary manner. The second objection is that it is impossible to calculate the retained dosage because the majority of the inoculation is rejected. Black noses were seen in King's rats 12 hr after exposure to coal dust. Begin has reported that when using sheep and exposing them to chrysotile asbestos, 90% of the fiber is rejected within the first 24 hr. The introduction of a huge bolus of dust in saline may completely overwhelm the animals' natural defenses.

Implantation into the Coelomic Cavities

Intrapleural. When I developed this method of exposure, it was designed to determine if mesotheliomas would occur following direct exposure of the mesothelial cells to fiber. It was highly successful. It was equally satisfactory in helping Mearle Stanton to develop his scheme illustrating the biological significance of exposure to fibers of specific length/diameter ratios. These findings were vital for predicting the occurrence of mesotheliomas.

Novel forms of fibrous dust, either natural or synthetic, were also used. We recognized from the first experiments that the value of this method of exposure was definitely limited to the ability to indicate which fibers should be regarded with suspicion. Later studies indicated that at least a 10% incidence of tumor was required before this interpretation could be made. These findings could not in any way be extrapolated to humans.

Intraperitoneal. Intraperitoneal inoculation is simpler to carry out than intrapleural inoculation. The peritoneal mesotheliomal layer is far more sensitive than the pleural; the incidence of tumors is far higher.

Inhalation

Exposure of animals to a specific concentration of dust, at a constant dosage, with an equal distribution of the fibers throughout the dusting chambers, is the most satisfactory method of exposure.

Recently there has been a return to exposing only the muzzle of the animals. This technique obviously requires less dust. However, the animals have to be restrained and must be moved to large cages at the end of each exposure period. These numerous handlings can only lead to a greater possibility of infection. The muzzle is never tightly fitting, so there is always a dust leak onto the animals' fur, which can also lead to cross-contamination.

II. Interpretation of the Results

The methods of quantifying the pulmonary reaction to particulate and fibrous dusts and the classification of tumors are described in Chapters 24 and 25, on experimental asbestosis and silicosis.

III. Reporting and Dissemination of Findings

This has been demonstrated far better than I can in the IARC Scientific Publication No. 79, *Statistical Methods in Cancer Research Volume III. The Design and Analysis of Long-Term Animal Experiments.*

24

Experimental Asbestosis

J. CHRISTOPHER WAGNER*

Weymouth, Dorset, England

Early experiments in asbestosis were limited to chrysotile and produced conflicting evidence on the toxicity of this dust. One of the first studies was a small dusting experiment with chrysotile, which was carried out in 1925 by Mavrogordato (quoted by Simson, 1928) and produced evidence of slight generalized fibrosis. Preliminary results from Gardner and Cummins (1931) demonstrated peribronchiolar fibrosis in guinea pigs dusted with chrysotile. Lesions were only produced with fibers smaller than 3 μm (even with higher dust concentrations) and these did not produce lesions in rabbits. Vorwald and co-workers (1951) confirmed these findings and, using an intratracheal method, produced similar lesions with amosite and crocidolite. These three authors emphasized that the lesions tended to regress when the animals were removed from the dusty atmosphere. King and colleagues (1946) and Smith et al. (1951) meanwhile published conflicting evidence from long and short fibers. By the 1950s there was little evidence of the toxicity of the amphibole dusts, so the relative danger of the various dusts was unknown. Furthermore, marked or moderate fibrosis, as seen in humans, had not been produced and disease did not tend to regress after removal from the dusty working conditions. These were the questions to be answered

---
*Medical Research Council, Retired.

761

when I was appointed Asbestosis Research Fellow at the South African Pneumoconi-osis Research Unit in 1954. I was asked specifically to investigate, using experimental animals and human material correlated with clinical and radiological material, whether all types of asbestos were responsible for the same disease patterns.

The initial experiments (Wagner, 1963) were inhalation studies using the Venturi method of exposure based on the investigations on silicosis undertaken previously in Johannesburg. Southern Africa was the only area where the three major fibers, chrysotile, crocidolite, and amosite, were mined; the dusts could be prepared from the crude fiber obtained from the mills or mines. Dust concen-tration in the chambers was initially controlled using konometers and later using thermal precipitators. The results of these initial experiments demonstrated that severe asbestosis could be produced with high dust concentrations used for 2 years. The lesions were similar to those in humans, with asbestos bodies seen in the animals exposed to amosite and crocidolite (only rarely in those exposed to chrysotile dusts).

The next phase of the experimental approach was determined by the discovery of mesotheliomas among people exposed to dust in the vicinity of the (blue asbestos) crocidolite mines (Wagner et al., 1960). We decided to implant the dust into the pleural cavity of the animals. This was initially undertaken by open thoracotomy, but the immediate postoperative mortality was high and a simpler method of direct inoculation into the pleural cavity was adopted (Wagner, 1962). In a small number of the rats that survived for 2 years mesotheliomas developed that were histologically and histochemically similar to those seen in humans. The situation was confused by the finding of pleural tumors in animals that had been inoculated with quartz dust as controls. It was eventually shown that the tumors that occurred following the silica treatment were histiocytic sarcomas and that these were associated with rats of the Wistar strain (Wagner et al., 1980b).

From these preliminary experiments it was demonstrated that it was possi-ble to produce asbestosis and mesotheliomas in experimental animals and that these lesions were similar to those seen in humans. If information was to be ob-tained to imply that there was a significant difference in the biological effects of exposure to the different types of asbestos, dusts, all the methods of study required refinement with statistical control and analysis.

The first improvement was in the experimental animal. The requirement was for a small animal in which it would be possible to produce asbestosis, car-cinoma of the lung, and mesotheliomas. The animals should be healthy and have a lifespan to allow for at least a 2 year survival period. The size of the animal should allow the exposure of sufficient animals to be suitable for statistical analysis. It was decided to obtain cesarean-derived barrier protected rats, so-called specific pathogen-free rat (SPF). This was an animal with clean lungs that did not develop rat bronchitis or bronchiectasis (as did many of the rats and mice in the original experiments). We were fortunate in being presented with a large number of these rats to start our intrapleural studies (and with a nucleus for a breeding colony when we had built suitable accommodation). The generous donors were the Experimental Pharmacology Division of ICI.

We also had to obtain suitable asbestos dusts for the intrapleural and inhalation studies. In some of the earlier experiments the source of the fibers was not specified, there was considerable variation in the nature of the fibers, and in several investigations we found that the asbestos had been obtained from bottles labelled "asbestos" left on laboratory shelves. The specific requirements were standardized samples of asbestos dusts that could be prepared for experimental studies in relatively large quantities to be made available to all workers wishing to undertake experiments either in vivo or in vitro. The fibers would then be fully characterized by both physical and mineralogical methods. In 1965 an international ad hoc committee on asbestos cancers was formed under the Union International Contre le Cancer (UICC). The dust physics and mineralogy sections of this committee were charged with the production of the dust samples. The fiber and its the cost of preparation were paid for by the International Asbestos Industries. A ton of the best commercial grade of the following fibres was donated: chrysotile "A" from Rhodesia (now Zimbabwe), chrysotile "B" from Canada, anthophyllite from Finland, crocidolite, and amosite from South Africa. All these samples were then prepared for inhalation studies in South Africa, and the fibers were characterized by scientists from South Africa and the United Kingdom (Timbrell et al., 1978). Material was then made available for experimental studies from the Pneumoconiosis Research Units in these two countries. The design was for these materials to be used as standard samples for comparison with other materials, but unfortunately most of the experimental studies have been confined to these specific materials.

Our third requirement was to produce reliable and repeatable methods for accurately exposing animals to a variety of specifically prepared asbestos dusts, using animals capable of developing asbestosis, carcinoma of the lung, and mesotheliomas. Suitable inhalation chambers for the studies and a specific dust dispenser that would produce a constant cloud of respirable asbestos dusts were needed. The preparation of this apparatus is discussed later in this chapter.

We hoped that if these objectives were met, we would have a means of comparing the biological effects of the various types of asbestos. Information from these studies should help to establish if exposure to all types of asbestos produced the same hazards. From this it was considered that it might be possible to suggest which asbestos minerals should be produced for commercial purposes.

I. Methods of Exposure

A. Inhalation Studies

Exposure Facilities

Inhalation experiments required exposure chambers large enough to accommodate an adequate number of animals to provide statistically viable results (Timbrell et al., 1970b). Animals were to be exposed to dust for 7 hr of each working day for periods of months and it was considered an advantage to house them in the chambers throughout the period as an extra safeguard of their SPF status. Chamber facilities were required for the introduction of food and water and also for the removal of feces and urine. The welfare of the animals required maintenance of a temperature of approximately 72 °F and relative humidity of approximately 50%. Since the chambers were to be located in a room supplied with filtered air (heated, when required, to this temperature) and in our temperate climate summer temperatures rarely significantly exceed this temperature, only dehumidification of the air supply was required. The safety of operators and the need to avoid cross-contamination required that the pressure within the chambers be less than that of the room they occupied. Cloud generators used for isometric dusts were unsuitable for fibrous dusts, so a new generator was required to disperse efficiently the previously prepared fibers, prior to feeding them into the chamber air supply. The aim was to generate clouds at a steady concentration high enough to produce disease but that was compatible with the industrial environment.

Cubic chambers with truncated top and bottom sections were constructed with a total volume of $1.4m^3$. Eight cages each capable of accommodating up to five rats were suspended across the middle section of the cube. An airlock was mounted on one side of the cube, with access provided by an internal door and gloves mounted on the front and rear panels of the chamber. Washout facilities were provided at the base of the chamber. Air was drawn through the chamber by applying suction near the base of the chamber. Air entering the chamber was first filtered, as an additional safeguard, and cooled to 4 °C, by passing it through a heat exchanger to provide adequately dried air to absorb the moisture released by the animals and thus maintain a satisfactory humidity within the chamber. Dust from the cloud generator was injected into the air inlet ducting just prior to the entry point, at the apex of the chamber, where it passed through a jet that mixed the dusty and clean air streams. This provided a uniform cloud over the whole area occupied by the animal cages.

A generator was devised that consisted of a small hardened steel chamber bowl housing a rotating, vertically mounted, blade. This blade wiped the face of a plug of previously prepared fibrous dust that was slowly advanced into the bowl through a side port. Dispersion was achieved by the motion of the rotating blade and the air supplied to the bowl. When adequately dispersed, the fibers were carried by the air in the chamber air supply duct. The cloud concentration could be adjusted by varying the fibrous dust plug advance rate and/or the volume of air drawn through the exposure chamber. The cloud concentrations were measured both gravimetrically and numerically, using instruments developed for use in industrial environments. In earlier experiments it was noted that the size distribution of coal dust found in the lungs of rats exposed to coal dust clouds was very similar to that recovered from the lungs of miners at postmortem examination. This indicated that on the basis of aerodynamic size, instruments evolved for measuring respirable dust concentrations in coal mines were appropriate for measuring these in rats. (The more complex configuration of the airways of the upper respiratory tract of the rat could be expected to reduce the penetration of the longer fibers.) This instrument, Casella Type 113A, sampling throughout each 7 hr exposure session, provided a measure of the mean daily concentration. Additional samples were obtained, by the approved method, on membrane filters and evaluated microscopically to provide fiber counts and size distributions corresponding to the respirable mass concentrations.

Without any experimental guidance, the decision to use a concentration of 10 mg/m^3 was a compromise between the need to use as high a concentration as possible to detect any adverse properties and yet not invoke spurious effects. At that time 10 mg/m^3 was the recommended threshold limit value for inactive dusts.

Five chambers were constructed to investigate each UICC asbestos reference sample.

Preparation of Samples for Inoculation Studies

In the inoculation experiments, fiber bypassed the lung defenses and was delivered to the test site. It was necessary to deposit only fibers in the appropriate aerodynamic size range (Wagner and Berry, 1969). The samples for these experiments were therefore prepared by generating a cloud and passing it through a horizontal elutriator (as fitted to the Casella Type 113A sampler) to remove from the collected sample all aerodynamically oversized particles. The collected respirable sample would include longer fibers, which, during inhalation, would be trapped in the upper nasal passages of rats but not humans.

Production and Measurement of Dusts

The dust clouds that were generated contained a wide size range of particles. The respirable dust concentration was measured gravimetrically using size-selective gravimetric samplers (Casella Type 113A). The dosage administered to the animal was calculated as the product of respirable gravimetric concentration and hours of exposure. In later studies, fiber counts were determined from membrane filter samples collected and evaluated according to recommendations for measurements made in factories. The amount of dust deposited and retained was determined chemically.

Improving the measurement of the amount and characteristics of the dust retained in the lungs depended on controlled maceration of the lung tissue and the use of a transmission electron microscope to measure the size and concentration of the fibers. For further characterization, energy-dispersive analysis of x-rays was required. The methods and results obtained by other authors are shown in Table 1. Most of these methods were designed to study particulate dusts and most have now been modified according to our investigations, providing the length of exposure and fiber dosage have been sufficient.

Treatment of Animals

Animals in the inhalation studies (Wagner et al., 1974) were exposed to the dusts for prescribed periods of time, usually 3 months, 6 months, 12 months, and 24 months. The remainder were allowed to live out their full lifespan. Animals were allocated randomly to the various treatments and selection for serial killings. At each of the time periods, an equal number of animals of each sex were removed from the dusting chambers. Of those removed from the dusting chambers a set number were killed, and another group was removed from the chambers and kept in cages in special rooms away from the unexposed stock. All the animals removed from the dusting chambers were killed at the end of the 2-year period to determine whether the disease had progressed in the period after they were removed from the dusting chambers. To remove dust particles from animals' fur, all animals were kept in the cages exposed only to ambient air for a period of 3 weeks.

The animals were initially killed by exposure to chloroform, but later carbon dioxide was used. Following exsanguination, the thorax was opened and the lungs were removed, inflated with air, and suspended in formal-saline by a lead weight attached to the trachea, which had been tied off after inflation. After fixation, whole lung slices were taken from each lung (middorsal, ventral, and through the bronchus). From these, histological sections were prepared and stained with hematoxylin and eosin and van Gieson's connective tissue stain. In various experiments portions of the right lung were taken for mineral analysis or electron microscopic study, and the left lung was kept for histological examination. In all animals a full necropsy was undertaken and all abnormal tissues examined histologically. Routine sections were taken for examination of the liver, spleen, and kidney. The lungs of the animals that had died spontaneously were inflated with formol-saline.

Histological Examination

The pulmonary histological findings in these studies was first interpreted on an arbitrary basis, usually a consensus of opinion in a single research establishment

Table 1 Inhalation Studies

Author/Year	Animal Species	Dosage and Fiber	Result	Method of Dosage
Lynch et al., 1957	AC/Fa hybrid mice	$150–300 \times 10^6$ particles/ml, 8–12hr/ day for 5 days/week for 17 months. Controls: Chrysotile:	80/222 pulmonary adenomas 58/127 pulmonary adenomas	Asbestos floats. Rotating metal agitators, dust blown toward animals.
Holt et al., 1964	Rats	5,000 particles/ml air exposed to dust for 100 hr over 30 days Dust from asbestos weaving shed	Slowly progressive fibrosis of lung killed at intervals of up to a year	Hammer mill with asbestos dust generator mill motor arranged to behave as centrifugal pump.
Gross et al., 1967	Rats	86 m/m³ 30 hr/week Controls: Chrysotile:	None 28/72 lung tumors 17 adeno-carcinomas, 4 squamous cell, 7 fibrosarcomas	Ball-milled fibers fed into hammer mill
Reeves et al., 1974	Rats	49 mg/³ 16 hr/week for 2 years Amosite: Chrysotile: Crocidolite:	0/31 tumors 5/40 pulmonary adenomas 2/31 squamous carcinoma	Ball-milled commercial samples. Hammer mill and fan systems.
Wagner et al., 1974	SPF Wistar rats	12 mg/m³, 7 hr/day, 5 days/week for 1 day, 3,6,12, or 24 months Amosite, anthophyllite, crocidolite, Rhodesian and Canadian chrysotile	All fibers produced asbestoses which progressed after removed from dust. Carcinoma of lung with all dusts; mesotheliomas with all dusts except Rhodesian chrysotile	UICC samples (Timbrell et al., 1978) See text for method
Davis et al., 1978	Rats SPF Han strain	Dosage calibrated by either equal fiber mass or equal fiber number Dusted as per Wagner for approx. 1 year	Chrysotile produced most fibrosis and lung tumors with longer fibers	UICC preparations. Chambers similar to Wagner

(Wagner et al., 1982; McConnell et al., 1984). An opportunity to standardize the diagnostic criteria for tumors and pulmonary fibrosis was provided by the European Insulation Manufacturers Association who, through their Joint European Medical Research Board, provided funds for a Working Party of Pathologists to consider the diagnostic criteria in December, 1981. The members of the Committee were Dr. L. Cobb (UK), Dr. D. L. Dungworth (USA), Prof. C. F. Holland (Holland), Dr. P. Kotin (USA), Prof. M. Kushner (USA), Dr. D. Lamb (UK), Dr. E. E. McConnell (USA), Prof. U. Mohr (FRG), Dr. D. M. Smith (USA), Dr. H.A. Solleveld (Holland), and myself.

The working party was organized to establish the criteria for the inhalation experiments with synthetic mineral fibers. These fibers did not produce severe fibrosis (McConnell et al., 1984). Therefore, the published illustrations only cover the early effects of exposure. In this chapter we have included a complete set of lesions from the normal to the very severe fibrosis.

The degree of pulmonary fibrosis was recorded in categories 1–8, with 1 being normal lung and 8 very severe fibrosis:

Stage 1:   normal lung, no lesions observed.

Stage 2:   minimal reaction, macrophages containing dust seen in the lumina of the terminal bronchioles.

Stage 3:   minimal tissue reaction, presence of patchy cuboidal epithelial change in the cells lining the alveoli arising from the respiratory bronchioles.

Stage 4:   collagen deposition in the walls of the terminal bronchioles. At this stage the deposition is very slight and affects a few isolated bronchioles.

Stage 5:   the collagen deposition has increased and there is evidence of linkage between the reaction around the individual bronchioles.

Stage 6:   this linkage has become more marked and not only are the respiratory bronchioles involved but fibrosis also has spread to involve the whole acinus.

Stage 7:   marked fibrosis with occasional areas of confluent fibrosis.

Stage 8:   confluent fibrosis with distorted and collapsed air spaces incorporated into the fibrous tissue.

Stage 4 was considered the earliest significant lesion, stage 5 the equivalent of early asbestosis, stage 6 moderate, and stages 7 and 8 as marked (or severe) (see Figs. 1–12).

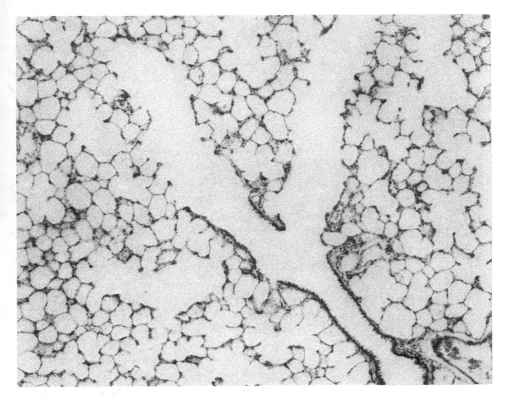

Figure 1 Fibrosis grade 1: normal: no lesion observed (×66).

B. Intrapleural Implantation

While we were developing the techniques for the inhalation studies, we decided to use a more direct method of exposure in our investigation of the association between exposure to asbestos dust and the development of diffuse mesotheliomas of the pleura. We planned to implant small quantities of various types of asbestos directly into the pleural cavity (Wagner, 1962). In our initial experiments we insufflated the dust directly into the right pleural cavity following open thoracotomy under general anesthetic. The survival of the animals was not encouraging. I then considered the techniques that had been used for establishing and maintaining the pneumothoraces for the treatment of pulmonary tuberculosis in the preantibiotic era.

We constructed a capillary monometer and attached this to a two-way tap. To one arm we attached a needle and to the other a tuberculin syringe containing the asbestos dust suspended in physiological saline. The animal was anesthetized

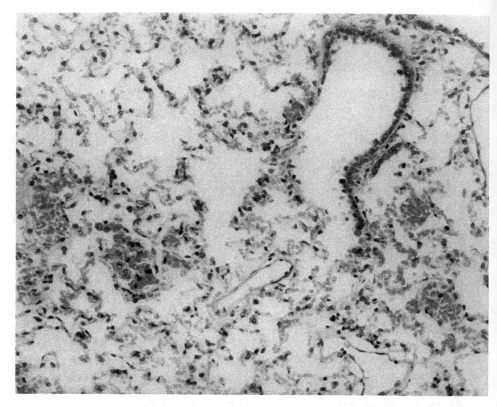

Figure 2 Fibrosis grade 2: cellular change is minimal: a few macrophages in the lumen of the terminal bronchioles and alveoli (×165).

with ether. The needle was then introduced into the thoracic cavity at the level of the fourth rib in the midaxillary line. Once the needle entered the thoracic cavity, a negative reading was obtained from the manometer, the tap was turned, and the suspension of dust was inoculated. The needle was then withdrawn and the animal placed in isolation until it recovered. Initially, rats, mice, and guinea pigs were used. The majority of the animals died of intercurrent infections. A few mesotheliomas occurred in the rats. These rats were of the Wistar strain and we could find no evidence that pleural mesotheliomas had been recorded in these animals.

In the next study the same technique was used on large groups of statistically randomized Wistar rats from both the SPF cesarean-derived barrier-protected strain (ICI Pharmaceutical Division, Alderly Edge, Cheshire, England) and an ordinary commercially bred colony of Wistar rats (Wagner and Berry, 1969).

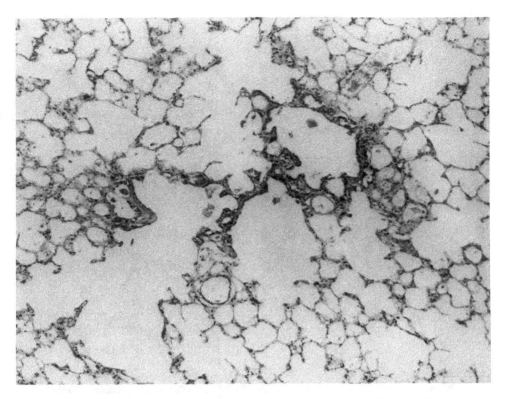

Figure 3 Fibrosis grade 3: cellular change, mild: presence of cuboidal epithelium lining the proximal alveoli. No collagen but reticulin fibers may be present in interstitium at the junctions of bronchiole and alveolus. Luminal macrophages are more conspicuous and mononuclear cells may be found in the interstitium (×66).

The ordinary rats had relatively clean lungs and the final results of the two studies were almost identical. The various asbestos and control dusts used were suspended in saline, and the containers autoclaved and submitted to ultrasonification immediately before inoculation. Each rat received 0.5 ml of a suspension of 50 mg dust/ml. It was observed that 0.1 ml of the solution remained in the needle and tap so that the actual dose was 20 mg of dust.

After their recovery the animals were caged in fours, each cage containing animals of a single sex. The SPF animals were kept in a special barrier-protected unit and their food was autoclaved. At a later stage irradiated cubes were used. The diet was standardized for all animals and water was given ad libitum. It was observed that the tumors occurred between 500 and 1000 days after the single inoculation. Therefore, the importance of keeping the animals alive and free of

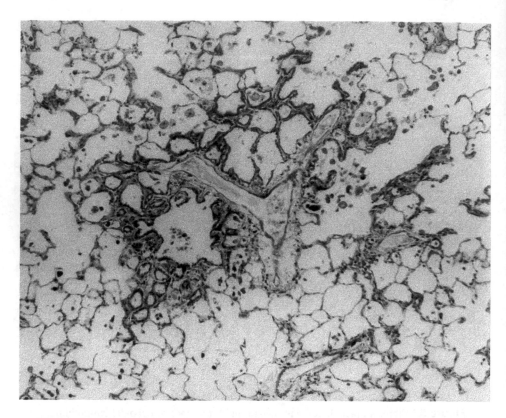

Figure 4 Fibrosis grade 4: minimal: minimal collagen deposition at level of terminal bronchiole and alveolus. Increased bronchiolization with mucoid debris (×66).

distress for as long as possible was recognized. The animals either lived their full lifespan or were killed if they showed any evidence of disease or discomfort. We found that a veterinary approved barbiturate, inoculated intraperitoneally, provided a simpler anesthetic than ether.

Stanton and Wrench (1972) adopted a different technique using open thoracotomy and the insertion of gelatine-coated fiberglas pledgets.

> In all experiments the surface of flat 45-mg pledgets composed of autoclaved, binder-coated, coarse fibrous glass of the type commonly used as insulation material, was coated with a standard dose of 40 mg test fibers that had been suspended in 1.5 ml of 10% gelatin by gentle continuous agitation. The gelatin-suspended coatings of test fiber were allowed to harden on the pledgets at 4 °C, so that the pledgets developed a rubbery consistency.

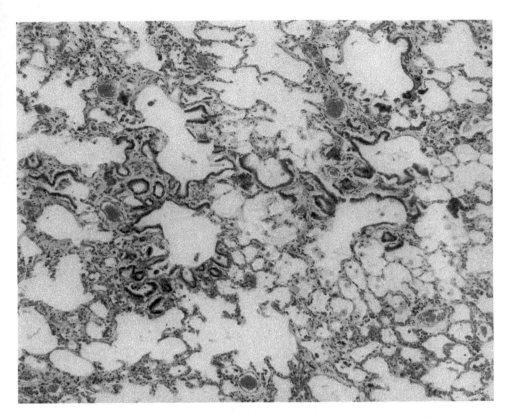

Figure 5 Fibrosis grade 5: fibrosis, mild: interlobular linking of lesion described in grade 4 and increased severity of fibrosis (×66).

Treatment consisted of a left-sided open thoracotomy under ether anesthesia and the direct placement of the pledget over the visceral pleura. Application in this manner had several advantages: The pledgets could be manipulated easily so that the test fibers on the surface of the pledget could be applied uniformly to a broad surface of the visceral pleura. Additionally, it permitted the application of fibers that could not pass readily through a hypodermic needle, ensuring that fibers were included that otherwise would have been fragmented or selectively filtered out by the injection procedures. Further, the coarse fibrous glass pledget acted as a partial barrier between the incision and the test material, thus retarding migration of the test fibers into the incision and surrounding tissues with the subsequent induction of tumours at extraneous sites (Stanton et al, 1977).

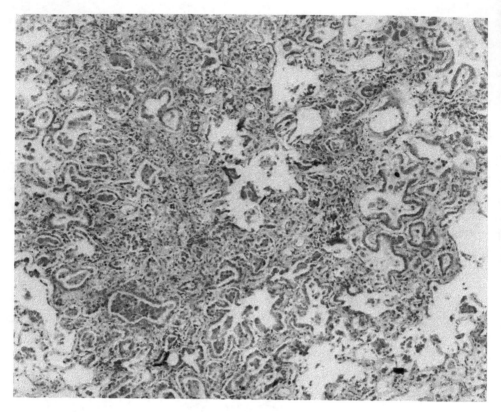

Figure 6   Fibrosis grade 6: moderate: early consolidation. Parenchymal decrease is apparent (×66).

C. Intraperitoneal Implantation

Both Pott (1978) and Davis (1974) used direct intraperitoneal inoculation of the dust, a technique that had earlier been favored by Schmal (1958).

D. Intratracheal Method

On the whole, the intratracheal method of introducing various forms of asbestos dust into rodents has not been a success, since the fibers fail to disperse and tend to obstruct the lumen of small bronchi and bronchioles. This method has been used to study cocarcinogenesis of chrysotile with benzo(a)pyrene in rats (Smith et al., 1970; Shabad et al., 1974). However, in sheep, Begin and his team (1983) have been successful in confining the inoculation to the tracheal lobe. This lobe was exposed to a dose of UICC asbestos fibers in phosphate-buffered saline by

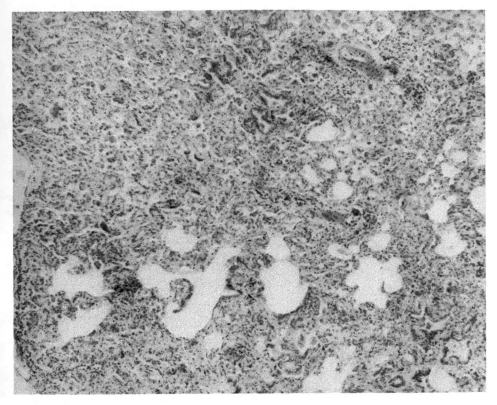

Figure 7 Fibrosis grade 7: severe: marked fibrosis and consolidation (×66).

slow infusion via bronchoscopic catheterization. They have used this technique to develop methods for studying bronchiolar lavage and fluid analysis, and correlating lung physiological and radiological findings with histological appearance. In most of the experiments described (Begin et al., 1986), sheep were killed at day 60 (see Table 2 for results).

E. Oral

Feeding experiments (Bonser and Clayson, 1967; Bolton and Davis, 1976; Wagner et al., 1977) did not produce tumors. A similar study carried out at N.I.E.H.S. National Institute of Environmental Health Sciences) and reported by McConnell et al. (1985) yielded similar results. Since the results of all these experiments were negative, the actual technique will not be discussed.

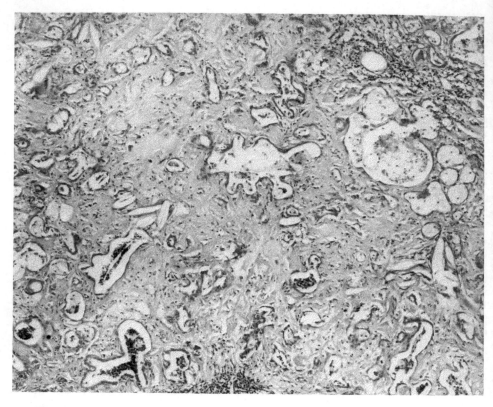

Figure 8   Fibrosis grade 8: severe: complete obstruction of most airways (×66).

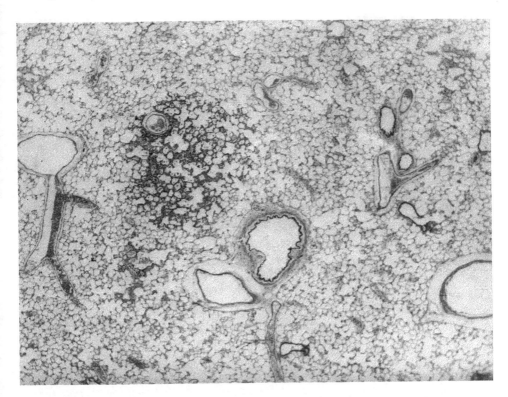

Figure 9 Bronchiolar alveolar hyperplasia: retention of alveolar architecture. Air spaces lined by cuboidal cells. Lack of cellular and nuclear atypia. There may or may not be significant amounts of fibrosis (×16.5).

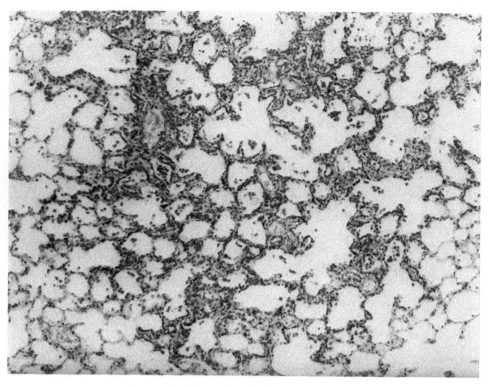

Figure 10 Benign alveolar hyperplasia (×66).

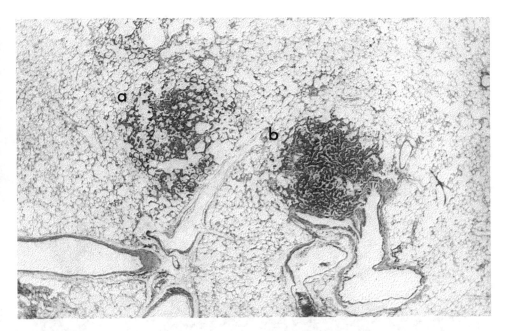

Figure 11 (a) Benign alveolar tumor: loss of preexisting architecture. Papillary formation into air spaces. Nuclear uniformity. There may or may not be metaplasia to other cell types. Compression but no invasion of adjacent tissue. (b) Malignant alveolar tumor differs by showing atypia of cells and early invasion of surrounding tissue (×16.5).

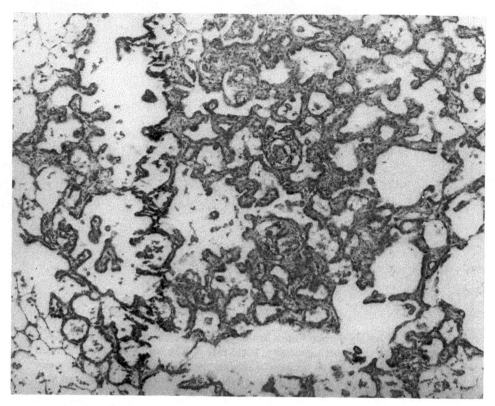

Figure 12   Benign alveolar tumor (×66).

Table 2 Intratracheal Experiments of Begin et al. in Sheep

Dosage, Fiber and Method	Results
Two groups each received monthly doses of UICC Canadian chrysotile by intratracheal instillation for 6/12 and then weekly for 6&12. Low-dosage: total = 328mg High-dosage: total = 2282mg Lung function and BAL studied every 2 months (Begin et al., 1983)	Initial alveolitis found at 20 months in the high-dosage group accompanied by significantly lower lung functions. and increased cellularity, LDH, proteases of BAL
Single dose of UICC Canadian chrysotile by bronchoscopic catheterization to tracheal lobe of 0, 1, 10, 25, 50 or 100 mg. BAL carried out sequentially to day 60 (Begin et al., 1986)	Albumen and fibronectin levels not significant Cellular counts, procollagen III, LDH, BG, and AK significantly raised for 100 mg only
100 mg UICC Canadian chrysotile compared with 100 mg short chrysotile (University of Sherbroke)[a]	Short fibers similar to saline controls
Comparison of lungs with and without alveolitis using 100 mg dose	Fiber counts significantly higher in those with alveolitis

[a]Rest of method is similar to Begin et al. (1986)
LDH, lactose dehydrogenase; BG, beta-glucuronidase; AK, alkaline phosphataease; BAL, bronchiolar lavage.

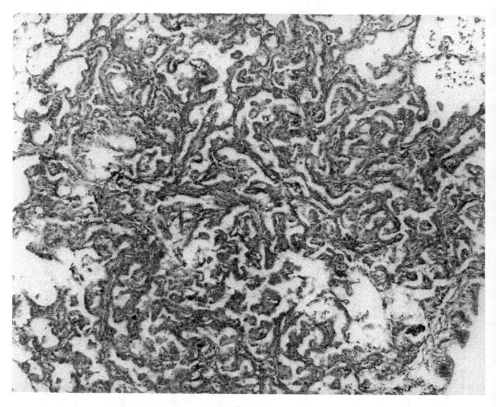

Figure 13   Malignant alveolar tumor (×66).

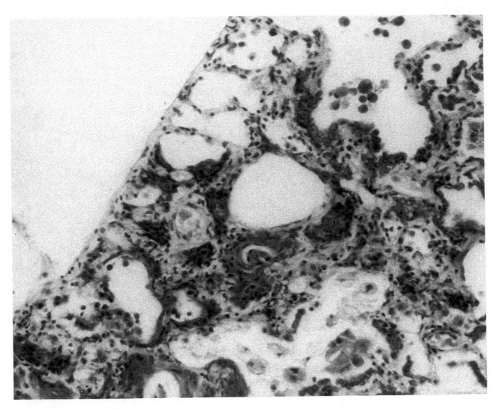

Figure 14 Squamous metaplasia (×165).

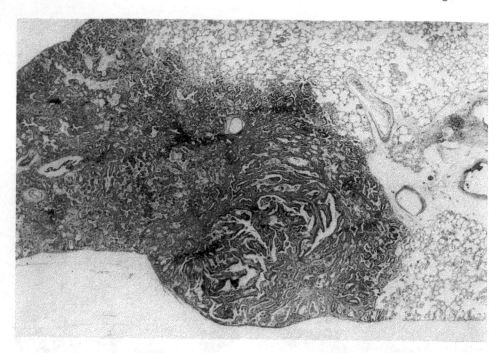

Figure 15 Malignant alveolar tumor: Adenocarcinoma. Expansion, compression, and invasion of adjacent tissues, atypia (nuclear and/or cellular), pleomorphism, and metastases. Metaplasia to other cell types may be present (×16.5).

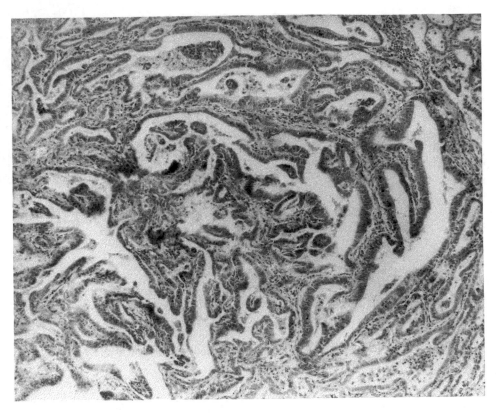

Figure 16 Adenocarcinoma (×66).

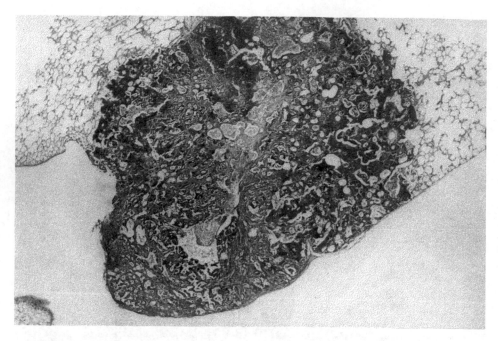

Figure 17 Malignant alveolar tumor: adenosquamous carcinoma (×16.5).

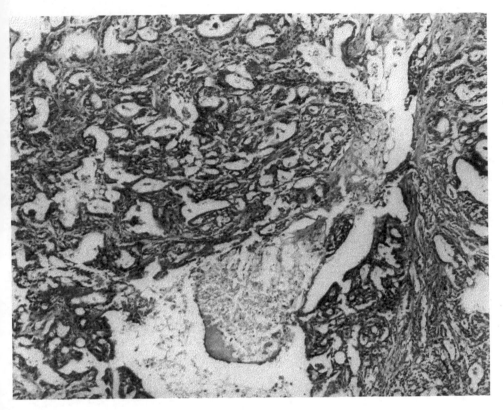

Figure 18 Adenosquamous carcinoma (×66).

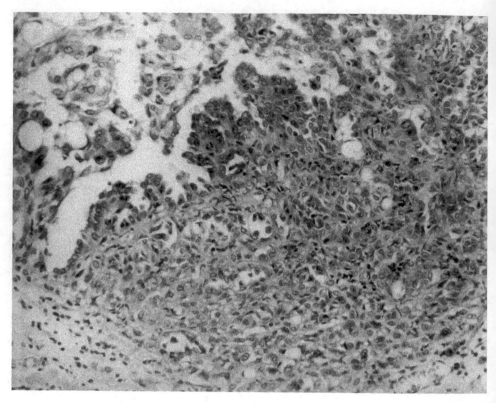

Figure 19   Mesothelioma, epithelial cell (×166).

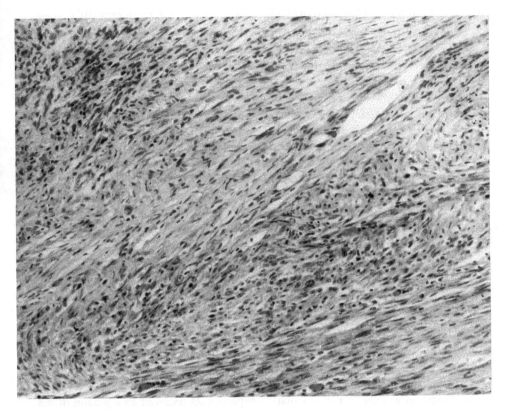

Figure 20   Mesothelioma, spindle cell (×166).

F.   Results of Inhalation Studies

The methods of diagnosing pulmonary thoracic tumors and the severity of pulmonary fibrosis have been defined and standardized by a working party of expert pathologists (McConnell et al., 1984). The correlation with human exposures has been difficult to assess. We have succeeded in producing the three major lesions that occur in humans following exposure to asbestos dust. Severe pulmonary fibrosis has been produced in vervet monkeys, rabbits, guinea pigs, and rats (Wagner, 1963). Lesions were most clearly illustrated with asbestos bodies in guinea pigs and monkeys. In rats, severe interstitial fibrosis occurred but asbestos bodies were rarely seen in the lung tissue. In contrast to this, asbestos bodies were frequently seen in pleural cavities of rats following inoculation of the amphibole types of asbestos.

In comparing the biological effects of exposure to the different UICC standard samples, there was initially good agreement. (Wagner et al., 1974). All the samples were involved in the production of asbestosis, pulmonary adenomatoses, adenomas, and carcinomas of the lungs. (In the later review by the Working Party, the term *adenomatoses* was dropped in favor of the more accurate *bronchiolar alveolar hyperplasia*.) Following all the exposures it could be shown that the fibrotic lesions progressed and the carcinomas, although peripheral, occurred in the more severely fibrosed lungs. These tumors were either bronchiolar or squamous in origin; the squamous tumors were associated with squamous metaplasia of the bronchiolar–alveolar epithelium. A few mesotheliomas were produced; four by both crocidolite and chrysotile B, two by amosite, and one by anthophyllite. Two mesotheliomas developed after a single day's exposure to dust and these should be discounted, since a 1% incidence was also found in the control subjects in later studies.

II. Statistical Methods

A. Design

Experiments were designed to conform to the basic principles of experimental design.

Controls

Experiments included one or more control treatments. Negative controls were not exposed to dust, but in inoculation experiments received an injection of the saline carrier solution. In the inhalation experiments (Wagner et al., 1974) the negative controls were caged throughout in racks in the main unit since it was not considered justifiable to use an inhalation chamber for this purpose. In some experiments (Wagner et al.,. 1977, 1979, 1980a) a positive control treatment was included, that is, a dust known from previous experiments to produce an effect. This allowed the effects for the new dusts under investigation to be related to the accumulated results from previous experiments.

Balance

Allocation of animals to treatments was balanced with respect to sex and age. Thus males and females were allocated separately; if the age range of the available rats covered more than about 2 weeks, they were organized into subgroups each covering a narrow age range and each subgroup was allocated separately. In some experiments (e.g., experiment 3 of Wagner et al., 1973) there were insufficient animals available at one time and rats were included from several breeding batches. Again, each batch was allocated separately.

Randomization

The available rats were allocated at random to cages using computer-generated random permutations. Cages were assigned to treatments in sequence; this procedure amounted to a random allocation of rats to the treatment groups within the constraints of allocating prespecified numbers to each treatment. The randomization was also stratified to take account of the factors to be balanced (see above).

Size of Experiment

Equal numbers of rats were usually allocated to each treatment, but in some experiments the different dusts were not of equal importance and more rats were allocated to the dusts considered more important. It was important to obtain precise data from the control groups, since all the treatments under investigation would be compared with the controls. It was therefore reasonable to allocate more rats to the controls. For example, in experiment 2 of Wagner and co-workers (1973), 16 rats were allocated to each of seven experimental dusts, 32 to the positive control dust, and 48 were used as negative controls. In the inhalation experiments the size of the chamber determined the number of rats that could be exposed to each dust.

End Point

In all experiments one of the end points was the occurrence of a tumor identified at necropsy following death, usually due to the tumor. For this endpoint it was shown (Berry, 1975) that the most efficient procedure was to allow each animal to live out its life. Thus in the injection experiments rats were only sacrificed if they appeared to be in distress. In the inhalation experiments there were two other end points: the amount of dust deposited in the lungs and fibrosis. These were required at intervals throughout, and following, the exposure period and so it was necessary to sacrifice some of the rats at various times. The remainder were allowed to live out their lives to provide data on tumor incidence (Wagner et al., 1973).

The majority of experiments had a simple design based on the principles described above. One experiment, however, required a more sophisticated approach. This determined if age at inoculation of asbestos influenced the resultant incidence of mesotheliomas (Berry and Wagner, 1976). The rats were allocated to two groups: rats in the first group were injected at the beginning of the experiment and those in the second group after a delay when they were older. If there was any effect of age at injection, the difference between the two groups would be expected to be greater the longer the delay before rats in the second group was injected. But the longer the delay, the more animals in the second group would die before sufficient time had elapsed for a mesothelioma to develop. The delay has to be a compromise between these two conflicting tendencies, and an optimum value was determined based on survival data from previous experiments and a specified size of the age effect it was required to be able to detect (Berry, 1975). It was reasoned intuitively that more rats should be included in the second group to compensate for the higher mortality due to other causes during the period when mesotheliomas would be occurring. The optimum relative size of the two groups was determined by similar means (Berry, 1975).

B. Analysis

Results were summarized by basic measures such as the proportion of rats developing a mesothelioma, mean survival, and mean fibrosis score. A treatment that results in rats developing mesotheliomas will show its effect both on the proportion of rats developing a mesothelioma and on mean survival time, which will be reduced as a consequence of the mortality due to mesothelioma. A method of combining all the information into a single index of mesothelioma incidence was required. It was also desirable to eliminate chance variations between treatment groups in mortality due to causes unrelated to the experimental treatments. One approach that eliminates natural mortality is the use of life table methods; these were employed in the early experiments (Wagner and Berry, 1969). This approach does not, however, produce a single index of mesothelioma incidence.

The approach adopted was to model the death rate of animals dying with a mesothelioma in terms of time since injection, using a model that had been published by Pike (see Berry and Wagner, 1976) shortly before the early experiments were ready for analysis. This model was based on the Weibull distribution and had the form

$$m = ck(t-w)^{k-1}$$

where m is the death rate of rats dying with a mesothelioma at time t after injection and c, k, and w are parameters. This model was found to fit the data well (Berry and Wagner, 1969). However, the estimates of the parameters k and w are negatively correlated and could not both be estimated precisely even with a great deal of data. It was found adequate to fix k as 3.25 and w—the lapse period before any mesotheliomas would occur—as 270 days (Wagner et al., 1973). The different treatments were then compared in terms of the parameter c.

The development of the proportional-hazards model and the conditional method of analysis by Cox (see Berry and Wagner, 1976) provided an alternative method of analysis. With this approach

$$m = bf_0(t)$$

where $f_0(t)$ is the time-dependent part of the incidence rate, which is assumed to be the same for all treatments, and the parameter b differs between treatments.

The relative values of b for the different treatment groups can be estimated without having to consider the form of $f_o(t)$ and this avoids the complications in the Weibull model of having to estimate k and w. Treatments are then compared in terms of b; this is an alternative to comparisons in terms of c, for fixed k and 2, in the Weibull model.

III. Future Work

The problem arose when the amount of dust (as estimated gravimetrically) in the lungs was compared. The severity of the fibrosis was similar for all exposure, yet the amount of both chrysotiles was minimal by weight when compared to the amphiboles. Similar results were obtained by Davis et al. (1978). It would suggest that chrysotile fibers have a similar fibrogenetic effect to the amphibole fibers on a similar mass basis. This is contrary to the findings of human environmental studies. This would confirm that in these inhalation studies the bulk of the chrysotile inhaled is not retained in the lung parenchyma. It is known that the chrysotile fibers shear into numerous fine fibrils in aqueous solutions. This is also seen when the chrysotile fibers are recovered from the lung tissue (Pooley, 1972). It has been calculated that one crocidolite fiber of equal length is 100 times the weight of a chrysotile fibril. There have been four suggestions as to the fate of the bulk of the chrysotile (Wagner, 1986): (1) because of the coiled, wavy configuration of the chrysotile fibers only a fraction of them reach the respiratory airways; (2) the chrysotile fibers are known to form gel when there is fibrillization in fluids; this gel is rejected in the larger airways and either expectorated or ingested; (3) the chrysotile is soluble in tissue fluids and the fibers rapidly disintegrate; (4) the fibers are inhaled into the respiratory portion of the lung where the fibrillization and rapid removal from the lung parenchyma take place. At the moment all these hypotheses have their supporters; solving the problem is a task for the future.

There is a general criticism that the working party's assessment of the various stages of fibrosis is too subjective and that a biochemical method would be more satisfactory. The earlier methods for detecting collagen increase are not sufficiently sensitive to discern the earlier lesions, but techniques are being developed to identify an increase in procollagen lysates. Positive identification of the mesotheliomas and carcinomas by enzyme and other specific markers would help to establish an objective diagnosis.

The inhalation experiments to establish the incidence of fibrosis and tumor formation are time consuming (the complete study takes about 3 years to complete) but there have been suggestions that the use of cocarcinogens might shorten this time. Few inhalation studies have been large enough to be acceptable for the code of Good Laboratory Practice of the Environmental Protection Agency

(EPA), although logistics for the exposure of 500 rats to each dust are being investigated, since expense would obviously limit such experiments in number and place.

IV. Results of the Implantation Studies

The initial intrapleural inoculation (Wagner, 1963) established that it was possible to produce mesotheliomas. The next stage was to see if it were possible to differentiate between the incidence of tumor production according to the type of asbestos fiber, using SPF rats (Wagner and Berry, 1969). The peak tumor incidence occurred at the 700th day after inoculation, with some animals surviving up to 1000 days. The result was a surprise, since the highest incidence of tumors (65%) occurred in the animals treated with the chrysotile samples; the crocidolite produced a 60% incidence of tumors. The experiment was repeated using a healthy strain of conventionally bred Wistar rats. The results were practically identical. The chrysotile used in this experiment was produced from a prototype plant at one of the Canadian mines that was trying to develop a wet weaving process in which the fiber was separated from the crushed ore by a sedimentation process, which left the actual fibers suspended in a detergent solution. The fiber was then sedimented out by the addition of a solution that neutralized the detergent. This resulted in a large number of very fine chrysotile fibers, the majority of which appeared to be straight under the transmission electron microscope. (In general, the chrysotile fibers have a wavy, coiled appearance, unlike the needle straight amphibole fiber [Timbrell et al., 1970a]). As stated previously, the electron microscopic investigation also showed that the chrysotile fibers in solution tended to be shredded into numerous very fine fibrils, whereas the longitudinal shearing characteristic of all asbestos fibers only produced two or three fine amphibole fibers. The possibility that brucite (or nemolite), a fibrous magnesium hydroxide found in the chrysotile deposits, may have been involved was considered. When inoculated the nemolite produced a high incidence of tumors (Wagner et al., 1976), but it was pointed out that by grinding the nemolite we had created a substance did not occur in the milling of the chrysotile ore. Studies of the minerals containing the chrysotile sample did not reveal nemolite. In the 20 years since these results were obtained, there has been considerable speculation as to why the Canadian chrysotile, which in human epidemiological studies produced few mesotheliomas, has been so active in experimental animals. Davis and his colleagues, in a more recent study (1986), obtained very similar results with dust from the final product of this fiber prepared for wet spinning.

Our inoculation studies (Wagner et al., 1976) were followed by others. It is not possible to discuss all the results, but the most relevant were those of Pott in Germany (1978), Stanton in the United States (Stanton et al., 1981), and Davis in the United Kingdom (1970). All these research groups undertook a series of investigations that became more and more precise as new techniques of analysis became available. The first stage of agreement was that by the implantation of various types of asbestos it was possible to produce mesotheliomas. (Stanton et al. in their later publication [1981] described the tumors as pleural sarcomas.) The second phase was to show that these tumors could be produced by numerous other mineral fibers. The third stage was the realization that the thinner the mineral fiber, the more likely it was to be associated with tumor formation. The final and most complex stage that has not been resolved is agreement on the dimensions of the fibers associated with the highest incidence of the tumours. Stanton's series of studies were the largest and more complex. He first suggested that the long fibers were not ingested by the macrophage and speculated about the surface area of free fibers. He observed that the long thin fibers seemed to escape the attention of these phagocytic cells. He next gave a precise range of fibers, with dimensions of less than 0.25 μm in diameter and greater than 8.0 μm in length, which would cause the highest rate of tumor production. He considered that the possible involvement could include a range with a maximum of 1.5 μm in diameter but he was never definite about the length. My interpretation of Stanton's conclusions is that the finer the mineral fiber, the more likely it is to cause mesothelioma, providing it is longer than 6 μm. The results of Davis's and our studies agreed with Stanton's but produced evidence that the maximum diameter would not be greater than 0.5 μm.

We believe that these views have been reinforced by our studies on the asbestos fiber tremolite (Wagner et al., 1982). Tremolite is an amphibole asbestos with fibers that have dimensions varying from large flakes to very fine fibrils. Following inoculation of the range of size of this material, we were only successful in producing mesotheliomas with the very fine material (diameter less than 0.3 μm). These findings, taken with the results of our collaborative investigations with epidemiologists and mineralogists, have been confirmed on a worldwide basis (Beck and Bignon, 1984). The very fine tremolite fibers are associated with mesotheliomas in humans, the fibers with a diameter of more than 1 μm are associated with pulmonary fibrosis and carcinoma of the lungs. The coarser fibers, which occur in many parts of the world as a contaminant of agricultural soils, are associated with the production of large calcified pleural plaques. The large flakelike particles that occur in some regions as a contaminant of talc do not cause lesions in humans or experimental animals.

V. Other Mineral Fibers

The methods and techniques developed for the study of asbestos have been adapted for the investigation of the biological effects of other mineral fibers. These have included synthetic mineral fibers both vitreous and crystalline, derived from glass, slag, volcanic rock, and ceramic materials, and the fibrous clay minerals: palygorskite, sepiolite, and anthophyllite. Most interesting so far, of these investigations has been the recognition of the zeolite fibre erionite as the cause of the devastating incidence of mesotheliomas in the Urgup region of Cappadocia in central Turkey. It has been possible to confirm by experimental study the sinister effects of exposure to the finer erionite fibers. This is the only dust in which we have been able to produce a 100% tumor incidence using both the inhalation and implantation techniques.

We have been fortunate in being able to incorporate the results of our experiments with findings from human epidemiological investigations. This collaboration between epidemiologists, clinicians, pathologists, and mineralogists has provided more valuable information in the study of the biological effects of the whole range of mineral fibers. These studies must continue in the future as new sources of natural and synthetic fibers are exploited. The in vitro studies on asbestosis are a rapidly developing field, which may yet overtake the experimental animal studies. The reader is referred to the proceedings of the three international workshops mentioned previously in Chapter 23.

Acknowledgments

In these experimental studies team work has been essential. I have therefore requested my colleagues Mr. J. W. Skidmore and Professor G. Berry to help in compiling this chapter. Mr. Skidmore prepared the dust clouds, constructed and calibrated the dusting chambers, monitored the dust dosage, and estimated the dust retained in the animals. Professor G. Berry was responsible for the statistical design of our experiments, from the animal breeding to the final assessment of the results.

References

Beck, E. G., and Bignon, J. (1984). *In Vitro Effects of Mineral Dusts*. Berlin, Springer Verlag.

Begin, R., Rola-Pleszczynski, M., Masse, S., Nadeau, D., and Drapeau, G. (1983). Assessment of progression of asbestosis in the sheep model by bronchoalveolar lavage and pulmonary function tests. *Thorax* **38**:449–457.

Begin, R., Masse, S., and Sebastien, P. (1986). Determinants of biological effects of chrysotile in the lung: exposure dose, fiber size, or chemistry and individual susceptibility. In: *Accomplishments in Cancer Research*. Edited by J. C. Wagner. Philadelphia, Lippincott, pp. 74–110.

Berry, G. (1975). Design of carcinogenesis experiments using the Weibull distribution. *Biometrika* **62**:321–328.

Berry, G., and Wagner, J. C. (1969). The application of a mathematical model describing the times of occurrence of mesotheliomas in rats following inoculation with asbestos. *Br. J. Cancer* **23**:582–586.

Berry, G., and Wagner, J. C. (1976). Effect of age at inoculation of asbestos on occurrence of mesotheliomas in rats. *Int. J. Cancer* **17**:477–483.

Bolton, R. E., and Davis, J. M. G. (1976). The short term effects of chronic asbestos ingestion in rats. *Ann. Occup. Hyg.* **19**:121–128.

Bonser, G. M., and Clayson, D. B. (1967). Feeding of blue asbestos to rats. *Br. Emp. Cancer Campaign Fes. Annu. Rep.* **45**:242.

Davis, J. M. G. (1970). The long term fibrogenic effects of chrysotile and crocidolite asbestos dust injected into the pleural cavity of experimental animals. *Br. J. Exp. Pathol.* **51**:617–627.

Davis, J. M. G. (1974). Histogenesis and fine structure of peritoneal tumours produced in animals by injections of asbestos. *J. Natl. Cancer Inst.* **52**:1823–1837.

Davis, J. M. G., Beckett, S. T., Bolton, R. E., Collings, P., and Middleton, A. P. (1978). Mass and number of fibres in the pathogenesis of asbestos-related lung disease in rats. *Br. J. Cancer* **37**:673–688.

Davis, J. M. G., Addison, J., Bolton, R. E., Donaldson, K., and Jones, A. D. (1986). Inhalation and injection studies in rats using dust samplers from chrysotile asbestos prepared by a wet dispersion process. *Br. J. Exp. Pathol.* **67**:113–129.

Gardner, L. U., and Cummins, D. E. (1931). Studies on experimental pneumoconiosis, VI. Inhalation of asbestos dust: Its effect upon primary tuberculosis infection. *J. Ind. Hyg.* **13**:65–81.

Gross, P., deTreville, R. T. P., Tolker, E. B., Kaschak, M., and Babyak, M. A. (1967). Experimental asbestosis: the development of lung cancer in rats with pulmonary deposits of chrysotile asbestos dust. *Arch. Environ. Health* **15**:343–355.

Holt, P. F., Mills, J., and Young, D. K. (1964). The early effects of chrysotile asbestos dust on the rat lung. *J. Pathol. Bacteriol* **87**:1–23.

King, E. J., Clegg, J. W., and Rae, V. M. (1946). The effect of asbestos, and of asbestos and aluminium on the lungs of rabbits. *Thorax* **1**:188–197.

Lynch, K. M., McIver, F. A., and Cain, J. R. (1957). Pulmonary tumors in mice exposed to asbestos dust. *Arch. Ind. Health* **15**:207–214.

McConnell, E. E., Wagner, J. C., Skidmore, J. W., and Moore, J. A. (1984). A comparative study of the fibrogenic and carcinogenic effects of UICC Canadian chrysotile B asbestos and glass microfibre (JM 100). In *Biological Effects of Man-Made Mineral Fibres*. Proceedings of a WHO/IARC Conference, Copenhagen. Vol. 2, pp. 234–250.

McConnell, E. E., Rutter, H., Ulland, B., and Moore, J. A. (1985). Chronic effects of dietary exposure to amosite asbestos and tremolite in F344 rats. Report from: The National Toxicology Program, Research Triangle Park, NC, NIEHS.

Pooley, F. D. (1972). Electron microscope characteristics of inhaled chrysotile asbestos fibre. *Br. J. Ind. Med.* **29**:146–153.

Pott, F. (1978). Some aspects of the dosimetry of the carcinogenic potency of asbestos and other fibrous dusts. *Staub-Reinhalt. Luft* **38**:486–490.

Reeves, A. L., Puro, H. E., Smith, R. G., and Vorwald, A. J. (1971). Experimental asbestos carcinogenesis. *Environ. Res.* **4**:496–511.

Schmal, D. (1958). Cancerogene wirkung von asbest bei implantation on ratten. (Cancerogenic effects of asbestos implantation in rats). *Z. Krebsforsch. (Berlin)* **62**(5):561–567.

Shabad, L. M., Pylev, L. N., Krivosheeva, L. V., Kulagina, T. F., and Nemenki, B. A. (1974). Experimental studies on asbestos carcinogenicity. *J. Natl. Cancer Inst.* **52**:1175–1187.

Simson, F. W. (1928). Pulmonary asbestosis in South Africa. *Br. Med. J.* **1**:885–887.

Smith, J. M., Wootton, I. D. P., and King, E. J. (1951). Experimental asbestosis in rats. The effect of particle size and of added alumina. *Thorax* **6**:127–136.

Smith, W. E., Miller, L., and Churg, J. (1970). An experimental model for study of cocarcinogenesis in the respiratory tract. In: *Morphology of Experimental Respiratory Carcinogenesis.* Edited by P. Nettesheim, M. G. Hanna, Jr., and J. W. Deatherage Jr. Oak Ridge, US Atomic Energy Commission, pp. 299–316.

Stanton, M. F., and Wrench, C. (1972). Mechanisms of mesothelioma induction with asbestos and glass fibre. *J. Natl. Cancer Inst.* **48**:797–816.

Stanton, M. F., Layard, M., Tegeris, A., Miller, E., May, M., Morgan, E., and Smith, A. (1981). Relation of particle dimension to carcinogenicity in amphibole asbestoses and other fibrous minerals. *J. Natl. Cancer Inst.* **67**:965–975.

Timbrell, V., Pooley, F., and Wagner, J. C. (1970a). Characteristics of respirable asbestos fibres in pneumoconiosis. In *Proceedings of the International Conference 1969.* Edited by M. A. Shapiro. Oxford, Oxford University Press, pp. 120–125.

Timbrell, V., Skidmore, J. W., Hyett, A. W., and Wagner, J. C. (1970b). Exposure chambers for inhalation experiments with standard reference samples of asbestos of the International Union Against Cancer (UICC). *Aerosol Sci.* **1**:215–223.

Timbrell, V., Gilson, J. C., Webster, I. (1978). UICC reference samples of asbestos, *Int. J. Cancer* **3**:406–408.

Wagner, J. C. (1962). Experimental production of mesothelial tumours of the pleura by implantation of dusts in laboratory animals. *Nature* **196**:180–181.

Wagner, J. C. (1963). Asbestosis in experimental animals. *Br. J. Ind. Med.* **20**:1–12.

Wagner, J. C. (1986). Mesothelioma and mineral fibers. Charles S. Mott Prize. *Cancer* **57**:1905–1911.

Wagner, J. C., and Berry, G. (1969). Mesotheliomas in rats following inoculation with asbestos. *Br. J. Cancer* **23**:567–581.

Wagner, J. C., Sleggs, C. A., and Marchand, P. (1960). Diffuse pleural mesotheliomas and asbestos exposure in the Northwestern Cape Province. *Br. J. Ind. Med.* **17**:260–271.

Wagner, J. C., Berry, G., and Timbrell, V. (1973). Mesotheliomata in rats after inoculation with asbestos and other materials. *Br. J. Cancer* **28**:173–85.

Wagner, J. C., Berry, G., Skidmore, J. W., and Timbrell, V. (1974). The effects of the inhalation of asbestos in rats. *Br. J. Cancer* **29**:252–269.

Wagner, J. C., Berry, G., and Skidmore, J. W. (1976). Studies of the carcinogenic effect of fibre glass of different diameters following intrapleural inoculation in experimental animals. In: U.S. Department of Health, Education and Welfare: Public Health Service: Center for Disease Control: National Institute for Occupational Safety and Health, Occupational exposure to fibrous glass. Proceedings of a symposium, College Park, Maryland, 26-27 June 1974. HEW Publication No. (NIOSH) 76-151. Washington, U.S.D.H.E.W., pp. 193–197.

Wagner, J. C., Berry, G., Cooke, T. J., Hill, R. J., Pooley, F. D., and Skidmore, J. W. (1977). Animal experiments with talc. In: *Inhaled Particles and Vapours,* Vo. IV. Edited by W. H. Walton. New York, Pergamon Press, pp. 647–653.

Wagner, J. C., Berry, G., Hill, R. J., Skidmore, J. W., and Pooley, F. D. (1979). An animal model for inhalation exposure to talc. In: *Dusts and Disease.* Edited by R. Lemen and J. M. Dement. Park Forest South, IL, Pathotox Publishers, pp. 389–392.

Wagner, J. C., Berry, G., Hill, R. J., Munday, D. E., and Skidmore, J. W. (1980a). Animal experiments with man-made mineral fibres. In: *Biological Effects of Mineral Fibres.* Edited by J. C. Wagner, Lyon, IARC Scientific Publications, No. 30, pp. 361–362.

Wagner, M. M. F., Wagner, J. C., Davies, R., and Griffiths, D. M. (1980b). Silica-induced malignant histiocytic lymphoma: incidence linked with strain of rat and type of silica. *Br. J. Cancer* **41**:908–917.

Wagner, J. C., Chamberlain, M., Brown, R. C., Berry, G., Pooley, F. D., Davies, R., and Griffiths, D. M. (1982). Biological effects of tremolite. *Br. J. Cancer* **45**:352–360.

25

Experimental Silicosis

MARGARET M. WAGNER * and **J. CHRISTOPHER WAGNER** *

Weymouth, Dorset, England

In 1913 it was still thought that any kind of dust was associated with increased risk of pulmonary tuberculosis.

> The first beams of dawn on this twilight are to be found in evidence placed before the Royal Commission on Metalliferous Mines and Quarries, and finally published in 1914. Statistical data were gathered which indicated that those old time occupational diseases, known as miners' phthisis, grinders' rot, masons' rot, potters' rot and knappers' rot, were in each case, the same disease, i.e., a specific form of dust phthisis associated with one kind of dust only: that composed of fine particles of silica (Collis, 1931).

The earliest experimental work, therefore, had to be directed towards elucidating the connection between silica and the tubercle bacillus. These bacilli were shown to grow more readily at a site damaged by silica (Gye and Kettle, 1922). Further work (Gardner, 1930) demonstrated that a strain of low virulance produces more tubercles of a worse nature when the animals are dusted with silica and that crystalline silica is far more dangerous than the silicates (with the exception of asbestos;

---

*Medical Research Council, Retired.

Miller and Sayer, 1934). Also, in 1934 Kettle's work revealed that a tubercle bacilli lesion promoted silicosis.

Pulmonary infection, in general, may exacerbate the effects of inhaled silica. More recent work with specific pathogen-free (SPF) rats resulted in only small fibrotic nodules (Chiappiani and Vilgiani, 1982). It is also noteworthy that in the SPF rats with their *relatively* clean lungs, alveolar proteinosis can be produced (Heppleston et al., 1974).

Work had also been progressing on methods of producing fibrosis. By 1950 King et al. were producing dense, round, collagenous nodules with a well-controlled standardized inhalation procedure, which has been little altered to this day. Kettle (1934) among others attempted to take a short cut by introducing suspensions of dust directly into the lungs, but both he and Gardner believed that "a single flooding of the lungs with an overwhelming dose of dust can have no relation to the continuous inhalation of small quantities." Nevertheless, the intratracheal method continues to find a use because there is no need for expensive equipment and fibrosis can be produced rapidly. Both of the above methods have also been applied to the study of lung carcinoma and silica (for review, see IARC, 1987). Other methods used in production of fibrosis have been intraperitoneal, which produces it in a matter of weeks or months (Miller and Sayers, 1934) in the coelomic cavity, while intravenous introduction will produce fibrosis of the liver (Svensson et al., 1956). Both crystalline and colloidal silica (in the 0.002 μm range) will produce silica shock (Gye and Purdy, 1922, 1924) when given by either of these two methods. The rationale for this was demonstrated by Harington and Sutton (1963).

The main use for intravenous or intracoelomic production of fibrosis must be for the comparison of dusts. Careful consideration must therefore be given to the results, which should not be applied inappropriately. For example, Wagner and Wagner (1972) produced malignant lymphoma of histiocytic type (probably of macrophage origin) by intrapleural inoculation. First, it is highly unlikely that a large dose of silica would ever reach this area. Second, there is no evidence that such a tumor is associated with silica exposure. Third, it was strongly associated with one strain of rat, and there is no evidence that a particular genetic type might have a *carcinoma* associated with silica. These experiments therefore do not supply evidence that silica is a carcinogen for the lung. On the other hand, Gye and Kettle (1922) and Kettle (1924), in some elegant experiments, successfully used the intact animals, giving subcutaneous injections of silica and intravenous injections of tubercle bacilli to show the collecting of these bacilli in the silicotic lesions. It cannot be stressed too strongly that physiological responses must be studied in the lung, preferably with inhalation studies.

I. Procedures

A. Type of Animal

Rats, guinea pigs, mice, and monkeys have all been used successfully in the production of fibrosis. The usefulness of having one type of animal (rat) that is not easily susceptible to tuberculosis, and another that is (guinea pig) was successfully exploited by Mavrogordato (1922), who showed that the rats in the gold mines were infected while those in the workers' compound were not. With the advent of SPF rats the difficulties with long-term experiments were solved. They must be used under suitable conditions and maintained as SPF rats.

B. Methods of Introduction

Inhalation

It is interesting that this method was first used in 1885 (Arnold) and adapted by Mavrogordato in 1914. The latter placed his guinea pigs in sponge bags (to prevent their urine "laying" the dust). The heads were held through an aperture. Dried and sterilized dusts were passed through filters, then added to the zinc-lined boxes so that there was a concentration of 30–45 mg/L. The animals were dusted for 24–36 hr, while a two-bladed fan kept the dust in a cloud. Gye and Kettle (1922) also used dusting chambers, but the next advance was achieved by Gardner (1932), in a well-known series of experiments. Animals were exposed in a room for 1, 2, or even 3 years, to a concentration of $30–80 \times 10^6$ particles, for 8 hr/day, 6 days/week. These particles were measured using the Greenburg-Smith (1932) impinger technique at the recommended magnification of 100 with both light and dark field illumination, both at the breathing zone and the airborne dust. Thus the amount and composition of the dusts were determined (Gardner, in Lanza, 1938). The animals always remained in the room, continuously exposed to the dust, which settled on their fur and bedding, which they constantly stirred up.

Major advances were made by Wright (1950, 1953) with further modifications in 1957. The basic principle was to keep the animals in a cabinet, which was not completely airtight but kept at a pressure slightly below that of atmospheric to minimize leakage. Care was taken not to clog the filters by having too high a dust concentration. The rate of air flow was controlled by reference to the ventilation rate of the animals (Guyton, 1947). Measurement of dust exposure was by a long-term continuous sampler consisting of a gravimetric thermal precipitator (Wright, 1953). These dust chambers eventually led to those constructed by Skidmore and Hyett under the supervision of Timbrell (1968). The chambers were also suitable for fibrous dusts, and Timbrell produced an apparatus for the dispersion of fibrous dusts, based on a well-known French coffee grinder (see Chapter 24).

Intratracheal Injections

The usual dosage given is 50 mg in 0.5 ml saline, under anesthesia. King (1953) and Saffiotti (1960), together with more recent workers, all produced fibrosis.

Subcutaneous and Intracutaneous Injections

The latter method, as described by Gardner (in Lanza, 1938), resulted in a central zone of ulceration, when 0.1 ml of a 0.2% solution was used with particles under 3 μm in diameter. Kettle (1926) used a similar dosage subcutaneously in a famous series of experiments to show the solubility of silica and that areas affected by silica are more susceptible to the tubercle bacillus (given intravenously).

Intravenous Injection

Gardner and Cummings (1933) gave 1–3 μm particles (total dose, 1g) in 20 doses over a period of 60 days into the ear vein of the rabbit. Gardner considered this a physiological method that did not produce nonspecific side effects. Rivers (1963) successfully developed the use of the tail vein of mice in a similar manner. Acute toxicity has been induced particularly by finely divided colloidal particles, 0.002 μm in diameter (Arienzo and Bresciano, 1968). Dale and King (1953) found that 25 mg of crystalline silica was lethal to mice.

Intraperitoneal Injections

Despite the "objectionable features" of injections, Gardner (in Lanza, 1938) agreed that Sayers and Miller's (1934) procedure could be used to produce fibrosis for simply, for example, a comparison of dusts. A 10% saline suspension (1-2 ml) of dust particles that will go through a 350 mesh screen was used by these authors.

Intrapleural

The essential technique is that described in Chapter 24.

C. Types of Silica

As well as the size of the particles, the state of the silica, whether crystalline, colloidal, or amorphous, must be considered (many comparative experiments are mentioned by Gardner in Lanza, 1938).

II. Pathological Findings

The histological assessment for severity of fibrosis was developed by Belt and King (1945).

Grade 1: loose reticulin fibrils with no collagen

Grade 2: compact reticulin without or with a small amount of collagen

Grade 3: Somewhat cellular but made up mostly of collagen

Grade 4: wholly composed of collagen fibrosis and completely acellular

Grade 5: Acellular, collagenous, and confluent

This grading schema was extended by Wagner (Brown et al., 1983).

References

Arienzo, R., and Bresciano, E. (1968). Experimental shock after silica and its prevention with heparin. *Boll. Soc. Ital. Biol. Sper.* **44**:1685–1687.

Belt, T. H., and King, E. J. (1945). Tissue reaction produced experimentally by selected dusts from South Wales Coalfields. In: *Chronic Pulmonary Disease in South Wales Coal Miners* III. Experimental Studies. London, Spec. Rep. Ser. Med. Res. Council, No. 250.

Brown, R. C., Munday, D. E., Sawitza, V. M., and Wagner, J. C. (1983). Angiotensin coverting enzyme in the serum of rats with experimental silicosis. *Br. J. Exp. Pathol.* **84**:286–296.

Arnold, J. (1885). Staub inhalation und staubmetastase.

Chiappino, G., and Vigliani, E. C. (1982). Role of infective, immunological and chronic irritative factors in the development of silicosis. *Br. J. Ind. Med.* **39**:253–258.

Collis, E. L., (1931). Silica. A critical review of the literature. *Bull. Hyg.* **6**(4):305–312.

Dale, J. C., and King, E. J. (1953). Acute toxicity of mineral dusts. *Arch. Ind. Hyg.* **7**:478–483.

Gardner, L. U. (1930). Proceedings, International Silicosis Conference, Johannesburg. (Quoted by Kettle, 1934).

Gardner, L. U. (1932). Inhalation of quartz dust. J. Ind. Hyg. **14**:18.

Gardner, L. U. (1938). Experimental pathology. In *Silicosis and Asbestosis*. Edited by A. Lanza, Oxford, Med. Publications, pp. 257–345.

Gardner, L. U., and Cummings, D. E. (1933). Reaction to fine and medium sized aluminium oxide particles. Silicotic cirrhosis of the liver. *Am. J. Pathol.* **9**:751.

Greenberg, L., and Bloomfield, J. J. (1932). Impinger dust sampling and apparatus as used by the U.S. Public Health Rep. **47**:654.

Guyton, A. C. (1947). Measurement of respiratory volume of laboratory animals. *Am. J. Physiol.* **150**:150–170.

Gye, W. E., and Kettle, E. H. (1922). Silicosis and miners' phthisis. *Br. J. Exp. Pathol.* **3**:241.

Gye, W. E., and Purdy, W. J. (1922). The poisonous properties of colloidal silica. *Br. J. Exp. Pathol.* **3**:75–85.

Harington, J. S., and Sutton, D. A. (1983). Studies on silica shock in the rabbit. The phenomenon, the apparent specificy of colloidal silica in producing bit, and protection against shock by various agents. *Med. Lav.* **54**:11:701–714.

Heppelston, A. G., Fletcher, K., and Wyatt, I. (1974). Changes in the composition of lung lipids and the turnover of dipalmitoyl lecithin in experimental alveolar lipoproteinosis induced by inhaled quartz. *Br. J. Exp. Pathol.* **55**:384–395.

IARC (1987). *Monographs on the Evaluation of Carcinogenic Risk of Chemicals to Humans. Silica and Some Silicates.* **42**:39–143. Lyon, France.

Kettle, E. H.(1924). The demonstration by the fixation abscess of the influence of silica in determining B tuberculosis infections. Br. J. Exp. Pathol. **5**:158.

Kettle, E. H. (1934). Experimental pneumoconiosis infective silicatosis. *J. Pathol. Bacteriol.* **5**(38):201–208.

King, E. J., Moharty, G. P., Harrison, C. V., and Nagelschmidt, G. (1953). The action of different forms of pure silica on the lungs of rats. *Br. J. Ind. Med.* **10**:9–17.

Mavrogordato, A. (1918). Experiments on the effects of dust inhalation. *J. Hyg.* **17**:439–459.

Mavrogordato, A. (1922). Studies in experimental silicosis and other pneumoconioses. South African Institute for Medical Research.

Miller, J. W., Sayers, R. R., and Yant, W. P. (1934). Response of peritoneal tissue to dusts introduced as foreign bodies. *J.A.M.A.* **103**:907.

Rivers, D., Morris, T. G., and Wise, M. E. (1963). The fibrogenicity of some respirable dusts measured in mice. *Br. J. Ind. Med.* **20**:13–23.

Saffiotti, V., Degna, A. T., and Mayer, L. (1960). The histogenesis of experimental silicosis II. Cellular and tissue reactions in the histogenesis of pulmonary lesions. *Med. Lav.* **51**:518–552.

Svensson, A., Glomme, J., and Bloom, G. (1956). On the toxicity of silica particles. *Arch. Ind. Health* **14**:482–486.

Timbrell, V., Hyett, A. W., and Skidmore, J. W. (1968).A simple dispenser for generating dust clouds from standard reference samples of asbestos. *Ann. Occup. Hyg.* **11**:273–281.

Wagner, M. M. F., and Wagner, J. C. (1972). Lymphomas in the Wistar rat after intrapleural inoculation of silica. *J. Natl. Cancer Inst.* **48**:81–91.

Wagner, M. M. F., Wagner, J. C., Davies, R., and Griffiths, D. M. (1980). Silica induced malignant histiocytic lymphoma incidence linked with strain of rat and type of silica. *Br. J. Cancer* **41**:908–917.

Wright, B. M. (1950). A new dust feed mechanism. *J. Sci. Instrum.* **27**:12–15.

Wright, B. M. (1953). Gravimetric thermal precipitator. *Science* **118**:195.

Wright, B. M. (1957). Experimental studies on the relative importance of concentration and duration of exposure to dust inhalation. *Br. J. Ind. Med.* **14**:219–228.

26

Methods in Experimental Pathology of the Pleura

NAI-SAN WANG

McGill University
Montreal, Quebec, Canada

I. Introduction

A. Development and Function of the Body Cavity

About day 20 of gestation a cavity begins to develop in the human embryo by the fusing of small clefts within the mesoderm (intraembryonic coelom) (Gray and Skandalakis, 1985). The coelom, the precursor of the body cavity, is lined by a layer of cuboidal to flattened mesothelial cells that show apical bushy and elongated microvilli and cytoplasmic fibrils at this early stage of development (Hesseldahl and Larsen, 1969; King and Wilson, 1983).

As the outer shell of the embryo becomes rigid by the formation of vertebrae, ribs, and muscles, the gut and other internal organs from endoderm and mesoderm also begin to bud into and fill the body cavity successively. The mesothelial cells with their bushy microvilli that trap hyaluronic-acid-rich glycoproteins, cover the surface of all organs, which grow into the body cavity. This arrangement allows the quickly enlarging organs to expand, move, and deform readily, yielding to each other, within the limited space of the body cavity (Wang, 1985a).

The body cavity is divided into pleural, pericardial, and peritoneal cavities when the formation of the diaphragm is complete. The pericardial and the right and left pleural cavities are completely separated from each other; each provides smooth, lubricated surfaces for the constant movements of the heart and lungs.

B. Pleural Cavity

The pleural cavity is a space between the visceral pleura, which covers the entire surface of the lung including the lobar septum, and the parietal pleura, which covers the inner surface of the thoracic cage, mediastinum, and diaphragm (Hayek, 1960). The visceral pleura reflects at the hilum of the lung to become the parietal pleura; the pleura and the pleural cavity resemble a flattened and sealed plastic bag inserted between the thoracic wall and also wrapped around the lung. The pleural cavity contains 0.1–0.2 ml/kg body weight of fluid in most mammals and is free of air (Miserocchi and Agostoni, 1971; Staub et al., 1985).

C. Regional Differences in the Pleural Cavity

Although all mesothelial cells and their substructures in the pleura are basically the same, substantial regional differences exist within a thorax and between animal species (Abertine et al., 1982; Mariassay and Wheeldon, 1983; Wang, 1974). These differences presumably reflect the substructures or adaptive changes of mesothelial cells to meet functional needs. Mesothelial cells over the rigid structures, such as ribs, are flattened with few microvilli and cellular organelles while those over loose structures, such as subcostal and lower mediastinal vascular and fatty tissues, are cuboidal with abundant microvilli and cytoplasmic organelles. Stimulated or reactive mesothelial cells from any pleural injury show the same abundance of microvilli and organelles, regardless of their location (Stoebner et al., 1970a; Wang, 1974, 1975).

The mesothelial cells outnumber other mesenchymal and endothelial cells in the pleura and could be the major source of connective tissue in the pleura (Rossi et al., 1981; Rennard et al., 1984, 1985).

D. Thickness and Blood Supply of the Pleura

The thickness and structure of the visceral pleura vary substantially between various species of animals (Bernaudin and Fleury, 1985). Those of the parietal pleura are, however, less well documented (Staub et al., 1985)

The visceral pleura has been classified into two groups: thin (mouse, cat, dog, and rabbit) and thick (cow, pig, and sheep) (McLaughlin et al., 1961). Staub et al. (1985) add whale as a mammal with a thick visceral pleura and human and horse as the third group with a pleura of intermediate thickness. The regional difference in the thickness of pleura is, however, substantial; in sheep it ranges from 85 μm (costal and diaphragmatic surfaces) to 25 μm (cephalad lobe). In dogs, some areas of pleura can be 25 μm thick while the average is 13 μm. The difference in the thickness of the pleura is due mainly to the amount of submesothelial connective tissue.

In humans the blood supply to the parietal pleura is all from the systemic circulation, but opinions are divided as to the blood supply of the visceral pleura. All agree that the visceral pleura of the mediastinal, some interlobar, and diaphragmatic surfaces are supplied by bronchial arteries (Verloop, 1948; Cudkowicz, 1979). Some believe that the rest of the visceral pleura are supplied by pulmonary artery (Hayek, 1960). Others (McLaughlin et al., 1966; Miller, 1947; Nagaishi, 1972; Staub et al., 1985) believe they are also supplied by the bronchial arteries, which reach the pleural surface probably via the lobar septum.

The thickness of the pleura is related to the type of blood supply. In sheep the bronchial arterial supply is most clear in the thicker part of the pleura (Staub et al., 1985). Bronchial arteries and lobar septa are not well developed in animals with thin pleura. Communications between pulmonary and bronchial vessels probably also exist (Verloop, 1948). In diseased conditions, systemic arteries from the chest wall frequently communicate with the vessels in the visceral pleura.

E. Fluid and Particle Exchange Through the Pleural Surface

Water, electrolytes, and particles smaller than 4 nm permeate freely through the mesothelial cell layer between cells (Casley-Smith, 1967; Cotran and Karnovsky, 1968). The intercellular junctions between mesothelial cells are varied in type and include tight (zonal occludens), intermediate, gap (nexus), and well-developed desmosome junctions (Legrand et al., 1971; Odor, 1954; Wang, 1983), although some of them are probably quite labile, as in the venous endothelium (Simionescu and Simionescu, 1977; Wang, 1985b). The rate of fluid permeability through the pleura is relatively slow with or without an intact layer of mesothelium (Leathes and Starling, 1895), and appears to be dependent on the endothelium (Kinasewitz et al., 1983).

Particles larger than 4 nm introduced in the peritoneal or pleural cavity, such as ferritin (11 nm) and colloidal gold, thorotrast, carbon (20–50 nm), and others, appear in the pinocytic vesicles and are transferred to the basal side of the mesothelial cell through the basal lamina (Baradi and Hope, 1964; Fedorko et al., 1971; Fukata, 1963; Odor, 1956; Wang, 1983), although this process appears to be slow and requires energy (Fedorko et al., 1971). Mesothelial cells can engulf particles as large as 1 μm (Wang, 1983) but this is probably the "forced phagocytosis" also observed in squamous and alveolar epithelial cells (Berry et al., 1978; Wolff and Konrad, 1972).

F. Fluid and Particle Exchange Through Pleurolymphatic Communication

Pleurolymphatic communication has been found only on some portions of the parietal pleura but not on the visceral pleura. At least four related structures have been described and are discussed here.

Stomas

Communications between the pleural cavity and the lymphatic or vascular channels were suspected as early as 1863 by von Recklinghausen. On scanning electron microscopy, pores or stomas measuring 2–12 μm have been found on the pleural surface of the lower mediastinum, the infracostal regions of the lower thorax, and the diaphragmatic surface, in rabbits, mice, and sheep (Wang, 1975; Albertine et al., 1982; Mariassay and Wheeldon, 1983). The stomas are particularly common on the peritoneal surface of the diaphragm (Leak and Rahil, 1978). Their number, size, and distribution in other species are not well documented.

Membrana Cribriformis

In the regions where stomas are found, collagen bundles form a netlike arrangement in the pleura. These netlike collagen bundles are covered by mesothelial cells on the pleural surface and lymphatic endothelial cells on the opposite surface to form the roof of a dilated lymphatic space called a lacuna (Kihara, 1950; Wang, 1975). The lacuna with stomas facing the pleural cavity and one-way lymphatic valves on the opposite side functions like a thumb pump, synchronizing with the respiratory movement to suck fluid and particles from the pleural cavity into the lacuna at inspiration and propel them into the lymphatic duct at expiration (Courtice and Simmons, 1949, 1954). The closely apposed mesothelial and lymphatic endothelial cells on the roof of the lacuna also transfer small particles readily through their cytoplasm (Wang, 1975).

Milky Spots (Kampmeier's Foci)

Small whitish flakes in the peritoneal cavity visible by the naked eye have been called Milchflecke by von Recklinghausen (1863) and tâches laiteuses by Ranvier (1873) (cited by Kampmeier, 1928). Similar small white spots have been found in the dorsal caudal portion of the mediastinum and occasionally in other parts of the pleural cavity in rats and humans (Cooray, 1949; Kampmeier, 1928). In dogs they have been found in the pleural tuft in the lower thoracic cavity (Lang, 1962; Lang and Liebich, 1976). The mesothelial cells over the Kampmeier's foci are cuboidal, loosely attached to each other, and appear activated with abundant cytoplasmic organelles. Beneath the mesothelial cells, lymphocytes, plasma cells, histiocytes, and pluripotential cells surround central lymphatic and vascular channels and probably play a focal immune-related function similar to that of the tonsils in the oropharynx (Kampmeier, 1928; Lang and Liebich, 1976; Wang, 1985a).

Mediastinal Crevices

Large cleftlike openings measuring up to 20–30 μm have been described in the lower mediastinum of mice (Kanazawa, 1985). Blood cells and macrophages can readily cross these clefts to reach lymphatic or vascular channels; presumably this is the way hemothorax is cleared.

G. Pathological Characteristics of the Pleural Cavity

The pleural cavity is better known for its disorders than for its normal functions because it is vulnerable to and readily shows alterations in many cardiopulmonary and other diseases. Although flattened mesothelial cells were recognized as early as 1827 by Bichat, the importance of this layer of cells was not realized at first. Instead, the pleural cavity was considered a peculiar interstitial space or a part of the lymphatic system (von Recklinghausen, 1863; Dybkowsky, 1866). Attention was, therefore, centered on the route and mechanisms of fluid and particles arriving, leaving, and accumulating in the peritoneal and pleural cavities. Most earlier studies were done in the peritoneal cavity. Some questions concerning the pleural cavity remain incompletely answered today (Staub et al., 1985).

Attention to the pathological appearance of the pleural cavity has since evolved through a few phases; from pleurisy, the inflammation and accumulation of fluid, blood, air and lymphs, to suppuration and fibrous obliteration; from the immune related or unexplained pleuritis to pneumoconiosis, especially asbestosis; and, finally, the development of mesothelioma and pleural responses in metastatic carcinoma.

Recent studies emphasize the cellular aspects of the pleural disease, especially the interactions between the mesothelial and reactive inflammatory and mesenchymal cells including mediators and other cellular products they produce and release. Genetic aspects of the pleural diseases have been only marginally explored.

II. Methods

The difficulty in studying a thin, fragile, and flattened cell layer over varied substructures using routine light microscopic technique is well recognized. Many methods to improve the preservation and ways of assessing the status and alterations of the mesothelial cells and their surrounding structures by light microscopy have been evaluated (Whitaker et al., 1980a).

A. Fixation and Processing of Pleura for Regular, Chemical, Enzymatic, and Immunochemical Light and Electron Microscopy

Fixatives

No specific fixatives have been favored for the mesothelial cell study. For most light microscopic (LM) studies, 4% formaldehyde in phosphate or other appropriately buffered solutions at neutral pH are adequate. For transmission and scanning electron microscopy (TEM, SEM), 3% glutaraldehyde in 0.1 N sodium cacodylate or similar fixatives appears to be satisfactory. To preserve mucoproteins containing hyaluronic acid, fixative in 100% alcohol or other nonaquous solvents are used. Depending on the investigator's other aims, such as immunohistochemistry, fixatives specifically designed to preserve the desired antigens or quick freezing followed by freeze-substitution or freeze-drying can be used.

Traditional Section Method for LM

The pleura with its surface mesothelial cells can be sampled fresh in strips, as in skin. As an alternative, a thin surface slice can be shaved off from the lung or the chest wall and fixed with a portion of the subpleural structure attached to maintain the original relationship before further sampling. Since normal mesothelial cells are damaged within minutes in air or by the slightest abrasion, sampling and fixation of pleura should be done promptly and carefully after the thoracic cavity is opened. The pleura can also be fixed in situ.

In Situ Fixation

The animals are anesthetized with pentobarbital (50 mg/kg) administered intraperitoneally. The trachea is cannulated with a polyethylene tube attached to a regular injection needle. A midline incision is then made in the upper abdomen. By pulling the liver and stomach caudally, the diaphragm is exposed and a small sharp incision is made on the membranous portion to collapse the lung without damaging it. The fixative is injected immediately into the pleural cavity through the small incision and then through the trachea to expand the lung to the desired pressure, usually at 20 cmH$_2$O.

The midline incision is then extended all the way up to the neck and the ventral thorax is freed of the skin, thoracic muscles, and upper limbs. The cervical spine is then cut at the level of the thoracic inlet. The thorax with trachea and the attached tube are lifted up and detached from the skin of the back, and finally completely freed from the rest of the body when the upper lumbar vertebrae are cut.

The separated thorax is immersed in the same fixative with the lung maintained at the desired pressure for 12–48 hr. The areas of interest on the visceral and parietal pleurae can be sampled systematically for LM, SEM, and TEM thereafter; the mesothelial surface should be kept wet all the time (Wang, 1975; Albertine, et al., 1982).

Rapid Freezing and Freeze Substitution

The animal is anesthesized as described above. While the animal is breathing spontaneously, the skin and muscle of the anterior and lateral thorax, including most of the intercostal muscles of the part of interest, are carefully removed by blunt dissection. The animal is then placed in a plastic box and liquid nitrogen is poured continuously over the thorax until the portion of interest is frozen solid (or when the whole thorax, in smaller animals, is completely immersed in liquid nitrogen). In larger animals, such as dogs or rabbits, the chest wall and the lung underneath can be sampled with a motor-driven trocar. A tissue cord 1–1.5 cm in diameter and up to 2 cm long can be obtained readily. In smaller animals the thorax can be processed as a whole or, after being sawed into coronal or sagittal slices, by the freeze substitution method for LM and TEM. Frozen sections 5 μm thick can also be prepared from the tissue cord or slice for histochemical and immunohistochemical study.

En Face Preparation of Mesothelial Cells

Clean regular histology slides are dipped into molten (45 °C) 3 % gelatin or 1 % agar for 5 sec and lifted and dried in a vertical position (Cheng and Berry, 1972; Watters and Buck, 1972; Whitaker et al., 1980a, b). A single layer of mesothelial cells can be detached by firmly pressing a coated glass slide onto a freshly exposed pleural surface for 10–20 seconds and then lifting the slide up gently (Hauchen technique).

Cell Deposit Method

A monolayer of mesothelial and other types of cells in the pleural effusion or from other forms of suspension can be obtained by the cytospin method or by regular centrifugation. One drop of the cells is then sprayed onto a gelatin- or agar-coated glass slide prepared as above. For SEM, 0.1 % poly-l-lysine-coated slides (Mazia et al., 1975) or nucleopore filter (0.2 μm) can be used.

Scanning and Transmission Electron Microscopy

If the pleura is sampled fresh, it should be flattened carefully, pinned around on a cork plate, and immersed face down in the fixative. Tissues for TEM are preferably processed within a few hours after fixation, but tissues for SEM, especially that prepared by the in situ method, are preferably kept in the fixative for more than 48 hr to harden the structure. The parietal pleura is usually processed with the rib attached to maintain, and also for identification of, the original relationship.

Processing procedures for TEM and SEM of the pleura are the same as for other tissues (Wang, 1974, 1975). The step of osmium tetroxide fixation is, however, not absolutely necessary for SEM. We use graded acetone for dehydration and embedded tissues in epoxy resin for TEM. Tissues for SEM are critical-point dried directly from acetone using CO_2. Areas of interest already documented by SEM can be cut out under the dissecting microscope and reprocessed for TEM (Wang, 1980).

B. Studying How Fluid or Particulate Matter Reaches, Accumulates, and Leaves the Pleural Cavity

Since the pleural cavity is sealed, fluid and particulate matter can only reach it indirectly through the lung, blood and lymphatic vessels, or adjacent soft tissues. The materials introduced by each route have been found by light and electron microscopy to reach and accumulate in the pleural cavity. The more commonly used method, however, is to introduce the tracing material directly into the pleural cavity and evaluate the time sequence and the route of clearance from the pleural cavity.

Intrapleural Injection

The easiest way to introduce tracer materials into the pleural cavity is through the intercostal space. The animal is anesthesized and placed in the lateral position. A skin incision is made parallel to the ribs between the anterior and posterior axillary line at the fifth or sixth intercostal space. The intercostal muscle is divided bluntly, a few fibers at a time, until the parietal pleura is visible. With a strong direct light, the moving lung under the parietal pleura can be seen clearly. The tracer is injected through a 22 gauge or smaller needle inserted almost horizontally between the visceral and parietal pleurae. With the needle hole upwards, each squirting of the tracer can be observed to disperse in all directions instantly.

A hole can also be drilled through the rib and a polyethelene tube inserted into the pleural cavity for the injection. The hole can be sealed with bone wax or a plastic plug after the injection. Intrapleural injection can also be done through the membranous portion of the diaphragm under direct vision following a laparotomy.

Inhalation Route

The methods for exposing different species of animals to aerosol inhalation have been well established and will not be detailed here (Timbrell et al., 1970). Particles can also be instilled directly into the trachea in a lightly anesthesized animal through a polyethylene or metal tube inserted orally into the trachea. The trachea can also be punctured with a needle through a small skin incision.

Intravenous and Subcutaneous Routes

Intravenous or subcutaneous introduction of foreign particles can be done readily in larger animals. In mice injection is usually carried out through the tail vein with a 30 gauge needle. Some of the intravenously injected dyes or particles can be recovered from the pleural cavity (Grober, 1901). Subcutaneously injected asbestos fibers also circulate through the blood stream (Kanazawa et al., 1970) and are trapped in the milky spots following peritoneal stimulation (Kanazawa et al., 1979).

Tracers

Many different sized tracers have been used to study the clearance from the serous cavity, mostly from the peritoneal cavity. These include autologous plasma labeled with Evans blue (Courtice and Simmonds, 1954), iron–dextran complex (Fukata, 1963), autologous and heterologous red blood cells (Courtice and Morris, 1953; Bettendorf, 1978; Wang, 1975), horseradish peroxidase (Kluge and Hovig, 1967; Cotran and Karnovsky, 1968), colloidal gold (Voth and Kohlhardt, 1962), saccharated iron oxide and barium sulfate (Cotran and Majno, 1967), lead and calcium salts (Cotran and Nicca, 1968), thorotrast and ferritin (Fedorko et al., 1971), carbon particles (Obata, 1978; Leak and Rahil, 1978; Wang, 1983), and various sizes of polystyrene and latex beads (Allen and Weatherford, 1959; Wang, 1983).

Collection of the Pleural Fluid

The content of the pleural cavity can be collected in anesthesized animal following exsanguination. Exsanguination is carried out first by exposing the abdominal aorta. In large animals the aorta is cannulated for exsanguination. In small animals both the aorta and vena cava are severed simultaneously and the blood is absorbed with gauze. Following exsanguination the thoracic cavity is opened by cutting the lateral border of the cartilaginous portion of the ribs along the sternum, making sure not to contaminate the pleural cavity with blood and tissue fluid. The pleural fluid is removed from the recesses carefully with a polyethylene tube attached to a syringe. Raising the head and upper chest facilitates the accumulation of fluid in the recesses. The collected fluid can be studied by the cell deposit as well as by biochemical and other methods. If the pleura is also intended for morphological studies, the fluid should be collected promptly and the cavity filled with a fixative immediately.

Route of Pleural Clearance

Fluids and particulate matter injected into the pleural cavity can be detected by gross and light microscopic examination to accumulate in the mediastinal lymphatic channels and nodes. Although technically difficult, quantitative estimation of the drainage can be done by ligating the thoracic duct at the diaphragmatic level and then cannulating the right lymphatic and thoracic ducts to collect the lymph. Thorough sampling of the mediastinal nodes estimates the amount of trapped particulate matters (Courtice and Simmonds, 1954).

C. Method of Isolating and Growing Pleural Mesothelial Cells

Neoplastic and nonneoplastic pleural mesothelial cells can be isolated from the pleural effusion for short-term culture (Maximow, 1927; Castor and Naylor, 1969). The fluid is centrifuged and deposited on a cover glass or in culture media. The population of the cells is mixed. Mesothelial cells can also be detached enzymatically from the visceral (Aronson and Cristofalo, 1981) or scraped off gently from the parietal pleura (Jaurand et al., 1981).

The method for obtaining a long-term mesothelial cell culture described by Jaurand et al. (1981) is as follows. Male Sprague-Dawley rats, 6–8 weeks old, are anesthesized and bled from the abdominal aorta. The thoracic cages are exposed by midline incision and the left and right cages are excised separately with heavy-duty scissors under sterile conditions. The excised thoracic cage is washed with the culture medium twice and mesothelial cells are obtained by scaping firmly between the ribs with a Demarres scarifier.

Parietal pleural mesothelial cells are initially placed in a multiwell dish. The culture medium used is NCTC 109 (Eutroph, Eurobio, France) supplemented with 10% fetal bovine serum, penicillin (100 U/ml) streptomycin (50 μg/ml), 1% glutamine, and sodium bicarbonate to adjust the pH to 7.3. Cultures are incubated in 5% CO_2 and 95% humidified air at 37 °C. The culture medium is changed at 48 hr and then weekly. The cultured cells can be processed in situ or after detachment with 0.25% trypsin, centrifuged, and processed as a cell block for SEM and TEM (Jaurand et al., 1985; Wang et al., 1987). They can also be tested for their phagocytic properties, cellular, enzyme and histochemistry, metabolism of benzo(3.4)pyrene, release of collagen, inflammatory mediators, and others.

D. Cytogenetic Studies

Cytogenetic studies of neoplastic and nonneoplastic cells can be done from the pleural effusion (Ayraud and Kermarec, 1968; Korsgaard, 1979; Mark, 1978), cultured mesothelial cell line (Jaurand et al., 1986), or solid tumors (Gibas et al., 1986).

Routine cytogenetic methods are followed. Colcemid (0.01 μg) is added to the cells 48–72 hr after plating for 1–2 hr. The cells can be grown in a plastic flask or on a cover glass. The cells are processed in situ or detached with 0.25% trypsin, swollen with 75 mM KCl, and fixed with 3:1 methanolacetic acid three times. The spread out chromosomes are usually G-banded (Jaurand et al., 1986).

Chromosomes prepared as above for light microscopy can be processed further for SEM. Following G-banding, the chromosomes are fixed with 3% cacodylate-buffered glutaraldehyde solution, postfixed in 1% osmium tetroxide, dehydrated with graded acetone, and critical-point dried using CO_2. The dried chromosomes are coated with a 20 nm layer of gold and examined by SEM (Wang et al., 1987).

E. Mineral Particle Analysis

Light Microscopy

The presence of mineral particles, fibers, and asbestos bodies in the pleura can be estimated by routine light microscopy on paraffin-embedded sections, using polarized light or with Prussian blue stain of the iron-containing coat of the asbestos body. Sub-light-microscopic particles, however, need to be assessed differently.

In Situ Particle Analysis by TEM and SEM

Mineral particles in tissue processed routinely for TEM or SEM can be identified by their morphological characteristics, selected area electron diffraction (SAED), and x-ray energy dispersive spectrometry (XEDS) (Abraham and Burnett, 1983; Vallyathan and Green, 1985).

The common and practical method is to use paraffin-embedded tissue prepared for routine LM. Sections 5–10 μm thick are placed directly on a carbon planchet. Following three changes of xylene to remove the paraffin, the tissues are washed with three changes of 100% acetone and air dried. The air-dried section is examined without coating in by SEM (Abraham and Burnett, 1983; Yao et al., 1984).

Particle Analysis Following Tissue Digestion

Ten to fifteen grams of tissue are minced and digested in 200 ml laboratory grade sodium hypochlorite for 2–3 days (Smith and Naylor, 1972; Sebastien et al., 1980). Most of the clear supernatant is then removed and the remaining solution with particles is mixed with 14 ml chloroform and 20 ml of 70% alcohol and agitated vigorously. Following centrifugation at 600 rpm for 10 min, the top two layers are discarded and the particles in the chloroform layer are placed on a carbon planchet or on nucleopore filter (pore size 0.2 μm) for SEM and XEDS (Kimizuka and Wang, 1985). The same material can also be deposited on a Formvar- and carbon-coated grid for TEM analysis. More elaborate procedures for the TEM study can also be used (Sebastien et al., 1980).

F. Studying Mesothelial Cell Damage, Repair, and Pleural Adhesion and Fibrosis

The mesothelial cells probably prevent pleural adhesion and fibrosis mainly through their fibrinolytic activity. The rapidity and completeness of mesothelial cell repair on the damaged pleural surface is therefore crucial.

Mesothelial cells are delicate and fragile and can be damaged by minimal wetting or drying. Intrapleural injection of air (Thorsrud, 1965), water (Ivanova and Puzyrev, 1977), blood (Ryan et al., 1971), phytohemagglutinin (PHA) (Mohr et al., 1970), endotoxin (Mohr et al., 1971), bovine serum albumin (BSA) (Mohr et al., 1972) ischemia (Ellis, 1962), foreign bodies, and many other agents has been used. The morphological changes are documented at regular intervals by LM, TEM, and SEM. The enhanced proliferative activities are documented using tritiated thymidine uptake and mitotic counts on the en face preparation or DNA measurements (Bertalanffy and Lau, 1962; Bryks and Bertalanffy, 1971).

The disappearance, restoration, or exaggeration of fibrinolytic activities of mesothelial cells during initial damage and subsequent regeneration is documented by the en face or monolayer preparations (Raftery, 1979; Ryan et al., 1973).

The predominant mechanisms leading to chronic pleural effusion and fibrosis involve persistent immune complexes and cellular immune responses. Intrapleural injection of carrageenan, Freud's adjuvant, purified protein derivatives (PPD), tuberculin, or insoluble agents such as asbestos has been used to study these mechanisms (Bignon and Gee, 1985; Whitaker et al., 1982a). Although morphological evaluation of cellular changes and deposition of immune complexes, as well as other amorphous and fibrous components, has been done on the tissue preparation, the documentation of cellular and humoral agents in the pleural fluid also appears to be very useful.

G. Studying Mesotheliomas and Metastatic Tumors of the Pleural Cavity

Exposure to asbestos fibers has been closely linked to the development of mesothelioma in humans (Wagner et al., 1960). Laboratory animals are, therefore, exposed to asbestos fibers through a variety of routes to study the mechanism of asbestos carcinogenesis. One intrapleural or intraperitoneal injection of a relatively large dose of asbestos fibers is usually used (Wagner, 1969). The animals are sacrificed either in intervals of 3–6 months through the lifespan or when tumors become clinically apparent.

Stanton and Wrench (1972) were the first to show that mesothelioma can be induced by glass fibers with physical properties similar to asbestos fibers. Many other types of mineral fibers have since been found to induce mesothelioma following an intrapleural or intraperitoneal injection (Monchaux al., 1985). As summarized by Monchaux et al. (1985), the study of asbestos and mineral carcinogenesis is directed toward determining a dose–effect relationship; the relationship between the physical and chemical properties of the agent and carcinogenesis; the host factors, including age, sex, and others; and synergistic or antagonistic actions of all factors involved.

Contrary to the intrapleural and intraperitoneal injection, inhalation of asbestos fibers needs to be repetitive for a long time and results in a low yield of mesothelioma in laboratory animals (Gross and de Treville, 1967). Retention and clearance of inhaled particles are mostly performed in the study of pulmonary fibrosis and lung cancers (Brody et al., 1981; Gross and de Treville, 1967). Mesotheliomas can also be induced in vitro by exposing asbestos or other carcinogens to cultured mesothelial cells. The transformed cells can be implanted in nude mice to show the cells' neoplastic nature.

Although metastatic tumors to the pleural cavity are very common, experimental designs to study their seeding, spread, and effects, particularly for the pleura, are relatively rare (Orr et al., 1986).

III. Interpretation of the Results

A. General Considerations

Processing Methods and Results

Mesothelial cells prepared by the traditional method for light microscopy are flattened and thin. Evaluation of the cytoplasmic contents or histochemical and immunohistochemical products is frequently difficult (Whitaker et al., 1980a). Mesothelial cells are also very fragile in the air (Ryan et al., 1971, 1973) and any delay in sampling or undue handling of the fresh specimen tends to damage the pleural surface and the results are frequently difficult to interpret.

For ultrastructural studies, in situ fixation is probably the best fixation procedure to follow especially when different parts of the pleural surface need to be sampled (Wang, 1975; Albertine et al., 1982). The thin mesothelial cell with the elongated bushy microvilli (Fig. 1), well-developed intercellular junctions, basal lamina, cytoplasmic fibrils, and organelles is well-characterized by TEM (Andrews and Porter, 1973; Cotran and Karnovsky, 1968; Fukata, 1963; Obata, 1978; Wang, 1983). The abundant surface microvilli are best appreciated on SEM (Fig. 2). Another advantage of the in situ preparation is that the lung and the pleura are maintained at a desired inflating condition following 48 hr of constant pressure fixation. Although shrinkage does occur during subsequent processing, the general architecture of pleura including the density of microvilli is maintained and is useful for morphometric analysis with appropriate correction factors.

Studies using SEM are also useful to evaluate surface healing of the pleura (Watters and Buck, 1972). Since evaluation of cell types and intracellular organelles is relatively difficult by SEM alone, a combined SEM and TEM study

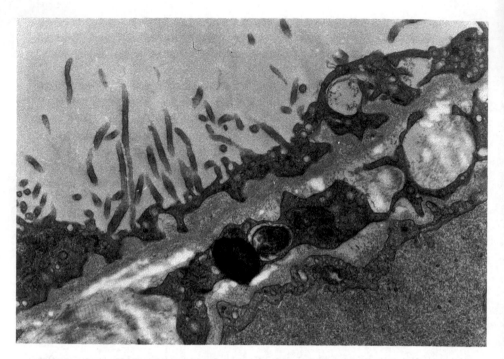

Figure 1 Elongated bushlike microvilli on the thin cytoplasm of a mesothelial cell are best preserved by the in situ fixation method. Cytoplasmic details of the mesothelial cell on the pleural surface, the fibroblast with dark-stained lipid droplets in the pleural wall, and the thin cytoplasm of the type I alveolar lining cell (bottom side) are usually well preserved by this method (rabbit, TEM × 21,000).

is necessary in most cases (Wang, 1974). Tissues prepared properly from the outset for SEM can be reprocessed for TEM with good results (Wang, 1980).

For light microscopy the ''en face'' or imprint technique obtains more than several hundred mesothelial cells in a single sheet and is ideal for evaluating the cell size, multinucleation, and mitotic incidence, and also for enzyme and histochemical studies (Whitaker et al., 1980a, b). The cells are, however, upside down showing the basal surface by SEM. The free cells deposited on the coated slide have the same advantage of the ''en face'' preparation and also show the original free cell surface, but the disadvantage is that the cell population is usually mixed. For enzyme and immunohistochemistry, routine phosphate-buffered 4% formaldehyde solution is usually adequate but the fixative and buffered solution can be modified for specific antigens (Singh et al., 1979).

The rapid freezing and freeze-substitution method has been used to estimate the thickness of the pleural fluid between the parietal and visceral pleura. Since

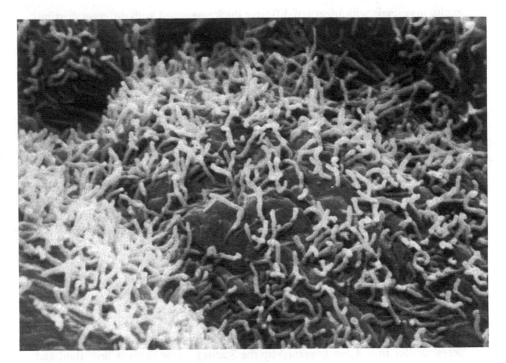

Figure 2 The abundant but unevenly distributed microvilli on the mesothelial cell is best appreciated by SEM (rabbit, SEM × 6,400).

complete removal of intercostal muscle is difficult and the depth of rapid freezing from the surface is quite limited (Weibel et al., 1982), preservation of fine structures of mesothelial cells is not ideal and estimation of the pleura fluid thickness is also not entirely certain. Whether this more laborious method can obtain a better result in enzyme or immunohistochemistry is unclear. For electron immunohistochemistry, rapid freezing followed by freeze drying and low-temperature embedding in low-viscosity resins appear promising (Chiovetti et al., 1987).

Species Differences

The implications of thin (mouse, cat, dog, rabbit), intermediate (human, horse), and thick (sheep, cow, pig, whale) pleurae for their function is not certain (Staub et al., 1985). In general, the thin pleura is supplied by the pulmonary circulation and the thick pleura by systemic circulation. The blood supply of the intermediate thickness of pleura is not completely agreed upon. Hemodynamics of the pleural cavity in animals with pleura of different thicknesses and with different distribution and sites of pleurolymphatic communications should be different (Lang, 1962; Staub et al., 1985): comparison of the results of experimental studies between different species should therefore take this into consideration.

Regional Differences

The concentration of microvilli, number and activity of mesothelial cells, and thickness of the pleura are influenced by their substructures and the degree of lung inflation. Additional factors, such as the more movable portion of the lung or those with specific functional modifications, also cause the regional difference (Albertine et al., 1982; Mariassay and Wheeldon, 1983; Wang, 1975). The pleura also becomes thicker with age (Dodson et al., 1983). The details of these regional differences have not been documented in most experimental animals except rabbits, mice, and sheep. To evaluate the pleural changes by a random biopsy can therefore by misleading.

B. Transport and Clearance of Fluid and Particulate Matter to and from the Pleural Cavity

Fluid and Small Particles

The mesothelial and endothelial layers are readily permeated by fluid and particles smaller than 4 nm governed by the Starling's law. This is how fluid and small-sized proteins normally reach the pleural cavity, regardless of the type of blood supply of the pleura (Staub et al., 1985). Some fluid and small particles are reabsorbed back into the low-pressure vascular system, probably mainly on the parietal pleura. Particles and proteins larger than 4 nm are, however, not removed efficiently either by the visceral or parietal pleurae (Fedorko et al., 1971; Wang, 1983). The only sensible passage through which larger particles might leave the pleural cavity, therefore, is pleurolymphatic communications, which have been suspected for a long time by LM (von Recklinghausen, 1863) and physiological (Courtice and Simmons, 1949) studies (see below).

In pulmonary edema induced by hyperoxia, [^{125}I]albumin, antihorseradish peroxidase IgG, and autologous [^{51}CR] red blood cells are found to be transferred from pulmonary interstitium through the visceral pleura and reabsorbed by the lymphatics of the parietal costal and diaphragmatic pleurae (Bernaudin et al., 1986). Since excessive vascular leakage and overflooding occur in the lung in this model, the expected direction of flow of the exudated protein and fluid is through the visceral pleura to the relatively negative pressured pleural cavity. This direction of flow may therefore not represent the physiological situation of protein and fluid clearance, although it certainly demonstrates a pathological pathway by which fluid and particulate accumulate in the pleural cavity are drained through the parietal lymphatics.

Pleurolymphatic Communications: Key Site of
Pleural Pathological Changes

Although pleurolymphatic communications have been found in similar locations in most species studied, their size and number differ between reports (Wang, 1975; Albertine et al., 1982; Mariassay et al., 1983; Kanazawa, 1985). A review of the findings suggests that the different results are not necessarily contradictory.

The pleural cavity is normally sterile and completely isolated (secluded) from the environment; few cells and a small amount of protein and fluid enter and leave the pleural cavity at a time. We have found very few pleurolymphatic communications (stomas) measuring 2–12 μm (Fig. 3) in healthy young adult rabbits (Wang, 1975). Stomas are found more readily on the peritoneal side (Leak and Rahil, 1978), which probably reflects the influence of the bowel contents, and also the presence of external communication in females, resulting in an increased necessity for clearance. We have suspected that stomas may be obliterated by fibrin at the acute stage of pleural inflammation to retain extravasated fluid and proteins in the pleural cavity. Later stomas, however, may increase both in number and size following fibrinolysis and lysosomal damage of the apposed mesothelial and lymphatic endothelial cells and collagen bundles in the lamina cribriformis to enhance pleural clearance (Wang, 1975). Contraction of mesothelial cells by inflammatory mediators, for instance, may also separate the cellular junctions to widen the stoma in the peritoneal cavity (Leak, 1983; Tsilibury and Wissig, 1983). The crevices measuring up to 10×30 μm (Fig. 4 and 5). described by Kanazawa (1985) may therefore represent an old and advanced destructive change of the lamina cribriformis. The relative rarity of crevices (only in old animals) supports this notion. The age and previous pulmonary and pleural insults to the animal should therefore be carefully evaluated before comparing the changes in the pleural cavity.

Milky spots (Kampmeier's foci), which are another form of pleurolymphatic communications, are also quite rare in healthy young adult animals but can be readily located in others, presumably due to chronic pleural irritations. They are inconstant changes (Seifert, 1921) and have varied and overlapping appearances (Kanazawa et al., 1979), which suggests that the reactive lymphoreticular cells around capillaries and lymphatic channels are constantly being modified.

Asbestos fibers, mineral particles, and other foreign substances can be transported to any area of the body, regardless of their primary site of introduction (Brody et al., 1981; Holt, 1983; Lee et al., 1981). Intravenously or subcutaneously injected dyes or mineral particles, however, accumulate in milky spots, especially if the peritoneum is irritated before the injection (Kanazawa et al., 1979). Intratracheally instilled asbestos fibers, especially chrysotile, however, alter the shape as well as the mitotic activities of mesothelial cells without reaching the pleura themselves. This suggests that inhaled agents may also generate indirect

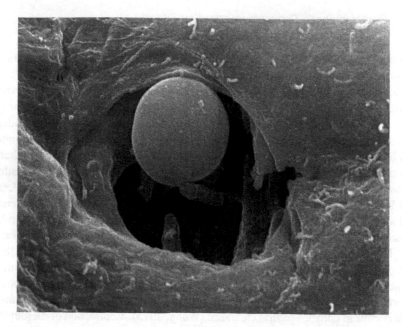

Figure 3 A pleurolymphatic communication at the infracostal region of the lower thorax. This stoma is approximately 12 μm in diameter and probably is at the upper limit for a normal rabbit. The round structure at the stoma is a red blood cell and the smaller rod-shaped structures beneath it are bacteria. Most proteins and particles in the pleural cavity are cleared through the stoma (rabbit, SEM × 4,800).

effects to alter the pleura (Bryks and Bertalauffy, 1971; Dodson and Ford, 1985): attraction or penetration of the inhaled agents into the pleura may follow (Holt, 1983; Sebastien et al., 1980). The local status of the pleura, especially the pleurolymphatic communications, therefore, is important in the attraction and accumulation of the circulating agents or agents introduced intrapulmonarily.

Irritating particles such as graphite or bacteria injected into the pleural cavity do not reach the mediastinal node as readily as nonirritant particles, such as homologous red blood cells, but are mainly retained at the sites of the pleurolymphatic communication (Cooray, 1949). This tendency to accumulate undesirable agents at certain locations blocks the rate of pleural clearance and is closely related to the development of pleural effusion, fibrosis, mesothelioma, and seeding of metastatic cancers. The details of these mechanisms require further documentation.

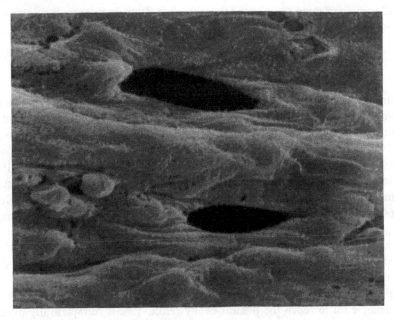

Figure 4 Crevices of approximately 30 × 10 μm are found in the lower mediastinum of an aged mouse. These are thought to be enlarged stomas following the rupture of surface lining cells and the supporting collagen bundles in the lamina cribriformis. One crevice in Figure 4 (× 1200) is enlarged in Figure 5 (× 2400; mouse, SEM).

Clearance of the Pleural Content and Influence of Respiratory Movements

The clearance of fluid and particles from the pleural cavity is much slower in anesthetized, spontaneously breathing animals than in awake stimulated animals (Courtice and Simmonds, 1949, 1954). Dybkowsky (1866) postulated that pleurolymphatic communicatins (stomas) and subpleural lymphatics (lacunae) are stretched to suck fluid and particles into the lacuna at inspiration. They are contracted or compressed at expiration to propel the contents of the lacuna through the one-way valve that opens towards the lymphatic channels. Miserocchi and colleagues (1981) have found that the end-expiratory pleural liquid pressure is always lower in the mediastinal than in the costal regions at the same thoracic level. This pressure difference is further exaggerated by increased ventilation and heart beats, resulting in a greater negative tidal swing (Miserocchi, 1985).

The lymphatic channels from the pleural cavity lead to mediastinal nodes and also mainly to the right lymphatic duct (Courtice and Simmonds, 1949). A small portion of them drain to the thoracic duct. Both ducts enter the veins (i.e., the systemic circulation).

C. Pathogenesis of the Pleural Plaques and Fibrosis

Pleural fibrosis following extensive damage of the pleural cavity, such as empyema, is simple scarring. The mechanisms of the insidious development of the pleural plaque and of focal adhesion or loculation of the pleural content are less well defined. The pleural plaques are most commonly found in patients with asbestos exposure and formation of asbestos bodies in the lung (Hourihane, 1964; Roberts and Ferrans, 1972). The LM studies, however, show that the plaque is mainly formed by dense collagen fibers with very few asbestos fibers. Kiviluoto (1960) suggested that the pleural plaque is formed by mechanical scratching of the parietal pleura by asbestos fibers that have partly penetrated and stuck out on the visceral pleura. Taskinen and colleagues (1973) proposed that small inhaled fibers may be drained to hilar nodes first and then spread retrogradely along and through the lymphatic channels into the intercostal space. As discussed above, inhaled fibers enter the bloodstream (Lee et al., 1981) and also penetrate the visceral pleura to reach the pleural cavity (Holt, 1983). Intrapleurally injected irritants also accumulate at the sites of pleurolymphatic communication (Cooray, 1949). The third and most likely explanation is, therefore, the collection of asbestos fibers at the pleurolymphatic communications either through the vascular channels or from the pleural cavity. Short fibers undetected by LM are particularly prone to this (Hillerdal, 1980; Sebastien et al., 1980). Mesothelial cells and fibroblasts are probably both stimulated by the fibers and start to synthesize collagen (Harvey and Amlot, 1983; Rennard et al., 1984; Rossi et al., 1981). The exact mechanisms of the mesothelial cell and fibroblast stimulation are not completely clear but immune-related reactions are suspected (Hillerdal, 1980).

Physicochemical properties of inhaled fibers are also important in pleural fibrosis and formation of the pleural plaque. Chrysotile is the common type of asbestos fibers found in the pleura (Sebastien et al., 1980). The population in Finnish endemic areas exposed to anthrophyllite shows a low incidence of pleural disease and mesothelioma while that of Turkish endemic area exposed to erionite shows a much higher incidence of pleural disease and mesothelioma (Hillerdal et al., 1984).

D. Normal and Activated Mesothelial Cells

Normal mesothelial cells are fragile (Ryan et al., 1973) but regenerated mesothelial cells are resilient and have altered cytoplasmic organelles and enzyme systems. Surface membrane and mitochondrium-associated enzymes such as 5'-nucleotidase, alkaline phosphatase, ATPase, cytochrome oxidase, and succinic dehydrogenase are present in normal and activated mesothelial cells but nonspecific esterases, acid hydrolases, and Golgi-linked enzymes such as thiamine pyrophosphatase are increased in the activated forms (Adnet et al., 1978; King and Wilson, 1983; Efrati and Nir, 1976; Marsan and Cayphas, 1974; Raftery, 1973b; Shanthaveerappa and Bourne, 1965; Whitaker et al., 1980b, 1982b). These enzyme patterns suggest that resting mesothelial cells are primarily involved in membrane transport but the active form is involved more in synthesis as well as digestive activities including prostaglandin synthesis (Coene et al., 1982) and fibrinolysis (Whitaker et al., 1982c; Merlo et al., 1983). Mesothelial cells at the pleurolymphatic communications are usually activated. How mesothelial cells transform from the resting to activated forms or vice versa is uncertain. The mesothelial cells are active in repair of the pleural surface.

E. Repair of the Pleural Surface

Although all investigators have identified mesothelial cells, macrophages, fibroblasts, and other reactive cells in the repair processes of the pleura, interpretations of the mode of healing of the pleural or peritoneal surface do not completely agree, except to note that healing is completed by replacement of mature mesothelial cells. New mesothelial cells are proposed to come from at least three sources: mature mesothelial cells from adjacent or opposing pleural surface (Watters and Buck, 1973; Johnson and Whitting, 1962); from floating serosal or free cells (Watters and Buck, 1972; Ryan et al., 1973; Curran and Clark, 1964); and from subserosal mesothelial cell precursors (Williams, 1955; Ellis et al., 1965; Raftery, 1973a, c). The cells are identified primarily by routine light microscopy but also by histochemistry of acid hydrolases (Mohr et al., 1971; Raftery, 1976; Whitaker et al., 1982b) and SEM and TEM (Watters and Buck, 1972; Raftery, 1973a).

Despite these detailed observations, the precise transition from the accumulation of reactive or inflammatory cells on the injured pleural surface to the final mature mesothelial cells on the healed pleural surface is still uncertain. A direct transformation of macrophages or mesenchymal cells into mesothelial cells or vice versa is unlikely. The question of the intermediate spindle cells and mesoblastic cells, however, is still unanswered. Mesothelial cells appear to derive ontogenically from spindle-shaped mesenchymal cells (Gray and Skandalakis, 1985) and neoplastic mesothelial cells classically are biphasic and may transform from epithelial to spindle forms or vice versa, repetitively with intermediate forms (Legrand and Pariente, 1974; Wagner et al., 1982), although this does not happen in cultured rat mesothelial cells (Jaurand et al., 1981). Moreover, mesoblastic cells, as described in the pleural effusion, are large, round or cuboidal cells with few organelles and can be recognized only by their association with more mature forms of mesothelial cells. To complicate the picture further, fibroblasts and myofibroblasts also proliferate between epithelial cells, in repair and also in neoplastic conditions, and spindle transformation of neoplastic cells occurs in almost all known epithelial neoplasms.

The problems of pleural repair, therefore, overlap with the neoplastic proliferation of mesothelial cells. Although the complete picture needs to be clarified further, more than one mechanism or cell may be involved in the mesothelial cell repair (Whitaker et al., 1982a).

F. Cytogenetic Study

Chromosomes remain stable in cultured mesothelial cells for up to 40 passages (Aronson and Cristofalo, 1981; Jaurand et al., 1985). Spontaneous transformation increases with age but can be enhanced by exposing the cultured cells to carcinogens. Although technical improvement is still needed, chromosomes prepared for cytogenetic study can be processed for SEM (Fig. 6) and show additional details. Interaction of asbestos fibers with cultured rat mesothelial cells and their chromosomes further sugests that SEM can be useful in the study of the interaction between chromosomes and carcinogens (Wang et al., 1987).

Karyotyping of the cells in the pleural fluid is most useful when the cellular population is relatively uniform. Desquamated mesothelial cells, lymphocytes, and macrophages are usually mixed with mesothelioma or metastatic tumor cells in the pleural fluid. Demonstration of polyploidy and presence of marker chromosomes are highly suggestive of neoplastic processes. Although chromosomal aberrations in mesothelioma appear to be very complex, repeated marker chromosomal aberrations have been found in chromosomes 3 and 6 and, to a lesser extent, 2, 7, and 13. Segmental losses preferentially affect chromosome 14 and 22 (Mark, 1978; Stenman et al., 1986). In another study 12 of 14 mesotheliomas show the presence of marker chromosome in numbers 1, 2, 3, 6, 9, 11, 17, and 22 (Gibas et al., 1986). These chromosomal alterations and the modification of oncogenes may reveal some clues to carcinogenesis, although many questions are yet to be answered.

G. Mineral Particle Analysis

Mineral particle analysis has been done mostly on lung tissue. Relatively few particles are found in the pleural plaque, although short chrysotile fibers appear to accumulate more in the pleura than in other sites in the lung (Sebastien et al., 1980). The LM identification of asbestos bodies appears to represent the total fiber load well, especially in heavily exposed lungs. Precise estimation and identification of mineral particles in the pleura and the lung are, however, best done by tissue digestion method with combined TEM, selected area electron diffraction, and x-ray energy dispersive spectrometry. In addition, SEM is particularly useful to evaluate the stereoscopic fine structures of mineral fibers as well as fiber and cell interactions (Kimizuka and Wang, 1985; Wang et al., 1987).

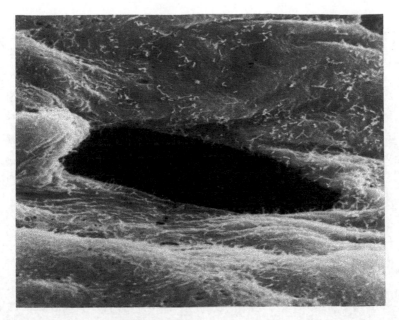

Figure 5   Crevices of approximately 30 × 10 μm are found in the lower mediastinum of an aged mouse. These are thought to be enlarged stomas following the rupture of surface lining cells and the supporting collagen bundles in the lamina cribriformis. One crevice in Figure 4 (× 1200) is enlarged in Figure 5 (× 2400; mouse, SEM).

H.   Mesothelioma and Metastatic Carcinoma

Numerous experimental animal studies have confirmed that mesothelioma can be induced by one intrapleural or intraperitoneal injection of a relatively large dose of fibers. Different types of asbestos and mineral fibers have been found to cause different incidences of mesothelioma. The physicochemical properties of fibers, concomitant exposures to polycyclic hydrocarbons such as benzo-alpha-pyrene, trace metals, cigarette smoke, ionizing radiation, and a variety of drugs and chemicals have been widely studied to evaluate their synergistic or antagonistic effects (Monchau et al., 1985).

The inhalational studies, however, highlight another important aspect of carcinogenesis. Fibers, even in an unrealistically high concentration in the nebulizing air, or when administered by intratracheal instillation, do not reach the pleura easily unless the pleura is already altered (Kanazawa et al., 1970). Metastatic cancer may also spread more easily to altered pleura, such as in the patient with breast cancer who has received irradiation of the chest wall.

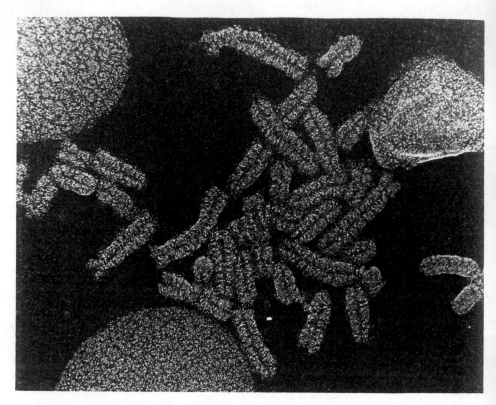

Figure 6 SEM of metaphase chromosomes originally prepared by the routine cytogenetic method. The interphase nuclei are round. Some chromosomal abnormalities may be better studied using this technique (rat pleural mesothelial cell in culture, SEM × 6,500).

Although chrysotile, talc, and other agents may act in more than one way to damage the pleura (Davis et al., 1978) and induce pleural fibrosis, epidemiologically they are less likely to enhance the development of mesothelioma than crocidolite, which is not found in the pleura (McDonald et al., 1971; Becklake, 1976). The apparently contradictory results from experimental and epidemiological studies need further clarification.

Problems with cell cultures involve clonal selection by the type of culture media, spontaneous transformation, and aging. In spite of these limitations, rat pleural mesothelial cells in culture appear to retain the characteristic bushlike microvilli and cytoplasmic fibrils (Jaurand et al., 1981). The latter demonstrate the immunohistochemical characteristics of cytokeratin (Lambre et al., 1983; Wu et al., 1982).

The bushlike elongated microvilli and abundant cytoplasmic cytokeratin of the mesothelial cell appear early in ontogeny and persist not only in the cultured but also in dysplastic and neoplastic mesothelial cells. These findings have been useful in the differential diagnosis of mesothelioma and metastatic pleural tumors, especially adenocarcinoma (Corson and Pinkus, 1982; Wang, 1973). The carcinoembryonic antigen found in most adenocarcinomas is also absent in most typical mesotheliomas (Wang et al., 1979; Whitaker and Shilkin, 1981).

IV. Summary and Perspectives

Substantial progresses have been made in the morphological study of the pleura by the en face preparation of mesothelial cells and in situ fixation of the pleura. The LM and EM studies have shown that the serous cavity and mesothelial cells, which develop early in ontogeny, are readily stretchable and the microvilli of the mesothelial cell trap slimy hyaluronic acid glycoproteins. These morphological findings explain how all organs in the pleural and peritoneal cavities can deform and displace or yield to each other easily during the rapid growth of the fetal period, and how the heart and lung expand and retract incessantly with minimal friction all through life.

Enzyme, cyto-, and immunohistochemistry, and radioautography, mainly in en face preparation and also in cell culture systems, have shown that mesothelial cells are primarily involved in membrane transport. However, they can also increase markedly the rate of fibrinolysis or activation of inflammatory mediators, such as prostacyclin, following stimulation to keep the pleural cavity open. They also increase the secretion of collagen and other matrix material at the repair stage. When stimulated or activated, the mesothelial cell shows increased microvilli and cytoplasmic organelles on TEM.

Although activated mesothelial cells participate in the repair of the pleural surface, what other inflammatory cells may contribute and from what the mesothelial cell may regenerate are not completely certain. Better cell markers are needed to identify and trace the evolution of the multitude of cells that participate in the repair of the pleural surface.

Endogenous fluid, proteins, cells, and cellular products, as well as foreign agents, reach the pleural cavity through vascular, lymphatic, and respiratory routes. In altered conditions the pleural cavity is frequently expanded; the permeability of the endothelial and mesothelial barriers is modified, primarily or secondarily, focally or diffusely, to increase or even attract circulating material into the pleural cavity.

Physiological studies suggest that clearance of the increased pleural content depends on pleurolymphatic communications. Combined LM, SEM, and TEM studies have demonstrated the distribution of stomas (2–12 μm) that connect the pleural cavity and the lymphatic channels. In disease, the stomas may increase in size and number, up to $10 \times 30 \mu m$, to facilitate clearance of the pleural cavity. As an alternative, lymphoreticular cells may proliferate and retain irritant particles at the pleurolymphatic communication to prevent them from reaching the mediastinum and the rest of the body. The collection of irritants in these foci and extravasated fluid and proteins in the pleural cavity may lead to pleural fibrosis, plaques, or mesothelioma.

Asbestos is an agent epidemiologically related to the development of mesothelioma. Different types of asbestos and even some nonasbestos fibers induce different incidence of mesothelioma. Although the mechanisms of carcinogenesis are not completely clarified, asbestos fibers alter metaphase chromosomes of cultured mesothelial cells; mesotheliomas also show complex but recurrent chromosomal changes. Genetic alterations in mesothelial cell hyperplasia, dysplasia, and the development of pleural fibrosis and mesothelioma need further clarification.

The propensity of the pleural cavity to collect fluid is very useful for biochemical, immunological, and physiological analyses of diseases of the pleura. The pleural content, however, is the mixed result of all biochemical and immunological events and physiological parameters and may not reflect immediate or localized changes in the pleural cavity. The morphological documentation of regional alterations, therefore, is most crucial and frequently also explains the results of other types of studies to enhance our understanding of pleural disease.

Acknowledgment

The author is most grateful to Mrs. Hassmig Minassian and Leng-Hwa Ling-Tsao for technical assistance, Mrs. Peggy Wang Hsu and C. Pollak for reference searches, and Miss Grace Pawelec for secretarial assistance. The chromosomal study was carried out in collaboration with Professor J. Bignon, Dr. M. C. Jaurand, and their staff.

References

Abraham, J. L., and Burnett, B. R. (1983). Quantitative analysis of inorganic particlate burden in situ in tissue sections. *Scanning Electron Microsc.* **II**:681–696.

Adnet, J. J., Petit, A., and Stoebner, P. (1978). Etude histo-enzymologique en microscopie optique du revétement pleural. *Bull. Eur. Physiopathol. Respir.* **14**:401–407.

Albertine, K. H., Wiener-Kronish, J. P., Ross, P. J., and Staub, N. C. (1982): Structure, blood supply, and lymphatic vessels of the sheep's visceral pleura. *Am. J. Anat.* **165**:277–294.

Allen, L., and Weatherford, T. (1959). Role of fenestrated basement membrane in lymphatic absorption from peritoneal cavity. *Am. J. Physiol.* **197**:551–554.

Andrews, P. M., and Porter, K. R. (1973). The ultrastructural morphology and possible functional significance of mesothelial microvilli. *Anat. Rec.* **177**:409–426.

Aronson, J. F., and Cristofalo, V. J. (1981). Culture of epithelial cells from the rat pleura. *In Vitro* **17**:61–70.

Ayraud, N., and Kermarec, J. (1968). Cytogenetic study of eight tumours of mesothelial origin. *Bull. Cancer* **55**:92–110.

Baradi, A. F., and Hope, J. (1964). Observation on ultrastructure of rabbit mesothelium. *Exp. Cell Res.* **34**:33–44.

Becklake, M. R. (1976). Asbestos-related diseases of the lung and other organs. Their epidemiology and implications for clinical practice. *Am. Rev. Respir. Dis.* **114**:187–227.

Bernaudin, J. F., and Fleury, J. (1985). Anatomy of the blood and lymphatic circulation of the pleural serosa. In *The Pleura in Health and Disease*. Edited by J. Chrétien, J. Bignon, and A. Hirsch. New York, Marcel Dekker, pp. 101–124.

Bernaudin, J. F., Theven, D., Pinchon, M. C., Brun-Pascaud, M., Bellon, B., and Pocidalo, J. J. (1986). Protein transfer in hyperoxic induced pleural effusion in the rat. *Exp. Lung Res.* **10**:23–38.

Berry, J. P., Henoc, P., and Galle, P. (1978). Phagocytosis by cells of the pulmonary alveoli: transformation of crystalline particles. *Am. J. Pathol.* **93**:27–44.

Bertalanffy, F. D., and Lau, C. (1962). Cell renewal. *Int. Rev. Cytol.* **13**:357–366.

Bettendorf, U. (1978). Lymph flow mechanisms of the subperitoneal diaphragmatic lymphatics. *Lymphology* **11**:111–116.

Bichat, M. F. X. (1827). *Traité des Membranes en General et des Diverses Membranes en Particulier*. Paris, Gabon.

Bignon, J., and Gee, J. B. L. (1985). Pleural fibrogenesis. In *the Pleura in Health and Disease*. Edited by J. Chrétien, J. Bignon, and A. Hirsch. New York, Marcel Dekker, pp. 417–444.

Brody, A. R., Hill, L. H., Adkins, B., and O'Connor, K. W. (1981). Chrysotile asbestos inhalation in rats: Deposition pattern and reaction of alveolar epithelial and pulmonary macrophages. *Am. Rev. Respir. Dis.* **123**:670–679.

Bryks, S., and Bertalanffy, F. D. (1971). Cytodynamic reactivity of the mesothelium. Pleural reaction to chrysotile asbestos. *Arch. Environ. Health* **23**:469–472.

Casley-Smith, J. R. (1967). An electron microscopical study of the passage of ions through the endothelium of lymphatic and blood capillaries and through the mesothelium. *Q. J. Exp. Physiol.* **52**:105–113.

Castor, W. C., and Naylor, B. (1969). Characteristics of normal and malignant human mesothelial cells studied in vitro. *Lab. Invest.* **20**:437–443.

Cheng, H., and Berry, M. (1972). A technique for the preparation of monolayers of mesothelium. *J. Histochem. Cytochem.* **20**:542–544.

Chiovetti, R., McGuffee, L. J., Little, S. A., Whealer-Clark, E., and Brass-Dale, J. (1987). Combined quick freezing, freeze-drying and embedding tissue at low temperature and in low viscosity resins. *J. Electron Microsc. Tech.* **5**:1–15.

Coene, M. C., Van Hove, C., Claeys, M., and Herman, A. G. (1982). Arachidonic acid metabolism by cultured mesothelial cells. *Biochim. Biophys. Acta* **710**:437–455.

Cooray, G. H. (1949). Defensive mechanisms in the mediastinum with special reference to the mechanics of pleural absorption. *J. Pathol. Bacteriol.* **61**:551–567.

Corson, J. M., and Pinkus, G. S. (1982). Mesothelioma: profile of keratin proteins and carcinoembryonic antigen. *Am. J. Pathol.* **108**:80–87.

Cotran, R. S., and Karnovsky, M. J. (1968). Ultrastructural studies on the permeability of the mesothelium to horseradish peroxidase. *J. Cell Biol.* **37**:123–137.

Cotran, R. S., and Majno, G. (1967). Studies on the intercellular junctions of mesothelium and endothelium. *Protoplasma* **63**:45–51.

Cotran, R. S., and Nicca, C. (1968). The intracellular localization of cations in mesothelium. A light and electron microscopic study. *Lab. Invest.* **18**:407–415.

Courtice, F. C., and Morris, B. (1953). The effect of diaphragmatic movement on the absorption of protein and of red cells from the pleural cavity. *Aust. J. Exp. Biol. Med. Sci.* **31**:227–238.

Courtice, F. C., and Simmonds, W. J. (1949). Absorption of fluids from the pleural cavities of rabbits and cats. *J. Physiol. (Lond.)* **109**:117–130.

Courtice, F. C., and Simmonds, W. J. (1954). Physiological significance of lymph drainage of the serous cavities and lungs. *Physiol. Rev.* **34**:419–448.

Cudkowicz, L. (1979). Bronchial arterial circulation in man. In *Normal Anatomy and Responses to Disease in Pulmonary Vascular Disease.* Edited by K. M. Moser. New York, Marcel Dekker, pp. 111–232.

Curran, R. C., and Clark, A. E. (1964). Phagocytosis and fibrogenesis in peritoneal implants in the rat. *J. Pathol. Bacteriol.* **88**:489–502.

Davis, J. M. G., Beckett, S. T., Bolton, R. E., Collings, P., and Middleton, A. P. (1978). Mass and number of fibers in the pathogenesis of asbestos related lung disease in rats. *Br. J. Cancer* **37**:673–688.

Dodson, R. F., and Ford, J. O. (1985). Early response of the visceral pleura following asbestos exposure: an ultrastructural study. *J. Toxicol. Environ. Health* **15**:1673–1686.

Dodson, R. F., O'Sullivan, M. F., Corn, C. J., Ford, J. O., and Hurst, G. A. (1983). The influence of inflation levels of the lung on the morphology of the visceral pleura. *Cytobios* **37**:171–179.

Dybkowsky (1866). Uber Aufsangung und Absonderung der Pleurawand. *Abhand Math. Phys. Clin. Ges. Wiss.* **18**:191–218.

Efrati, P., and Nir, E. (1976). Morphological and cytochemical investigation of human mesothelial cells from pleura and peritoneal effusions: a light microscopic study. *Isr. J. Med Sci.* **12**:662–673.

Ellis, H. (1962). The aetiology of post-operative abdominal adhesions. An experimental study. *Br. J. Surg.* **50**:10–16.

Ellis, M., Harrison, W., and Hugh, T. B. (1965). The healing of peritoneum under normal and pathological conditions. *Br. J. Surg.* **52**:471–476.

Fedorko, M. E., Hirsch, J. G., and Fried, B. (1971). Studies on transport of macromolecules and small particles across mesothelial cells of the mouse omentum. II. Kinetic features and metabolic requirements. *Exp. Cell Res.* **69**:313–323.

Fukata, H. (1963). Electron microscopic study on normal rat peritoneal mesothelium and its changes in absorption of particulate iron dextran complex. *Acta Pathol. Jpn.* **13**:309–325.

Gibas, Z., Li, F. P., Antman, K. H., Bernals, S., Stahel, R., and Sandberg, A. A. (1986). Chromosomal changes in malignant mesothelioma. *Cancer Genet. Cytogenet.* **20**:191–201.

Gray, S. W., and Skandalakis, J. E. (1985). Development of the pleura. In *The Pleural in Health and Disease*. Edited by J. Chrétien, J. Bignon, and A. Hirsch. New York, Marcel Dekker, pp. 3–19.

Grober, J. A. (1901). Die Resorptionkraft der Pleura. *Beitr. Pathol. Anat. Allg. Pathol.* **30**:269–347.

Gross, P., and de Treville, R. T. P. (1967). Experimental asbestosis: studies on the progressiveness of the pulmonary fibrosis caused by chrysotile dust. *Arch. Environ. Health* **15**:638–649.

Harvey, W., and Amlot, P. L. (1983). Collagen production by human mesothelial cells in vitro. *J. Pathol.* **139**:337–347.

Hayek, V. H. (1960). The parietal pleura (pleura parietalis) and the visceral pleura (pleura pulmonalis). In *The Human Lung*. New York, Hafner, pp. 34–49.

Hesseldahl, H., and Larsen, J. F. (1969). Ultrastructure of human yolk sac: endoderm, mesenchyme, tubules and mesothelium. *Am. J. Anat.* **126**:315–336.

Hillerdal, G. (1980). The pathogenesis of pleural plaques and pulmonary asbestosis: possibilities and impossibilities. *Eur. J. Respir. Dis.* **61**:129–138.

Hillerdal, G., Zitting, A., Van Assendleft, A. H. W., and Kuusela, T. (1984). Rarity of mineral fiber pleurisy among persons exposed to Finnish anthophyllite and with low risk of mesothelioma. *Thorax* **39**:608–611.

Holt, P. F. (1983). Translocation of inhaled dust to the pleura. *Environ. Res.* **31**:212–220.

Hourihane, D. O'B. (1964). The pathology of mesotheliomata and an analysis of their association with asbestos exposure. *Thorax* **19**:268–278.

Ivanova, V. F., and Puzyrev, A. A. (1977). Autoradiographic study of mesothelial proliferation in white mice under experimental conditions. *Arkh. Anat. Gistol. Embriol.* **72**:10–17.

Jaurand, M. C., Bernandi, J. F., Renier, A., Kaplan, H., and Bignon, J. (1981). Rat mesothelial cells in culture. *In Vitro* **17**:98–105.

Jaurand, M. C., Pinchon, M. C., and Bignon, J. (1985). Mesothelial cells in vitro. In *The Pleura in Health and Disease*. Edited by J. Chrétien, J. Bignon, and A. Hirsch. New York, Marcel Dekker, pp. 43–67.

Jaurand, M. C., Kheuang, L., Magne, L., and Bignon, J. (1986). Chromosomal changes induced by chrysotile fibers of Benzo(3-4)pyrene in rat pleural mesotheial cells. *Mutat. Res.* **169**:141–148.

Johnson, F. R., and Whitting, H. W. (1962). Repair of parietal peritoneum. *Br. J. Surg.* **49**:653–660.

Kampmeier, O. F. (1928). Concerning certain mesothelial thickenings and vascular plexuses of the mediastinal pleura, associated with histiocyte and fat-cell production, in the human newborn. *Anat. Rec.* **39**:201–214.

Kanazawa, K. (1985). Exchanges through the pleura, cells, and particles. In *The Pleura in Health and Disease*. Edited by J. Chrétien, J. Bignon, and A. Hirsch. New York, Marcel Dekker, pp. 195–231.

Kanazawa, K., Birbeck, M. S. C., Carter, R. L., and Roe, F. J. C. (1970). Migration of asbestos fibers from subcutaneous sites in mice. *Br. J. Cancer* **24**:96–106.

Kanazawa, K., Roe, F. J. C., and Yamamoto T. (1979). Milky spots (taches laiteuses) as structures which trap asbestos in mesothelial layers and their significance in the pathogenesis of mesothelial neoplasia. *Int. J. Cancer* **23**:858–865.

Kiviluoto, R. (1960). Pleural calcificatin as a roentgenologic sign of non-occupational endemic anthophyllite-asbestos. *Acta Radiol. (Suppl).* **194**:1–67.

Kihara, T. (1950). The extravascular fluid passway system. *Ketsuekigaku Togikai Hokoku* **3**:118.

Kimizuka, G., and Wang, N. S. (1985). Method of direct particle deposition on a carbon planchet to study mineral dust in human lung by scanning electron microscopy. *J. Electron Microsc. Tech.* **2**:209–215.

Kinasewitz, G. T., Groome, L. J., Marshal, R. P., and Diana, J. N. (1983). Permeability of the canine visceral pleural. *J. Appl. Physiol.* **55**:121–130.

King, B. F., and Wilson, J. M. (1983). A fine structural and cytochemicalstudy of the Rhesus monkey yolk sac: endoderm and mesothelium. *Anat. Rec.* 205:143–158.

Kluge, T., and Hovig, T. (1967). The ultrastructure of human and rat pericardium. I. Parietal and visceral mesothelium. *Acta Pathol. Microbiol. Scand.* 71:529–546.

Korsgaard, R. (1979). Chromosome analysis of malignant human effusions in vivo. *Scand. J. Respir. Dis.* (Suppl) 105:1–100.

Lambre, C. R., Jaurand, M. C., Renier, A., and Bignon, J. (1983). Immunoenzymatic study of the keratin content of mesothelial cells in vitro. In *Immunoenzymatic Techniques*. Edited by S. Avraveas, P. Druet, R. Masseyeff, and G. Feldman. Amsterdam, Elsevier, pp. 81–84.

Lang, J. (1962). Über eigenartig Kapillarkonvolute der Pleura parietalis. I. *Z. Zellforsch.* 58:487–523.

Lang, J., and Liebich, H. G. (1976). Über eigenartig Kapillarkonvolute der Pleura parietalis. III. Elektronenmikroskopische Untersuchungen. *Z. Mikrosk. Anat. Forsch.* 90:1074–1094.

Leak, L. V. (1983). Interaction of mesothelium to intraperitoneal stimulation. I. Aggregation of peritoneal cells. *Lab. Invest.* 48:479–491.

Leak, L. V., and Rahil, K. (1978). Permeability of the diaphragmatic mesothelium: the ultrastructural basis for "stomata." *Am. J. Anat.* 151:557–594.

Leathes, J. B., and Starling, E. H. (1895). On the absorption of salt solutions from the pleural cavities. *J. Physiol.* 18:106–116.

Lee, K. P., Barras, C. E., Griffith, F. D., and Wartz, R. S. (1981). Pulmonary response and transmigration of inorganic fibers by inhalation exposure. *Am. J. Pathol.* 102:314–323.

Legrand, M., and Pariente, R. (1974). Etude au microscope électronique de 18 mésothéliomes pleuraux. *Pathol. Biol.* 22:409–420.

Legrand, M., Pariente, R., and André J. (1971). Ultrastructure de la pleure parietale humaine. *Presse Méd.* 55:2515–2520.

Mariassay, A. T., and Wheeldon, E. B. (1983). The pleura: a combined light microscopic, scanning and transmission electron microscopic study in the sheep. I. Normal pleura. *Exp. Lung Res.* 4:293–314.

Mark, H. (1978). Three chromosomal abnormalities observed in cells of two malignant mesotheliomas studied by banding techniques. *Acta Cytol.* (Baltimore) 22:398–401.

Marsan, C., and Cayphas, J. (1974). The aid of some histochemical stains in the identification of mesothelial cells: preliminary results. *Acta Cytol.* 18:252–258.

Maximow, A. (1927). Ueber das Mesothel (Deckzellen der serosen Haute) und die Zellen der serosen Exudate. Untersuchungen an entzundetem Gewebe und an Gewebskulturen. *Arch. Exp. Zellforsch.* 4:1–42.

Mazia, D., Schatten, G., and Sale, W. (1975). Adhesion of cells to surfaces coated with polylysine. Applications to electron microscopy. *J. Cell Biol.* **66**:198–200.

McDonald, J. C., McDonald, A. D., Gibbs, G. W., Siemiatycki, J., and Rossiter, C. E. (1971). Mortality in the chrysotile asbestos mines and mills of Quebec. *Arch. Environ. Health* **22**:677–686.

McLaughlin, R. F., Tyler, W. S., and Canada, R. O. (1961). A study of the subgross pulmonary anatomy in various mammals. *Am. J. Anat.* **108**:149–165

McLaughlin, R. F., Tyler, W. S., and Canada, R. O. (1966). Subgross anatomy of the rabbit, rat, and guinea pig with additional notes on the human lung. *Am. Rev. Respir. Dis.* **94**:380–387.

Merlo, G., Fausone, G., and Castagna, B. (1983). Fibrinolytic activity of mesothelial lining of the displaced peritoneum. *Am. J. Med. Sci.* **286**:12–14.

Miller, W. S. (1947). *The Lung*, 2nd ed. Springfield, IL, Charles C. Thomas, pp. 89–118.

Miserocchi, G. (1985). Pleural liquid pressure. In *The Pleura in Health and Disease*. Edited by J. Chrétien, J. Bignon, and A. Hirsch. New York, Marcel Dekker, pp. 131–168.

Miserocchi, G., and Agostoni, E. (1971). Contents of the pleural space. *J. Appl. Physiol.* **30**:208–213.

Miserocchi, G., Nakamura, T., Mariani, E., and Negrini, D. (1981). Pleural liquid pressure over the interlobular, mediastinal, and diaphragmatic surfaces of the lung. *Respir. Physiol.* **46**:61–69.

Mohr, W., Beneke, G., and Murr, L. (1970). Transformation von Zellen der Peritoneal und Pleurahohle durch Phytohamagglutinin. *Beitr. Pathol.* **142**:90–113.

Mohr, W., Beneke, G., and Murr, L. (1971). Proliferation der Zellsysteme in cavum peritonei. I. *Beitr. Pathol.* **143**:345–359.

Mohr, WS., Beneke, G., and Murr, L. (1972). Proliferation der Zellsysteme im Cavum peritonei. II. *Beitr. Pathol.* **145**:381–394.

Monchaux, G., Bignon, J., Lafume, J., and Hirsch, A. (1985). Experimental pleural carcinogenesis induced by mineral fibers. In *The Pleura in Health and Disease*. Edited by J. Chrétien, J. Bignon, and A. Hirsch. New York, Marcel Dekker, pp. 551–570.

Nagaishi, C. (1972). *Functional Anatomy and Histology of the Lung*. Baltimore, University Park Press, pp. 79–179.

Obata, H. (1978). Differences in normal structure and reaction to adjuvant between the costal and the visceral pleura. *Arch. Histol. Jpn.* **41**:65–86.

Odor, D. L. (1954). Observations of the rat mesothelium with the electron and phase microscopes. *Am. J. Anat.* **95**:433–466.

Orr, F. W., Adamson, I. Y., and Young, L. (1986). Quantification of metastatic tumor growth in bleomycin-injured lungs. *Clin. Exp. Metastasis* **4**:105–116.

Raftery, A. T. (1973a). Regeneration of parietal and visceral peritoneum: an electron microscopical study. *J. Anat.* **115**:365–373.

Raftery, A. T. (1973b). An enzyme histochemical study of mesothelial cells in rodents. *J. Anat.* **115**:365–373.

Raftery, A. T. (1973c). Regeneration of parietal and visceral peritoneum. A light microscopical study. *Br. J. Surg.* **60**:293–299.

Raftery, A. T. (1976). Regeneration of parietal and visceral peritoneum: an enzyme histochemical study. *J. Anat.* **121**:589–597.

Raftery, A. T. (1979). Regeneration of peritoneum: a fibrinolytic study. *J. Anat.* **129**:659–664.

Rennard, S. I., Jaurand, M. C., Bignon, J., Kawanami, O., Ferrans, V. J., Davison, J., and Crystal, R. G. (1984). Role of pleural mesothelial cells in the production of the submesothelial connective tissue matrix of the lung. *Am. Rev. Respir. Dis.* **130**:267–274.

Rennard, S. I., Jaurand, M. C., Bignon, J., Ferrans, V. J., and Crystal, R. G. (1985). Connective tissue matrix of the pleura. In *The Pleura in Health and Disease*. Edited by J. Chrétien, J. Bignon, and A. Hirsch. New York, Marcel Dekker, pp. 69–85.

Roberts, W. C., and Ferrans, V. J. (1972). Pure collagen plaques on the diaphragm and pleura. *Chest* **61**:357–360.

Rossi, G. A., Hunninghake, G. W., Szapiel, S. V., and Crystal, R. G. (1981). Pulmonary responses to environmental agents and drugs are influenced by immune response and non-immune response genes. *Am. Rev. Respir. Dis.* **123**(Part 2):144.

Ryan, G. B., Grobety, J., and Majno, G. (1971). Postoperative peritoneal adhesions. *Am. J. Pathol.* **65**:117–148.

Ryan, G. B., Grobety, J., and Majno, G. (1973). Mesothelial injury and recovery. *Am. J. Pathol.* **71**:93–102.

Sebastien, P., Janson, X., Gaudichet, A., Hirsch, A., and Bignon, J. (1980). Asbestos retention in human respiratory tissues: comparative measurments in lung parenchyma and in parietal pleura. In *Biological Effects of Mineral Fibers*. Edited by J. C. Wagner. Lyon, France, IARC Sci. Pub. (No. 30), Vol. 1, pp. 237–246.

Seifert, E. (1921). Zur Biologie des Menschlichen Grossen Netzes. *Arch. Klin. Chirurg. Bd.* **116**:510–517.

Shanthaveerappa, T. R., and Bourne, G. H. (1965). Histochemical studies on the localization of oxidative and dephosphorylating enzymes and esterases in the peritoneal mesothelial cells. *Histochemie* **51**:331–338.

Simionescu, M., and Simionescu, N. (1977). Organization of cell junctions in the peritoneal mesothelium. *J. Cell. Biol.* **74**:98–110.

Singh, G., Whiteside, T. L., and Dekker, A. (1979). Immunodiagnosis of mesothelioma. Use of antimesothelial cell serum in an indirect immunofluorescence assay. *Cancer* **43**:2288–2296.

Smith, Y. S., and Naylor, B. (1972). A method for extracting ferruginous bodies from sputum and pulmonary tissue. *Am. J. Clin. Pathol.* **58**:250–254.

Stanton, M. F., and Wrench, C. (1972). Mechanism of mesothelioma induction with asbestos and fibrous glass. *J. Natl. Cancer Inst.* **48**:797–821.

Staub, N. C., Wiener-Kronish, J. P., and Albertine, K. H. (1985). Transport through the pleura, physiology of normal liquid and solute exchange in the pleural space. In *The Pleura in Health and Disease.* Edited by J. Chrétien, J. Bignon, and A. Hirsch. New York, Marcel Dekker, pp. 169–193.

Stenman, G., Olofsson, K., Mansson, T., Hagmar, B., and Mark, J. (1986). Chromosomes and chromosomal evolution in human mesotheliomas as reflected in sequential analyses of two cases. *Hereditas* **105**:233–239.

Stoebner, P., Miech, G., Sengel, A., and Witz, J. P. (1970a). Notions d'ultrastructure pleurale I. L'hyperplasie mésothéliale. *Presse Med.* **78**:1179–1184.

Taskinen, E., Ahlman, K., and Wiikeri, M. (1973). A current hypothesis of the lymphatic transport of inspired dust to the parietal pleura. *Chest* **64**:193–196.

Thorsrud, G. K. (1965). Pleural reactions to irritants. An experimental study with special reference to pleural adhesions and concrescence in relation to pleural turnover of fluid. *Acta Chir. Scand. (Suppl.)* **355**:1–74.

Timbrell, V., Skidmore, J. W., Hyett, A. W., and Wagner, J. C. (1970). Exposure chambers for inhalation experiments with standard reference samples of asbestos of the international union against cancer (UICC). *Aeros. Sci.* **1**:215–223.

Tsilibury, E. C., and Wissig, S. L. (1983). Lymphatic absorption from the peritoneal cavity: regulation of patency of mesothelial stoma. *Microvasc. Res.* **25**:22–39.

Vallyathan, V., and Green, F. H. Y. (1985). The role of analytical techniques in the diagnosis of asbestos associated disease. *CRC Crit. Rev. Clin. Lab. Sci.* **22**:1–42.

Verloop, M. C. (1948). The arteriae bronchiales and their anastomoses with the arteria pulmonalis in the human lung: a micro-anatomical study. *Acta Anat. (Basel)* **5**:171–205.

von Recklinghausen, F. (1863). Zur Fettresorption. *Virchows Arch. (Pathol. Anat.)* **26**:172–208.

Voth, D., and Kohlhardt, M. (1962). Untersuchungen zur Histomorphologie und Zytologie des menschlichen Mesothels. *Z. Zellforsch.* **58**:546.

Wagner, J. C. (1969). Experimental production of mesothelial tumors of the pleura in implantation of dusts in laboratory animals. *Nature* (London) **196**:180–181.

Wagner, J. C., Sleggs, C. A., and Marchand, P. (1960). Diffuse pleural mesotheioma and asbestos exposure in the north western Cape province. *Br. J. Ind. Med.* **17**:260–271.

Wagner, J. C., Johnson, N. F., Brown, D. G., and Wagner, M. M. F. (1982). Histology and ultrastructure of serially transplanted rat mesotheliomas. *Br. J. Cancer* **46**:294–299.

Wang, N. S. (1973). Electron microscopy in the diagnosis of pleural mesotheliomas. *Cancer* **31**:1046–1054.

Wang, N. S. (1974). The regional difference of pleural mesothelial cells in rabbits. *Am. Rev. Respir. Dis.* **110**:623–633.

Wang, N. S. (1975). The preformed stomas connecting the pleural cavity and the lymphatics in the parietal pleura. *Am. Rev. Respir. Dis.* **111**:12–20.

Wang, N. S. (1980). The surface appearance of some lung tumors, mesotheliomas and their precursor lesions. *Scanning Electron Microsc.* **III**:79–88.

Wang, N. S. (1983). Morphologic data of pleura—normal conditions. In *Diseases of the Pleura*. Edited by J. Chrétien and A. Hirsch. New York, Masson USA, pp. 1–24.

Wang, N. S. (1985a). Anatomy and physiology of the pleural space. *Clin. Chest Med.* **6**:3–16.

Wang, N. S. (1985b). Mesothelial cells in situ. In *The Pleura in Health and Disease*. Edited by J. Chrétien, J. Bignon, and A. Hirsch. New York, Marcel Dekker, pp. 23–42.

Wang, N. S., Huang, S. N., and Gold, P. (1979). Absence of carcinoembryonic antigen-like material in mesothelioma: an immunohistochemical differentiation from other lung cancers. *Cancer* **44**:937–943.

Wang, N. S., Jaurand, M. C., Magne, L., Kheuang, L., Pinchon, M. C., and Bignon, J. (1987). The interactions between asbestos fibers and metaphase chromosomes of rat pleural mesothelial cells in culture. *Am. J. Pathol.* **126**:343–349.

Watters, W. B., and Buck, R. C. (1972). Scanning electron microscopy of mesothelial regeneration in the rats. *Lab. Invest.* **26**:604–609.

Watters, W. B., and Buck, R. C. (1973). Mitotic activity of peritoneum in contact with a regenerating area of peritoneum. *Virchows Arch. Cell Pathol.* **13**:48–54.

Weibel, E. R., Limacher, W., and Bachofen, H. (1982). Electron microscopy of rapidly frozen lungs: Evaluation on the basis of standard criteria. *J. Appl. Physiol.* **53**:516–527.

Whitaker, D., and Shilkin, K. B. (1981). Carcinoembryonic antigen in tissue diagnosis of malignant mesothelioma. *Lancet* **1**:1369.

Whitaker, D., Papadimitriou, J. M., and Walters, M. N-I. (1980a). The mesothelium; techniques for investigating the origin, nature and behaviour of mesothelial cells. *J. Pathol.* **132**:263–271.

Whitaker, D., Papadimitriou, J. M., and Walters, M. N-I. (1980b). The mesothelium: a histochemical study of resting mesothelial cells. *J. Pathol.* **132**:273–284.

Whitaker, D., papadimitriou, J. M., and Walters, M. N-I. (1982a). The mesothelium and its reactions: a review. *Crit. Rev. Toxicol.* **10**:81–144.

Whitaker, D., Papadimitriou, J. M., and Walters, M. N-I. (1982b). The mesothelium: a cytochemical study of "activated" mesothelial cells. *J. Pathol.* **136**:169–179.

Whitaker, D., Papadimitriou, J. M., and Walters, M. N-I. (1982c). The mesothelium: its fibrinolytic properties. *J. Pathol.* **136**:291–299.

Williams, D. C. (1955). The peritoneum: plea for change in attitude towards this membrane. *Br. J. Surg.* **42**:401–405.

Wolff, K., and Konrad, K. (1972). Phagocytosis of latex beads by epidermal keratinocytes in vivo. *J. Ultrastruct. Res.* **39**:262–280.

Wu, Y. J., Parker, L. M., Binder, N. E., Beckett, M. A., Sinard, J. H., Griffiths, C. T., and Rheinwald, J. G. (1982). The mesothelial keratins: a new family of cytoskeletal proteins identified in cultured mesothelial cells and non-keratinizing epithelia. *Cell* **31**:693–703.

Yao, Y. T., Wang, N. S., Michel, R. P., and Poulsen, R. S. (1984). Mineral dusts in lungs with scar or scar cancer. *Cancer* **54**:1814–1823.

27

Methods in Experimental Pathology of Pulmonary Vasculature

PAUL DAVIES

University of Pittsburgh
School of Medicine
Pittsburgh, Pennsylvania

LYNNE M. REID

Harvard Medical School and
The Children's Hospital
Boston, Massachusetts

DAPHNE deMELLO

St. Louis University and
Cardinal Glennon Children's Hospital
St. Louis, Missouri

I. Introduction

At each heartbeat, half the cardiac output passes through the lungs. The lungs, which are a low-pressure system with a resting volume only one-twentieth that of the body, accepts the same blood volume as the rest of the body. Because atheroma, although common in the large systemic arteries, is rare in the lung it was considered that the lungs had a special metabolic protection, but Brenner (1935) showed that when the pulmonary artery pressure was raised, the lungs' arteries were susceptible to atheroma.

At present, it seems more useful to emphasize the differences between the systemic and pulmonary systems. While the cast of characters, the cells of the vascular walls, are similar in each, behavior of the pulmonary bed or even of its cells in culture cannot be predicted from the behavior of the systemic system. Even within the pulmonary system the cells at different levels can be strikingly different from each other.

The blood vessels of the systemic bed were much studied in the 19th century, as in Bright's classic studies of hypertension and renal disease. In these studies the vessels were fixed in the distended state under controlled conditions so that comparisons could be made of size, structure, and wall thickness. For reasons not clear, these techniques fell out of use. Brenner, in his classic papers, regretted that his retrospective study could not be carried out on injected, distended material since so much additional information would then have been available. Short (1956) reintroduced the technique of injection to the lung and, as befitted a radiologist, he chose a barium sulfate–gelatin mixture that provided a radiographic image while, fortunately for the pathologist, it also permitted microscopic examination.

For a critical consideration of structure, it cannot be emphasized too strongly that distention under controlled conditions is necessary to allow comparison and interpretation of vessel size and wall thickness. One considers intuitively that the thickness of the wall of an artery, if left undisturbed, will reflect the tone and degree of constriction. Thus, depending on the constriction of the medial coat, an artery is smaller with a thicker wall, or larger with a relatively thin wall. When an artery is distended, structure and wall thickness can be related to each other, but to relate size to the position of the artery in the pattern of branching, it is necessary to "landmark" it in some way. This can most conveniently be done by using the accompanying airway as a point of reference.

A. Normal Structure

Double Arterial Supply, Venous Drainage, and Lymphatics

The arterial supply to the lung includes a contribution from both the right and left ventricle; blood drains to the right as well as to the left atrium. The largest volume of blood passes through the main pulmonary artery to pulmonary vein route, that is, from the right ventricle to the left atrium.

The pulmonary artery branches and does not supply a capillary bed until it reaches the alveoli. The pulmonary artery branches more often than the airway, giving rise to two types of branch. The first type, the so-called conventional branch, runs with the pulmonary artery and ensures that the airway is always accompanied by a pulmonary artery branch. These arteries enter the acinus at its center and are distributed within the alveolar region also, according to the distribution of the airways. The other type of branch, the so-called supernumerary artery, is in fact more common; at a given level the supernumerary artery is smaller than the conventional one. These vessels supply the capillary bed of alveoli close to the bronchoarterial sheath and so the periphery of the acinus. The supernumerary branches represent an additional, or collateral, supply to the respiratory region. For a given acinus, the supernumerary branches have arisen proximally to the conventional ones. Thus, if an axial pulmonary artery is blocked, collateral supply can be provided through the more proximal supernumerary branches.

The systemic or bronchial artery supply to the lungs develops from the ventral branches of the dorsal aorta, and two major arteries supply each lung (Kasai and Chiba, 1979; Miller, 1947). They originate either from the aorta directly or from the intercostal arteries (Pump, 1963). Since they send branches to several mediastinal structures, anastomoses from these sites help to overcome obstruction in a major branch. Entering the lung at the hilum, they divide at the main stem bronchus and follow the bronchial tree within the peribronchial connective tissue sheath. As far as the terminal bronchioli, bronchial artery branches provide a peribronchial as well as a submucosal plexus. Beyond the terminal bronchiolus, they anastomose with the microvessels from the pulmonary artery (Tobin, 1952). An important consideration is that the bronchial arteries also supply this hilum, lymph nodes, vasa vasorum, walls of the pulmonary arteries and veins, and a perihilar region of visceral pleura about the size of an adult hand. Bronchopulmonary artery branches that directly end in pulmonary alveolar microvessels were described by the Wagenvoorts (1967). Such arteries, while common in neonates, are less apparent in older children and adults (Wagenvoort et al., 1964).

The true bronchial veins that drain the hilar region empty into the azygos system. The rest of the bronchial artery distribution drains to pulmonary veins. That is, the intrapulmonary capillary bed, whether around the airway or in the alveolar wall, drains to the pulmonary veins. The drainage from the airway walls, whether bronchial or bronchiolar, represents a degree of venous admixture even in the normal patient.

Precapillary anastomoses are present between the pulmonary artery and bronchial artery in the wall of the airways. These are precapillary in position and, under certain functional states, also larger, but under normal conditions, because of their structure, these vessels are closed. If pressure drops to zero in either one of the systems or under conditions of disease such as inflammation or tumors, in which there is an increased flow to the lung, mediators seem capable of relaxing these channels and shunts develop between the two systems. Under conditions of disease, new structural channels can develop through granulation tissue.

In this chapter the pulmonary artery and vein will be the main focus because these are those more commonly affected in disease and have certainly been more frequently studied. The bronchial artery can also be important and will be mentioned when appropriate. Lymphatics are part of the vascular system of lung; although they are not found in the alveolar wall, they are present in the alveolar region within the connective tissue framework, that is, in the connective tissue septa and in peribronchial, periarterial, and perivenous connective tissue sheaths. Under conditions of disease, they achieve greater importance since they drain the increased fluid produced by edema, whether secondary to a rise in filtration or to increased permeability in inflammation.

Anatomical Structure and Levels Along the Vascular Pulmonary Loop

In the lung, pulmonary artery blood traverses the loop of artery, capillary, and vein. Whether this is along a short loop, as through alveoli close to the hilum, or a long one, for example, through alveoli against the diaphragmatic pleura, regional differences along the artery or vein are apparent but additional segments can also be identified. The elastic and then transitional structure of an artery gives way to a muscular wall (Fig. 1). Each of these structures includes both elastic fibers and muscle in its wall but, as conventionally described, the muscular artery is one in which a medial muscle coat lies between a well-developed internal and external elastic lamina. In the transitional and elastic arteries an increasing number of central laminae are found between the two main elastic laminae. The muscular arteries are the resistance arteries. They have the thickest walls of these arteries.

Beyond these arteries is a special precapillary unit (Fig. 2). Here the medial muscle coat is replaced by a spiral of muscle. As judged by light microscopic examination, the media becomes incomplete so that muscle is seen in only part of the wall: the partially muscular artery segment. In arteries larger than capillaries the muscle coat is no longer apparent, they are nonmuscular arteries. Electron microscopic examination reveals thin cytoplasmic processes in the nonmuscular parts of these walls. The details of this arrangement are described later. The cells can apparently rapidly hypertrophy and multiply under challenge by injury or stimuli.

Downstream from the capillary, along a postcapillary segment, a mirror image of this region is present: a nonmuscular and then partially muscular venous segment before the fully muscular veins are reached.

For these various anatomical levels in the vascular loop, not only is the structure different but reactivity to pharmacological agents also differs. Increasingly, therefore, as in response to injury, vessel structure must be related to segmental regional arrangement. With growth, characteristic changes occur at these various levels; the levels also vary in their susceptibility to injury. A striking example is that under conditions of hypoxia the precapillary unit changes more strikingly, although the alveolar oxygen tension will affect both systems and the relative difference from the normal would be greater in the vein than in the artery. On the other hand, the injury of hyperoxia quickly produces damage to the endothelium of both the pre- and postcapillary units.

Bronchial Circulation

The bronchial circulation has been studied in several laboratory animals (Abdalla and King, 1976,1977; Albertine et al., 1982; Balding et al., 1964; Berry et al., 1931; Collier, 1979; Magno and Fishman, 1982; May, 1970; McLaughlin et al., 1961; Notkovich, 1957; Parry and Yates, 1979; Verloop, 1949) and

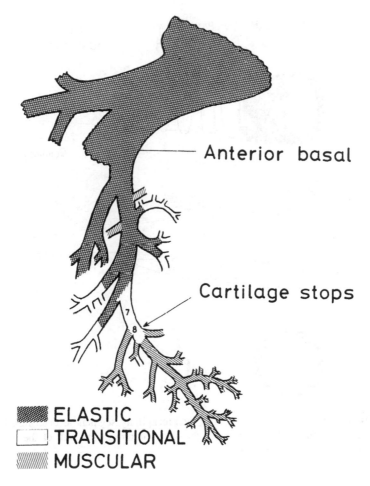

Anterior basal

Cartilage stops

7
8

▓▓ ELASTIC
☐ TRANSITIONAL
▨ MUSCULAR

Figure 1 Adult human lung: diagram of anterior basal arterial pathway shows the structural types of large artery.

physiological studies of the bronchial circulation in animals have been recently reviewed (Deffebach, 1987).

Most lung diseases alter bronchial blood flow. Because the bronchial tree is the first pulmonary line of defense from the external environment, inhaled or regurgitated material first lodges in the bronchial mucosa. This is exclusively supplied by the bronchial circulation, which therefore plays an important role in airway disease. The bronchial circulation, which normally forms 1% of the cardiac output, can increase up to one-third the cardiac output in disease states

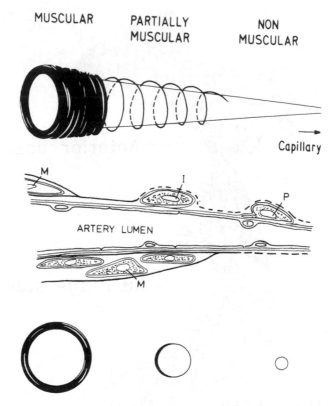

MUSCULAR   PARTIALLY   NON
                   MUSCULAR   MUSCULAR

Capillary

M   I   P

ARTERY LUMEN

M

Figure 2 The structure of the wall of the precapillary segment as determined by light microscopic methods. Diagram shows the ultrastructure of the wall of the intraacinar, precapillary segment of the rat.

that call for an increased circulation, such as inflammation, scar tissue, hypoxia, intravascular thrombosis, emboli, edema, and congenital heart disease (Allanby et al., 1950; Aziz et al., 1977; Lacina et al., 1983; Liao et al., 1985; Weibel, 1960). Primary or secondary lung neoplasms are supplied by the bronchial circulation (Cudkowicz, 1967; Milne, 1967). It seems that new tissue prefers the higher oxygenation of the bronchial system. Bronchial artery to pulmonary artery anastomosis occurs in the wall of large airways. These are usually closed but they open up when flow in the pulmonary arterial system is reduced or when bronchial blood flow increases in disease. In many disease states, the bronchial circulation has not received the attention it deserves and descriptions of the structural alterations are few.

Bronchopulmonary anastomotic blood flow increases acutely after experimental pulmonary arterial obstruction (Jindal et al., 1984). If the bronchial blood flow is simultaneously restricted artificially, pulmonary infarcts consistently occur (Parker and Smith, 1957). In rats, Weibel (1960) demonstrated dilation of the bronchial vessels 2 days after pulmonary artery ligation, followed by vascular proliferation at 5 days and development of new vessels through pleural adhesions at 10 days.

An angiographic study of patients with chronic bronchitis and emphysema has shown distended, enlarged bronchial arteries (Boushy et al., 1969). Likewise, Liebow (1953) found an expanded bronchial venous circulation in these patients. In pneumonia and acute lung injuries such as adult respiratory distress syndrome (ARDS), the bronchial arteries and bronchopulmonary anastomoses are moderately dilated (Barkin et al., 1986; Cudkowicz, 1952; Lakshminarayan et al., 1983; Mathes et al., 1932; Noszkov et al., 1979). Bronchial artery dilation is also seen in chronic inflammatory processes such as bronchiectasis (Cudkowicz, 1968; Liebow et al., 1949), tuberculosis (Cudkowicz, 1965; Wagner et al., 1964; Wood and Miller, 1938), and interstitial fibrosis (Cudkowicz, 1979; Muller and Bordt, 1980; Turner-Warwick, 1963).

The bronchial circulation probably also plays a role in airway reactivity and influences heat and water loss, thus playing a role in the bronchospasm of asthma and in the delivery of aerosol bronchodilators to the more distal airways.

The advent of heart and lung transplantation highlights the significance of the bronchial circulation. Although at transplantation routine bronchial circulation is not reestablished directly, preservation or reestablishment of the bronchial circulation by wrapping omentum around the anastomotic site or anastomosis of transplant bronchial vessels to the recipient's intercostals results in better healing and fewer complications at the bronchial anastomosis (Fell et al., 1985; Lima et al., 1982; Mills et al., 1970; Morgan et al., 1982; Pinsker et al., 1980, 1984).

Lymphatics

The discovery of the lymphatic system came much later than that of the other vascular components. As recently as this century there was skepticism as to its existence (Kinmonth, 1972). This is not surprising, since the vessels are small, colorless, often collapsed, and hence difficult to see. It is perhaps small wonder that while studies monitoring lung lymph flow in lung injury abound, studies of lymphatic structural alterations are few.

After a somewhat diffident start by Hippocrates (460–377 B.C.), with the first description of axillary lymph nodes as "glands that everyone has in the armpit," and in subsequent eras several controversies raged, for example, about the distinction of lymph channels from veins, the work of Pecquet (1651), Bartholinus (1653), and Rudbeck (1653) eventually established the existence of the lymphatic system. The work of Ludwig, Starling, and, more recently Drinker and Yoffey and Courtice (1970) went a long way toward documenting the normal physiological function of the lymphatics. The embryonic origin of lymphatics from sacs that bud from veins was established by Florence Sabin (1911).

A major stumbling block to progress in the study of lymphatics in different diseases has been the lack of techniques for their visualization. Injection of a variety of chemicals, dyes, and organisms has been utilized (Bartels, 1909; Drinker et al., 1935; Fischer, 1935; Gerota, 1896; Hudack and McMaster, 1933). The history of the lymphatic system and methods for visualization have been recently reviewed by Kanter (1987).

Modern lymphangiography has succeeded in allowing good clinical visualization of the lymphatics by the use of a water-soluble, radiopaque contrast medium (Kinmonth, 1954; Trapnell, 1963).

In the lung two sets of lymphatics are described: superficial and deep (Muller, 1947; Pick and Howden, 1977). The superficial lymphatics are beneath the pleura and form a rich plexus covering the outer surface of the lung. The deep lymphatics accompany the blood vessels and run along the bronchi. They are not present in the alveolar wall. Both sets of lymphatics connect with bronchial glands (lymph nodes) at the root of the lungs. Efferent vessels from these glands drain into the thoracic duct on the left side and into the right lymphatic duct on the right. There is some crossover drainage, especially by the left lower lobe lymphatics.

Although inconspicuous in health, pulmonary lymphatics are dilated in certain disease states, particularly when there is pulmonary venous obstruction, as in congenital anomalous pulmonary venous connection or severe mitral stenosis. Drake et al. (1987) have recently described a mechanical model of the lung interstitial–lymphatic system and shown that decreases in lymph vessel resistance are essential if lymph flow is to increase substantially in the initial stages of edema. Dilated lymphatics reflect a reduced resistance, and it is suggested that dilatation results from increased tension in anchoring filaments surrounding the lymphatic walls as the lungs become edematous and tissue matrix swells (Casey-Smith, 1980). The mechanical effect is to pull the lymph vessels open.

Studies of pleural lymphatic albumin concentration in sheep have recently indicated that when sheep are suspended head up in the vertical position, a gradient in microvascular hydrostatic pressure occurs from the apex to the base of the lung (Albertine et al., 1987). Studies of pulmonary lymphatic clearance of [^{99}TC]DTPA from air spaces during lung inflation and lung injury indicate that new pathways can open through airway epithelium that provide direct access between air spaces and lymphatics (Peterson and Gray, 1987). These studies underscore our incomplete understanding of how the pulmonary lymphatic system functions.

II. Methods

A. Preparation of the Vasculature

General Considerations

If the lung's vasculature is to be studied the following requirements need to be identified, since they will dictate methods:

1. The system of interest: pulmonary or bronchial, and whether arterial, venous or capillary.

2. Technique of examination: light microscopy, electron microscopy, immunohistochemistry. Light microscopy is adequate for morphometric study of all vessels except the capillaries, but offers insufficient resolution for the microcirculation or for detailed examination of the vessel wall. On the other hand, ultrastructural studies preclude injecting casting material such as barium sulfate–gelatin that is helpful in quantitative analysis of the vasculature. For immunohistochemical examination, only short periods of fixation are usually desirable.

3. If the study is quantitative, the vasculature should be fixed under controlled conditions. This involves consideration of several additional factors.

Smooth Muscle Tone

Unless the study sets out expressly to examine vascular reactivity, the presence of vascular smooth muscle tone can lead to variability in the vessel dimension and add to the difficulty in quantifying vessel structure. Tone can be diminished or even eliminated by perfusing a vasodilator such as papaverine sulfate or a metabolic poison such as cyanide. Its effects can largely be overcome by distending the vessel so that the smooth muscle cells are stretched to lengths at which any contractile force they generate is minimal.

Vessel Distention

For quantitation, the vessels should be fixed under consistent conditions of pressure and flow. In practice, this is difficult to achieve and, in any case, begs the question of what pressure. For studies of the normal circulation, physiological PaP could be used, but a choice still has to be made between systolic, diastolic, or mean pressure and a normal pressure would not be likely to perfuse the restricted beds of hypertensive lungs. Conversely, pressures sufficient to perfuse hypertensive beds could overdistend normal vessels.

Although input pressure may be rigorously monitored, pressure dissipation within the lung means that the transmural pressure in distal vessels will not be the same and will vary in different parts of the lung, causing variable degrees of distention. One way of avoiding this is to inject a continuous column of liquid under a maintained pressure that is transmitted to all filled vessels. This method is used in preparing vascular casts for which various materials have been used, including resin (Singhal et al., 1973) and silicone rubber (Yen et al., 1983). After polymerization, tissue is digested away.

If the casting material is capable of being infiltrated in an embedding medium, the filled vessels can be studied microscopically. This method has been successfully used by Reid and associates who modified the technique of Short that involved barium gelatin mixture injected into the pulmonary artery (Elliott and Reid, 1965; Short, 1956). Horsfield (1978) used this technique for studying vessels not reached by the resin cast.

In the technique of Reid et al., a solution of gelatin at 60 °C containing a suspension of barium sulfate is injected into the vasculature at a pressure of 100 cmH$_2$O generated by an air pump. This is sufficiently high to fill patent beds from the most hypertensive subjects and, by distending vessels to maximal or near maximal levels, counteracts the effects of tone. The composition of the mixture prevents it from passing through the capillary bed. It is thus an excellent marker for the arterial or venous sides of the circulation, pulmonary or bronchial, depending on the vessel into which it was introduced. After the vascular bed is injected, the lung is fixed by intratracheal instillation at a transpulmonary pressure sufficient to inflate it to maximal volume (e.g., 25 cmH$_2$O). The gelatin quickly polymerizes. One major advantage is that the barium sulfate contained in the mixture allows the injected bed to be radiographed and the angiograms are useful in analyzing the degree of filling and showing the branching pattern and the sites of blockage.

Tissue samples are embedded either in paraffin wax or methacrylate. The mixture does not infiltrate well with epoxy resins, however, so the method is not suitable for electron microscopic studies.

Fixation

If the vasculature is not injected with a casting material, or if the study demands transmission electron microscopy, vessels are distended with fixative. When conditions closer to normal pressure or flow are desired or when the capillary bed and air–blood barrier are of interest, the fixative is perfused through the pulmonary vasculature (Bachoven et al., 1982). After catheterization, the vascular bed is flushed with a saline solution warmed to body temperature to prevent cold-induced muscle spasm and containing an agent to reduce or eliminate smooth muscle tone. To prevent edema formation, the perfusate should also contain high-molecular-weight compounds that increase oncotic pressure. When the lung is free of blood, the perfusate is changed to fixative solution. Requisite perfusion pressure is established in various ways: by setting the supply reservoir at a given height, attaching it to an air supply at given pressure, or adjusting flow through a peristaltic pump (Bachofen et al., 1982; Hayat, 1981). Once the fixative acts on the wall of the vessels, vascular compliance falls, and flow needs to be reduced to maintain pressure at a low level. For this reason, conditions that combine normal pressure and flow are impossible to achieve. Edema is monitored by suspending the lung from a balance that indicates changes in weight.

The composition of the fixative determines its osmolality. The effects of osmolality on cell preservation and on the proportional composition of the medial muscle layer have been examined in systemic arteries (Mathieu-Costello and Fronek, 1985). A slightly hypertonic solution was found to give the best ultrastructural preservation.

During perfusion the lung is inflated at a given transpulmonary pressure. The balance between inflow perfusion pressure, outflow pressure at the left atrium, and the pressure in the air spaces will control whether the lung is fixed under zone I, II, or III conditions. The requirements are usually for a well-inflated lung that has a fully recruited microcirculation, (i.e., zone III). This can be achieved by raising outflow pressure at the left atrium slightly above airway pressure.

To ensure near maximal distention of pulmonary arteries comparable to that given by barium/gelatin injection, Meyrick and Reid (1979a) ligated the pulmonary veins and injected the glutaraldehyde fixative into the pulmonary artery at a pressure of 100 cmH$_2$O. At the same time, fixative was introduced into the airways at a pressure of 25 cmH$_2$O. Because the blood had not previously been flushed from the system, the red blood cells tended to pack the veins. This method has been successfully used in a number of studies that have combined light microscopic morphometry with ultrastructural studies.

B. Macroscopic Morphometry

Casts

Like the airways, the pulmonary vasculature forms an irregular dichotomous branching system. The branches can be counted or measured in orders from the center outwards or from the periphery inwards. Order number can be increased at each branch, proceeding in either direction, or, in the Strahler system, the most distal branch is the first order and only when two branches of the same order meet is the order number increased.

Angiograms

When the vasculature has been injected with barium–gelatin the angiograms can be used to measure the luminal diameter of axial pathways. Two approaches are possible: measuring the diameter of vessel branches classified by order (counted outwards from an easily identifiable central point) or, more frequently, measuring the diameter at levels that represent given fractions of the total pathway length.

C. Microscopic Morphometry

General Considerations

A morphometric study has several requirements:

1. An index of level within the branching system.

2. Characterization of vessel structure.

3. A measure of vessel size, usually external diameter, but luminal diameter may also be important when estimating the amount of vascular restriction produced by a given change.

4. A measure of the thickness of the vessel wall, and of its three constituent layers: intima, media, and adventitia. Since the media is the contractile layer, it is the most frequently selected for measurement, but in pathological conditions change in the adventitia or intima also occur.

Since these measurements rely on delineating the elastic laminae, sections for vascular morphometry should be stained with an appropriate elastic stain.

Distended Vessels

If the vessels are fully distended, such as after injection with barium–gelatin, linear measurements can be made with an eyepiece graticule. This approach has the advantage of being fast, allowing more vessels to be measured per unit time. Wax or methacrylate sections stained with an elastic stain (e.g., Miller's elastic stain, counterstained with Van Gieson) are viewed by light microscopy. Each artery profile is characterized in four ways (Fig. 3):

1. Level in the branching system is determined by identifying the accompanying airway: preacinar (bronchiolus, terminal bronchiolus) and intraacinar (respiratory bronchiolus, alveolar duct, and alveolar wall).

2. Structure is classified as muscular, partially muscular, or nonmuscular.

3. External diameter is measured as the diameter of the circle formed by the external elastic lamina; if the vessel is sectioned obliquely this is the shorter of its two profile axes.

4. If structure is muscular or partially muscular, medial thickness (MT) is measured as the distance between the internal and external elastic laminae. Medial thickness is expressed as a percentage of external diameter (ED): $2 \times MT \times 100/ED$. Note that some other workers have not used the factor of 2 in calculating this value. Luminal diameter is $ED - (2 \times MT)$.

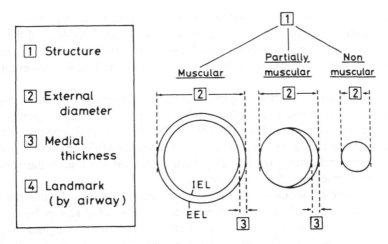

Figure 3 Morphometry of distal pulmonary arteries in tissue section.

For partially muscular arteries, 1 × MT can be used in the equation, but using 2 × MT makes it possible to compare the medial coat in partially and fully muscular arteries.

Constricted/Collapsed Vessels

If the method of fixation does not overcome constriction due to smooth muscle tone, the above methods will not yield consistent results and may give artificially high values for medial thickness. In undistended pulmonary arteries of human subjects, Fernie and Lamb (1985a, b) and others have suggested using medial area as an index of size. The theoretical basis for this is that if vessel length is unchanged, the absolute area of the media will be the same regardless of its shape. Medial area can be measured using a computerized digitizer tablet. Area must be related to an index of size independent of constriction. The length of the internal elastic lamina, usually the most easily resolved of the two laminae, can be used to provide this. It is measured planimetrically. This length can be considered the circumference of the "lumen" of the fully distended vessel, whose area and diameter can then be calculated. This is based on an additional assumption that the length of the elastic lamina does not change with constriction or dilatation of the vessel. This seems not to be so. In the mesenteric arteries of the rat, Lee and co-workers (1983b) showed that the internal lamina was shorter and thicker during constriction than when the artery was relaxed. The differences were apparent in the normotensive Wistar-Kyoto strain, but not in a spontaneously hypertensive strain, which suggested that the physical properties of the lamina were changed in the hypertensive rats.

Fernie and Lamb found that the square root of medial area plotted against length of internal lamina gave a straight line regression. The slope of the line was different in uninjected and injected arteries and indicated that in injected arteries the internal lamina stretched up to 1.5 times the length in undistended arteries.

In the absence of a digitizing tablet, medial and other areas can be estimated by point-counting methods, but this is labor-intensive and limits the number of vessels that can be examined in a given time. This is not to say that point-counting and other stereological techniques have not been successfully applied to studies of the vessel wall, but generally, they have been used to estimate the volume contribution of cells and extracellular matrix.

The directed orientation of the blood vessels and of the cells that line the walls means that totally different results can be obtained if an equally rigidly oriented stereological test lattice is superimposed on a vessel in longitudinal versus transverse sections. For this reason, the orientation of sectioning should be consistent.

The more isotropic distribution of the normal microcirculation allows it to be sampled at random. Stereological methods have been used to determine capillary luminal volume and surface area, endothelial volume, and other variables. These methods have found considerable application in toxicology studies.

Measuring Smooth Muscle Hypertrophy Versus Hyperplasia

Medial hypertrophy can result from hypertrophy of the smooth muscle cells, their hyperplasia, or a combination of both, and usually the amount of extracellular matrix is increased.

Although hyperplasia can be assessed by autoradiography following a pulse of tritiated thymidine, the period of cell proliferation may be short and easily missed. Other ways to answer the question have been explored in systemic, but not in pulmonary, vessels. In mesenteric arteries from spontaneously hypertensive rats, Lee and co-workers (1983 a, b) used stereological point- and intersection-counting to determine the volume to surface ratio (V/S) of the smooth muscle cells. They argued that hypertrophy would increase V/S. If hyperplasia had occurred without a change in cell size, V/S would remain the same, but there would be more cell layers within the media. If a combination of hypertrophy and hyperplasia occurred, both V/S and number would increase.

In the same arterial system and animals model, Mulvany and associates (1985) used a different approach. They cut eight serial cross-sections of an artery. Each was photographed by light microscopy at high magnification so that, within the wall, the smooth muscle cells and their nuclei could be easily seen. Two parallel lines approximately perpendicular to the vessel wall were superimposed on each photograph. The lines contained a roughly oblong area of wall, which over the eight sections reconstructed a known volume. By identifying the counting nuclei in the first section (n1), following the nuclei in subsequent sections, and counting these same nuclei in the last (n8), these workers could determine the number of nuclei within the containing volume (n1-n8).

Both methods came to the same conclusion: in the hypertensive rat hyperplasia of smooth muscle cells is the predominant mechanism contributing to medial thickening.

Scanning Electron Microscopy

Scanning electron microscopy has proved useful in qualitative descriptions of vessels. It has been used to study the morphology and orientation of endothelial cells viewed en face, and the orientiation of elastic fibers in the elastic laminae of the pulmonary artery (Smith, 1976).

D. Special Identification of Wall Cells and Components

Medial Smooth Muscle Cell

Ultrastructural techniques have been used to identify smooth muscle cells by their possession of a contractile apparatus. In the systemic literature, increasing emphasis is being placed on the immunohistochemical demonstration of smooth-muscle-specific contractile proteins such as α-actin or intermediate filament proteins such as desmin.

Intermediate Cell and Pericyte

By conventional light or electron microscopic examination, these cells are identified by their location rather than by any special characteristics. In the pulmonary circulation their similarity to smooth muscle cells was demonstrated immunohistochemically by their staining with antibodies to smooth muscle myosin (Meyrick et al., 1981). Later, in systemic vessels, the expression by pericytes of smooth muscle characteristics was demonstrated for actin (Herman and D'Amore, 1985), tropomyosin (Joyce et al., 1985), and guanosine triphosphate (GTP)-dependent protein kinase (Joyce et al., 1984).

Endothelial Cell

In the normal vessel the endothelial cell has a unique location, so it is unlikely to be confused with another cell type. Weibel-Palade bodies (Weibel and Palade, 1964) in the endothelial cytoplasm are characteristic, but are not found at all levels of the pulmonary vasculature and are uncommon in the rat (Meyrick and Reid, 1979).

Where there is intimal damage or proliferation, the identification of endothelial characteristics may be more difficult to achieve by light microscopy. Immunohistochemical staining for Factor VIIR.Ag is one marker. Current interest centers on the expression of leukocyte adhesion molecules (Worthen et al., 1987).

Adventitial Fibroblast

In muscular arteries or larger veins, the adventitial fibroblast can be recognized by its location outside the external elastic lamina. In nonmuscular vessels its identification is more difficult. Since the adventitial fibroblast actively produces collagen and elastin, its expression of mRNA for these matrix components can be studied by in situ hybridization, using cDNA or cRNA probes (Mecham et al., 1987). Fibroblasts also produce a variety of growth factors: mRNA for one of them, somatomedin C, has been localized in the fetal lung by in situ hybridization in the adventitial fibroblasts of large pulmonary arteries (Han et al., 1987).

Extracellular Matrix

In the past, histochemical stains were used to localize collagen and elastin in the vessel wall, and these stains will continue to be used for routine purposes. For more specific identification of collagen and elastin subtypes, fibronectin, and other matrix components, immunohistochemical staining can be used (Furthmayr, 1982).

Glycosaminoglycan species can be identified by cationic stains used in combination with specific enzyme degradation (Wight, 1980). Identification of specific molecules is achieved by immunohistochemical staining using poly- or monoclonal antibodies (Caterson et al., 1985).

E. Models of Injury

To gain insight into pathogenetic mechanims involved in different disease processes, an animal model offers the investigator a way to study an organ system under artificially manipulated environmental conditions. Animal models are of two types: spontaneously occurring and experimentally induced. Although spontaneous models have the obvious advantage of providing new leads for the study of human disease, for many diseases spontaneous models do not exist. Artifactual models are important for the analysis of pathological changes and their pathogenesis but are limited in that they do not necessarily identify the primary abnormality or cause of a given human disease.

The pulmonary vasculature responds in a number of ways to various injurious stimuli: constriction, inflammation that includes increased permeability resulting in edema, and structural remodeling. Studies of the speed and magnitude of these responses and their distribution along the pulmonary vascular pathway extend our knowledge of human disease. In this section only a few models of injury have been chosen to illustrate methods of producing injury and then of detecting and measuring its effect on the pulmonary vasculature.

Edema

There are still many unanswered questions regarding the effects of edema on pulmonary vessels and pulmonary vascular resistance. The hemodynamic mechanisms causing pulmonary edema have been much studied (Bhattacharya et al., 1980, 1984; Crandall et al., 1983; Greenfield and Harrison, 1971; Hogg, 1978; Hurley, 1978; Ngeow and Mitzner, 1983; Wang et al., 1985; West et al., 1965), and the associated structural changes very little. Analysis of the distribution and amount of edema fluid in the different lung compartments is important, since it can explain the effects of edema on lung mechanics and vascular resistance and elucidate the reasons for the site of accumulation of edema fluid. Edema is caused in two ways: by an increase in either hydrostatic pressure or permeability of the vessel wall. Different experimental methods are used to produce each type.

Hydrostatic Edema

Michel et al. (1986) produced hydrostatic edema in dogs with a fluid overload. They raised pulmonary artery wedge pressure by inflating a balloon in the proximal aorta to obstruct flow partially without lowering distal aortic pressure excessively. The ureters were ligated and the development of edema was monitored clinically. On sections of snap-frozen lung, quantitative morphometric analysis of the area of edema cuffs around vessels, airways, and bronchovascular bundles was performed.

Permeability Edema

Inflammation causes edema by increasing permeability. For example, Michel et al. (1984) produced permeability edema in dogs by intravenous administration of alpha-naphthylthiourea (ANTU). Within 6 hr a dose of 27 mg/kg produced edema, and its development was monitored clinically. In addition, in this model (Michel et al., 1983) the effect on the amount of interstitial edema of three methods of fixation—airway instillation, vascular perfusion, and snap freezing—was compared between lobes processed by each of the three methods. In the experience of Michel et al. (1983) and that of others (Mazzone et al., 1978; Mazzone, 1981; Nicolaysen et al., 1975; Staub and Storey, 1962; Staub et al., 1967), snap freezing did not result in an increase in the amount of interstitial edema and also adequately preserved alveolar edema for visualization by light microscopy. Therefore this would be the method of choice for measuring the absolute amount of edema in the lung. Michel et al. (1983) found that fixation through the airways decreases the amount of alveolar edema, perhaps by washing it from the alveoli into the interstitium because of a pressure gradient in that direction. Perfusion fixation, on the other hand, was found to preserve alveolar fluid well but increased interstitial fluid. Compared with snap-frozen lungs, the absolute amount of edema in lungs fixed either by airway instillation or perfusion increased, but the proportions of interstitial edema around arteries and veins was not altered. Thus the latter methods could be effectively used for studying preferential edema accumulation.

Inflammation

Two processes are involved in the vascular injury of inflammation in the lung, edema and cellular infiltration.

Edema results from increased permeability of the vessel wall, which can be simulated by the administration of chemical agents (for its measurement, see above). In pulmonary microvascular injury, several studies (Craddock et al., 1979; Flick et al., 1981; Heflin and Brigham, 1981; Henson et al., 1982; Hinson et al., 1983; Irvin et al., 1982; Johnson and Malik, 1980; Larsen et al., 1981) have established a pathogenetic role for inflammatory cells, in particular leukocytes. We discuss here only the use of endotoxin to produce a model of vascular injury through inflammation.

Endotoxin Injury

To produce a model of acute microvascular injury, endotoxin is administered as a bolus. When administered over a period of time, whether intermittently or continuously, endotoxin produces vascular remodeling and, in the case of the latter, pulmonary hypertension. Only the acute use of endotoxin to produce a model of inflammation is described here.

Meyrick and Brigham (1983) gave a single infusion of *E. coli* endotoxin (1.25 µg/kg) over 30 min to anesthetized, catheterized sheep. They documented the acute injury it produced with hemodynamic monitoring and sequential lung biopsy at varying intervals over the next 4 hr. To correlate structural and functional changes, they counted the total number of granulocytes and alveoli in 10 microscopic fields of the lung biopsies. Pulmonary artery pressure increased early, peaked at 30 min, and returned to baseline levels by 60 min after the start of the infusion. Within 15 min of the start of infusion, a marked increase in the number of peripheral lung granulocytes and mononuclear cells was found; this increased progressively over 4 hr. At first there was margination of granulocytes within alveolar capillaries and small arteries and veins, later granulocytes also appeared within alveolar walls and the perivascular sheaths. On ultrastructural examination, disruption and fragmentation of granulocytes and damage to capillary endothelial cells, pericytes, smooth muscle cells, fibroblasts, and contractile interstitial cells was seen. This model shows species variation, since in the rat endotoxin infusion does not cause an acute early rise in pulmonary artery pressure (Kirton et al., 1988).

Structural Remodeling

Hypoxia and hyperoxia both cause remodeling of the pulmonary microvasculature but the initial injury and the subsequent response of the pulmonary vasculature differ between the two.

Hypoxia

In hypoxic injury, the basic process is not necrotizing; the remodeling essentially represents a hypertrophic response probably from change in the metabolism of the smooth muscle cells. Acute hypoxia causes vasoconstriction that reverses rapidly on return to air. Chronic hypoxia produces a steady rise in pulmonary artery pressure that persists on return to air, and within 3 days of starting exposure is associated with structural changes (Hislop and Reid, 1976). Vasoconstriction is not necessary for this structural remodeling. These changes are well established after 10 days of hypoxia (Rabinovitch et al., 1979). While in most species some animals do not respond to hypoxia (Hu et al., 1988), the rat is a useful choice for this model since it usually gives a hypoxic, constrictive response and its small size permits use of a conveniently sized chamber. The response to both normobaric and hypobaric hypoxia is similar, although the latter is easier and more practical to use, since application of a vacuum to the experimental chamber produces hypoxia at the desired level.

Hyperoxia

The injury of hyperoxia at adequate levels is a necrotizing and obliterative process that results in a markedly restricted pulmonary vascular bed. Remodeling occurs in the residual patent bed. Practical considerations include the level of hyperoxia, since a high rate of mortality occurs with exposure to 100% O_2. Exposure to 80–87% O_2 achieves both prolonged survival of the exposed rats and progression of cell necrosis to the structural changes of obliteration of microvessel lumens and interstitial fibrosis. Structural remodeling of both preacinar and intraacinar pulmonary arteries was described by Jones et al. (1984) in rats within 7 days of exposure to 90% O_2.

In both hypoxia and hyperoxia, the changes of structural remodeling involve all three muscle coats (intima, media, and adventitia), which can be quantified by conventional morphometric techniques (see discussion of microscopic morphometry above).

Persistent Pulmonary Hypertension of the Newborn

All forms of persistent pulmonary hypertension of the newborn (PPHN) are characterized by failure of normal postnatal adaptation that causes a drop in pressure by dilatation and an increase in compliance of resistance arteries. In some cases, there is abnormal intrauterine structural development of the pulmonary vasculature; in others a functional failure occurs perinatally. Fatal cases have characteristic structural alterations including a medial muscle coat in intraacinar arteries that are normally nonmuscular, abnormally small pre- and intraacinar arteries, and dense collagen sheaths around arteries of all sizes.

While in many cases of PPHN a cause can be identified (see discussion later), for patients with the "idiopathic" type no causative agent has yet been identified. Two proposed causes for this variety were hypoxia or fetal ductal constriction because of maternal ingestion of indomethacin or other cyclooxygenase inhibitors. Attempts to produce experimental models of the structural and functional features of idiopathic PPHN by either of these causes have not been successful.

F. Structural Assessment of Vascular Reactivity

Morphometric Methods

The location of the vasoconstrictive response to hypoxia was investigated structurally in the cat by Kato and Staub (1966) using rapid freezing methods to preserve the tissue. In cross-sections of peripheral pulmonary arteries, internal diameter was measured. This value was significantly decreased in lungs ventilated with an hypoxic gas mixture.

Davies and co-workers (1984, 1985) placed explants of adult rat lung in short-term organ culture. After adding norepinephrine or epinephrine for a given period of time, the explants were fixed and embedded in Epon. Sections 1 μm thick were stained in hematoxylin and basic fuchsin to delineate the elastic laminae. They used a light microscope fitted with a camera lucida to draw on paper the internal lamina of muscular arteries or the single lamina of nonmuscular vessels. A digitizing tablet was used to measure the length of the lamina and enclosed "lunimal area" (Fig. 4). A ratio was calculated of actual luminal area divided by computed area of the idealized circle formed by the lamina. If the ratio was 1, the vessel was theoretically fully distended, while values below this indicated greater degrees of constriction or retraction. By comparing the mean ratio of vessels incubated with and without the constrictive agent, its effects could be determined. In this way, the reactivity was determined for muscular arteries and nonmuscular vessels in normal rats and for the corresponding vessels and newly muscularized arteries in rats made hypertensive by chronic hypoxia.

Myography

The mechanical and contractile properties of pulmonary arteries have been investigated by excising them from the lung and mounting them isometrically on

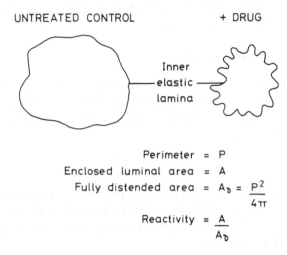

Figure 4 continued below:

UNTREATED CONTROL + DRUG

Inner elastic lamina

Perimeter = P
Enclosed luminal area = A
Fully distended area = $A_D = \dfrac{P^2}{4\pi}$

Reactivity = $\dfrac{A}{A_D}$

Figure 4 Method used to determine the degree of luminal closure of vessels incubated in culture and treated with a vasoconstrictor. After being fixed and sectioned, the length of the internal elastic lamina is measured planimetrically.

a myograph. Force generated is normalized for wall area by dividing by morphometric measurements taken on the vessels after histological processing (Coflesky et al., 1987). Because this method depends on the ability to dissect out vessels undamaged from the lung, it is technically difficult and therefore not applicable to vessels below about 150 μm in external diameter. In the systemic circulation, vessels smaller than this have been investigated by microperfusion, but in the lung this is difficult because frequency of branching prevents one from obtaining sufficiently long segments. In subpleural vessels, micropuncture has been used to determine intraluminal pressure and monitor the site of hypoxic vasoconstriction (Nagasaka et al., 1984).

G. Innervation

Various techniques are available to study the innervation of the vasculature within the lung. They are capable of differentiating between adrenergic, cholinergic, and peptidergic pathways and have been described in a recent review (McLean, 1986).

H. Cell Culture

The ability to obtain vascular cells in culture enables the investigator to study normal function and the response to conditions that trigger pathological changes in the lung. Space does not allow detailed description of culture methods here.

Instead, we will examine briefly the methods used to obtain, in culture, each of the constituent cells of the vessel wall.

Medial Smooth Muscle Cell

This cell can easily be obtained in culture from explants of the pulmonary artery. Most experimentors have used the large central artery (Davies et al., 1987), but differences between cells obtained from this and more distal arteries have not been investigated.

Pericyte

Pericytes have been cultured from rat lung by the digestion of minced tissue with collagenase (Davies et al., 1987). The cell population obtained is largely a mixture of endothelial cells and pericytes. In a serum-supplemented medium, pericytes overgrow the endothelial cells.

Endothelial Cells

These cells can be obtained from the large central pulmonary arteries by stripping the intima (Ryan et al., 1980) or by collagenase treatment (Ryan et al., 1978). Because it is generally agreed that the cells from the large arteries are not representative of the microcirculation, various methods of obtaining the latter have been tried. One technique that also obtains the pericyte uses collagenase digestion of minced tissue (Davies et al., 1987). To prevent overgrowth by pericytes, the endothelial cells are plated on gelatin and grown in medium supplemented with plasma and endothelial growth factor(s). Another technique avoids the use of proteases and claims to obtain cells from vessels of given dimensions. Microspheres of selected diameter are injected into the vasculature and lodge in vessels of the same internal diameter. Endothelial cells migrate onto the spheres and are obtained from the lung by reverse perfusion. In culture, cells can be allowed to migrate off the spheres (Ryan et al., 1982). Alternatively, they can be grown and passaged on the spheres, thus avoiding trypsinization procedures. This technique may still only be suitable for cells from the larger arteries. Spheres appropriate to the microcirculation are ingested by the endothelial cells (Ryan, 1987).

Adventitial Fibroblast

These cells grow out from explants of the wall of central arteries. They have not been obtained from smaller vessels.

III. Interpretation

A. Macroscopic Morphometry

The pulmonary vasculature is an irregularly branching system. Nevertheless, it roughly conforms to the mathematics of a regularly branching system, so that the diameter (d_z) of a vessel at generation z is given by:

$$d_z = d_o \times (3/\frac{1}{2})^z$$

where do is the diameter of the main pulmonary artery.

This can be interpreted as indicating that the vasculature is designed to facilitate mass flow (Weibel, 1984).

Casts

In the human, a resin cast of the pulmonary arteries was used to measure the number, diameter (luminal), and length of branches classified by Strahler order (Singhal et al., 1973). Arteries were uniformly filled only down to 800 μm diameter; the dimensions of more distal arteries were estimated by extrapolation. A total of 17 orders was found to be present. A later study (Horsfield, 1978) used both a resin cast and gelatin injection to measure arteries down to 10 μm diameter. In both studies, diameter or length plotted semilogarithmically against order gave a positive straight line. Number of branches plotted semilogarithmically against order also gave a straight line, but of negative slope. The antilog of the slope is the branching ratio: the number of branches in an order divided by the number in the subsequent order. For orders 3–17, branching ratio was about 3. For orders 1–3, representing the most distal arteries with diameters 10–34 μm, branching ratio was somewhat higher (about 3.6), indicating a greater number of small distributing branches just before the capillaries. The fact that the branching ratio for the whole system was greater than that for airways (2.8) is due to the presence of supernumerary arteries. Strahler ordering of this kind, however, does not provide an adequate picture of the arterial system because it fails to indicate the abrupt stepdown in size that is a feature of lateral and supernumerary branches and that is accompanied by an equally abrupt change in wall structure.

For the pulmonary venous system of the human, Horsfield and Gordon (1981) prepared resin casts of autopsy lungs from two women and one man. They used Strahler orders to classify and measure all branches down to 1 mm diameter and a sample of branches down to 0.2 mm. The dimensions of the smallest branches were estimated to be similar to those of the arteries measured in another study. Fifteen orders of veins were estimated to be present. Branch ratio was 3.301. Branch diameter and length plotted semilogarithmically against order could be described by single straight regression lines. Total volume of the pulmonary veins was estimated as 74 ml.

In the cat, silicone rubber casts were made of the pulmonary veins and measured down to about 100 μm diameter (Yen et al., 1983). For smaller veins, the authors made measurements on thick (1 mm) histological sections of uninjected lung tissue. They identified a total of 11 Strahler orders. Although the description is similar to that of the human pulmonary veins, the plot of log branch length against order gave two regression lines: the one for orders 1–3 was steeper than that for orders 4–10, indicating that at the periphery there is a more rapid increase in the length of more proximal branches.

B. Microscopic Morphometry

Elliott and Reid (1965) found that from the hilum to the alveolar wall two types of pulmonary artery can be recognized (Fig. 5):

1. Conventional arteries that accompany the airways and for proximal arteries are contained within the same connective tissue sheath.

2. Supernumerary arteries that arise from a conventional artery to take a short route to supply adjacent lung tissue that is distal in the conventional pattern of airway branching. They are narrower than conventional arteries at the same level and are more numerous distally.

Structure

Central arteries are elastic in that they have many elastic laminae in addition to the internal and external. Wagenvoort and Wagenvoort (1977) applied the term *elastic* to any artery that had additional laminae, but Meyrick and Reid (1979) followed Brenner's (1935) original description and restricted the term to arteries with more than seven central laminae. They used the term *transitional* for arteries with four to six central laminae, and *muscular* to arteries with fewer than four.

On light microscopic examination, a muscular artery possessing only an external and internal elastic lamina is identified within the acinus of the lung. Each is followed by a short segment in which the medial layer is not circumferentially continuous: a partially muscular artery. This is followed by a segment lacking a media: a nonmuscular artery (Fig. 6). Because the term *arteriole* has no agreed structural definition, Reid and colleagues have called all vessels proximal to the capillary bed *arteries*.

In the rat, Meyrick et al. (1978) have described, in addition, a thick-walled oblique muscle segment in which the usual artery wall has an additional layer of muscle with an oblique pitch.

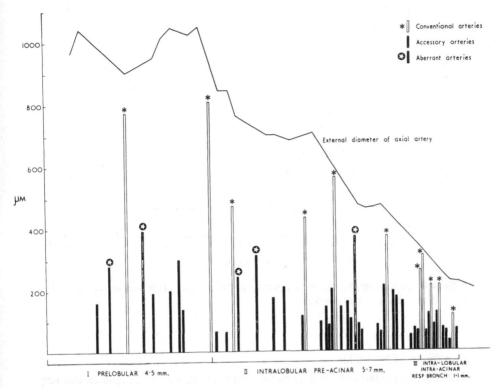

Figure 5 Adult human lung: the external diameter and level of artery branches along the axial artery and their classification into conventional and supernumerary (accessory and aberrant) types.

External Diameter

In the adult human lung the following simple generalization holds. Arteries larger than 1000 μm external diameter are elastic. Between 1000 and 150 μm they are muscular. For arteries smaller than this, structure may be muscular, partially muscular, or nonmuscular.

Medial Thickness

When medial thickness expressed as a percentage of external diameter is plotted against diameter, a characteristic curve is obtained that is steepest over the smallest range of diameters and slopes gently downwards for the remaining sizes. In the smallest muscular arteries the media thickness is only one cell, so that its hypertrophy or hyperplasia has a greater effect on percentage medial thickness than in proximal arteries.

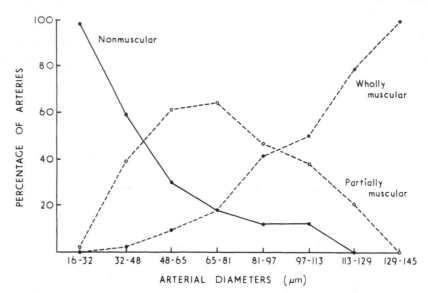

Figure 6 Adult human lung: the frequency distribution of intraacinar arteries classified by structure and external diameter.

Density and Distribution

The fact that the pulmonary arteries run parallel with the airways means that they lie centrally within the acinus. Veins, in contrast, lie at the periphery of the acinus. In the normal lung, the pressure drop has occurred by the end of the capillaries. The veins must therefore transport blood at low pressure and low resistance. For this reason, there are more veins than arteries.

In a given area of lung section, the number of intraacinar arteries is always less than that of alveoli; the ratio of alveoli to arteries or veins is always greater than 1. An intraacinar artery supplies or a vein drains several alveoli. In cat lung, the same conclusion was reached by Sobin et al (1980) using silicone-injected vascular tissue. Singhal et al. (1973), however, assumed that the number of terminal, precapillary arteries was approximately equal to that of the alveoli. Horsfield and Gordon (1981) even suggested that the number was greater, but these were assumptions not made on the basis of direct quantitation.

The density of small vessels in the lung determines the length of capillary traversed between artery and vein. This is reflected physiologically as mean transit time. In the cat lung, Sobin et al. (1980) attempted to determine this length by injecting silicone elastomer into the arterial *and* venous systems of the same lung, but not into the capillaries. Randomly oriented histological sections were prepared and a number of criteria used to differentiate small arteries and veins. Since one of these, however, was the presence, in the arteries, of smooth muscle cells, the method did not identify the smallest pre- and postcapillary vessels, which are nonmuscular. The borders of arterial and venous "zones" were outlined on photomicrographs and stereological assumptions were used to calculate the mean length tranversed by the blood between artery and vein. This was found to be 556 μm. It was interpreted by the authors as the length of capillary, but their inability to identify terminal artery and vein means that this includes pre- and postcapillary vessels and is therefore too long. In rapidly frozen lungs, a more direct method gave values of 240 μm in guinea pigs, 256 μm in rabbits, and 273 μm in dogs (Miyamoto and Moll, 1971). In rat lung, computerized reconstruction of the acinus gave a value of between 205 and 268 μm (Mercer and Crapo, 1987).

Elastic Laminae

Planimetric measurements made on the constricted/retracted muscular arteries smaller than 200 μm in distended diameter in rat lung explants produced an unexpected finding. The length of the internal elastic lamina was sometimes longer than that of the external. When the data were analyzed by least squares regression, a straight line could be drawn, indicating that for larger arteries this was indeed the case, but in smaller arteries the converse applied: the internal lamina was the shorter (Fig. 7). This suggested a crossover point at which the lengths were the same, calculated as an external diameter of 49 μm. When muscular arteries from rats exposed to hypoxia for 14 days were examined in the same way, the relationship still applied, but now the regression line was shifted significantly downwards so that the crossover point occurred at an external diameter of 72 μm. This presumably reflected the presence of medial hypertrophy that has been accommodated by the internal lamina becoming shorter (as lumen was narrowed). In none of our other studies of hypoxia did we find that the external became longer.

For newly muscularized arteries the relationship is different: the regression line is not superimposed even on the line for small muscular arteries. In these remodeled vessels, the new internal lamina is much shorter than the original single lamina, now the external.

The fact that in larger muscular arteries the internal lamina is longer means that it is crenated. These crenations extend as folds along the length of the artery. Their function is unknown, but they probably have a considerable influence on blood flow and on the relationship of the endothelial cell to components of the blood. Recent planimetric studies on the arteries of rat lungs perfused at pressures higher than normal (30 mmHg) showed that in arteries distended at high pressure the length of the internal lamina was always shorter than that of the external. This indicates that the external lamina is sensitive to transmural pressure and responds to high pressure by stretching. Cells lying close to the external lamina

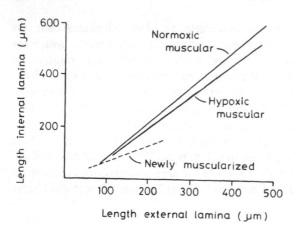

Figure 7 The relationship between the planimetric lengths of internal and external elastic laminae of muscular arteries in rat lung explants. In the hypoxic group, rats were exposed for 14 days to 10% hypobaric hypoxia.

are also presumed to be stretched, whereas cells lying close to the intima are less affected by stretch and more by vessel closure (Davies, personal observation).

C. Details of the Wall of Intraacinar Vessels

Artery

In the rat, ultrastructural studies showed that in the nonmuscular parts of the wall of the partially muscular artery, the cell corresponding to the smooth muscle cell of the media was poorly differentiated, lacking myofilaments and dense bodies. Meyrick and Reid (1979) called these intermediate cells because they seemed transitional between the pericyte and the vascular smooth muscle cell. The level at which they were formed was between nonmuscular and muscular segments. In nonmuscular arteries, the media is present, but the cell corresponding to the smooth muscle cell is a pericyte.

More recent work has extended these studies by using microdissection to relate branching pattern to structure (Davies et al., 1986). In the artery running with the terminal bronchiolus, the wall is several muscle layers thick. Once the bronchiolus becomes respiratory, the number of layers falls abruptly to one or two. This is the structure of the artery accompanying first generation of alveolar ducts. At the level of the second-generation ducts, the internal elastic lamina becomes discontinuous and the smooth muscle cells lose their fully differentiated phenotype to resemble intermediate cells. At the level of the third (terminal) generation duct, the intermediate cell layer is no longer continuous. In the smallest precapillary arteries, the subendothelial cells are pericytes (Fig. 2).

Vein

Microdissection has not been applied to the venous system but other studies indicate that the sequence of structural change is similar. In the veins the media is thinner than in arteries and, in the dog lung, Michel (1982) found that the internal elastic lamina is poorly represented. Medial smooth muscle cells are more loosely arranged than in arteries, but more often form myoendothelial junctions.

D. Special Identification of Wall Cells and Components

Medial Smooth Muscle Cells

On ultrastructural examination, smooth muscle cells can be identified by plentiful cytoplasmic microfilaments (actin, 5–6 nm diameter) and less frequent myosin filaments (15 nm diameter); caveolae appear along ad- and abluminal membranes. Dense bodies occur within the contractile apparatus and dense plaques are distributed along the membrane. The plaques are regions of adhesion between the cell and matrix. One study has described a preferential distribution of plaques along the abluminal membrane and speculated that this was the distribution necessary to bring about luminal closure (Gabella, 1983).

In systemic arteries smooth muscle cells communicate by gap junctions, but these have not been reported in the pulmonary circulation. In systemic arteries the myoendothelial junction also appears to be a gap junction (Davies, 1986), but in the intraacinar arteries of the rat, contacts between smooth muscle and endothelial cells appear more like areas of adhesion, containing intercellular matrix (Davies et al., 1986)

Intermediate Cells and Pericytes

The intermediate cell and pericyte lack the fully differentiated characteristics of the smooth muscle cell. They are closely apposed to the endothelium and its basement membrane. In places, this membrane is interrupted by a process of the pericyte, which forms a "peg and socket" junction with the endothelial cell. Although gap junctions have been described between pericytes and endothelial cells in brain microvessels (Larson et al., 1987), none have so far been detected in the lung.

Because, during hypertensive remodeling, the intermediate cell and pericyte acquire the morphological characteristics of smooth muscle cells, considerable interest has been shown in their normal expression of smooth muscle contractile proteins. In the rat lung, the precursor cells lining the walls of nonmuscular arteries gave positive immunohistochemical staining when tested with antibody against smooth muscle myosin (Meyrick et al., 1981). In the lungs of rats exposed to hypoxia, staining was increased in newly muscularized arteries and, indeed, was enhanced in the media of muscular preacinar arteries.

E. Models of Injury

Perhaps no injury or adaptation by lung is without an effect on its vessels. Here we have selected a few examples to illustrate interpretation of injury, its pathophysiology, or pathogenesis, and always with a focus on structure. Hypoxia offers an example of vasoconstriction, metabolic perturbation, and structural remodeling that includes even a change in cell phenotype. Inflammation represents the humoral and cellular response to a variety of injuries and includes a variety of changes calling for structural interpretation. Adult respiratory failure (ARF) and the adult respiratory distress syndrome (ARDS) exemplify acute alveolar damage and its progression. Hyperoxia is doubly linked with this since it contributes to the lung appearance of fatal cases. While it is a life-saving treatment, it also causes significant injury. It is included here because it provides an example of both a necrotizing lesion in its early phases and structural adaptations in the healing phase. It illustrates several important biological concepts of injury, adaptation, and repair, and is relevant to disease of all ages: adult, child, and newborn. Bronchopulmonary dysplasia is discussed as an example here of both hyperoxic injury and its impairment of lung growth.

Edema

In dog lungs with hydrostatic edema, Michel et al. (1986) studied the distribution of fluid in the interstitium around arteries, veins, airways, and connected bronchovascular bundles. From photomicrographs of these structures, the individual areas of the vessels and airways in addition to total bronchovascular area and the edema cuff area around each structure were obtained with a digitizer. To calculate edema ratios (ER), the area of the vessel at the external border of the media (B) was subtracted from A, the area of the vessel with its edema cuff. The ER was computed from the equation A-B/B. Larger vessels had higher edema ratios than smaller vessels, and arteries had larger edema ratios than veins. This means the arteries with bronchi and bronchioli in their bronchovascular bundle had higher edema ratios than those running with respiratory bronchioli, despite the anatomical continuity of the entire interstitium of the lung (Low, 1978). Thus, it was shown that with hydrostatic edema there is an apparent compartmentalization, with a preferential pattern of distribution of edema in the absence of anatomical barriers. Two factors have been invoked to explain this pattern of distribution. One is anatomical: large arteries have a larger amount of surrounding loose connective tissue than veins and small arteries, and large airways are more loosely attached to the surrounding parenchyma than small airways. The larger amount of connective tissue and looser parenchymal attachment permits more fluid accumulation around larger arteries and airways, respectively. Another factor that could influence fluid accumulation is the interstitial pressure. This suggestion is supported by evidence indicating that interstitial pressure is more negative around vessels than airways (Permutt, 1965) and more negative at the hilum than at the periphery of the lung (Bhattacharya et al., 1984; Taylor et al., 1982). Thus, interstitial compliance reflects both the structural properties of the interstitium and the pressure gradients within it, and this combination could explain the selective pattern of distribution of edema.

Low interstitial compliance within tight perialveolar tissue has been invoked to explain the small quantities of perialveolar fluid in lung edema (Pritchard, 1982). Based on experimental studies of the movement of protein from alveoli to interstitium, Gee and Staub (1977) and Staub (1983) proposed that in severe edema fluid flows from the peribronchovascular connective tissue into the lumen of small airways, filling the alveoli in a retrograde manner. The probable site of this retrograde filling of alveoli is more proximal than the respiratory bronchiolus because of the paucity of fluid accumulation around small airways, as demonstrated by Michel et al. (1986).

The relative distribution of fluid in this hydrostatic edema in dogs is similar to that in the permeability edema produced by ANTU (Michel et al., 1984), but in hydrostatic edema the cuffs were smaller than in permeability edema. This difference could reflect the shorter time that elapsed before fixation after the development of hydrostatic edema than after ANTU administration. Another explanation may be the higher intravascular pressure during hydrostatic edema, which lowers the interdependence between airways, arteries, and parenchyma and results in smaller amounts of interstitial fluid within bronchovascular bundles and around vessels (Parker et al., 1981). Other substances, such as oleic acid (Erhart and Hofman, 1981; Malo et al., 1984) and nitrogen dioxide (Vassilyadi and Michel, 1985) produce more alveolar flooding and less interstitial edema than in hydrostatic edema, and these differences are most likely related to the severity of edema and rapidity of onset.

The effect of edema on pulmonary vessels and vascular resistance is not clear (Bhattacharya et al., 1980; Hogg, 1978; Ngeow and Mitzner, 1983; Wang et al., 1985). Pulmonary vascular resistance is increased in edema and while several mechanisms have been proposed for increased resistance, including compression of vessels (Hogg, 1978; Ngeow and Mitzner, 1983; West et al., 1965), increased hematocrit, and intravascular endothelial abnormalities (Wang et al., 1985), the sites at which such an increase might occur were not analyzed until the recent work by Michel and associates (1987). They compared the cross-sectional areas and diameters of small airways and arteries in the lungs of normal dogs and those in which edema had been produced by fluid overload (hydrostatic edema) and by ANTU (permeability edema). In edema of either type, as transpulmonary pressure increased (the lungs became stiffer), airway and arterial areas and diameters did not decrease. These studies show that interstitial edema fails to compress small arteries either because of the relative rigidity of their walls or the high intravascular pressure or both. The interstitial edema encroaches instead on the relatively more compliant alveoli in the surrounding lung. Mechanisms other than compression of small arteries by interstitial edema must be invoked to explain the increased pulmonary vascular resistance observed in edema.

Elevation of pulmonary hydrostatic pressure (Uhley et al., 1967) and isobaric reduction of plasma oncotic pressure (Zarins et al., 1978) are both associated with a marked increase in pulmonary lymph flow. Following prolonged periods of increased pulmonary microvascular pressure or decreased plasma oncotic pressure, the pulmonary lymphatic capacity is exceeded and fluid accumulates in extraalveolar connective tissue (Staub et al., 1967; West et al., 1965). Alveolar wall edema and alveolar space flooding subsequently occur. Significant ultrastructural differences are seen in the dog lung whether isolated or in vivo, between edema produced by increasing microvascular pressure or decreasing oncotic pressure (Defouw, 1982; Defouw and Chinard, 1983; Defouw et al., 1983, 1985). In cells of the air–blood barrier of isolated dog lungs, increased micropinocytotic vesicles were seen after the development of alveolar wall edema and alveolar lumen flooding. In vivo, micropinocytotic vesicles were not seen, and neither alveolar wall nor alveolar lumen edema developed. Instead, the extravascular fluid accumulated in connective tissue spaces away from the alveoli. This indicates that alveolar wall fluid accumulation does not occur until pulmonary lymphatic capacity is exhausted, and that the increased vesicles in the isolated lungs represent a defense against excessive interstitial fluid accumulation by providing a way to return the filtrate to the blood. It is postulated that persistent edema would lead to coalescence of these vesicles, cytoplasmic vacuolation, and cellular disruption.

Inflammation

Permeability edema and cellular infiltration are the two main components of inflammation (for interpretation of edema, see above).

Cellular Infiltration Produced by Acute Endotoxin

Meyrick and Brigham (1983) studied the acute injury following a single infusion of *E. coli* endotoxin to sheep. Within 15 min of the start of the infusion, margination of granulocytes and mononuclear cells was seen in alveolar capillaries and arteries and veins less than 60 μm in external diameter. By 30 min there was marked thickening of the alveolar walls because of the increased cellularity. Within 2 hr edema was noted within perivascular sheaths, interlobular septa, and intraalveolar spaces. At this time a mixed cellular infiltrate (mononuclear and granulocytic) was present in the edematous perivascular sheaths, and granulocytes admixed with mucus appeared within the lumen of terminal bronchioli. In absolute terms, at 15 min the granulocyte increase was threefold over the baseline, at 60 min a fourfold increase, and at 4 hr a sixfold increase was apparent. Ultrastructural study revealed that granulocytes and lymphocytes were sequestered in equal numbers after the endotoxin infusion. By 30 min, more specific granules were free than at 15 min, and disrupted granulocytes were noted. By 60 min the electron density of capillary endothelium and type I pneumocytes was greater than normal and the number of pinocytotic vesicles was increased. After 2 hr in some capillaries the endothelial cell layer was disrupted. The infusion caused an early rise in pulmonary artery pressure that peaked at 30 min after the start of the infusion and had returned to baseline by 60 min. Lymph flow increased within 15 min and remained elevated throughout the study. Lymph to plasma protein concentration ratio decreased below baseline at 30 min. By 2½ hr it had increased above baseline, indicating increased vascular permeability.

Precisely how endotoxin mediates these vascular changes is not clear. Both the complement and coagulation systems as well as several cell types—granulocytes, monocytes, macrophages, platelets and mast cells—have been implicated (Morrison and Ulevitch, 1978). The hemodynamic response and specific cellular infiltrate varies with species and type of endotoxin.

In the model described here, margination of leukocytes in the microcirculation coincides with the early rise in pulmonary artery pressure, whereas interstitial infiltration and cellular damage precedes an increase in vascular permeability. Granulocytes appear to contribute more to the later physiological changes than to the initial ones, since in granulocyte-depleted sheep the initial pulmonary hypertension still occurs, whereas the later increase in vascular permeability is reduced (Heflin and Brigham, 1981). While the effect of granulocyte depletion on the endothelial injury has not been investigated, this attenuated second-phase vascular permeability response suggests that it should be less. In vitro, activated granulocytes have been shown to cause endothelial cell damage (Sacks et al., 1978). Granulocytes seem to be responsible for the increased airway resistance associated with endotoxemia (Snapper et al., 1981). This change in lung mechanics coincides with the appearance of mucus and granulocytes in the airway (Snapper et al., 1981) and is not seen in granulocyte-depleted sheep (Hinson et al., 1982, 1983).

Lymphocytes also have a role in mediating the response to endotoxin. The accumulation of lymphocytes in the microcirculation is similar to that of granulocytes (Meyrick and Brigham, 1983); the early rise in pulmonary artery pressure is attenuated by chronic thoracic duct drainage that depletes T lymphocytes (Bohs et al., 1979). In sheep, administration of 6-methylprednisolone before endotoxemia attenuates the early phase and inhibits the late phase effect, probably because of steroid modification of lymphocyte function, although steroids may also affect platelet and granulocyte function (Brigham et al., 1981).

Increased levels of prostaglandin metabolites in lung lymph following endotoxemia (Frolich et al., 1980; Ogletree et al., 1982; Watkins et al., 1982) suggest that prostaglandins also play a role in the response to endotoxin.

Structural Remodeling

Each coat of the vessel wall consists of a single cell type, and each coat is involved in the response to injury. The response is mainly a hypertrophic one. The endothelial cells undergo hyperplasia or hypertrophy and increase their production of elastin; the smooth muscle cells show hypertrophy and hyperplasia and an increase in extracellular matrix; the adventitial fibroblasts proliferate and increase deposition of collagen. Such increased activity could be caused by mediators that act as growth factors, or the response could represent removal of inhibition, as if endothelial cells were to decrease their production of a pericyte-inhibiting factor (EPIF).

Hypoxia leads to a metabolic or hypertrophic process that encroaches on vessel lumen; hyperoxia causes necrosis and obliteration of part of the vascular bed and remodeling in the much reduced residual patent bed. Thus both restrict the vascular bed and provide examples of relatively high blood flow. A common example of high flow is offered by congenital heart disease, with a left to right shunt when total flow is increased through the pulmonary vascular bed. With obliteration of part of the bed as a response to injury, a relative increase in load is placed on the patent vascular bed. Thus in pulmonary hypertension localized high flow occurs. In hypoxia, although the vessels are patent, the restriction of cross-sectional area results in a generalized increase in velocity, especially in exercise.

Hypoxia

Media

In large and small arteries in the rat structural changes reflecting altered cell activity are seen within a few hours of hypoxic exposure (Meyrick and Reid, 1979b, 1980a; Rabinovitch et al., 1979). In the main pulmonary artery, the media increases in mass and thickness due to an increase in extracellular matrix (Meyrick and Reid, 1980a) and smooth muscle cell hypertrophy that affects the organelles of protein synthesis more than the contractile elements is seen. The fully muscularized low-resistance arteries at the periphery also show only a modest increase in labeling with radioactive thymidine (Meyrick and Reid, 1979b).

Adventitia

In preacinar arteries during the first 3 days of hypoxia, the adventitia increases in mass and thickness. By 1 week it is three times the normal thickness. The fibroblasts show a striking increase in labeling index: a sixfold increase at 24 hr and an eightfold increase by 3 days (Meyrick and Reid, 1979b). This activity is probably a metabolic effect of hypoxia rather than one that is pressure driven, since the initial hypoxic vasoconstriction causes only a small rise in pressure. A smaller response is also seen in the adventitial fibroblasts of the small muscularized resistance arteries at the periphery, where labeling increased to twice baseline levels (Meyrick and Reid, 1979b).

Intima

In the large preacinar arteries, the endothelial cells undergo hypertrophy and their basal surface becomes separated from the elastic lamina by edema, although the cell edge remains tethered. Labeling indices show a doubling in mitotic activity throughout the study period (14 days is the maximum time studied) (Meyrick and Reid, 1979b). In the recovery phase, elastin is produced by the endothelial cell so that when the fluid disappears, elastin remains between the cell and basement membrane. In partially muscular and nonmuscular peripheral arteries the endothelial cell does not show a peak of mitosis until day 7 of hypoxic exposure, and even then it affects only 8% of the population and is short lived. In these vessels, the pericyte and intermediate cell population (presursor smooth muscle cells) undergo hypertrophy and hyperplasia and by day 5 nearly 20% of these cells are labeled (Meyrick and Reid, 1979b). The pericyte and endothelial cell are close neighbors and even share the same basement membrane, yet they show this striking difference in activity. The pericyte produces its own basement membrane and an elastic lamina that separates it from the endothelial cell. These precursor cells undergo a phenotypic change and differentiate into a smooth muscle cell.

The endothelial cell from the normal lung microcirculation produces a substrate that inhibits pericyte multiplication. The pericyte proliferation during hypoxia suggests that this inhibition is no longer present or that the cell has lost its sensitivity to it.

Hyperoxia

Media

In the rat, exposure to 90% O_2 for 7 days increases the medial thickness of preacinar arteries, whether muscular or partially muscular in structure (Jones et al., 1984). This increase is due to both hypertrophy of smooth muscle cells and increase in extracellular matrix and is even more striking after 4 weeks of exposure (Jones et al., 1985). In intraacinar arteries the medial thickness of both muscular and partially muscular arteries also increases (Jones et al., 1984). The proportion of muscularized intraacinar arteries increases (Jones et al., 1984, 1985) and of nonmuscularized arteries decreases, indicating structural remodeling of artery walls. The new muscle probably arises from a phenotypic change by the precursor cells (intermediate cells and pericytes) within the intima (see below).

Adventitia

Hyperoxia increases the adventitia that occurs at all arterial levels, from proliferation of fibroblasts and increase in collagen. Fibroblasts in the aveolar wall also show hyperplasia. The alveolar wall shows an interstitial fibrosis (Pappas et al.,

1983), and the small intraalveolar wall arteries develop cuirasses of collagen (Jones et al., 1985). Fibrosis is perhaps the most significant of the lung's responses to hyperoxia. Bronchopulmonary dysplasia following oxygen therapy in premature neonates is an example (see below).

Intima

The endothelial cell of the intima is the site of the first identified structural injury of hyperoxia (Dobuler et al., 1982). Injury is presumably due to the intracellular generation of toxic oxygen metabolites. The cellular mechanisms of oxygen injury and defense have been extensively reviewed (Borg et al., 1978; Cadet and Teoule, 1978; Deneke and Fanburg, 1980; Frank, 1985; Frank and Massaro, 1980; Freeman and Crapo, 1982; Fridovich, 1978; McCord, 1974; McCord and Fridovich, 1978; Michelson, 1977; Svingen et al., 1978). Within 60 hr of hyperoxic exposure in rats (Crapo et al., 1980), about one-third of the endothelial cells in pulmonary capillaries have been destroyed, and their mass and total capillary surface area are proportionately decreased. The rate of endothelial cell loss is more gradual with exposure to 85% O_2 (Crapo et al., 1980). Platelets and polymorphonuclear leukocytes accumulate within capillaries, probably augmenting endothelial cell injury and death by their release of free radicals (Barry and Crapo, 1985). Ultrastructural examination of these microvessels reveals intimal edema with separation of the endothelial cells from their basement membrane (Jones et al., 1985).

Edema, cellular debris, and thrombosis all contribute to widespread occlusion of capillaries and microvessels, which are well illustrated by the macroscopically white and avascular lung after 1 week of hyperoxia (Jones et a., 1984).

The damage to the vascular endothelium is not uniform, however, and in some vessels the endothelial cells undergo hypertrophy to the point of causing luminal closure (Jones et al., 1985). In the pre- and postcapillary microvessels ($<$ 100 μm), the endothelial cells remain largely intact and undergo hypertrophy. In these vessels, intermediate cells and pericytes that lie in close apposition to the endothelial cells undergo hypertrophy. With chronic hyperoxia these cells also undergo a phenotypic change to smooth muscle cells, thereby gradually converting nonmuscularized arteries to partially muscular arteries and partially muscular to fully muscular arteries. This process of remodeling only takes place in the residual patent vascular bed (Jones et al., 1985). The overall effect of this process is to restrict further the vascular bed by wall encroachment on the lumen.

Bronchopulmonary Dysplasia

The effect of exposure to high concentrations of oxygen (hyperoxia) in the newborn period has an adverse effect on lung growth, especially its vessels. The degree and reversibility of such alterations have significant clinical implications, because treatment with oxygen and supported ventilation constitutes standard therapy for the respiratory distress syndrome, a state resulting from deficiency of surfactant in premature infants. Northway et al. (1967) described the chronic pulmonary disease bronchopulmonary dysplasia (BPD), which resulted from the use of high oxygen and intermittent positive-pressure respirators in these infants. Following the stage of acute injury and cell loss, characterized by hyaline membranes lining distal air passages, hyperemia, collapse, and lymphatic dilatation, the period of regeneration occurs. Although hyaline membranes persist in this stage, histological studies show repair of alveolar epithelium, dilatation of alveoli, and squamous metaplasia of epithelium lining the small airways. The next stage represents a transition to chronic disease and is characterized by fewer hyaline membranes, more epithelial regeneration and metaplasia, and more severe emphysematous change. In addition, hypertrophy of peribronchiolar smooth muscle, perimucosal fibrosis, patchy areas of collapse, large numbers of intraalveolar macrophages, and thickening of alveolar septal walls and basement membranes are also seen. Vascular structural changes that include lesions typical of pulmonary hypertension were studied morphometrically by Tomashefski et al. (1984). Axial arterial pathways, while maintaining lengths equivalent to those of age-matched fetuses, had internal diameters that were either excessively wide or diffusely narrow. Percentage medial thickness of muscular arteries was lower than fetal values, which suggested that postnatal vascular adaptation had occurred. The proportion of muscularized intraacinar arteries was increased, indicating peripheral extension of smooth muscle. Others have described the pulmonary vasculature in BPD and also reported a variety of findings ranging from minimal change (Anderson and Strickland, 1971) to medial hypertrophy (Tagnizadeh and Reynolds, 1976), endothelial proliferation and adventitial fibrosis (Bonikos et al. 1976), and reduction of capillaries (Larroche and Nessman, 1971). The variability of vascular structural alteration seen in BPD probably reflects the variety of factors that determine injury, recovery, and repair. The degree of prematurity, the severity of hyaline membrane development, injury by high oxygen and ventilation, residual scarring, and growth in the lung capable of growth each contributes to the evolving picture.

Ultrastructural studies of the effects of oxygen therapy on the neonatal lung (Anderson et al., 1973) have shown generalized capillary endothelial damage, interstitial edema, proliferation of type II alveolar cells, and incorporation of hyaline membranes into septal walls. The mechanisms of oxygen toxicity have been extensively reviewed (Deneke and Fanburg, 1980; Frank and Massaro, 1980; Frank, 1985; Freeman and Crapo, 1982). It is generally believed that in tissues a variety of biochemical processes involving the utilization of oxygen result in different end products, including hydrogen peroxide, superoxide radicals, hydroxyl radicals, and singlet excited oxygen; each of these intermediate metabolites is capable of causing tissue damage (Fridovich, 1978; McCord and Fridovich, 1978).

The intracellular production results in chain reactions causing lipid peroxidation, depolymerization of mucopolysaccharides, protein sulfhydryl oxidation, and cross-linking, eventually leading to enzyme inactivation and nucleic acid damage (Borg et al., 1978; Cadel and Teoule, 1978; McCord, 1974; Michelson, 1977; Svingen et al., 1978).

The body's natural defense against oxygen damage includes groups of enzymes and antioxidants such as catalases, peroxidases, glutathione, ascorbate, and alpha-tocopherol. Premature infants have been shown to have lower levels of antioxidants in the first few days of life and hence are at greater risk for oxygen toxicity (McCarthy et al., 1984).

High Flow

Congenital heart disease, with a left to right shunt, offers an example of pulmonary vascular structural injury and remodeling secondary to high flow (Aziz et al., 1977; Berry et al., 1981; Edwards, 1957; Ferencz, 1960; Friedli et al., 1974; Hallidie-Smith, 1968; Haworth and Macartney, 1980; Haworth and Reid, 1977a, b, c, 1978; Haworth et al., 1977; Heath and Edwards, 1958; Heath et al., 1958; Meyrick and Reid, 1980; Naeye, 1962; Newfeld et al., 1974; Rabinovitch et al., 1978, 1981; Robertson, 1968; Wagenvoort, 1973). Because the vascular changes occur here when the lung is still growing, high flow also modifies vascular growth. The normal density of pulmonary arteries fails to develop with overall restriction of the vascular bed.

Heath and Edwards (1958) classified the injury of pulmonary vascular occlusive disease into six grades based on increasing damage:

Grade I   :   Medial hypertrophy

Grade II   :   Medial hypertrophy and intimal proliferation

Grade III :   Progressive fibrous occlusion

Grade IV   :   Complex dilatation lesions

Grade V   :   Dilatation lesions and pulmonary hemosiderosis

Grade VI   :   Fibrinoid medial necrosis

Grade I

An increase in thickness of the media of muscular arteries is an early change in the vascular remodeling of high flow, and it occurs fairly quickly in response to high flow. An increase in medial thickness is present at birth in babies born with atresia or critical stenosis of the aortic valve and an intact ventricular septum, in whom prenatal blood volume in lung was probably increased (Haworth and Reid, 1977b; Naeye, 1962). In such infants, the entire cardiac output is ejected into the pulmonary artery and the body is perfused by the ductus arteriosus. The medial coat of the preacinar pulmonary arteries is markedly thickened in these children.

Grade II

The initial intimal lesion in high flow is a cellular proliferation in small muscular arteries. The proliferation can be so marked as to occlude the lumen; when this change is widespread, the capacity of the vascular bed is significantly reduced.

Grade III

With persistence of high flow, there is intimal fibrosis in vessels smaller than 300 μm in diameter. It begins beneath the endothelium, causing concentric narrowing; it progresses centrifugally. This is important because it represents an "irreversible" structural alteration (Heath and Edwards, 1958); there will be incomplete regression of pulmonary vascular disease after repair or cessation of high flow. Lesions of this severity are rarely seen in the first 2 years of life and since correction or palliation of most cardiac defects is now possible in early childhood, this form of structural change is becoming uncommon. This lesion must be differentiated microscopically from fibrosis secondary to organization of a thromboembolus, which is usually eccentric. As high flow continues, there is splitting and reduplication of the internal elastic lamina, and the fibrotic process extends proximally to involve the origin of the affected arteries and also the artery of origin. Muscular arteries larger than 600 μm in diameter are rarely affected by this process.

Grades IV and V

In the more severe forms of pulmonary hypertension, complex lesions involve perhaps all three vessel coats. In small and medium-sized muscular arteries, both proximal and distal to fibrotic intimal occlusion, there occurs a marked thinning of the media and dilation of the vessel wall to form microaneurysmal sacs. This change probably represents disuse atrophy in sites distal to obstruction, and decompensation due to the elevated resistance in sites proximal to the obstruction. From the walls of the sacs, a vast network of thin-walled vessels and capillaries arise that are distributed within the alveolar walls, thus providing collateral pulmonary blood flow. Thrombi and papilliferous endothelial proliferation are frequently seen within the sacs, giving rise to a "plexiform lesion." At this stage, at least focally, the lungs are highly vascular. Hemosiderin deposition is also seen at this time; whether this is due to hemorrhage or chelation with iron of glycoprotein in macrophages is not clear.

Grade VI

With severe pulmonary hypertension, fibrinoid necrosis in the media results from necrosis of the medial smooth muscle cells and fibrin deposition. If this is accompanied by an inflammatory reponse, polymorphonuclear leukocytes are present and are seen perhaps in the adventitia and adjacent alveoli as well. It is believed that intense vasoconstriction (vasospasm) contributes to the necrosis (Harris and Heath, 1962; Wagenvoort and Wagenvoort, 1977).

The first three grades reflect progressive changes and a labile pulmonary vascular bed, whereas the higher grades are not necessarily seen in progression but probably represent a less labile pulmonary vasculature. Although this classification was useful in assessing structural damage in older patients, it failed to encompass the modified vascular growth in the child's developing lung. We have (Haworth and Reid, 1977a, b, c, 1978; Haworth et al., 1977; Rabinovitch et al., 1978, 81) devised a scheme that also incorporates these developmental alterations.

Grade Ia

This describes the presence of muscle in the walls of distal (intraacinar) arteries that in the fetus and child are free of muscle (Haworth and Reid, 1977a, b; Rabinovitch et al., 1978, 1981). In the adult, some intraacinar arteries are normally muscularized but a greater proportion of small intraacinar (i.e., alveolar wall) arteries develop a muscle coat in high flow.

Grade Ib

This is characterized by an increase in medial thickness of normally muscularized arteries. These changes—increased medial thickness of muscular arteries and extension of muscle into smaller, more peripheral arteries than normal—were also seen in experimentally produced high flow in pigs following aortopulmonary anastomosis (Rendas et al., 1979).

Grade Ic

This feature is of special significance because it correlates with an elevated pulmonary vascular resistance and often points to irreversible functional impairment of the pulmonary vascular bed. It is characterized by a reduction in the density of small peripheral arteries, and in the growing lung signifies a failure of peripheral arterial branching to keep pace with alveolar multiplication. It is identified by an increased alveolar arterial ratio per unit area of lung. This change chronologically follows Grade Ia and Ib changes (Rabinovitch et al., 1978, 1981) and is generally seen in severe pulmonary vascular occlusive disease (grade III and upward on the Heath-Edwards [1958] scale). Obliterative injury could also result in this.

While in adults the distribution of pulmonary vascular change can vary throughout the lung, with the lower lobes being generally more severely affected than the upper lobes (Harisson, 1958), in young children extensive postmortem studies have not demonstrated regional differences in structural change in the absence of differences in perfusion (Haworth and Reid, 1978).

Structural Remodeling in Persistent Pulmonary Artery Hypertension of the Newborn, Idiopathic Variety

The drop in pulmonary artery resistance and pressure that occurs at birth results from adaptation based on diffuse dilation as well as an increased distensibility of the special resistance precapillary segment. At birth in the newborn human and in most species studied, the intraacinar arteries are nonmuscular. In a number of types of PPHN in which a cause can be identified (Haworth and Reid, 1977a, b; Murphy et al., 1984; Rabinovitch and Reid, 1980), precocious muscularization of intraacinar arteries has occurred in utero. In the idiopathic variety, this structural change is also found (Haworth and Reid, 1976; Murphy et al., 1981) and we discuss this here.

Clinical

In the idiopathic variety of PPHN, which at present needs to be considered a separate disease, fatal cases show abnormal arterial muscularization that has developed in utero (Haworth and Reid, 1976; Murphy et al., 1981). Morphometric analysis of the lungs of those infants who die soon after birth show a normal number of bronchial and arterial generations and alveoli, normal arterial density per acinus, and normal pulmonary veins. The structure of the pulmonary arteries is abnormal: preacinar arteries have a thickened media and reduced luminal cross-sectional area, normally nonmuscular intraacinar arteries have a muscle coat, and arteries of all sizes have a thick collagen adventitial sheath. In fatal cases, such a remodeled vascular bed has not adapted postnatally or has done so only transiently, and these infants have elevated postnatal pulmonary arterial pressure and resistance, which produce shunting of blood from pulmonary to systemic circuits through the foramen ovale or ductus arteriosus. It is likely that these structural alterations also make the vascular bed more reactive to vasoconstrictive stimuli (Goldberg et al., 1971) and less responsive to stimuli for vasodilation (Meyrick and Reid, 1980b).

Animal Studies

Two conditions had been widely suggested as causes of precocious muscularization in utero: hypoxia and early closure of the ductus arteriosus.

Morphometric analysis of the pulmonary arterial bed was performed in two species: newborn rat and guinea pig whose mothers were made hypoxic during pregnancy (Geggel et al., 1986; Murphy et al., 1986). In either model the structural changes of fatal human PPHN were not found, and in the guinea pigs functional studies showed a normal drop in pulmonary artery pressure at birth. These animals showed the normal adaptation at birth, including the increase in external diameter and drop in wall thickness typical of the low-resistance arteries. Rendas et al. (1978) studies these changes in the newborn pig, using morphometric analysis and hemodynamic assessments and found that wall thickness of arteries

15–200 μm diameter falls to adult values within 3 days after birth. These perinatal changes are similar to those described in the human (Hislop and Reid, 1974) and are closely related to changes in cardiorespiratory function (Avery and Cook, 1961; Rudolph, 1974). The changes in the pig, however, occur at a much faster rate; the human changes in childhood and adolescence are telescoped into the first 3 months of life.

We studied the effects of hypoxia on this postnatal vascular adaptation in new born rabbits (deMello et al., 1988). Pups were delivered by cesarean section into either 10% hypobaric hypoxia or normobaric normoxia and sacrificed at 1, 6, 24, 48, and 72 hr after birth. To remove tone, the pulmonary vasculature was perfused with saline containing 10 mM KCN, then with glutaraldehyde at 30 mmHg. The airways were distended with fixative at 10 mmHg. Morphometric analysis revealed that at 1 hr in normoxic animals, preacinar artery percentage medial thickness (%MT) had fallen below the zero hour value, while external diameter remained unchanged, indicating a change in medial mass. A further drop in %MT accompanied by an increase in external diameter occurred at 72 hr. Hypoxic pups did not survive beyond 48 hr, but in the first 48 hr their vascular structure did not differ from those born into normoxia, indicating that in the first 48 hr of life hypoxia did not change this perinatal structural adaptation by resistance arteries.

Indomethacin

Reports of human newborn infants manifesting perinatal pulmonary hypertension following maternal ingestion of acetylsalicylic acid or indomethacin (Csaba et al., 1978; Levin et al., 1978; Manchester et al., 1976) led to the suspicion that fetal ductal constriction (because of cyclooxygenase inhibition) was the cause of PPHN. An epidemiological survey of thousands of pregnant women with rheumatoid arthritis who required large doses of acetylsalicylic acid revealed no deleterious effect on the fetal circulation and no difference in perinatal mortality (Shapiro et al., 1976). We attempted to produce an animal model of PPHN by administering indomethacin to pregnant guinea pigs (deMello et al., 1987). During preliminary experiments to determine dosage, placental crossing of indomethacin and guinea pig ductal sensitivity to the drug was apparent at higher dosages as manifested by abortion of fetuses with closed ducti. For the study, the dosage selected was only slightly lower than that causing abortion with ductal closure. Hemodynamic studies in the "treated" newborns revealed no pulmonary hypertension. Morphometric analysis did not reveal the structural alterations of PPHN (Haworth and Reid, 1976; Murphy et al., 1981), although the lungs of "treated" animals were sufficiently different to indicate that indomethacin had produced an effect on the fetus: preacinar arteries had an increased external diameter and increased muscle mass and there was an increase in alveolar multiplication; alveoli were smaller and alveolar number per unit area was higher.

Whether these changes represent the effect of increased blood flow or a direct effect of indomethacin on the lung and vascular bed is not clear.

From these studies it appears that neither fetal hypoxia or fetal ductal constriction is the cause of the precocious muscularization associated with idiopathic PPHN. Other possible mechanisms of pulmonary vascular control need to be investigated.

F. Reactivity

Morphometric Methods

Short-term organ culture of rat lung explants was used to demonstrate the reactivity of pulmonary arteries that were too small to be tested by more conventional methods (Davies et al., 1984, 1985). Explants were incubated for up to 30 min in medium containing vasoconstrictor (either norepinephrine or epinephrine) and then fixed and embedded in Epon. Planimetric measurements of arteries and nonmuscular vessels in section were used to calculate an areal ratio in which unity represented full distention and lower values greater degrees of constriction/retraction. Mean ratios from treated explants were compared with those from control explants from the same animals. For muscular arteries, the ratio was significantly reduced, indicating vasoconstriction (Fig. 8). Although this was expected, the study showed that for nonmuscular vessels the ratio was also reduced, although to a smaller degree than for muscular arteries (Figs. 9, 10). The study was not able to differentiate between nonmuscular arteries and veins, but the variance in the data was small, suggesting that arteries and veins show a similar response. Thus, vessels that lack a layer of medial smooth muscle cells are, nevertheless, capable of constriction. This can be ascribed to the presence of subendothelial effector cells: the intermediate cell and pericyte.

The same methods were used to study arteries from rats exposed for 14 days to hypoxia. In the muscular arteries the degree of constriction produced by incubation with epinephrine was the same as in arteries from normal animals. Nonmuscular arteries also constricted to the same degree as in normals. Newly muscularized arteries, however, constricted less than nonmuscular vessels of similar size and level, which are the vessels from which they derive.

Myography

Isometric force generation has been studied in the extra- and intrapulmonary arteries of rats exposed for 21 days to 87% oxygen (Coflefsky et al., 1987). Resting tension circumference curves showed increased stiffness in the wall of these arteries compared with controls, and active tension induced by KCl was lower despite morphometrically demonstrated medial hypertrophy. In the larger arteries, resting wall stress was calculated by dividing resting tension by total wall thickness: this

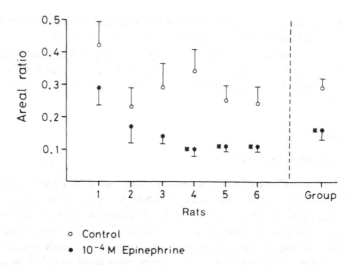

Figure 8 The reactivity of muscular arteries in rat lung explants treated in culture with epinephrine. Areal ratio was determined by the method described in Figure 4; lower values indicate luminal closure. Open circles, untreated arteries; closed circles, treated arteries *from the same lung,* nine arteries in each (from Davies et al., 1985, with permission).

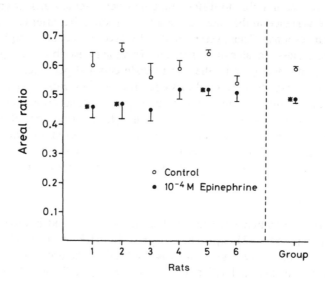

Figure 9 The reactivity of nonmuscular vessels in rat lung explants treated with epinephrine (from Davies et al., 1985, with permission).

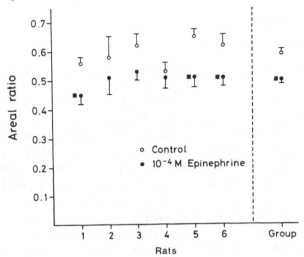

Figure 10 The reactivity of nonmuscular vessels in lung explants from rats exposed to 10% hypobaric hypoxia for 14 days (from Davies et al., 1985, with permission).

was appreciably less in the hyperoxic group than in controls. Active medial stress was calculated by dividing active tension by medial thickness: once again it was reduced in the hyperoxic group. These results indicated that increased constrictive activity is unlikely to be important in the onset or pathogenesis of hyperoxic pulmonary hypertension. The authors did not identify reasons for the decreased stress generation, but speculated that a shift from contractile to synthetic phenotype in the medial smooth muscle cells is one explanation that fits with the deposition of extra collagen that increases wall stiffness.

G. Innervation

The pattern of vascular innervation has been extensively studied in several species and described in a recent review (McLean, 1986). The relevance of adrenergic innervation to hypoxic hypertension has also been noted. Relatively little hypertension is seen at altitude in the dog, guinea pig, sheep, and llama whereas in the horse, rat, pig, and cow, severe hypertension develops. In the first group, pulmonary arteries less than about 70 μm in diameter have adrenergic innervation. In the second group adrenergic fibers are rare or absent. In normotensive animals there is an inverse correlation between innervation and medial thickness. This association has prompted speculation that adrenergic nerves inhibit smooth muscle trophic effects or that adrenergic vasomotor control maintains smooth muscle tone, preventing excessive stretch produced by large transmural pressures.

H. Cell Culture

Medial Smooth Muscle Cells

In primary culture and after one or two passages, these cells can be characterized ultrastructurally using criteria similar to those applied in vivo. Thus, the cells have abundant organized filaments with dense bodies. After repeated passaging, however, the cells lose this "contractile" phenotype, and acquire a less differentiated "synthetic" phenotype with abundant rough endoplasmic reticulum.

Rat pulmonary artery smooth muscle cells in culture stain positively for smooth muscle and nonmuscle myosin. Expression of nonmuscle contractile proteins is characteristic of smooth muscle cells in culture and of cells in the wall of atherosclerotic systemic vessels, although not of normal cells in vivo.

Pulmonary artery smooth muscle cells produce elastin in culture. Cells from newborn calves exposed to hypoxia produced more elastin than cells from control animals (Mecham et al., 1987). Conditioned medium from these smooth muscle cells obtained from hypoxic cases stimulated elastin production by adventitial fibroblasts.

Pericytes

Pericytes in culture, examined by electron microscopy, are poorly differentiated, with rough endoplasmic reticulum and no myofilaments or dense bodies. They resemble smooth muscle cells, however, in staining positively with antibody against smooth muscle and nonmuscle myosin (Davies et al., 1987).

Endothelial Cells

As in vivo, endothelial cells cells stain positively for factor VIIIR.Ag, although microvascular cells from the rat lung stain poorly with antibody against human factor VIIIR.Ag.

Endothelial cells from rat lung stain positively with antibody against nonmuscle myosin, but not for smooth muscle myosin (Davies et al., 1987).

Calf pulmonary artery endothelial cells in culture produce collagen type III and VIII (Macarak, 1984) and proteoglycans (Humphries et al., 1986). After exposure in vitro to 3% oxygen, the cells secreted less sulfated proteoglycan (mainly heparan sulfate) into the medium, but secreted more after exposure to 80% hyperoxia.

Adventitial Fibroblast

The adventitial fibroblast has been obtained in culture from the main pulmonary artery of the newborn calf (Mecham et al., 1987). When medium conditioned by smooth muscle cells obtained from calves chronically exposed to hypoxia was added to cultures of adventitial fibroblasts, it caused some increase in the production of elastin in comparison with medium conditioned by smooth muscle cells from control calves. This was due to the presence of one or more elastogenic peptides of low molecular weight.

References

Abdalla, M. A., and King, A. S. (1976). The functional anatomy of the bronchial system of the domestic fowl. *J. Anat.* **121**:537–50.

Abdalla, M. A., and King, A. S. (1977). The avian bronchial arteries: species variations. *J. Anat.* **123**:697–704.

Albertine, K. H., Schultz, E. L., Wiener-Kronish, J. P., and Staub, N. C. (1987). Regional differences in pleural lymphatic albumin concentration in sheep. *Am. J. Physiol.* **252**:H64–H70.

Albertine, K. M., Wiener-Kronish, J. P., Roos, P. J., and Staub, N. C. (1982). Structure, blood supply, and lymphatic vessels of the sheep's visceral pleura. *Am. J. Anat.* **165**:277–94.

Allanby, K. D., Brinton, W. D., Capbell, M., and Gardner, T. (1950). Pulmonary atresia and the collateral circulation of the lungs. *Guys Hosp. Rep.* **99**:1–40.

Anderson, W. R., and Strickland, M. B. (1971). Pulmonary complications of oxygen therapy in the neonate, postmortem study of bronchopulmonary dysplasia with emphasis on fibroproliferative obliterative bronchitis and bronchiolitis. *Arch. Pathol. Lab. Med.* **91**:506–514.

Anderson, W. R., Strickland, M. B., Tsai, S. H., and Haglin, J. J. (1973). Light microscopic and ultrastructural study of the adverse effects of oxygen therapy on the neonate lung. *Am. J. Pathol.* **73**:327–348.

Avery, M. E., and Cook, C. D. (1961). Volume–pressure relationships of lungs and thorax in fetal, newborn and adult goats. *J. Appl. Physiol.* **16**:1034–1038.

Aziz, K. A., Paul, M. H., and Rowe, R. D. (1977). Bronchopulmonary circulation in d-transposition of the great arteries: possible role in genesis of accelerated pulmonary disease. *Am. J. Cardiol.* **39**:432–438.

Bachofen, H., Ammann, A., Wagensteed, D., and Weibel, E. R. (1982). Perfusion fixation of lungs for structure–function analysis: credits and limitations. *J. Appl. Physiol.* **53**:528–533.

Balding, J. D. Jr, Ogilvie, R. W., Hoffman, C. L., and Knisely, W. M. (1964). The gross morphology of the arterial supply to the trachea, primary bronchi and esophagus of the rabbit. *Anat. Rec.* **148**:611–4.

Barkin, P., Jung, W., Pappagianopoulos, P., Balkas, C., Lamborghini, D., Burke, J., and Hales, C. (1986). The role of the bronchial circulation in production of pulmonary edema in dogs exposed to acrolein in smoke (abstract). *Am. Rev. Respir. Dis.* (Suppl.) **133**:A270.

Barry, B. E., and Crapo, J. D. (1985). Patterns of accumulation of platelets and neutrophils in rat lungs during exposure to 100% and 85% oxygen. *Am. Rev. Respir. Dis.* **132**:548–555.

Bartels, P. (1909). *Das Lymphgefassystem.* Jena, Gustav Fischer Verlag, pp 22–24.

Bartholinus, T. (1653). Vasa Lymphatica, nuperhafniae in animantibus inventa et hepatis exsequiae, 195 (Copenhagen).

Berry, C. L., Greenwald, S. E., and Haworth, S. G. (1981). Mechanical properties of the pulmonary vessels in the normal and in congenital heart disease. In: *Paediatric Cardiology 4.* Edited by M. J. Godman. Edinburgh, Churchill Livingstone, pp. 64–70.

Berry, J., Brailsford, J. F., and Daly, I deB. (1931). Bronchial vascular system in dog. *Proc. R. Soc. Lond.* **109**:214–28.

Bhattacharya, J., Nakahara, K., and Staub, N. C. (1980). Effect of edema on pulmonary blood flow in the isolated perfused dog lung lobe. *J. Appl. Physiol.* **48**:444–449.

Bhattacharya, J., Gropper, M. A., and Staub, N. C. (1984). Interstitial fluid pressure gradient measured by micropuncture in excised dog lung. *J. Appl. Physiol.* **56**:271–277.

Bohs, C. T., Fish, J. C., Miller, T. H., and Traber, D. L. (1979). Pulmonary vascular response to endotoxin in normal and lymphocyte depleted sheep. *Circ. Shock* **6**:13–21.

Bonikos, D. S., Bensch, K. G., Northway, W. H. Jr., and Edwards, D. K. (1976). Bronchopulmonary dysplasia: the pulmonary pathologic sequel of necrotizing bronchiolitis and pulmonary fibrosis. *Hum. Pathol.* **7**:643–666.

Borg, D. C., Schaich, K. M., Elmore, J. J. Jr., and Bell, J. A. (1978). Cytoxic reactions of free radical species of oxygen. *Photochem. Photobiol.* **28**:887–908.

Boushy, S. F., North, L. B., and Trice, J. A. (1969). The bronchial arteries in chronic obstructive pulmonary disease. *Am. J. Med.* **46**:506–15.

Brenner, O. (1935). Pathology of the vessels of the pulmonary circulation. *Arch. Intern. Med.* **56**:211–237.

Brigham, K. L., Bowers, R. E., and McKeen, C. R. (1981). Methylprednisolone prevention of increased lung vascular permeability following endotoxemia in sheep. *J. Clin. Invest.* **67**:1103–1110.

Cadet, J., and Teoule, R. (1978). Comparative study of oxidation of nucleic acid components by hydroxyl radicals, singlet oxygen and superoxide anion radicals. *Photochem. Photobiol.* **28**:661–667.

Casey-Smith, J. R. (1980). Are the initial lymphatics normally pulled open by the anchoring filaments? *Lymphology* **13**:120–129.

Caterson, B., Christner, J. E., Baker, J. R., and Couchman, J. R. (1985). Production and characterization of monoclonal antibodies directed against connective tissue proteoglycans. *Fed. Proc.* **44**:386–393.

Coflefsky, J., Jones, R., Reid, L., and Evans, J. N. (1987). Mechanical properties and structure of isolated pulmonary arteries remodeled by chronic hyperoxia. *Am. Rev. Respir. Dis.* **136**:388–394.

Collier, P. S. (1979). Bronchial arteries of the dog. A pharmacoanatomical study. *J. Anat.* **128**:887–8.

Craddock, P. R., Fehr, J., Brigham, K., Kronenberg, R. S., and Jacob, H. S. (1979). Complement and leukocyte mediated pulmonary dysfunction in hemodialysis. *N. Engl. J. Med.* **296**:769–774.

Crandall, E. D., Staub, N. C., Goldberg, H. S., and Effros, R. M. (1983). Recent developments in pulmonary edema. *Ann. Intern. Med.* **99**:808–822.

Crapo, J. D., Barry, B. E., Foscue, H. A., and Shelburne, J. (1980). Structural and biochemical changes in rat lungs occurring during exposures to lethal and adaptive doses of oxygen. *Am. Rev. Respir. Dis.* **122**:123–143.

Csaba, I. F., Sulyok, E., and Ertl, T. (1978). Relationship of maternal treatment with indomethacin to persistence of fetal circulation syndrome. *J. Pediatr.* **92**:484.

Cudkowicz, L. (1952). Some observations of the bronchial arteries in lobar pneumonia and pulmonary infarction. *Br. J. Tuberc.* **46**:99–102.

Cudkowicz, L. (1965). Cardiorespiratory studies in patients with pulmonary tuberculosis. *Can. Med. Assoc. J.* **92**:111–115.

Cudkowicz, L. (1967). Cardiorespiratory studies in patients with lung tumors. *Dis. chest* **51**:427–32.

Cudkowicz, L. (1968). *The Human Bronchial Circulation in Health and Disease.* Baltimore, William & Wilkins.

Cudkowicz, L. (1979). Bronchial arterial circulation in man: normal anatomy and responses to disease. In *Pulmonary Vascular Diseases*. Edited by K. M. Moser. New York, Marcel Dekker, pp. 111–232.

Davies, P., Maddalo, F., and Reid, L. (1984). The response of microvessels in rat lung explants to incubation with norepinephrine. *Exp. Lung. Res.* **7**:93–100.

Davies, P., Maddalo, F., and Reid, L. (1985). Effects of chronic hypoxia on structure and reactivity of rat lung microvessels. *J. Appl. Physiol.* **58**:795–801.

Davies, P., Burke, G., and Reid, L. (1986). The structure of the wall of the rat intraacinar pulmonary artery: an electron microscopic study of microdissected preparations. *Microvasc. Res.* **32**:50–63.

Davies, P., Smith, B. T., Maddalo, F. B., Langleben, D., Tobias, D., Fujiwara, K., and Reid, L. (1987). Characteristics of lung pericytes in culture including their growth inhibition by endothelial substrate. *Microvasc. Res.* **33**:300–314.

Davies, P. F. (1986). Biology of disease. Vascular cell interactions with special reference to the pathogenesis of atherosclerosis. *Lab. Invest.* **55**:5–24.

Deffebach, M. E., Charan, N. B., Lakshiminarayan, S., and Butler, J. (1987). The bronchial circulation, small, but a vital attribute of the lung. *Am. Rev. Respir. Dis.* **135**:463–481.

Defouw, D. O. (1982). Ultrastructure of the pulmonary alveolar septa after hemodynamic edema. *Ann. N. Y. Acad. Sci.* **384**:45–53.

Defouw, D. O., and Chinard, F. P. (1983). Variations in cellular attentuation and vesicle numerical densities in capillary endothelium and type I epithelium of isolated, perfused dog lungs after acute severe edema formation. *Microvasc. Res.* **26**:15–26.

Defouw, D. O., Cua, W. O., and Chinard, F. P. (1983). Morphometric and physiological studies of alveolar microvessels in dog lungs in vivo after sustained increases in pulmonary microvascular pressures and after sustained decreases in plasma oncotic pressures. *Microvasc. Res.* **25**:56–67.

Defouw, D. O., Ritter, A. B., and Chinard, F. P. (1985). Alveolar microvessels in isolated perfused dog lungs: structural and functional studies after production of moderate and severe hydrodynamic edema. *Exp. Lung. Res.* **8**:67–79.

deMello, D. E., Murphy, J., Aronovitz, M., Davies, P., and Reid, L. M. (1987). Effects of indomethacin in utero on the pulmonary vasculature of the newborn guinea pig. *Pediatr. Res.* **22**:693–697.

deMello, D. E., Hu, L. M., Gashi-Luci, L., Davies, P., and Reid, L. M. (1988). The effect of hypoxia on postnatal vascular structure in newborn rabbits. (Abstr.) *F.A.S.E.B.*

Deneke, S. M., and Fanburg, B. L. (1980). Normobaric oxygen toxicity of the lung. *N. Engl. J. Med.* **303**:76–86.

Dobuler, N. K., Catnavas, J. D., and Gillis, C. N. (1982). Early detection of oxygen-induced liver injury in conscious rabbit. Reduced in vivo activity of angiotensin converting enzyme and removal of 5-hydroxytryptamine. *Am. Rev. Respir. Dis.* **126**:534–539.

Drake, R. E., Lauie, G. A., Allen, S. J., Katz, J., and Gabel, J. (1987). A model of the lung interstitial-lymphatic system. *Microvasc. Res.* **34**:96–107.

Drinker, C. K., Enders, J. F., Shaffer, M. F., and Leigh, O. C. (1935). The emigration of pneumococci type III from the blood into the thoracic duct of rabbits, and the survival of these organisms in the lymph following intravenous injection of specific antiserum. *J. Exp. Med.* **62**:849–860.

Edwards, J. E. (1957). Functional pathology of the pulmonary vascular tree in congenital cardiac disease. *Circulation* **15**:164–196.

Elliott, F. M., and Reid, L. (1965). Some new facts about the pulmonary artery and its branching pattern. *Clin. Radiol.* **16**:193–198.

Erhart, I. C., and Hofman, W. F. (1981). Oleic acid dose-related edema in isolated canine lung perfused at constant pressure. *J. Appl. Physiol.* **50**:1115–1120.

Fell, S. C., Mollenkopf, F. P., and Montefusco, C., Mitsudo, S., Kamholz, S. L., Goldsmith, J., and Veith, F. (1985). Revascularization of ischemic bronchial anastomoses by an intercostal pedicle flap. *J. Thorac. Cardiovasc. Surg.* **90**:172-178.

Ferencz, C. (1960). The pulmonary vascular bed in tetralogy of Fallot. II. Changes following systemic–pulmonary arterial anastomosis. *Bull. Johns Hopkins Hosp.* **106**:100-118.

Fernie, J. M., and Lamb, D. (1985a). A new method for measuring intimal component of pulmonary arteries. *J. Clin. Pathol.* **38**:1374-1379.

Fernie, J. M., and Lamb, D. (1985b). A new method for quantitating the medial component of pulmonary arteries. The measurements. *Arch. Pathol. Lab. Med.* **109**:156-162.

Fischer, E. (1935). Lymphgefassuntersuchungen an Serosen Hauten mit Luftfullungmethoden. *Verh. Dtsch. Patho. Ges.* **28**:223.

Flick, M., Perel, A., and Staub, N. (1981). Leukocytes are required for increased lung microvascular permeability after microembolization in sheep. *Circ. Res.* **48**:344-351.

Frank, L. (1985). Oxidant injury to pulmonary endothelium. In *The Pulmonary Circulation in Acute Lung Injury*. Edited by S. I. Said. Mt. Kisco, NY, Futura, pp. 283-305.

Frank, L., and Massaro, D. (1980). Oxygen toxicity. *Am. J. Med.* **69**:117-126.

Freeman, B. A., and Crapo, J. D. (1982). Biology of disease: free radicals and tissue injury. *Lab. Invest.* **47**:412-426.

Fridovich, I. (1978). The biology of oxygen radicals. *Science* **201**:875-80.

Friedli, B., Kidd, B. S. L., Mustard, W. T., and Keith, J. D. (1974). Ventricular septal defect with increased pulmonary vascular resistance. Late results of surgical closure. *Am. J. Cardiol.* **33**:403-409.

Frolich, J. C., Ogletree, M. L., Peskar, B. A., and Brigham, K. L. (1980). Pulmonary hypertension correlated to thromboxane synthesis. *Adv. Prostaglandin Thromboxane Res.* **7**:745-750.

Furthmayr, H. (1982). *Immunochemistry of the Extracellular Matrix.* Vol. 1 and II. Boca Raton, CRC Press.

Gabella, G. (1983). Asymmetric distribution of dense bands in muscle cells of mammalian arterioles. *J. Ultrastruct. Res.* **84**:24-33.

Gee, M. H., and Staub, N. C. (1977). The role of bulk fluid flow in protein permeability of the dog lung alveolar membrane. *J. Appl. Physiol.* **42**:144-149.

Geggel, R. L., Aronovitz, M. J., and Reid, L. M. (1986). Effects of chronic in utero hypoxemia on rat neonatal pulmonary arterial structure. *J. Pediatr.* **108**:756-759.

Gerota (1896). Zur Technik der Lymphgefassinjektion. *Anat. Anzeiger.* **12**:216-224.

Goldberg, S. J., Levy, R. A., Siassi, B., and Betten, J. (1971). The effects of maternal hypoxia and hyperoxia upon the neonatal pulmonary vasculature. *Pediatrics* **48**:528–533.

Greenfield, L. J., and Harrison Jr, L. H. (1971). Pulmonary vascular response to graded hydrostatic edema in the perfused lung. *J. Surg. Res.* **11**:410–414.

Hallidie-Smith, K. A. (1968). The long-term results of closure of ventricular septal defect with pulmonary vascular disease. *Am. Heart J.* **76**:591–595.

Han, V. K. M., D'Ercole, J. D., and Lund, P. K. (1987). Cellular localization of somatomedin (insulin-like growth factor) messenger RNA in the human fetus. *Science* **236**:193–197.

Harisson, C. V. (1958). The pathology of the pulmonary vessels in pulmonary hypertension. *Br. J. Radiol.* **31**:217–226.

Harris, P., and Heath, D. (1962). *The Human Pulmonary Circulation: Its Form and Function in Health and Disease*. London, E. and S. Livingston.

Haworth, S. G., and Macartney, F. J. (1980). Growth and development of the pulmonary circulation in pulmonary atresia with ventricular septal defect and major aorto-pulmonary collateral arteries. *Br. Heart J.* **44**:14–24.

Haworth, S. D., and Reid, L. (1976). Persistent fetal circulation: newly recognized structural features. *J. Pediatr.* **88**:614–620.

Haworth, S. G., and Reid, L. (1977a). Structural study of pulmonary circulation and of heart in total anomalous pulmonary venous return in early infancy. *Br. Heart J.* **39**:80–92.

Haworth, S. G., and Reid, L. (1977b). A quantitative structural study of pulmonary circulation in the newborn with aortic atresia, stenosis or coarctation. *Thorax* **32**:121–128.

Haworth, S. G., and Reid, L. (1977c). Quantitative structural study of the pulmonary circulation in the newborn with pulmonary atresia. *Thorax* **32**:129–133.

Haworth, S. G., and Reid, L. (1978). A morphometric study of regional variation in lung structure in infants with pulmonary hypertension and congenital cardiac defect: a justification of lung biopsy. *Br. Heart J.* **40**:825–831.

Haworth, S. G., Sauer, U., Buhlmeyer, K., Reid, L. (1977). Development of the pulmonary circulation in ventricular septal defect: A quantitative structural study. *Am. J. Cardiol.* **40**:781–788.

Hayat, M. A. (1981). *Fixation for Electron Microscopy*. New York, Academic Press, p. 245.

Heath, D., and Edwards, J. E. (1958). The pathology of hypertensive pulmonary vascular disease. A description of six grades of structural changes in the pulmonary arteries with special reference to congenital cardiac septal defects. *Circulation* **18**:533–547.

Heath, D., Helmholtz, H. F., Burchell, H. B., DuShane, J. W., Kirklin, J. W., Edwards, J. E. (1958). Relation between structural changes in the small pulmonary arteries and the immediate reversibility of pulmonary hypertension following closure of ventricular and atrial septal defects. *Circulation* **18**:1167–1174.

Heflin, A. C., and Brigham, K. L. (1981). Prevention by granulocyte depletion of increasd lung vascular permeability of sheep lung following endotoxemia. *J. Clin. Invest.* **68**:1253–1260.

Henson, P. M., Larsen, G. L., Webster, R. O., Mitchell, B. C., Goins, A. J., and Henson, J. E. (1982). Pulmonary microvascular alterations and injury induced by complement fragments: synergistic effect of complement activation, neutrophil sequestration, and prostaglandins. *Ann. N. Y. Acad. Sci.* **384**:287–300.

Herman, I. M., and D'Amore, P. A. (1985). Microvascular pericytes contain muscle and nonmuscle actins. *J. Cell Biol.* **101**:43–52.

Hinson, J. M., Jr., Brigham, K. L., Hutchison, A. A., and Snapper, J. R. (1982). Granulocytes participate in the early changes in lung mechanics caused by endotoxemia. (Abstr.) *Am. Rev. Respir. Dis.* **125**:275.

Hinson, J. M. Jr., Hutchinson, A. A., Ogletree, M. L., Brigham, K. L., and Snapper, J. R. (1983). Effect of granulocyte depletion on altered lung mechanics after endotoxemia in sheep. *J. Appl. Physiol.* **55**(1):92–99.

Hislop, A., and Reid, L. (1974). Growth and development of the respiratory system. In *Scientific Foundations of Pediatrics*. Edited by J. A. Davis and J. Dobbing. London, Heinemann Medical Books, pp. 214–254.

Hislop, A., and Reid, L. (1976). New findings in pulmonary arteries of rats with hypoxia-induced pulmonary hypertension. *Br. J. Exp. Pathol.* **57**:542–554.

Hogg, J. C. (1978). Effect of pulmonary edema on distribution of blood flow in the lung. In *Lung Water and Solute Exchange*, Vol 7. Edited by N. C. Staub. New York, Marcel Dekker, pp. 167–182.

Horsfield, K. (1978). Morphometry of the small pulmonary arteries in man. *Circ. Res.* **42**:593–597.

Horsfield, K., and Gordon, W. I. (1981). Morphometry of pulmonary veins in man. *Lung* **159**:211–218.

Hu, L.-M., Geggel, R., Davies, P., and Reid, L. (1988). The effect of heparin on the hemodynamic and structural response in the rat to acute and chronic hypoxia. Submitted for publication. *Br. J. Exp. Patho.*

Hudack, S. S., and McMaster, P. D. (1933). The lymphatic participation in human cutaneous phenomena. *J. Exp. Med.* **57**:751–774.

Humphries, D. E., Lee, S.-L., Fanburg, B. S., and Silbert, J. E. (1986). Effects of hypoxia and hyperoxia on proteoglycan production by bovine pulmonary artery endothelial cells. *J. Cell Physiol.* **126**:249–253.

Hurley, J. V. (1978). Current views on the mechanisms of pulmonary edema. *J. Pathol.* **125**:59–79.

Irvin, C. G., Henson, P. M., and Berend, N. (1982). Acute effect of airways inflammation on airway function and reactivity (abstr.) *Fed. Proc.* **41**:1358.

Jindal, S. K., Lakshminarayan, S., Kirk, W., and Butler, J. (1984). Acute increase in anastomotic bronchial blood flow after pulmonary artery obstruction. *J. Appl. Physiol.* **57**:424–8.

Johnson, A., and Malik, A. (1980). Effect of granulocytopenia on extravascular lung water content after microembolization. *Am. Rev. Respir. Dis.* **122**:561–566.

Jones, R., Zapol, W. M., and Reid, L. (1984). Pulmonary artery remodeling and pulmonary hypertension after exposure to hyperoxia for 7 days. *Am. J. Pathol.* **117**:273–285.

Jones, R., Zapol, W. M., and Reid, L. (1985). Oxygen toxicity and restructuring of pulmonary arteries. A morphometric study. The response to 4 weeks exposure to hyperoxia and return to breathing room air. *Am. J. Pathol.* **121**:212–223.

Joyce, N. C., Decamilli, P., and Boyles, J. (1984). Pericytes, like vascular smooth muscle cells, are immunocytochemically positive for cGMP-dependent protein kinase. *Microvasc. Res.* **28**:206–219.

Joyce, N., Haire, M., and Palade, G. (1985). Contractile proteins in pericytes. II. Immunocytochemical evidence for the presence of two isomyosins in graded concentrations. *J. Cell Biol.* **100**:1387–1395.

Kanter, M. A. (1987). The lymphatic system: an historical perspective. *Plast. Reconstr. Surg.* **79**:131–139.

Kasai, T., and Chiba, S. (1979). Macroscopic anatomy of the bronchial arteries. *Anat. Anz.* **145**:166–181.

Kato, M., and Staub, N. C. (1966). Response of small pulmonary arteries to unilobar hypoxia and hypercapnia. *Circ. Res.* **19**:426–440.

Kinmonth, J. B. (1954). Lymphangiography in clinical surgery and particularly in the treatment of lymphoedema. *Ann. R. Coll. Surg. Engl.* **15**:300–315.

Kinmonth, J. B. (1972). *The Lymphatics: Diseases, Lymphography, and Surgery.* Baltimore, Williams & Wilkins, p.v.

Kirton, O. C., Jones, R. C., and Carvalho, A. (1988). Thromboxane A_2 release but not pulmonary artery hypertension after bolus *Escherichia coli* endotoxin infusion in the rat. Submitted for publication. *Circ. Shock.*

Lacina, S. S., Hamilton, W. T., Thilenius, O. G., Bharati, S., Lev, M., and Arcilla, R. A. (1983). Angiographic evidence of absent ductus arteriosus in severe right ventricular outflow obstruction. *Pediatr. Cardiol.* **4**:5–11.

Lakshminarayan, S., Jindal, S. K., Kirk, W., and Butler, J. (1983). Bronchial blood flow increases in inflammatory lung edema (abstr.). *Physiologist* **26**:A56.

Larroche, J., and Nessman, C. (1971). La maladie des membranes hyalines: evolution, cicarrisation, sequelles; etude histologique. *Arch. Fr. Pediatr.* **28**:113–132.

Larsen, G. L., Mitchell, B. C., and Henson, P. M. (1981). The pulmonary response of C5 sufficient and deficient mice to immunocomplexes. *Am. Rev. Respir. Dis.* **123**:434–439.

Larson, D. M., Carson, M. P., and Haudenschild, C. C. (1987). Junctional transfer of small molecules in cultured bovine brain microvascular endothelial cells and pericytes. *Microvasc. Res.* **34**:184–199.

Lee, R. M. K. W., Forrest, J. B., Garfield, R. E., and Daniel, E. E. (1983a). Ultrastructural changes in mesenteric arteries from spontaneously hypertensive rats. A morphometric study. *Blood Vessels* **20**:72–91.

Lee, R. M. K. W., Forrest, J. B., Garfield, R. E., and Daniel, E. E. (1983b). Comparison of blood vessel wall dimensions in normotensive and hypertensive rats by histometric and morphometric methods. *Blood Vessels* **20**:245–254.

Levin, D. L., Fixler, D. E., Morriss, F. C., and Tyson, J. (1978). Morphologic analysis of the pulmonary vascular bed in infants exposed in utero to prostaglandin synthetase inhibitors. *J. Pediatr.* **92**:478–483.

Liao, P. K., Edwards, W. D., Julsrud, P. R., Puga, F. J., Danielson, G. K., and Feldt, R. H. (1985). Pulmonary blood supply in patients with pulmonary atresia and ventricular septal defects. *J. Am. Coll. Cardiol.* **6**:1343–50.

Liebow, A. A. (1953). The bronchopulmonary venous collateral circulation with special reference to emphysema. *Am. J. Pathol.* **29**:251–89.

Liebow, A. A., Hales, M. R., and Lindskog, G. E. (1949). Enlargement of the bronchial arteries and their anastomoses with the pulmonary arteries in bronchiectasis. *Am. J. Pathol.* **25**:211–31.

Lima, O., Goldberg, M., Peters, W. J., Ayabe, H., Townsend, E., and Cooper, J. D. (1982). Bronchial omentoplexy in canine lung transplantation. *J. Thorac. Cardiovasc. Surg.* **83**:418–21.

Low, F. N. (1978). Lung interstitium: development, morphology, fluid content. In *Lung Water and Solute Exchange*. Edited by N. C. Staub. New York, Marcel Dekker, p. 17.

Macarak, E. J. (1984). Collagen synthesis by cloned pulmonary artery endothelial cells. *J. Cell Physiol.* **119**:175–182.

Magno, M. G., Fishman, A. P. (1982). Origin, distribution, and blood flow of bronchial circulation in anesthetized sheep. *J. Appl. Physiol.* **53**:272–279.

Malo, J., Ali, J., and Wood, L. D. H. (1984). How does positive end-expiratory pressure reduce intra-pulmonary shunt in canine pulmonary edema? *J. Appl. Physiol.* **57**:1002–1010.

Manchester, D., Margolis, H. S., and Sheldon, R. E. (1976). Possible association between maternal indomethacin therapy and primary pulmonary hypertension of the newborn. *Am. J. Obstet. Gynecol.* **126**:467–469.

Mathes, M. E., Holman, E., and Reichert, F. L. (1932). A study of the bronchial, pulmonary, and lymphatic circulations of the lung under various pathologic conditions experimentally produced. *J. Thorac. Surg.* **1**:339–62.

Mathieu-Costello, O., and Fronek, K. (1985). Morphometry of the amount of smooth muscle cells in the media of various rabbit arteries. *J. Ultrastruct. Res.* **91**:1–12.

May, N. D. S. (1970). *Anatomy of the Sheep*, 3rd ed. St. Lucia, Australia, University of Queensland Press.

Mazzone, R. W. (1981). Influence of pulmonary edema on capillary morphology (abstr.). *Fed. Proc.* **41**:404.

Mazzone, R. W., Durand, C. M., and West, J. B. (1978). Electron microscopy of lung rapidly frozen under controlled physiological conditions. *J. Appl. Physiol.* **45**:325–333.

McCarthy, K., Bhogal, M., Nardi, M., and Hart, D. (1984). Pathogenic factors in bronchopulmonary dysplasia. *Pediatr. Res.* **18**:483–488.

McCord, J. M. (1974). Free radicals and inflammation: protection of synovial fluid by superoxide dismutase. *Science* **185**:529–31.

McCord, J. M., and Fridovich, I. (1978). The biology and pathology of oxygen radicals. *Ann. Intern. Med.* **89**:122–7.

McLaughlin, R. F., Tyler, W. S., and Canada, R. O. (1961). A study of the subgross pulmonary anatomy in various mammals. *Am. J. Anat.* **108**:149–65.

McLean, J. R., (1986). Pulmonary vascular innervation. In *Abnormal Pulmonary Circulation*. Edited by E. H. Bergofsky. New York, Churchill Livingstone, pp. 27–81.

Mecham, R. P., Whitehouse, L. A., Wrenn, D. S., Parks, W. C., Griffin, G. L., Senior, R. M., Crouch, E. C., Stenmark, K. R., and Voelkel, N. F. (1987). Smooth muscle-mediated connective tissue remodeling in pulmonary hypertension. *Science* **237**:423–426.

Mercer, R. R., and Crapo, J. D. (1987). Determination of the pulmonary capillary length within the acinus of the rat (abstr.). *Fed. Proc.* **46**:520.

Meyrick, B., and Reid, L. (1979a). Ultrastructural features of the distended pulmonary arteries of the normal rat. *Anat. Rec.* **193**:71–97.

Meyrick, B., and Reid, L. (1979b). Hypoxia and incorporation of ^{3}H-thymidine by cells of the rat pulmonary arteries and alveolar wall. *Am. J. Pathol.* **96**:51–70.

Meyrick, B., and Reid, L. (1980a). Hypoxia induced structural changes in the media and adventitia of the rat hilar pulmonary artery and their regression. *Am. J. Pathol.* **100**:151–178.

Meyrick, B., and Reid, L. (1980b). Ultrastructural findings in lung biopsy material from children with congenital heart defects. *Am. J. Pathol.* **101**:527–542.

Meyrick, B., and Brigham, K. (1983). Acute effects of Escherichia coli Endotoxin on the pulmonary microcirculation of anesthetized sheep. Structure: function relationships. *Lab. Invest.* **48**:458–470.

Meyrick, B., Hislop, A., and Reid, L. (1978). Pulmonary arteries of the normal rat: the thick walled oblique muscle segment. *J. Anat.* **125**:209–221.

Meyrick, B., Fujiwara, K., and Reid, L. (1981). Smooth muscle myosin in precursor and mature smooth muscle cells in normal pulmonary arteries and the effect of hypoxia. *Exp. Lung Res.* **2**:303–313.

Michel, R. (1982). Arteries and veins of the normal dog lung: qualitative and quantitative structural differences. *Am. J. Anat.* **164**:227–241.

Michel, R. P., Hakim, T. S., Smith, T. T., and Poulsen, R. S. (1983). Quantitative morphology of permeability lung edema in dogs induced by alpha-naphthylthiourea. *Lab. Invest.* **149**:412–418.

Michel, R. P., Smith, T. T., and Poulsen, R. S. (1984). Distribution of fluid in bronchovascular bundles with permeability lung edema induced by alpha-naphthylthiourea in dogs. A morphometric study. *Lab. Invest.* **51**:97–103.

Michel, R. P., Meterissian, S., and Poulsen, R. S. (1986). Morphometry of the distribution of hydrostatic pulmonary edema in dogs. *Br. J. Exp. Pathol.* **67**:865–877.

Michel, R. P., Zocchi, L., Rossi, A., Cardinal, G. A., Ploy-Song-Sang, Y., Poulsen, R. S., Milic-Emili, J., and Staub, C. (1987). Does interstitial lung edema compress airways and arteries? A morphometric study. *J. Appl. Phys.* **62**(1):108–115.

Michelson, A. M. (1977). Toxicity of superoxide radical anions. In *Superoxide and Superoxide Dismutases*. Edited by A. M. Michelson, J. M. McCord, and I. Fridovich. London, Academic Press, pp. 245–55.

Miller, W. S. (1947). *The Lung*. Springfield, IL, Charles C. Thomas.

Mills, N. L., Boyd, A. D., and Gheranpong, C. (1970). The significance of bronchial circulation in lung transplantation. *J. Thorac. Cardiovasc. Surg.* **60**:866–78.

Milne, E. N. C. (1967). Circulation of primary and metastatic pulmonary neoplasms. A post mortem microarteriographic study. *Am. J. Roentgenol.* **100**:603–18.

Miyamoto, Y., and Moll, W. A. (1971). Measurements of dimensions and pathway of red blood cells in rapidly frozen lungs *in situ*. *Respir. Physiol.* **12**:141–156.

Morgan, E., Lima, O., Goldberg, M., Ferdman, A., Luk, S. K., and Cooper, J. D. (1982). Successful revascularization of totally ischemic bronchial autografts with omental pedicle flaps in dogs. *J. Thorac. Cardiovasc. Surg.* **84**:204–10.

Morrison, D. C., and Ulevitch, R. J. (1978). The effects of bacterial endotoxins on host mediation systems. *Am. J. Pathol.* **93**:527–617.

Mueller, W. S. (1947). *The Lung*, 2nd. edition. Springfield, IL, Charles C. Thomas.

Muller, K. M., and Bordt, J. (1980). Der bronchial terienkreislauf unter krankhaften verhaltnissen. *Pra. Pneumol.* **34**:324–31.

Mulvany, M. J., Brandrup, U., and Gundersen, H. J. G. (1985). Evidence for hyperplasia in mesenteric resistance vessels of spontaneously hypertensive rats using a three-dimensional disector. *Circ. Res.* **57**:794–800.

Murphy, J. D., Rabinovitch, M., Goldstein, J. D., and Reid, L. M. (1981). The structural basis of persistent pulmonary hypertension of the newborn infant. *J. Pediatr.* **98**:962–96.

Murphy, J. D., Vawter, G. F., and Reid, L. M. (1984). Pulmonary vascular disease in fatal meconium aspiration. *J. Pediatr.* **104**:758–762.

Murphy, J. D., Aronvitz, M. J., and Reid, L. M. (1986). Effects of chronic in utero hypoxia on the pulmonary vasculature on the newborn guinea pig. *Pediatr. Res.* **20**:292–295.

Naeye, R. L. (1962). Perinatal vascular changes associated with underdevelopment of the left heart. *Am. J. Pathol.* **41**:287–293.

Nagasaka, Y., Bhattacharya, F., Nanjo, S., Groper, M. A., and Staub, N. C. (1984). Micropuncture measurements of lung microvascular pressure profile during hypoxia in cats. *Circ. Res.* **54**:90–95.

Newfeld, E. A., Paul, M. H., Muster, A. J., and Idriss, F. S. (1974). Pulmonary vascular disease in complete transposition of the great arteries: a study of 200 patients. *Am. J. Cardiol.* **34**:75–82.

Ngeow, Y. K., and Mitzner, W. (1983). Pulmonary hemodynamics and gas exchange properties during progressive edema. *J. Appl. Physiol.* **55**:1154–1159.

Nicolaysen, G., Nicholaysen, A., and Staub, N. C. (1975). A quantitative radioautographic comparison of albumin concentration in different sized lymph vessels in normal mouse lungs. *Microvasc. Res.* **10**:138–152.

Northway, W. H., Rosan, R. C., and Porter, D. Y. (1967). Pulmonary disease following respirator therapy of hyaline membrane disease. *N. Engl. J. Med.* **276**:357–368.

Noszkov, A., Gulacsy, I., and Kiss, T. (1979). Angiography of the bronchial and pulmonary arteries in chronic non-specific lung diseases. *Acta Chir. Acad. Sci. Hung.* **20**:225–34.

Notkovich, H. (1957). The anatomy of the bronchial arteries of the dog. *J. Thorac Surg.* **33**:242–53.

Ogletree, M. L., Oates, J. A., Brigham, K. L., and Hubbard, W. C. (1982). Evidence for pulmonary release of 5-hydroxyeicosatetraenoic acid (5-HETE) during endotoxemia in unanesthetized sheep. *Prostaglandins* **23**:459–468.

Pappas, C. T. E., Ohara, H., Beausch, K. G., and Northway, W. H. (1983). Effect of prolonged exposure to 80% oxygen on the lung of the newborn mouse. *Lab. Invest.* **48**:735–748.

Parker, B. M., and Smith, J. R. (1957). Studies of experimental pulmonary embolism and infarction and the development of collateral circulation in the affected lung lobe. *J. Lab. Clin. Med.* **49**:850-7.

Parker, J. C., Allison, R. C., and Taylor, A. E. (1981). Edema affects intra-alveolar fluid pressures and interdependence in dog lungs. *J. Appl. Physiol.* **51**:911-921.

Parry, K., and Yates, M. S. (1979). Observations on the avian pulmonary and bronchial circulation using labeled microspheres. *Respir. Physiol.* **28**:131-40.

Pecquet, J. (1651). Experimenta nova anatomica quibus incognitum chyli receptaculum, et ab eo per thoracem in ramos usque subclavis vasa lactea defergunter. Paris, Carmoisy S. & G.

Permutt, S. (1965). Effect of interstitial pressure of the lung on pulmonary circulation. *Med. Thorac.* **22**:118-131.

Peterson, B. T., and Gray, L. D. (1987). Pulmonary lymphatic clearance of [99]m Tc-DTPA from air spaces during lung inflation and lung injury. *J. Appl. Physiol.* **63**(3):1136-1141.

Pick, T. P., and Howden, R. (1977). *Anatomy, Descriptive and Surgical*, 15th Edition. Edited by H. Gray. New York, Bounty Books.

Pinsker, K. L., Montefusco, C., Kamholz, Hagstrom, J. W. C., Gliedman, M. L., and Veith, F. J. (1980). Improved bronchial anastomotic healing secondary to maintenance or restoration of bronchial arterial circulation by microsurgical techniques. *Surg. Forum* **31**:230-2.

Pinsker, K. L., Veith, F. J., Kamholz, S. L., Montefusco, C., Emerson, E., and Hagstrom, J. W. C. (1984). Influence of bronchial circulation and corticosteroid therapy on bronchial anastomotic healing. *J. Thorac. Cardiovasc. Surg.* **87**:439-44.

Pritchard, J. S. (1982). *Edema of the Lung.* Springfield, IL, Charles C. Thomas, p. 114.

Pump, K. K. (1963). The bronchial arteries and their anastomoses in the human lung. *Dis. Chest.* **43**:245-55.

Rabinovitch, M., and Reid, L. M. (1980). Quantitative structural analysis of the pulmonary vascular bed in congenital heart defects. *Pediatr. Cardiovasc. Dis.* **11**:149-169.

Rabinovitch, M., Haworth, S. G., Cataneda, A. R., Nadas, A. S., and Reid, L. M. (1978). Lung biopsy in congential heart disease: a morphometric approach to pulmonary vascular diseases. *Circulation* **58**:1107-1122.

Rabinovitch, M., Gamble, W., Nadas, A. S., Miettinen, O. S., and Reid, L. (1979). Rat pulmonary circulation after chronic hypoxia: hemodynamic and structural features. *Am. J. Physiol.* **236**:H818-H827.

Rabinovitch, M., Castaneda, A. R., and Reid, L. (1981). Lung biopsy with frozen section as a diagnostic aid in patients with congenital heart defects. *Am. J. Cardiol.* **47**:77-84.

Rendas, A., Branthwaite, M., and Reid, L. M. (1978). Growth of pulmonary circulation in normal pig—structural analysis and cardiopulmonary function. *J. Appl. Physiol.* **45**:806–817.

Rendas, A., Lennox, S., and Reid, L. (1979). Aorto-pulmonary shunts in growing pigs. *J. Thorac. Cardiovasc. Surg.* **77**:109–118.

Robertson, B. (1968). The intrapulmonary arterial pattern in normal infancy and in transposition of the great arteries. *Acta Pediatr. Scand. (Suppl)* **184**:7–36.

Rudbeck, O. (1653). Novo exercitatio anatomica, exhibens ductus hepaticos, aquosus et vasa glandularum serosa. Arosiae Hexud e Lauringerus. English translation (1942), *Bull. Hist. Med.* **11**:304.

Rudolph, A. M.(1974). *Congenital Diseases of the Heart*. Chicago, IL, Yearbook, pp. 17–48.

Ryan, U. S. (1987). Endothelial cell activation responses. In *Pulmonary Endothelium in Health and Disease*. Edited by U. S. Ryan. New York, Marcel Dekker, pp. 3–33.

Ryan, U. S., Clements, E., Habliston, D., and Ryan, J. W. (1978). Isolation and culture of pulmonary artery endothelial cells. *Tissue Cell* **10**:535–554.

Ryan, U. S., Mortara, M., and Whitaker, C. (1980). Methods for microcarrier culture of bovine pulmonary artery endothelial cells avoiding the use of enzymes. *Tissue Cell* **12**:619–635.

Ryan, U. S., White, L. A., Lopez, M., and Ryan, J. W. (1982). Use of microcarriers to isolate and culture pulmonary microvascular endothelium. *Tissue Cell* **14**:597–606.

Sabin, F. R. (1911). A critical study of the evidence presented in several recent articles on the development of the lymphatic system. *Anat. Rec.* **5**:417–446.

Sacks, T., Moldon, C. F., Craddock, P. R., Bowers, T. K., and Jacob, H. S. (1978). Oxygen radicals mediate endothelial cell damage by complement-stimulated granulocytes: An in vitro model of immune vascular damage. *J. Clin. Invest.* **61**:1161–1167.

Shapiro, S., Siskind, V., Monson, R. R., Heinonen, O. P., Kaufman, D. W., and Slone, D. (1976). Perinatal mortality and birth-weight in relation to aspirin taken during pregnancy. *Lancet* **1**:1375–1376.

Short, D. S. (1956). Post-mortem pulmonary arteriography with special reference to the study of pulmonary hypertension. *J. Fac. Radiol.* **8**:118–131.

Singhal, S., Henderson, R., Horsfield, K., Harding, K., and Cumming, G. (1973). MOrphometry of the human pulmonary arterial tree. *Circ. Res.* **33**:190–197.

Smith, P. (1976). A comparison of the orientation of elastin fibers in the elastic laminae of the pulmonary trunk and aorta of rabbits using the scanning electron microscope. *Lab. Invest.* **35**:525–529.

Snapper, J. R., Ogletree, M. L., Hutchison, A. A., and Brigham, K. L. (1981). Meclofenamate prevents increased resistance of the lung (R_L) following endotoxemia in unanesthetized sheep. *Am. Rev. Respir. Dis.* **123**:200–204.

Sobin, S. S., Fung, Y.-C., Lindal, R. G., Tremer, H. M., and Clark, L. (1980). Topology of pulmonary arterioles, capillaries, and venules in the cat. *Microvasc. Res.* **19**:217–233.

Staub, N. C. (1983). Alveolar flooding and clearance. *Am. Rev. Respir. Dis.* **127**:S44–S51.

Staub, N. C., and Storey, W. F. (1962). Relation between morphological and physiological events in lung studied by rapid freezing. *J. Appl. Physiol.* **17**:381–390.

Staub, N. C., Nagano, H., and Pearce, M. L. (1967). Pulmonary edema in dogs, especially the sequence of flud accumulation in lungs. *J. Appl. Physiol.* **22**:227–240.

Svingen, B. A., O'Neal, F. O., and Aust, S. D. (1978). The role of superoxide and singlet oxygen in lipid peroxidation. *Photochem. Photobiol.* **28**:803–9.

Tagnizadeh, A., and Reynolds, E. O. R. (1976). Pathogenesis of bronchopulmonary dysplasia following hyaline membrane disease. *Am. J. Pathol.* **82**:241–257.

Taylor, A. E., Parker, J. C., Kvietys, P. R., and Perry, M. A. (1982). The pulmonary interstitium in capillary exchange. *Ann. N. Y. Acad. Sci.* **384**:146–165.

Tobin, C. E. (1952). The bronchial arteries and their connections with other vessels in the human lung. *Surg. Gynecol. Obstet.* **95**:741–50.

Tomashefski, J. F., Oppermann, H. C., Vawter, G. F., and Reid, L. M. (1984). Bronchopulmonary dysplasia: a morphologic study with emphasis on the pulmonary vasculature. *Pediatr. Pathol.* **2**:469–487.

Trapnell, D. H. (1963). The peripheral lymphatics. *Br. J. Radiol.* **36**:660–672.

Turner-Warwick, M. (1963). Precapillary system pulmonary anastomoses. *Thorax* **18**:225–37.

Uhley, H. N., Leeds, S. E., Sampson, J. J., and Friedman, M. (1967). Right duct lymph flow in experimental heart failure following acute elevation of left atrial pressure. *Circ. Res.* **20**:306–310.

Vassilyadi, M., and Michel, R. P. (1985). Sequence of fluid accumulation in nitrogen dioxide (NO_2)-induced lung edema in dogs: a morphometric study (abstr.). *Physiologist* **28**:350.

Verloop, M. C. (1949). On the arteriae bronchiales and their anastomoses, with the arteriae pulmonales in some rodents; a micro-anatomical study. *Acta Anat.* **7**:1–32.

Wagenvoort, C. A. (1973). Hypertensive pulmonary vascular disease complicating congenital heart disease: a review. *Cardiovasc. Circ.* **5**:43–60.

Wagenvoort, C. A., and Wagenvoort, N. (1967). Arterial anastomoses, bronchopulmonary arteries, and pulmobronchial arteries in perinatal lungs. *Lab. Invest.* **16**:13–24.

Wagenvoort, C. A., and Wagenvoort, N. (1977). *Pathology of Pulmonary Hypertension*. New York, John Wiley, pp. 17–55, 119–161, 217–231.

Wagenvoort, C. A., Heath, D., and Edwards, J. E. (1964). *The Pathology of the Pulmonary Vasculature*. Springfield, IL, Charles C. Thomas.

Wagner, H. N. Jr., Sabiston, D. C. Jr., Iio, M., McAfee, J. G., Meyer, J. K., and Langan, J. K. (1964). Regional pulmonary blood flow in man by radioisotope scanning. *J.A.M.A.* **187**:601–3.

Wang, C. G., Hakim, T. S., Michel, R. P., and Chang, H. K. (1985). Segmental pulmonary vascular resistance in progressive hydrostatic and permeability edema. *J. Appl. Physiol.* **59**(1):242–247.

Watkins, W. D., Hutlemeier, P. C., Kong, D., and Peterson, M. B. (1982). Thromboxane and pulmonary hypertension following *E. coli* endotoxin infusion in sheep: effect of an imidazole derivative. *Prostaglandins* **23**:273–285.

Weibel, E. R. (1960). Early stages in the development of collateral circulation to the lung in the rat. *Circ Res.* **8**:353–76.

Weibel, E. R. (1984). *The Pathway of Oxygen*. Cambridge, MA, Harvard University Press, p. 289.

Weibel, E. R., and Palade, G. E. (1964). New cytoplasmic components in arterial endothelia. *J. Cell Biol.* **23**:101–112.

West, J. B., Dollery, C. T., and Heard, B. E. (1965). Increased pulmonary vascular resistance in the dependent zone of the isolated dog lung caused by perivascular edema. *Circ. Res.* **17**:191–206.

Wight, T. N. (1980). Vessel proteoglycans and thrombogenesis. In *Progress in Hemostasis and Thrombosis*. Edited by T. Spact. New York, Grune & Stratton, p. 1.

Wood, D. A., and Miller, M. (1938). The role of the dual pulmonary circulation in various pathologic conditions of the lungs. *J. Thorac. Surg.* **7**:649–70.

Worthen, G. S., Lien, D. C., Tonnesen, M. G., and Hensen, P. M. (1987). Interaction of leukocytes with the pulmonary endothelium. In *Pulmonⁱry Endothelium in Health and Disease*. Edited by U. S. Ryan. New York, Marcel Dekker, pp. 123–160.

Yen, R. T., Zhuang, F. Y., Fung, Y. C., Ho, H. H., Tremer, H., and Sobin, S. S. (1983). Morphometry of cat pulmonary venous tree. *J. Appl. Physiol.* **55**:236–242.

Yoffey, J. M., and Courtice, F. C. (1970). *Lymphatics, Lymph and the Lymphomyeloid Complex*. New York, Academic Press.

Zarins, C. K., Rice, C. L., Peters, R. M., and Virgilio, R. W. (1978). Lymph and pulmonary response to isobaric reduction in plasma oncotic pressures in baboons. *Circ. Res.* **43**:925–930.

28

Experimental Lung Carcinogenesis by Intratracheal Instillation

ULRICH MOHR and SHINJI TAKENAKA*

Medizinische Hochschule Hannover
Hannover, West Germany

I. Introduction

Intratracheal (IT) instillation techniques began to be used for lung carcinogenesis studies in the 1950s (Cember et al., 1959; Kotschetkowa and Awrunina, 1957). The methods were first described in detail by Della Porta et al. (1958) and Te Dunga and Preston (1958). Further development by Saffiotti et al. (1968) established IT instillation as a common technique for lung carcinogenesis studies.

Early studies using this technique were aimed mainly at inducing a model of human bronchogenic carcinoma in laboratory animals. Strong carcinogens combined with inert particles, such as benzo(a)pyrene and ferric oxide particles (Saffiotti et al., 1968), were best selected for this purpose. With the increasing interest in environmental protection and occupational hygiene, the IT instillation technique has been used in risk assessment tests as an alternative method to inhalation. This technique is available for almost all substances except gaseous ones.

Details of the IT instillation technique and a survey of major carcinogenicity studies in which the technique has been used will be provided in this chapter.

---

*Present affiliation: Gesellschaft für Strahlen-und Umweltforschung, Neuherberg, West Germany

Since inhalation is a method comparable to IT instillation, studies using inhalation exposure will also be reviewed.

II. Method

Saffiotti et al. (1968) established the technique using Syrian golden hamsters. Although there are some technical variations in mice, rats, and hamsters (Ho and Furst, 1973; Brain et al., 1976; Steinhoff et al., 1986), IT instillation is generally performed based on Saffiotti's method.

Before instillation, the hamsters are anesthetized with 0.4 ml of 1% solution of sodium pentobarbital injected intraperitoneally through the abdominal wall, which has previously been disinfected with a few drops of 70% ethanol. This short-acting barbiturate is selected to avoid the edema and other changes in the lung often associated with ether anesthesia, and to permit rapid resumption of the normal respiratory activity. As soon as the animal is anesthetized, it is placed on its back, head uppermost, on a slanted board. Its mouth is kept open by attaching the lower incisor teeth to a wire hook while the upper incisors are retained by a tight rubber band. Then the selected volume of agitated suspension is drawn out of the flask using a 0.25 ml tuberculin syringe fitted with a blunt 19-gauge needle about 60 mm long, bent to an angle of 135 degrees about 45 mm from the tip. A direct-focusing headlight, worn by the operator, provides a clear view of the pharynx when the tongue of the hamster is gently pulled outward and laterally with a forceps. When necessary, the oral cavity is cleaned of mucus with a small cotton swab. The blunt tip of the needle is inserted dorsally under the epiglottis to uncover the vocal cords and is then gently pushed between these into the tracheal lumen. Very light but definite bumping against the tracheal cartilage rings indicates that the needle is properly inserted. The tip of the needle is pushed almost to the tracheal bifurcation, the suspension gently injected, and the needle withdrawn. Inspection of the pharynx is continued for a short time and the hamster is kept on the board for a minute or two to ensure that none of the suspension is regurgitated. Following the injection, the animals show brief apnea, after which they rapidly resume regular respiration. With technical help for the anesthesia, each treatment takes less than 1 min. The apparatus used to prepare and inject the suspensions is sterilized before use.

For rats or mice, needles of about 80 mm or 30 mm, respectively, should be used. Three to 4% halothane (Hoechst, Frankfurt, FRG) can be used for anesthesia in hamsters, rats, and mice.

Test substances are prepared as follows (Steinhoff et al., 1986). Substances soluble in water, such as sodium dichromate, are administered in 0.9% NaCl solution. Slightly water-soluble substances, such as benzo(a)pyrene, are administered in 0.9% NaCl and 2% Tween 60 (Merck-Schuchardt, Hohenbrunn, FRG). Insoluble substances, such as dimethylcarbamyl chloride, are administered in peanut oil (DAB 7).

Maximal volumes to be administered are as follows: for rats, 0.3 ml; for Syrian golden hamsters, 0.15 ml; and for mice, 0.1 ml. Materials can be given up to twice weekly.

III. Dosimetry

The main feature of the IT instillation technique is the exclusion of test substances from the upper respiratory tract. This characteristic obliterates the normal relationship between inhaled dose and local tissue dose, which is an important factor in animals or humans naturally exposed to airborne substances. The advantage is that it allows massive exposure of the pulmonary parenchyma to test materials so that the chances of inducing tumors are maximized and associated pathogenetic studies can be undertaken. The disadvantages are that because normal dosimetry factors are bypassed there can be no quantitative risk estimates relevant to inhalation exposure, and there is no opportunity to investigate the subtle interplay of factors that operate during "real-life" inhalation exposures with relatively low concentrations of test substances.

The distribution pattern of particles in the alveolar region clearly differs between studies using instillation and inhalation (Brain et al., 1976). Substances administered by IT instillation have a nonuniform distribution, with preferential deposition in the dependent portion of the lung.

Substances deposited in the alveolar region are mainly phagocytosed by alveolar macrophages, which are then eliminated slowly via the mucociliary escalator of the tracheobronchial tract. Substances are also cleared via lymphatics and blood vessels, even though this occurs inefficiently.

Depending on the dosage and physicochemical properties, some substances administered remain in the alveolar region for extended periods of time. Clearance mechanisms and rates appear similar for both IT instillation and inhalation, as long as low dosages are administered (Finch et al., 1987).

IV. Carcinogenicity

A. Polynuclear Aromatic Compounds

As described in the Introduction, animal experiments using IT instillation techniques began with a class of strong carcinogens, the polynuclear aromatic compounds (PAC)s. These included dibenzanthracene (DBA), benzo(a)pyrene; B(a)P, and 3-methylcholanthrene (3-MC).

The first induction of lung tumors in rats was reported by Niskanen (1949) after direct injection of DBA into the trachea through skin incision. Later, Pylev (1961) and Shabad (1962) also demonstrated squamous cell carcinomas in rats by IT instillation of DBA.

DMBA suspended in 1% gelatin induced squamous cell carcinomas and adenocarcinomas of the trachea and bronchi in hamsters when given by IT instillation (Della Porta et al., 1958). Nettesheim and Hammons (1971) and Schreiber et al. (1972) demonstrated bronchiolar–alveolar squamous cell tumors in mice and rats by IT instillation of 3-MC.

B(a)P, like other PACs, occurs frequently in the environment. It occurs mainly as a result of incomplete combustion. Pylev (1961) and Shabad (1962) produced lung tumors by IT instillation of B(a)P into rats. Herrold and Dunham (1962), Gross et al. (1965), and Saffiotti et al. (1968) demonstrated bronchogenic lung tumors in hamsters by IT instillation of B(a)P.

In the experiment by Saffiotti et al. (1968), the combination of B(a)P and ferric oxide induced a high incidence of lung tumors, mainly bronchogenic squamous cell carcinomas, in Syrian golden hamsters. Owing to the high incidence of tumors and well-established similarities to human bronchogenic cancer, this technique has since been used by many researchers.

Schreiber et al. (1974) investigated sequential cytological changes in developing respiratory tract tumors induced in hamsters by B(a)P–ferric oxide. They demonstrated that cytological abnormalities, indistinguishable from those found in cigarette smokers and uranium miners, could be induced by the application of B(a)P–ferric oxide in hamsters; as in humans, the sequence from mild atypia to moderate and marked atypia leading to cancer was observed.

IT instillations of ^{210}Po adsorbed onto hematite carrier particles produced adenosquamous carcinomas in the peripheral lung of hamsters, while B(a)P–hematite produced mainly epidermoid carcinomas in the major bronchi and trachea (Saffiotti et al., 1968; Little and O'Toole, 1974). The mechanism of differences in the histological appearance and apparent sites of origin of induced tumors was investigated by Kennedy and Little (1974), who examined the transport and localization of these carcinogens in the lung. They found that B(a)P induced central tumors because a significant amount of the carcinogen reached the upper airway epithelial cells, while ^{210}Po induced peripheral tumors because the major radiation dose is delivered to the alveolar region.

Hamsters and rats showed striking species differences in the response of their respiratory tracts to the IT instillation of B(a)P–ferric oxide. Whereas in hamsters squamous metaplasia of the trachea and large bronchi were the major changes, squamous cell tumors of bronchiolar–alveolar origin developed in rats within a few weeks after carcinogen application. A species difference was detected in the distribution of B(a)P in the tracheal tissues; the carcinogen was found in the epithelium of hamsters but not in the epithelium of rats, suggesting a species difference in penetration of the carcinogen from the lumen into the tracheal tissues (Schreiber et al., 1975).

In many mice strains, pulmonary adenoma occurs with a high spontaneous incidence. Because of the morphological and biological characteristics, the site of origin, and the spontaneous occurrence of this type of tumor, the pulmonary adenoma in mice is generally not considered to be an adequate model for studies designed to elucidate the pathogenesis of human bronchogenic carcinoma (Nettesheim and Hammons, 1971). However, Nettesheim and Hammons (1971), Ho and Furst (1973), and Yoshimoto et al. (1973) demonstrated induction of squamous cell carcinomas by IT instillation of 3-MC or B(a)P in strains with low incidences of spontaneous pulmonary adenomas.

The effects of vitamin A on chemical carcinogenesis were examined using IT instillation of PACs. Saffiotti et al. (1967) demonstrated inhibitory effects of vitamin A on the induction of tracheobronchial squamous metaplasia and squamous cell tumors in Syrian golden hamsters by IT instillation of B(a)P–hematite. Harris et al. (1972) demonstrated squamous metaplasia with defective basement membranes, enlarged nuclei with cytoplasmic invagination, and pleomorphic nuclei in the trachea of hamsters fed a vitamin-A-deficient diet after IT instillation of B(a)P–ferric oxide. Nettesheim et al. (1976) examined the effects of retinyl acetate on the development of metaplastic lung changes induced by 3-MCA in F344 rats and found that retinyl acetate had a significant inhibitory effect on the postinitiation phase of preneoplastic lung nodules. Later, Dogra et al. (1985), using IT instillation techniques, demonstrated the effects of vitamin A deficiency not only on the postinitiation phase but also on the initiation phase in B(a)P–ferric-oxide-induced lung tumorigenesis in Wistar rats.

In recent years, PACs, especially B(a)P, have been used in studies of cocarcinogenic or syncarcinogenic effects of environmental or occupational hazards. Ishinishi et al. (1977) and Pershagen et al. (1984) found indications of a synergism between arsenic trioxide and B(a)P in rats and hamsters. Shabad et al. (1974) also found indication of a synergism between chrysotile asbestos and B(a)P. Little et al. (1978) investigated the interaction between instilled B(a)P and polonium 210 radiation in the induction of lung cancer in Syrian golden hamsters. They observed additive effects after simultaneous administration. A significant synergistic interaction between the two agents also occurred when B(a)P exposure followed 4 months after ^{210}Po exposure. Metivier et al. (1984) also demonstrated multiplicative effects of IT-instilled B(a)P and inhaled plutonium oxide on lung carcinogenesis in rats. Kobayashi and Okamoto (1974) examined the carcinogenic effects of lead oxide and/or B(a)P in Syrian golden hamsters by IT instillation. They found bronchiolar–alveolar adenomas and adenocarcinomas in the lungs of hamsters given B(a)P mixed with lead oxide, whereas no tumors occurred in the other groups. Heinrich et al. (1986) examined the syncarcinogenic effects of diesel exhaust after administration of PACs or N-nitroso compounds in rats, hamsters, or mice. They found synergistic effects only in rats.

B. *N*-Nitroso Compounds

Although N-nitroso compounds are well known to induce lung tumors when given parenterally, orally, or by inhalation, the use of N-nitrosamines for lung carcinogenicity studies using IT instillation has been limited (Dotenwill et al., 1962; Moiseev and Benemansky, 1975; IARC, graphs, 1978). Herrold and Dunham (1963), however, clearly demonstrated that IT instillation of N-nitrosodiethylamine

(NDEA) induced multiple squamous cell papillomas of the trachea and bronchi and carcinomas of the ethmoid region of the nasal cavity in Syrian golden hamsters. Ferron et al. (1972) also confirmed the development of tumors in Syrian golden hamsters after IT instillation of NDEA.

C. Radionuclides

Little and O'Toole (1974) investigated the lung carcinogenicity of polonium 210 combined with ferric oxide when given by IT instillation to Syrian golden hamsters. They found high yields of respiratory cancers in hamsters receiving 225–4500 rads of polonium 210 given by IT instillation. ^{210}Po-induced tumors were almost exclusively adenosquamous carcinomas that arose peripherally (Little and O'Toole, 1974). A serial sacrifice study on the pathogenesis of ^{210}Po-induced lung tumors in Syrian golden hamsters revealed rapid progression from hyperplasia of bronchiolarlike cells in proximal alveolar regions to development of malignant tumors (Kennedy et al. 1978). Kennedy et al. (1978) demonstrated no significant difference in ultimate tumor incidence following administration of polonium 210 with or without ferric oxide, although there were different distribution patterns of polonium 210 in the lungs of the two groups.

Another source of alpha-radiation, ^{239}Pu, also induced lung tumors in rats (Yerokhin et al., 1971). Temple et al. (1960) reported an increased incidence of lung adenomas in mice exposed to $[^{239}Pu]O_2$ or to the beta emitter, $[^{106}Ru]O_2$.

In a series of experiments Cember and Watson (1958) and (1959) demonstrated the lung carcinogenicity of beta and gamma rays in rats by IT instillation. The administration of 375 μCi$[Ba^{35}]SO_4$, in the form of 1.45 μm diameter $BaSO_4$ particles, induced severe squamous metaplasia and squamous cell carcinomas in rats after IT instillation. $[^{144}Ce]Fl_3$ at dosages of 5–50 μCi also induced squamous cell carcinomas in all groups of experimental rats. Soluble ^{144}Ce, as $[^{144}Ce]Cl_3$ solution, was also found to be carcinogenic in the rat lung after IT instillation of dosages of 10–30 μmCi per rat. Histological findings included squamous cell carcinomas and few adenocarcinomas, undifferentiated carcinomas, and sarcomas (Cember and Stemmer, 1964).

$[^{103}Ru]O_2$, $[^{106}Ru]Cl$, $[^{59}Fe_2]O_3$, ^{198}Au, and $[Cr^{32}]PO_4$ colloid were reported to be weakly carcinogenic in rats after IT instillation (Annalys of ICRP, 1980).

Alpha and beta emitter have also been shown to be carcinogenic following inhalation in rats (Annals of ICRP, 1980; Hahn, 1985), mice (Hahn et al., 1980; Lundgren et al., 1981), and Syrian golden hamsters (Cross et al., 1981; Lundgren et al., 1982, 1983; Thomas et al., 1981).

D. Arsein

Arsenic trioxide (As) induced tumors of the respiratory tract, including lungs, after IT instillation in Syrian golden hamsters, although the incidence was relatively low (Pershagen et al., 1984). At the microscopic level, adenomas, adenocarcinomas, and anaplastic carcinomas occurred. As described previously, synergistic effects of As on B(a)P carcinogenicity were also demonstrated (Ishinishi et al., 1977; Pershagen et al., 1984).

On the other hand, two reports on inhalation experiments showed no lung carcinogenicity of arsein in mice (Berteau et al., 1978) or rats (Glaser et al., 1986).

E. Beryllium

Beryllium compounds, such as beryllium–aluminium alloy, beryllium chloride, beryllium fluoride, and beryllium sulfate are known to be carcinogenic in the rat lung after inhalation (Schepers et al., 1957; Reeves et al., 1967; Wagner et al., 1969; Litvinov et al., 1975). Intratracheal instillation of beryllium metal, passivated beryllium metal (99% Be, containing 0.26% chromium), beryllium–aluminium alloy, and beryllium hydrooxide also induced lung tumors in rats (Groth et al., 1980).

Beryllium is one of the substances that resulted in lung carcinogenicity in rats after both IT instillation and inhalation.

In hamsters, no lung tumors were observed after inhalation of beryl or bertrandite ore dusts at 15 mg/m^3 for 16 hr/day, 5 days/week for up to 17 months (Wagner et al., 1969).

F. Chromium

Although there is sufficient evidence for the carcinogenicity of calcium chromate and some relatively insoluble chromium compounds parenterally administered in rats, very few data are available for the carcinogenicity of IT-instilled or inhaled chromium (IARC, 1980). Recently, Steinhoff et al. (1986) found that sodium dichromate showed weak but significant lung carcinogenicity in rats after IT instillation.

G. Cadmium

Similarly to chromium compounds, cadmium is well known to be carcinogenic when it is injected parenterally (IARC, 1976). However, carcinogenic data from IT instillation or inhalation are limited. Sanders and Mahaffy (1983) examined the carcinogenic effects of low doses of cadmium oxide after single or multiple IT instillation. They found that cadmium oxide was not carcinogenic in the lung of rats. Recently, Pott et al. (1986) demonstrated a high incidence of lung tumors in rats by IT instillation of cadmium sulfide (a total dose, 0.63–50 mg). In the same series of experiments, Pott et al. examined carcinogenic effects of instilled cadmium chloride (maximal total dose, 135 µg/rat), and cadmium oxide (maximal total dose, 135 µg/rat) in rats. They found 6.4–7.5% incidences of lung tumor in about 40 rats. Adenomas, adenocarcinomas, and squamous cell carcinomas occurred in rats exposed to these cadmium compounds.

Takenaka et al. (1983) demonstrated dosage-dependent tumor induction in rats after inhalation of low level of cadmium chloride.

Cadmium is therefore one of the substances that induced lung carcinogenicity in rats after both IT instillation and inhalation.

H. Nickel

Although parenteral administration of nickel and nickel compounds produced tumors locally or systemically, it has been reported that IT instillation of nickel powder, nickel subsulfide, or nickel oxide does not induce any significant increases in tumor incidence in mice, rats or hamsters (IARC, 1976; Fisher et al., 1986). Recently, however, Pott et al (1986) demonstrated a significant increase in the occurrence of lung tumors in rats after IT instillation of nickel oxide (total dosage, 50 or 150 mg/raty), nickel powder (total dosage, 6 or 9 mg/rat), or nickel subsulfide (total dosage, 0.95 mg or 3.75 mg/rat). Adenomas, adenocarcinomas, squamous cell carcinomas, and adenosquamous carcinomas were found. In Syrian golden hamsters, Muhle et al. confirmed no significant increase in lung tumors after IT instillation of nickel powder or nickel subsulfide.

Inhalation of nickel subsulfide showed carcinogenic effects in the rat lung (Ottolenghi et al., 1974). It has also been suggested that nickel carbonyl induces lung tumors in rats (Sunderman et al., 1959). Rats exposed to nickel powder (Hueper, 1958; Hueper and Payne, 1962) or nickel oxide (Takenaka et al., 1985) developed no lung tumors. Wehner et al. (1975b) observed no lung tumors in Syrian golden hamsters following inhalation of nickel oxide.

In conclusion, nickel is one of the substances that induce lung tumors in rats after both IT instillation and inhalation. However, its lung carcinogenicity in hamsters has never been demonstrated.

I. Silica

Silica, which has been known for a long time as a causative agent of silicosis, has been recently found to be carcinogenic to the lung of rats. Holland et al. (1983) showed lung tumors in 6 of 36 rats after 10 IT instillations of 7 mg/rat of quartz (Minu-U-Sil). In the experiment by Groth et al. (1986), rats developed lung tumors after a single IT instillation of 20 mg of quartz (Min-U-Sil or Novaculite). All 30 lung tumors in rats that received Min-U-Sil were adenocarcinomas. Twenty of 21 lung tumors in rats that received Novaculite were also adenocarcinomas, and 1 was a squamous cell carcinoma. After inhalation of 12 mg/m^3 or 50 mg/m^3 quartz (Min-U-Sil) for 24 months, rats developed lung tumors in the studies by Holland et al. (1986) and Dagle et al. (1986).

On the other hand, neither IT instillation nor inhalation of Silica has been shown to cause lung tumors in hamsters (Holland et al., 1983; Niemeier et al., 1986; Stenbaeck et al., 1986).

J. Asbestos

Gross et al. (1967) demonstrated 3 adenocarcinomas in 19 rats given chrysotile asbestos by IT instillation (total dosages, 14–21 mg/rat). In the same series of experiments, they also found lung tumors in rats exposed to chrysotile by inhalation.

Pylev (1980) produced malignant and benign lung tumors in Syrian golden hamsters after IT instillation of chrysotile (10 mg twice monthly) or amphibole fibers (5 mg twice monthly). In his comparison of pretumorous lesions in rats and hamsters, both species showed focal proliferation of the bronchial and alveolar epithelium; in rats these proliferative foci were usually solid, with squamous metaplasia. No squamous metaplasia was observed in hamsters.

A slight increase in the tumor rate was observed in hamster lungs on inhalation of chrysotile asbestos, although only adenomas occurred (Wehner et al., 1975a).

Recently, Pott et al. (1986) demonstrated lung carcinogenicity of crocidolite asbestos in rats after IT instillation (total dosage 10 mg/rat). Adenocarcinomas, squamous cell carcinomas, and adenosquamous carcinomas occurred.

V. Conclusion

A. Comparison Between IT Instillation and Inhalation

For IT instillation, the ease of exposure, low cost, and ability to administer multiple large doses in a short period of time (Finch et al., 1987) are major advantages over inhalation. In addition, it is probably one of the safer methods for exposing animals to very hazardous materials in terms of handling and contamination (Phalen, 1984). However, the distribution of instilled doses in the respiratory tissue is artificial. IT instillation tends to lead to less uniform deposition than inhalation and, due to gravitational settling of the instilled material, tends to favor the dependent portions of the lung (Brain et al., 1976; Phalen, 1984). Other problems are possible mechanical damage to the trachea by the intubation tube and potential alterations in the pulmonary fluid balance (Watson et al., 1960; Finch et al., 1987).

Although data directly comparing the lung carcinogenicity of inhaled and instilled substances are limited, a definite tendency can be noticed. Substances listed in this chapter as carcinogenic to the lung after IT instillation did not always show lung carcinogenicity after inhalation, whereas substances that were carcinogenic to the lung after inhalation always showed lung carcinogenicity after IT instillation. Moreover, only strong carcinogens, such as radionuclides, were able to induce lung tumors after both IT instillation and inhalation in three species—hamsters, rats, and mice—although hamsters and mice developed low incidences of lung tumors after inhalation. Other substances reviewed in this article showed lung carcinogenicity only in rats after both IT instillation and inhalation. In hamsters, some substances, such as B(a)P or arsein, showed lung carcinogenicity after IT instillation but not after inhalation. Lung carcinogenicity of substances administered by IT instillation or inhalation in rats and hamsters is listed in Table 1.

B. Species Differences Between Rats and Hamsters

Data have recently accumulated on the species-dependent differences in the lung carcinogenicity of various compounds for rats and hamsters. The results obtained by IT instillation of B(a)P (Schreiber et al., 1975), radioactive materials (Hahn, 1985), and silica (Saffiotti, 1986) indicated that the Syrian golden hamsters were far less susceptible than the rats to the lung carcinogen. In addition, the two species showed clearly different localization of the induced lesions and eventual tumors. The IT instillation of B(a)P–ferric oxide in hamsters produced squamous metaplasia of the trachea and large bronchi; in contrast, the same treatment resulted in squamous cell nodules of bronchiolar–alveolar origin in rats (Schreiber et al., 1975). Similar species differences between hamsters and rats were also observed in the results of inhalation exposure to silica (Saffiotti, 1986) or diesel exhaust (Heinrich et al., 1986).

   One of the major factors responsible for such species differences would be the different susceptibility of cells at risk in the two species. This difference can be attributed to various aspects of cellular activity. The rat tracheal cells may be better protected from carcinogen penetration than the hamster tracheal epithelial cells because of the overwhelming abundance of mucus production in the rat trachea compared with the hamster trachea (Schreiber et al., 1975; Plopper, 1983). Differences in ability to metabolize indirect-acting carcinogens were also demonstrated in hamsters in rats. B(a)P-directed monooxygenase levels were two to three times greater in microsomes from lungs of methylcholanthrene-preinduced rats than in identical preparations from hamster lungs (Prough et al., 1979). In contrast, organ-cultured tracheas of hamsters produced a larger amount of B(a)P metabolites and more binding of them to DNA than those of rats (Mass and Kaufman, 1983). The results of these studies on metabolism corroborate well the incidence, types, and localization of the tumors and preneoplastic changes induced by IT instillation of B(a)P in both species (Schreiber et al., 1975). Another factor that is probably responsible for the species differences would be the species-

Table 1 Lung Carcinogenicity of Substances Administered by IT Instillation or Inhalation in Rats and Hamsters

	Rat		Hamster	
	Instillation	Inhalation	Instillation	Inhalation
Radionuclides	+	+	+	+
N-nitroso and its compounds	?	+	+	+
Asbestos	+	+	+	+ −
PACs	+	+	+	−
Nickel and its compounds	+	+	−	−
Silica	+	+	−	−
Arsein and its compounds	?	−	+	−
Beryllium and its compounds	+	+	?	−
Cadmium and its compounds	+	+	?	?
Chromium and its compounds	+	+ −	?	?

+, Positive; −, negative; + −, weakly positive; ?, not examined
Data on inhalation studies from Mohr and Dungworth (1988).

dependent variation in the dosimetry of the instilled carcinogen. The inhalation experiment using monodisperse ^{169}Yb-labeled alumino-silicate aerosols revealed that the bronchial deposition of particles with diameters smaller than 2.09 aerodynamic diameters (μm) was detected more constantly in hamsters than in rats, while the lung deposition varied between the two species depending on the particle size (Raabe et al., 1977). In the same study, tracheal deposition of particles larger than 1.04 aerodynamic diameter (μm) was greater in hamsters than in rats, while that of smaller particles showed a reversed relationship. In the hamster tracheal tissues examined by fluorescence microscopy after IT instillation of B(a)P–ferric oxide, penetration of the carcinogen into the epithelium was clearly indicated (Schreiber et al., 1975). In rats, however, this was not the case despite significant amounts of the substances being detected in the tracheal lumen. Clearance of substances deposited in alveolar regions will also affect the dosimetric profile. Alveolar macrophages, which play a large part in alveolar clearance, showed some species differences in their function. The uptake of insoluble gold particles by the alveolar macrophages was faster in hamsters than in rats (Brain and Mensah, 1983). Saffiotti (1986) examined mechanisms of the species difference in the lung carcinogenicity of silica. He found that the rats treated with silica showed necrosis of macrophages, a complex reticuloendothelial reaction with continuous recruitment and necrosis of macrophages, and fibrotic changes. In Syrian golden hamsters, however, neither necrosis of macrophages nor any of the subsequent fibrogenic processes occurred after IT instillation of silica.

These comparisons show that IT instillation of substances tends to induce lung tumors at higher incidences than does inhalation. Histological types of tumors induced by IT instillation were comparable to those induced by inhalation. These results confirm the usefulness of the IT instillation technique in the study of lung carcinogenesis. Because of the advantages already described, the IT instillation technique will continue to be frequently used in tests to detect substances' potential to cause lung tumors.

References

Annals of the ICRP (1980). *Biological Effects of Inhaled Radionuclides*. New York, Pergamon.

Berteau, P. E., Flom, J. O., Dimmick, R. L., and Boyd, A. R. (1978). Long-term study of potential carcinogenicity of inorganic aerosols to mice. *Toxicol. Appl. Pharmacol.* **45**:323.

Brain, J. D., and Mensah, G. A. (1983). Comparative toxicology of the respiratory tract. *Am. Rev. Respir. Dis.* **128**:S87–S90.

Brain, J. D., Knudson, D. E., Sorokin, S. P., and Davis, M. A. (1976). Pulmonary distribution of particles given by intratracheal instillation or by aerosol inhalation. *Environ. Res.* **11**:13–33.

Cember, H., and Stemmer, K. (1964). Lung cancer from radioactive cerium chloride. *Health Phys.* **10**:43–48.

Cember, H., and Watson, J. A. (1958). Bronchogenic carcinoma from radioactive barium sulfate. *Arch. Ind. Health* **17**:230–235.

Cember, H., Watson, J. A., and Spritzer, A. A. (1959). Bronchogenic carcinoma from radioactive cerium fluoride. *Arch. Ind. Health* **19**:14–23.

Cross, F. T., Palmer, R. F., Busch, R. H., Filipy, R. E., and Stuart, B. O. (1981). Development of lesions in Syrian golden hamsters following exposure to radon daughters and uranium ore dust. *Health Phys.* **41**:135–153.

Dagle, G. E., Wehner, A. P., Clark, M. L., and Buschbom, R. L. (1986). Chronic inhalation exposure of rats to quartz. In *Silica, Silicosis and Cancer*. Edited by D. F. Goldsmith, D. M. Winn, and M. Shy. Westport, Greenwood Press, pp. 255–266.

Della Porta, G., Kolb, L., and Shubik, P. (1958). Induction of tracheobronchial carcinomas in the Syrian golden hamster. *Cancer Res.* **18**:592–597.

Dogra, S. C., Khanduja, K. L., and Gupta, M. P. (1985). The effect of vitamin A deficiency on the initiation and postinitiation phases of benzo(a)pyrene-induced lung tumorigenesis in rats. *Br. J. Cancer* **52**:931–935.

Dotenwill, W., Mohr, U., and Zagel, M. (1962). Ueber die unterschiedliche Lungen-carcinogene Wirkung des Diaethylnitrosamine bei Hamster und Ratte. *Z. Krebsforsch.* **64**:499–502.

Ferron, V. J., Emmelot, P., and Vossenaar, T. (1972). Lower respiratory tract tumors in Syrian golden hamsters after intratracheal instillation of diethylnitrosamine alone and with ferric oxide. *Eur. J. Cancer* **8**:445–449.

Finch, G. L., Fisher, G. L., and Hayes, T. L. (1987). The pulmonary effects and clearance of intratracheally instilled Ni3S2 and TiO2 in mice. *Environ. Res.* **42**:83–93.

Fisher, G. L., Chrisp, C. E., and McNeill, D. A. (1986). Lifetime effects of intratracheally instilled nickel subsulfide on B6C3F1 mice. *Environ. Res.* **40**:313–320.

Glaser, U., Hochrainer, D., Oldiges, H., and Takenaka, S. (1985). Long-term inhalation studies with NiO and As2O3 aerosols in Wistar rats. Proceedings of the International Conference on Health Hazards and Biological Effects of Welding Fumes and Gases, Copenhagen.

Gross, P., Tolker, E., Babyak, M. A., and Kaschak, M. (1965). Experimental lung cancer in hamsters. *Arch. Environ. Health* **11**:59–65.

Gross, P., deTreville, R. T. P., Tolker, E. B., Kashak, M., and Babyak, M. A. (1967). Experimental asbestosis. The development of lung cancer in rats with pulmonary deposits of chrysotile asbestos dust. *Arch. Environ. Health* **15**:343–355.

Groth, D. H., Kommineni, C., and MacKay, G. R. (1980). Carcinogenicity of berillium hydroxide and alloys. *Environ. Res.* **21**:63–84.

Groth, D. H., Stettler, L. E., Platek, S. F., Lal, J. B., and Burg, J. R. (1986). Lung tumors in rats treated with quartz by intratracheal instillation. In *Silica, Silicosis and Cancer.* Edited by D. F. Goldsmith, D. M. Winn, and M. Shy. Westport, Greenwood Press, pp. 243–253.

Hahn, F. F. (1985). Radiation-induced squamous cell carcinoma, lung of rodents. In *Monographs on Pathology of Laboratory Animals. Respiratory System.* Edited by T. C. Jones, U. Mohr, and R. D. Hunt. Berlin, Springer-Verlag, pp. 127–137.

Hahn, F. F., Lundgren, D. L., and McClellan, R. O. (1980). Repeated inhalation exposure of mice to 144CeO2. II. Biological effects. *Radiat. Res.* **82**:123–137.

Harris, C. C., Sporn, M. B., Kaufman, D. G., Smith, J. M., Jackson, F. E., and Saffiotti, U. (1972). Histogenesis of squamous metaplasia in the hamster tracheal epithelium caused by vitamin A deficiency or Benzo(a)pyrene-ferric oxide. *J. Natl. Cancer Inst.* **48**:743–761.

Heinrich, U., Muhle, H., Takenaka, S., Ernst, H., Fuhst, R., Mohr, U., Pott, F., and Stoeber, W. (1986). Chronic effects on the respiratory tract of hamsters, mice and rats after long-term inhalation of high concentrations of filtered and unfiltered diesel engine emissions. *J. Appl. Toxicol.* **6**:383–395.

Herrold, K. M., and Dunham, L. J. (1962). Induction of carcinoma and papilloma of the tracheobronchial mucosa of the Syrian hamster by intratracheal instillation of benzo(a)pyrene. *J. Natl. Cancer Inst.* **28**:467–491.

Herrold, K. M., and Dunham, L J. (1963). Induction of tumors in the Syrian hamster with diethylnitrosamine (N-nitrosodiethylamine). *Cancer Res.* **23**:733–777.

Ho, W., and Furst, A. (1973). Intratracheal instillation method for mouse lungs. *Oncology* **27**:385–393.

Holland, L. M., Gonzales, M., Wilson, J. S., and Tillery, M. I. (1983). Pulmonary effects of shale dusts in experimental animals. In *Health Issues Related to Metal and Nonmetallic Mining.* Edited by W. L. Wagner, W. N. Rom, and J. A. Merchant. Boston, Butterworth, pp. 485–496.

Holland, L. M., Wilson, J. S., Tillery, M. I., and Smith, D. M. (1986). Lung cancer in rats exposed to fibrogenic dusts. In *Silica, Silicosis and Cancer.* Edited by D. F. Goldsmith, D. M. Winn, and M. Shy. Westport, Greenwood Press, pp. 267–279.

Hueper, W. C. (1958). Experimental studies in metal carcinogenesis. IX. Pulmonary lesions in guinea pigs and rats exposed to prolonged inhalation of powdered metallic nickel. *Arch. Pathol.* **65**:600–607.

Hueper, W. C., and Payne, W. W. (1962). Experimental studies in metal carcinogenesis: chromium, nickel, iron and arsenic. *Arch. Environ. Health* **5**:445–462.

IARC (1976). IARC monographs on the evaluation of carcinogenic risk of chemicals to man. Vol. 11, Lyon.

IARC (1978). IARC monographs on the evaluation of carcinogenic risk of chemicals to man. Vol. 17, Lyon.

IARC (1980). IARC monographs on the evaluation of carcinogenic risk of chemicals to man. Vol. 23, Lyon.

Ishinishi, N., Kodama, Y., Nobutomo, K., and Hisanaga, A. (1977). Preliminary experimental study on carcinogenicity of arsenic trioxide in rat lung. *Environ. Health Perspect.* **19**:191–196.

Kennedy, A. R., and Little, J. B. (1974). The transport and localization of benzo(a)pyrene–hematite and hematite–210 Po in the hamster lung following intratracheal instillation. *Cancer Res.* **34**:1344–1352.

Kennedy, A. R., McGandy, R. B., and Little, J. B. (1978). Serial sacrifice study of pathogenesis of 210Po-induced lung tumors in Syrian golden hamsters. *Cancer Res.* **38**:1127–1135.

Kobayashi, N., and Okamoto, T. (1974). Effects of lead oxide on the induction of lung tumors in Syrian hamsters. *J. Natl. Cancer Inst.* **52**:1605–1610.

Kotschetkowa, T. A., and Awrunina, G. (1957). Veraenderungen in den Lungen und in anderen Organen bei intratrachealer Einfuehrung einiger Radio-Isotope (24Na, 32P, 198Au). *Arch. Gewerbepathol. Gewerbehyg.* **16**:24–33.

Little, J. B., McGandy, R. B., and Kennedy, A. R. (1978). Interactions between polonium-210 alfa-radiation, benzo(a)pyrene, and 0.9% NaCl solution instillations in the induction of experimental lung cancer. *Cancer Res.* **38**:1929–1935.

Little, J. B., and O'Toole, W. F. (1974). Respiratory tract tumors in hamsters induced by benzo(a)pyrene and 210Po alfa-radiation. *Cancer Res.* **34**:3026–3039.

Litvinov, N. N., Bugryshev, P. E., and Kazenashev, V. F. (1975). Toxic properties of some soluble beryllium compounds (based on experimental morphological investigations). *Gig. Tr. Prof. Zabol.* **7**:34–37.

Lundgren, D. L., Hahn, F. F., and McClellan, R. O. (1981). Toxicity of 90Y in relatively insoluble fused aluminosilicate particles when inhaled by mice. *Radiat. Res.* **88**:510–523.

Lundgren, D. L., Hahn, F. F., and McClellan, R. O. (1982). Effects of single and repeated inhalation exposure of Syrian hamsters to aerosols of 144CeO2. *Radiat. Res.* **90**:374–394.

Lundgren, D. L., Hahn, F. F., Reber, A. H., and McClellan, R. O. (1983). Effects of the single or repeated inhalation exposure of Syrian hamsters to aerosols of 239PuO2. *Int. J. Radiat. Biol.* **43**:1–18.

Mass, M. J., and Kaufman, D. G. (1983). Species differences in the activation of benzo(a)pyrene in the tracheal epithelium of rats and hamsters. *Basic Life Sci.* **24**:331–351.

Metivier, H., Wahrendorf, J., and Masse, R. (1984). Multiplicative effect of inhaled pultonium oxide and benzo(a)pyrene on lung carcinogenesis in rats. *Br. J. Cancer* **50**:215–221.

Mohr, U., and Dungworth, D. L. (1988). Relevance to humans of experimentally-induced pulmonary tumors in rats and hamsters. In *The Design and Interpretation of Inhalation Studies and Their Use in Risk Assessment*. Edited by U. Mohr and D. L. Dungworth. Heidelberg, Springer-Verlag. pp. 209–232.

Moiseev, G. E., and Benemansky, V. V. (1975). On carcinogenic activity of low concentrates of nitrosodimethylamine in inhalation. *Vop. Onkol.* **21**:107–109.

Muhle, H., Bellmann, B., and Takenaka, S. Chronic effect of i.t. instilled nickel containing particles in hamsters. In *Nickel and Human Health: Current Perspective. Advance in Environmental Science & Technology*. Nebraska. John Willy Press: (in press).

Nettesheim, P., and Hammons, A. S. (1971). Induction of squamous cell carcinoma in the respiratory tract of mice. *J. Natl. Cancer Inst.* **47**:697–701.

Nettesheim, P., Virginia Cone, M., and Snyder, C. (1976). The influence of retinyl acetate on the postinitiation phase of preneoplastic lung nodules in rats. *Cancer Res.* **36**:996–1002.

Neimeier, R. W., Mulligan, L. T., and Rowland, J. (1986). Cocarcinogenicityof foundry silica sand in hamsters. In *Silica, Silicosis and Cancer*. Edited by D. F. Goldsmith, D. M. Winn, and M. Shy. Westport, Greenwood Press, pp. 215–227.

Niskanen, K. O. (1949). Observation on metaplasia of the bronchial epithelium and its relation to carcinoma of the lung. *Acta Pathol. Microbiol. Scand. Suppl.* **80**:1–80.

Ottolenghi, A. D., Haseman, J. K., Payne, W. W., Falk, H. L., and MacFarland, H. N. (1974). Inhalation studies of nickel sulfide in pulmonary carcinogenesis of rats. *J. Natl. Cancer Inst.* **54**:1165–1170.

Pershagen, G., Nordberg, G., and Bjoerklund, N.-E. (1984). Carcinomas of the respiratory tract in hamsters given arsenic trioxide and/or benzo(a)pyrene by the pulmonary route. *Environ. Res.* **34**:227–241.

Phalen, R. F. (1984). *Inhalation Studies: Foundations and Techniques*. Boca Raton, FL, CRC Press.

Plopper, C. G. (1983). Comparative morphologic features of bronchiolar epithelial cells; the Clara cell. *Am. Rev. Respir. Dis.* **128**:S37–S41.

Pott, F., Ziem, U., Reiffer, F. J., Huth, F., Ernst, H., and Mohr, U. (1987). Carcinogenicity studies on fibers, metal compounds, and some other dusts in rats. *Exp. Pathol.* **32**:129–152.

Prough, R. A., Patrizi, V. W., Okita, R. T., Masters, B. S. S., and Jakobsson, S. W. (1979). Characteristics of benzo(a)pyrene metabolism by kidney, liver, and lung microsomal fractions from rodents and humans. *Cancer Res.* **39**:1199–1206.

Pylev, L. N. (1961). Experimental induction of lung cancer in rats by intra-tracheal administration of 9, 10-dimethyl-1,2-benz-anthracene. *Bull. Exp. Biol. Med.* **52**:99–102.

Pylev, L. N. (1980). Pretumorous lesions and lung and pleural tumors induced by asbestos in rats, Syrian golden hamsters and Macaca Mulatta (Rhesus) monkeys. In *Biological Effects of Mineral Fibers*, Vol. I. Edited by J. C. Wagner. Lyon, IARC Scientific Publication 30, pp. 343–355.

Raabe, O. G., Yen, H., Newton, G. J., Phalen, R. F., and Velasquez, D. J. (1977). Deposition of inhaled monodisperse aerosols in small rodents. In *Inhaled Particles* IV. Edited by W. H. Walton. New York, Pergamon Press, pp. 3–21.

Reeves, A. L., Deitch, D., and Vorwald, A. J. (1967). Beryllium carcinogenesis. I. Inhalation exposures of rats to beryllium sulfate aerosol. *Cancer Res.* **27**:439–445.

Saffiotti, U. (1986). The pathology induced by silica in relation to fibrogenesis and carcinogenesis. In *Silica, Silicosis and Cancer*. Edited by D. F. Goldsmith, D. M. Winn, and M. Shy. Connecticut, Greenwood Press, pp. 287–37.

Saffiotti, U., Montesano, R., Sellakumar, A. R., and Borg, S. A. (1967). Experimental cancer of the lung. Inhibition by vitamin A of the induction of tracheobronchial squamous metaplasia and squamous cell tumors. *Cancer* 20:857–864.

Saffiotti, U., Cefis, F., and Kolb, L. H. (1968). A model for the experimental induction of bronchogenic carcinoma. *Cancer Res.* 28:104–124.

Sanders, C. L., and Mahaffey, J. A. (1984). Carcinogenicity of single and multiple intratracheal instillations of cadmium oxide in the rat. *Environ. Res.* 33:227–233.

Schepers, G. W. H., Durkan, T. M., Delehant, A. B., and Creedon, F. T. (1957). The biological action of inhaled beryllium sulfate. *Arch. Ind. Health* 15:32–58.

Schreiber, H., Nettesheim, P., and Martin, D. H. (1972). Rapid development of bronchiolo-alveolar squamous cell tumors in rats after intratracheal injection of 3-methylcholanthrene. *J. Natl. Cancer Inst.* 49:541–554.

Schreiber, H., Saccomanno, G., Martin, D. H., and Brennan, L. (1974). Sequential cytological changes during development of respiratory tract tumors induced in hamsters by benzo(a)pyrene-ferric oxide. *Cancer Res.* 34:689–698.

Schreiber, H., Martin, D. H., and Pazmino, N. (1975). Species difference in the effect of benzo(a)pyrene-ferric oxide on the respiratory tract of rats and hamsters. *Cancer Res.* 35:1654–1661.

Shabad, L. M. (1962). Experimental cancer of the lung. *J. Natl. Cancer Inst.* 28:1305–1332.

Shabad, L. M., and Pylev, L. N. (1970). Morphological lesions in rat lungs induced by polycyclic hydrocarbons. In *Morphology of Experimental Respiratory Carcinogenesis*. Edited by P. Nettesheim, M. G. Hanna, and J. Deatherage. Oak Ridge, TN, Division of Technical Information, U. S. Atomic Energy Commission (USAEC Symposium Series, Monograph No. 21), pp. 227–242.

Shabad, L. M., Pylev, L. N., Krivosheeva, L. V., Kulagina, T. F., and Nemenko, B. A. (1974). Experimental studies on asbestos carcinogenicty. *J. Natl. Cancer Inst.* 52:1175–1187.

Steinhoff, D., Gad, S. C., Hatfield, G. K., and Mohr, U. (1986). Carcinogenicity study with sodium dichromate in rats. *Exp. Pathol.* 30:129–141.

Stenbaeck, F., Wasenius, V.-M., and Rowland, J. (1986). Alveolar and interstitial changes in silica-associated lung tumors in Syrian hamsters. In *Silica, Silicosis and Cancer*. Edited by D. F. Greenwood, D. M. Winn, and M. Shy. Westport, Greenwood Press, pp. 199–213.

Sunderman, F. W., Donnelly, A. J., West, B., and Kincaid, J. F. (1959). Nickel poisoning. IX. Carcinogenesis in rats exposed to nickel carbonyl. *Arch. Ind. Health* 20:36–41.

Takenaka, S., Oldiges, H., Koenig, H., Hochrainer, D., and Oberdoerster, G. (1983). Carcinogenicity of cadmium chloride aerosols in W rats. *J. Natl. Cancer Inst.* **70**:367–373.

Takenaka, S., Hochrainer, D., and Oldiges, H. (1985). Alveolar proteinosis induced in rats by long-term inhalation of nickel oxide. In *Progress in Nickel Toxicology*. Edited by S. S. Brown and F. W. Sunderman, Jr. Oxford, Blackwell Scientific Publishers, pp. 89–92.

Temple, L. A., Marks, S., and Bair, W. J. (1960). Tumors in mice after pulmonary deposition of radioactive particles. *Int. J. Radiat. Biol.* **2**:143–156.

Te Punga, W. A., and Preston, N. W. (1958). Intratracheal infection of mice with haemophilus pertussis. *J. Pathol. Bacteriol.* **6**:275–283.

Thomas, R. G., Drake, G. A., London, J. E., Anderson, E. C., Prine, J. R., and Smith, D. M. (1981). Pulmonary tumors in Syrian hamsters following inhalation of 239PuO2. *Int. J. Radiat. Biol.* **40**:605–611.

Wagner, W. D., Groth, D. H., Holtz, J. L., Madden, G. E., and Stokinger, H. E. (1969). Comparative chronic inhalation toxicity of beryllium ores, bertrandite and beryl, with production of pulmonary tumors by beryl. *Toxicol. Appl. Pharmcol.* **15**:10–29.

Watson, J. A., Sprintzer, A. A., Auld, J. A., and Guetthof, M. A. (1960). Deposition and clearance following inhalation and intratracheal injection of particles. *Arch. Environ. Health* **19**:51–58.

Wehner, A. P., Busch, R. H., Olson, R. J., and Craig, D. K. (1975a). Chronic inhalation of asbestos and cigarette smoke by hamsters. *Environ. Res.* **10**:368–383.

Wehner, A. P., Busch, R. H., Olson, R. J., and Craig, D. K. (1975b). Chronic inhalation of nickel oxide and cigarette smoke by hamsters. *Am. Ind. Hyg. Assoc. J.* **36**:801–810.

Yerokhin, R. A., Koshurnikova, N. A., Lemberg, V. K., Nifatov, A. P., and Puzyrev, A. A. (1971). Remote after effects of radiation damage. Atomizdat: 315–333 (in Russian). Translated in AEC-tr-7883: 344–363.

Yoshimoto, T., Inoue, T., Iizuka, H., Nishikawa, H., Sakatani, M., Ogura, T., Hirao, F., and Yamamura, Y. (1980). Differential induction of squamous cell carcinomas and adenocarcinomas in mouse lung by intratracheal instillation of benzo(a)pyrene and charcoal powder. *Cancer Res.* **40**:4301–4307.

29

Experimental Lung Carcinogenesis by Intravenous Administration

ULRICH MOHR and SHINJI TAKENAKA[*]

Medizinische Hochschule Hannover
Hannover, West Germany

I. Introduction

In experimental lung carcinogenesis studies, systemic administration of carcinogens, including intravenous (IV) injection, has been used ever since this type of experiment first appeared in the literature. Andervont and Shimkin (1940) demonstrated the lung carcinogenicity of some heterocyclic compounds, such as dibenz(a,h)acridine, by IV administration in mice.

In association with the frequent occurrence of lung metastases and demonstration of lung tumors by IV injection of tumor cell suspension, IV administration of chemical carcinogens has attracted much attention in experimental lung carcinogenesis.

Chemical substances administered intravenously are mainly cleared by the reticuloendothelial systems of the liver, spleen, and bone marrow. In mice, rats, rabbits, and dogs, 0.5–4% of the intravenously injected doses are later found in the lung (Dobson, 1957). If substances are potent enough to affect lung tissues

---

*Present affiliation: Gesellschaft für Strahlen-und Umweltforschung, Neuherberg, West Germany

alone, it must be possible to induce lung tumors using substances with organ specificity. In reality, many substances show carcinogenicity not only in the lung but also in other organs. *N*-nitroso compounds especially tend to induce multiple tumors after parenteral administration. This attribute causes serious complications when these compounds are used as experimental tools (Pour et al., 1976).

In this chapter we review the results of lung carcinogenesis experiments, since almost no problems have been encountered with the techniques applied.

II. Carcinogenicity

A. Polynuclear Aromatic Compounds

As mentioned above, the first observation of lung carcinogenicity of polynuclear aromatic compounds (PACs) administered intravenously was reported by Andervont and Shimkin (1940), and Shimkin (1940). The IV or subcutaneous (SC) administration of eight PACs: 20-methylcholanthrene; 1,2,5,6-dibenzanthracene (a,h); 3,4,5,6,-dibenzcarbazole (7H-dibenzo(c,g)carbazole); 3,4-benzpyrene (benzo(a)pyrene); 15,16-benzdehydrocholanthrene; 1,2,5,6,-dibenzacridine (a,h); 2-methyl-3,4-benzphenanthrene; or 4′-methyl-3,4-benzpyrene led to a significant increase in lung tumor incidences in comparison with the untreated control.

The majority of PACs have been examined for their carcinogenic effects by oral administration or skin application only. Yet some were examined by SC and/or intraperitoneal (IP) administration, and found to have the potential to induce sarcomas at the site of injection (IARC, 1972, 1983).

Benzo(a)pyrene was extensively investigated for its lung carcinogenicity after SC (Pietra et al., 1961; Toth and Shubik, 1967; Grant et al., 1968) and intratracheal administration (Pylev; 1961, Shabad, 1962, Herrold and Dunham, 1962; Gross et al., 1965; Saffiotti et al., 1968). Following IV administration of benzo(a)pyrene, however, an increased incidence of mammary tumors was only reported in rats (Pataki and Huggins, 1969).

Stanton and Blackwell (1961) examined the effects of pulmonary infarction on lung carcinogenicity of 3-methylcholanthrene by IV administration. They found increased lung tumor incidences in rats when 3-methylcholanthrene was mixed with the infarct-inducing vesicles and injected intravenously.

B. *N*-Nitroso Compounds

Although *N*-nitroso compounds have been examined systematically, the number of lung carcinogenicity studies using IV administration is limited.

Repeated IV injections of *N*-nitroso-n-methylurethane (NMUT) to rats produced malignant lung tumors, diagnosed as alveolar cell and squamous cell carcinomas (Druckrey et al., 1962; Thomas and Schmaehl, 1963). Itano et al. (1972) also demonstrated papillary and adenomatous lung tumors in rabbits after IV administration of NMUT.

N-nitroso-n-methylurea (NMU) was examined for its carcinogenicity by IV administration in many animal species, including mouse, rat, hamster, gerbil, rabbit, dog, and monkey. In mice treated with IV administration of NMU, tumors developed in the lung as well as in the stomach, liver, lymphoid organs, and brain (Denlinger et al., 1974). The IV administration of NMU to other species induced tumors at multiple sites, especially in the nervous tissues (Druckrey et al., 1967; IARC monographs, 1978).

Stekar (1977) and Stekar and Gimmy (1980) noted the differing organotropism of NMUT and NMU in rats. NMUT, which, on IV administration, selectively induces lung tumors, reacts strongly with cysteine in vitro, resulting in the evolution of N_2 (Schoental, 1961). On the other hand, NMU neither reacts with this thiol nor induces lung tumors (Wheeler and Bowdon, 1972). Wheeler and Bowdon postulated that all those nitrosamides that react with thiols in vitro under concomitant liberation of elementary nitrogen should be capable of inducing lung cancer upon IV administration. *N*-nitroso-*N*-methylpropionamide and *N*-nitroso-*N*-methyl-*N'*-nitroguanidine were selected to test the validity of the hypothesis, since they were shown to react rapidly with cysteine, simultaneously liberating N_2 (Schultz and McCalla, 1969; Lawley and Thatcher 1970). After IV administration, both compounds selectively induced lung tumors with a high yield.

Habs et al. (1978) compared the carcinogenic effects of *N*-nitroso-acetoxymethylamine after SC, IV, or intrarectal administration in rats. The compound is hydrolized to the unstable intermediate hydroxymethyl-methyl-nitrosamine by unspecific esterase; therefore it should be a locally acting carcinogen. They found high incidences of lung and heart tumors and a lower incidence of kidney and earduct tumors after IV and SC administration. As expected, local tumors, such as subcuticals tumors after SC administration and colon tumors after intrarectal administration, were also detected. The compound showed systemic as well as local carcinogenicity.

N-nitrosomethylbenzylamine produced esophageal tumors after oral or subcutaneous administration in rats (Druckrey et al., 1967; Scheinberg et al., 1977; Stinson et al., 1978). In mice, multiple intraperitoneal injection of the compound induced a high incidence of forestomach carcinomas and lung adenomas, but no esophageal tumors, although chronic oral administration resulted in a 100% tumor incidence in both the esophagus and forestomach (Sander and Scheinberg, 1973). Concerning the relationship between alkylation of DNA and carcinogenic activity (Magee and Farber, 1962; Preussmann and Stewart, 1984), Wiestler et al. (1984) compared the degree of DNA methylation in targeted and nontargeted tissues of rats and mice that received IV, IP, or oral administration of *N*-nitroso-methylbenzylamine. They found that methylation of DNA purines in rats on oral or IV administration was most extensive in the esophagus, followed by the liver, lung, and forestomach. In mice, DNA methylation after intraperitoneal administration of the compound was highest in the liver, followed by the lung and forestomach. Oral administration of the compound, however, led to very high concentrations of alkylated DNA bases in both the esophagus and forestomach. From the results, they concluded that the induction of esophageal, forestomach, and lung tumors in rats and mice by *N*-nitrosomethylbenzylamine resulted from a preferential bioactivation of the compound in the target tissue. The incidence and location of the tumors induced by the compound correlated with the initial extent of DNA alkylation. However, the concentration of methylated purine bases required to induce neoplastic transformation differed between animal species and tissues.

The toxicity and carcinogenicity of derivatives of *N*-nitroso-(2-chloroethyl)urea, which showed chemotherapeutic activity in rats after IV administration, were investigated (Eisenbrand and Habs, 1980). They compared clinically established compounds and newly developed analogs with high cytostatic activity, and found different incidences of lung, nervous system, and forestomach tumors depending on the chemical properties. 1,3-bis(2-chloroethyl)-1-nitrosourea (BCNU), a clinically established compound, showed the highest toxicity and carcinogenicity in the lung.

In the absence of any clear data showing the clinical superiority of BCNU over the other analogs (Zeller et al., 1978; Zeller and Eisenbrand, 1981), and in view of its toxicity/carcinogenicity, they recommended replacing it with analogs having reduced toxicity and carcinogenicity. This animal experiment using the IV administration method has direct application to clinical medicine.

Other *N*-nitroso compounds, such as *N*-nitrosodi-*n*-butylamine, *N*-nitrosodiethylamine, *N*-nitrosodimethylamine, *N*-nitrosodi-*n*-propylamine and its derivatives and N-nitrosofolic acid, resulted in lung carcinogenicity when given by SC and/or IP administration (IARC, 1974). However, evidence of unequivocal lung carcinogenicity of those compounds has not been demonstrated after IV administration.

C. Miscellaneous Substances

Whereas some heavy metals, such as chromium or nickel compounds, were demonstrated to be carcinogenic in rat lungs after inhalation/intratracheal instillation, there is no clear evidence of lung carcinogenicity following IV administration of heavy metals (Hueper, 1955; Lau et al., 1972; IARC, 1976, 1980).

A solution of mustard gas was reported to increase lung tumor incidence in mice after IV administration (Heston, 1950; IARC, 1975).

III. Conclusion

In *N*-nitroso carcinogenesis, some studies using IV administration have contributed to a clearer understanding of the mechanism of carcinogenesis. Eisenbrand and Habs (1980) showed a direct application of the IV administration method to clinical medicine.

Nevertheless, this review confirms that only a limited number of studies demonstrated lung carcinogenicity of IV-administered substances. To be able to induce lung tumors, chemical compounds or their metabolites should accumulate in the lung in a sufficient amount depending on their reactivity within the tissues. In routinely studied mammalian species, such as mice, rats, rabbits, and dogs, however, the lungs rarely play a major role in removing particulates from the blood (Warner et al., 1986). This may be the main reason why lung tumors occurred so infrequently after IV administration. Recently, some domestic animals, such as calves, goats, sheep, and pigs, were found to have a unique reticuloendothelial system in the lung. Warner et al. (1986) showed that in calves and goats 35–100% (depending on particle type) of IV injected particles were sequestered in the lung by pulmonary intravascular macrophages. Although intravascular macrophages have not yet been demonstrated in human lungs, experiments using domestic animals, such as calves, would explore a new possibility in the study of experimental lung carcinogenesis by IV administration.

Many studies have demonstrated frequent uptake of tumor cells in the lung of various species following IV administration of tumor cell suspension (Wallace et al., 1978; Weiss, 1980; Glaves and Weiss, 1981; Orr et al., 1985). However, there are such large differences in size and plysicochemical properties that tumor cells and chemical carcinogens may not be directly comparable in their behavior.

References

Andervont, H. B., and Shimkin, M. B. (1940). Biological testing of carcinogens. II. Pulmonary tumor induction technique. *J. Natl. Cancer Inst.* 1:225–239.

Denlinger, R. H., Koestner, A., and Wechsler, W. (1974). Induction of neurogenic tumors in C3HeB/FeJ mice by nitrosourea derivatives: observations by light microscopy, tissue culture, and electron microscopy. *Int. J. Cancer* 13:559–571.

Dobson, E. L. (1957). Factors controlling phagocytosis. In *Physiopathology of the Reticuloendothelial System*. Edited by B. N. Halpern. Springfield, IL, Charles C. Thomas, p. 80.

Druckrey, H., Afkham, J., and Blum, G. (1962). Erzeugung von Lungenkrebs durch Nitrosomethylurethan bei intravenoeser Gabe an Ratten. *Naturwissenschaften* 49:451–452.

Druckrey, H., Preussman, S., Ivankovic, S., and Schmaehl, D. (1967). Organotrope carcinogene Wirkung bei 65 verschiedenen N-Nitroso-Verbindungen an BR-Ratten. *Z. Krebsforsch.* 69:103–201.

Eisenbrand, G., and Habs, M. (1980). Chronic toxicity and carcinogenicity of cytostatic N-Nitroso-(2-chloroethyl)ureas after repeated intravenous application to rats. In *Mechanismus of Toxicity and Hazard Evaluation*. Edited by B. Holmstedt, R. Lauwerys, M. Mercier, and M. Roberfroid. Amsterdam. Elsevier/North-Holland Biomedical Press, pp. 273–278.

Glaves, D., and Weiss, L. (1981). Metastasis and the reticulo-endothelial system. II. Effects of triamcinolone acetonide on organ retention of malignant cells in endotoxin-treated mice. *Int. J. Cancer* 27:475–479.

Grant, G. A., Carter, R. L., Roe, F. J. C., and Pike, M. C. (1968). Effects of the neonatal injection of a carcinogen on the induction of tumors by the subsequent application to the skin of the same carcinogen. *Br. J. Cancer* 22:346–358.

Gross, P., Tolker, E., Babyak, M. A., and Kaschak, M. (1965). Experimental lung cancer in hamsters. *Arch. Environ. Health* 11:59–65.

Habs, M., Schmaehl, D., and Wiessler, M. (1978). Carcinogenicity of acetoxymethyl-methyl-nitrosamine after subcutaneous, intravenous and intrarectal application in rats. *Z. Krebsforsch.* 91:217–221.

Herrold, K. M., and Dunham, L. J. (1962). Induction of carcinoma and papilloma of the tracheobronchial mucosa of the Syrian hamster by intratracheal instillation of Benzo(a)pyrene. *J. Natl. Cancer Inst.* 28:467–491.

Heston, W. E. (1950). Carcinogenic action of the mustards. *Cancer Res.* 10:224 (abstract).

Hueper, W. C. (1955). Experimental studies in metal carcinogenesis. VII. Tissue reactions to parenterally introduced powdered metallic chromium and chromium ore. *J. Natl. Cancer Inst.* 16:447–462.

IARC (1972). IARC monographs on the evaluation of carcinogenic risk of chemicals to man. Vol. 3, Lyon.

IARC (1974). IARC monographs on the evaluation of carcinogenic risk of chemicals to man. Vol. 4, Lyon.

IARC (1975). IARC monographs on the evaluation of carcinogenic risk of chemicals to man. Vol. 9, Lyon.

IARC (1976). IARC monographs on the evaluation of carcinogenic risk of chemicals to man. Vol. 11, Lyon.

IARC (1978). IARC monographs on the evaluation of carcinogenic risk of chemicals to man. Vol. 17, Lyon.

IARC (1980). IARC monographs on the evaluation of carcinogenic risk of chemicals to man. Vol. 23, Lyon.

IARC (1983). IARC monographs on the evaluation of carcinogenic risk of chemicals to man. Vol. 32, Lyon.

Itano, T., Takimoto, R., Konishi, T., and Kagawa, T. (1972). Occurrence of lung cancer of rabbit induced by administration of N-methyl-N-nitrosourethane. *Igaku No Ayumi* 81:826.

Lau, T., Hackett, R. L., and Sunderman, F. Jr. (1972). The carcinogenicity of intravenous nickel carbonyl in rats. *Proc. Am. Assoc. Cancer Res.* 13:13.

Lawley, P. D., and Thatcher, G. L. (1970). Methylation of deoxyribonucleic acid in cultured mammalian cells by N-methyl-N'-nitrosoguanidine. *Biochem. J.* 116:693–707.

Magee, P. N., and Farber, E. (1962). Toxic liver injury and carcinogenesis. Methylation of rat-liver nucleic acids by dimethylnitrosamine in vivo. *Biochem. J.* **83**:114–124.

Orr, F. W., Adamson, I. Y. R., and Young, L. (1985). Pulmonary inflammation generates chemotactic activity for tumor cells and promotes lung metastasis. *Am. Rev. Respir. Dis.* **131**:607–611.

Pataki, J., and Huggins, C. (1969). Molecular site of substituents of benz(a)anthracenes related to carcinogenicity. *Cancer Res.* **29**:506–509.

Pietra, G., Rappaport, H., and Shubik, P. (1961). The effects of carcinogenic chemicals in newborn mice. *Cancer* **14**:308–317.

Pour, P., Stanton, M. F., Kuschner, M., Laskin, S., and Shabad, L. M. (1976). Tumors of the respiratory tract. In *Pathology of Tumors in Laboratory Animals*, Vol. 1. Editor-in-Chief, V. S. Turusov. Lyon. IARC Scientific Publication 23, pp. 1–40.

Preussmann, R., and Stewart, B. W. (1984) N-nitroso carcinogens. In *Chemical carcinogens*, Vol. 2. Edited by C. E. Searle. ACS Monograph 182. Washington, D.C., American Chemical Society, pp. 643–828.

Pylev, L. N. (1961). Experimental induction of lung cancer in rats by intra-tracheal administration of 9,10-dimethyl-1,2-benz-anthracene. *Bull. Exp. Biol. Med.* **52**:99–102.

Saffiotti, U., Cefis, F., and Kolb, L. H. (1968). A model for the experimental induction of bronchogenic carcinoma. *Cancer Res.* **28**:104–124.

Sander, J., and Schweinsberg, F. (1973). Tumorinduktion bei Maeusen durch N-Methylbenzyl-nitrosamin in niedriger Dosierung. *Z. Krebsforsch.* **79**:157–161.

Scheinberg, F., Schott-Kollat, P., and Buerkle, G. (1977). Veraenderung der Toxizitaet und Carcinogenitaet von N-Methyl-N-nitrosobenzylamin durch Methylsubstitution am Phenylrest bei Ratten. *Z. Krebsforsch.* **88**:231–236.

Schoental, R. (1961). Interaction of the carcinogenic N-methyl-N-nitrosourethane with sulphydryl groups. *Nature* **192**:670.

Schultz, U., and McCalla, D. R. (1969). Reactions of cysteine with N-methyl-N-nitroso-p-toluenesulfonamide and N-methyl-N'-nitroso-N-nitrosoguanidine. *Can. J. Chem.* **47**:2021–2027.

Shabad, L. M. (1962). Experimental cancer of the lung. *J. Natl. Cancer Inst.* **28**:1305–1332.

Shimkin, M. B. (1940). Biological testing of carcinogens. I. Subcutaneous-injection technique. *J. Natl. Cancer Inst.* **1**:211–223.

Stanton, M. F., and Blackwell, R. (1961). Induction of epidermoid carcinoma in lungs of rats: a "new" method based upon deposition of methylcholanthrene in areas of pulmonary infarction. *J. Natl. Cancer Inst.* **27**:375–407.

Stekar, J. (1977). Pulmotropic carcinogenic activity of N-methyl-N-nitrosopropioamide. *Eur. J. Cancer* **13**:1183–1189.

Stekar, J., and Gimmy, J. (1980). Induction of lung tumors in rats by i.v. injection of N-methyl-N'-nitro-N-nitrosoguanidine. *Eur. J. Cancer* **16**:395–400.

Stinson, S. F., Squire, R. A., and Sporn, M. B. (1978). Pathology of esophageal neoplasms and associated proliferative lesions induced in rats by N-methyl-N-benzylnitrosamine. *J. Natl. Cancer Inst.* **61**:1471–1475.

Thomas, C., and Schaehl, D. (1963). Zur Morphologie der durch intravenoese Injektion von Nitrosomethylurethan erzeugten Lungen-tumoren bei der Ratte. *Z. Krebsforsch.* **65**:294–302.

Toth, B., and Shubik, P. (1967). Carcinogenesis in AKR mice injected at birth with benzo(a)pyrene and dimethylnitrosamine. *Cancer Res.* **27**:43–51.

Wallace, A. C., Chew, E.-C., and Jones, D. S. (1978). Arrest and extravasation of cancer cells in the lung. In *Pulmonary Metastasis*. Edited by L. Weiss and H. A. Gilbert. Boston, G. K. Hall, pp. 26–42.

Warner, A. E., Barry, B. E., and Brain, J. D. (1986). Pulmonary intravascular macrophages in sheep. Morphology and function of a novel constituent of the mononuclear phagocyte system. *Lab. Invest.* **55**:276–288.

Weiss, L. (1980). Cancer cell traffic from the lungs to the liver: an example of metastatic inefficiency. *Int. J. Cancer* **25**:385–392.

Weiss, L., Ward, P. M., and Holmes, J. C. (1983). Liver-to-lung traffic of cancer cells. *Int. J. Cancer* **32**:79–83.

Wheeler, G. P., and Bowdon, B. J. (1972). Comparison of the effects of cysteine upon the decomposition of nitrosoureas and of 1-methyl-3-nitro-1-nitrosoguanidine. *Biochem. Pharmacol.* **21**:265–267.

Wiestler, O. D., Uozumi, A., and Kleihues, P. (1984). DNA methylation by N-nitrosomethylbenzylamine in target and non-target tissues of laboratory rodents. Comparison with carcinogenicity. In *N-nitroso Compounds: Occurrence, Biological Effects and Relevance to Human Cancer*. Edited by I. K. O'Neill, R. C. Von Borstel, C. T. Miller, J. Long, and H. Bartsch. Lyon, IARC Scientific Publication 57, pp. 595–601.

Zeller, W. J., and Eisenbrand, G. (1981). Examination of newly synthesized 2-chloroethylnitrosoureas on rat leukemia L 5222. *Oncology* **38**:39–42.

Zeller, W. J., Eisenbrand, G., and Fiebing, H. H. (1978). Chemotherapeutic activity of new 2-chloroethylnitrosoureas in rat L5222 leukemia: comparison of bifunctional and water-soluble derivatives with 1,3-bis(2-chloroethyl)-1-nitrosourea. *J. Natl. Cancer Inst.* **60**:345–348.

AUTHOR INDEX

Italic numbers give the page on which the complete reference is listed.

Y

SUBJECT INDEX

A

Abnormal lung growth, factors producing, 451–452
Acute alveolar injury (AAI), 641–733
  caused by *N*-nitroso-*N*-methylure-thane, 682–703
    lung mechanics, 684–685
    qualitative alterations in surfac-tants, 697–703
    surfactant alterations, 697
    volume-pressure and morphome-tric observations, 685–697
  experimental models, 662
  hyperoxic lung injury, 670–682
    alveolar epithelial cells, 679–680
    interstitial cells, 680–682
    morphologic changes, 675–678
  induced by oleic acid, 703–719
  morphologic features of AAI, 643–661
  neutrophil-mediated oxidant lung injury, 663–670
  mediators of lung injury, 668–669
  neutrophil-mediated inflamma-tion, 663–665
  neutrophils in acute lung in-jury, 665–667

[Acute alveolar injury (AAI)]
  in vitro systems, 667–668
  in vivo experimental models, 667–668
Adaptive growth in lung develop-ment, 452–460
Adsorptive nonspecific probes, 374–376
Adsorptive specific probes, 376–377
Age, lung development and, 411
Air drying as method of lung fixa-tion, 25
Airway casting methods, 265–277
  development of airway casting, 265
  future developments, 274–275
  methods of casting, 266–273
    injecting the casting medium, 271–272
    macerating the tissues, 272–273
    materials for hollow casts, 268–269
    materials for negative casts, 266–268
    preparation of the lungs for casting, 269–270
    region of airways demonstrated by casting, 273
  negative casts, 274

N

O